Application of Plant Growth Promoting Microorganism and Plant Growth Regulators in Agricultural Production and Research

Application of Plant Growth Promoting Microorganism and Plant Growth Regulators in Agricultural Production and Research

Editor

Naeem Khan

MDPI • Basel • Beijing • Wuhan • Barcelona • Belgrade • Manchester • Tokyo • Cluj • Tianjin

Editor
Naeem Khan
Agronomy Department
University of Florida
Gainesville
United States

Editorial Office
MDPI
St. Alban-Anlage 66
4052 Basel, Switzerland

This is a reprint of articles from the Special Issue published online in the open access journal *Agronomy* (ISSN 2073-4395) (available at: www.mdpi.com/journal/agronomy/special_issues/pgr).

For citation purposes, cite each article independently as indicated on the article page online and as indicated below:

LastName, A.A.; LastName, B.B.; LastName, C.C. Article Title. *Journal Name* **Year**, *Volume Number*, Page Range.

ISBN 978-3-0365-1442-0 (Hbk)
ISBN 978-3-0365-1441-3 (PDF)

Contents

About the Editor

Naeem Khan

Dr. Naeem Khan (gold medalist) is a senior biological scientist at the Department of Agronomy, University of Florida, USA. Presently, he is working on plant breeding, plant metabolites, abiotic stresses, plant hormones, plant growth regulation, plant protection, plant–microbe interactions, and soil microbiology. As in the proceeding climate change scenario, abiotic stresses have been categorized as the main stressors affecting plant growth and productivity. His research addresses this problem in detail using physiological, microbiological, and molecular mechanisms. He has more than 12 years of research expertise and has authored 76 peer-reviewed research papers, 2 books, and 7 book chapters.

agronomy

Editorial

Application of Plant Growth Promoting Microorganism and Plant Growth Regulators in Agricultural Production and Research

Naeem Khan 

Department of Agronomy, Institute of Food and Agricultural Sciences, University of Florida, Gainesville, FL 32611, USA; naeemkhan@ufl.edu

Citation: Khan, N. Application of Plant Growth Promoting Microorganism and Plant Growth Regulators in Agricultural Production and Research. *Agronomy* **2021**, *11*, https://doi.org/10.3390/agronomy11030524

Received: 23 February 2021
Accepted: 10 March 2021
Published: 11 March 2021

Publisher's Note: MDPI stays neutral with regard to jurisdictional claims in published maps and institutional affiliations.

Plant growth-promoting microorganisms (PGPM) are rhizosphere microorganisms that colonize the root environment. Some of them are beneficial rhizobacteria while others are fungi that efficiently colonize roots and rhizosphere [1,2]. These microorganisms are capable of improving agriculture production and can also be used as biofertilizers under stressful environmental conditions. Continuous yield losses due to abiotic stresses are one of the important reasons for socio economic imbalance. As abiotic stresses decrease the synthesis of photosynthetic pigments, plant biomass and yield and negatively impact physiological and biochemical mechanisms, and eventually reduce plant growth and yield. The yield damages due to abiotic stresses vary from 50–82% [3]. The modern cultivation methods play an important role for good agricultural and horticultural practices. These methods include the use of cover crops, living mulches, PGPM, plant growth regulators (PGR) and other biostimulants that can protect the soil degradation and phytopathogens and improve the tolerance of plants to stress [4]. One of the utmost common stress tolerance plans in plants is the overproduction of diverse types of low molecular weight and non-toxic compatible organic solutes. They protect plants from unfavourable environmental conditions by different means like, adjustment of osmotic stress, detoxifying reactive oxygen species, membrane stabilization and protecting the structure of enzymes and proteins [5].

It has been suggested that tolerance mechanisms, such as leaf hydration, increased intrinsic water use efficiency, reduced oxidative damage or improved nutritional status, can explain the contribution of PGPR to the stress resistance of host plants. The use of PGPR and other symbiotic microorganisms, may play an important role in developing strategies to assist water conservation in plants. More precisely, the soil-borne *Pseudomonads* and *Paecilomyces variotii* have received a particular attention because of their metabolic flexibility, excellent root-colonising ability and competence to produce a wide range of enzymes and metabolites that benefit the plant in water conservation and enable them to endure diverse biotic and abiotic stresses [3,6].

Rhizosphere microorganisms as well as plant secondary metabolites are well-known for their role in improving growth patterns of roots as they result in rhizosheaths formation around the roots and protecting them from desiccation, pollutant degradation, maintenance of primary cellular functions and from antimicrobial activity of various predators. Many mechanisms have been described for the action of PGPR [7]. Some strains produce metabolites such as hydrogen cyanide, 2,4-diacetylphloroglucinol; antibiotics, and volatile compounds that motivate plant growth. Other strains are responsible for siderophores production and thus play a critical role in sequestering iron for plants, delay senescence, improve biological control, and produce phytohormones which influence plant physiological processes. Some inoculants enter inside root and establish endophytic populations with compliance to the niche and paybacks to the host plants while some enhance surface area of root, thus attract nutrients uptake, and in turn, tempt plant productivity [8,9]. The application of PGPR alone or in combination with chitosan play an important role in combating salinity stress by maintaining higher chlorophyll content, chlorophyll fluorescence, and antioxidant enzymes activity [10]. Noshin et al. [11] isolated the halo-tolera[nt] bacterial species and evaluated their ability to improve seed germination, plant grow[th] and phytohormones content in plants grown under saline stress conditions. Similar[ly] Ahmad et al. [12] noted the synergistic effects of PGPR and biochar on the growth a[nd] yield of maize grown under semi-arid climate. The beneficial effects of the fungal strai[n] *aggressivum* f. *europaeum* Tae52481 and *T. saturnisporum* Ca1606 were also evident on t[he] growth and yield of pepper and tomatoes [13].

One of the major benefits of PGPR is to produce effective antibacterial compoun[ds] that can be used against certain plant pathogens and pests. Moreover, PGPR arbitrat[es] biological control not directly by eliciting induced systemic resistance against a lar[ge] number of plant diseases [14]. Allelopathic rhizosphere bacteria also improve the whe[at] growth as they act as biocontrol agents to control the weeds in wheat growing areas [1[5]. Whereas, He et al. [16] reported the nematicidal activities of *A. japonicas* against root kn[ot] nematodes. These microorganisms are also an essential part of the soil phosphorus (P) cyc[le] as they are concerned in a series of processes that have an effect on the transformation [of] soil P. Particularly, soil microorganisms are efficient in the release of P from inorganic a[nd] organic pools of total soil P by the process of mineralization and solubilisation. Shorta[ge] and fixation of P in alkaline calcareous soils initiate a decrease in crop production. T[he] impact of rock phosphate and chemical fertilizers were evaluated in a two year fie[ld] experiment both individually and in combination with PGPR on the growth and yield [of] wheat and on physico-chemical properties of soil. The study revealed substantial increas[es] in wheat growth and yield treated with *Pseudomonas* sp. + poultry litter. Whereas, all oth[er] treatments i.e., rock phosphate + poultry litter + *Proteus* sp.; rock phosphate and poult[ry] litter; half dose inorganic P from Single Super Phosphate-SSP with 18% P_2O_5 and poult[ry] litter alone were useful for maintaining the soil biological and biochemical properties [1[7]. It was also reported that mechanical pot seedling transplanting together with deep nitrog[en] (N) fertilization significantly improves the yield and antioxidant enzymes activities in ri[ce] thus may also play an important role in improving the stress tolerance in test plants [18].

PGPR also assists in phytoremediation and microbial based phytoremediation is o[ne] of the utmost developing and environmentally friendly methods used for the purificati[on] of pollutants from the soil. The PGPR *S. aureus* K1 revealed to regulate the plant growth a[nd] antioxidant enzyme activities by decreasing oxidative stress and chromium (Cr) toxicity [by] converting Cr^{6+} to Cr^{3+} and the accumulation of Cr^{6+} was significantly reduced in whe[at] plants inoculated with *S. aureus* K1. This shows that the application of *S. aureus* K1 could [be] an effective approach to lessen the Cr toxicity in wheat and other crop plants [19]. Vario[us] free-living rhizosphere bacteria that promote the growth of plants can be applied in hea[vy] metal polluted soils to alleviate lethal effects of heavy-metals on the flora. These benefic[ial] microbes either entirely inhibit metal ions by inhabiting different metabolic activities [or] enhance the plant tolerance mechanism to high concentration of heavy metals [20].

Plant growth regulators (PGR) are chemical compounds that play a significant ro[le] in plant growth and yield. They are involved in plant's intercellular communication a[nd] particularly present in the actively growing tissues of plants [21]. PGR associated with t[he] control of cell division, root formation, embryogenesis, fruit development and ripening, a[nd] tolerance to biotic and abiotic stresses [22]. Plant growth regulators are designated in t[he] literature as taking a significant part in acquiring crop management of modern agricultu[re] in conditions of abiotic and biotic stressors. Plant growth regulators may improve t[he] antioxidant activity in plants. Foliar application of GA3 significantly improved both ro[ot] length and diameter, root and foliage fresh weights/plant, and root and foliage yields/[ha] increased with the incremental level of nitrogen and/or GA_3 concentration [23]. Oxalic ac[id] (OA) is an important calcium regulator and plays an important role in fruit yield and qual[ity]. In this special issue Benítez García, et al. [24] studied the PGR present in the seaweeds a[nd] evaluated their plant growth promoting abilities. Whereas, García-Pastor et al. [25] point[ed] out the effects of preharvest oxaloacetic acid (OA) treatments on pomegranate trees. Th[ey] reported an increase in the respiration rate, fruit size, fruit quality and crop yield. The C[a] treatment was also augmenter to sugars and organic acids content, as well as to bioactive compounds and antioxidant activity. They authors also noted a stimulation in the fruit ripening process, increase in the number of fruits with improved quality. Nawaz et al. [26] studied the effects of seed priming with SA on the growth, pigmentation and mineral concentrations of maize (*Zea mays* L.) grown under B toxicity. The findings suggested that the exogenously applied SA moderates the reaction of plants grown under the boron toxicity, and therefore could be used as a plant growth regulator to motivate plant growth and augment mineral nutrient uptake under B-stressed conditions. Ali et al. [27] studied the effects of α-Tocopherol foliar spray on the growth, photosynthetic pigments, nutrient uptake, and drought tolerance in maize. They reported that α-Tocopherol is important in improving water stress tolerance in maize, and its foliar application was found to be effective in decreasing the adverse effects of water-stress on growth by modulating the metabolic activities of plants.

Funding: This research received no external funding.

Data Availability Statement: All the data generated or analyzed during this study are included in this published editorial.

Conflicts of Interest: The author declares no conflict of interest.

References

ALKahtani, M.D.F.; Fouda, A.; Attia, K.A.; Al-Otaibi, F.; Eid, A.M.; Ewais, E.E.-D.; Hijri, M.; St-Arnaud, M.; Hassan, S.E.-D.; Khan, N.; et al. Isolation and Characterization of Plant Growth Promoting Endophytic Bacteria from Desert Plants and Their Application as Bioinoculants for Sustainable Agriculture. *Agronomy* **2020**, *10*, 1325. [CrossRef]

Asghar, W.; Kondo, S.; Iguchi, R.; Mahmood, A.; Kataoka, R. Agricultural Utilization of Unused Resources: Liquid Food Waste Material as a New Source of Plant Growth-Promoting Microbes. *Agronomy* **2020**, *10*, 954. [CrossRef]

Khan, N.; Ali, S.; Tariq, H.; Latif, S.; Yasmin, H.; Mehmood, A.; Shahid, M.A. Water Conservation and Plant Survival Strategies of Rhizobacteria under Drought Stress. *Agronomy* **2020**, *10*, 1683. [CrossRef]

Patkowska, E.; Mielniczuk, E.; Jamiołkowska, A.; Skwaryło-Bednarz, B.; Błażewicz-Woźniak, M. The Influence of *Trichoderma harzianum* Rifai T-22 and Other Biostimulants on Rhizosphere Beneficial Microorganisms of Carrot. *Agronomy* **2020**, *10*, 1637. [CrossRef]

Naamala, J.; Smith, D.L. Relevance of Plant Growth Promoting Microorganisms and Their Derived Compounds, in the Face of Climate Change. *Agronomy* **2020**, *10*, 1179. [CrossRef]

Moreno-Gavíra, A.; Diánez, F.; Sánchez-Montesinos, B.; Santos, M. *Paecilomyces variotii* as A Plant-Growth Promoter in Horticulture. *Agronomy* **2020**, *10*, 597. [CrossRef]

Shah, A.; Smith, D.L. Flavonoids in Agriculture: Chemistry and Roles in, Biotic and Abiotic Stress Responses, and Microbial Associations. *Agronomy* **2020**, *10*, 1209. [CrossRef]

Ali, S.; Hayat, K.; Iqbal, A.; Xie, L. Implications of Abscisic Acid in the Drought Stress Tolerance of Plants. *Agronomy* **2020**, *10*, 1323. [CrossRef]

Shahid, M.A.; Sarkhosh, A.; Khan, N.; Balal, R.M.; Ali, S.; Rossi, L.; Gómez, C.; Mattson, N.; Nasim, W.; Garcia-Sanchez, F. Insights into the Physiological and Biochemical Impacts of Salt Stress on Plant Growth and Development. *Agronomy* **2020**, *10*, 938. [CrossRef]

ALKahtani, M.D.F.; Attia, K.A.; Hafez, Y.M.; Khan, N.; Eid, A.M.; Ali, M.A.M.; Abdelaal, K.A.A. Chlorophyll Fluorescence Parameters and Antioxidant Defense System Can Display Salt Tolerance of Salt Acclimated Sweet Pepper Plants Treated with Chitosan and Plant Growth Promoting Rhizobacteria. *Agronomy* **2020**, *10*, 1180. [CrossRef]

Ilyas, N.; Mazhar, R.; Yasmin, H.; Khan, W.; Iqbal, S.; Enshasy, H.E.; Dailin, D.J. Rhizobacteria Isolated from Saline Soil Induce Systemic Tolerance in Wheat (*Triticum aestivum* L.) against Salinity Stress. *Agronomy* **2020**, *10*, 989. [CrossRef]

Ahmad, M.; Wang, X.; Hilger, T.H.; Luqman, M.; Nazli, F.; Hussain, A.; Zahir, Z.A.; Latif, M.; Saeed, Q.; Malik, H.A.; et al. Evaluating Biochar Microbe Synergies for Improved Growth, Yield of Maize, and Post-Harvest Soil Characteristics in a Semi-Arid Climate. *Agronomy* **2020**, *10*, 1055. [CrossRef]

Sánchez-Montesinos, B.; Diánez, F.; Moreno-Gavíra, A.; Gea, F.J.; Santos, M. Role of *Trichoderma aggressivum* f. *europaeum* as Plant-Growth Promoter in Horticulture. *Agronomy* **2020**, *10*, 1004. [CrossRef]

Kousar, B.; Bano, A.; Khan, N. PGPR Modulation of Secondary Metabolites in Tomato Infested with *Spodoptera litura*. *Agronomy* **2020**, *10*, 778. [CrossRef]

Abbas, T.; Zahir, Z.A.; Naveed, M.; Abbas, S.; Alwahibi, M.S.; Elshikh, M.S.; Mustafa, A. Large Scale Screening of Rhizospheric Allelopathic Bacteria and Their Potential for the Biocontrol of Wheat-Associated Weeds. *Agronomy* **2020**, *10*, 1469. [CrossRef]

He, Q.; Wang, D.; Li, B.; Maqsood, A.; Wu, H. Nematicidal Evaluation and Active Compounds Isolation of *Aspergillus japonicus* ZW1 against Root-Knot Nematodes *Meloidogyne incognita*. *Agronomy* **2020**, *10*, 1222. [CrossRef]

17. Billah, M.; Khan, M.; Bano, A.; Nisa, S.; Hussain, A.; Dawar, K.M.; Munir, A.; Khan, N. Rock Phosphate-Enriched Compost in Combination with Rhizobacteria; A Cost-Effective Source for Better Soil Health and Wheat (*Triticum aestivum*) Productivity. *Agronomy* **2020**, *10*, 1390. [CrossRef]

18. Li, L.; Zhang, Z.; Tian, H.; Mo, Z.; Ashraf, U.; Duan, M.; Wang, Z.; Wang, S.; Tang, X.; Pan, S. Roles of Nitrogen Deep Placement on Grain Yield, Nitrogen Use Efficiency, and Antioxidant Enzyme Activities in Mechanical Pot-Seedling Transplanting Rice. *Agronomy* **2020**, *10*, 1252. [CrossRef]

19. Zeng, F.; Zahoor, M.; Waseem, M.; Anayat, A.; Rizwan, M.; Ahmad, A.; Yasmeen, T.; Ali, S.; El-Sheikh, M.A.; Alyemeni, M.N.; et al. Influence of Metal-Resistant *Staphylococcus aureus* Strain K1 on the Alleviation of Chromium Stress in Wheat. *Agronomy* **2020**, *10*, 1354. [CrossRef]

20. Jabborova, D.; Wirth, S.; Kannepalli, A.; Narimanov, A.; Desouky, S.; Davranov, K.; Sayyed, R.Z.; El Enshasy, H.; Malek, R.A.; Syed, A.; et al. Co-Inoculation of Rhizobacteria and Biochar Application Improves Growth and Nutrientsin Soybean and Enriches Soil Nutrients and Enzymes. *Agronomy* **2020**, *10*, 1142. [CrossRef]

21. Daniel, E.C.; Fabio, G. An Assessment of Seaweed Extracts: Innovation for Sustainable Agriculture. *Agronomy* **2020**, *10*, 1433. [CrossRef]

22. Cappellari, L.D.R.; Chiappero, J.; Palermo, T.B.; Giordano, W.; Banchio, E. Volatile Organic Compounds from Rhizobacteria Increase the Biosynthesis of Secondary Metabolites and Improve the Antioxidant Status in *Mentha piperita* L. Grown under Salt Stress. *Agronomy* **2020**, *10*, 1094. [CrossRef]

23. Leilah, A.A.A.; Khan, N. Interactive Effects of Gibberellic Acid and Nitrogen Fertilization on the Growth, Yield, and Quality of Sugar Beet. *Agronomy* **2021**, *11*, 137. [CrossRef]

24. Benítez García, I.; Dueñas Ledezma, A.K.; Martínez Montaño, E.; Salazar Leyva, J.A.; Carrera, E.; Osuna Ruiz, I. Identification and Quantification of Plant Growth Regulators and Antioxidant Compounds in Aqueous Extracts of *Padina durvillaei* and *Ulva lactuca*. *Agronomy* **2020**, *10*, 866. [CrossRef]

25. García-Pastor, M.E.; Giménez, M.J.; Valverde, J.M.; Guillén, F.; Castillo, S.; Martínez-Romero, D.; Serrano, M.; Valero, D.; Zapata, P.J. Preharvest Application of Oxalic Acid Improved Pomegranate Fruit Yield, Quality, and Bioactive Compounds at Harvest in a Concentration-Dependent Manner. *Agronomy* **2020**, *10*, 1522. [CrossRef]

26. Nawaz, M.; Ishaq, S.; Ishaq, H.; Khan, N.; Iqbal, N.; Ali, S.; Rizwan, M.; Alsahli, A.A.; Alyemeni, M.N. Salicylic Acid Improves Boron Toxicity Tolerance by Modulating the Physio-Biochemical Characteristics of Maize (*Zea mays* L.) at an Early Growth Stage. *Agronomy* **2020**, *10*, 2013. [CrossRef]

27. Ali, Q.; Tariq Javed, M.; Haider, M.Z.; Habib, N.; Rizwan, M.; Perveen, R.; Ali, S.; Nasser Alyemeni, M.; El-Serehy, H.A.; Al-Misned, F.A. α-Tocopherol Foliar Spray and Translocation Mediates Growth, Photosynthetic Pigments, Nutrient Uptake, and Oxidative Defense in Maize (*Zea mays* L.) under Drought Stress. *Agronomy* **2020**, *10*, 1235. [CrossRef]

 **agronomy**

Review

Water Conservation and Plant Survival Strategies of Rhizobacteria under Drought Stress

Naeem Khan [1,*], Shahid Ali [2], Haleema Tariq [3], Sadia Latif [4,5], Humaira Yasmin [6], Asif Mehmood [7] and Muhammad Adnan Shahid [8]

1 Department of Agronomy, Institute of Food and Agricultural Sciences, University of Florida, Gainesville, FL 32611, USA
2 Plant Epigenetic and Development, Northeast Forestry University Harbin, Harbin 150040, China; Shahidsafi926@gmail.com
3 Department of Biosciences, University of Wah, Wah Cantt 47040, Pakistan; haleema.tariq2020@gmail.com
4 Department of Plant Sciences, Quaid-i-Azam University, Islamabad 45320, Pakistan; slatif@bs.qau.edu.pk
5 Department of Biology and Environmental Science, Allama Iqbal Open University, Islamabad 44000, Pakistan
6 Department of Bio-Sciences, COMSATS University, Islamabad 45550, Pakistan; humaira.yasmin@comsats.edu.pk
7 Institute of Biological Sciences, Sarhad University of Science and Information Technology, Peshawar 25000, Pakistan; asif.ibs@suit.edu.pk
8 Department of Agriculture, Nutrition and Food Systems, University of New Hampshire, Durham, NH 03824, USA; muhammad.shahid@unh.edu
* Correspondence: naeemkhan@ufl.edu or naeemkhan001@gmail.com

Received: 18 September 2020; Accepted: 28 October 2020; Published: 30 October 2020

Abstract: Drylands are stressful environment for plants growth and production. Plant growth-promoting rhizobacteria (PGPR) acts as a rampart against the adverse impacts of drought stress in drylands and enhances plant growth and is helpful in agricultural sustainability. PGPR improves drought tolerance by implicating physio-chemical modifications called rhizobacterial-induced drought endurance and resilience (RIDER). The RIDER response includes; alterations of phytohormonal levels, metabolic adjustments, production of bacterial exopolysaccharides (EPS), biofilm formation, and antioxidant resistance, including the accumulation of many suitable organic solutes such as carbohydrates, amino acids, and polyamines. Modulation of moisture status by these PGPRs is one of the primary mechanisms regulating plant growth, but studies on their effect on plant survival are scarce in sandy/desert soil. It was found that inoculated plants showed high tolerance to water-deficient conditions by delaying dehydration and maintaining the plant's water status at an optimal level. PGPR inoculated plants had a high recovery rate after rewatering interms of similar biomass at flowering compared to non-stressed plants. These rhizobacteria enhance plant tolerance and also elicit induced systemic resistance of plants to water scarcity. PGPR also improves the root growth and root architecture, thereby improving nutrient and water uptake. PGPR promoted accumulation of stress-responsive plant metabolites such as amino acids, sugars, and sugar alcohols. These metabolites play a substantial role in regulating plant growth and development and strengthen the plant's defensive system against various biotic and abiotic stresses, in particular drought stress.

Keywords: PGPR; RIDER; drylands; water conservation

---

1. Introduction

Desertification, drought, and land degradation are major challenges to sustainable crop production throughout the world especially in developed countries. Water scarcity mainly due to low annual precipitation is very damaging for plant growth, and ultimately sustainable crop production. However, there is an inordinate need to use these areas even with marginal productivity due to damage to basic farmlands. Therefore, there is more interest in producing crops using low or marginal yields of soil (e.g., sandy soil) [1]. However, sandy soil has high temperatures and suffers severe drought. Stress losses can range from 50% to 80%, depending on the stress period and type of plant species [2]. Drought stress in desert areas affects plant water potential, restricts the normal plant performance, [3], and alters the plant physiological and morphological characteristics [4,5]. Drought stress-induced plant growth was studied in wheat [6], barley [7], rice, and corn [8]. Moisture content and plant biomass are common growth factors impacted by drought stress in these plants [9]. Besides, drought stress stimulus negatively impacts the nutrient uptake and translocation as the soil nutrients are transferred to the roots via water.

Consequently, drought stress reduces the absorption of nutrient and mass-flux of water-soluble nutrients, for example, calcium, nitrate, sulfate, silicon, and magnesium [10]. Drought stress enhances formation of free radicals that damage plant defence system resulting in an increase in reactive oxygen species (ROS), such as superoxide radicals, hydroxyl radicals, and hydrogen peroxide induces oxidative stress. ROS can cause tissue damage, to membrane corrosion, proteins and nucleic acids by causing their lipid peroxidation [11–13].

Water stress is responsible for high economic losses in arid and semi-arid regions. It disturbs plant–water relations at cellular and whole plant levels, resulting in specific and non-specific responses [14]. Plant reaction to water stress is a complex process that tends to include polyamine formation and a collection of novel proteins with relatively unknown functions. Drought decreases the photosynthesis supply of carbon dioxide, which may contribute to ROS production from misguided electrons in the camera system [15,16]. It also creates free radicals during abiotic tension. ROS, such as radical superoxide (O_2^-), radical hydroxyl (OH), and hydrogen peroxide, enhance the damaging effect of lipid peroxidation throughout the membrane [17]. Plants have an antioxidant defensive system which prevents cellular membranes and DNA from ROS-induced oxidative damage by converting ROS into non-toxic forms such as water and oxygen [18,19].

Inoculation of plants with growth-promoting microorganisms can improve water retention strategies and drought tolerance of plants grown in arid or semi-arid regions [20]. These useful microbes inhabit the rhizosphere/endogenous rhizosphere of the plant through various direct-indirect mechanisms and promote plant growth (Figure 1). The rhizosphere is a thin layer of soil surrounding the roots of the plant and is a very critical and active area of root activity and metabolism [21–23]. A significant number of microorganisms coexist in the rhizosphere, such as bacteria, fungi, protozoa, and algae, but mostly different types of bacteria. Plants release organic compounds through exudate to select the bacteria that contribute most to the plant's health under stressful conditions [24]. The beneficial relationships of plant-microbes in the rhizosphere are the key determinants in water conservation, soil productivity, and plant health. Plant growth-promoting rhizobacteria (PGPR) affect growth, yield, and nutrient uptake through a series of mechanisms. Some strains (e.g., *Azospirillum brasilense*, *Aeromonas punctata*, Bacillus megaterium, Pseudomonas fluorescens, Serratia marcescens) directly modulate plant physiology by stimulating the production of plant hormones, while others upturn minerals and nitrogen in the soil as a means of increasing growth under water-deficient conditions [25–28].

The current review comprehensively covers major research to evaluate the effectiveness of PGPR in alleviating crop water stress and to find effective PGPR to help crops in maintaining water status under drought conditions. The aim of the present review is to provide insights into the role of phytohormones, plant metabolites, exopolysaccharides (EPS), and 1-aminocyclopropane-1-caroboylic acid (ACC) deaminase activity in stress tolerance of plants in response to PGPR inoculation. This review identifies the challenges of drought stress and involvement of PGPR in the mitigation of drought stress in plants for sustainable production.

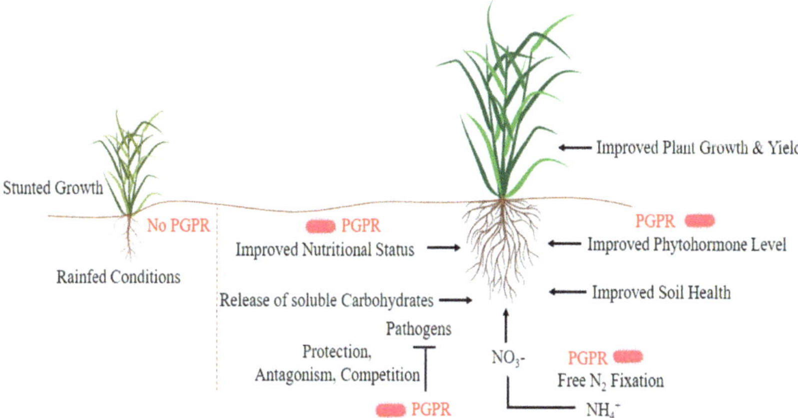

Figure 1. Plant growth-promoting strategies of plant growth-promoting rhizobacteria (PGPR) under drought stress.

2. Plant Survival Strategies under Drought Stress

A species may have a complementary set of survival strategies enabling it to survive under small and unpredictable distribution of rain [29]. Desert plants may have no water for many years. Plants exhibit different responses when sensing abiotic stimuli, which are related to specific stress-tolerance mechanisms [30–32]. A series of epidermis waxes protect plants from excessive moisture loss and provide protection against various pathogenic antagonistic activities [33]. In addition, osmoprotectants like proline accumulation aid in sustaining the plant's water potential, and promotes the plant's extraction of water from the soil [34]. Changes in primary metabolism are considered to be the most obvious of all metabolic reactions and comprise changes in the level of sugar/sugar alcohol, amino acid, and tricarboxylic acid cycle intermediates, exhibiting a common tendency for ecological stress reactions. However, changes in secondary metabolism are exact to particular stress and are precise to the type of plant species [35,36].

Some of the metabolic compounds that are associated with abiotic stresses and act as protectants include the sorbitol, polyols, mannitol, sucrose, fructan, proline, and ectoine [37]. Other small molecules such as carotenoids, ascorbic acid, tocopherols and anthocyanins also protect plants from being subjected to oxidative injury and protect plants by eliminating stress-induced ROS in plants. The production of phytoalexins and initiation of phenylpropanoid pathways and lignin biosynthesis are related to plant defence mechanisms [38–40]. Plant molecules such as salicylic acid, jasmonic acid, methyl salicylate, and methyl jasmonate are formed under stress. They can also act as signalling molecules that trigger defences against various biotic and abiotic stresses in crop plants [41]. In recent years, metabolomics has been used for various purposes, such as (1) assessing the effect of various stresses in plants; (2) pursuing the contribution of specific compounds in a specific biosynthetic or secondary deprivation pathway and (3) organizing various plant samples [42]. Stability, defence, and signalling of metabolites can be used to measure the degree of plant lenience to diverse abiotic stresses [43,44]. Extensive research is carried out to develop policies against drought stress by growing drought tolerant crops, improving crop calendars and resource management practices [45].

3. Water Conservation Strategies of Plant Growth-Promoting Rhizobacteria (PGPR)

Using PGPR and commensal microorganisms, in particular arbuscular mycorrhiza (AM) fungi, may help to develop strategies to improve water retention potential in plants. More specifically, *Pseudomonas* sp. in the soil is of specific importance due to its versatility in catabolism, exceptional root colonization capacity, and the capability to produce a variety of enzymes and metabolites that contribute to abiotic stress tolerance in plants [46]. Relatively few pathways have been found to clarify the improved tolerance of *Pseudomonas*-treated plants to environmental stresses. Tolerance mechanisms such as increased osmotic adjustment and hydration of the leaves decreased oxidative damage, enhanced nutritional status, or increased the efficiency of intrinsic water usage have been suggested to elaborate the contribution of PGPR in improving stress tolerance [47].

Soil microbes have developed complex survival methods in desiccated soil. For example, bacteria have been documented to alter their membrane structure to improve their survival during the phases of low external water potential [48]. Increased water content in bacterial colonies can increase nutrient utilization [49]. In particular, the release of soluble carbohydrates in rhizosphere in PGPR-treated plants are higher, which can improving the survival rate of microorganisms under water deficit conditions. Some PGPR, such as *Azospirillum*, have the capacity to preserve water by developing cyst formation around the roots and by synthesizing polyhydroxybutyrate and melanin [50]. Likewise, extracellular bacterial polysaccharides will form organic mineral sheaths around cells along with surrounding mineral particles, which contributes to an improvement in development of macro-aggregates as an additional indirect consequence [51]. On the other hand, excessive drought stress decreases the amounts of water-soluble carbon and carbohydrates in rhizosphere of plants inoculated with *Glomus intraradices*, suggesting that mycorrhizal fungi serve as an effective sink for photosynthates and that these carbon fractions contribute to the stabilization of soil aggregates to a lesser extent. As a result, increased soil accumulation can be expected to increase water absorption by plants, thereby improving plant growth [52].

The free-moving soil bacteria sustain associations with the plant roots, thereby helping plant defence against various stresses, including drought, heavy metals toxicity, pathogens, and salinity [53–55]. Some PGPR such as *Azospirillum* sp., *Pseudomonas fluorescens* and *Azotobacter* sp. are widely used for increased crop yield [56,57]. The increased hydration caused by the PGPR strain could be due to improved water efficiency and/or enzymatic reduction of the concentration of plant ethylene, thus diminishing the inhibitory effect of ethylene on seedling root biomass [58]. The co-inoculation of arbuscular mycorrhizal fungi (*Glomus intraradicers* or *Glomus mosseae*) and PGPR *Pseudomonas mendocina* with *Lactuca sativa* L. improved antioxidant catalase under extreme drought conditions, indicating that inoculants can be used to mitigate oxidative damage induced by drought [59,60]. Kohler et al. [61] demonstrated that when PGPR, *P. mendocina*, and arbuscular mycorrhizal fungi were inoculated, antioxidant catalase activity was higher in lettuce plants under severe drought conditions. The aforementioned PGPR species were also found to be useful in reducing drought-induced oxidative damage in plants (Figure 2). Interestingly, the plant growth promoting bacteria (PGPB) strain *Pseudomonas fluorescens* Pf1 augmented enzymatic activities of catalase and peroxidase in green grams under water stress. Similarly, it can also serve as a storing compound for protein synthesis. Starch biosynthesis is reduced under stressed conditions, and proline accumulation is used as a carbon and nitrogen source for plant survival [62–65]. Other common mechanisms of maintaining water status by plants in response to PGPR under water stress are as follows:

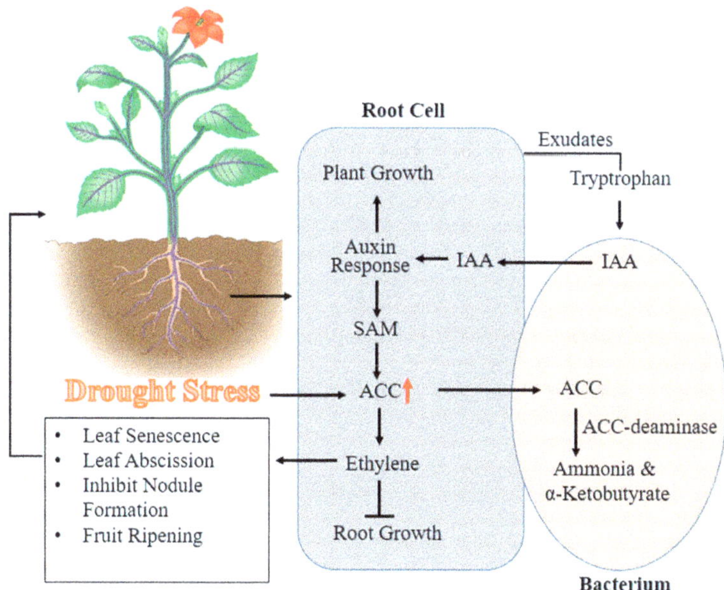

Figure 2. Water conservation and drought stress alleviating mechanisms employed by PGPR. SAM-Shoot apical meristem.

3.1. Modifications in Phytohormones Content

Plant growth-promoting rhizobacteria have developed various phytohormones, such as abscisic acid (ABA), ethylene, gibberellins, auxins, cytokinins and salicylic acid. Such hormones stimulate plant growth either directly or through certain secondary bacterial metabolites [66]. These plant hormones maintain plant water status water deficient conditions and are important for plant growth and disease prevention. *Acetobacter, Bacillus, Herbaspirillum* and *Rhizobium* species render gibberellins (Figure 2) [67]. Indole-3-acetic acid (IAA) is generated by some species of *Micrococcus, Pseudomonas* and *Staphylococcus*. Furthermore, cytokinins are produced by *Azotobacter chroococcum* that helps plants in sustaining proper moisture under extreme drought conditions [68]. The plant stress hormone abscisic acid (ABA), mediates plant stress tolerance by regulating several stress response genes and is responsible for maintaining proper moisture level in cells under drought conditions [69,70]. It has been previously reported that about 80% of the microorganisms extracted from the rhizosphere of different crops are recorded to be able to synthesize and release auxin as a secondary metabolite [71]. The rhizosphere bacteria-secreted IAA can interfere with plant growth and development, since receiving IAA from the soil bacteria may change the endogenous level of Plant IAA [72]. The IAA also serves as a signalling molecule that influences gene expression in a variety of microorganisms. The previous studies confirmed that phytohormones work as bi-directional communication between microbes and plant. For example, under nitrogen or phosphate starvation, the strigolactones exuded from the root, which attracts AM fungi, and downregulated their biosynthesis upon colonization [73]. Auxin and ABA have concentration-dependent positive effects on AM development while salicylic acid (SA), ethylene (ET), and gibrillic acid (GA) inhibit the root nodule and AM symbiosis. To understand the underlying complexity, it is essential to complement the genetics with system biology approaches, including hormone profiling, metabolomics, global network analysis, and computational molecular modeling of various processes in plants and soil. IAA is produced by many plant-associated microbes, including PGPR, nitrogen-fixing symbionts, and pathogens, which assist in interactions between plant-microbes [74,75]. The pathogenic bacteria that produce IAA, when grown in culture including

Erwinia herbicola, *Xanthomonas campestris*, *Erwinia Chrysanthemi*, and several *Pseudomonas syringae* pathovars [76–79].

In plant-associated microbes, the IAA regulates the expression of genes that promote the interaction with plants. IAA induced large-scale changes in the transcriptome of PGPR, *A. brasilense*, which upregulate the gene expression involved in IAA biosynthesis and genes involved in the metabolism, respiration, and transportation [80]. These findings suggest that IAA promotes physiological and metabolic adjustment for growth in the rhizosphere [81]. Furthermore, IAA induces the expression of genes predicted to be involved in the Type VI secretion system (T6SS). In PGPR and other plant-associated bacteria, the role of the Type VI secretion is not well understood but may help the bacteria by injecting toxins into other microbes in the vicinity. Also, the exogenous IAA enhanced the expression of genes involved in stress responses. In *Escherichia coli* and *Bradyhizobium japonicum* the IAA treatments enhanced the cell viability when the bacteria was grown in stressful conditions, including oxidative stress, heat shock, and osmotic shock, and furthermore promoted biofilm production [82]. Consequently, IAA plays a very significant function in the relationship between rhizobacteria and plants [83]. Bacterial IAA can have enhanced the root length and surface area, making soil nutrients and water easier for plants to obtain. In addition, rhizobacteria IAA can relax plant cell walls, thereby increasing the root secretions and thus providing more nutrients to promote the growth of rhizosphere bacteria [84–87].

Agrobacterium tumefaciens strains STM196 isolated from the rhizosphere of rapeseed rape [88] have been shown to improve resistance to moderate water deficit, and alter plant physiology and delay developmental transition in *Arabidopsis thaliana* [89]. In addition, previous in vitro experiments have shown that STM196 modifies root architecture and hormonal signalling [90]. Importantly, STM196 not only improved plant longevity but also improved recovery of growth in living plants in post-stress and increased biomass production during flowering [91]. In *A. thaliana*, it was interpreted that the inculcation of *Bacillus subtilis* augmented the photosynthetic rate by reducing the concentration of ABA [92]. In the common bean, co-inoculation of *Rhizobium tropici* and *Paenibacillus polymyxa* has been shown to reduce the ABA content in water-deficient conditions [93]. The caronatine is exuded from *Pseudomonas syringe*, which inhibits the signalling pathways of ABA and prevents stomatal closure [94]. *Arabidopsis thaliana* inoculated with *A. brasilense* showed contrast results by increasing ABA content two-fold [95], and this increase in the content of ABA plays an important role in water conservation and alleviation of drought effects [96].

In addition to this, *B. subtilis* produces numerous polyamines which promote plant growth and development under water stress. The different types of polyamines, including spermine, spermidine, cadaverine and putrescine, are natural small-molecular-weight compound that modify physiological and biochemical attributes in plants and improve plant growth and development under drought environments [97]. Polyamine functions in the regulation of plant growth and water conservation. It has been reported to play a significant role both in promoting active growth and the division of cells into young tissues of the plants [98]. Polyamines promote the growth and cell differentiation in plant roots and provide insight into morphological variations [21,99]. Furthermore, they also play a major role in sustaining optimal ionic and pH environments, cell differentiation, organ development and secondary metabolite production under stress [100]. Polyamine has previously been documented to assist in stress tolerance either by regulating ROS homeostasis or by regulating antioxidant processes or by suppressing ROS production [101,102].

3.2. PGPR Mediated Metabolites Involved in Drought Stress Tolerance

Genetically engineered rhizobacteria, which overproduce trehalose in their cells, thereby allow plants to retain their water status and may increase survival of plants under extreme water-deficient conditions, in particular by increasing leaf water contents or by causing an accumulation of trehalose in the soil [103]. Some sugars including galactinol and raffinose act as osmoprotectants, which are synthesized in response to water-deficient conditions, mannitol scraps ROS, hydroxyl radicals and stabilize the structure of the enzyme [104,105]. The osmolytes prevent the creation of intramolecular hydrogen bonds in macromolecules by forming hydrogen bonds with them. Trehalose stabilizes the formation of membranes and macromolecules during water stress conditions and allows plants to retain water under harsh conditions. During different environmental pressures, the aggregation of different osmolytes such as glycinebetaines, proline, ectoine, etc. has been reported [106]. Similarly, Khan et al. [107] reported a different accumulation of metabolites such as L-proline, L-arginine, L-histidine, L-isoleucine, and tryptophan in chickpea cultivars grown under drought conditions (Figure 3). Furthermore, other metabolites like alanine, choline, phenylalanine, tyrosine, glucosamine, guanine gamma-aminobutyric acid, and aspartic acid had reduced accumulation under drought conditions. Amino acids, such as branch chain alanine, valine, leucine also increased in samples of *Triticeae* species (IG132864, TR39477 and Bolal) under water-deficient conditions [108,109]. Urano et al. [110] also described the increased accretion of these branch chain amino acids in *A. thaliana*. Less and Galili [111] have suggested that activities of enzymes of amino acid catabolism were rapidly increased under drought stress.

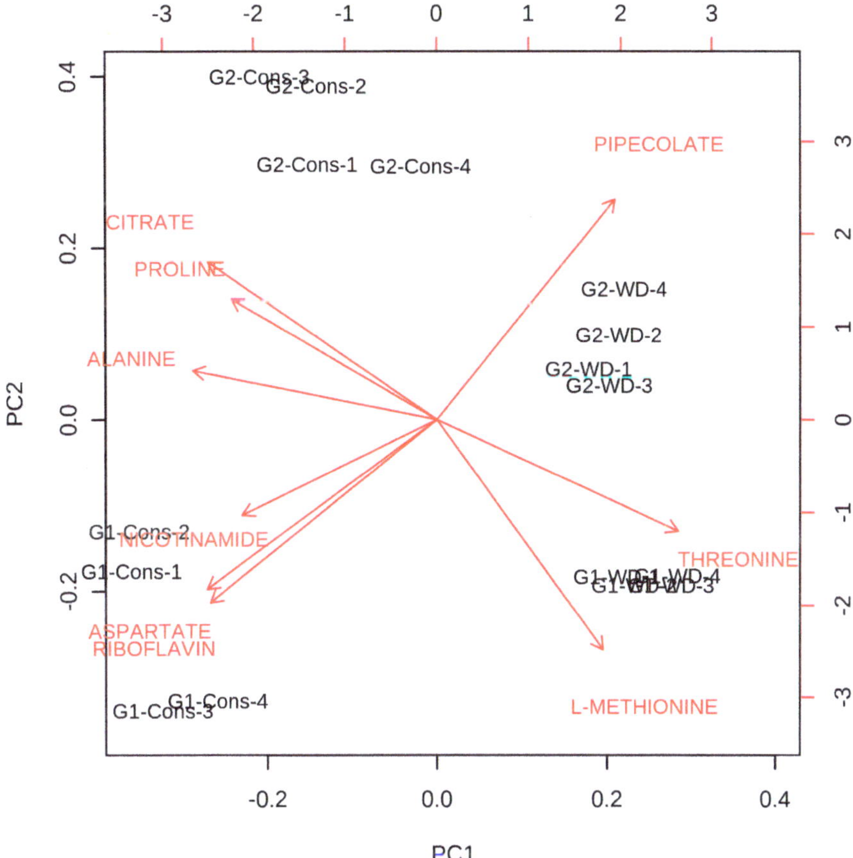

Figure 3. A principal component analysis (PCA) based biplot showing association among different metabolites induced by PGPR in chickpea leaves grown under *consortium* (Cons) and water deficit (WD) conditions. Samples with *consortium* and water deficit treatments did not overlap with each other showing that both the treatments have different levels of metabolites. G1-Drought Sensitive genotype; G2-Drought Tolerant genotype.

On the other hand, sugar and its compounds such as fructose, mannitol, galactose, mannose and other non-reducing sugars and oligosaccharides provide a hydration shield around drought-sensitive proteins which can provide an initial defensive condition against further water depletion [112]. It has been reported previously that trehalose, glycinebetaine, carnitine, glutamate, proline, mannitol, polyols, fructans, sorbitol, oligosaccharides, and inorganic ions such as K^+, sucrose, etc. are used as osmolytes to suppress cellularosmotic shock under water stress. The osmolyte accumulation prevents macromolecules by stabilizing the tertiary structure of proteins and by scavenging ROS [113,114].

3.3. Biofilm and Exopolysaccharides (EPS) Production by PGPR under Water Deficiency

Biofilms are microbial communities in which differentiated cell populations are encapsulated by bacterial made extracellular matrices [115]. Most microorganisms are capable of forming biofilms in natural, clinical and industrial environments (Figure 4). Rhizobacteria are often found to form micro-colony or biofilm-like structures at the roots of plants [116,117]. Biofilms enhance soil accumulation, improve water status and enhance microbial biomass, thereby stimulating root exudates under pressure. Therefore, the production of a viscous extracellular matrix layer at the rhizosphere has a strong selectivity advantage, especially under stress conditions [118]. The matrix may also contribute to the mechanical stability of the biofilm and interact with other macromolecules and solutes of low molecular weight to provide several microenvironments within the biofilm [119–121].

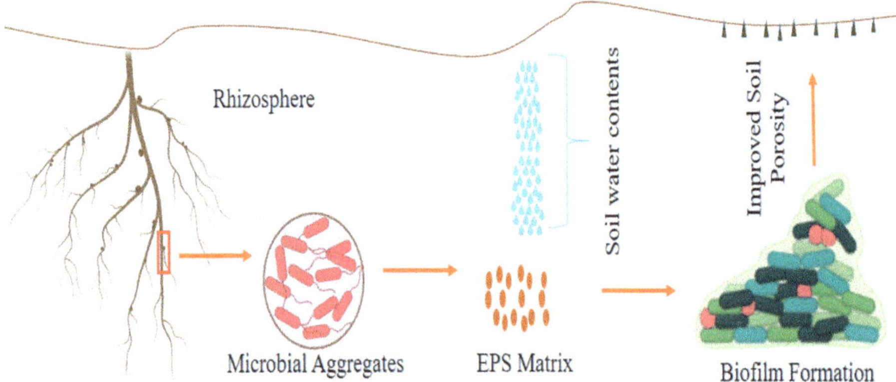

Figure 4. Microbial aggregates result in the formation of exopolysaccharides (EPS) and biofilm that provide protection to roots under abiotic stresses and improve the soil moisture content and soil porosity.

Studies have shown that the formation of PGPR biofilms has a pivotal role in defending plants under water deficient conditions. For example, *P. polymyxa* has been shown to colonize plant root tips, form biofilm-like structures and protect plants from water stress and pathogen infection [122]. As another example, a high mucus mutant of *P. fluorescens* CHAO strain indicating enhanced biofilm formation is believed to be beneficial for water budget and exhibits significantly enhanced carrot root colonization compared to its wild-type parent [123]. Khan et al. [124] reported that PGPR *Planomicrobium chinense*, *Bacillus cereus* and *P. fluorescens* alone or in combination resulted in biofilm formation in inoculated chickpea plants grown under sandy soil condition and protected the roots of plants from the adverse effects of unfavourable conditions. Besides this, *B. subtilis* strain ATCC 6051 is capable of forming biofilm-like structures on the roots of *Arabidopsis* plants and protecting *Arabidopsis* from high temperatures and infection with *P. syringae* [125–127].

Microbial EPS is essential for the production of biofilms and cell aggregates that help to protect cells from harsh conditions and may protect the substantial amount of heavy metals (Figure 4) [128]. Many studies demonstrated the importance of microbial EPS in water conservation under water-deficient conditions [129]. Furthermore, the bacterial EPS has also been found to be important in the bioremediation of wastes from water [130]. Exopolysaccharides produced by PGPR improved soil moisture content, plant biomass and leaf area of maize plant grown under stress condition [131]. Alami et al. [132] reported that the EPS-producing rhizobacteria control the aggregation of root-adhering soils. EPS-producing rhizobacteria will dramatically increase the number of soil macropores and soil aggregation in the rhizosphere, contributing to increased supply of water and fertilizer to inoculated plants [133–135]. EPS-producing rhizobacteria also bind cations such as Na$^+$, with a rise in the population density of EPS-producing bacteria in the root zone expected to reduce the Na$^+$ amount available for plant absorption, thereby alleviating salt stress in plants grown in saline environments [136]. The EPS-producing bacterial population responded to adverse environmental conditions by contributing to soil aggregation and increased the retention of water in the root rhizosphere under water stress conditions [137]. The efficacy of inoculation with rhizobacteria, *P. mendocina*, has been documented for both soil stabilization and soil fertility enhancement under non-saline conditions [138]. The function of alginates in bacterial adhesion and biofilm formation for *Pseudomonas aeruginosa* has been examined in depth. It has been shown that alginate increases the binding and anchoring of *P. aeruginosa* strains on many surfaces and is the key constituent in the biofilm matrix [139,140].

3.4. 1-Aminocyclopropane-1-Carboxylic Acid (ACC) Deaminase Activity of PGPR to Combat Water Deficit Stress

One of the key pathways used by PGPR to promote water conservation by plant growth and development during a water shortage is the reduction of ethylene levels by 1-aminocyclopropane-1-carboxylic acid (ACC), the immediate precursor of ethylene in plants (Figure 2) [141]. The enzyme hydrolyzes ACC to α-ketobutyrate and ammonia [142]. Plants that are inoculated with PGPR containing ACC deaminase have been found to be substantially more resistant to the deleterious effects of stress ethylene that is synthesized as a result of adverse environments such as heavy metals toxicity [143], presence of phytopathogens [144], hypoxia [145], high salinity and drought stress [146]. It has been stated in most of these cases that the PGPR containing ACC deaminase significantly decreases the ACC activity in the stressed plants, thus reducing the level of ethylene biosynthesis and consequently protecting the plants from their deleterious effects. The usage of plant growth-promoting bacteria containing ACC deaminase is helpful to facilitate plant growth and water conservation in water-deficient conditions [147].

The impact of inoculation with ACC deaminase-containing rhizobacteria on water use efficiency was clearly evident in crop plants. *Pseudomonas fluorescens* biotype G (ACC-5) was found to be more promising at low humidity (25% field capacity), because of its maximum water use efficiency compared to the respective uninoculated controls. Similarly, inoculation was shown to be efficient in increasing the productivity of water usage when measured on a dry weight basis [148]. The inoculation of ACC deaminase PGPR *Achromobacter piechaudii* ARV8 with pepper and tomato seedling exposed to transient water stress significantly increased the fresh and dry weights of the plants [149]. Recently, beneficial effects of ACC deaminase-producing bacterium *Variovorax* sp. in improving the nodulation and growth of spiny brooms grown in arid regions of Tunisia have been observed [150]. Likewise, ACC deaminase-containing *Variovorax paradoxus* has also caused physiological modifications in *Pisum sativum* L. grown under moisture deficient condition [151]. Positive effects of ACC deaminase bacteria on shoots and roots biomass, transpiration rate and leaf area of plants were also observed in short-term experiments (Table 1). In long-term studies, plants inoculated with ACC deaminase bacteria provided more seed yield (25–41%), number of seeds, and accumulation of seed nitrogen than uninoculated plants, and helped preserve water status [152].

Table 1. Impacts of 1-aminocyclopropane-1-carboxylic acid (ACC) deaminase-producing bacteria on plant growth and drought stress tolerance.

ACC Producing PGPR	Host Plants	Results	References
A. piechaudii ARV8	*Solanum lycopersicum* L.	Enhanced plant biomass and decrease in ethylene levels.	[153]
A. piechaudii ARV8	*Pisum sativum* L.	Improve root-shoot ratio under low soil moisture content.	[154]
Pseudomonas sp.	*Pisum sativum* L.	Improve the plant growth and yield and reduce the triple response of ethylene.	[155]
P. fluorescens	*Pisum sativum* L.	Positive impacts on plant growth under severe drought stress.	[156]
V. paradoxus 5C-2	*Pisum sativum* L.	Induce the abscisic acid (ABA) signalling in plants and Improve the soil nutrient content.	[157]
Rhizosphere bacteria containing ACC-deaminase	*Triticum aestivum*	Enhance root-shoot length and improve the water and nutrient uptakes.	[158]
Bacillus 23-B + *Pseudomonas* sp. 6-P + *Mesorhizobium ciceri*		Improve seed germination and root length in chickpea under moisture stress.	[159]
Bacillus licheniformis K11	*Pepper nigrum*	Enhance the expression of stress related genes e.g., Cadh, VA, sHSP, and CaPR-10.	[160]
Citricoccus zhacoinesis B-4	*Allium cepa*	Promote plant growth and germination index.	[161]
Ochrobactrum pseudogrignonense RJ12, *Pseudomonas* sp. RJ15 and *B. subtilis* RJ46	*Vigna mungo* L. *Pisum sativum* L.	Show positive impacts on seed germination, improve root and shoot length and regulate ethylene level.	[162]

4. Development of Root System by PGPR

Rhizobium helps plants in maintaining a favourable water status in their tissues under water-deficient conditions by enhancing root development (Figure 1). Rhizosphere bacteria that promote plant growth colonize roots and maintain symbiotic interactions to promote plant growth and provide protections against stresses [163]. Different strains of PGPRs are well known for their positive impact on plant growth and also helping in water stress such as *Azospirillum* sp., *Azotobacter* sp., and *P. fluorescens* [164]. Root biomass was large in plants inoculated with *Phyllobacterium brassicacearum* STM196, and also increased the water absorption by changing root architecture. Studies conducted under reproductive conditions have shown that STM196 increases lateral root length [165], as well as the density and length of root hairs [166]. STM196 resulted in a greater contact surface with the soil causing higher water flow from roots to the shoot. Some rhizosphere bacteria help plants in maintaining a desirable moisture level in their tissues under water-deficient conditions by improving the root development [167]. Inoculation with PGPR strains improved plant growth by strengthening the root architecture, consequently increasing nutrient uptake [168,169].

Some PGPRs like *A. brasilense*, *B. japonicum*, *B. cereus*, *Paenibacillus illinoisensis*, *P. fluorescens* promote root development and alter root structure by producing plant hormones such as IAA, resulting in increased root surface area and increased number of root tips [170]. This root stimulation can help plants fight pathogens. It has also been suggested that PGPR increases plant uptake of water and mineral ions by proton pump ATPase stimulation, despite the lack of experimental evidence [171].

5. Improving Nutrient Availability and Maintenance of Soil Quality

Drought stress affects soil biological, physical and chemical activities. It not only decreases plant nutrient supply but also has harmful impacts on plant and soil health due to a rise in soil temperature [172]. Under drought conditions, the nutrient availability to the plant is seriously impaired; however, the usage of appropriate microorganisms restores nutrient bioavailability in drought-stressed conditions. Several PGPR have been isolated in recent decades and have been suggested for use in sustainable agriculture under water deficit conditions [173]. Plant growth-promoting bacteria found to be very effective in substantially increasing soil nutrients content thus improving crop yields [174]. PGPRs of the genera *Arthrobacter, Azotobacter, Azospirillum, Bacillus, Enterobacter, Pseudomonas, Serratia* and *Streptomyces* were largely documented for this purpose [175]. PGPR may have a beneficial impact on plant growth and development, possibly because of a nutrient mobilization in the soil, nitrogen fixation and excretion of different plant hormones (Figure 2). Using PGPR may help in reducing the use of chemical fertilizers or increase the nutrient-use efficiencies, particularly for low-mobility nutrients such as iron and phosphorus [176,177]. In particular, iron forms insoluble hydroxides in high pH soils, and supply of bioavailable Fe to plants is reduced. It was previously reported that the production of siderophores by genus *Pseudomonas* enhanced the solubility of Fe and contributed to the overall iron requirements of plants, especially in calcareous soils [178].

Soil quality is critical for the improvement of sustainable farming under extreme drought conditions. It encourages penetration of water, provides optimal habitat for soil organisms and an optimal aeration to roots and soil organisms, and helps in preventing soil erosion [179]. Microorganisms have been observed to have a direct impact on soil properties and quality, since they may associate with other microorganisms in the rhizosphere (Figure 4) [180–182]. Soil microorganisms play an important role in controlling the processes of decomposition of organic matter, and the supply of plant nutrients, such as N, P and K. Microbial inoculants are well known as an essential component of advanced nutrient management contributing to sustainable agriculture [183]. In addition, microbial inoculants can be used to improve crop production as an economic input; fertilizer doses can be reduced, and nutrient use efficiencies can be increased [184].

6. Changes in Plant Functional Traits

It is well established that extreme deficient water stress affects plant growth, water status, and is responsible for a decrease in photosynthetic ability, especially through stomatal closure and leaf senescence. Non-destructive experiments based on chlorophyll fluorescence imaging were commonly used to decipher the effect of different water potentials on plant physiology, but only rarely used at high performance [185]. Measurements of chlorophyll fluorescence were performed at high throughput to unravel the influence of rhizobacteria on the responses of plants to drought (Figure 5). There are numerous photosynthetic parameters that exist; dark-adapted Fv/Fm represents the optimal efficiency of photosystem II (PSII) and is thus one of the most commonly used parameters for analyzing the physiological modulations in leaf. Most commonly, the mean Fv/Fm of the photosynthetic organ or the whole plant is used to describe the stressor response [186]. The major decrease in mean Fv/Fm during a prolonged water deficit is commonly correlated with high leaf senescence. With a higher proportion of leaf senescence, STM196-inoculated plants may persist and thus provide higher resistance to photosynthetic damage from the leaves [187–189].

Inoculated plants thus demonstrated reduced mortality rate after the establishment of water tension. Leaf senescence reflects a common way to conserve resources. It allows translocation to reproductive organs and decreases water intake of older and less-productive leaves [190]. Therefore, leaf senescence is an adaptive trait which will help plant survival under stressful conditions. The increase in chlorophyll content may contribute to the improvement in plant photosynthetic efficiency triggered by PGPR. Rice plants inoculated with arbuscular mycorrhizal fungus under water tension showed a positive association between water budget and PSII efficiency [191]. In *A. thaliana*, the inoculation of the *Burkholderia phytofirmans* PsJn strain enhances the senescent leave at flowering under well-watered environments. It is widely confirmed that rhizobacteria increase the content of leaf water which results in increased plant resistance under drought stress [192–194].

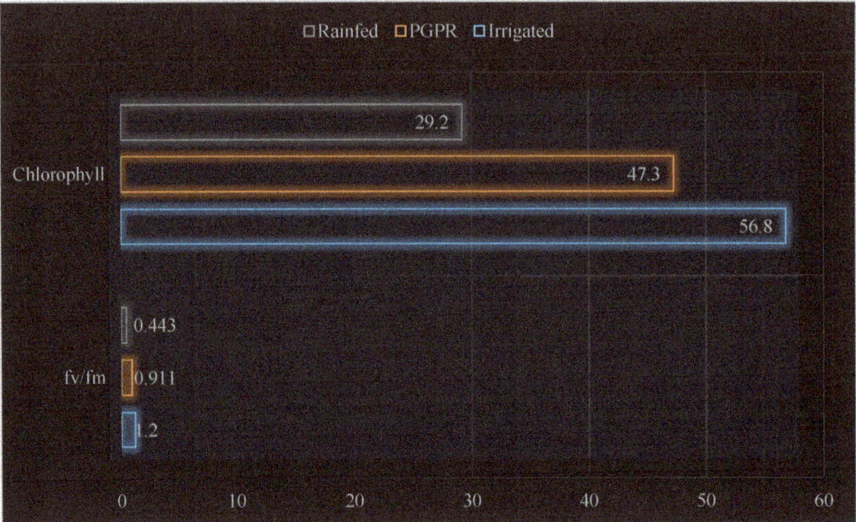

Figure 5. Effects of PGPR on leaf chlorophyll content and photochemical efficiency of chickpea plants grown under moisture stress conditions.

7. Molecular Mechanisms to Mitigate Drought Stress Induced by PGPR

Plant responses to environmental stresses are complex mechanisms which involve modulation in the expression of stress-related genes [195,196]. These genes support stress management by inducing two different types of protein, either functional proteins that act directly, such as mRNA binding proteins, chaperones, LEA proteins, and osmotic regulators, or regulatory proteins that regulate transcription and signalling pathways [197]. Plants recognise abiotic stresses by specific receptors in the cell walls or intracellularly, which involve various sensing system. ABA plays a significant role in abiotic stress responses by influences in the expression of various genes to mediate systemic stress tolerance [198–200]. Additionally, a variety of compounds serve as systemic signals to alleviate stress within the plants, for example, small RNAs (sRNA), peptides and metabolites [198].

Stress tolerance can be enhanced by treating plants with several PGPR stains which up-regulate stress tolerance inducing genes. The rice plants treated with *P. fluoresces* induced multiples differential gene expression, for example, ERD15 (Early response to dehydration15), COC1, Hsp20 and bZIP1 (chaperones in ABA signalling pathway), PKDP (protein kinase), and COX1 (regulate energy and carbohydrate metabolism). *Arabidopsis thaliana* treated with *Pseudomonas* strains promote the expression of ACO, ACS (ethylene biosynthesis), ADC, CPA, AIH, SPMS, SPDS and SAMDC (polyamine biosynthesis), VSP1 (ethylene-responsive gene), Pdf1.2 (JA marker genes) and PR1 (SA regulated gene). Similarly in *Lycium barium* it increased the expression of LbKT1, LbSKOR (encoding potassium channels) and RAB18 (ABA-responsive gene) in drought conditions [201–203]. Pepper plant inoculated with *Bacillus* sp. enhanced 1.5 fold increased the expression of sHSP (small heat-shock proteins), VA, and Cadhn. Under drought conditions, the inoculation of *A. brasilense* NO40 and *Bacillus amyloliquefaciens* 5113 alleviates the deleterious effects in the leaves of wheat plants by upregulation of APX1, HSP17.8 and SAMS1 stress-responsive genes [204]. This overexpression of genes increased the ascorbate-glutathione redox cycle, which helps to overcome the adverse effect of water stress.

Recent approaches to system biology and omics analysis of transcripts, proteins and metabolites have improved our knowledge of molecular responses in stressed plants and plant–microbe interactions [205,206]. PGPR-induced physiological and metabolic alternation is anticipated to be driven by molecular alteration that has culminated in protein and post-translation modifications. The combination of proteomics and metabolomics profiling for stressed, non-stressed and PGPR-treated plants will also help to classify metabolic and molecular modulations involved under stress conditions in beneficial plant–microbe interactions and help to elucidate the essence of the defence. Under drought stress, the proteomic and metabolic studies and PGPR responses have been reported in many plants [207,208]. However, the impact of beneficial microbes on gene expression and metabolite aggregation in PGPR-treated crop plants remains poorly investigated.

8. Conclusions and Future Perspectives

Many plant-associated bacteria are well known for their ability to promote plant growth and improve water-use efficiency and tolerance to various abiotic stresses. These PGPR species improve the water conservation status in many plants and are capable of overproducing biofilms, exopolysaccharides and trehalose, in their cells and improve the root system and soil fertility status. They help plants to improve their root system and maintain its proper cellular moisture status, resultantly improving plant survival under severe water-deficient conditions. Phytohormones are an important component of plant growth and development under drought stress. The PGPR inoculation changes the levels of plant hormones and other metabolites which help in plant adaptations through their response to the plant water balance, nutrient uptake and translocation, gas exchange, and the movement of photosynthates between tissues. In addition, polyamines are also found to be highly associated in enhancing the water balances and promote the growth of the plants.

In future research, studies can be focused on how PGPR can alter metabolic profiling in plants under water deficiency and on examining further the gene expression or protein changes that are directly involved in the production of these metabolites. It is also important to unravel the complex genetic network and metabolic-interacting events which mediate the host–microbe interactions.

Author Contributions: Conceived of the presented idea, N.K.; developed the theory and performed the computations, N.K. and S.A.; writing—original draft preparation, N.K., S.A., A.M., M.A.S. writing—review and editing, N.K., S.A., H.T., S.L., H.Y. and S.A.; supervision, N.K.; verified the numerical results, N.K., S.A., H.T., S.L., M.A.S., H.Y. and A.M. All authors have read and agreed to the published version of the manuscript.

Funding: This research received no external funding.

Conflicts of Interest: The authors declare no conflict of interest.

References

1. Campbell, A. The use of wild food plants, and drought in Botswana. *J. Arid. Environ.* **1986**, *11*, 81–91. [CrossRef]
2. Christensen, J.H.; Christensen, O.B. A summary of the PRUDENCE model projections of changes in European climate by the end of this century. *Clim. Chang.* **2007**, *81*, 7–30. [CrossRef]
3. Hsiao, T.C.; Xu, L. Sensitivity of growth of roots versus leaves to water stress: Biophysical analysis and relation to water transport. *J. Exp. Bot.* **2000**, *51*, 1595–1616. [CrossRef] [PubMed]
4. Vessey, J.K. Plant growth promoting rhizobacteria as biofertilizers. *Plant Soil* **2003**, *255*, 571–586. [CrossRef]
5. Zhao, J.; Ren, W.; Zhi, D.; Wang, L.; Xia, G. Arabidopsis DREB1A/CBF3 bestowed transgenic tall fescue increased tolerance to drought stress. *Plant Cell Rep.* **2007**, *26*, 1521–1528. [CrossRef]
6. Rampino, P.; Pataleo, S.; Gerardi, C.; Mita, G.; Perrotta, C. Drought stress response in wheat: Physiological and molecular analysis of resistant and sensitive genotypes. *Plant Cell Environ.* **2006**, *29*, 2143–2152. [CrossRef]
7. Khan, N.; Bano, A. Effects of exogenously applied salicylic acid and putrescine alone and in combination with rhizobacteria on the phytoremediation of heavy metals and chickpea growth in sandy soil. *Int. J. Phytoremediat.* **2018**, *16*, 405–414. [CrossRef] [PubMed]
8. Lafitte, H.R.; Yongsheng, G.; Yan, S.; Li, Z.-K. Whole plant responses, key processes, and adaptation to drought stress: The case of rice. *J. Exp. Bot.* **2006**, *58*, 169–175. [CrossRef]
9. Kamara, A.Y.; Ekeleme, F.; Chikoye, D.; Omoigui, L.O. Planting Date and Cultivar Effects on Grain Yield in Dryland Corn Production. *Agron. J.* **2009**, *101*, 91–98. [CrossRef]

10. Atouei, M.T.; Pourbabaee, A.A.; Shorafa, M. Alleviation of Salinity Stress on Some Growth Parameters of Wheat by Exopolysaccharide-Producing Bacteria. *Iran. J. Sci. Technol. Trans. A Sci.* **2019**, *43*, 2725–2733. [CrossRef]

11. Dimkpa, C.; Merten, D.; Svatoš, A.; Büchel, G.; Kothe, E. Siderophores mediate reduced and increased uptake of cadmium byStreptomyces tendaeF4 and sunflower (Helianthus annuus), respectively. *J. Appl. Microbiol.* **2009**, *107*, 1687–1696. [CrossRef]

12. Farooq, M.; Hussain, M.; Siddique, K.H.M. Drought Stress in Wheat during Flowering and Grain-filling Periods. *Crit. Rev. Plant Sci.* **2014**, *33*, 331–349. [CrossRef]

13. Nair, P.R. Agroecosystem management in the 21st century: It is time for a paradigm shift. *J. Trop. Agric.* **2008**, *46*, 1–12.

14. Guo, P.; Baum, M.; Grando, S.; Ceccarelli, S.; Bai, G.; Li, R.; Von Korff, M.; Varshney, R.K.; Graner, A.; Valkoun, J. Differentially expressed genes between drought-tolerant and drought-sensitive barley genotypes in response to drought stress during the reproductive stage. *J. Exp. Bot.* **2009**, *60*, 3531–3544. [CrossRef]

15. Venuprasad, R.; Lafitte, H.R.; Atlin, G.N. Response to Direct Selection for Grain Yield under Drought Stress in Rice. *Crop. Sci.* **2007**, *47*, 285–293. [CrossRef]

16. Gaffney, J.; Schussler, J.; Löffler, C.; Cai, W.; Paszkiewicz, S.; Messina, C.D.; Groeteke, J.; Keaschall, J.; Cooper, M. Industry-Scale Evaluation of Maize Hybrids Selected for Increased Yield in Drought-Stress Conditions of the US Corn Belt. *Crop. Sci.* **2015**, *55*, 1608–1618. [CrossRef]

17. Akram, R.; Natasha; Fahad, S.; Hashmi, M.Z.; Wahid, A.; Adnan, M.; Mubeen, M.; Khan, N.; Rehmani, M.I.A.; Awais, M.; et al. Trends of electronic waste pollution and its impact on the global environment and ecosystem. *Environ. Sci. Pollut. Res.* **2019**, *26*, 16923–16938. [CrossRef]

18. Denef, K.; Roobroeck, D.; Wadu, M.C.M.; Lootens, P.; Boeckx, P. Microbial community composition and rhizodeposit-carbon assimilation in differently managed temperate grassland soils. *Soil. Biol. Biochem.* **2009**, *41*, 144–153. [CrossRef]

19. Nasim, W.; Amin, A.; Fahad, S.; Awais, M.; Khan, N.; Mubeen, M.; Wahid, A.; Rehman, M.H.; Ihsan, M.Z.; Ahmad, S.; et al. Future risk assessment by estimating historical heat wave trends with projected heat accumulation using SimCLIM climate model in Pakistan. *Atmos. Res.* **2018**, *205*, 118–133. [CrossRef]

20. Ngumbi, E.; Kloepper, J. Bacterial-mediated drought tolerance: Current and future prospects. *Appl. Soil Ecol.* **2016**, *105*, 109–125. [CrossRef]

21. Gontia-Mishra, I.; Sapre, S.; Sharma, A.; Tiwari, S. Amelioration of drought tolerance in wheat by the interaction of plant growth-promoting rhizobacteria. *Plant Biol.* **2016**, *18*, 992–1000. [CrossRef]

22. Yasmin, F.; Othman, R.; Sijam, K.; Saad, M.S. Effect of PGPR inoculation on growth and yield of sweetpotato. *J. Biol. Sci.* **2007**, *7*, 421–424.

23. Bresson, J.; Varoquaux, F.; Bontpart, T.; Touraine, B.; Vile, D. The PGPR strain P hyllobacterium brassicacearum STM 196 induces a reproductive delay and physiological changes that result in improved drought tolerance in A rabidopsis. *New Phytol.* **2013**, *200*, 558–569. [CrossRef]

24. Barnawal, D.; Singh, R.; Singh, R.P. Role of Plant Growth Promoting Rhizobacteria in Drought Tolerance. In *PGPR Amelioration in Sustainable Agriculture*; Elsevier BV: Amsterdam, The Netherlands, 2019; pp. 107–128.

25. Yadav, S.K. Cold stress tolerance mechanisms in plants. A review. *Agron. Sustain. Dev.* **2010**, *30*, 515–527. [CrossRef]

26. Kaushal, M.; Wani, S.P. Plant-growth-promoting rhizobacteria: Drought stress alleviators to ameliorate crop production in drylands. *Ann. Microbiol.* **2015**, *66*, 35–42. [CrossRef]

27. Wang, C.; Guo, Y.; Wang, C.; Liu, H.; Niu, D.; Wang, Y.; Guo, J. Enhancement of tomato (Lycopersicon esculentum) tolerance to drought stress by plant-growth-promoting rhizobacterium (PGPR) Bacillus cereus AR156. *J. Agric. Biotechnol.* **2012**, *20*, 1097–1105.

28. Barnawal, D.; Bharti, N.; Pandey, S.S.; Pandey, A.; Chanotiya, C.S.; Kalra, A. Plant growth-promoting rhizobacteria enhance wheat salt and drought stress tolerance by altering endogenous phytohormone levels and TaCTR1/TaDREB2 expression. *Physiol. Plant.* **2017**, *161*, 502–514. [CrossRef]

29. Liu, F.; Xing, S.; Ma, H.; Du, Z.; Ma, B. Cytokinin-producing, plant growth-promoting rhizobacteria that confer resistance to drought stress in Platycladus orientalis container seedlings. *Appl. Microbiol. Biotechnol.* **2013**, *97*, 9155–9164. [CrossRef]

30. Vaishnav, A.; Choudhary, D.K. Regulation of Drought-Responsive Gene Expression in Glycine max L. Merrill is Mediated Through Pseudomonas simiae Strain AU. *J. Plant Growth Regul.* **2018**, *38*, 333–342. [CrossRef]

31. Ansari, F.A.; Ahmad, I. Alleviating Drought Stress of Crops through PGPR: Mechanism and Application. In *Microbial Interventions in Agriculture and Environment*; Springer Science and Business Media LLC: Berlin/Heidelberg, Germany, 2019; pp. 341–358.

32. Zheng, W.; Zeng, S.; Bais, H.; Lamanna, J.M.; Hussey, D.S.; Jacobson, D.L.; Jin, Y. Plant Growth-Promoting Rhizobacteria (PGPR) Reduce Evaporation and Increase Soil Water Retention. *Water Resour. Res.* **2018**, *54*, 3673–3687. [CrossRef]

33. Kumar, A.; Patel, J.S.; Meena, V.S.; Srivastava, R. Recent advances of PGPR based approaches for stress tolerance in plants for sustainable agriculture. *Biocatal. Agric. Biotechnol.* **2019**, *20*, 101271. [CrossRef]

34. Armada, E.; Roldán, A.; Azcon, R. Differential Activity of Autochthonous Bacteria in Controlling Drought Stress in Native Lavandula and Salvia Plants Species Under Drought Conditions in Natural Arid Soil. *Microb. Ecol.* **2013**, *67*, 410–420. [CrossRef]

35. Venkateswarlu, B.; Shanker, A.K. Climate change and agriculture: Adaptation and mitigation stategies. *Indian J. Agron.* **2009**, *54*, 226–230.

36. Niu, X.; Song, L.; Xiao, Y.; Ge, W. Drought-Tolerant Plant Growth-Promoting Rhizobacteria Associated with Foxtail Millet in a Semi-arid Agroecosystem and Their Potential in Alleviating Drought Stress. *Front. Microbiol.* **2018**, *8*, 2580. [CrossRef]

37. Augé, R.M. Water relations, drought and vesicular-arbuscular mycorrhizal symbiosis. *Mycorrhiza* **2001**, *11*, 3–42. [CrossRef]

38. Piccoli, P.; Bottini, R. Abiotic Stress Tolerance Induced by Endophytic PGPR. In *Soil Biology*; Springer Science and Business Media LLC: Berlin/Heidelberg, Germany, 2013; pp. 151–163.

39. Panda, R.; Das, M.; Nayak, S. Estimation and optimization of exopolysaccharide production from rice rhizospheric soil and its interaction with soil carbon pools. *Rhizosphere* **2020**, *14*, 100206. [CrossRef]

40. Calvo-Polanco, M.; Sánchez-Romera, B.; Aroca, R.; Asins, M.J.; Declerck, S.; Dodd, I.C.; Martinez-Andujar, C.; Albacete, A.; Ruiz-Lozano, J.M. Exploring the use of recombinant inbred lines in combination with beneficial microbial inoculants (AM fungus and PGPR) to improve drought stress tolerance in tomato. *Environ. Exp. Bot.* **2016**, *131*, 47–57. [CrossRef]

41. Naseem, H.; Ahsan, M.; Shahid, M.A.; Khan, N. Exopolysaccharides producing rhizobacteria and their role in plant growth and drought tolerance. *J. Basic Microbiol.* **2018**, *58*, 1009–1022. [CrossRef]

42. Joshi, R.; Wani, S.H.; Singh, B.; Bohra, A.; Dar, Z.A.; Lone, A.A.; Pareek, A.; Singla-Pareek, S.L. Transcription Factors and Plants Response to Drought Stress: Current Understanding and Future Directions. *Front. Plant Sci.* **2016**, *7*, 1029. [CrossRef]

43. Jaleel, C.A.; Manivannan, P.A.R.A.M.A.S.I.V.A.M.; Wahid, A.; Farooq, M.; Al-Juburi, H.J.; Somasundaram, R.A.M.A.M.U.R.T.H.Y.; Panneerselvam, R. Drought stress in plants: A review on morphological characteristics and pigments composition. *Int. J. Agric. Biol.* **2009**, *11*, 100–105.

44. Lucy, M. *Management Strategies for Balance Herbicide in Chickpeas*; GRDC: Canberra, Australia, 2004.

45. Zhao, P.; Liu, P.; Shao, J.; Li, C.; Wang, B.; Guo, X.; Yan, B.; Xia, Y.; Peng, M. Analysis of different strategies adapted by two cassava cultivars in response to drought stress: Ensuring survival or continuing growth. *J. Exp. Bot.* **2014**, *66*, 1477–1488. [CrossRef] [PubMed]

46. Pallai, R.; Hynes, R.K.; Verma, B.; Nelson, L.M. Phytohormone production and colonization of canola (Brassica napusL.) roots byPseudomonas fluorescens6-8 under gnotobiotic conditions. *Can. J. Microbiol.* **2012**, *58*, 170–178. [CrossRef] [PubMed]

47. Kasim, W.A.; Osman, M.E.; Omar, M.N.; El-Daim, I.A.A.; Bejai, S.; Meijer, J. Control of Drought Stress in Wheat Using Plant-Growth-Promoting Bacteria. *J. Plant Growth Regul.* **2012**, *32*, 122–130. [CrossRef]

48. Todaka, D.; Shinozaki, K.; Yamaguchi-Shinozaki, K. Recent advances in the dissection of drought-stress regulatory networks and strategies for development of drought-tolerant transgenic rice plants. *Front. Plant Sci.* **2015**, *6*, 84. [CrossRef]

49. Szota, C.; Farrell, C.; Williams, N.S.G.; Arndt, S.K.; Fletcher, T.D. Drought avoiding plants with low water use can achieve high rainfall retention without jeopardising survival on green roofs. *Sci. Total. Environ.* **2017**, *603*, 340–351. [CrossRef] [PubMed]

50. Dos Reis, S.P.; Marques, D.N.; Lima, A.M.; De Souza, C.R.B. Plant Molecular Adaptations and Strategies Under Drought Stress. In *Drought Stress Tolerance in Plants*; Springer Science and Business Media LLC: Berlin/Heidelberg, Germany, 2016; Volume 2, pp. 91–122.

51. García, J.E.; Maroniche, G.; Creus, C.; Suarez-Rodríguez, R.; Ramirez-Trujillo, J.A.; Groppa, M.D. In vitro PGPR properties and osmotic tolerance of different Azospirillum native strains and their effects on growth of maize under drought stress. *Microbiol. Res.* **2017**, *202*, 21–29. [CrossRef]

52. Schellenbaum, L.; Muller, J.; Boller, T.; Wiemken, A.; Schuepp, H. Effects of drought on non-mycorrhizal and mycorrhizal maize: Changes in the pools of non-structural carbohydrates, in the activities of invertase and trehalase, and in the pools of amino acids and imino acids. *New Phytol.* **1998**, *138*, 59–66. [CrossRef]

53. Khan, N.; Bano, A. Rhizobacteria and Abiotic Stress Management. In *Plant Growth Promoting Rhizobacteria for Sustainable Stress Management*; Springer: Singapore, 2019; pp. 65–80.

54. Ikram, M.; Ali, N.; Guljan, F.; Khan, N. Endophytic fungal diversity and their interaction with plants for agriculture sustainability under stressful condition. *Recent Pat. Food Nutr. Agric.* **2019**, *10*, 1. [CrossRef]

55. Shulaev, V. Metabolomics technology and bioinformatics. *Brief. Bioinform.* **2006**, *7*, 128–139. [CrossRef]

56. Uppalapati, S.R.; Ishiga, Y.; Doraiswamy, V.; Bedair, M.; Mittal, S.; Chen, J.; Chen, R. Loss of abaxial leaf epicuticular wax in Medicago truncatula irg1/palm1 mutants results in reduced spore differentiation of anthracnose and nonhost rust pathogens. *Plant Cell* **2012**, *24*, 353–370. [CrossRef]

57. Lamizadeh, E.; Enayatizamir, N.; Motamedi, H. Isolation and Identification of Plant Growth-Promoting Rhizobacteria (PGPR) from the Rhizosphere of Sugarcane in Saline and Non-Saline Soil. *Int. J. Curr. Microbiol. Appl. Sci.* **2016**, *5*, 1072–1083. [CrossRef]

58. Pagnani, G.; Galieni, A.; Stagnari, F.; Pellegrini, M.; Del Gallo, M.; Pisante, M. Open field inoculation with PGPR as a strategy to manage fertilization of ancient Triticum genotypes. *Biol. Fertil. Soils* **2019**, *56*, 1–14. [CrossRef]

59. Rezaei-Chiyaneh, E.; Amirnia, R.; Machiani, M.A.; Javanmard, A.; Maggi, F.; Morshedloo, M.R. Intercropping fennel (*Foeniculum vulgare* L.) with common bean (*Phaseolus vulgaris* L.) as affected by PGPR inoculation: A strategy for improving yield, essential oil and fatty acid composition. *Sci. Hortic.* **2020**, *261*, 108951. [CrossRef]

60. Rubin, R.L.; Van Groenigen, K.J.; Hungate, B.A. Plant growth promoting rhizobacteria are more effective under drought: A meta-analysis. *Plant Soil* **2017**, *416*, 309–323. [CrossRef]

61. Kohler, J.; Hernández, J.-A.; Caravaca, F.; Roldán, A. Plant-growth-promoting rhizobacteria and arbuscular mycorrhizal fungi modify alleviation biochemical mechanisms in water-stressed plants. *Funct. Plant Biol.* **2008**, *35*, 141–151. [CrossRef]

62. Odoh, C.K. Plant Growth Promoting Rhizobacteria (PGPR): A Bioprotectant bioinoculant for Sustainable Agrobiology. A Review. *Int. J. Adv. Res. Biol. Sci.* **2017**, *4*, 123–142. [CrossRef]

63. Swamy, M.K.; Akhtar, M.S.; Sinniah, U.R. Response of PGPR and AM Fungi Toward Growth and Secondary Metabolite Production in Medicinal and Aromatic Plants. In *Plant, Soil and Microbes*; Springer Science and Business Media LLC: Berlin/Heidelberg, Germany, 2016; pp. 145–168.

64. Rouphael, Y.; Franken, P.; Schneider, C.; Schwarz, D.; Giovannetti, M.; Agnolucci, M.; De Pascale, S.; Bonini, P.; Colla, G. Arbuscular mycorrhizal fungi act as biostimulants in horticultural crops. *Sci. Hortic.* **2015**, *196*, 91–108. [CrossRef]

65. Kenneth, O.C.; Nwadibe, E.C.; Kalu, A.U.; Unah, U.V. Plant Growth Promoting Rhizobacteria (PGPR): A Novel Agent for Sustainable Food Production. *Am. J. Agric. Biol. Sci.* **2019**, *14*, 35–54. [CrossRef]

66. Prasad, M.; Srinivasan, R.; Chaudhary, M.; Choudhary, M.; Jat, L.K. Plant Growth Promoting Rhizobacteria (PGPR) for Sustainable Agriculture. In *PGPR Amelioration in Sustainable Agriculture*; Elsevier BV: Amsterdam, The Netherlands, 2019; pp. 129–157.

67. Yang, J.; Kloepper, J.W.; Ryu, C.-M. Rhizosphere bacteria help plants tolerate abiotic stress. *Trends Plant Sci.* **2009**, *14*, 1–4. [CrossRef]

68. Pavlova, A.; Leontieva, M.; Smirnova, T.; Kolomeitseva, G.; Netrusov, A.; Tsavkelova, E. Colonization strategy of the endophytic plant growth-promoting strains of *Pseudomonas fluorescens* and *Klebsiella oxytoca* on the seeds, seedlings and roots of the epiphytic orchid, Dendrobium nobileLindl. *J. Appl. Microbiol.* **2017**, *123*, 217–232. [CrossRef]

69. Singh, J.S.; Singh, D.P. Plant Growth Promoting Rhizobacteria (PGPR): Microbes in Sustainable Agriculture. In *Management of Microbial Resources in the Environment*; Springer Science and Business Media LLC: Berlin/Heidelberg, Germany, 2013; pp. 361–385.

70. Gerhardt, K.E.; MacNeill, G.J.; Gerwing, P.D.; Greenberg, B.M. Phytoremediation of Salt-Impacted Soils and Use of Plant Growth-Promoting Rhizobacteria (PGPR) to Enhance Phytoremediation. In *Phytoremediation*; Springer Science and Business Media LLC: Berlin/Heidelberg, Germany, 2017; Volume 5, pp. 19–51.

71. Agrawal, R.; Satlewal, A. Characterization of Plant Growth-Promoting Rhizobacteria (PGPR): A Perspective of Conventional Versus Recent Techniques. In *Applied Bioremediation and Phytoremediation*; Springer Science and Business Media LLC: Berlin/Heidelberg, Germany, 2015; Volume 44, pp. 471–485.

72. Saravanakumar, D.; Samiyappan, R. ACC deaminase from Pseudomonas fluorescens mediated saline resistance in groundnut (Arachis hypogea) plants. *J. Appl. Microbiol.* **2007**, *102*, 1283–1292. [CrossRef]

73. Lynch, J.M.; Whipps, J.M. Substrate flow in the rhizosphere. *Plant Soil* **1990**, *129*, 1–10. [CrossRef]

74. Rezazadeh, S.; Aghayari, F.; Paknejad, F.; Rezaee, M. The physiological and biochemical responses of directly seeded and transplanted maize (Zea mays L.) supplied with plant growth-promoting rhizobacteria (PGPR) under water stress. *Plant Physiol.* **2019**, *10*, 3009–3021.

75. Khan, N.; Bano, A. Role of PGPR in the Phytoremediation of Heavy Metals and Crop Growth under Municipal Wastewater Irrigation. In *Phytoremediation*; Springer: Cham, Switzerland, 2018; pp. 135–149.

76. Castillo, P.; Molina, R.; Andrade, A.; Vigliocco, A.; Alemano, S.; Cassán, F. Phytohormones and Other Plant Growth Regulators Produced by PGPR: The Genus Azospirillum. In *Handbook for Azospirillum*; Springer Science and Business Media LLC: Berlin/Heidelberg, Germany, 2015; pp. 115–138.

77. Kumar, A.; Bahadur, I.; Maurya, B.R.; Raghuwanshi, R.; Meena, V.S.; Singh, D.K.; Dixit, J. Does a plant growth-promoting rhizobacteria enhance agricultural sustainability. *J. Pure Appl. Microbiol.* **2015**, *9*, 715–724.

78. Tsukanova, K.A.; Meyer, J.J.M.; Bibikova, T.N. Effect of plant growth-promoting Rhizobacteria on plant hormone homeostasis. *South Afr. J. Bot.* **2017**, *113*, 91–102. [CrossRef]

79. Kang, S.-M.; Khan, A.L.; Waqas, M.; You, Y.-H.; Kim, J.-H.; Kim, J.-G.; Hamayun, M.; Lee, I.-J. Plant growth-promoting rhizobacteria reduce adverse effects of salinity and osmotic stress by regulating phytohormones and antioxidants in Cucumis sativus. *J. Plant Interact.* **2014**, *9*, 673–682. [CrossRef]

80. Maheshwari, D.K.; Dheeman, S.; Agarwal, M. Phytohormone-Producing PGPR for Sustainable Agriculture. In *Bacterial Metabolites in Sustainable Agroecosystem*; Springer Science and Business Media LLC: Berlin/Heidelberg, Germany, 2015; pp. 159–182.

81. Fard, M.D.; Habibi, D.; Fard, F.D. Effect of plant growth promoting rhizobacteria and foliar application of amino acids and silicic acid on antioxidant enzyme activity of wheat under drought stress. *Chem. Eng.* **2011**, *23*, 80–85.

82. Kurepin, L.V.; Zaman, M.; Pharis, R.P. Phytohormonal basis for the plant growth promoting action of naturally occurring biostimulators. *J. Sci. Food Agric.* **2014**, *94*, 1715–1722. [CrossRef]

83. Bhattacharyya, P.N.; Jha, D.K. Plant growth-promoting rhizobacteria (PGPR): Emergence in agriculture. *World J. Microbiol. Biotechnol.* **2011**, *28*, 1327–1350. [CrossRef] [PubMed]

84. Yoneyama, K.; Xie, X.; Kim, H.I.; Kisugi, T.; Nomura, T.; Sekimoto, H.; Yokota, T.; Yoneyama, K. How do nitrogen and phosphorus deficiencies affect strigolactone production and exudation? *Planta* **2012**, *235*, 1197–1207. [CrossRef]

85. Yin, C.; Park, J.-J.; Gang, D.R.; Hulbert, S.H. Characterization of a Tryptophan 2-Monooxygenase Gene from Puccinia graminis f. sp. tritici Involved in Auxin Biosynthesis and Rust Pathogenicity. *Mol. Plant-Microbe Interact.* **2014**, *27*, 227–235. [CrossRef]

86. Duca, D.; Lorv, J.; Patten, C.L.; Rose, D.; Glick, B.R. Indole-3-acetic acid in plant–microbe interactions. *Antonie van Leeuwenhoek* **2014**, *106*, 85–125. [CrossRef]

87. Barash, I.; Manulis-Sasson, S. Recent Evolution of Bacterial Pathogens: The Gall-FormingPantoea agglomeransCase. *Annu. Rev. Phytopathol.* **2009**, *47*, 133–152. [CrossRef]

88. Crépin, A.; Barbey, C.; Beury-Cirou, A.; Hélias, V.; Taupin, L.; Reverchon, S.; Nasser, W.; Faure, D.; Dufour, A.; Orange, N.; et al. Quorum Sensing Signaling Molecules Produced by Reference and Emerging Soft-Rot Bacteria (Dickeya and Pectobacterium spp.) *PLoS ONE* **2012**, *7*, e35176. [CrossRef]

89. Aragón, I.M.; Pérez-Martínez, I.; Moreno-Pérez, A.; Cerezo, M.; Ramos, C. New insights into the role of indole-3-acetic acid in the virulence ofPseudomonas savastanoipv.savastanoi. *FEMS Microbiol. Lett.* **2014**, *356*, 184–192. [CrossRef]

90. McClerklin, S.; Lee, S.G.; Harper, C.P.; Nwumeh, R.; Jez, J.M.; Kunkel, B.N. Pseudomonas syringae DC3000-derived auxin contributes to virulence on Arabidopsis. *PLoS Path.* **2018**, *14*, e1006811. [CrossRef]

91. Broek, A.V.; Gysegom, P.; Ona, O.; Hendrickx, N.; Prinsen, E.; Van Impe, J.; Vanderleyden, J. Transcriptional Analysis of the Azospirillum brasilense Indole-3-Pyruvate Decarboxylase Gene and Identification of a cis-Acting Sequence Involved in Auxin Responsive Expression. *Mol. Plant-Microbe Interact.* **2005**, *18*, 311–323. [CrossRef]

92. Van Puyvelde, S.; Cloots, L.; Engelen, K.; Das, F.; Marchal, K.; Vanderleyden, J.; Spaepen, S. Transcriptome Analysis of the Rhizosphere Bacterium Azospirillum brasilense Reveals an Extensive Auxin Response. *Microb. Ecol.* **2011**, *61*, 723–728. [CrossRef]

93. Donati, A.J.; Lee, H.-I.; Leveau, J.H.J.; Chang, W.-S. Effects of Indole-3-Acetic Acid on the Transcriptional Activities and Stress Tolerance of Bradyrhizobium japonicum. *PLoS ONE* **2013**, *8*, e76559. [CrossRef]

94. Vandana, U.K.; Singha, B.; Gulzar, A.; Mazumder, P. Molecular mechanisms in plant growth promoting bacteria (PGPR) to resist environmental stress in plants. In *Molecular Aspects of Plant Beneficial Microbes in Agriculture*; Elsevier BV: Amsterdam, The Netherlands, 2020; pp. 221–233.

95. Ghosh, D.; Gupta, A.; Mohapatra, S. Dynamics of endogenous hormone regulation in plants by phytohormone secreting rhizobacteria under water-stress. *Symbiosis* **2018**, *77*, 265–278. [CrossRef]

96. Galván, M.; Eduardo, A.; Cortés-Patiño, S.; Romero-Perdomo, F.; Uribe-Vélez, D.; Bashan, Y.; Bonilla, R.R. Proline accumulation and glutathione reductase activity induced by drought-tolerant rhizobacteria as potential mechanisms to alleviate drought stress in Guinea grass. *Appl. Soil Ecol.* **2020**, *147*, 103367. [CrossRef]

97. Shivakumar, S.; Bhaktavatchalu, S. Role of Plant Growth-Promoting Rhizobacteria (PGPR) in the Improvement of Vegetable Crop Production Under Stress Conditions. In *Microbial Strategies for Vegetable Production*; Springer Science and Business Media LLC: Berlin/Heidelberg, Germany, 2017; pp. 81–97.

98. Puga-Freitas, R.; Blouin, M. A review of the effects of soil organisms on plant hormone signalling pathways. *Environ. Exp. Bot.* **2015**, *114*, 104–116. [CrossRef]

99. Ali, S.; Khan, N.; Nouroz, F.; Erum, S.; Nasim, W.; Shahid, M.A. In vitro effects of GA3 on morphogenesis of CIP potato explants and acclimatization of plantlets in field. *Invitro. Cell. Dev. Biol.-Plant* **2018**, *54*, 104–111. [CrossRef]

100. Naseri, R.; Maleki, A.; Naserirad, H.; Shebibi, S.; Omidian, A. Effect of plant growth promoting rhizobacteria (PGPR) on reduction nitrogen fertilizer application in rapeseed (*Brassica napus* L.). *Middle-East J. Sci. Res.* **2013**, *14*, 213–220.

101. Baxter, H.L.; Stewart, C.N. Effects of altered lignin biosynthesis on phenylpropanoid metabolism and plant stress. *Biofuels* **2013**, *4*, 635–650. [CrossRef]

102. Prasad, R.; Kumar, M.; Varma, A. Role of PGPR in Soil Fertility and Plant Health. In *Applied Bioremediation and Phytoremediation*; Springer Science and Business Media LLC: Berlin/Heidelberg, Germany, 2014; pp. 247–260.

103. Ranty, B.; Aldon, D.; Cotelle, V.; Galaud, J.-P.; Thuleau, P.; Mazars, C. Calcium Sensors as Key Hubs in Plant Responses to Biotic and Abiotic Stresses. *Front. Plant Sci.* **2016**, *7*, 327. [CrossRef]

104. Sgroy, V.; Cassán, F.; Masciarelli, O.; Del Papa, M.F.; Lagares, A.; Luna, V. Isolation and characterization of endophytic plant growth-promoting (PGPB) or stress homeostasis-regulating (PSHB) bacteria associated to the halophyte Prosopis strombulifera. *Appl. Microbiol. Biotechnol.* **2009**, *85*, 371–381. [CrossRef]

105. Mazumdar, D.; Saha, S.P.; Ghosh, S. Isolation, screening and application of a potent PGPR for enhancing growth of Chickpea as affected by nitrogen level. *Int. J. Veg. Sci.* **2019**, *26*, 333–350. [CrossRef]

106. Ali, S.; Charles, T.C.; Glick, B.R. Amelioration of high salinity stress damage by plant growth-promoting bacterial endophytes that contain ACC deaminase. *Plant Physiol. Biochem.* **2014**, *80*, 160–167. [CrossRef] [PubMed]

107. Khan, N.; Bano, A. Exopolysaccharide producing rhizobacteria and their impact on growth and drought tolerance of wheat grown under rainfed conditions. *PLoS ONE* **2019**, *14*, e0222302. [CrossRef]

108. Atieno, M.; Herrmann, L.; Nguyen, H.T.; Phan, H.T.; Nguyen, N.K.; Srean, P.; Than, M.M.; Zhiyong, R.; Tittabutr, P.; Shutsrirung, A.; et al. Assessment of biofertilizer use for sustainable agriculture in the Great Mekong Region. *J. Environ. Manag.* **2020**, *275*, 111300. [CrossRef] [PubMed]

109. Alkahtani, M.D.F.; Attia, K.A.; Hafez, Y.M.; Khan, N.; Eid, A.M.; Ali, M.A.M.; Abdelaal, K.A.A. Chlorophyll Fluorescence Parameters and Antioxidant Defense System Can Display Salt Tolerance of Salt Acclimated Sweet Pepper Plants Treated with Chitosan and Plant Growth Promoting Rhizobacteria. *Agronomy* **2020**, *10*, 1180. [CrossRef]

110. Urano, K.; Yoshiba, Y.; Nanjo, T.; Igarashi, Y.; Seki, M.; Sekiguchi, F.; Yamaguchi-Shinozaki, K. Characterization of Arabidopsis genes involved in biosynthesis of polyamines in abiotic stress responses and developmental stages. *Plant Cell Environ.* **2003**, *26*, 1917–1926. [CrossRef]

111. Rahdari, P.; Tavakoli, S.; Hosseini, S.M. Studying of salinity stress effect on germination, proline, sugar, protein, lipid and chlorophyll content in purslane (*Portulaca oleracea* L.) leaves. *J. Stress Physiol. Biochem.* **2012**, *8*, 182–193.

112. Gusain, Y.S.; Singh, U.S.; Sharma, A.K. Bacterial mediated amelioration of drought stress in drought tolerant and susceptible cultivars of rice (*Oryza sativa* L.). *Afr. J. Biotechnol.* **2015**, *14*, 764–773.

113. Vurukonda, S.S.K.P.; Vardharajula, S.; Shrivastava, M.; SkZ, A. Enhancement of drought stress tolerance in crops by plant growth promoting rhizobacteria. *Microbiol. Res.* **2016**, *184*, 13–24. [CrossRef]

114. Kumari, S.; Vaishnav, A.; Jain, S.; Varma, A.; Choudhary, D.K. Bacterial-mediated induction of systemic tolerance to salinity with expression of stress alleviating enzymes in soybean (Glycine max L. Merrill). *J. Plant Growth Regul.* **2015**, *34*, 558–573. [CrossRef]

115. Gou, W.; Tian, L.; Ruan, Z.; Zheng, P.E.N.G.; Chen, F.U.C.A.I.; Zhang, L.; Shi, W. Accumulation of choline and glycinebetaine and drought stress tolerance induced in maize (Zea mays) by three plant growth promoting rhizobacteria (PGPR) strains. *Pak J. Bot.* **2015**, *47*, 581–586.

116. Vigani, G.; Rolli, E.; Marasco, R.; Dell'Orto, M.; Michoud, G.; Soussi, A.; Daffonchio, D. Root bacterial endophytes confer drought resistance and enhance expression and activity of a vacuolar H+-pumping pyrophosphatase in pepper plants. *Environ. Microbiol.* **2019**, *21*, 3212–3228. [CrossRef]

117. SkZ, A.; Vardharajula, S.; Vurukonda, S.S.K.P. Transcriptomic profiling of maize (Zea mays L.) seedlings in response to Pseudomonas putida stain FBKV2 inoculation under drought stress. *Ann. Microbiol.* **2018**, *68*, 331–349. [CrossRef]

118. Lugtenberg, B.; Kamilova, F. Plant-growth-promoting rhizobacteria. *Annu. Rev. Microbiol.* **2009**, *63*, 541–556. [CrossRef] [PubMed]

119. Vílchez, J.I.; García-Fontana, C.; Román-Naranjo, D.; González-López, J.; Manzanera, M. Plant drought tolerance enhancement by trehalose production of desiccation-tolerant microorganisms. *Front. Microbiol.* **2016**, *7*, 1577. [CrossRef] [PubMed]

120. Basra, S.M.; Lovatt, C.J. Exogenous Applications of Moringa Leaf Extract and Cytokinins Improve Plant Growth, Yield, and Fruit Quality of Cherry Tomato. *HortTechnology* **2016**, *26*, 327–337. [CrossRef]

121. Less, H.; Egalili, G. Principal Transcriptional Programs Regulating Plant Amino Acid Metabolism in Response to Abiotic Stresses. *Plant Physiol.* **2008**, *147*, 316–330. [CrossRef] [PubMed]

122. Jaleel, C.A.; Manivannan, P.; Sankar, B.; Kishorekumar, A.; Gopi, R.; Somasundaram, R.; Panneerselvam, R. *Pseudomonas fluorescens* enhances biomass yield and ajmalicine production in Catharanthus roseus under water deficit stress. *Colloids Surfaces B Biointerfaces* **2007**, *60*, 7–11. [CrossRef]

123. Saakre, M.; Baburao, T.M.; Salim, A.P.; Ffancies, R.M.; Achuthan, V.P.; Thomas, G.; Sivarajan, S.R. Identification and Characterization of Genes Responsible for Drought Tolerance in Rice Mediated by Pseudomonas fluorescens. *Rice Sci.* **2017**, *24*, 291–298. [CrossRef]

124. Khan, N.; Ali, S.; Shahid, M.A.; Kharabian-Masouleh, A. Advances in detection of stress tolerance in plants through metabolomics approaches. *Plant Omics* **2017**, *10*, 153–163. [CrossRef]

125. Furlan, A.L.; Bianucci, E.; Castro, S.; Dietz, K.-J. Metabolic features involved in drought stress tolerance mechanisms in peanut nodules and their contribution to biological nitrogen fixation. *Plant Sci.* **2017**, *263*, 12–22. [CrossRef]

126. Bounedjah, O.; Hamon, L.; Savarin, P.; Desforges, B.; Curmi, P.A.; Pastré, D. Macromolecular Crowding Regulates Assembly of mRNA Stress Granules after Osmotic Stress new role for compatible osmolytes. *J. Biol. Chem.* **2012**, *287*, 2446–2458. [CrossRef]

127. Selvakumar, G.; Panneerselvam, P.; Ganeshamurthy, A.N. Bacterial Mediated Alleviation of Abiotic Stress in Crops. In *Bacteria in Agrobiology: Stress Management*; Springer: Berlin/Heidelberg, Germany, 2012; pp. 205–224.

128. Su, Z.; Yacob, A.; Wen, J.; Roerink, G.; He, Y.; Gao, B.; Boogaard, H.; Van Diepen, C. Assessing relative soil moisture with remote sensing data: Theory, experimental validation, and application to drought monitoring over the North China Plain. *Phys. Chem. Earth Parts A/B/C* **2003**, *28*, 89–101. [CrossRef]

129. Ansari, F.A.; Ahmad, I. Biofilm Development, Plant Growth Promoting Traits and Rhizosphere Colonization by Pseudomonas entomophlla FAP1. A Promising PGPR. *Adv. Microbiol.* **2018**, *8*, 235–251. [CrossRef]

130. Khan, N.; Bano, A. Growth and Yield of Field Crops Grown Under Drought Stress Condition Is Influenced by the Application of PGPR. In *Field Crops: Sustainable Management by PGPR*; Springer: Cham, Switzerland, 2019; pp. 337–349.

131. Singh, A.; Chauhan, P.S. Ecological Significance of Soil-Associated Plant Growth-Promoting Biofilm-Forming Microbes for Stress Management. In *Biofilms in Plant and Soil Health*; Wiley: Hoboken, NJ, USA, 2017; pp. 291–326.

132. Alami, Y.; Achouak, W.; Marol, C.; Heulin, T. Rhizosphere Soil Aggregation and Plant Growth Promotion of Sunflowers by an Exopolysaccharide-Producing Rhizobiumsp. Strain Isolated from Sunflower Roots. *Appl. Environ. Microbiol.* **2000**, *66*, 3393–3398. [CrossRef] [PubMed]

133. Bramhachari, P.V.; Nagaraju, G.P.; Kariali, E. Current Perspectives on Rhizobacterial-EPS interactions in Alleviation of Stress Responses: Novel Strategies for Sustainable Agricultural Productivity. In *Role of Rhizospheric Microbes in Soil*; Springer Science and Business Media LLC: Berlin/Heidelberg, Germany, 2018; pp. 33–55.

134. Khan, N.; Bano, A. Modulation of phytoremediation and plant growth by the treatment with PGPR, Ag nanoparticle and untreated municipal wastewater. *Int. J. Phytoremediat.* **2016**, *18*, 1258–1269. [CrossRef]

135. Gontia-Mishra, I.; Sapre, S.; Kachare, S.; Tiwari, S. Molecular diversity of 1-aminocyclopropane-1-carboxylate (ACC) deaminase producing PGPR from wheat (Triticum aestivum L.) rhizosphere. *Plant Soil* **2016**, *414*, 213–227. [CrossRef]

136. Timmusk, S.; Timmusk, K.; Behers, L. Rhizobacterial plant drought stress tolerance enhancement: Towards sustainable water resource management and food security. *J. Food Secur.* **2013**, *1*, 6–9.

137. Bogino, P.; Abod, A.; Nievas, F.; Giordano, W. Water-Limiting Conditions Alter the Structure and Biofilm-Forming Ability of Bacterial Multispecies Communities in the Alfalfa Rhizosphere. *PLoS ONE* **2013**, *8*, e79614. [CrossRef]

138. Kovács, Á.T.; Dragoš, A. Evolved Biofilm: Review on the Experimental Evolution Studies of Bacillus subtilis Pellicles. *J. Mol. Biol.* **2019**, *431*, 4749–4759. [CrossRef]

139. Gauri, S.S.; Mandal, S.M.; Pati, B.R. Impact of Azotobacter exopolysaccharides on sustainable agriculture. *Appl. Microbiol. Biotechnol.* **2012**, *95*, 331–338. [CrossRef]

140. Banerjee, A.; Sarkar, S.; Cuadros-Orellana, S.; Bandopadhyay, R. Exopolysaccharides and Biofilms in Mitigating Salinity Stress: The Biotechnological Potential of Halophilic and Soil-Inhabiting PGPR Microorganisms. In *Microorganisms in Saline Environments: Strategies and Functions*; Springer: Cham, Switzerland, 2019; pp. 133–153.

141. Din, B.U.; Sarfraz, S.; Xia, Y.; Kamran, M.A.; Javed, M.T.; Sultan, T.; Munis, M.F.H.; Chaudhary, H.J. Mechanistic elucidation of germination potential and growth of wheat inoculated with exopolysaccharide and ACC- deaminase producing Bacillus strains under induced salinity stress. *Ecotoxicol. Environ. Saf.* **2019**, *183*, 109466. [CrossRef]

142. Arshad, M.; Shaharoona, B.; Mahmood, T. Inoculation with Pseudomonas spp. Containing ACC-Deaminase Partially Eliminates the Effects of Drought Stress on Growth, Yield, and Ripening of Pea (Pisum sativum L.). *Pedosphere* **2008**, *18*, 611–620. [CrossRef]

143. Gupta, G.; Snehi, S.K.; Singh, V. Role of PGPR in Biofilm Formations and Its Importance in Plant Health. *Biofilms Plant Soil Health* **2017**, *27*, 27–42. [CrossRef]

144. Chen, L.; Dodd, I.C.; Theobald, J.C.; Belimov, A.A.; Davies, W.J. The rhizobacterium Variovorax paradoxus 5C-2, containing ACC deaminase, promotes growth and development of Arabidopsis thaliana via an ethylene-dependent pathway. *J. Exp. Bot.* **2013**, *64*, 1565–1573. [CrossRef]

145. Zahir, Z.A.; Munir, A.; Asghar, H.N.; Shaharoona, B.; Arshad, M. Effectiveness of rhizobacteria containing ACC deaminase for growth promotion of peas (Pisum sativum) under drought conditions. *J. Microbiol. Biotechnol.* **2008**, *18*, 958–963.

146. Belimov, A.A.; Dodd, I.C.; Hontzeas, N.; Theobald, J.C.; Safronova, V.I.; Davies, W.J. Rhizosphere bacteria containing 1-aminocyclopropane-1-carboxylate deaminase increase yield of plants grown in drying soil via both local and systemic hormone signalling. *New Phytol.* **2009**, *181*, 413–423. [CrossRef]

147. Sharma, P.; Khanna, V.; Kumari, P. Efficacy of aminocyclopropane-1-carboxylic acid (ACC)-deaminase-producing rhizobacteria in ameliorating water stress in chickpea under axenic conditions. *Afr. J. Microbiol. Res.* **2013**, *7*, 5749–5757.

148. Ghosh, D.; Gupta, A.; Mohapatra, S. A comparative analysis of exopolysaccharide and phytohormone secretions by four drought-tolerant rhizobacterial strains and their impact on osmotic-stress mitigation in Arabidopsis thaliana. *World J. Microbiol. Biotechnol.* **2019**, *35*, 90. [CrossRef]

149. Kousar, B.; Bano, A.; Khan, N. PGPR Modulation of Secondary Metabolites in Tomato Infested with Spodoptera litura. *Agronomy* **2020**, *10*, 778. [CrossRef]

150. Bessadok, K.; Navarro-Torre, S.; Pajuelo, E.; Mateos-Naranjo, E.; Redondo-Gómez, S.; Caviedes, M.Á.; Fterich, A.; Mars, M.; Rodríguez-Llorente, I.D. The ACC-Deaminase Producing Bacterium Variovorax sp. CT7.15 as a Tool for Improving Calicotome villosa Nodulation and Growth in Arid Regions of Tunisia. *Microorganisms* **2020**, *8*, 541. [CrossRef]
151. Arora, N.K.; Fatima, T.; Mishra, J.; Mishra, I.; Verma, S.; Verma, R.; Verma, M.; Bhattacharya, A.; Verma, P.; Mishra, P.; et al. Halo-tolerant plant growth promoting rhizobacteria for improving productivity and remediation of saline soils. *J. Adv. Res.* **2020**, *26*, 69–82. [CrossRef]
152. Grover, M.; Ali, S.Z.; Sandhya, V.; Rasul, A.; Venkateswarlu, B. Role of microorganisms in adaptation of agriculture crops to abiotic stresses. *World J. Microbiol. Biotechnol.* **2011**, *27*, 1231–1240. [CrossRef]
153. Upadhyay, S.K.; Singh, J.; Singh, D. Exopolysaccharide-Producing Plant Growth-Promoting Rhizobacteria under Salinity Condition. *Pedosphere* **2011**, *21*, 214–222. [CrossRef]
154. Scandalios, J.G. Regulation and Properties of Plant Catalases. In *Causes of Photooxidative Stress and Amelioration of Defense Systems in Plants*; CRC Press: Boca Raton, FL, USA, 2019; pp. 275–316.
155. Murphy, D.V.; Stockdale, E.A.; Brookes, P.C.; Goulding, K. Impact of Microorganisms on Chemical Transformations in Soil. In *Soil Biological Fertility*; Springer Science and Business Media LLC: Berlin/Heidelberg, Germany, 2007; pp. 37–59.
156. Abawi, G.; Widmer, T. Impact of soil health management practices on soilborne pathogens, nematodes and root diseases of vegetable crops. *Appl. Soil Ecol.* **2000**, *15*, 37–47. [CrossRef]
157. Bowler, C.; Montagu, M.V.; Inze, D. Superoxide dismutase and stress tolerance. *Ann. Rev. Plant Physiol. Mol. Biol.* **1992**, *43*, 83–116. [CrossRef]
158. Shahid, M.A.; Sarkhosh, A.; Khan, N.; Balal, R.M.; Ali, S.; Rossi, L.; Gómez, C.; Mattson, N.; Nasim, W.; Garcia-Sanchez, F. Insights into the Physiological and Biochemical Impacts of Salt Stress on Plant Growth and Development. *Agronomy* **2020**, *10*, 938. [CrossRef]
159. Deng, Y.; Chen, C.; Zhao, Z.; Zhao, J.; Jacq, A.; Huang, X.; Yang, Y. The RNA Chaperone Hfq Is Involved in Colony Morphology, Nutrient Utilization and Oxidative and Envelope Stress Response in Vibrio alginolyticus. *PLoS ONE* **2016**, *11*, e0163689. [CrossRef]
160. Guterman, L. *Distortions to Agricultural Incentives in Colombia*; Agricultural Distortions Working Paper Series No. 1856-2016-152637; Word Bank, December 2007. Available online: https://ideas.repec.org/p/ags/wbadwp/48392.html (accessed on 5 September 2020).
161. Mayak, S.; Tirosh, T.; Glick, B.R. Plant growth-promoting bacteria that confer resistance to water stress in tomatoes and peppers. *Plant Sci.* **2004**, *166*, 525–530. [CrossRef]
162. Qureshi, M.A.; Iqbal, A.; Akhtar, N.; Shakir, M.A.; Khan, A. Co-inoculation of phosphate solubilizing bacteria and rhizobia in the presence of L-tryptophan for the promotion of mash bean (*Vigna mungo* L.). *Soil Environ.* **2012**, *31*, 47–54.
163. Saikia, J.; Sarma, R.K.; Dhandia, R.; Yadav, A.; Bharali, R.; Gupta, V.K.; Saikia, R. Alleviation of drought stress in pulse crops with ACC deaminase producing rhizobacteria isolated from acidic soil of Northeast India. *Sci. Rep.* **2018**, *8*, 1–16. [CrossRef]
164. Zhu, M.-L.; Wu, X.-Q.; Wang, Y.-H.; Dai, Y. Role of Biofilm Formation by Bacillus pumilus HR10 in Biocontrol against Pine Seedling Damping-Off Disease Caused by Rhizoctonia solani. *Forest* **2020**, *11*, 652. [CrossRef]
165. Rana, K.L.; Kour, D.; Yadav, A.N.; Yadav, N.; Saxena, A.K. Agriculturally important microbial biofilms: Biodiversity, ecological significances, and biotechnological applications. In *New and Future Developments in Microbial Biotechnology and Bioengineering: Microbial Biofilms*; Elsevier BV: Amsterdam, The Netherlands, 2020; pp. 221–265.
166. Cohen, A.C.; Bottini, R.; Piccoli, P. Role of Abscisic Acid Producing PGPR in Sustainable Agriculture. In *Sustainable Development and Biodiversity*; Springer Science and Business Media LLC: Berlin/Heidelberg, Germany, 2015; pp. 259–282.
167. Kalam, S.; Basu, A.; Ankati, S. Plant Root-Associated Biofilms in Bioremediation. *Biofilms Plant Soil Health* **2017**, *337*, 337–355. [CrossRef]
168. Sarma, B.K.; Yadav, S.K.; Singh, D.P.; Singh, H.B. Rhizobacteria mediated induced systemic tolerance in plants: Prospects for abiotic stress management. In *Bacteria in Agrobiology: Stress Management*; Springer: Berlin/Heidelberg, Germany, 2012; pp. 225–238.

169. Primo, E.D.; Ruiz, F.; Masciarelli, O.; Giordano, W. Biofilm Formation and Biosurfactant Activity in Plant-Associated Bacteria. In *Sustainable Development and Biodiversity*; Springer Science and Business Media LLC: Berlin/Heidelberg, Germany, 2015; Volume 12, pp. 337–349.

170. Kour, D.; Rana, K.L.; Kaur, T.; Yadav, N.; Yadav, A.N.; Rastegari, A.A.; Saxena, A.K. Microbial biofilms: Functional annotation and potential applications in agriculture and allied sectors. In *New and Future Developments in Microbial Biotechnology and Bioengineering: Microbial Biofilms*; Elsevier BV: Amsterdam, The Netherlands, 2020; pp. 283–301.

171. Khan, N.; Zandi, P.; Ali, S.; Mehmood, A.; Shahid, M.A.; Yang, J. Impact of Salicylic Acid and PGPR on the Drought Tolerance and Phytoremediation Potential of Helianthus annus. *Front. Microbiol.* **2018**, *9*, 2507. [CrossRef]

172. Gupta, S.; Pandey, S. Unravelling the biochemistry and genetics of ACC deaminase-An enzyme alleviating the biotic and abiotic stress in plants. *Plant Gene* **2019**, *18*, 100175. [CrossRef]

173. Selvakumar, G.; Bhatt, R.M.; Upreti, K.K.; Bindu, G.H.; Shweta, K. Citricoccus zhacaiensis B-4 (MTCC 12119) a novel osmotolerant plant growth promoting actinobacterium enhances onion (Allium cepa L.) seed germination under osmotic stress conditions. *World J. Microbiol. Biotechnol.* **2015**, *31*, 833–839. [CrossRef]

174. Kim, S.-G.; Kim, S.-Y.; Park, C.-M. A membrane-associated NAC transcription factor regulates salt-responsive flowering via FLOWERING LOCUS T in Arabidopsis. *Planta* **2007**, *226*, 647–654. [CrossRef]

175. Marulanda, A.; Porcel, R.; Barea, J.M.; Azcón, R. Drought Tolerance and Antioxidant Activities in Lavender Plants Colonized by Native Drought-tolerant or Drought-sensitive Glomus Species. *Microb. Ecol.* **2007**, *54*, 543–552. [CrossRef]

176. Vu, B.; Chen, M.; Crawford, R.J.; Ivanova, E.P. Bacterial Extracellular Polysaccharides Involved in Biofilm Formation. *Molecules* **2009**, *14*, 2535–2554. [CrossRef]

177. Barriuso, J.; Ramos-Solano, B.; Lucas, J.A.; Lobo, A.P.; Garca-Villaraco, A.; Maero, F.J.G.; García-Villaraco, A.; Gutierrez-Mañero, F.J. Ecology, Genetic Diversity and Screening Strategies of Plant Growth Promoting Rhizobacteria (PGPR). *Plant-Bact. Interact.* **2008**, *4*, 1–17. [CrossRef]

178. Chen, S.; Zhao, H.; Zou, C.; Li, Y.; Chen, Y.; Wang, Z.; Jiang, Y.; Liu, A.; Zhao, P.; Wang, M.; et al. Combined Inoculation with Multiple Arbuscular Mycorrhizal Fungi Improves Growth, Nutrient Uptake and Photosynthesis in Cucumber Seedlings. *Front. Microbiol.* **2017**, *8*, 2516. [CrossRef]

179. Colinet, H.; Larvor, V.; Laparie, M.; Renault, D. Exploring the plastic response to cold acclimation through metabolomics. *Funct. Ecol.* **2012**, *26*, 711–722. [CrossRef]

180. Glick, B.R. The enhancement of plant growth by free-living bacteria. *Can. J. Microbiol.* **1995**, *41*, 109–117. [CrossRef]

181. Kloepper, J.W.; Lifshitz, R.; Zablotowicz, R.M. Free-living bacterial inocula for enhancing crop productivity. *Trends Biotechnol.* **1989**, *7*, 39–44. [CrossRef]

182. Rahdari, P.; Hoseini, S.M. Drought stress: A review. *Int. J. Agron. Plant Prod.* **2012**, *3*, 443–446.

183. Shafeek, M.R.; Helmy, Y.I.; Omar, N.M. Use of some bio-stimulants for improving the growth, yield and bulb quality of onion plants (Allium cepa L.) under sandy soil conditions. *Middle East J. Appl. Sci.* **2015**, *5*, 68–75.

184. Nelson, L.M. Plant Growth Promoting Rhizobacteria (PGPR): Prospects for New Inoculants. *Crop Manag.* **2004**, *3*, 1–7. [CrossRef]

185. Kumar, A.; Prakash, A.; Johri, B.N. Bacillus as PGPR in Crop Ecosystem. In *Bacteria in Agrobiology: Crop Ecosystems*; Springer Science and Business Media LLC: Berlin/Heidelberg, Germany, 2011; pp. 37–59.

186. Singh, G.G.S. Plant Growth Promoting Rhizobacteria (PGPR): Current and Future Prospects for Development of Sustainable Agriculture. *J. Microb. Biochem. Technol.* **2015**, *7*, 096–102. [CrossRef]

187. Khan, N.; Ebano, A. Role of plant growth promoting rhizobacteria and Ag-nano particle in the bioremediation of heavy metals and maize growth under municipal wastewater irrigation. *Int. J. Phytoremediat.* **2015**, *18*, 211–221. [CrossRef]

188. Arora, N.K.; Tewari, S.; Singh, S.; Lal, N.; Maheshwari, D.K. PGPR for protection of plant health under saline conditions. In *Bacteria in Agrobiology: Stress Management*; Springer: Berlin/Heidelberg, Germany, 2012; pp. 239–258.

189. Sharma, I.P.; Chandra, S.; Kumar, N.; Chandra, D. PGPR: Heart of Soil and Their Role in Soil Fertility. In *Agriculturally Important Microbes for Sustainable Agriculture*; Springer Science and Business Media LLC: Berlin/Heidelberg, Germany, 2017; pp. 51–67.

190. Costa, O.Y.A.; Raaijmakers, J.M.; Kuramae, E.E. Microbial Extracellular Polymeric Substances: Ecological Function and Impact on Soil Aggregation. *Front. Microbiol.* **2018**, *9*, 1636. [CrossRef]

191. Chenu, C. Clay- or sand-polysaccharide associations as models for the interface between micro-organisms and soil: Water related properties and microstructure. *Geoderma* **1993**, *56*, 143–156. [CrossRef]

192. Cheynier, V.; Comte, G.; Davies, K.M.; Lattanzio, V.; Martens, S. Plant phenolics: Recent advances on their biosynthesis, genetics, and ecophysiology. *Plant Physiol. Biochem.* **2013**, *72*, 1–20. [CrossRef]

193. Durán, P.; Acuña, J.; Armada, E.; López-Castillo, O.; Cornejo, P.; Mora, M.; Azcón, R. Inoculation with selenobacteria and arbuscular mycorrhizal fungi to enhance selenium content in lettuce plants and improve tolerance against drought stress. *J. Soil Sci. Plant Nutr.* **2016**, *16*, 211–225. [CrossRef]

194. Nakashima, K.; Ito, Y.; Yamaguchi-Shinozaki, K. Transcriptional Regulatory Networks in Response to Abiotic Stresses in Arabidopsis and Grasses: Figure 1. *Plant Physiol.* **2009**, *149*, 88–95. [CrossRef] [PubMed]

195. Zhu, J.-K. Abiotic Stress Signaling and Responses in Plants. *Cell* **2016**, *167*, 313–324. [CrossRef] [PubMed]

196. Wang, M.; Li, P.; Li, C.; Pan, Y.; Jiang, X.; Zhu, D.; Zhao, Q.; Yu, J. SiLEA14, a novel atypical LEA protein, confers abiotic stress resistance in foxtail millet. *BMC Plant Biol.* **2014**, *14*, 290. [CrossRef] [PubMed]

197. Takahashi, F.; Shinozaki, K. Long-distance signaling in plant stress response. *Curr. Opin. Plant Biol.* **2019**, *47*, 106–111. [CrossRef]

198. Ali, S.; Hayat, K.; Iqbal, A.; Xie, L. Implications of Abscisic Acid in the Drought Stress Tolerance of Plants. *Agronomy* **2020**, *10*, 1323. [CrossRef]

199. Khan, N.; Bano, A.; Zandi, P. Effects of exogenously applied plant growth regulators in combination with PGPR on the physiology and root growth of chickpea (Cicer arietinum) and their role in drought tolerance. *J. Plant Interact.* **2018**, *1*, 239–247. [CrossRef]

200. Zong, N.; Wang, H.; Li, Z.; Ma, L.; Xie, L.; Pang, J.; Fan, Y.; Zhao, J. Maize NCP1 negatively regulates drought and ABA responses through interacting with and inhibiting the activity of transcription factor ABP9. *Plant Mol. Biol.* **2020**, *102*, 339–357. [CrossRef]

201. Wang, D.; Pan, Y.; Zhao, X.; Zhu, L.; Fu, B.; Li, Z. Genome-wide temporal-spatial gene expression profiling of drought responsiveness in rice. *BMC Genom.* **2011**, *12*, 149. [CrossRef]

202. Kasual, M. Microbes in cahoots with plants: MIST to hit the jackpot of agricultural productivity during drought. *Int. J. Mol. Sci.* **2019**, *20*, 1769.

203. Wang, Y.; Ohara, Y.; Nakayashiki, H.; Tosa, Y.; Mayama, S. Microarray Analysis of the Gene Expression Profile Induced by the Endophytic Plant Growth-Promoting Rhizobacteria, Pseudomonas fluorescens FPT9601-T5 in Arabidopsis. *Mol. Plant-Microbe Interact.* **2005**, *18*, 385–396. [CrossRef]

204. Wu, Y.; Ma, L.; Liu, Q.; Vestergård, M.; Topalovic, O.; Wang, Q.; Feng, Y. The plant-growth promoting bacteria promote cadmium uptake by inducing a hormonal crosstalk and lateral root formation in a hyperaccumulator plant Sedum alfredii. *J. Hazard. Mater.* **2020**, *395*, 122661. [CrossRef] [PubMed]

205. Urano, K.; Kurihara, Y.; Seki, M.; Shinozaki, K. 'Omics' analyses of regulatory networks in plant abiotic stress responses. *Curr. Opin. Plant Biol.* **2010**, *13*, 132–138. [CrossRef] [PubMed]

206. Mhlongo, M.I.; Piater, L.A.; Madala, N.E.; Labuschagne, N.; Dubery, I.A. The Chemistry of Plant–Microbe Interactions in the Rhizosphere and the Potential for Metabolomics to Reveal Signaling Related to Defense Priming and Induced Systemic Resistance. *Front. Plant Sci.* **2018**, *9*, 112. [CrossRef] [PubMed]

207. Meena, K.K.; Sorty, A.M.; Bitla, U.M.; Choudhary, K.; Gupta, P.; Pareek, A.; Singh, D.P.; Prabha, R.; Sahu, P.K.; Gupta, V.K.; et al. Abiotic Stress Responses and Microbe-Mediated Mitigation in Plants: The Omics Strategies. *Front. Plant Sci.* **2017**, *8*, 172. [CrossRef]

208. Khan, N.; Bano, A.; Curá, J.A. Role of Beneficial Microorganisms and Salicylic Acid in Improving Rainfed Agriculture and Future Food Safety. *Microorganisms* **2020**, *8*, 1018. [CrossRef]

Publisher's Note: MDPI stays neutral with regard to jurisdictional claims in published maps and institutional affiliations.

Article

α-Tocopherol Foliar Spray and Translocation Mediates Growth, Photosynthetic Pigments, Nutrient Uptake, and Oxidative Defense in Maize (*Zea mays* L.) under Drought Stress

Qasim Ali [1,*], Muhammad Tariq Javed [1], Muhammad Zulqurnain Haider [1], Noman Habib [1], Muhammad Rizwan [2], Rashida Perveen [3], Shafaqat Ali [2,4,*], Mohammed Nasser Alyemeni [5], Hamed A. El-Serehy [6] and Fahad A. Al-Misned [6]

[1] Department of Botany, Government College University, New Campus, Jhang Road, Faisalabad 38000, Pakistan; mtariqjaved@gcuf.edu.pk (M.T.J.); dr_mzhaider@yahoo.com (M.Z.H.); nomi4442003@yahoo.com (N.H.)

[2] Department of Environmental Sciences and Engineering, Government College University, Allama Iqbal Road, Faisalabad 38000, Pakistan; mrazi1532@yahoo.com

[3] Department of Physics, University of Agriculture, Faisalabad 38040, Pakistan; 2007ag942@uaf.edu.pk

[4] Department of Biological Sciences and Technology, China Medical University, Taichung 40402, Taiwan

[5] Botany and Microbiology Department, College of Science, King Saud University, Riyadh 11451, Saudi Arabia; mnyemeni@ksu.edu.sa

[6] Department of Zoology, College of Science, King Saud University, Riyadh 11451, Saudi Arabia; helserehy@ksu.edu.sa (H.A.E.-S.); almisned@ksu.edu.sa (F.A.A.-M.)

[*] Correspondence: drqasimali@gcuf.edu.pk (Q.A.); shafaqataligill@gcuf.edu.pk or shafaqataligill@yahoo.com (S.A.)

Received: 21 June 2020; Accepted: 10 August 2020; Published: 21 August 2020

Abstract: A pot experiment was conducted to assess the induction of drought tolerance in maize by foliar-applied α-tocopherol at early growth stage. Experiment was comprised two maize cultivars (Agaiti-2002 and EV-1098), two water stress levels (70% and 100% field capacity), and two α-tocopherol levels (0 mmol and 50 mmol) as foliar spray. Experiment was arranged in a completely randomized design in factorial arrangement with three replications of each treatment. α-tocopherol was applied foliary at the early vegetative stage. Water stress reduced the growth of maize plants with an increase in lipid peroxidation in both maize cultivars. Contents of non-enzymatic antioxidants and activities of antioxidant enzymes increased in studied plant parts under drought, while the nutrient uptake was decreased. Foliary-applied α-tocopherol improved the growth of both maize cultivars, associated with improvements in photosynthetic pigment, water relations, antioxidative mechanism, and better nutrient acquisition in root and shoot along with tocopherol contents and a decrease in lipid peroxidation. Furthermore, the increase of tocopherol levels in roots after α-Toc foliar application confers its basipetal translocation. In conclusion, the findings confer the role of foliar-applied α-tocopherol in the induction of drought tolerance of maize associated with tissue specific improvements in antioxidative defense mechanism through its translocation.

Keywords: α-Tocopherol; antioxidants; drought; nutrient dynamics; tissue specific response

1. Introduction

Among different environmental adversities, water shortage is of major focus, which has hampered the production of global agricultural systems [1,2]. At a global level, about 45% of all land is prevailed by drought [3]. On the other hand, an estimated increase in the world population will be about

2.5 billion in the next 25 years, which will exert huge pressure on agriculture to fulfill world food demand and on the available freshwater resources. From the last two decades, Pakistan has also faced the problem of agricultural productivity to fulfill the food demand of the sixth largest population in the world. With an agriculture-based economy, Pakistan is predominantly categorized as arid country lying within the geographic coordinates of 23.38°–30.25° N latitude and 61.78°–74.30° E longitude, with a total land area of 796,096 km^2 [4]. The interannual rainfall variability makes the arid region (covering 75% land area of Pakistan) more susceptible to drought risks. Approximately 34.15 Mha of land area is in agriculture use, and uncultivated land is 23.60 Mha. About 25% of the cultivated land is rainfed, which plays a vital role in the country's economy [5]. Due to the major contribution of the agriculture sector in Pakistan's economy, Pakistan is more susceptible to drought risks [6]. In recent decades, unexpected and rapid changes in climate have severely affected socioeconomic and environmental conditions in Pakistan [6]. The major cause of drought stress is a decrease in soil water contents in combination with evaporation due to over-changing atmospheric conditions [7]. Shortage of water induces drastic changes in plants' physio-biochemical and molecular properties that ultimately affects all growth stages of a plant's life cycle, including the final yield [8,9]. At present (and in the near future), the maintenance of crop productivity for a large population under limited water supply is a challenge for the researchers working in the agriculture sector.

To survive under water deficit conditions, plants have manipulated metabolic defensive systems/mechanisms, which are species- and genotype-specific [10–12]. Disturbance in plant water status is the important effect of water shortage that triggers various other metabolic processes to survive under water stress [10,11,13,14]. It results in reduced growth and final grain yield due to perturbations in photosynthesis by disturbances in the biosynthesis of photosynthetic pigments and impaired nutrient uptake [15,16]. Water deficit conditions cause sub-optimal plant photosynthetic efficiency due to limited CO_2 diffusion into the leaves due to less stomatal opening or reduced Rubisco activity [17,18]. To cope with a stressful environment, the plant mineral uptake mechanism plays a significant role in improving resistance [19,20]. Generally, under water deficit conditions, mineral uptake and transport reduces due to a decrease in the nutrient diffusion rate [16,21]. Among different nutrients, potassium (K^+), nitrogen (N), calcium (Ca^{2+}), phosphorus (P), and magnesium (Mg^{2+}) have prime importance due to their vital functions in plant physio-biochemical processes [14,20,22].

The stress tolerance in crop plants that results in better yield is growth-stage and species-specific [23,24]. The seedling stage is of prime importance in potentially contributing to better seed yield. Uniform crop stand leads to better yield, which depends on better seedling growth [25,26]. Furthermore, it was found that at early seedling stages, crop cultivars with better antioxidative potential are more drought tolerant than cultivars with less antioxidative activity [27] because the disturbances in different physiological mechanisms results in another secondary stress (oxidative stress) by excessive production of reactive oxygen species (ROS).

Stress-induced oxidative stress due to production of ROS ($O_2{}^-$, H_2O_2, OH^-, and O*) is a common phenomenon in all organisms [28]. Over-production of ROS damages membrane lipids [28], thereby increasing malondialdehyde (MDA) accumulation due to limited activity of antioxidative defense mechanisms [29]. Under stressful environments, the levels of MDA are parallel with antioxidant enzyme activities, which are the indices to assess the status of the extent of damage due to the overproduction of ROS [30]. Other than the levels of antioxidant enzyme to counteract ROS damage, plants also have non-enzymatic antioxidative defense mechanisms such as the production of ascorbic acid, phenolic acid, carotenoids, tocopherols, etc. [31]. Furthermore, it is well known that the antioxidative defensive phenomenon is inter-species, cultivar, and growth-stage-specific. However, most of the higher yielding genotypes are not drought tolerant when considering stress tolerance mechanisms [32].

Furthermore, some high-yield crop cultivars are deficit with regard to such anti-stress mechanisms [14,28,33]. For the induction of drought tolerance, different approaches have been adopted, including the exogenous application of secondary growth metabolic compounds [34–37]. Exogenous application such as the foliar spray of different secondary metabolites of which the plant is in deficit is considered as an effective means among others for stress tolerance induction [38,39]. It is well known that foliar application of such compounds is translocatable to different plant parts. Furthermore, after their translocation to different plant parts, they play a potential role in the induction of drought tolerance. Along with modulating metabolic activities, plants also control their own metabolisms [34,40]. Among different secondary metabolic compounds, the tocopherols are lipophilic in nature and scavenge ROS, with the ability to recycle themselves and, as a result, reduce lipid peroxidation. Tocopherols belong to a family of eight members including α, β, γ, and δ tocopherols, along with their respective precursors (tocotrienols) that have high antioxidative activity and protect plants from stress through different metabolic processes [41]. Among these, α-Toc is largely known as vitamin E, with large antioxidant potential in comparison with other family members, but the production of α-Toc to reduce oxidative damage is cultivar-specific [42]. However, α-Toc exogenous application was found to be helpful for stress tolerance induction. For example, in wheat, exogenous application of α-Toc improved salt-stress tolerance [43]. In flax, genotypes foliar-applied tocopherol significantly improved salt stress tolerance [44]. Most of the studies presented are regarding salt tolerance induction and the application of α-Toc on adult-stage plants, and there is a lack of knowledge regarding its exogenous use at other growth stages. However, the discovery of the proper plant stage for better drought-stress induction through exogenous use of this compound is of prime importance [45].

Furthermore, there are missing gaps in understanding the proper physiological mechanism for the induction of stress tolerance at different growth stages by the exogenous use of organic compounds like that of α-Toc, also considering its translocation to specific plant parts. Therefore, the current work was aimed to quantify to which extent the foliar applied α-Toc could modulate growth in water-stressed maize plants and when it should be applied in the early growth stage. The goal of the study was to draw parallels among tissue-specific alterations in endogenous tocopherol levels, antioxidative defense mechanisms, and nutrient mobility patterns after α-Toc foliar application in maize plants grown in a drought-stressed rhizosphere. The research outcomes are helpful for optimizing strategies for growing maize with limited irrigation and in semi-arid and arid regions for better growth and production.

Maize (*Zea mays* L.) is the third most commonly produced cereal, after wheat and rice. It has a potential to grow in a wide range of environmental conditions and has gained great economic priority due to its potential nutritional quality all over the world, including in Pakistan [46]. In Pakistan, 1.016 million hectares are under maize cultivation, and 35% of the total cultivated area is rainfed, which is now facing problems in getting better production under dry environmental spells; this situation has further become more severe due to the present change in environmental conditions. Maize kernels are not only good and cheap source of carbohydrates but are also a rich source of carotenoids, proteins, and edible oil. However, due to changes in rainfall patterns along with the shortage of fresh water for irrigation, its production is under threat, along with that of other crops.

2. Materials and Methods

The present experiment was arranged in the research area of the Department of Botany, Government College University Faisalabad, Pakistan, (latitude 30°30 N, longitude 73°10 E, and altitude 213 m) under natural environmental conditions during August–September 2018. To avoid disturbances due to rain, the experimental area was covered with a polyethylene sheet. The design of the experiment was completely randomized in factorial arrangement, with three replications of each treatment. The experiment consisted of two drought levels (control and 70% field capacity), two highly yielding maize genotypes (EV-1098 and Agaiti-2002), and two levels of α-Toc (0 mmol and 50 mmol) in solution form applied as foliar spray with three replications of each treatment. The 70% field capacity used in the present study was selected following some earlier studies [47,48]. These two maize cultivars selected for study are used frequently in breeding programs to produce high-yielding hybrid genotypes. The experimental unit was comprised a total 24 equal-size plastic pots (28 cm × 30 cm), each filled with 10 kg soil. The soil was fully irrigated with canal water before seed sowing. When the soil was at field capacity, seeds of both maize genotypes were hand sown. Before sowing, the soil was prepared well by hand digging. The seeds of both maize genotypes were purchased from Maize and Millet Research Institute, Yousafwala Sahiwal, Pakistan. Ten healthy seeds were sown in each pot. After five days of the completion of seed germination, five seedlings per pot were maintained by thinning. The water stress treatment was started just after the thinning of the seedlings by controlling the irrigation of half of the pots at 70% field capacity, and the other half of the pots were treated as control plants and irrigated to maintain 100% field capacity. Average mean daily length was 13/11 h, mean minimum and maximum day/night temperatures were 38 ± 3/30 ± 3 °C and 25 ± 2.5/20 ± 2.5 °C, respectively, the mean relative humidity during whole experiment (at daytime) was 50%. During the whole experimental period the averaged photosynthetically available radiation (PAR) measured at noon was varied from 794 μmolm^{-2} s^{-1} to 1154 μmolm^{-2} s^{-1}. Soil moisture content was maintained on daily basis and using a tensiometer, (Irrometer, Model RT-12 inch Riverside, CA, USA). Ten days after thinning, the seedlings were supplied exogenously as foliar spray with 0 mmol and 50 mmol solution of α-toc. Foliar spray of α-Toc solution was done in evening before sunset for the maximum absorption of the solution in leaf. The spray of α-Toc solution was made only once during the whole experimental period. An aliquot of 50 mL solution of each of α-Toc level was applied manually per replicate as foliar spray that costs only \$0.015 USD for six plants and \$65 USD per acre. The solution was prepared by dissolving the required measured quantity in minimal amount of ethanol, and then the final volume was maintained with distilled water. The 0 mmol treatment without α-Toc was considered as control treatment. Before foliar spray, 0.1% of Tween-20 was added as the surfactant to the finally prepared solution for the maximum absorption of the solution. The data for varying attributes was calculated after 15 days of α-Toc foliar spray. Fresh leaf material was taken in liquid nitrogen and stored at −80 °C for different biochemical studies.

2.1. Soil Analysis

The soil used was sandy loam with a saturation percentage of 47.5, average pH, and the ECe of the soil solution was 7.63 ds.m^{-1} and 0.045 ds.m^{-1}, respectively, organic matter (1.21%), with the available P (0.051 mg kg^{-1}), K (30 mg kg^{-1}), and total N (6.1 mg kg^{-1}). The soil solution had soluble CO_3^{2-} (traces), HCO_3^- (5.01 meq L^{-1}), Cl^- (8.49 meq L^{-1}), SO_4^{-2} (2.01 meq L^{-1}), Na (3.01 meq L^{-1}), $Ca^{2+}+Mg^{2+}$ (13.91 meq L^{-1}), and SAR (0.079 meq L^{-1}).

2.2. Estimation of Different Growth Parameters

Two plants per replicate were uprooted and washed with distilled water for the estimation of different growth attributes. After calculating root and shoot lengths, number of leaves, leaf area, and fresh masses of roots and shoots, the same plants was then oven-dried using an electric oven at 70 °C for 48 h, and their dry masses were calculated.

2.3. Estimation of Leaf Photosynthetic Pigments

For the estimation of leaf chlorophyll (Chl.) *a*, *b*, total Chl, and Chl *a/b*, we followed the method described by Arnon [49]. The content of carotenoids (Car) was estimated following Kirk and Allen [50]. The extraction of the pigments was done using 80% acetone. Briefly, fresh leaf material (0.1 g) was chopped and put in 10 mL acetone for overnight at 4 °C and the absorbance of the extract was read at 663, 645, and 480 nm using a spectrophotometer (Hitachi U-2001, Tokyo, Japan). The quantities were computed using the specific formulae:

$$\text{Chl. } a = [12.7 \text{ (OD 663)} - 2.69 \text{ (OD 645)}] \times V/1000 \times W \tag{1}$$

$$\text{Chl. } b = [22.9 \text{ (OD 645)} - 4.68 \text{ (OD 663)}] \times V/1000 \times W \tag{2}$$

$$\text{Total Chl.} = [20.2 \, (\Delta A645) - 8.02(\Delta A663)] \times v/w \times 1/1000 \tag{3}$$

$$A\ carotenoid\ (\mu g/g\ FW) = \Delta A480 + (0.114 \times \Delta A663) - (0.638 \times \Delta A645) \tag{4}$$

$$Car = A\ Car./Em\ 100\% \times 100 \tag{5}$$

$$Emission = Em\ 100\% = 2500 \tag{6}$$

$$\Delta A = absorbance\ at\ respective\ wavelength \tag{7}$$

$$V = volume\ of\ the\ extract\ (mL) \tag{8}$$

$$W = weight\ of\ the\ fresh\ leaf\ tissue\ (g) \tag{9}$$

2.4. Leaf Relative Water Content (LRWC)

For the estimation of LRWC, the second one from top was used. In first step, after excising the leaf, the fresh weight was measured and tagged with a specific mark. Then, the leaf was soaked in dH_2O for 4 h. Then, the leaf was taken out of the water, it absorbed the extra surface water, and we measured its weight again and termed the result the turgid weight. The same leaf was then oven-dried at 75 °C for 48 h and again weighed and termed this the dry weight of leaf. Then LRWC was estimated using the formula from the obtained data

$$LRWC\ (\%) = \frac{Fresh\ weight\ of\ leaf\ -\ dry\ weight\ of\ leaf}{Turgid\ weight\ of\ leaf\ -\ dry\ weight\ of\ leaf} \times 100 \tag{10}$$

2.5. Leaf Relative Membrane Permeability

We followed the method described by Yang et al. [51] to find out the leaf relative membrane permeability (LRMP). The known amount (0.5 g) of excised leaf was cut into small pieces (approximately 1 cm) and put in test tubes having 20 mL of deionized dH_2O. After vortexing well for 5 s, the EC of the assayed material was measured and termed as EC0. The test tubes containing leaf were then kept at 4 °C for 24 h, and the EC1 was measured. These test tubes containing leaf material were then autoclaved for 30 min at 120 °C and assayed the EC2. The LRMP was measured using the following equation:

$$RMP\ (\%) = \frac{EC1\ -\ EC0}{EC2\ -\ EC0} \times 100 \tag{11}$$

2.6. Estimation of Leaf Malondialdehyde Content

Content of malondialdehyde (MDA) was measured using the method given by Cakmak and Horst [52] as the measure of lipid peroxidation. The trichloroacetic acid (TCA) method was used for the estimation of MDA content. One gram of freshly taken leaf material was ground in TCA (10% solution). The supernatant (0.5 mL) was obtained from the homogenized material and mixed with 3 mL of thiobarbituric acid (TBA), prepared in 20% TCA. Test tubes having the triturate were kept at 95 °C for 50 min and then cooled immediately in chilled water. After centrifugation (10,000× g) of mixture for 10 min, the absorbance of colored part was read at 600 nm and 532 nm. The content of MDA was calculated using the following formula:

$$MDA\ (nmol) = \Delta\ (A532\ nm\ -\ A\ 600\ nm)/1.56 \times 105 \tag{12}$$

Absorption coefficient for the calculation of MDA is 156 $mmol^{-1}\ cm^{-1}$.

2.7. Extraction of Antioxidant Enzymes and Total Soluble Proteins from Different Plant Parts

For the extraction of antioxidant enzymes and total soluble proteins (TSP) from each plant part (root, stem, leaf), fresh material was ground (0.5 g) in chilled (10 mL) 50-mM phosphate buffer (pH 7.8). The mixture was then centrifuged at 10,000× g for 20 min at 4 °C. The supernatant so obtained was then used for the estimation of total soluble proteins (TSP) and estimation of antioxidative enzymes activities.

2.7.1. Estimation of Total Soluble Proteins in Different Plant Parts

TSP in the buffer extracts was estimated following the method of Bradford [53]. The absorbance of the triturate was measured at 595 nm, and the quantities of the TSP in samples were computed using a series of protein standards (200–1400 mg/kg) prepared from analytical-grade bovine serum albumin (BSA).

2.7.2. Estimation of the Activities of Superoxide Dismutase, Peroxidase, and Catalase in Different Plant Parts

Activity of superoxide dismutase (SOD) was estimated using the method of Giannopolitis and Ries [54]. The method works based on the principle of photochemical reduction inhibition of nitroblue tetrazolium (NBT), which was used, and absorbance was read at 560 nm using an UV-visible spectrophotometer. However, the method of Chance and Maehly [55] was followed to measure the peroxidase (POD) and catalase (CAT) activities.

2.8. Determination of Non-Enzymatic Antioxidants in Different Plant Parts

Ascorbic acid (ASA) content in different plant parts was determined following Mukherjee and Choudhuri [56] after extraction in TCA. The flavonoid contents in different plant parts were determined following the methods ascribed by Karadeniz et al. [57]. However, the total tocopherol content in different plant parts was assayed following the method of Backer et al. [58]. The contents of ASA, flavonoids, and tocopherol were measured quantitively using the standard curves prepared with known concentration of analytical grade ASA, rutin, and α-toc, respectively, obtained from Sigma-Aldrich Chemie GmbH - Schnelldorf, Germany.

2.9. Determination Mineral Nutrients

2.9.1. Estimation of K^+, Ca^{2+}, and Mg^{2+} in Different Plan Parts

For the estimation of mineral elements in different plant parts, 0.1 g dry material was digested using a 2 mL digestion mixture (prepared from H_2O_2, H_2SO_4, $LiSO_4$, and Se metal). The final volume was maintained 50 mL using a volumetric flask. Flame photometer was used for determination of the contents of K^+ and Ca^{2+}, while of Mg^{2+}, contents were estimated using an Atomic Absorption Spectrophotometer (Hitachi, Model 7JO-8024, Tokyo, Japan).

2.9.2. Determination of N and P

The nitrogen (N) content from the digested material was determined following the method described by Bremner and Keeney [59]. The phosphorus (P) content from the digested material was estimated using Barton's reagent by spectrophotometrically, and quantity was estimated spectroscopically.

2.10. Statistical Analysis

Microsoft Excel software 2010, US was used for the estimation of means and standard errors from the collected. To find the significant differences among treatments, analysis of variance (ANOVA) was performed using Co-Stat window version 6.3, Cohorts, Berkeley, California, USA. To compare means for significant differences among treatments at 5% levels, Tukey's test (HSD-test) was performed. Correlations and PCA analysis were performed of the studied parameters using the XLSTAT software, version 2014.5, New York, USA and the significance among the generated values of each attribute was found using the Spearman's correlation table.

3. Results

3.1. Different Growth Attributes and Content of Leaf Photosynthetic Pigments of Water-Stressed Maize Plants Foliar-Applied Alpha Tocopherol

Data for different morphological and growth attributes as presented in Table 1, which shows that water shortage imposed significant adverse impacts on the lengths of shoots and roots, the number of leaves, and the total leaf area of both maize cultivars (Table 2). Foliar application of α-Toc significantly reduced the adverse impacts of water shortage on these growth attributes for both cultivars, and both wheat genotypes showed similar increasing response in this regard. However, root length and root fresh weights remained unaffected due to foliar spray of alpha tocopherols.

Reduced water supply significantly decreased the roots and shoots fresh and dry masses of both maize genotypes (Tables 1 and 2). Foliar spray of α-Toc significantly reduced the adverse effects of water stress on these growth attributes. A similar increase in the root and shoot fresh and dry biomasses was found in both genotypes due to foliary-supplied α-Toc, both under stressed and non-stressed conditions.

Leaf Chl. *a*, Chl. *b*, and total Chl. contents decreased significantly of both maize cultivars when grown under limited water supply. Both maize genotypes showed similar decreasing trend in leaf Chl. *a*, Chl. *b*, and total Chl. contents under drought stress. Significant increasing the effect of foliary-supplied α-Toc was recorded on the contents of leaf Chl. *a*, Chl. *b*, and total Chl. of both maize cultivars both under non-stressed and stressed conditions (Tables 2 and 3).

Chl. *a/b* ratio was also significantly affected due to drought stress in both maize genotypes. An improvement in Chl. *a/b* was recorded in cv. EV-1098, but the opposite was true for cv. Agaiti-2002. α-Toc foliar spray significantly improved the leaf Chl. *a/b* only in cv. Agaiti-2002 under conditions of limited water supply. However, the carotenoids content in different plant parts increased significantly due to water shortage in both maize genotypes (Tables 2 and 3), but this increase was cultivar and plant-part-specific. A significantly higher increase in carotenoids was found in leaf and root of cv. Agaiti-2002 in comparison to cv. EV-1098, but in relation with stem carotenoids content, this cultivar-specific difference was not found under drought stress. Foliar spray of α-Toc further enhanced the content of carotenoids in all studied plant parts. Significantly more increase was recorded in the leaf and root of cv. Agaiti-2002 in comparison to cv. EV-1098. However, this improvement in stem carotenoids due to α-Toc foliar application was same in both genotypes. Similar increasing trend in carotenoids under normal irrigation in all studied plant parts was also found in both genotypes due to α-Toc foliar application (Tables 2 and 3).

Table 1. Influence of foliar-applied alpha tocopherols on different growth and morphological attributes of maize cultivars grown under different water regimes (mean $\pm$ SE; $n = 3$). SL = shoot length; RL = root length; NOL = number of leaves; PLA = plant leaf area; SFW = shoot fresh weight; SDW = shoot dry weight; RFW = root fresh weight; RDW = root dry weight; α-toc = alpha tocopherol.

	α-Toc	Cultivars	SL (cm)	RL (cm)	NOL/plant	PLA (cm^2)	SFW (g)	SDW (g)	RFW (g)	RDW (g)
Control	0 mmol	Agaiti-2002	12.23 $\pm$ 1.10 [a]	7.77 $\pm$ 0.48 [a]	3.75 $\pm$ 0.20 [b]	69.10 $\pm$ 5.73 [bc]	2.92 $\pm$ 0.215 [b]	0.37 $\pm$ 0.04 [a]	1.18 $\pm$ 0.15 [a]	0.13 $\pm$ 0.019 [b]
		EV-1098	12.03 $\pm$ 0.67 [a]	6.70 $\pm$ 0.50 [ab]	3.85 $\pm$ 0.20 [b]	62.40 $\pm$ 4.04 [cd]	2.76 $\pm$ 0.224 [bc]	0.32 $\pm$ 0.03 [abc]	1.18 $\pm$ 0.11 [a]	0.13 $\pm$ 0.022 [b]
	50 mmol	Agaiti-2002	13.33 $\pm$ 0.82 [a]	6.20 $\pm$ 0.35 [bc]	4.25 $\pm$ 0.30 [a]	79.98 $\pm$ 5.08 [a]	3.47 $\pm$ 0.402 [a]	0.40 $\pm$ 0.03 [a]	1.39 $\pm$ 0.14 [a]	0.16 $\pm$ 0.014 [a]
		EV-1098	12.85 $\pm$ 1.11 [a]	7.50 $\pm$ 0.79 [a]	4.50 $\pm$ 0.35 [a]	73.60 $\pm$ 0.63 [ab]	3.74 $\pm$ 0.140 [a]	0.38 $\pm$ 0.03 [a]	1.35 $\pm$ 0.15 [a]	0.17 $\pm$ 0.024 [a]
Drought	0 mmol	Agaiti-2002	8.27 $\pm$ 1.11 [d]	4.67 $\pm$ 0.49 [cd]	3.25 $\pm$ 0.13 [c]	45.50 $\pm$ 1.06 [f]	2.28 $\pm$ 0.145 [c]	0.23 $\pm$ 0.03 [bd]	0.74 $\pm$ 0.13 [c]	0.09 $\pm$ 0.021 [d]
		EV-1098	8.45 $\pm$ 0.66 [d]	4.00 $\pm$ 0.39 [d]	3.00 $\pm$ 0.35 [c]	47.50 $\pm$ 1.13 [ef]	2.37 $\pm$ 0.135 [c]	0.24 $\pm$ 0.01 [cd]	0.78 $\pm$ 0.12 [c]	0.10 $\pm$ 0.011 [cd]
	50 mmol	Agaiti-2002	10.30 $\pm$ 1.51 [b]	5.15 $\pm$ 0.59 [e]	3.67 $\pm$ 0.20 [b]	55.00 $\pm$ 1.74 [de]	2.56 $\pm$ 0.136 [b]	0.27 $\pm$ 0.01 [cd]	0.93 $\pm$ 0.16 [b]	0.11 $\pm$ 0.007 [bcd]
		EV-1098	9.77 $\pm$ 1.16 [c]	4.73 $\pm$ 0.37 [cd]	3.75 $\pm$ 0.25 [b]	54.50 $\pm$ 1.07 [def]	2.72 $\pm$ 0.139 [b]	0.28 $\pm$ 0.03 [bcd]	0.99 $\pm$ 0.12 [b]	0.12 $\pm$ 0.018 [bc]
	LSD 5%		1.311	1.11	0.414	4.26	0.50	0.081	0.15	0.025

Values in column with same alphabets in superscript do not differ significantly.

Table 2. Mean squares from analysis of variance of the data for the studied attributes of water stressed maize plants foliar-applied with α-Toc at seedling stage.

SOV	d.f	SL	RL	NOL	LA	SFW	SDW	RFW	RDW
WS	1	66.7 ***	34.08 **	2.37 **	2440 ***	2.85 ***	0.045 **	1.06 ***	0.011 ***
Toc	1	8.13 **	0.02 ns	2.01 **	41.5 ***	1.67 **	0.003*	0.23 ns	0.004 ***
CV	1	0.53 ns	0.42 ns	0.01 ns	37.5 *	0.04 ns	0.007 ns	0.002 ns	5.04×10^{-4} ns
WS * Toc	1	0.54 ns	0.88 ns	0.02 ns	8.16 ns	0.34 ns	0.004 ns	6×10^{-4} ns	3.37×10^{-4} ns
WS * CV	1	0.16 ns	0.66 ns	0.17 ns	121 ***	0.001 ns	0.014 ns	0.010 ns	1.04×10^{-4} ns
Toc * CV	1	0.67 ns	1.92 ns	0.09 ns	0.17 ns	0.08 ns	0.002 ns	2.7×10^{-4} ns	1.04×10^{-4} ns
WS * Toc * CV	1	0.10 ns	2.66 ns	0.04 ns	0.17 ns	0.06 ns	0.002 ns	0.002 ns	1.04×10^{-4} ns
Error	16	0.78	3.31	0.17	7.08	0.13	0.003	0.065	2.17×10^{-4}
SOV	d.f	Protein L	Protein R	Potein S	LRWC	LRMP	Chl. *a*	Chl. *b*	Chl *a/b*
WS	1	46728 ***	6834 ***	2281 ***	3314 ***	517 ***	0.15 ***	0.043 ***	1.270 ***
Toc	1	11051 ***	3978 ***	253 **	172 **	50.0*	0.10 ***	0.015 ***	0.130 ns
CV	1	651.04 **	1395 **	13.5 ns	118 *	0.18 ns	0.01 ns	2.8×10^{-4} ns	0.004 ns
WS * Toc	1	35.04 ns	234 ***	181.5 **	157 ***	76.5 *	2.4×10^{-6} ns	0.004 **	0.290 *
WS * CV	1	1717 ***	513 ***	37.5 ns	22.6ns	0.80 ns	4.8×10^{-4} ns	2.2×10^{-5} ns	0.003 ns
Toc * CV	1	1785 ***	18.4 ns	37.5 ns	1.25 ns	16.1 ns	0.002 ns	5.8×10^{-5} ns	0.010 ns
WS * Toc * CV	1	3.37 ns	408 ***	1.5 ns	6.56 ns	0.19 ns	1.14×10^{-54} ns	0.002 *	0.199 ns
Error	16	62.83	18	19.5	17.06	10.8	0.002	3.1×10^{-4}	0.051

Table 2. *Cont.*

SOV	d.f	Total Chl.	POD R	POD S	POD L	CAT R	CAT S	CAT L	SOD S
WS	1	0.351 ***	2223 ***	260 ***	1932 ***	1220 ***	329.8 ***	2011 ***	270 ***
Toc	1	0.190 ***	198 ***	37.6 ***	146 ***	359 ***	12.01 ns	268 ***	4.74 ns
CV	1	0.008 ns	135 **	149 *	157 **	45.8 **	126.1 **	414 ***	71.4 *
WS * Toc	1	0.004 ns	30.4 ns	1.7×10^{-4} ns	0.89 ns	18.6 ns	1.054 ns	0.219 *	1.1 ns
WS * CV	1	7×10^{-4} ns	9.4 ns	1.50 ns	1.21 ns	3.2 ns	12.07 ns	19.65 *	27.4 ns
Toc * CV	1	0.001 ns	9.3 ns	1.53 ns	1.21 ns	3.2 ns	1.054 ns	0.22 ns	0.39 ns
WS * Toc * CV	1	0.003 ns	3.4 ns	5.93 ns	0.08 ns	3.5 ns	1.032 ns	2.85 ns	0.018 ns
Error	16	0.002	9.5	7.10	12.1	6.9	8.286	8.91	8.485
SOV	d.f	SOD L	SOD R	MDA S	MDA L	MDA R	TOC S	TOC L	TOC R
WS	1	1229 ***	7222 ***	3174 ***	1890 ***	3313 ***	1493 ***	1666 ***	1741 ***
Toc	1	86.98 *	176 **	73.5 **	828 ***	793 ***	83.7 **	1148 ***	489 ***
CV	1	81.40 *	218 **	4.0 *	84.3 ***	384 ***	9.0 ns	60.11 *	55.2 *
WS * Toc	1	107 **	24.7 ns	1.5 ns	108 ***	726 ***	3.24 ns	170 **	1.30 ns
WS * CV	1	0.06 ns	16.1 ns	6.0 ns	18.3 ns	337 ***	0.30 ns	2.66 ns	1.71 ns
Toc * CV	1	62.3 *	0.05 ns	1.5 ns	30.3 *	253 ***	0.42 ns	0.16 ns	1.71 ns
WS * Toc * CV	1	28.8 ns	0.80 ns	1.5 ns	18.3ns	216 ***	3.24 ns	0.66 ns	0.01 ns
Error	16	10.9	15.65	4.9	5.12	8.62	5.52	12.0	11.3
SOV	d.f	AsA S	AsA L	AsA R	Car S	Car L	Car R	Flav S	Flav L
WS	1	14259 ***	23814 ***	26498 ***	5953 ***	10688 ***	5730 ***	45.37 **	84.38 ***
Toc	1	1053.***	14113 ***	4858 ***	1722 ***	4776 ***	273 ns	0.37 ns	18.38 *
CV	1	630 ***	4056 ***	8720 ***	67.3 ns	412.7 *	894 *	9.37 ns	9.375 ns
WS * Toc	1	108 *	937 ***	881 ***	146 ns	265	8.77 ns	0.38 ns	3.35 ns
WS * CV	1	63.3 ns	0.01 ns	930 ***	2.53 ns	39.9	6.20 ns	0.37 ns	0.37 ns
Toc * CV	1	30.4 ns	1.5 ns	12.7 ns	0.26 ns	27.9	8.19 ns	0.38 ns	0.38 ns
WS * Toc * CV	1	30.3 ns	253 *	304 **	0.25 ns	9.9	3.04 ns	0.37 ns	0.37 ns
Error	16	8.62	31.75	29.7	59.4	64.9	160.6	4.5	3.5
SOV	d.f	Flav R	K S	K L	K R	Ca S	Ca L	Ca R	Mg S
WS	1	135.4 ***	261 **	403 ***	396 ***	56.8 ***	172 ***	32.08 ***	0.014 ns
Toc	1	30.3 *	53.2 ns	68.2 *	45.7 ns	1.67 ns	15.6 ***	14.7 **	0.144 ns
CV	1	9.37 ns	2.83 ns	1.82 ns	0.55 ns	2.66 *	4.61 *	0.26 ns	0.768 ns

Table 2. *Cont.*

WS * Toc	1	3.375 ns	0.15 ns	2.95 ns	0.40 ns	6.36 **	0.90 ns	0.011 ns	1.264 *
WS * CV	1	3.375 ns	7.67 ns	0.12 ns	0.22 ns	2.11 ns	1.14 ns	0.206 ns	9×10^{-4} ns
Toc * CV	1	0.375 ns	8.89 ns	11.5 ns	2.01 ns	0.77 ns	0.98 ns	0.061 ns	1.92 **
WS * Toc * CV	1	0.375 ns	0.81 ns	0.26 ns	1.02 ns	0.85 ns	0.14 ns	0.160 ns	0.069 ns
Error	16	4.125	20.1 ns	11.9	14.2	0.57	0.79	1.035	0.221
SOV	**d.f**	**Mg L**	**Mg R**	**P S**	**P L**	**P R**	**N S**	**N L**	**N R**
WS	1	0.29 ns	3.93 ***	10.5 ***	11.7 ***	5.15 *	260 **	321.84 ***	368.8 ***
Toc	1	0.28 ns	0.02 ns	2.45 *	1.15 *	1.56 ns	52.3 ns	45.49 ns	109.4 *
CV	1	2.86 ***	0.01 ns	0.40 ns	0.02 ns	0.36 ns	1.53 ns	26.44 ns	36.03 ns
WS * Toc	1	0.056 ns	0.20 ns	0.34 ns	0.295 ns	0.43 ns	0.58 ns	3.37 ns	1.89 ns
WS * CV	1	2.95 ***	0.16 ns	0.32 ns	0.039 ns	0.001 ns	1.89 ns	3.26 ns	3.07 ns
Toc * CV	1	0.43 ns	0.04 ns	0.30 ns	0.005 ns	0.36 ns	1.73 ns	2.72 ns	3.50 ns
WS * Toc * CV	1	1.23 **	0.12 ns	1.11 ns	0.324 ns	0.35 ns	1.81 ns	1.52 ns	2.14 ns
Error	16	0.11	0.13	0.51	0.203	0.69	21.4ns	11.65	19.3

*, ** and *** = Significant at 0.5, 0.1 and 0.01 levels respectively; ns = non-significant.

Table 3. Influence of foliar-applied alpha tocopherol on photosynthetic pigments of maize cultivars grown under different water regimes (mean ± SE; *n* = 3). Chl. *a* = leaf chlorophyll *a*; Chl. *b* = leaf chlorophyll *b*; Chl *a/b* = chlorophyll *a/b* ratio; Total Chl. = total chlorophyll; Leaf Car = leaf carotenoids; Root Car = root carotenoids; Stem Car = stem carotenoids.

Stress	α-Toc	Cultivars	Chl. *a* (mg/g FW)	Chl. *b* (mg/g FW)	Chl *a/b*	Total Chl. (mg/g FW)	Leaf Car (µg/g FW)	Root Car (µg/g FW)	Stem Car (µg/g FW)
Control	0 mmol	Agaiti-2002	1.52 ± 0.11 [ab]	0.36 ± 0.01 [bc]	4.53 ± 0.16 [a]	1.88 ± 0.12 [ab]	102.16 ± 1.78 [e]	99.26 ± 9.60 [de]	82.70 ± 5.97 [e]
		EV-1098	1.47 ± 0.18 [bc]	0.38 ± 0.02 [bc]	3.87 ± 0.25 [d]	1.85 ± 0.17 [b]	93.52 ± 4.41 [f]	84.16 ± 12.26 [f]	80.41 ± 6.24 [e]
	50 mmol	Agaiti-2002	1.63 ± 0.06 [a]	0.47 ± 0.01 [a]	4.40 ± 0.14 [ab]	2.00 ± 0.16 [ab]	120.27 ± 13.12 [d]	105.34 ± 10.99 [d]	105.00 ± 2.83 [d]
		EV-1098	1.60 ± 0.14 [ab]	0.44 ± 0.01 [ab]	3.64 ± 0.33 [ed]	2.04 ± 0.18 [a]	118.23 ± 4.89 [d]	94.79 ± 6.55 [e]	101.89 ± 4.40 [d]
Drought	0 mmol	Agaiti-2002	1.34 ± 0.04 [cd]	0.33 ± 0.02 [c]	3.62 ± 0.09 [e]	1.67 ± 0.05 [b]	139.98 ± 10.40 [b]	129.65 ± 9.31 [ab]	120.00 ± 5.83 [bc]
		EV-1098	1.31 ± 0.12 [bc]	0.31 ± 0.02 [c]	4.28 ± 0.05 [bc]	1.62 ± 0.10 [c]	127.25 ± 5.61 [c]	118.75 ± 2.98 [c]	116.00 ± 5.75 [c]
	50 mmol	Agaiti-2002	1.46 ± 0.12 [bcd]	0.34 ± 0.02 [c]	4.29 ± 0.16 [bc]	1.80 ± 0.10 [b]	176.08 ± 2.74 [a]	134.73 ± 11.52 [a]	132.00 ± 6.54 [a]
		EV-1098	1.46 ± 0.09 [bcd]	0.35 ± 0.02 [c]	4.17 ± 0.16 [c]	1.81 ± 0.12 [b]	163.33 ± 5.61 [b]	124.26 ± 9.60b [c]	128.00 ± 4.89 [ab]
	LSD 5%		0.15	0.12	0.64	0.52	8.74	11.81	8.15

Values in column with same alphabets in superscript do not differ significantly.

3.2. Leaf Relative Water Content, Leaf Relative Membrane Permeability, Total Soluble Proteins, and H_2O_2 Contents of Leaf Photosynthetic Pigments of Maize Plants Foliar-Applied with Alpha Tocopherol

Data presented in Table 3 reveals that the imposition of water stress decreased the LRWC of both genotypes, and a slightly higher decrease in LRWC was found in cv. EV-1098 in comparison to cv. Agaiti-2002. The foliar application of α-Toc significantly increased the LRWC of both genotypes, and this increase was found only under drought-stressed conditions; both cultivars showed a similar increasing trend in this regard (Tables 2 and 4).

Leaf relative membrane permeability (LRMP) increased significantly under water deficit conditions, and this increase was similar in both maize cultivars. Exogenous application of α-Toc as foliar spray was found to be effective in decreasing the LRMP in both maize cultivars under water-stressed conditions, and both maize cultivars showed similar responses in this regard (Tables 2 and 4).

Drought stress exerted a tissue-specific increment in leaf, root, and stem TSP contents of both genotypes when grown without foliar application of α-Toc. In leaf and root, this improvement in TSP was higher in cv. Agaiti-2002 in comparison to cv. EV-1098, but in relation to stem TSP, both cultivars showed the same increasing trend. Exogenous application of α-Toc further improved TSP accumulation in all studied plant parts in both maize cultivars under stressed and non-stressed conditions. Alpha-toc-induced this improvement in TSP contents was significantly more prominent in leaves of cv. Agaiti-2002 in comparison to cv. EV-1098 under limited water supply, but a similar increasing trend was recorded in root and stem (Tables 2 and 4).

Under stressful conditions, the extent of oxidative damage is measured in terms of MDA contents. The data presented shows that MDA contents in all studied plant parts of both the cultivars increased significantly under limited water supply. α-Toc foliar-application significantly reduced the MDA accumulation in all studied plant parts, and a more prominent reduction was found in leaves in comparison to other plant parts in both maize genotypes (Tables 2 and 4).

Table 4. Influence of foliar application of α-Toc on leaf relative water content, leaf relative membrane permeability, tissue specific total soluble proteins and malondialdehyde content of maize cultivars grown under different water regimes (mean ± SE; $n = 3$). LRWC = leaf relative water content; RMP = leaf relative membrane permeability; Leaf TSP = leaf total soluble proteins; Root TSP = root total soluble proteins; Stem TSP = stem total soluble proteins; Leaf MDA = leaf malondialdehyde; Root MDA = root malondialdehyde; Stem MDA = stem malondialdehyde.

Stress	α-Toc	Cultivars	LRWC (%)	RMP (%)	Leaf TSP (mg/kg FW)	Root TSP (mg/kg FW)	Stem TSP (mg/kg FW)	Leaf MDA (nmol/g FW)	Root MDA (nmol/g FW)	Stem MDA (nmol/g FW)
Control	0 mmol	Agaiti-2002	87.54 ± 1.71 [a]	32.33 ± 3.92 [e]	210.00 ± 29.89 [f]	144.33 ± 7.50 [f]	73.00 ± 15.15 [c]	68.10 ± 1.19 [bc]	66.18 ± 6.40 [c]	62.02 ± 4.48 [cd]
		EV-1098	81.73 ± 5.00 [b]	33.96 ± 2.33 [de]	233.33 ± 23.34 [e]	160.33 ± 4.58 [e]	72.00 ± 9.36 [c]	62.35 ± 2.94 [c]	66.10 ± 8.17 [c]	60.31 ± 5.18 [cd]
	50 mmol	Agaiti-2002	88.36 ± 1.62 [a]	34.83 ± 2.76 [d]	267.33 ± 24.22 [d]	186.67 ± 2.57 [cd]	83.67 ± 9.01 [b]	60.18 ± 8.75 [c]	65.23 ± 7.33 [c]	60.35 ± 2.12 [c]
		EV-1098	81.38 ± 5.36 [b]	33.55 ± 3.19 [de]	257.00 ± 18.26 [e]	182.00 ± 3.45 [d]	86.67 ± 14.81 [b]	55.49 ± 3.26 [bc]	66.53 ± 4.37 [c]	56.42 ± 3.29 [c]
Drought	0 mmol	Agaiti-2002	58.23 ± 3.52 [d]	45.32 ± 0.50 [b]	312.33 ± 31.67 [c]	183.67 ± 10.05 [d]	101.67 ± 16.96 [a]	90.65 ± 6.94 [a]	99.83 ± 6.21 [a]	85.59 ± 4.37 [ab]
		EV-1098	51.34 ± 3.09 [e]	47.88 ± 3.78 [a]	304.67 ± 25.01 [c]	201.33 ± 5.64 [b]	94.33 ± 13.58 [ab]	84.84 ± 3.74 [a]	102.84 ± 1.98 [a]	84.04 ± 4.32 [b]
	50 mmol	Agaiti-2002	67.87 ± 0.62 [c]	40.29 ± 2.08 [c]	376.67 ± 22.30 [a]	196.00 ± 9.98 [bc]	99.00 ± 9.93 [a]	70.72 ± 1.83 [b]	64.3 ± 7.68 [b]	81.92 ± 4.91 [ab]
		EV-1098	66.17 ± 2.02 [c]	39.08 ± 2.44 [c]	330.67 ± 33.06 [b]	227.67 ± 5.81 [a]	98.00 ± 10.98 [a]	72.22 ± 3.74 [b]	92.50 ± 6.40 [b]	80.60 ± 3.67 [a]
	LSD 5%		4.06	3.49	24.44	12.64	10.26	5.82	7.88	6.12

Values in column with same alphabets in superscript do not differ significantly.

3.3. Root, Stem, and Leaf Total Tocopherol (Figure 1A–C); Ascorbic Acid (Figure 1D–F); and Total Flavonoid Contents (Figure 1G–I) of Maize Plants Foliar-Applied with α-Toc

Imposition of water stress significantly increased the accumulation of total Toc contents in the studied plant parts of both maize cultivars. This accumulation in total-Toc content in all studied plant parts was increased further due to the foliar application of α-Toc. This increased accumulation in internal total-Toc in all studied plant parts due to its foliar application was more in root and leaf in comparison to stem in both genotypes under both non-stressed and stressed conditions. α-Toc applied this increase in all studied plant parts and was similar in both maize cultivars (Figure 1A–C).

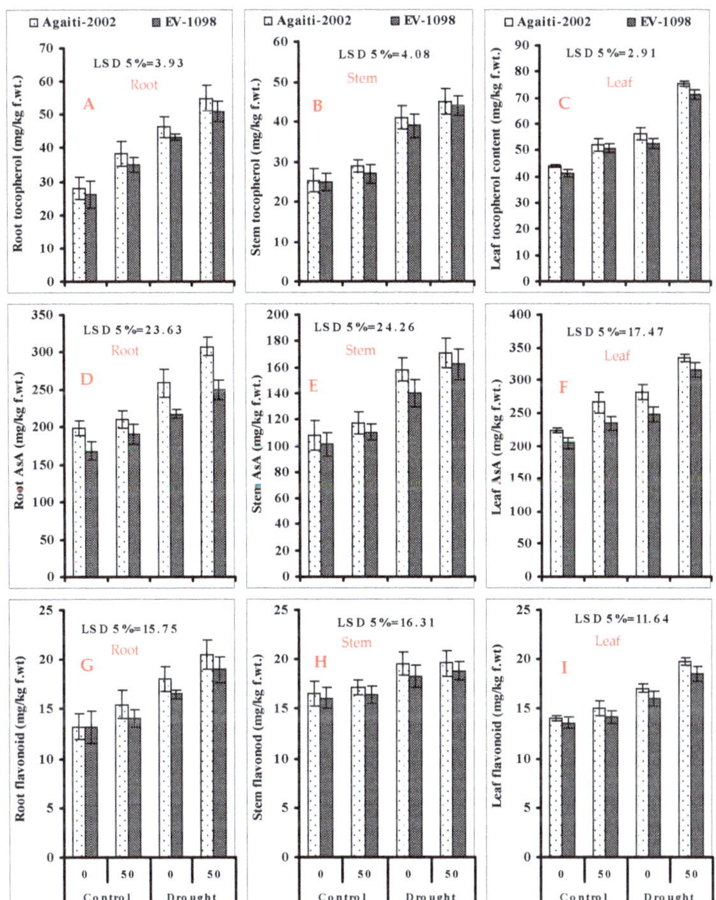

Figure 1. Root, stem, and leaf total-Toc (**A–C**), AsA (**D–F**), and total flavonoids (**G–I**) of maize plants foliar-applied with α-Toc when grown under water deficit conditions (mean ± SE; *n* = 4); AsA = ascorbic acid; 0 and 50 = mmol solution of α-Tocopherol for foliar spray.

AsA and flavonoid contents in different studied plant parts also increased significantly in both genotypes under water deficit conditions, and this improvement in AsA and flavonoid accumulation was more in root and leaf in cv. Agaiti-2002 in comparison to cv. EV-1098 (Figure 1; Table 2). Exogenous application of α-Toc as foliar spray further enhanced the AsA accumulation in all studied plant parts of both maize genotypes; accumulation was higher in cv. Agaiti-2002, both under stressed and non-stressed conditions. However, improvement in flavonoids was found only in the leaf and root of both maize genotypes when grown under water deficit conditions; this improved accumulation in flavonoids was not found in stem flavonoids (Figure 1D–I).

3.4. Activities of CAT (Figure 2A–C), SOD (Figure 2D–F) and POD (2G–I) in Root, Stem, and Leaf of Maize Plants Foliar-Applied with α-Toc

Activities of CAT and SOD in all studied plant parts increased significantly in both genotypes when grown under limited water supply, and comparatively more improvement was found in root and leaf of cv. Agaiti-2002 in comparison with stem. Alpha-toc application further enhanced the CAT and SOD activities in root and leaf in both genotypes, but such improvement in CAT and SOD activities was not found in the stem of both genotypes. In leaf, significantly more improvement in CAT activity due to α-Toc application was recorded in cv. Agaiti-2002 as compared with cv. EV-1098; however, in relation with SOD activity in root and leaf, cv. Agaiti-2002 was superior in comparison to cv. EV-1098 due to α-Toc application (Figure 2A–F; Table 2).

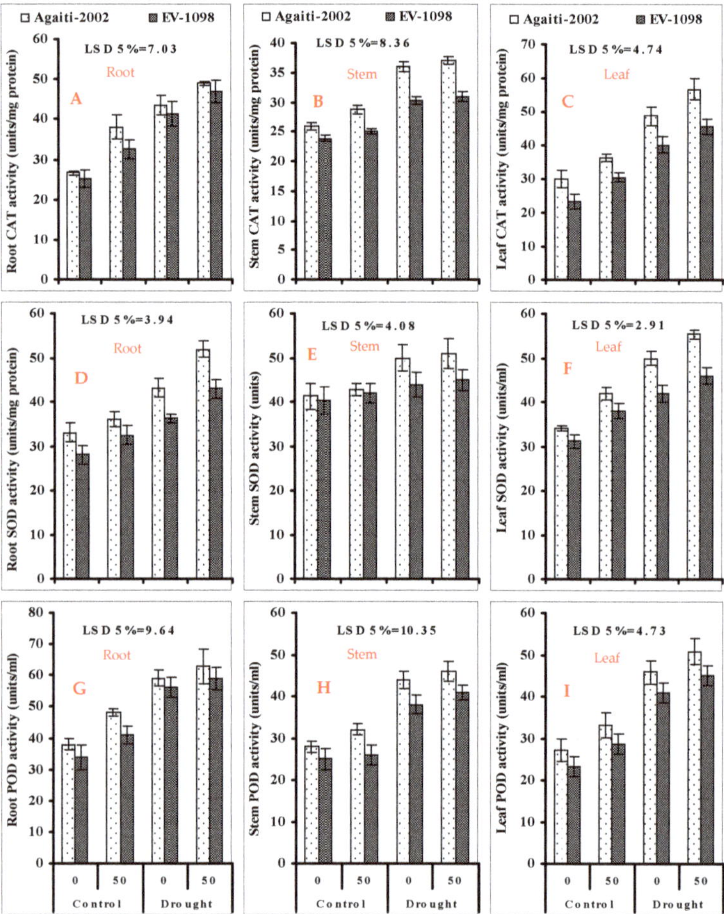

Figure 2. Activities of CAT (**A–C**), SOD (**D–F**), and POD (**G–I**) in root, stem and leaf, respectively, of drought-stressed maize plants applied with α-Toc as foliar spray when grown under water deficit conditions (mean ± SE; *n* = 4). CAT = catalase; SOD = superoxide dismutase; POD = peroxidase 0 and 50 = mmol solution of α-Tocopherol for foliar spray.

Like CAT and SOD activities, POD activity was also improved significantly under limited water supply in both genotypes in all studied plant parts. Foliar spray of α-Toc further enhanced the POD activity in root and leaf of both genotypes under non-stressed and stressed conditions, but this improvement in POD activity was not found in stem of both genotypes. A non-significant difference between genotypes was found in this regard (Figure 2G–I; Table 2).

3.5. Contents of K, Ca, Mg, N, and P in Different Parts of Maize Plants Foliary-Applied with α-Toc When Grown under Different Water Regimes

Drought stress significantly altered the tissue-specific acquisition patterns of macro-nutrients of both the studied cultivars when grown without α-Toc application (Tables 2 and 5). Potassium contents of leaf, root, and stem were reduced significantly grown under water stress without foliar spray of α-Toc. Foliar application of α-Toc increased the potassium content in specific organs under non-stressed and stressed conditions, and the impact was significant for leaf and root K of cv. Agaiti-2002 under water stress. The leaf, root and stem Ca and Mg uptake was also significantly improved after foliar application of α-Toc in both maize genotypes under non-stressed and stressed conditions, which was impaired due to limited water supply. This prominent difference in the uptake of K^+, Ca^{2+}, and Mg^{2+} due to α-Toc foliar application was similar in both maize genotypes under non-stressed and stressed conditions.

Like other nutrients, drought stress also negatively affected the P and N uptake in leaf, root, and stem of both the cultivars and this impact was more prominent on leaf and stem N. Exogenous application of α-Toc helped both the cultivars to maintain their N and P nutrition of root, leaf and stem under non-stressed and stressed conditions. Regarding the N contents in studied plant parts, comparatively more improvement in N uptake due to α-Toc foliar spray was found in leaf and root than stem (Tables 2 and 5).

Table 5. Influence of foliar-applied α-Toc on tissue specific organic and inorganic minerals of two maize (*Zea mays* L.) cultivars under non-stress and water-stressed conditions (mean ± SE; n = 3).

Stress	α-Toc	Cultivars	K^+ Leaf (mg g^{-1} DW)	K^+ Root (mg g^{-1} DW)	K^+ stem (mg g^{-1} DW)	Ca^{2+} leaf (mg g^{-1} DW)	Ca^{2+} Root (mg g^{-1} DW)	Ca^{2+} stem (mg g^{-1} DW)	Mg^{2+} leaf (mg g^{-1} DW)	Mg^{2+} root (mg g^{-1} DW)	Mg^{2+} stem (mg g^{-1} DW)
Control	0 mmol	Agaiti-2002	39.30 ± 3.18 [a]	41.35 ± 2.99 [ab]	33.09 ± 3.20 [bc]	11.35 ± 0.20 [d]	10.34 ± 0.75 [c]	10.03 ± 0.61 [b]	3.41 ± 0.06 [b]	4.13 ± 0.30 [a]	3.31 ± 0.32 [ab]
		EV-1098	41.17 ± 1.47 [ab]	40.21 ± 2.12 [b]	31.99 ± 2.38 [c]	12.92 ± 0.61 [c]	10.05 ± 0.03 [c]	9.02 ± 0.43 [c]	2.73 ± 0.15 [d]	4.02 ± 0.12 [a]	3.14 ± 0.31 [b]
	50 mmol	Agaiti-2002	43.15 ± 2.19 [a]	42.90 ± 1.41 [a]	35.12 ± 1.67 [ab]	14.36 ± 1.46 [b]	10.99 ± 0.18 [b]	11.71 ± 1.22 [a]	4.01 ± 0.44 [a]	4.29 ± 0.14 [a]	3.51 ± 0.33 [a]
		EV-1098	43.67 ± 2.45 [a]	42.93 ± 1.77 [a]	36.00 ± 1.21 [a]	15.33 ± 0.40 [a]	12.23 ± 0.31 [a]	10.40 ± 0.36 [b]	3.27 ± 0.16 [bc]	4.09 ± 0.22 [a]	3.45 ± 0.07 [a]
Drought	0 mmol	Agaiti-2002	30.33 ± 3.47 [e]	35.67 ± 2.82 [cd]	25.00 ± 1.85 [ef]	6.48 ± 0.59 [g]	7.33 ± 0.40 [e]	5.74 ± 0.71 [g]	2.03 ± 0.35 [e]	2.64 ± 0.29 [b]	2.38 ± 0.22 [d]
		EV-1098	32.33 ± 1.76 [de]	33.00 ± 2.79 [e]	24.33 ± 1.76 [f]	7.67 ± 0.40 [f]	7.67 ± 0.81 [f]	6.87 ± 0.72 [f]	2.24 ± 0.19 [e]	2.34 ± 0.29 [b]	2.82 ± 0.26 [c]
	50 mmol	Agaiti-2002	36.00 ± 1.85 [c]	37.64 ± 2.29 [c]	27.33 ± 1.76 [d]	8.45 ± 0.30 [e]	9.15 ± 0.82 [d]	8.33 ± 0.40 [e]	2.54 ± 0.09 [d]	2.66 ± 0.33 [b]	2.49 ± 0.31 [d]
		EV-1098	34.81 ± 2.19 [cd]	34.67 ± 2.45 [de]	27.00 ± 3.89 [de]	8.33 ± 0.40 [e]	9.96 ± 0.66 [c]	8.88 ± 0.13 [d]	3.11 ± 0.19 [c]	2.71 ± 0.25 [b]	2.47 ± 0.20 [d]
	LSD 5%		2.99	2.20	2.26	0.50	0.47	0.40	0.29	0.41	0.22

Stress	α-Toc	Cultivars	N Leaf (mg g^{-1} DW)	N Root (mg g^{-1} DW)	N Stem (mg g^{-1} DW)	P Leaf (mg g^{-1} DW)	P Root (mg g^{-1} DW)	P Stem (mg g^{-1} DW)
Control	0 mmol	Agaiti-2002	38.63 ± 2.28 [b]	41.35 ± 2.99 [ab]	33.09 ± 3.20 [b]	5.20 ± 0.33 [b]	5.91 ± 0.43 [b]	4.50 ± 0.46 [c]
		EV-1098	41.17 ± 1.47 [b]	40.21 ± 2.12 [ab]	32.72 ± 2.93 [b]	5.55 ± 0.25 [b]	5.74 ± 0.27 [b]	4.01 ± 0.28 [de]
	50 mmol	Agaiti-2002	45.82 ± 2.77 [a]	42.90 ± 1.41 [a]	38.16 ± 1.90 [a]	6.06 ± 0.34 [a]	6.13 ± 0.20 [ab]	5.02 ± 0.22 [b]
		EV-1098	47.00 ± 2.52 [a]	43.93 ± 2.72 [a]	39.06 ± 2.73 [a]	6.01 ± 0.23 [a]	6.28 ± 0.31 [a]	5.28 ± 0.49 [a]
Drought	0 mmol	Agaiti-2002	25.33 ± 3.46 [d]	33.33 ± 2.45 [c]	26.00 ± 2.27 [c]	4.33 ± 0.50 [cd]	4.62 ± 0.42 [d]	3.84 ± 0.22 [e]
		EV-1098	23.33 ± 2.64 [d]	34.41 ± 3.00 [e]	23.00 ± 3.20 [d]	4.06 ± 0.27 [d]	4.86 ± 0.50 [cd]	3.58 ± 0.25 [f]
	50 mmol	Agaiti-2002	30.00 ± 2.10 [c]	36.61 ± 3.27 [b]	31.00 ± 1.85 [b]	4.29 ± 0.27 [cd]	5.02 ± 0.47 [c]	4.08 ± 0.23 [d]
		EV-1098	32.67 ± 0.81 [c]	37.67 ± 1.45 [b]	27.67 ± 2.02 [c]	4.54 ± 0.22 [c]	5.05 ± 0.53 [c]	3.82 ± 0.11 [e]
	LSD 5%		2.95	4.00	2.80	0.39	0.32	0.22

Values in column with same alphabets in superscript do not differ significantly.

3.6. PCA Analysis and Spearman's Correlation Coefficient (r^2) Values Extracted from XLSTAT Software of All the Studied Attributes of Maize Plants Foliar-Applied with α-Toc

PCA and correlations coefficients among studied attributes revealed a significant positive correlation of total-Toc contents in leaf, root, and stem with morphological and growth attributes, levels of antioxidants, and uptake of mineral nutrients (K, Ca, Mg, N, and P) in all studied tissues of maize. A positive correlation of leaf and stem Toc was found with leaf area (0.768 *** and 0.664 **) and fresh weights (0.921 *** and 0.661 ***), respectively, that depicts the role of Toc in the improved growth under drought stress. Positive correlation was also recorded of shoot dry weight with Toc levels in studied plant tissues such as in leaf (0.578 **) and root (0.643 ***), respectively. Significantly positive correlation was found of Toc levels in the root with LRWC (0.721 ***). CAT, POD, and SOD activities in different plant parts like leaf (0.966 ***, 0.961 *** and 0.936 ***) and stem (0.863 ***, 0.872 *** and 0.859 ***), respectively, were also positively correlated with plant Toc levels. Tocopherol contents were also positively correlated with potassium and calcium contents in leaf (0.553 ** and 0.606 **, 0.569 ** and 0.633 ***), root (0.555 ** and 0.675 ***, 0.674 ** and 0.461 *), and stem (0.470 * and 0.673 ***, 0.749 *** and 0.437 *), respectively. Furthermore, a positive correlation was also recorded between nitrogen and phosphorus contents with Toc levels in studied plant tissues such as in leaf (0.610 ** and 0.613 **, 0.539 **, and 0.683 ***), root (0.669 ***, 0.494 * and 0.488 * and 0.729 ***, 0.430 * and 0.620 **), and stem (0.626 *** and 0.601 **, 0.536 **, and 0.688 ***), respectively. Figure 3 shows the PCA analysis of varying studied attributes that confirmed correlation studies. Of the extracted components, F1 has a major contribution (67.43%) that has divided the studied attributes in different groups. Of them, the major group encircled has parameters that are positively correlated include Pr L, RFW, RDW, N R, S L, SDW, K L, Ca L, P S, P L, K S, and LRWC, and L A, Ca R, Ca S, Mg L, Pr R, P R, N R, and N L contributed maximally in determining the variance. The F2 component has less variance (17.70%). Both components have a total variance of 80.13% (Figure 3; Table 6).

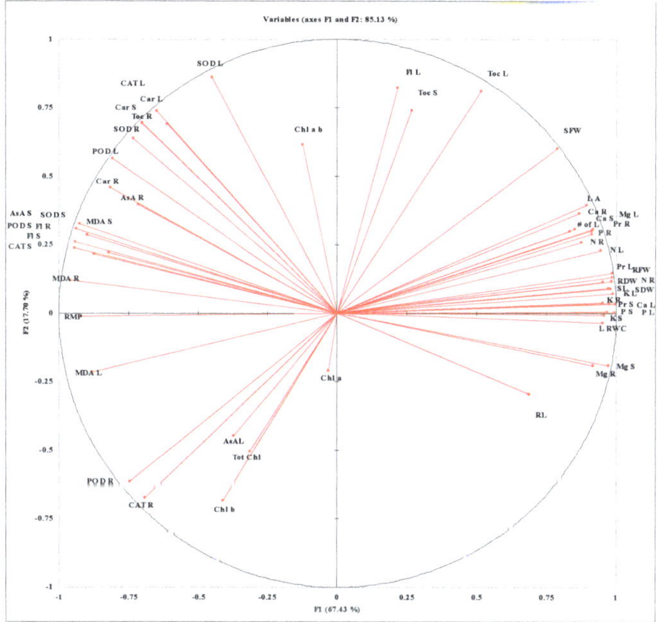

Figure 3. Principle component analysis of tocopherol levels in different plant tissues of maize with studied growth and physio-biochemical attributes, and nutrient accumulation.

Table 6. Spearman correlation coefficient values (r^2) of Toc levels in different plant parts of maize with growth, biochemical attributes, and nutrient uptake.

	Toc L	Toc R	Toc S
Toc L	1.000		
Toc R	0.204 ns	1.00	
Toc S	0.847 ***	0.302 ns	1.00
SL	0.534 **	−0.644 ***	0.287 ns
RL	0.192 ns	−0.661 ***	0.122 ns
NL	0.797 ***	−0.353 ns	0.557 **
L A	0.768 ***	−0.371 ns	0.664 ***
SFW	0.921 ***	−0.124 ns	0.661 ***
SDW	0.578 **	0.643 ***	0.368 ns
RFW	0.627 ***	−0.597 **	0.381 ns
RDW	0.780 ***	−0.420*	0.531 ns
L RWC	0.391 ns	0.721 ***	0.164 ns
Chl. a	−0.088 ns	−0.059 ns	0.266 ns
Chl. b	−0.787 ***	−0.181 ns	−0.423*
Chl. a/b	0.553 **	0.538 **	0.740 ***
Tot Chl.	−0.624 ***	−0.133 ns	−0.239 ns
RMP	−0.416*	0.664 ***	−0.084 ns
MDA L	−0.534 **	−0.463 *	−0.202 ns
MDA R	−0.342 ns	−0.751 ***	−0.052 ns
MDA S	−0.240 ns	−0.845 ***	−0.037 ns
Protien L	0.561 **	0.591 **	0.392 ns
Protein R	0.695 ***	0.324 ns	0.314 ns
Protein S	0.468 *	0.666 ***	0.159 ns
AsA L	−0.434 *	−0.115 ns	−0.149 ns
AsA R	−0.180 ns	0.709 ***	−0.023 ns
AsA S	−0.196 ns	0.797 ***	0.145 ns
Car L	0.191 ns	0.925 ***	0.339 ns
Car R	−0.078 ns	0.870 ***	0.190 ns
Car S	0.218 ns	0.974 ***	0.411 ns
Flav L	0.669 ***	0.443*	0.426*
Flav R	−0.348 ns	0.679 ***	−0.269 ns
Flav S	−0.211 ns	0.733 ***	0.047 ns
CAT L	0.275 ns	0.966 ***	0.434 *
CAT R	−0.884 ***	0.041 ns	−0.757 ***
CAT S	−0.283 ns	0.863 ***	−0.114 ns
POD L	0.049 ns	0.961 ***	0.255 ns
POD R	−0.831 ***	0.116 ns	−0.633 ***
POD S	−0.263 ns	0.872 ***	−0.107 ns
SOD L	0.438 *	0.936 ***	0.430 *
SOD R	0.173 ns	0.951 ***	0.273 ns
SOD S	−0.220 ns	0.859 ***	0.047 ns
K L	0.553 **	0.606 **	0.282 ns
K R	0.555 **	0.675 ***	0.333 ns
K S	0.470 *	0.673 ***	0.319 ns
Ca L	0.569 **	0.633 ***	0.339 ns
Ca R	0.674 ***	0.461 *	0.416 *
Ca S	0.749 ***	0.437 *	0.469 *
Mg L	0.656 ***	0.439 *	0.470 *
Mg R	0.372 ns	0.761 ***	0.201 ns
Mg S	0.335 ns	0.833 ***	0.151 ns
N L	0.610 **	0.613 **	0.368 ns
N R	0.669 ***	0.494 *	0.488 *
N R	0.626 ***	0.601 **	0.337 ns
P L	0.539 **	0.683 ***	0.383 ns
P R	0.729 ***	0.430*	0.620 **
P S	0.536 **	0.688 ***	0.281 ns

4. Discussion

The exogenous application of water-soluble antioxidants have been widely investigated to improve stress tolerance, but plant growth modulations by foliar application of lipophilic antioxidants like α-Toc has been little studied, probably due to limited information regarding their application, absorption, and translocation within the plant. Kumar et al. [60] and Ali et al. [61] reported that the exogenously applied α-Toc can partly alleviate the deleterious impacts of heat and water stress in wheat. In another study, it was found that the exogenous application of α-Toc effectively decreased the adverse effects of salt stress in flax cultivars [44]. In most of the earlier studies, the α-Toc was applied at adult growth stages. However, the seedling stage (among other growth stages) is considered important due to its involvement in better seed yield by establishing better crop stand [14]. In view of the available information in literature, the present experiment was planned with the objective to study the involvement of α-Toc in the improvement of water stress tolerance in relation to the growth modulations of maize depending upon tissue specific partitioning of macro-nutrients and antioxidants in relation with its own translocation/synthesis in specific terms. For this purpose, the response of selected maize genotypes (Agaiti-2002 and EV-1098) was examined under water stress at an early growth stage with and without foliar spray of α-Toc.

4.1. Tocopherol Content in Different Plant Parts

Foliar spray of α-Toc significantly increased the leaf tocopherol levels under non-stressed and stressed conditions, which pointed out the existence of an appropriate mechanism for the uptake of α-Toc in the leaves of maize. The increments in root tocopherol contents exhibited a similar pattern, as did the leaves, after foliar application, which suggests an efficient basipetal translocation of α-Toc in maize. Our findings are in agreement with Kumar et al. [60], who reported an elevation in the endogenous levels of α-Toc in heat-stressed wheat plants after its exogenous application. Furthermore, it has been reported that the exogenous application of these organic compounds, along with altering the cellular metabolic activities, also controls the plant's own metabolism. In the present study, the improvement in the internal levels of α-Toc by its exogenous application might also be due to its involvement in regulating plant metabolism [34,36,40].

4.2. Growth, Water Relations, and Photosynthetic Pigments

Seedling growth of maize plants was adversely affected in plants grown without foliar application of α-Toc under water stress, which is in line with the findings that drought-caused growth reduction is a clear phenomenon in crop plants [8,14]. Similarly, in the present study, a drought-induced decrease was recorded in root and shoot lengths, root and shoot fresh and dry weights, leaf area, and number of leaves of both maize genotypes. Growth is dependent on physiological factors, including the content of plant photosynthetic pigments and water relations that directly influences the leaf photosynthetic rate by affecting the capacity of light capturing and assimilation process [61,62]. Different plant species and even cultivars in the same species have different potentials to tolerate the adverse conditions regarding these attributes [63].

In the present study, water-stress-induced reduction in biomass is associated with reduced photosynthetic pigment along with disturbed plant water relations, and this reduction was less in cv. Agaiti-2002, showing its better tolerance to drought [34]. The foliar spray of α-Toc substantially elevated the plant's endogenous levels and resulted in significant growth improvement under stressed and non-stressed conditions. Increments in plant biomass production is positively associated with the improvement in plant water relations and biosynthesis of biosynthetic pigments such as chlorophyll and carotenoids under the influence of α-Toc foliar application. The increment in plant water status might probably be due to impact of α-Toc on H ATPase system showing its role in cellular osmotic adjustment, due to a necessary part of cellular membranes. This involvement of alpha tocopherol in cellular osmotic adjustment confers its role in maintaining the cellular water relations under stressful conditions. Similar might be in present study where foliar application of alpha tocopherol improved the leaf relative water content of water stressed maize plants. This improvement in plant water relations further confers its role in improving the leaf net photosynthetic efficiency because plant better water content is necessary to regulate stomatal regulation for better photosynthesis [62]. Furthermore, it is found that α-Toc, being a part of cellular membranes, plays a significant role in decreasing the degradation of photosynthetic pigments in a stressful environment [64]. Tocopherols also protect D_1 protein [65] and chloroplastic membranes from damaging effects when grown under stressful conditions.

In the present study, foliar-applied α-Toc under drought stress further enhanced its internal levels in parallel with the improvement in leaf photosynthetic pigments, which might be due to the significant role of alpha tocopherol in reducing the adverse effects on leaf photosynthetic pigments, resulting in improved photosynthetic efficiency along with better plant water relations that resulted in better plant biomass production. In an earlier study, it was found by Sakr and El-Metwally [43] and El-Quesni [66] in wheat and *Hibiscus rosa sinensis*, respectively, that exogenous application of α-Toc enhanced plant biomass production, which might be due to the role of α-Toc in the accumulation of total carbohydrates and protein biosynthesis, confirming its role in photosynthesis and assimilation [67]; this can be correlated with present findings, where higher biomass production was associated with α-Toc levels in different parts that improved plant water relations and net photosynthesis as a result of better net assimilation with improved biomass production. Furthermore, this study reveals the increased plant dry weights due to foliar application of α-Toc, which points toward the improved photosynthetic activity and assimilation with the establishment of new binding sites [68] after its exogenous application.

Furthermore, in the present study, both maize cultivars maintained an optimum level of their carotenoid contents even under drought and α-Toc supplementation, which further enhanced the plant carotenoid contents, especially in leaf and root. These observations point out that α-Toc-induced improvement in the growth of maize plants might be due to an improvement in the contents of accessory pigments as additional support to different photosynthetic attributes. In an earlier study, it was found that, in different wheat cultivars [43] and *Vicia faba* [69], foliar-applied α-Toc improved the leaf carotenoid concentration in association with its enhanced growth. Without α-Toc application, a decrease in leaf water contents was found in maize plants, which is a well-known phenomenon in all plants. α-Toc foliar application significantly increased the leaf water content of water-stressed maize plants, showing its protective role in drought-stressed plants, which might be due to its role in the management of cellular turgor potential through imparting its role in cellular osmotic adjustment by enhancing biosynthesis of osmolytes [7], resulting in better growth by providing an environment for increased cell division and provide an environment for better photosynthesis.

4.3. Lipid Peroxidation and Antioxidative Defence Mechanism

An increase in the levels of ROS under stressful environment is a general phenomenon due to O_2 excitation to form singlet oxygen or its conversion to hydroxyl radicals (OH^-), hydrogen peroxide (H_2O_2), or superoxide (O^{-2}) due to the transfer of excited electrons, respectively [34,70], with restricted e^- transfer at different steps in photosynthesis and respiration under reduced metabolic activities. These overly produced ROS directly affect different cellular membranes through lipid peroxidation. As a defense for the protection of the cellular membranes and other components from the deleterious and damaging effects of overproduced ROS, plants have evolved well-developed mechanisms for the antioxidation of ROS, i.e., comprised of non-enzymatic (AsA, phenolics, carotenoids, flavonoids, tocopherol, etc.) and enzymatic (SOD, POD, CAT, APX) components [14,34,71]. This antioxidative system works well in combination. In the present study, the α-Toc-treated plants suffered significantly lower oxidative damage, especially in root and leaf, as depicted by the lower MDA contents in these plant parts relative to untreated ones (as reported earlier for wheat) [60]. Drought stress significantly increased oxidative stress in maize plants; this is obvious from increased levels of MDA, a product of lipid peroxidation. Damage to biological membranes due to oxidative stress is a general phenomenon that generally increases in specific environments [14,45]. In an earlier study, significantly lower oxidative stress was recorded in α-Toc applied plants as obvious from lower membrane permeability which is in line with its role in quenching lipid peroxyl radicals, responsible for propagating lipid peroxidation [69,72,73]. It was reported that during early growth stages, α-Toc played a significant role in counteracting the adverse effects of membrane lipid peroxidation. Furthermore, being lipophilic, α-Toc has a significant role in membrane stabilization [74] and also protects them from ROS [75]. Furthermore, α-Toc directly scavenges singlet oxygen [76], giving rise an intermediate tocopherol quinone, which again yields α-Toc in chloroplasts, thereby conferring the recycling for oxidized tocopherols [77]. Reports exist that α-Tocopherol is also an excellent quencher and scavenger of singlet oxygen by controlling the lifetime of ROS. By resonance energy transfer, one α-Toc molecule can neutralize up to 120 molecules of singlet oxygen [78]. The activities of antioxidants such as SOD, POD, and CAT were found to be higher in leaves and roots of maize plant after α-Toc treatment, which suggested their antioxidative role to be stimulated in the presence of α-Toc.

Higher activity rates of these enzymes were found in leaves and roots where more accumulation of α-Toc was found in comparison with stem, showing the supportive role of α-Toc in the activities of antioxidative enzymes. Furthermore, the higher levels of non-enzymatic antioxidant in root and leaf as compared to stem (such as AsA, phenolics, and flavonoids) are also associated with high content of α-Toc in these plant parts. These findings show that α-Toc application after its translocation to the studied plant parts played a significant role in increasing the activities of antioxidative enzymes and the levels of non-enzymatic antioxidant compounds and thus played an imperative role in protecting cellular membranes by boosting the plant's own mechanism. It was found by Fahrenholtz et al. [79] that α-Toc acts as an antioxidative defense mechanism in plants. It was also found that α-Toc minimizes the oxidative changes in the cellular membrane in a significant way with other antioxidants [80–82].

4.4. Uptake of Mineral Nutrients

Drought-induced growth reduction can also be attributed to disturbances in the uptake of mineral nutrients along with other physiological attributes. It is well known that disturbance or reductions in the leaf uptake of mineral nutrient in plants is probably due to nutrient availability, partitioning, and transport, which is negatively affected under drought conditions. Plant mineral nutrients status played a major role in determining drought tolerance [83]. In the present study, the PCA analysis and the correlations studied suggest that an improvement in the levels of α-Toc contents in different plant parts induced by its foliar application increased the uptake of mineral nutrients (K, Ca, N, and P). Mineral nutrients effectively decrease the harsh effects of water stress by various mechanisms [22]. For example, it has been found that better uptake of mineral nutrients like Ca^{2+}, N, and K^+ reduces the deleterious effects of over produced ROS by increasing the concentration of antioxidants like CAT, POD, and SOD [22]. It has been reported that P, K, and Mg improve root growth, which results in improved water intake conferring the drought tolerance. It can be interpreted that optimum nutrient levels maintained after α-Toc application confer drought tolerance induction in maize plants in parallel with improved growth. This is more likely because leaf water contents were significantly improved by foliar spray of α-Toc. The supportive role of α-Toc after its application in the absorption of nutrients from the soil in stressful environment has been found extensively [44,78], and it is reported that α-Toc induced increase uptake of nutrients due to α-Toc being an antioxidant, along with membrane permeability. Furthermore, previous studies found that α-Toc induced an increase in growth, water relation, and nutrient uptake associated with improved stem and leaf anatomy, which further improved translocation to different plant parts. Therefore, the studies confirm that, as in the present study, α-Toc application might improve the uptake and translocation of different nutrients from the soil solution to the roots and then to different plant parts, resulting in better assimilation and growth.

5. Conclusions

It can be concluded that endogenous levels of α-Toc have an important role in enhancing water stress tolerance of maize cultivars, and its foliar application is found to be effective in reducing water-stress-induced adversative effects on growth by modulating different metabolic activities. Our results confirmed that α-Tocopherol application resulted in membrane protection through increased activities of antioxidative enzymes (CAT, POD, and SOD) and the content of non-enzymatic antioxidants with improved water relations. The correlations and PCA analysis revealed that the increase in α-Tocopherol contents in different plant parts after its foliar application increased the uptake of mineral nutrients (K^+, Ca^{2+}, N, and P). Optimum water content and nutrients, along with better antioxidant potential, ultimately resulted in drought tolerance in both maize cultivars that increased growth. In relation to translocation-dependent effects, it was found that α-Toc followed basipetal translocation, concentrating mainly in the roots rather than the shoot after its foliar application. Therefore, analysis of the impact of foliar application of α-Toc on seed yield and nutritional quality of arable crops under stressful environment should be the subject of future studies.

Author Contributions: Conceptualization, Q.A., M.R., S.A., M.N.A., H.A.E.-S., and F.A.A.-M.; Data curation, Q.A., M.T.J., M.Z.H., and R.P.; Formal analysis, Q.A., M.T.J., N.H., and R.P.; Funding acquisition, Q.A., M.N.A., H.A.E.-S., and F.A.A.-M.; Investigation, N.H. and R.P.; Methodology, M.T.J., M.Z.H., and R.P.; Resources, M.R., S.A., M.N.A., H.A.E.-S., and F.A.A.-M.; Software, M.R., S.A., M.N.A., H.A.E.-S., and F.A.A.-M.; Supervision, Q.A. and M.T.J.; Validation, M.T.J., M.Z.H., and N.H.; Visualization, M.Z.H. and N.H.; Writing—original draft, Q.A. and F.A.A.-M.; Writing—review and editing, M.R., S.A., M.N.A., and H.A.E.-S. All authors have read and agreed to the published version of the manuscript.

Funding: The authors would like to extend their sincere appreciation to the Researchers Supporting Project Number (RSP-2020/19), King Saud University, Riyadh, Saudi Arabia.

Acknowledgments: The authors are grateful to the Higher Education Commission (HEC) Islamabad, Pakistan for its support.

Abbreviations

Toc L	leaf tocopherol
Toc R	root tocopherol
Toc S	stem tocopherol
SL	shoot length
RL	root length
NL	number of leaves
LA	leaf area
SFW	shoot fresh weight
SDW	shoot dry weight
RFW	root fresh weight
RDW	root dry weight
L RWC	leaf relative water content
Chl. a	chlorophyll a
Chl. b	chlorophyll b
Chl. a/b	chlorophyll a/b ratio
Tot Chl.	total chlorophyll
RMP	relative membrane permeability
MDA L	MDA leaf
MDA R	MDA root
MDA S	MDA stem
Protien L	protein leaf
Protein R	protein root
Protein S	protein stem

AsAL	ascorbic acid leaf
AsA R	ascorbic acid root
AsA S	ascorbic acid stem
Car L	carotenoids leaf
Car R	carotenoids root
Car S	carotenoids stem
Flav L	flavonoids leaf
Flav R	flavonoids root
Flav S	flavonoids stem
CAT L	catalase leaf
CAT R	catalase root
CAT S	catalase stem
POD L	peroxidase leaf
POD R	peroxidase root
POD S	peroxidase stem
SOD L	superoxide dismutase leaf
SOD R	superoxide dismutase root
SOD S	superoxide dismutase stem
K L	potassium leaf
K R	potassium root
K S	potassium stem
Ca L	calcium leaf
Ca R	calcium root
Ca S	calcium stem
Mg L	magnesium leaf
Mg R	magnesium root
Mg S	magnesium stem
N L	nitrogen leaf
N R	nitrogen root
N S	nitrogen stem
P L	phosphorus leaf
P R	phosphorus root
P S	phosphorus stem

References

1. Gobin, A. Impact of Heat and Drought Stress on Arable Crop Production in Belgium. *Nat. Hazards Earth Syst. Sci.* **2012**, *12*, 1911–1922. [CrossRef]
2. Khan, N.; Bano, A. Rhizobacteria and Abiotic Stress Management. In *Plant Growth Promoting Rhizobacteria for Sustainable Stress Management*; Springer: Singapore, 2019; pp. 65–80.
3. Cairns, J.E.; Sanchez, C.; Vargas, M.; Ordoñez, R.; Araus, J.L. Dissecting Maize Productivity: Ideotypes Associated with Grain Yield Under Drought Stress and Well-Watered Conditions. *J. Integr. Plant Biol.* **2012**, *54*, 1007–1020. [CrossRef] [PubMed]
4. Adnan, S.; Ullah, K.; Gao, S.; Khosa, A.H.; Wang, Z. Shifting of Agro Climatic Zones, their Drought Vulnerability, and Precipitation and Temperature Trends in Pakistan. *Int. J. Climatol.* **2017**, *37*, 529–543. [CrossRef]
5. Kazmi, D.H.; Li, J.; Rasul, G.; Tong, J.; Ali, G.; Cheema, S.B.; Liu, L.; Gemmer, M.; Fisherr, T. Statistical Downscaling and Future Scenario Generation of Temperatures for Pakistan Region. *Theor. Appl. Climatol.* **2015**, *120*, 311–350. [CrossRef]
6. Hina, S.; Saleem, F. Historical Analysis (1981–2017) of Drought Severity and Magnitude over a Predominantly Arid Region of Pakistan. *Clim. Res.* **2019**, *78*, 189–204. [CrossRef]
7. Um, M.; Kim, Y.; Park, D.; Jung, K.; Wang, Z.; Kim, M.M.; Shin, H. Impacts of Potential Evapotranspiration on Drought Phenomena in Different Regions and Climate Zones. *Sci. Total Environ.* **2020**, *703*, 135590. [CrossRef]

8. Ali, Q.; Anwar, F.; Ashraf, M.; Saari, N.; Perveen, R. Ameliorating Effects of Exogenously Applied Proline on Seed Composition, Seed Oil Quality and Oil Antioxidant Activity of Maize (*Zea mays* L.) Under Drought Stress. *Int. J. Mol. Sci.* **2013**, *14*, 818–835. [CrossRef]

9. Khan, N.; Bano, A.; Rahman, M.A.; Guo, J.; Kang, Z.; Babar, M.A. Comparative physiological and metabolic analysis reveals a complex mechanism involved in drought tolerance in chickpea (*Cicer arietinum* L.) induced by PGPR and PGRs. *Sci. Rep.* **2019**, *9*, 1–19. [CrossRef]

10. Rana, V.; Singh, D.; Dhiman, R.; Chaudhary, H.K. Evaluation of Drought Tolerance Among Elite Indian Bread Wheat Cultivars. *Cereal. Res. Commun.* **2014**, *42*, 91–101. [CrossRef]

11. Chaves, M.M.; Maroco, J.P.; Pereira, J.S. Understanding Plant Responses to Drought From Genes to the Whole Plant. *Physiol. Mol. Biol. Plants* **2003**, *30*, 239–264. [CrossRef]

12. Khan, N.; Bano, A.; Rahman, M.A.; Rathinasabapathi, B.; Babar, M.A. UPLC-HRMS-based untargeted metabolic profiling reveals changes in chickpea (Cicer arietinum) metabolome following long-term drought stress. *Plant Cell Environ.* **2019**, *42*, 115–132. [CrossRef] [PubMed]

13. Demirevska, K.; Simova-Stoilova, L.; Fedina, I.; Georgieva, K.; Kunert, K. Response of *Oryza cystatin* L. Transformed Tobacco Plants to Drought, Heat and Light Stress. *J. Agron. Crop Sci.* **2010**, *196*, 90–99. [CrossRef]

14. Ali, Q.; Javed, M.T.; Noman, A.; Haider, M.Z.; Waseem, M.; Iqbal, N.; Perveen, R. Assessment of Drought Tolerance in Mung Bean Cultivars/Lines as Depicted by the Activities of Germination Enzymes, Seedling's Antioxidative Potential and Nutrient Acquisition. *Arch. Agron. Soil. Sci.* **2018**, *64*, 84–102. [CrossRef]

15. Chen, J.; Junying, D. Effect of Drought on Photosynthesis and Grain Yield of Corn Hybrids With Different Drought Tolerance. *Acta-Agron Sin.* **1996**, *22*, 757–762.

16. Khan, N.; Bano, A.; Curá, J.A. Role of Beneficial Microorganisms and Salicylic Acid in Improving Rainfed Agriculture and Future Food Safety. *Microorganisms* **2020**, *8*, 1018. [CrossRef]

17. Lawlor, D.W.; Cornic, G. Photosynthetic Carbon Assimilation and Associated Metabolism in Relation to Water Deficits in Higher Plants. *Plant Cell Environ.* **2002**, *25*, 275–294. [CrossRef]

18. Rapacz, M.; Kościelniak, J.; Jurczyk, B.; Adamska, A.; Wójcik, M. Different Patterns of Physiological and Molecular Response to Drought in Seedlings of Malt- and Feed-Type Barleys (*Hordeum vulgare*). *J. Agron. Crop Sci.* **2010**, *196*, 9–19. [CrossRef]

19. Marschner, P. *Mineral Nutrition of Higher Plants*; Academic Press: London, UK, 2011.

20. Mukhtar, I.; Shahid, M.A.; Khan, M.W.; Balal, R.M.; Iqbal, M.M.; Naz, T.; Zubair, M.; Ali, H.H. Improving salinity tolerance in chili by exogenous application of calcium and sulphur. *Soil Environ.* **2016**, *1*, 35.

21. Heidari, M.; Karami, V. Effects of Different Mycorrhiza Species on Grain Yield, Nutrient Uptake and Oil Content of Sunflower Under Water Stress. *J. Saudi Soc. Agric. Sci.* **2014**, *13*, 9–13. [CrossRef]

22. Waraich, E.A.; Ahmad, R.; Ashraf, M.Y.; Ahmad, S.M. Improving Agricultural Water Use Efficiency by Nutrient Management in Crop Plants. *Acta Agric. Scand. B Soil Plant Sci.* **2011**, *61*, 291–304. [CrossRef]

23. Fahad, S.; Bajwa, A.A.; Nazir, U.; Anjum, S.A.; Farooq, A.; Zohaib, A.; Ihsan, M.Z. Crop Production Under Drought and Heat Stress: Plant Responses and Management Options. *Front. Plant Sci.* **2017**, *8*, 1147. [CrossRef] [PubMed]

24. Hussain, H.A.; Hussain, S.; Khaliq, A.; Ashraf, U.; Anjum, S.A.; Men, S.; Wang, L. Chilling and Drought Stresses in Crop Plants: Implications, Cross Talk, and Potential Management Opportunities. *Front. Plant Sci.* **2018**, *9*, 393. [CrossRef]

25. Rajala, A.; Niskanen, M.; Isolahti, M.; Peltonen-Sainio, P. Seed Quality Effects on Seedling Emergence, Plant Stand Establishment and Grain Yield in Two-Row Barley. *Agric. Food Sci.* **2011**, *20*, 228–234. [CrossRef]

26. Singla, S.; Grover, K.; Angadi, S.V.; Schutte, B.; VanLeeuwen, D. Guar Stand Establishment, Physiology, and Yield Responses to Planting Date in Southern New Mexico. *Agron. J.* **2016**, *8*, 2289–2300. [CrossRef]

27. Srivalli, B.; Sharma, G.; Khanna-Chopra, R. Antioxidative Defense System in an Upland Rice Cultivar Subjected to Increasing Intensity of Water Stress Followed by Recovery. *Physiol. Plant.* **2003**, *119*, 503–512. [CrossRef]

28. Ashraf, M. Inducing Drought Tolerance in Plants: Recent Advances. *Biotechnol. Adv.* **2010**, *28*, 169–183. [CrossRef]

29. Zhang, L.X.; Li, S.X.; Zhang, H.; Liang, Z.S. Nitrogen Rates and Water Stress Effects on Production, Lipid Peroxidation and Antioxidative Enzyme Activities in Two Maize (*Zea mays* L.) genotypes. *J. Agron. Crop Sci.* **2007**, *193*, 387–397. [CrossRef]

30. Farooq, M.; Wahid, A.; Lee, D.J.; Cheema, S.A.; Aziz, T. Comparative Time Course Action of the Foliar Applied Glycinebetaine, Salicylic Acid, Nitrous Oxide, Brassinosteroids and Spermine in Improving Drought Resistance of Rice. *J. Agron. Crop Sci.* **2010**, *196*, 336–345. [CrossRef]
31. Posmyk, M.M.; Kontek, R.; Janas, K.M. Antioxidant Enzymes Activity and Phenolic Compounds Content in Red Cabbage Seedlings Exposed to Copper Stress. *Ecotoxicol. Environ. Saf.* **2009**, *72*, 596–602. [CrossRef]
32. Farooq, M.; Basra, S.M.A.; Wahid, A.; Cheema, Z.A.; Cheema, M.A.; Khaliq, A. Physiological Role of Exogenously Applied Glycinebetaine to Improve Drought Tolerance in Fine Grain Aromatic Rice (*Oryza sativa* L.). *J. Agron. Crop. Sci.* **2008**, *194*, 325–333. [CrossRef]
33. Khan, S.; Hasan, M.; Bari, A.; Khan, F. Climate Classification of Pakistan. In Proceedings of the Balwois Conference, Ohrid, Macedonia, 25–29 May 2010; pp. 1–47.
34. Ali, Q.; Ashraf, M. Induction of Drought Tolerance in Maize (*Zea mays* L.) Due to Exogenous Application of Trehalose: Growth, Photosynthesis, Water Relations and Oxidative Defence Mechanism. *J. Agron. Crop Sci.* **2011**, *197*, 258–271. [CrossRef]
35. Tayebi-Meigooni, A.; Awang, Y.; Biggs, A.R.; Mohamad, R.; Madani, B.; Ghasemzadeh, A. Mitigation of Salt-Induced Oxidative Damage in Chinese Kale (*Brassica alboglabra* L.) Using Ascorbic Acid. *Acta Agric. Scand. Sect. B-Soil Plant Sci.* **2014**, *64*, 13–23.
36. Noman, A.; Ali, Q.; Maqsood, J.; Iqbal, N.; Javed, M.T.; Rasool, N.; Naseem, J. Deciphering Physio-Biochemical, Yield, and Nutritional Quality Attributes of Water-Stressed Radish (*Raphanus sativus* L.) Plants Grown From Zn-Lys Primed Seeds. *Chemosphere* **2018**, *195*, 175–189. [CrossRef] [PubMed]
37. Khan, N.; Bano, A.; Zandi, P. Effects of exogenously applied plant growth regulators in combination with PGPR on the physiology and root growth of chickpea (Cicer arietinum) and their role in drought tolerance. *J. Plant Interact.* **2018**, *1*, 239–247. [CrossRef]
38. Khan, N.; Zandi, P.; Ali, S.; Mehmood, A.; Adnan Shahid, M.; Yang, J. Impact of salicylic acid and PGPR on the drought tolerance and phytoremediation potential of Helianthus annus. *Front. Microbiol.* **2018**, *9*, 2507. [CrossRef]
39. Khan, N.; Bano, A.; Ali, S.; Babar, M.A. Crosstalk amongst phytohormones from planta and PGPR under biotic and abiotic stresses. *Plant Growth Regul.* **2020**, *90*, 189–203. [CrossRef]
40. Jamil, S.; Ali, Q.; Iqbal, M.; Javed, M.T.; Iftikhar, W.; Shahzad, F.; Perveen, R. Modulations in Plant Water Relations and Tissue-Specific Osmoregulation by Foliar-Applied Ascorbic Acid and the Induction of Salt Tolerance in Maize Plants. *Braz. J. Bot.* **2015**, *38*, 527–538. [CrossRef]
41. Hasegawa, P.M.; Bressan, R.A.; Zhu, J.K.; Bohnert, H.J. Plant Cellular and Molecular Responses to High Salinity. *Annu. Rev. Plant Biol.* **2000**, *51*, 463–499. [CrossRef]
42. Fritsche, S.; Wang, X.; Jung, C. Recent Advances in Our Understanding of Tocopherol Biosynthesis in Plants: An Overview of Key Genes, Functions, and Breeding of Vitamin E Improved Crops. *Antioxidants* **2017**, *6*, 99. [CrossRef]
43. Sakr, M.T.; El-Metwally, M.A. Alleviation of the Harmful Effects of Soil Salt Stress on Growth, Yield and Endogenous Antioxidant Content of Wheat Plant by Application of Antioxidants. *Pak. J. Biol. Sci.* **2009**, *12*, 624–630. [CrossRef]
44. Sadak, S.M.; Dawood, S. Role of Ascorbic Acid and α Tocopherol in Alleviating Salinity Stress on Flax Plant (*Linum usitatissimum* L.). *J. Stress Physiol. Biochem.* **2014**, *10*, 93–111.
45. Gill, S.S.; Tuteja, N. Reactive Oxygen Species and Antioxidant Machinery in Abiotic Stress Tolerance in Crop Plants. *Plant Physiol. Biochem.* **2010**, *48*, 909–930. [CrossRef]
46. Saleem, A.; Saleem, U.; Subhani, G.M. Correlation and Path Coefficient Analysis in Maize (*Zea mays* L.). *J. Agric. Sci.* **2007**, *45*, 177–183.
47. Aslam, M.; Zamir, M.S.I.; Anjum, S.A.; Khan, I.; Tanveer, M. An Investigation into Morphological and Physiological Approaches to Screen Maize (*Zea mays* L.) Hybrids for Drought Tolerance. *Cereal Res. Commun.* **2015**, *43*, 41–51. [CrossRef]
48. Anjum, S.A.; Ashraf, U.; Tanveer, M.; Khan, I.; Hussain, S.; Shahzad, B.; Zohaib, A.; Abbas, F.; Saleem, M.F.; Ali, I.; et al. Drought Induced Changes in Growth, Osmolyte Accumulation and Antioxidant Metabolism of Three Maize Hybrids. Front. *Plant Sci.* **2017**, *8*, 69. [CrossRef]
49. Arnon, D.I. Copper Enzyme in Isolated Chloroplast Polyphenol Oxidase in *Beta vulgaris. Plant Physiol.* **1949**, *24*, 1–15.

50. Kirk, J.T.O.; Allen, R.L. Dependence of Chloroplast Pigment Synthesis on Protein Synthesis: Effect of Actidione. *Biochem. Biophys. Res. Commun.* **1965**, *21*, 523–530. [CrossRef]
51. Yang, G.; Rhodes, D.; Joly, R.J. Effects of High Temperature on Membrane Stability and Chlorophyll Fluorescence in Glycinebetaine-Deficient and Glycinebetaine-Containing Maize Lines. *Physiol. Mol. Biol. Plants* **1996**, *23*, 437–443. [CrossRef]
52. Cakmak, I.; Horst, W.J. Effect of Aluminium on Lipid Peroxidation, Superoxide Dismutase, Catalase, and Peroxidase Activities in Root Tips of Soybean (*Glycine max*). *Physiol. Plant.* **1991**, *83*, 463–468. [CrossRef]
53. Bradford, M.M.A. Rapid and Sensitive Method for the Quantitation of Microgram Quantities of Protein Utilizing the Principle of Protein-Dye Binding. *Anal. Biochem.* **1976**, *72*, 248–254. [CrossRef]
54. Giannopolitis, C.N.; Ries, S.K. Superoxide Occurrence in Higher Plants. *Plant Physiol.* **1977**, *59*, 309–314. [CrossRef] [PubMed]
55. Chance, B.; Maehly, A.C. Assay of Catalase and Peroxidase. *Methods Enzymol.* **1955**, *2*, 764–775.
56. Mukherjee, S.P.; Choudhuri, M.A. Implications of Water Stress-Induced Changes in the Levels of Endogenous Ascorbic Acid and Hydrogen Peroxide in Vigna Seedlings. *Physiol. Plant.* **1983**, *58*, 166–170. [CrossRef]
57. Karadeniz, F.; Burdurlu, H.S.; Koca, N.; Soyer, Y. Antioxidant Activity of Selected Fruits and Vegetables Grown in Turkey. *Turk. J. Agric. For.* **2005**, *29*, 297–303.
58. Backer, H.; Frank, O.; De Angelis, B.; Feingold, S. Plasma Tocopherol in Man at Various Times After Ingesting Free or Acetylated Tocopherol. *Nutr. Rep. Int.* **1980**, *21*, 531–536.
59. Bremner, J.M.; Keeney, D.R. Steam Distillation Methods for Determination of Ammonium, Nitrate and Nitrite. *Anal. Chim. Acta* **1965**, *32*, 485–495. [CrossRef]
60. Kumar, S.; Singh, R.; Nayyar, H. α-Tocopherol Application Modulates the Response of Wheat (*Triticum aestivum* L.) Seedlings to Elevated Temperatures by Mitigation of Stress Injury and Enhancement of Antioxidants. *J. Plant Growth Regul.* **2012**, *32*, 307–314. [CrossRef]
61. Ali, Q.; Ali, S.; Iqbal, N.; Javed, M.T.; Rizwan, M.; Khaliq, R.; Wijaya, L. Alpha-Tocopherol Fertigation Confers Growth Physio-Biochemical and Qualitative Yield Enhancement in Field Grown Water Deficit Wheat (*Triticum aestivum* L.). *Sci. Rep.* **2019**, *9*, 1–15. [CrossRef]
62. Taize, L.; Zeiger, E.; Moller, I.M.; Murphy, A. *Plant Physiology and Development*, 6th ed.; Sinauer Associates, Inc.: Sunderland, MA, USA, 2015.
63. Foryer, C.; Noctor, G. Oxygen Processing in Photosynthesis: Regulation and Signaling. *New Phytol.* **2000**, *146*, 359–388. [CrossRef]
64. Munné-Bosch, S.; Alegre, L. The Function of Tocopherols and Tocotrienols in Plants. *Crit. Rev. Plant Sci.* **2002**, *21*, 31–57. [CrossRef]
65. Trebst, A.; Depka, B.; Holländer-Czytko, H. A Specific Role for Tocopherol and of Chemical Singlet Oxygen Quenchers in the Maintenance of Photosystem II Structure and Function in *Chlamydomonas reinhardtii*. *FEBS Lett.* **2002**, *516*, 156–160. [CrossRef]
66. El-Quesni, F.E.M.; Abd El-Aziz, N.G.; Kandil, M.M. Some Studies on the Effect of Ascorbic Acid and a-Tocopherol on the Growth and Some Chemical Composition of *Hibiscus rosa sinensis* L. at Nubaria. *Ozean J. Appl. Sci.* **2009**, *2*, 159–167.
67. Sadak, M.S.; Rady, M.M.; Badr, N.M.; Gaballah, M.S. Increasing Sun Flower Salt Tolerance Using Nicotinamide and a-Tocopherol. *Int. J. Acad. Res.* **2010**, *2*, 263–270.
68. Javed, M.T.; Greger, M. Cadmium Triggers *Elodea canadensis* to Change the Surrounding Water pH and Thereby Cd uptake. *Int. J. Phytoremediation* **2010**, *13*, 95–106. [CrossRef] [PubMed]
69. Orabi, S.A.; Abdelhamid, M.T. Protective Role of α-Tocopherol on Two *Vicia faba* Cultivars Against Seawater-Induced Lipid Peroxidation by Enhancing Capacity of Anti-Oxidative System. *J. Saudi Soc. Agric. Sci.* **2016**, *15*, 145–154. [CrossRef]
70. Shigeoka, S.; Ishikawa, T.; Tamoi, M.; Miyagawa, Y.; Takeda, T.; Yabuta, Y.; Yoshimura, K. Regulation and Function of Ascorbate Peroxidase Isoenzymes. *J. Exp. Bot.* **2002**, *53*, 1305–1319. [CrossRef]
71. Ali, Q.; Haider, M.Z.; Iftikhar, W.; Jamil, S.; Javed, M.T.; Noman, A.; Perveen, R. Drought Tolerance Potential of *Vigna mungo* L. Lines as Deciphered by Modulated Growth, Antioxidant Defense, and Nutrient Acquisition Patterns. *Braz. J. Bot.* **2016**, *39*, 801–812. [CrossRef]
72. Maeda, H.; Sakuragi, Y.; Bryant, D.A.; DellaPenna, D. Tocopherols Protect *Synechocystis* sp. Strain PCC 6803 From Lipid Peroxidation. *Plant Physiol.* **2005**, *138*, 1422–1435. [CrossRef]

73. Sattler, S.E.; Cahoon, E.B.; Coughlan, S.J.; DellaPenna, D. Characterization of Tocopherol Cyclases From Higher Plants and Cyanobacteria. Evolutionary Implications for Tocopherol Synthesis and Function. *Plant Physiol.* **2003**, *132*, 2184–2195. [CrossRef]

74. Wang, X.; Quinn, P.J. Preferential Interaction of α-Tocopherol With Phosphatidylcholines in Mixed Aqueous Dispersions of Phosphatidylcholine and Phosphatidylethanolamine. *Eur. J. Biochem.* **2000**, *267*, 6362–6368. [CrossRef]

75. Maeda, H.; DellaPenna, D. Tocopherol Functions in Photosynthetic Organisms. *Curr. Opin. Plant Biol.* **2007**, *10*, 260–265. [CrossRef]

76. Fukuzawa, K.; Matsuura, K.; Tokumura, A.; Suzuki, A.; Terao, J. Kinetics and Dynamics of Singlet Oxygen Scavenging by α-Tocopherol in Phospholipid Model Membranes. *Free Radic. Biol. Med.* **1997**, *22*, 923–930. [CrossRef]

77. Kobayashi, N.; DellaPenna, D. Tocopherol Metabolism, Oxidation and Recycling Under High Light Stress in Arabidopsis. *Plant J.* **2008**, *55*, 607–618. [CrossRef]

78. Fahrenholtz, S.R.; Doleiden, F.H.; Trozzolo, A.M.; Lamola, A.A. On the Quenching of Singlet Oxygen by α-Tocopherol. *J. Photochem. Photobiol. Biol.* **1974**, *20*, 505–509. [CrossRef]

79. Semida, W.M.; Taha, R.S.; Abdelhamid, M.T.; Rady, M.M. Foliar-Applied α-Tocopherol Enhances Salt-Tolerance in *Vicia faba* L. Plants Grown Under Saline Conditions. *S. Afr. J. Bot.* **2014**, *95*, 24–31. [CrossRef]

80. Khattab, H. Role of Glutathione and Polyadenylic Acid on the Oxidative Defense Systems of Two Different Cultivars of Canola Seedlings Grown Under Saline Conditions. *Aust. J. Basic Appl. Sci.* **2007**, *1*, 323–334.

81. Pourcel, L.; Routaboul, J.M.; Cheynier, V.; Lepiniec, L.; Debeaujon, I. Flavonoid Oxidation in Plants: From Biochemical Properties to Physiological Functions. *Trends Plant Sci.* **2007**, *12*, 29–36. [CrossRef]

82. Shao, H.B.; Chu, L.Y.; Shao, M.A.; Jaleel, C.A.; Hong-mei, M. Higher Plant Antioxidants and Redox Signaling Under Environmental Stresses. *C. R. Biol.* **2008**, *331*, 433–441. [CrossRef]

83. Hu, Y.; Schmidhalter, U. Drought and Salinity: A Comparison of Their Effects on Mineral Nutrition of Plants. *J. Plant Nutr. Soil Sci.* **2005**, *168*, 541–549. [CrossRef]

 *agronomy*

Article

Role of *Trichoderma aggressivum* f. *europaeum* as Plant-Growth Promoter in Horticulture

Brenda Sánchez-Montesinos [1], Fernando Diánez [1,*] , Alejandro Moreno-Gavíra [1], Francisco J. Gea [2] and Mila Santos [1]

[1] Departamento de Agronomía, Escuela Superior de Ingeniería, Universidad de Almería, 04120 Almería, Spain; brensam@hotmail.com (B.S.-M.); alejanmoga@gmail.com (A.M.-G.); msantos@ual.es (M.S.)
[2] Centro de Investigación, Experimentación y Servicios del Champiñón (CIES), Quintanar del Rey, 16220 Cuenca, Spain; fjgea.cies@dipucuenca.es
* Correspondence: fdianez@ual.es; Tel.: +34-628188359

Received: 4 June 2020; Accepted: 8 July 2020; Published: 13 July 2020

Abstract: The main objective of this study was to determine the capacity of *Trichoderma aggressivum* f. *europaeum* to promote pepper and tomato seedling growth compared to that of *T. saturnisporum*, a species recently characterised as a biostimulant. Consequently, in vitro seed germination and seedling growth tests were performed under commercial plant nursery conditions. Additionally, the effects of different doses and a mixture of both species on seedling growth under plant nursery and subsequently under greenhouse conditions were determined. Furthermore, mass production of spores was determined in different substrates, and their siderophore and indole acetic acid production and phosphate (P) solubilisation capacity were also determined. Direct application of *Trichoderma aggressivum* f. *europaeum* to seeds in vitro neither increases the percentage of pepper and tomato seed germination nor improves their vigour index. However, substrate irrigation using different doses under commercial plant nursery conditions increases the quality of tomato and pepper seedlings. Tomato roots increased by 66.66% at doses of 10^6 spores per plant. Applying *T. aggressivum* f. *europaeum* or *T. saturnisporum* under plant nursery conditions added value to seedlings because their growth-promoting effect is maintained under greenhouse conditions up to three months after transplantation. The combined application of the two species had no beneficial effect in relation to that of the control. The present study demonstrates the biostimulant capacity of *T. aggressivum* f. *europaeum* in pepper and tomato plants under commercial plant nursery and greenhouse conditions.

Keywords: *Trichoderma*; plant growth promotion; tomato; pepper; biostimulant

---

1. Introduction

The success of applying *Trichoderma* in agriculture results from the multiple benefits that it generates in plants. Thus, the genus *Trichoderma* is characterised by its strong competitive and reproductive potential, presenting high survival rates under unfavourable or abiotic stress conditions, such as salinity [1], water stress [2], or the presence of various toxic chemicals, including fungicides [3], among others. Similarly, *Trichoderma* exhibits high efficiency in the promotion of nutrient uptake [4], the capacity to modify the rhizosphere and root structure in which the fungus is established [5,6], high aggressiveness against plant pathogenic fungi, efficiency in the promotion of plant growth [7–12], and the ability to induce plant defence mechanisms, among many additional benefits [8,9,13]. The properties of *Trichoderma* have generated considerable research interest in these fungi for use in agriculture, and a large number of commercial products have been developed using different *Trichoderma* species [10,14]. Many formulations contain mixtures of different species that provide a wider range of direct and indirect beneficial effects for the plants. Numerous studies have reported the benefits of *Trichoderma* application for plant growth and even increased production yield. Thus, applying *Trichoderma* species, to both soil and seeds, allows the multiplication of the fungus in conjunction with the developing root system [15]. Its ability to colonise plant roots from the appressorium-like structure directly enhances seed vigour [16] and germination and promotes seedling growth [10,11,17]; thereby, suggesting that these fungi should be applied from the plant nursery stage in the case of horticultural, ornamental, or forest species, which would allow the early colonisation of the roots by *Trichoderma*, before transplanting the seedlings in the field.

It has been reported that plant growth is enhanced in association with *Trichoderma* species similar to that of other plant-growth-promoting microorganisms (PGPMs), but the effects are greater with *Trichoderma* when plants are under biotic, abiotic, or physiological stress conditions [9,18–21]. Recently, *T. aggressivum* f. *europaeum* has been described as a melon seedling growth promoter under saline stress conditions, in addition to its capacity to control *Pythium ultimum*, decreasing the severity of the disease in seedlings [1]. *Trichoderma aggressivum* Samuels & W. Gams is the causal agent of the green mould disease, which causes economic losses in the cultivation of white button mushrooms (*Agaricus bisporus* (J.E. Lange) Imbach) worldwide. There are two subspecies, *T. aggressivum* f. *aggressivum* and *T. aggressivum* f. *europaeum* found in North America and Europe, respectively [22]. *Trichoderma aggressivum*, a fast-growing filamentous fungus, colonises compost and casings used as growth substrates in mushroom cultivation and produces dense white mycelial colonies that change colour to green after sporulation [23]. This aggressive competitor is known to produce metabolites that are toxic to *A. bisporus* [24,25]. In areas colonised by *T. aggressivum*, fruit body formation is retarded, and fruit bodies may be of poor quality because of damage or discolouration [23]. Numerous *Trichoderma* species have been isolated from *Agaricus* compost and *Pleurotus* substrates, such as *T. harzianum*, *T. longibrachiatum*, *Trichoderma ghanense*, *T. asperellu*, and *T. atroviride*, although its aggressiveness has not been determined [26]. Sánchez-Montesinos et al. [1] demonstrated its high mycelial growth and sporulation on roots. Thus, *T. aggressivum* f. *europaeum* is a potential biofertilizer for different crops. In our study, the growth-promoting capacity of this species has been analysed in comparison to that of *T. saturnisporum* Ca1606, which was recently characterised as a biocontrol agent and a seedling growth promoter for different horticultural plants [11,16,27]. Since the effectiveness of microorganisms as growth promoters will depend on the crop, dose and application method, among many other factors, further studies on *T. aggressivum* f. *europaeum* are needed to determine its efficacy.

Consequently, in the present study, *T. aggressivum* f. *europaeum* Tae52481 and *T. saturnisporum* Ca1606, were tested to evaluate: (a) the effects of direct application to seeds of a fungus suspension on root colonisation of tomatoes and peppers and subsequent plant vigour; (b) the promotion of growth and quality of pepper and tomato seedlings under a conventional production system; and (c) the effects of applying different doses and the synergistic effect of both isolates on tomato seedlings and on their subsequent transplantation under greenhouse conditions.

2. Materials and Methods

2.1. Fungal Isolates

Trichoderma saturnisporum Ca1606 (TS), already known for their plant growth promotion properties, were extracted from suppressive soils. TS was cultivated on potato dextrose agar (PDA) for 7 days at 25 °C in lightless conditions. The growth results measured were used to establish a comparison value.

For this study *Trichoderma aggressivum* f. *europaeum* Tae52481 (TA) were isolated from samples of substrate used for *Agaricus bisporus* cultivation at mushroom farms. These fungal spore samples were similarly cultivated on potato dextrose agar (PDA) for 7 days at 25 °C in dark conditions. The corresponding growth results were recorded. The spore suspensions for both samples were prepared using sterile distilled water. A concentration of 1×10^7 spores/mL was achieved with a Neubauer haemocytometer.

2.2. Analysis of Plant Growth-Promoting Attributes

In accordance with the method of Louden et al. [28], by the transference of fungal mycelial discs (5 mm) of active culture onto Chrome-Azurol S (CAS) agar medium, siderophore production was determined. At 24, 48 and 72 h the diameter of the siderophore colony indicative orange halos on blue were measured.

Indole-3-acetic acid (IAA) production was estimated according to the procedure described by Diánez et al. [16]. Five independent replicates of TA and TS were analysed. This process is described as follows. A glucose peptone broth (GPB) of 50 mL, amended with or without L-tryptophan (Sigma-Aldrich) at a concentration of 100 mg L^{-1} was prepared. Flasks containing this broth inoculated with TA and TS were incubated on an orbital shaker at 150 rpm in dark conditions for 7 days at 25 °C. Subsequently the supernatants from each flask, having first being centrifuged for 30 min at 12,000× *g* and filtered through sterile Millipore membranes (pore size 0.22 μm), were collected into sterile test tubes. In order to determine the quantity of IAA, optical density tests were carried out and compared to a standard IAA curve. For both the TA and TS, 3 mL of the culture supernatant and 2 mL (0.5 mol L^{-1} $FeCl_3$ + 98 mL of 35% $HClO_4$) Salkowski reagent were combined and left for 30 min. The intensity of the resulting red pigmentation density was measured at 530 nm using a scanning spectrophotometer for each of the samples.

To determine the quantitative estimation of phosphate solubilisation, a modified version of the Lima–Rivera procedure [29] was followed. Then, 250 mL capacity flasks containing 50 mL National Botanical Research Institute's phosphate (NBRIP) broth, inoculated with two 5 mm pure *Trichoderma* isolates agar disks were agitated at 100 rpm and incubated at 26 °C for 3, 5, 7, 10 and 15 days. As a control the procedure was carried out on uninoculated flasks containing the same NBRIP broth. The experiments were conducted in triplicate.

Using the Fiske and Subbarow method [30] phosphate concentrations in culture supernatants were estimated as equivalent phosphate (μg mL^{-1}), mean values expressed and pH analysed. The total P (phosphate) in the flasks was 10 mg mL^{-1}.

2.3. Mass Production of TA and TS on Solid Substrates

A mixture of two kinds of substrates, one containing buckwheat husk (BH) and oat (O), the other containing BH and rice (R) were tested for the mass multiplication of TA and TS [31]. Different proportions of BH-O (90–10%, 80–20% and 70–30% *v/v*) and BH-R (90–10%, 80–20% and 70–30% *v/v*) were submerged in 30% *v/v* of water for 24 h. Each mixture was sterilised for 1 h at 125 °C twice on consecutive days. Each mixture was placed on a tray and aseptically inoculated by spraying with 5 mL of spore suspension containing 1×10^7 spores mL^{-1} of each isolate. The trays were kept at 25 °C in the dark for 15 days. In total, three samples (2 g) of the fungus-colonised substrate were removed from the trays in each treatment. The samples were successively diluted in sterile distilled water + 0.01% Tween 20® and the number of conidia g^{-1} of the solid substrate was quantified for each replicate using a Neubauer haemocytometer. There were three replications per treatment. The collected spores were used in the different experiments conducted in this study.

2.4. Analysis of Effects of TA and TS on Seed Germination under Laboratory Conditions

Three treatments (control, TA and TS) and four repetitions following a random block experimental design were implemented in this study. For each repetition of the three treatments 50 seeds of tomato (*Solanum lycopersicum* 'Red Cherry') and pepper (*Capsicum annuum* 'Largo de Reus') were germinated on two sheets of sterile distilled water moistened Whatman No. 1 filter paper in (150 mm) Petri dishes. These seeds were first surface sterilized for 5 min with 1.5% sodium hypochlorite (NaOCl), rinsed twice with sterile distilled water and dried under laminar airflow on sterile paper [16]. Germination was achieved by treating the seeds with 50 μL of spore suspension (1×10^5 spores mL^{-1}) of TA, TS or 50 μL of sterile water (control). The trays were placed in a lightless incubator at 25 ± 1 °C, 7 days for tomato and 10 days for pepper seeds. For each Petri dish treated with one of the three solutions (control, TA and TS), percent germination, root length and shoot length of tomato and pepper seeds were recorded. A Seed Vigour Index (SVI) was calculated as follows: SVI (length) = seed germination% (mean root length + mean shoot length) [32].

2.5. Analysis of Promoter Effects of TA and TS on Pepper and Tomato Seedlings: Experiment 1

The following experiment was conducted in autumn using a completely randomised design at a commercial nursery (Almería, Spain). Pepper (*Capsicum annuum* 'Largo de Reus') and tomato (*Solanum lycopersicum* 'Red Cherry') seeds were sown in 96-cell commercial peat mix filled nursery polystyrene planting trays (70 mL volume) and covered with vermiculite. Trays were relocated to a greenhouse and rinsed with sterile distilled water (control), or a 5 mL (TA or TS) spore suspension per cell at 10^5 spores per plant, after a 2 day (tomato) or 4 day (pepper) period in a germination room (relative humidity (RH) = 95%; 25 °C). Four trays of seedlings for each treatment were cultivated under standard nursery culture conditions (18–28 °C; 75.4 ± 6.7% RH). Then, 20 plants per treatment and control were randomly selected from the four replications at 45 days after sowing across the four replications. Different growth parameters: number of leaves, stem length, stem base diameter, total leaf area and root dry weights, as well as leaf area using the WINDIAS 3.1 of the plants, were measured. The formula: DQI = TDW/((LS/D) + SDW/RDW)) where TDW is the total dry weight (g), LS is stem length (cm), D is stem diameter (mm), SDW and RDW are stem and root dry weight (g), respectively; they were employed to determine the Dickson Quality Index (DQI) [33].

2.6. Analysis of Effects of Applying Different Doses of TA and TS to Tomatoes: Experiment 2

The experimental procedure followed for experiment 2 was similar to that described for experiment 1, although conducted in winter. Again, propagated in substrate appropriately irrigated according to climate and crop necessity under commercial plant nursery conditions and supplemented with a commercial complex nutrient fertiliser, 96 tomato seedlings per replicate of four were treated with three solutions of spore suspension, each with 5 mL of TA, TS conidia and TA + TS (M) (TA D1, TS D1 and M D1: 10^5 spores mL^{-1}; TA D2, TS D2 and M D2: 10^6 spores mL^{-1}; and TA D3, TS D3 and M D3: 10^7 spores mL^{-1}). After 30 days of sowing, twenty plants from each of the three treatment batches and control were randomly selected for harvest. The plants were measured, and data were recorded for the same parameters described in experiment 1. In mid-February a further 25 plants were transplanted into a sandy soil and analysed in mid-May.

In all tests, roots inoculated with *Trichoderma* isolates were collected at the end of the tests. Roots were surface sterilized in 0.1% sodium hypochlorite and washed with sterilised water. Root fragments were placed in PDA medium to determine root colonisation by the fungal isolate.

2.7. Statistical Analysis

The experimental results are presented as the means and standard error (± SE) for the different replicates. Mean separation was carried out using Fisher's Least Significant Difference (LSD) test. The data were tested by one-way analysis of variance (ANOVA) or Student's *t*-test with significance defined as *p*-values less than 0.05 ($p < 0.05$). Statgraphics Centurion 18 Software was utilised for statistical analysis.

3. Results

3.1. Mass Production of Trichoderma Isolates on Solid Substrates

The results are outlined in Table 1. Both isolates grew and sporulated well in all mixtures tested. The proportion of 70 + 30% for buckwheat husk and oats (Figure 1), respectively, and 80 + 20% for buckwheat husk and rice, resulted in significantly higher spore production for both species, followed by

90 + 10% and 70 + 30% of BH + 10% R (Table 1). The lowest spore production rate was observed for 80 BH + 20% O.

Table 1. Mass production of spores on solid substrates (CFU g^{-1}).

Treatments	*T. aggressivum* **f.** *europaeum*	*T. saturnisporum*
90% BH + 10% O	$6.65 \cdot 10^8 \pm 3.04 \cdot 10^7$ c	$6.48 \cdot 10^8 \pm 2.84 \cdot 10^7$ c
80% BH + 20% O	$5.63 \cdot 10^8 \pm 3.20 \cdot 10^7$ d	$5.17 \cdot 10^8 \pm 6.60 \cdot 10^7$ d
70% BH + 30% O	$1.04 \cdot 10^9 \pm 1.44 \cdot 10^7$ a	$9.98 \cdot 10^8 \pm 5.69 \cdot 10^7$ a
90% BH + 10% R	$8.32 \cdot 10^8 \pm 1.61 \cdot 10^7$ b	$7.88 \cdot 10^8 \pm 6.45 \cdot 10^7$ b
80% BH + 20% R	$1.04 \cdot 10^9 \pm 1.04 \cdot 10^7$ a	$1.02 \cdot 10^8 \pm 6.26 \cdot 10^7$ a
70% BH + 30% R	$8.00 \cdot 10^8 \pm 5.00 \cdot 10^7$ b	$7.12 \cdot 10^8 \pm 4.25 \cdot 10^7$ bc
p-value	0.0000	0.0000

BH: buckwheat husk; O: oat; R: rice; CFU: colony forming unit. Data were analysed by ANOVA and treatment means were compared according to Fisher's Least Significant Difference (LSD) statistical procedure (*F*-test at $p < 0.05$). Different letters indicate significant differences according to the one-way ANOVA test ($p = 0.05$).

Figure 1. Mass production of (**A**) *Trichoderma aggressivum* f. *europaeum* and (**B**) *Trichoderma saturnisporum* on 70 + 30% for buckwheat husk and oats.

3.2. Siderophore Production, IAA and P Solubilisation

TA and TS siderophore production was observed in the formation of an orange-coloured zone around the fungal colonies at 24 and 48 h, and the production of TA was higher, in both cases. No increase in the diameter of the halo (mm) was detected at 72 h in any isolate (Table 2).

Table 2. Siderophores and IAA production by *Trichoderma* isolates.

	Radius of Siderophores Production (mm)			IAA (mg mL^{-1})	
Treatment	24 h	48 h	72 h	+Trp	−Trp
p-value	0.0000	0.0000	-	0.0068	0.0304
T. aggressivum	9.73 ± 0.89	18.50 ± 1.70	-	0.145 ± 0.011	0.085 ± 0.009
T. saturnisporum	5.45 ± 0.31	9.82 ± 0.56	-	0.199 ± 0.014	0.129 ± 0.021

Values are average of five replications; values after ± represent standard deviation. IAA: indole-3-acetic acid; +Trp: with L-tryptophan; −Trp: without L-tryptophan.

Although both *Trichoderma* strains exhibited an in vitro ability to produce IAA in medium supplemented with and without 100 mg L^{-1} tryptophan during a 7-day period, the production of *T. saturnisporum* was higher under both conditions tested (Table 2). In both TA and TS, IAA production increased in the medium supplemented with tryptophan.

The effects of TA and TS on the soluble phosphate concentration are shown in Figure 2. The initial concentration of P in the medium was used to quantify the concentration of P solubilised by both isolates.

As shown in Figure 2, P solubilisation was significant from the fifth day of incubation, with no significant differences between the two isolates. Furthermore, no change in the pH of the medium was detected, which remained at approximately 6.5–7.

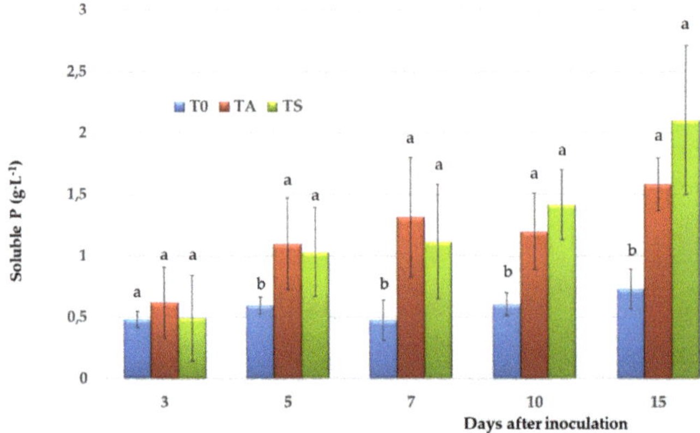

Figure 2. Effects of *Trichoderma aggressivum* f. *europaeum* (TA) and *Trichoderma saturnisporum* (TS) on phosphate solubilisation in National Botanical Research Institute's phosphate (NBRIP) broth containing tribasic calcium phosphate (10 g). T0: NBRIP broth without *Trichoderma* isolates. The results are shown as the average of the three replicates, in g L^{-1}. Mean standard deviation is expressed in the error bar ($n = 3$). For each isolate, columns marked with different letters indicate a significant difference at $p < 0.05$.

3.3. Effects of TA and TS Treatment on Germination and Vigour Index

The results from the in vitro application of TA and TS spores to tomato and pepper seeds are outlined in Table 3. No significant effects on pepper and tomato seed germination percentages were observed in either treatment. However, the application of both *Trichoderma* isolates led to a decrease in the radicle and hypocotyl length (growth) parameters and significantly decreased the SVI in peppers. The tomato seed vigour index was not affected by TA or TS treatment ($p = 0.1918$).

Table 3. Effects of *T. aggressivum* f. *europaeum* and *T. saturnisporum* on tomato and pepper seed germination 7 and 10 days after treatment, respectively.

Treatment	% Germination	Root Length (cm)	Shoot Length (cm)	Seed Vigour Index
		Pepper		
p-value	0.5420	0.0126	0.0010	0.0030
T. aggressivum	83 ± 6.83a	0.64 ± 0.18b	1.66 ± 0.55b	138.14 ± 35.48b
T. saturnisporum	80 ± 3.26a	0.91 ± 0.11b	1.29 ± 0.17b	176.50 ± 18.14b
Control	78 ± 4.61a	2.16 ± 1.01a	1.98 ± 0.56a	320.32 ± 83.36a
		Tomato		
p-value	0.5268	0.0020	0.3154	0.1918
T. aggressivum	89 ± 6.83a	4.53 ± 0.31a	2.97 ± 0.47a	671.80 ± 112.91a
T. saturnisporum	92 ± 7.30a	3.25 ± 0.37b	2.55 ± 0.27a	536.29 ± 76.26a
Control	85 ± 8.32a	3.95 ± 0.28b	2.81 ± 0.29a	580.13 ± 96.37a

Different letters indicate significant differences according to the one-way ANOVA test ($p = 0.05$).

3.4. Effects of Trichoderma Inoculation on Tomato and Pepper Seedlings

The effects of TA and TS application on morphological parameters and DQI are shown in Table 4. Unlike the results from the direct application of both *Trichoderma* to the seeds, the application to the substrate increased the study parameters compared to that of the control, and the results were better in peppers than in tomatoes, with better quality seedlings, according to the DQI values (Figure 3). There were no significant differences after the application of TA and TS in both horticultural plants. The increased percentage assessed in pepper seedlings for each species (TA/TS) was 8%/8.5% for stem length, 12.32/~0.01 for stem diameter, 7.77/5.5 for leaf number, 22.22/25 for shoot dry weight, 36.36/63.63 for root dry weight and 13.83/13.74 for leaf area, respectively. In tomato seedlings, the percentages were 9/6 for stem length, 0.5/1.5 for stem diameter, 6/8.8 for leaf number, 12.5/5.3 for shoot dry weight, 0/–6.6 for root dry weight and 8/9.2 for leaf area. No significant differences in DQI were found in tomato seedlings for any treatment applied with respect to that of the control.

Figure 3. (**A**) Differential growth of pepper seedlings with *T. aggressivum* f. *europaeum* (TA), compared to control. (**B**) Tomato plants grown under field transplantation conditions (60 days).

3.5. Effects of Dose of Application of T. aggressivum f. europaeum and T. saturnisporum

Since no significant plant growth-promoting results were found in tomato seedlings, the effects of applying three doses of both species separately, as well as jointly, were determined. The results are outlined in Table 5, wherein values significantly higher than that of the control are highlighted in green, and negative values in red, for better visualisation.

The increase in the dose of both species improved seedling quality, increasing all study parameters in TA D2 and TS D3 treatments, with respect to that of the control. In treatment TA D2, stem length increased 14.37%, plant diameter 9.4%, leaf number 21.58%, shoot dry weight 16.66% and root dry weight 66.66%. In treatment TS D3, stem length increased 39.05%, plant diameter 15.22%, leaf number 11.55%, shoot dry weight 12.5% and root dry weight 33.33%. Although most treatments favoured the development of seedling shoots, no favourable results were found in roots; therefore, the seedling quality was not improved. The combination or mixture of the two species for the three doses tested did not improve the results compared to that of their separate application.

The results of the study parameters after transplantation of the seedlings into the soil are outlined in Table 5 (Figure 3). Three treatments, TA D1, TA D2 and TS D2, led to a good relationship between tomato shoots and roots, with significantly higher plant quality, compared to that of the control, without a new application of *Trichoderma*. Thus, shoot dry weight increased 43.20%, 22.84% and 29.58% and root dry weight increased 29.94%, 39.32% and 31.51% after the TA D1, TA D2 and TS D2 treatment, respectively. The establishment of the endophytic fungus at the root (Figure 4) enabled its effects to persist after transplantation.

Table 4. Morphological parameters and quality index of pepper and tomato seedlings treated with *T. aggressivum* f. *europaeum* and *T. saturnisporum* at 45 days after sowing.

Treatment	Length of Stem (cm)	Diameter (mm)	Number of Leaves	Aerial Dry Weight (g)	Root Dry Weight (g)	Leaf Area (mm^2)	DQI
			Pepper				
p-value	0.0000	0.0000	0.0092	0.0000	0.0000	0.0001	0.0000
T. aggressivum	29.18 ± 1.36a	4.10 ± 0.24a	7.22 ± 0.76a	0.44 ± 0.06a	0.15 ± 0.30b	86.38 ± 12.58a	0.06 ± 0.01a
T. saturnisporum	29.31 ± 1.75a	3.68 ± 0.22b	7.07 ± 0.82a	0.45 ± 0.06a	0.18 ± 0.09a	86.31 ± 13.22a	0.06 ± 0.01a
Control	27.01 ± 2.07b	3.65 ± 0.23b	6.70 ± 0.72b	0.36 ± 0.04b	0.11 ± 0.03c	75.88 ± 11.15b	0.04 ± 0.01b
			Tomato				
p-value	0.0000	0.4387	0.0031	0.0232	0.0245	0.0295	0.0793
T. aggressivum	27.93 ± 1.99a	3.86 ± 1.84a	4.45 ± 0.50a	0.63 ± 0.09a	0.15 ± 0.02ab	74.63 ± 10.99a	0.06 ± 0.01a
T. saturnisporum	27.17 ± 1.62a	3.90 ± 0.24a	4.57 ± 0.50a	0.59 ± 0.10ab	0.14 ± 0.02b	75,52 ± 13.40a	0.06 ± 0.01a
Control	25.61 ± 2.07b	3.84 ± 0.26a	4.20 ± 0.46b	0.56 ± 0.10b	0.15 ± 0.02a	69.12 ± 6.96b	0.06 ± 0.00a

Different letters indicate significant differences according to the one-way ANOVA test (p = 0.05). DQI: Dickson Quality Index.

Table 5. Morphological parameters and DQI of tomato seedlings and plants treated with different doses (10^5, 10^6 and 10^7 spores per plant; D1, D2 and D3, respectively) of *T. aggressieum* f. *europaeum* (TA), *T. saturnisporum* (TS) and mix (M) of two species.

	Tomato Seedling					
Treatment	Length of Stem (mm)	Diameter (mm)	N° Leaves	Aerial Dry Weight (g)	Root Dry Weight (g)	DQI
p-value	0.0000	0.0000	0.0000	0.0000	0.0000	0.0000
TA D1	135.21 ± 28.79de	3.91 ± 0.46ab	4 ± 0.65b	0.27 ± 0.09a	0.03 ± 0.02cd	0.025 ± 0.01de
TA D2	149.80 ± 32.58d	3.81 ± 0.34b	4 ± 0.51b	0.28 ± 0.07a	0.05 ± 0.018a	0.034 ± 0.01a
TA D3	127.02 ± 25.41e	3.76 ± 0.46b	3.98 ± 0.64b	0.27 ± 0.10a	0.04 ± 0.02ab	0.030 ± 0.01ab
TS D1	141.25 ± 19.83d	3.55 ± 0.49c	3.65 ± 0.60cd	0.23 ± 0.10b	0.03 ± 0.01cd	0.024 ± 0.01de
TS D2	188.44 ± 30.29a	4.07 ± 0.33a	4.25 ± 0.73a	0.28 ± 0.07a	0.03 ± 0.01d	0.021 ± 0.01e
TS D3	174.83 ± 27.78b	4.01 ± 0.45a	3.67 ± 0.63cd	0.27 ± 0.09a	0.04 ± 0.02a	0.030 ± 0.01ab
M D1	157.25 ± 29.14c	3.5 ± 0.39c	3.56 ± 0.62de	0.27 ± 0.061a	0.04 ± 0.01bc	0.026 ± 0.00bcd
M D2	99.08 ± 21.02b	3.13 ± 0.23b	3.35 ± 0.39bc	0.28 ± 0.08a	0.04 ± 0.01bcd	0.029 ± 0.00bcd
M D3	98.54 ± 21.37f	3.11 ± 0.45d	3.33 ± 0.63ef	0.18 ± 0.08c	0.04 ± 0.01ab	0.029 ± 0.00abc
Control	125.73 ± 22.3e	3.48 ± 0.31c	3.29 ± 0.62f	0.24 ± 0.07b	0.03 ± 0.01cd	0.025 ± 0.01cde

	Tomato Plants					
Treatment	Length of Stem (cm)	Diameter (mm)	Internodes	Aerial Dry Weight (g)	Root Dry Weight (g)	DQI
p-value	0.0000	0.5379	0.0159	0.0015	0.2769	0.5373
TA D1	101.40 ± 15.06ab	12.11 ± 0.90a	15.00 ± 1.15ab	61.19 ± 11.27a	4.99 ± 0.68ab	3.23 ± 0.24a
TA D2	102.80 ± 9.47a	12.11 ± 1.24a	14.10 ± 1.29ab	52.49 ± 11.26	5.35 ± 0.57a	3.20 ± 0.46a
TA D3	92.20 ± 9.82bc	11.26 ± 1.17ab	14.40 ± 0.84abc	54.74 ± 13.09abc	4.70 ± 1.10abc	3.01 ± 0.58ab
TS D1	77.55 ± 7.76e	11.74 ± 1.18ab	13.90 ± 1.37bc	49.26 ± 14.78bcde	4.40 ± 1.04bc	2.91 ± 0.56ab
TS D2	88.75 ± 10.63cd	11.92 ± 1.57ab	15.20 ± 1.03a	55.37 ± 8.24ab	5.05 ± 1.23ab	3.19 ± 0.42a
TS D3	81.00 ± 12.39de	11.58 ± 1.37ab	13.20 ± 2.57c	51.71 ± 7.97abcd	4.34 ± 1.17bc	3.01 ± 0.94ab
M D1	79.05 ± 13.20e	10.87 ± 1.38b	13.90 ± 1.29bc	47.00 ± 11.38bcde	4.14 ± 1.11bc	2.73 ± 0.64ab
M D2	76.18 ± 11.91e	11.61 ± 0.95ab	13.20 ± 1.23c	45.30 ± 9.79cde	4.23 ± 0.99bc	2.85 ± 0.59ab
M D3	78.45 ± 8.08e	11.22 ± 1.18ab	14.30 ± 1.16abc	40.91 ± 11.56e	4.28 ± 1.37bc	2.72 ± 0.86ab
Control	77.96 ± 7.29e	11.52 ± 2.03ab	14.25 ± 1.16abc	42.73 ± 6.46de	3.84 ± 0.62c	2.60 ± 0.43b

Different letters indicate significant differences according to the one-way ANOVA test ($p = 0.05$). Green: favourable; Red: unfavourable; Orange: no effect compared to control. DQI: Dickson Quality Index.

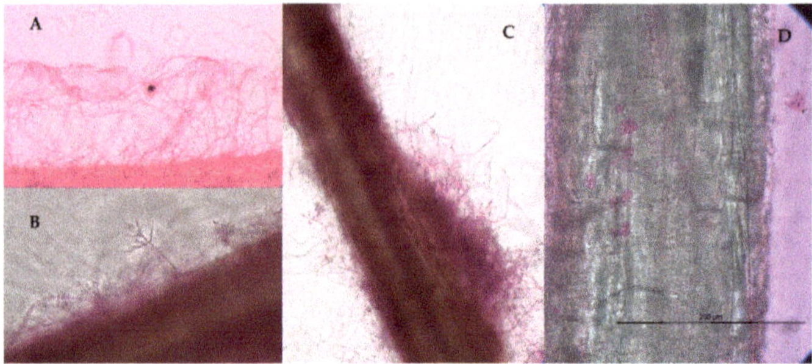

Figure 4. (**A**) Colonization of pepper and tomato roots by *T. aggressivum* f. *europaeum*. (**C**) Mycelium in pepper root (100×). (**B**) Conidiophores and mycelium in tomato root (100×). (**D**) Chlamydospores in pepper root (200×).

4. Discussion

Numerous *Trichoderma* species have been described as plant-growth promoters, including *T. harzianum, T. longipile, T. tomentosum, T. viride, T. koningii, T. asperellum, T. aureoviride* and *T. saturnisporum*, among others [34]. This ability to promote growth depends on several factors, including the existence of isolates of the same species that may or may not promote plant growth, or for example, the crop and/or variety to which the species is applied [34]. Similarly, the use of a mixture of species has been extensively studied and commercialised to increase this activity [10]. In this study, the plant growth-promoting capacity of a new species, *T. aggressivum* f. *europaeum*, which is characterised by its rapid growth and sporulation, was analysed and compared to that of *T. saturnisporum*, a species characterised as a plant-growth promoter by Diánez et al. [16,18]. Although Allaga et al. [35] recommend not using species that produce green mould disease, these species do not create any problems in horticultural crops or pose any danger to mushroom crops, as long as they are applied in different geographical areas. Additionally, mushrooms are produced in closed locations and under completely different conditions. Furthermore, plant remains under horticultural production are not used to prepare substrates for mushroom cultivation, as shown in many commercial species; neither are plant remains that have been studied with plant-growth promoters, which may also cause green mould disease, such as *T. harzianum* [36] or *T. longibrachiatum* [37].

The first objective was to obtain viable spores with high yield on low-cost substrates. This product was used for additional tests, which demonstrated that the nutritional composition of the substrates used did not affect the biostimulant capacity of either *Trichoderma* species. Lane [38] determined that the nutrients provided in the medium could affect the biocontrol or biostimulant capacity of the agent. Different substrates have been used for *Trichoderma* spore production, including barley straw [39], wheat, rice, corn kernels [40] or a mixture of substrates, such as wheat straw, bran, cassava, potato starch and sugar beets [41,42], among others. In general, in our study, high yields, expressed as colony forming unit (CFU) g^{-1}, were assessed in all substrate mixtures tested; the yields increased both in 80% buckwheat husk + 20% rice and 70% buckwheat husks + 30% oats. Although in laboratory tests, extraction could be performed without a problem in all mixtures, in the extractor tank, mixtures containing rice adhered to the walls and pipes, complicating the subsequent extraction and filtration processes. For this reason, to develop low-cost production methods for industrial scale-up, rice was rejected as a constituent of the production substrate for TA and TS. A high siderophore and IAA production and P solubilisation by TA and TS compared to other *Trichoderma* species or isolates were demonstrated in our study. These three components play key roles in plant biostimulation by increasing nutrient availability to plants, such as for hormone production [43,44]. However, the direct relationship between IAA production and plant-growth promotion is not yet clear because numerous species can produce IAA, but they do not promote plant growth [45]. Hoyos et al. [45] concluded that IAA production is not a species-dependent quality of *Trichoderma* and found no direct correlation between biostimulation and IAA and siderophore production or P solubilisation. In turn, Vinale et al. [46] highlighted the effects of siderophore (harzianic acid) production on the germination of tomato seeds and the improved growth of the seedlings even under iron-deficient conditions. Similarly, Qi and Zhao [47] demonstrated that applying *T. asperellum* enhanced cucumber growth by inducing physiological protection under saline stress, and its siderophores played a key role in mitigating the negative effects of salinity.

Many *Trichoderma* species can produce IAA, and high IAA secretion in the presence of tryptophan indicates the importance of tryptophan as a precursor for IAA production [48,49]. Gravel et al. [50] reported that IAA production induced by L-tryptophan increased the fresh weight of tomato shoots and roots. Our results indicate that TA and TS produce much higher amounts of IAA than those assessed by other authors. Accordingly, Saber et al. [48] described IAA production of *T. harzianum* isolates that were 10 times lower than that of *T. aggressivum* f. *europaeum* and *T. saturnisporum* assessed in this study. Bader et al. [51] reported that IAA production ranged from 13.38 to 21.14 μg mL^{-1} in *T. brevicompactum*, *T. gamsii* and *T. harzianum*. Diánez et al. [16] described a highly similar IAA production for *T. saturnisporum*; therefore, the in vitro production capacity of IAA was preserved despite maintaining the isolate in the laboratory for 10 years. Similarly, phosphate solubilisation by *Trichoderma* species has been described both in vitro and in vivo [52–54]. Recently, Tandon et al. [55] evaluated P solubilisation of different *Trichoderma koningiopsis* isolates under abiotic stress conditions and determined a range from 1.6 to 71 μg mL^{-1}. Bononi et al. [12] found that *Trichoderma* isolated from soils of the Amazon rainforest demonstrated a high potential for phosphate solubilisation, which ranged from 51.7 to 90.3% 10 days after inoculation. Despite their high P solubilisation capacity, some of these isolates inhibited the germination of soybean seeds. In our study, the P solubilisation range of both isolates was lowest on the tenth day of incubation, at 5.9% and 6.16% for TA and TS, respectively.

Applying PGPMs to seeds makes it possible to use a lower concentration of spores while ensuring that the PGPMs are readily accessible at germination and during early developmental plant stages, stimulating healthy and rapid establishment, and consequently, maximising crop production [43]. However, the direct application of different *Trichoderma* isolates or species to seeds (bioprimming) has not always had beneficial effects. In this study, the seed germination rate was not affected by *T. aggressivum* f. *europaeum* or *T. saturnisporum* application. Similar results were found by Azarmi et al. [18] after applying *T. harzianum* isolates to tomato seeds. Hajieghrari et al. [56] demonstrated that direct exposure of corn seeds to *Trichoderma* spores decreased the percentage of seed germination, as well as radicle and shoot length. However, You et al. [57] demonstrated that *T. harzianum* and *T. koningiopsis* isolates significantly enhanced the tomato seed vigour index when they were used to treat tomato seeds. Our results demonstrated that direct *T. aggressivum* f. *europaeum* and *T. saturnisporum* application decreased seed vigour, significantly so in peppers but not in tomatoes. However, the application of either species under commercial plant nursery conditions, via substrate irrigation, similarly enhanced pepper seedling quality significantly, albeit again non-significantly for tomatoes. Optimising the application dose for each species is a factor that should be considered, among other factors, to enable companies and producers to adopt this technology with higher security [58]. Increasing the dose of *T. aggressivum* f. *europaeum* and *T. saturnisporum* applied to tomato seedlings increased most of the study parameters, as well as the DQI value in treatments TA D2, TA D3 and TS D3. The endophytic establishment of *Trichoderma* in plant nurseries may ensure its colonisation once transplanted. As such, in the TA D2 treatment, tomato plants continued to show better quality, without any additional application of *Trichoderma*, and plant quality improved in other treatments with *Trichoderma* applied separately. The poorest results were obtained for mixtures of both species, with no improvement in study parameters for any dose tested, and even a reduction of 21.62%, 10.63% and 25% in stem length, diameter and shoot dry weight of tomato seedlings in the MD3 treatment, respectively. Similar results were found by Liu et al. [59], who reported that the combination of three species, *T. afroharzianum*, *T. pseudoharzianum* and *T. asperelloides*, decreased the biocontrol and growth-promoting effects in comparison to the application of each species separately.

Although major reductions in the use of chemical fertilisers without production losses is currently difficult in many farming systems, their gradual decrease accompanied by the use of biostimulants or biofertilizers is a tool that can optimise the use of chemical inputs while reducing environmental pollution and food crop contamination.

5. Conclusions

The present study demonstrated, for the first time, the biostimulant capacity of *T. aggressivum* f. *europaeum* in pepper and tomato plants under commercial plant nursery and greenhouse conditions, with similar results to those of *T. saturnisporum*.

6. Patents

This isolate was patented with a Spanish patent number ES2706099: New strain of *T. aggressivum* f. *europaeum*, compositions and applications.

Author Contributions: F.D. and M.S. conceived and designed the experiments; B.S.-M., F.J.G., and A.M.-G. performed the experiments; M.S. analysed the data; F.D. wrote the paper. All authors have read and agreed to the published version of the manuscript.

Funding: This research received no external funding.

Acknowledgments: The present work benefited from the input of the project RTC-2017-6486-2 and was supported by the Spanish Ministry of Science, Innovation and Universities.

Conflicts of Interest: The authors declare no conflicts of interest.

References

1. Sánchez-Montesinos, B.; Diánez, F.; Moreno-Gavíra, A.; Gea, F.J.; Santos, M. Plant growth promotion and biocontrol of *Pythium ultimum* by saline tolerant *Trichoderma* isolates under salinity stress. *Int. J. Environ. Res. Public Health* **2019**, *16*, 2053. [CrossRef] [PubMed]

2. Khoshmanzar, E.; Aliasgharzad, N.; Neyshabouri, M.R.; Khoshru, B.; Arzanlou, M.; Asgari Lajayer, B. Effects of *Trichoderma* isolates on tomato growth and inducing its tolerance to water-deficit stress. *Int. J. Environ. Sci. Technol.* **2020**, *17*, 869–878. [CrossRef]

3. Adnan, M.; Islam, W.; Shabbir, A.; Khan, K.A.; Ghramh, H.A.; Huang, Z.; Chen, H.Y.H.; Lu, G. Plant defense against fungal pathogens by antagonistic fungi with *Trichoderma* in focus. *Microb. Pathog.* **2019**, *129*, 7–18. [CrossRef] [PubMed]

4. Halifu, S.; Deng, X.; Song, X.; Song, R. Efects of Two *Trichoderma* Strains on Plant Growth, Rhizosphere Soil Nutrients, and Fungal Community of *Pinus sylvestris* var. mongolica Annual Seedlings. *Forests* **2019**, *10*, 758.

5. Vargas, W.A.; Crutcher, F.K.; Kenerley, C.M. Functional characterization of a plant-like sucrose transporter + from the beneficial fungus *Trichoderma virens*. Regulation of the symbiotic association with plants by sucrose metabolism inside the fungal cells. *New Phytol.* **2010**, *189*, 777–789. [CrossRef]

6. Garnica-Vergara, A.; Barrera-Ortiz, S.; Muñoz-Parra, E.; Raya-Gonzalez, J.; Mendez-Bravo, A.; Macias-Rodriguez, L.; Ruiz-Herrera, L.F.; Lopez-Bucio, J. The volatile 6-pentyl-2H-pyran-2-one from *Trichoderma* atroviride regulates *Arabidopsis thaliana* root morphogenesis via auxin signaling and ethylene insensitive 2 functioning. *New Phytol.* **2016**, *209*, 1496–1512. [CrossRef]

7. Harman, G.E.; Howell, C.R.; Viterbo, A.; Chet, I.; Lorito, M. *Trichoderma* species -opportunistic, avirulent plant symbionts. *Nat. Rev. Microbiol.* **2004**, *2*, 43–56. [CrossRef]

8. Shoresh, M.; Harman, G.E.; Mastouri, F. Induced systemic resistance and plant responses to fungal biocontrol agents. *Annu. Rev. Phytopathol.* **2010**, *48*, 21–43. [CrossRef]

9. Hermosa, R.; Viterbo, A.; Chet, I.; Monte, E. Plant-beneficial effects of *Trichoderma* and of its genes. *Microbiology* **2012**, *158*, 17–25. [CrossRef]

10. López-Bucio, J.; Pelagio-Flores, R.; Herrera-Estrella, A. *Trichoderma* as biostimulant: Exploiting the multilevel properties of a plant beneficial fungus. *Sci. Hortic.* **2015**, *196*, 109–123. [CrossRef]

11. Diánez, F.; Santos, M.; Carretero, F.; Marín, F. Biostimulant activity of *Trichoderma saturnisporum* in melon (Cucumis melo). *Hortscience* **2018**, *53*, 810–815.

12. Bononi, L.; Chiaramonte, J.B.; Pansa, C.C.; Moitinho, M.A.; Melo, I.S. Phosphorus-solubilizing *Trichoderma* pp. from Amazon soils improve soybean plant growth. *Sci. Rep.* **2020**, *10*, 2858. [CrossRef] [PubMed]

13. Yuan, M.; Huang, Y.; Ge, W.; Jia, Z.; Song, S.; Zhang, L.; Huang, Y. Involvement of jasmonic acid, ethylene and salicylic acid signaling pathways behind the systemic resistance induced by *Trichoderma longibrachiatum* H9 in cucumber. *BMC Genom.* **2019**, *20*, 144–157. [CrossRef] [PubMed]

14. Sanjeev, K.; Manibhushan, T.; Archana, R. *Trichoderma*: Mass production, formulation, quality control, delivery and its scope in commercialization in India for the management of plant diseases. *Afr. J. Agric. Res.* **2014**, *9*, 3838–3852.

15. Khan, M.R.; Mohiddin, F.A. Trichoderma: Its Multifarious Utility in Crop Improvement. In *Crop Improvement through Microbial Biotechnology. New and Future Developments in Microbial Biotechnology and Bioengineering*; Aligarh Muslim University: Aligarh, India; Srinagar, India, 2018; Volume 13, pp. 263–291.

16. Diánez, F.; Santos, M.; Carretero, F.; Marín, F. *Trichoderma saturnisporum*, a new biological control agent. *J. Sci. Food Agric.* **2016**, *96*, 1934–1944. [CrossRef]

17. Rocha, I.; Ma, Y.; Souza-Alonso, P.; Vosátka, M.; Freitas, H.; Oliveira, R.S. Seed coating: A tool for delivering beneficial microbes to agricultural crops. *Front. Plant Sci.* **2019**, *10*, 1357. [CrossRef]

18. Azarmi, R.; Hajieghrari, B.; Giglou, A. Effect of *Trichoderma* isolates on tomato seedling growth response and nutrient uptake. *Afr. J. Biotechnol.* **2011**, *10*, 5850–5855.

19. Druzhinina, I.S.; Seidl-Seiboth, V.; Herrera-Estrella, A.; Horwitz, B.A.; Kenerley, C.M.; Monte, E.; Mukherjee, P.K.; Zeilinger, S.; Grigoriev, I.V.; Kubicek, C.P. *Trichoderma: The* genomics of opportunistic success. *Nat. Rev. Microbiol.* **2011**, *9*, 749–759. [CrossRef]

20. Bimenya, G.S.; Kaviri, D.; Mbona, N.; Byarugaba, W. Monitoring the severity of iodine deficiency disorders in Uganda. *Afr. Health Sci.* **2002**, *2*, 63–68.

21. Berg, G. Plant–microbe interactions promoting plant growth and health: Perspectives for controlled use of microorganisms in agriculture. *Appl. Microbiol. Biot.* **2009**, *84*, 11–18. [CrossRef]

22. Samuels, G.J.; Dodd, S.L.; Gams, W.; Castlebury, L.A.; Petrini, O. *Trichoderma* species associated with the green mold epidemic of commercially grown *Agaricus bisporus. Mycologia* **2002**, *94*, 147–170. [CrossRef]

23. Largeteau, M.L.; Savoie, J.M. Microbially induced diseases of *Agaricus bisporus*: Biochemical mechanisms and impact on commercial mushroom production. *App. Microbiol. Biotechnol.* **2010**, *86*, 63–73. [CrossRef]

24. Krupke, O.A.; Castle, A.J.; Rinker, D.L. The North American mushroom competitor, *Trichoderma aggressivum* f. *aggressivum*, produces antifungal compounds in mushroom compost that inhibit mycelial growth of the commercial mushroom *Agaricus bisporus. Mycol. Res.* **2003**, *102*, 1467–1475.

25. Guthrie, J.L.; Castle, A.J. Chitinase production during interaction of *Trichoderma aggressivum* and *Agaricus bisporus. Can. J. Microbiol.* **2006**, *96*, 961–967. [CrossRef]

26. Hatvani, L.; Antal, Z.; Manczinger, L.; Szekeres, A.; Druzhinina, I.S.; Kubicek, C.P.; Nagy, A.; Nagy, E.; Vagvolgyi, C.; Kredics, L. Green mold diseases of *Agaricus* and *Pleurotus* spp. are caused by related but phylogenetically different *Trichoderma* species. *Phytopathology* **2007**, *97*, 532–537. [CrossRef]

27. Sharma, V.; Shanmugam, V. Unraveling the multilevel aspects of least explored plant beneficial *Trichoderma saturnisporum* isolate GITX-Panog. *Eur. J. Plant Pathol.* **2018**, *152*, 169–183. [CrossRef]

28. Louden, B.C.; Harmann, D.; Lynne, A.M. Use of Blue Agar CAS Assay for Siderophore Detection. *J. Microbiol. Biol. Educ.* **2011**, *12*, 51–53. [CrossRef] [PubMed]

29. Lima-Rivera, D.L.; Lopez-Lima, D.; Desgarennes, D.; Velázquez-Rodríguez, A.S.; Carrion, G. Phosphate solubilization by fungi with nematicidal potential. *J. Soil Sci. Plant Nutr.* **2016**, *16*, 507–524. [CrossRef]

30. Fiske, C.H.; Subbarow, Y. The colorimetry determination of phosphorous. *J. Biol. Chem.* **1925**, *66*, 375–400.

31. Moreno-Gavíra, A.; Diánez, F.; Sánchez-Montesinos, B.; Santos, M. *Paecilomyces variotii* as a plant-growth promoter in horticulture. *Agronomy* **2020**, *10*, 597. [CrossRef]

32. Murali, M.; Amruthesh, K.N.; Sudisha, J.S.; Niranjana, R.; Shetty, H.S. Screening for plant growth promoting fungi and their ability for growth promotion and induction of resistance in pearl millet against downy mildew disease. *J. Phytol.* **2012**, *4*, 30–36.

33. Dickson, A.; Leaf, A.L.; Hosner, J.F. Quality appraisal of white spruce and white pine seedling stock in nurseries. *Forest Chron.* **1960**, *36*, 10–13. [CrossRef]

34. Stewart, A.; Hill, R. Applications of Trichoderma in Plant growth promotion. In *Biology and Applications of Trichoderma*; Mukherjee, P.K., Horwitz, B.A., Singh, U.S., Mukherjee, M., Schmoll, M., Eds.; CABI: Wallingford, CT, USA, 2013; pp. 415–425.

35. Allaga, H.; Bóka, B.; Poór, P.; Nagy, V.D.; Szuzs, A.; Stankovics, I.; Takó, M.; Manczinger, L.; Vágvölgyi, C.; Kredics, L.; et al. Composite Bioinoculant Based on the Combined Application of Beneficial Bacteria and Fungi. *Agronomy* **2020**, *10*, 220. [CrossRef]

36. Wang, G.; Cao, X.; Ma, X.; Guo, M.; Liu, C.; Yan, L.; Bian, Y. Diversity and effect of *Trichoderma* spp. associated with green mold disease on *Lentinula edodes* in China. *Microbiologyopen* **2016**, *5*, 709–718. [CrossRef] [PubMed]

37. Zhang, T.; Lu, M.Z.; Zhang, C.L.; Xu, J.Z. First Report of *Trichoderma longibrachiatum* Causing Green Mold Disease on *Ganoderma lingzhi*. *Plant Dis.* **2019**, *103*, 156. [CrossRef]

38. Lane, B.S.; Trinci, A.P.J.; Gillespie, A.T. Influence of culture conditions on the virulence of conidia and blastospores of *Beauveria bassiana* to the green leafhopper, *Nephotettix virescens*. *Mycol. Res.* **1991**, *95*, 829–833. [CrossRef]

39. Serna-Díaz, M.G.; Mercado-Flores, Y.; Jiménez-González, A.; Anducho-Reyes, M.A.; Medina-Marín, J.; Seck Tuoh-Mora, J.C.; Téllez-Jurado, A. Use of barley straw as a support for the production of conidiospores of *Trichoderma harzianum*. *Biotechnol. Rep.* **2020**, *26*, e00445. [CrossRef]

40. Flodman, H.R.; Noureddini, H. Effects of intermittent mechanical mixing on solid-state fermentation of wet corn distiller grain with *Trichoderma reesei*. *Biochem. Eng. J.* **2013**, *81*, 24–28. [CrossRef]

41. Deschamps, F.; Giuliano, C.; Asther, M.; Huet, M.C.; Roussos, S. Cellulase production by *Trichoderma harzianum* in static and mixed solid-state fermentation reactors under nonaseptic conditions. *Biotechnol. Bioeng.* **1985**, *27*, 1385–1388. [CrossRef]

42. Michel-Aceves, A.C.; Otero-Sánchez, M.A.; Martínez-Rojero, R.D.; Rodríguez-Morán, N.L.; Ariza-Flores, R.; Barrios-Ayala, A. Producción masiva de *Trichoderma harzianum* Rifai en diferentes sustratos orgánicos. *Rev. Chapingo Ser. Hortic.* **2008**, *14*, 185–191. [CrossRef]

43. Zhang, F.G.; Yuan, J.; Yang, X.M.; Cui, Y.Q.; Chen, L.H.; Ran, W.; Shen, Q.R. Putative *Trichoderma harzianum* mutant promotes cucumber growth by enhanced production of indoleacetic acid and plant colonization. *Plant Soil* **2013**, *368*, 433–444. [CrossRef]

44. Contreras-Cornejo, H.A.; Macías-Rodríguez, L.; Cortés-Penagos, C.; López-Bucio, J. *Trichoderma virens*, a plant beneficial fungus, enhances biomass production and promotes lateral root growth through an auxin-dependent mechanism in Arabidopsis. *Plant Phys.* **2009**, *149*, 1579–1592. [CrossRef] [PubMed]

45. Hoyos-Carvajal, L.; Orduz, S.; Bissett, J. Growth stimulation in bean (*Phaseolus vulgaris* L.) by *Trichoderma*. *Biol. Control.* **2009**, *51*, 409–416. [CrossRef]

46. Vinale, V.; Nigro, M.; Sivasithamparam, K.; Flematti, G.; Ghisalberti, E.L.; Ruocco, M.; Varlese, R.; Marra, R.; Lanzuise, S.; Eid, A.; et al. Harzianic acid: A novel siderophore from *Trichoderma harzianum*. *FEMS Microbiol. Lett.* **2013**, *347*, 123–129. [CrossRef] [PubMed]

47. Qi, W.; Zhao, L. Study of the siderophore-producing *Trichoderma asperellum* Q1 on cucumber growth promotion under salt stress. *J. Basic Microbiol.* **2013**, *53*, 355–364. [CrossRef]

48. Saber, W.I.A.; Ghoneem, K.; Rashad, Y.M.; AlAskar, A.A. *Trichoderma Harzianum* WKY1: An indole acetic acid producer for growth improvement and anthracnose disease control in sorghum. *Biocontrol. Sci. Technol.* **2017**, *27*, 654–676. [CrossRef]

49. Vinale, F.; Sivasithamparam, K.; Ghisalberti, E.L.; Woo, S.L.; Nigro, M.; Marra, R.; Lombardi, N.; Pascale, A.; Ruocco, M.; Lanzuise, S.; et al. Trichoderma secondary metabolites active on plants and fungal pathogens. *Open Mycol. J.* **2014**, *8*, 127–139. [CrossRef]

50. Gravel, V.; Antoun, V.; Tweddell, R.J. Growth stimulation and fruit yield improvement of greenhouse tomato plants by inoculation with *Pseudomonas putida* or *Trichoderma atroviride*: Possible role ofindoleacetic acid (IAA). *Soil Biol. Biochem.* **2007**, *39*, 1968–1977. [CrossRef]

51. Bader, A.N.; Salerno, G.L.; Covacevich, F.; Consolo, V.F. Native *Trichoderma harzianum* strains from Argentina produce indole-3 acetic acid and phosphorus solubilization, promote growth and control wilt disease on tomato (*Solanum lycopersicum* L.). *J. King Saud Univ.-Sci.* **2020**, *32*, 867–873. [CrossRef]

52. Altomare, C.; Norvell, W.A.; Bjorkman, T.G.; Harman, E. Solubilization of phosphates and micronutrients by the plant-growth-promoting and biocontrol fungus *Trichoderma harzianum* Rifai. *Appl. Environ. Microbiol.* **1999**, *65*, 2926–2933. [CrossRef]

53. Kapri, A.; Tiwari, L. Phosphate solubilization potential and phosphatase activity of rhizospheric *Trichoderma* spp. *Braz. J. Microbiol.* **2010**, *41*, 787–795. [CrossRef]

54. Tandon, A.; Fatima, T.; Gautam, A.; Yadav, U.; Srivastava, S.; Singh, P.C. Effect of *Trichoderma koningiopsis* on chickpea rhizosphere activities under different fertilization regimes. *Open J. Soil Sci.* **2018**, *8*, 261–275. [CrossRef]

55. Tandon, A.; Fatima, T.; Shukla, D.; Tripathi, P.; Srivastava, S.; Singh, P.C. Phosphate solubilization by *Trichoderma koningiopsis* (NBRI-PR5) under abiotic stress conditions. *J. King Saud Univ.-Sci.* **2020**, *32*, 791–798. [CrossRef]

56. Hajieghrari, B. Effects of some Iranian *Trichoderma* isolates on maize seed germination and seedling vigor. *Afr. J. Biotechnol.* **2010**, *9*, 4342–4347.

57. You, J.; Zhang, J.; Wua, M.; Yang, L.; Chen, W.; Li, G. Multiple criteria-based screening of *Trichoderma* isolates for biological control of *Botrytis cinerea* on tomato. *Biol. Control* **2016**, *101*, 31–38. [CrossRef]

58. Diniz, K.A.; Silva, P.A.; Oliveira, J.A.; Evangelista, J.R.E. Sweet pepper seed responses to inoculation with microorganisms and coating with micronutrients, aminoacids and plant growth regulators. *Sci. Agric.* **2009**, *66*, 293–297. [CrossRef]

59. Liu, B.; Jib, S.; Zhang, H.; Wang, Y.; Liu, Z. Isolation of *Trichoderma* in the rhizosphere soil of *Syringa oblata* from Harbin and their biocontrol and growth promotion function. *Microbiol. Res.* **2020**, *235*, 126445. [CrossRef]

Article

Paecilomyces variotii as A Plant-Growth Promoter in Horticulture

Alejandro Moreno-Gavíra, Fernando Diánez, Brenda Sánchez-Montesinos and Mila Santos *

Departamento de Agronomía, Escuela Superior de Ingeniería, Universidad de Almería, 04120 Almería, Spain; alejanmoga@gmail.com (A.M.-G.); fdianez@ual.es (F.D.); brensam@hotmail.com (B.S.-M.)
* Correspondence: msantos@ual.es; Tel.: +34-950-015511

Received: 1 April 2020; Accepted: 19 April 2020; Published: 22 April 2020

Abstract: In the present study, *P. variotii*, an endophytic fungus isolated from plant roots from the Cabo de Gata Natural Park (Parque Nacional Cabo de Gata—Spain), was tested to determine the effect on the growth promotion of tomato and pepper seeds and seedlings. For these purposes, germination trials in a laboratory and two experiments in a commercial nursery and greenhouse conditions were performed. The *P. variotii* isolate has shown a high ability to produce siderophores and IAA, but low ability to solubilize P. High values for germination percentage, seedling vigor, root and shoot length were obtained by *P. variotii* on tomato and pepper against control. *P. variotii* applications resulted in improved most of the growth parameters evaluated, for both horticultural crops, with the best results in the development of pepper seedlings. The application of a higher dose of *P. variotii* improved most of the morphological parameters and the Dickson quality index (DQI) value in tomato in seedlings and plants. The establishment of the endophytic fungus at the root enabled its biostimulant effects to persist after transplantation without any additional application. Few studies have analyzed this species as a biostimulant. The positive results from the tests showed its high potential. The application of this isolate can be of enormous benefit to horticultural crops for its high reproductive and establishment capacity.

Keywords: *Paecilomyces*; PGPF; tomato; pepper; plant probiotic microorganisms

---

1. Introduction

The genus *Paecilomyces* includes more than 100 species known for their multiple activities and habitat heterogeneity [1]. Among them, *Byssochlamys spectabilis* (Udagawa and Shoji Suzuki) Houbraken and Samson, formerly known as *Paecilomyces variotii* Bainier, is an ascomycete characterized by its ability to produce secondary metabolites, which belong to different chemical groups with wide biological activity [2–5]. This species has been described as a biological control agent (BCA) against nematodes [6,7], trematode eggs [8] and phytopathogenic fungi, such as *Biscogniauxia mediterranea*, *Fusarium moniliforme* and *Phytophthora cinnamomi* [9], *Pyricularia oryzae* [10], *Fusarium graminearum* [11] and *Magnaphorte oryzae* [12], among others, that function through their raw extracts, secondary bioactive metabolites or their mycelia. *P. variotii* produces metabolites with herbicidal [13] and insecticidal [14] activity and has been reported to control infections caused by pathogenic bacteria in fish [15] and humans [9]. In turn, this fungal species has even been shown to degrade aromatic compounds [16,17], in addition to removing ammonium from synthetic media and reducing ammonia emissions from chicken manure. [18]. However, it is also associated with many types of human infections in immunosuppressed patients [19]. Nevertheless, only studies related to the possible activity of these metabolites as a hormone-like substance or a promoter of phytohormone production by plant hosts have been published to date [20]. Very few references describe the application of *P. variotii* as a plant-growth promoter. The biocontrol agent ZhiNengCong (ZNC), which is an extract of *P. variotii*, is used in China [21]. ZNC is a highly effective plant elicitor that promotes plant growth by inducing auxin accumulation in root tips with low concentrations [21].

The use of plant probiotic microorganisms (PPMs) is an effective alternative to the use of chemical fertilizers [22–24]. The most studied PPMs are plant growth-promoting bacteria (PGPB), although there are numerous examples of plant growth-promoting fungi (PGPF), which increase crop yield [25–27]. Thus, the most relevant are those that establish endosymbiotic relationships, such as arbuscular mycorrhizal fungi, which solubilize nutrients, such as phosphorous, and micronutrients absorbed by plants [28–30]. *Trichoderma* is one of the most studied genera as PGPF [31], although many others fungi have demonstrated their potential growth-promoting capacity, such as *Penicillium oxalicum* [32], *Penicillium simplicissimum* [33], *Fusarium oxysporum* [34], *Fusarium equiseti* [35], *Alternaria* sp. [36], *Aspergillus* spp. [37] and *Phoma* [38], among others.

In the present study, *P. variotii*, an endophytic fungus isolated from plant roots from the Cabo de Gata Natural Park (Parque Nacional Cabo de Gata—Spain), was tested to evaluate: (a) the effects of seed priming with a fungus suspension on root colonization and tomato and pepper plant vigor; (b) the promotion of growth and quality of pepper and tomato seedlings under a conventional production system and (c) the effects of applying different doses to tomato seedlings and their subsequent transplantation in a greenhouse.

2. Materials and Methods

2.1. Isolation of P. variotii from Plant Roots

Twenty roots of different species of autochthonous plants from the Cabo de Gata Natural Park (CGNP; Almería, Spain) were collected for the isolation of fungal organisms in 2017. Collected samples were cleaned under running tap water to remove debris before use, air dried and processed for isolation of endophytic fungi. To remove epiphytic and surface-adhering microbes, the roots were cut into small, 2–3 cm long, pieces, were surface-sterilized with 2% sodium hypochlorite for 3 min and washed three times with sterile distilled water. The surface-sterilized samples were allowed to dry on sterile paper towels. Ten fragments from each root were placed onto potato dextrose agar (PDA, Difco) supplemented with 50 μg mL^{-1} chloramphenicol to suppress bacterial growth. After incubation at 25 °C for 7 d, individual hyphal tips of the developing fungal colonies were removed, placed on PDA medium and incubated for 5–7 d.

Colony morphology of the pure cultured isolates on PDA and conidiophore morphology were examined and identified by light microscopy, and all selected isolates were stored for further studies. Only one isolate, whose identification under a microscope was consistent with the genus *Paecilomyces* (Figure 1), was selected for this study (*P. variotii* CDG33). Molecular identification of the selected fungi was conducted following the procedure described by Diánez et al. [26]. The sequence was analyzed using a BLAST search in the GenBank database of the National Centre for Biotechnology Information (NCBI, http://blast.ncbi.nlm.nih.gov/Blast.cgi) and aligned to the nearest neighbors. The sequence has not been deposited in the GenBank database because the isolate is subject to patent.

The culture of *P. variotii* has been deposited in the CECT (Spanish Type Culture Collection, Valencia, Spain) with the collection number CECT 20957. This strain was selected for the experiments based on the results of a preliminary assay (data not shown).

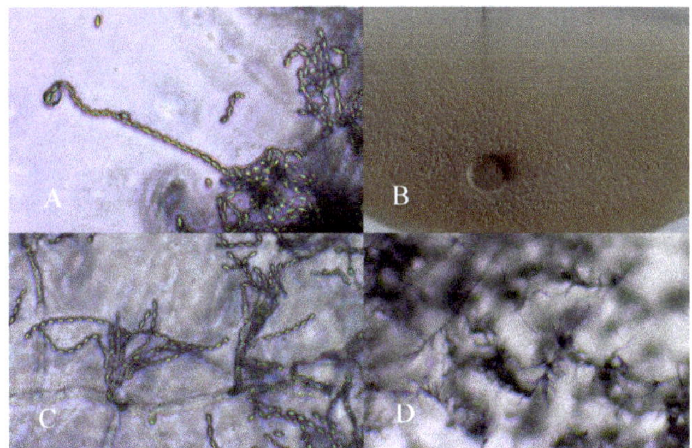

Figure 1. Conidiophore of *Paecilomyces variotii* (**A**: 400×; **C**: 200×; **D**: 100×) and an aspect of colony morphology (**B**) in the potato dextrose agar (PDA) medium.

2.2. Analysis of Plant Growth-Promoting Attributes

Siderophore production was determined on the chrome-azurol S (CAS) medium following the method of Schwyn and Neilands [39] and Louden et al. [40]. Fungal mycelial discs (5 mm) of active culture were transferred to CAS medium and orange halos around the colonies on blue were indicative of siderophore production. The diameter of the orange halo was measured at 24, 48 and 72 h.

Indole-3-acetic acid (IAA) production was estimated according to the procedure described by Diánez et al. [26]. *P. variotii* was grown in 50 mL of glucose peptone broth (GPB) amended with or without L-tryptophan (Sigma-Aldrich) at a concentration of 100 mg L^{-1}. The flasks were inoculated and incubated on an orbital shaker at 150 rpm at 25 °C in the dark for 7 d. After incubation, the suspension from each flask was centrifuged for 30 min at 12,000× g. The supernatant was filtered through sterile Millipore membranes (pore size 0.22 μm) and collected in sterile tubes. The culture supernatants (3 mL) were pipetted into test tubes, and 2 mL Salkowski reagent (2 mL of 0.5 mol L^{-1} FeCl$_3$ + 98 mL of 35% HClO$_4$) was added to it. The tubes containing the mixture were left for 30 min for red color development. The intensity of the color was determined by measuring the optical density at 530 nm using a scanning spectrophotometer. The quantity of IAA was determined by comparison with a standard curve for IAA. Five independent replicates of *P. variotii* were analyzed.

The qualitative evaluation of the phosphorus solubilized by *P. variotii* was performed using NBRIP and PVK media supplemented with 2% agar (Difco Laboratories, Detroit, MI, USA). Phosphate solubilization was detected by the formation of transparent zones surrounding fungal colonies in both media [41]. For the quantitative estimation of phosphate solubilization, a modified version of the procedure by Lima-Rivera [42] was followed. Flasks (250 mL capacity) containing 50 mL of NBRIP broth were inoculated with two disks of agar (5 mm diameter) that had been taken from pure cultures of *P. variotii*. Uninoculated flasks were used as a control (three replicates). Incubation was conducted at 26 °C at a shaking speed of 100 rpm for 3, 5, 7, 10 and 15 d. Supernatants of each culture were analyzed for pH and phosphate concentration. Phosphates in culture supernatants were estimated using the Fiske and Subbarow method [43] and expressed as equivalent phosphate (μg mL^{-1}). The experiments were conducted in triplicate and values were expressed as the mean. The total P in flasks was 10 mg mL^{-1}.

2.3. Mass Production of P. variotii on Solid Substrates

A mixture of two kinds of substrates, buckwheat husk (BH) and oat (O), were tested for the mass multiplication of *P. variotii*. Different percentages (90–10%, 80–20% and 70–30% *v/v* BH-O) of both substrates were submerged in different percentages of water (10%, 20% and 30% *v/v*) for 24 h. Each mixture was sterilized for 1 h at 125 °C twice on consecutive days. Each mixture was placed on a tray and aseptically inoculated by spraying with 5 mL of spore suspension containing 4×10^6 spores mL^{-1} of *P. variotii*. The trays were kept at 25 °C in the dark for 10 d. In total, three samples (2 g) of the fungus-colonized substrate were removed from the trays in each treatment. The samples were successively diluted in sterile distilled water + 0.01% Tween 20® and the number of conidia g^{-1} of the solid substrate was quantified for each replicate using a Neubauer hemocytometer. There were three replications per treatment. The collected spores were used in the different experiments conducted in this study.

2.4. Analysis of Effects of P. variotii on Seed Germination under Laboratory Conditions

Seeds of tomatoes (*Solanum lycopersicum* 'Red Cherry') and pepper (*Capsicum annuum* 'Largo de Reus') were used in this study. The trial used a random block experimental design with two treatments (control and *P. variotii*) and four repetitions. Each repetition included 50 seeds that were germinated in Petri dishes (150 mm diameter) containing two sheets of Whatman No. 1 filter paper that were moistened with sterile distilled water. The seeds were surface-sterilized sterilized with 1.5% sodium hypochlorite (NaOCl) for 5 min, rinsed twice with sterile distilled water and dried under laminar airflow on sterile paper [26]. Treatments were performed by pipetting 50 µL of *P. variotii* spore suspension (1×10^5 spores mL^{-1}) or 50 µL of water (control) on each seed; all boxes were placed in an incubator (25 ± 1 °C in the dark). Root length (mm) was measured from the tip of the primary root to the base of the hypocotyl. After 7 and 10 d, for tomatoes and peppers, respectively, percent germination, root length and shoot length were recorded and a seed vigor index (SVI) was calculated as follows: SVI (length) = seed germination% (mean root length + mean shoot length) [44].

2.5. Analysis of Promoter Effects of P. variotii on Pepper and Tomato Seedlings: Experiment 1

This experiment was performed in nursery polystyrene planting trays, each with 96 cells (70 mL volume), at a commercial nursery (Almería Province, Spain). Pepper and tomato seeds red cherry and Largo de Reus, respectively, were sown into commercial peat mix and covered with vermiculite. After 2 d (tomato) and 4 d (pepper) in a germination room (relative humidity (RH) = 95%; 25 °C), trays were located in a greenhouse and rinsed with water (control) or a 5 mL spore suspension per cell at 10^5 spores per plant. Seedlings were grown using standard nursery culture conditions (18–28 °C; 75.4% ± 6.7% RH) and four trays were used for each treatment. At 45 d after sowing, 20 plants per treatment and control were randomly selected from the four replications and measured for different growth parameters: number of leaves, stem length, stem base diameter, total leaf area and aerial and root dry weights. Leaf area was measured using the WINDIAS 3.1. (Delta-T Devices Ltd., Cambridge, UK, 2009) leaf area processing software. The Dickson quality index [45] was determined using the formula Dickson quality index (DQI) = TDW/((LS/D) + SDW/RDW)), where TDW is the total dry weight (g), LS is the stem length (cm), D is the stem diameter (mm) and SDW and RDM are the stem and root dry weight (g), respectively. The experiments were conducted in autumn using a completely randomized design.

2.6. Analysis of Effects of Applying Different Doses of P. variotii to Tomatoes: Experiment 2

For this experiment, the procedure described in experiment 1 was followed. Three doses of *P. variotii* conidia (PaeD1: 10^4 spores mL^{-1}, PaeD2: 10^5 spores mL^{-1} and PaeD3: 10^6 spores mL^{-1}) were applied to tomato seedlings growing under commercial plant nursery conditions with irrigation of the substrate, by adding 5 mL of spore suspension to each plant. The test was conducted in winter and four replicates were performed with 96 plants per replicate. The seedlings were harvested 30 d after sowing. Twenty plants per treatment and control were randomly selected from the four replicates and measured for the same parameters as described above. Another 25 plants from each treatment were transplanted into sandy soil in mid-February and analyzed in mid-May. The experiment was performed under greenhouse conditions (Figure 2). Water requirements were established according to climatic conditions and crop needs. Plants were fertilized with a commercial complex nutrient fertilizer.

Figure 2. Tomato plants grown under commercial seedlings (**A**) and field assay, 30 (**B**) and 60 (**C**) days after transplanting.

In all tests (experiments 1 and 2), tomato and pepper roots inoculated with *P. variotii* were collected at the end of the tests. Roots were surface-sterilized in 0.1% sodium hypochlorite and washed with sterilized water. Lastly, 2 cm root fragments were placed in PDA medium to determine root colonization by the fungal isolate.

2.7. Statistical Analysis

Statistical analysis was conducted using Statgraphics Centurion XVIII Software. The data were tested by a one-way analysis of variance (ANOVA) or Student's *t*-test with significance defined as p values less than 0.05 ($p < 0.05$). The experimental results are presented as the means and standard error (±SE) for the different replicates. Mean separation was performed using Fisher's least significant difference (LSD) test.

3. Results

A total of 42 fungal isolates were obtained from the analyses of 20 root samples taken from plants native to the CGNP. All fungal isolates were obtained in pure cultures using standard techniques. The isolates were identified as filamentous fungi belonging to the phylum Ascomycota. The fungal isolates were identified to the genus level. *Aspergillus*, *Trichoderma* sp., *Rustroemia* sp. and *Penicillium* sp., were the most characteristic genera. They differed in color, type of conidiophore or presence of microsclerotia-like structures. Therefore were considered to be different. The presence of the genus *Paecilomyces* was especially relevant, and this genus was thus selected for this study (isolate CDG33). The amplified sequences of isolate of *Paecilomyces* were compared with available DNA sequences using BLAST, having 99% homology with *Paecilomyces variotii* accession number JX282326.1.

3.1. Mass Production of P. variotii on Solid Substrates

The fungus multiplied well in all mixtures tested. Among the different treatments, whose composition varied in water content and the percentage of oats or buckwheat husk that were tested for mass multiplication of *P. variotii*, the proportion of 70% + 30% for buckwheat husk and oats, respectively, with 20% water content, resulted in significantly higher spore production (Figure 3), followed by 90% + 10% with 10% water content (Table 1). The lowest spore production rate was observed at a percentage of 80% + 20% with 10% water (Figure 3).

Figure 3. Mass production of *P. variotii* spores at a proportion of 70% + 30% buckwheat husk and oats, respectively, +20% water (**A**,**B**) and at 80% + 20% with 10% water (**C**).

Table 1. Effects of solid media on mass multiplication of *P. variotii* at varying substrates percentages (*v/v*) and water contents (*v/v*).

Treatments					
% (*v/v*) water					
10		20		30	
BH-O%	CFU·g^{-1}	BH-O%	CFU g^{-1}	BH-O%	CFU g^{-1}
90-10	$1.13 \times 10^8 \pm 1.2 \times 10^7$ b	90-10	$8.5 \times 10^7 \pm 7.09 \times 10^6$ ef	90-10	$9.14 \times 10^7 \pm 9.73 \times 10^6$ de
80-20	$7.0 \times 10^7 \pm 1.32 \times 10^7$ g	80-20	$1.05 \times 10^8 \pm 1.09 \times 10^7$ bc	80-20	$1.0 \times 10^8 \pm 5.11 \times 10^6$ d
70-30	$8.04 \times 10^7 \pm 1.14 \times 10^7$ f	70-30	$1.24 \times 10^8 \pm 1.06 \times 10^7$ a	70-30	$1.07 \times 10^8 \pm 9.54 \times 10^6$ bc

BH: buckwheat husk; O: oats. Data were analyzed by ANOVA and treatment means were compared according to Fisher's least significant difference (LSD) statistical procedure (*F*-test at $p < 0.05$).

3.2. Plant Growth-Promoting Characteristics of P. variotii: Siderophore Production, IAA and P Solubilization

The formation of an orange-colored zone around the fungal colonies was observed, which indicated siderophore production by *P. variotii* (Figure 4A). The diameter of the halo (mm) was estimated at 3.88 ± 0.33, 5.55 ± 0.22 and 8.83 ± 1.29 for 24, 48 and 72 h, respectively.

IAA was produced by *P. variotii* in medium supplemented with 100 mg L^{-1} tryptophan during a 7 d period, and the final concentration was 0.049 ± 0.001 mg mL^{-1}. The final concentration of IAA was 0.03 ± 0.001 mg mL^{-1} in medium supplemented without tryptophan.

No halo of P solubilization by *P. variotii* was detected in any of the media used (NBRIP and PVK media supplemented with 2% agar). The effect of *P. variotii* on the soluble phosphate concentration is shown in Figure 4. The initial concentration of P in the medium was used to quantify the concentration of P solubilized by *P. variotii*. As shown in Figure 4B, no P solubilization was detected during up to 15 d of incubation, assessing a soluble P of 2.01 ± 0.68 g L^{-1} versus 0.74 ± 0.25 in the control ($p = 0.0243$). In turn, no change in the pH of the medium was detected, which remained at approximately 7.

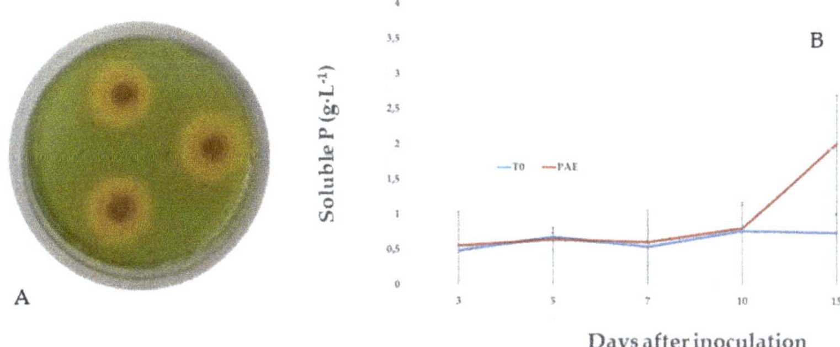

Figure 4. Formation of orange-colored halos owing to production of siderophores by *P. variotii* (**A**). Effects of *P. variotii* on phosphate solubilization (**B**) in NBRIP broth containing tribasic calcium phosphate (10 g). Mean standard deviation is expressed in error bar. The results are shown as the average of the three replicates, in g L^{-1}.

3.3. P. variotii Inoculation Effects on Tomato and Pepper Seed Germination

Table 2 outlines the results of the in vitro application of *P. variotii* to tomato and pepper seeds. Direct inoculation of seeds by *P. variotii* isolate spores had a significant effect ($p < 0.05$) on the percent seed germination, root and shoot length, and SVI in tomatoes. The increase in root and shoot length was 18.23% and 17.85%, respectively. However, pepper seeds treated with *P. variotii* showed no improvement in SVI ($p > 0.05$).

Table 2. Effects of *P. variotii* isolate on tomato and pepper seed germination 7 and 10 d after treatment, respectively. pp

Treatment	% Germination	Root Length (cm)	Shoot Length (cm)	Seed Vigor Index
		Pepper		
p -value	0.5369	0.6162	0.0010	0.9975
P. variotii	80 ± 9.79	2.37 ± 1.30	1.66 ± 0.55	323.00 ± 127.07
T0	76 ± 7.30	2.26 ± 1.24	1.98 ± 0.56	322.93 ± 119.36
		Tomato		
p-value	0.0020	0.0013	0.0036	0.0000
P.variotii	89 ± 3.82	4.41 ± 1.33	3.30 ± 1.15	686.70 ± 210.40
T0	75 ± 3.82	3.73 ± 1.32	2.80 ± 1.02	498.70 ± 164.28

T0: control without *P. variotii*. Data were analyzed using Student's *t*-test ($p < 0.05$).

3.4. Promoter Effects of P. variotii Isolates on Tomato and Pepper Seedlings: Experiment 1

The effects of *P. variotii* application on morphological parameters and DQI are shown in Table 3. Most of the growth parameters evaluated in tomato and pepper were improved by *P. variotii*, when compared with those of the experimental control, with the best results in the development of pepper seedlings.

These increases were statistically significant for most parameters. The increases assessed in pepper and tomato seedlings (P/T) were 9.7%/6.9% for stem length, 4.9%/0.8% for stem diameter, 10.6%/6.0% for leaf number, 18.2–6.7% for root dry weight, 16.7%/10.7% for aerial dry weight and 10.1%/7.5% for leaf area. In tomatoes, *P. variotii* applications resulted in a decrease in root dry weight, albeit without significant differences from that of the control. *P. variotii* application improved plant quality, albeit without significant differences for tomato plants ($p = 0.2059$). In all tests (experiments 1 and 2), *P. variotii* was observed in tomato and pepper roots analyzed in PDA medium.

Table 3. Morphological parameters and quality index of pepper and tomato seedlings treated with *P. variotii* isolate at 45 d after sowing.

Treatment	Length of Stem (cm)	Diameter (mm)	Number of Leaves	Areal Dry Weight (g)	Root Dry Weight (g)	Leaf Area mm^2	DQI
			Pepper				
p-value	*0.0000*	*0.0008*	*0.0000*	*0.0000*	*0.0022*	*0.0020*	*0.0054*
P. variotii	29.63 ± 1.75	3.83 ± 0.21	7.39 ± 0.54	0.42 ± 0.05	0.13 ± 0.02	83.58 ± 11.27	0.051 ± 0.00
T0	27.01 ± 2.07	3.65 ± 0.23	6.68 ± 0.70	0.36 ± 0.04	0.11 ± 0.03	75.88 ± 11.15	0.046 ± 0.00
			Tomato				
p-value	*0.0002*	*0.0039*	*0.0330*	*0.0042*	*0.1443*	*0.0397*	*0.2059*
P. variotii	27.37 ± 2.06	3.81 ± 0.27	4.40 ± 0.54	0.62 ± 0.07	0.14 ± 0.02	74.45 ± 11.85	0.067 ± 0.01
T0	25.61 ± 2.07	3.78 ± 0.22	4.15 ± 0.48	0.56 ± 0.10	0.15 ± 0.02	69.28 ± 10.17	0.063 ± 0.01

T0: control without *P. variotii*. Data were analyzed using Student's *t*-test ($p < 0.05$).

3.5. Effects of Dose of Application of P. variotii Isolates on Tomato Seedlings and Transplanted Plants: Experiment 2

The effects of applying three doses on the development of tomato seedlings under commercial plant nursery conditions and subsequent transplantation into soil under greenhouse conditions are shown in Table 4. As shown in Table 4, applying different doses of *P. variotii* promoted seedling shoot development, significantly decreasing, in some cases, root development. The best results were assessed for dose 3, which resulted in a 30.2% increase in stem length, 15.7% in stem diameter, 19.5% in leaf number and 46.2% in aerial dry weight. However, root dry weight decreased 25.0%. The decrease in root length can cause stress when transplanting tomato seedlings, thereby decreasing plant quality. The DQI [45] expresses the global aptitude of a plant to successfully overcome the transplantation phase, based on overall plant development, while considering the balance between plant shoots and roots. Higher values of this index indicate higher seeding quality. In this case, despite the decrease in root length, a higher value of this index was observed for dose 3, with a significant difference for all doses tested, but not with the control treatment.

Table 4. Morphological parameters and Dickson quality index (DQI) of tomato seedlings (30 d after sowing) and plants (90 days after transplanting) treated with different doses (10^4, 10^5 and 10^6 spores per plant, (D1, D2 and D3, respectively) of *P. variotii*.

	Tomato seedling					
Treatment	Length of stem (cm)	Diameter (mm)	N° leaves	Areal dry Weight (g)	Root dry Weight (g)	DQI
p-value	*0.0000*	*0.0000*	*0.0002*	*0.0000*	*0.0028*	*0.0481*
Pae D1	16.12 ± 2.11a	3.72 ± 0.38b	3.58 ± 0.53b	0.27 ± 0.09a	0.039 ± 0.01a	0.045 ± 0.01b
Pae D2	14.46 ± 2.58b	3.50 ± 0.38c	3.71 ± 0.84ab	0.25 ± 0.08a	0.035 ± 0.01a	0.046 ± 0.01b
Pae D3	16.27 ± 3.05a	3.99 ± 0.28a	3.93 ± 0.78a	0.38 ± 0.09a	0.027 ± 0.01b	0.054 ± 0.01a
T0	12.5 ± 2.22b	3.45 ± 0.36c	3.29 ± 0.61c	0.25 ± 0.08b	0.036 ± 0.01a	0.045 ± 0.01b
	Tomato Plants					
Treatment	Length of stem (cm)	Diameter (mm)	Internodes	Areal dry Weight (g)	Root dry Weight (g)	DQI
p-value	*0.1876*	*0,1244*	*0.9759*	*0.0048*	*0.7796*	*0.0283*
Pae D1	82.75 ± 12.97ab	11.57 ± 1.08ab	14.20 ± 1.93a	50.67 ± 10.80a	4.22 ± 1.20a	24.84 ± 7.68a
Pae D2	87.20 ± 9.42a	11.59 ± 1.02ab	14.30 ± 1.25a	49.19 ± 10.21a	4.26 ± 1.64a	24.23 ± 8.16a
Pae D3	77.60 ± 7.75b	11.84 ± 1.08a	14.10 ± 1.28a	49.38 ± 10.53a	4.20 ± 1.37a	25.78 ± 7.66a
T0	84.00 ± 7.10ab	10.58 ± 0.91b	14.00 ± 1.63a	39.13 ± 7.99b	3.73 ± 0.91a	16.84 ± 3.79b

T0: control without *P. variotii*. Different letters indicate significant differences according to the one-way ANOVA test ($p = 0.05$). Treatment means were compared according to the Fisher's LSD statistical procedure (*F*-test at $p < 0.05$).

Once transplanted in soil, the findings showed that the three doses favored several morphological parameters studied in the development of tomato plants in relation to those of control plants, with higher values for dry weights and DQI but without significant differences for the three doses tested. The establishment of the endophytic fungus at the root (Figure 5), therefore, enabled its effects to persist after transplantation.

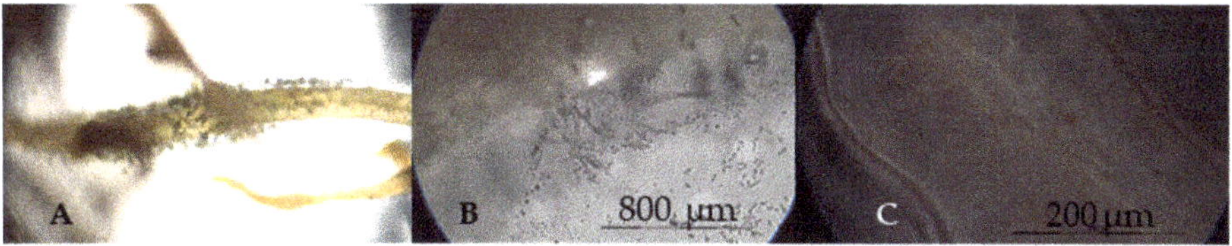

Figure 5. *P. variotii* mycelium and conidiophores colonizing tomato (**A**: 100×; **B**: 200×) and pepper roots (**C**: 400×).

4. Discussion and Conclusions

In this study, fungal isolates from surface-sterilized root segments of native plants and soil were collected from the CGNP. This park is located in an arid-to-semiarid Mediterranean region where the predominant fungi genera are *Penicillium* and *Aspergillus*, as well as isolates from the group termed Dark Septate Endophytic (DSE) fungi, which colonize plant roots. The DSE fungi isolated herein formed dark brownish microsclerotia-like structures. The role of DSE fungi in nature has been considered to be similar to that of mycorrhizal fungi [46,47]. These isolates will be analyzed in future studies. Among all isolates, the plant growth-promoting capacity of the only isolate from *P. variotii* was analyzed in this study.

Nevertheless, *P. variotii* was selected given the sparse literature on this species as a plant-growth promoter. The ability of *P. variotii* to stimulate plant growth is poorly studied, while for other species of the same genus, the growth-promoting effect is generally associated with an improvement in the plant status for its nematicide effect for the control of diseases caused by different nematode species [48].

A key requirement for selling microorganisms with a biostimulant capacity is that they have a high-spore production capacity in substrates with a low-production cost. Nevertheless, grains whose agitation leads to a rapid release of spores must be selected to obtain the formulation. For this reason, buckwheat husk and oats were selected for high performance in the spore production of *P. variotii*. This production occurred in the entire substrate, not only on the surface. Generally, commercial spore production methods often only use cereal grains, rice or other starch-based substrates [49]. The nutritional composition of the medium in which the growth promotor multiplies can affect its biocontrol or biostimulant capacity [50]. Based on our results, *P. variotii* can stimulate plant development under the production conditions described herein.

The *P. variotii* isolate has shown the ability to produce siderophores and IAA. Biostimulation is generally associated with increased nutrient availability, similar to biofertilization [51], but it is also caused by multiple other factors, such as mechanisms including enhancement of plant systemic resistance [52]. Thus, siderophore production plays a key role by enhancing the Fe uptake of plants and can be considered an ecofriendly alternative to the use of chemical fertilizers and pesticides in the agricultural sector [53]. Vala et al. [54] describe the production of both hydroxamate and carboxylate-type siderophores by *P. variotii* isolated from the surface of mangrove plants. Terrestrial *P. variotii* has previously been shown to secrete the trihydroxamate siderophore ferrirubin [55]. The *P. variotii* isolate has shown a relatively high IAA production in relation to its genus, and tryptophan incorporation slightly increased in vitro IAA biosynthesis by *P. variotii*. Other studies also reported that Trp considerably stimulated microbial IAA yield in vitro [56–58]. Ali [59] described IAA production by *P. variotii* lower than 2 ng mL^{-1} produced in Czapek medium supplemented with tryptophan. In our study, we described a much higher production for the same species. Waqas [60] described IAA production by *P. formosus* at 34.07 ± 3.92 µg mL^{-1}, a result very similar to the findings of this study, and consequently, a plant growth-promoting effect. Nevertheless, Waqas [61] analyzed the variability of IAA production as a function of the culture medium used and conversely found that *P. variotii* had virtually undetectable phosphatase activity. These results differed from the high phosphatase production by a *P. variotii* isolate from the medicinal plant *Caralluma acutangula* [59]. In soil, P-solubilizing fungi constitute approximately 0.1–0.5% of the total fungal population [62]. The use of microbial inoculants (biofertilizers) possessing P-solubilizing activities in crop productivity is considered as an alternative to further application of mineral P fertilizers [63]. Endophyte co-inoculation of plants exponentially improved the phosphatase activity of soil compared to that of the non-inoculated plants under stress conditions [64]. Etesami et al. [65] showed the relationship among ACC deaminase activity, IAA production, siderophore production and phosphate solubilization of bacterial strains and their effect on root elongation of rice seedlings. In this study, this correlation could not be determined since an isolate of *P. variotii* was analyzed.

The application of beneficial microorganisms (biopriming) may not only help to improve germination and vigor parameters but also relieve a wide range of physiological, abiotic and biotic stresses in both seeds and seedlings [66]. Nevertheless, biopriming can potentially lead to a more resistant plant after transplanting. This can depend on numerous factors, such as plant species, microorganisms, applied dose and substrates, among others. Thus, in this study, the *P. variotii* isolate enhanced germination and root vigor in tomato seeds. However, the results with pepper plants were different, with no significant differences from that of the controls and even with a 19.27% decrease in shoot length ($p < 0.05$). This may be because germination of pepper seeds is often more heterogeneous. Cochran [67] determined that the germination percentage and the accumulation of dry matter in the large seeds of bell pepper were higher in relation to the small seeds. *P. variotii* enhanced several tomato and pepper seedling parameters. The application doses for different crops must be studied for optimal outcomes. In this study, the application of a higher dose (10^6 spores mL^{-1} versus 10^5 spores mL^{-1}) improved the different morphological parameters value in tomato seedlings under commercial plant nursery conditions. After transplanting in the greenhouse, the biostimulant effects persisted 4 months after applying the three doses without any additional application. Similar benefits have been observed as a result of the application of other species of the genus *Paecilomyces*. Waqas [68] described improved soybean seedling germination and SVI when applying the endophyte *Paecilomyces* sp. The application of *P. variotii* extracts significantly increased cherry radish yield, dry matter accumulation, the root–shoot ratio and quality. This extract has a very high biological activity, with a low cost, which has a great application prospect [69]. Similar studies conducted by Anis [70] showed increased sunflower seedling vigor under biotic stress conditions when applying *P. variotii* and *Macrophomina phaseolina* spores, in vitro, with no favorable results when conducting the tests in pots. In turn, Maitlo [71] assessed chickpea plant biostimulation when inoculating them with *P. lilacinus* and *F. oxysporum* f. sp. *ciceris*, which also reduced chickpea wilt.

In intensive horticulture under plastic, the benefits of the application of biostimulants or biofungicides based on microorganisms, are in question, because farmers perceive low efficacy of these products as disease controllers when compared with the rapid response presented by chemical fertilizers or fungicide. The current changes in legislation regarding the reduction of active ingredients and the commercialization of biostimulants and biopesticides, together with the need to increase the sustainability of agriculture in terms of public health and the environment, require the use of PPMs as a key element in horticulture.

To the best of our knowledge, the present study demonstrated for the first time the biostimulant capacity of *P. variotii* in pepper and tomato plants under commercial plant nursery and greenhouse conditions.

5. Patents

This isolate was patented with a Spanish patent number ES2684858A1: New strain of *Paecilomyces variotii*, compositions and applications.

Author Contributions: F.D. and M.S. conceived and designed the experiments; A.M.-G. and B.S.-M. performed the experiments; F.D. analyzed the data; M.S. wrote the paper. All authors have read and agreed to the published version of the manuscript.

Funding: This research received no external funding.

Acknowledgments: The present work benefited from the input of the project RTC-2017-6486-2 was supported by the Spanish Ministry of Science, Innovation and Universities.

Conflicts of Interest: The authors declare that there is no conflict of interests regarding the publication of this manuscript.

References

1. Nguyen, T.T.T.; Paul, N.C.; Lee, H.B. Characterization of Paecilomyces variotii and Talaromyces amestolkiae in Korea Based on the Morphological Characteristics and Multigene Phylogenetic Analyses. *Mycobiology* **2016**, *44*, 248–259. [CrossRef] [PubMed]

2. Mioso, R.; Marante, F.J.T.; De Laguna, I.H.B. The Chemical Diversity of the Ascomycete Fungus Paecilomyces variotii. *Appl. Biochem. Biotechnol.* **2015**, *177*, 781–791. [CrossRef] [PubMed]

3. De Laguna, I.H.B.; Marante, F.J.T.; Mioso, R. Enzymes and bioproducts produced by the ascomycete fungus Paecilomyces variotii. *J. Appl. Microbiol.* **2015**, *119*, 1455–1466. [CrossRef] [PubMed]

4. Akinfala, T.O.; Houbraken, J.; Sulyok, M.; Adedeji, A.R.; Odebode, A.C.; Krska, R.; Ezekiel, C.N. Moulds and their secondary metabolites associated with the fermentation and storage of two cocoa bean hybrids in Nigeria. *Int. J. Food Microbiol.* **2020**, *316*, 108490. [CrossRef] [PubMed]

5. Lòpez-Fernàndez, S.; Campisano, A.; Schulz, B.J.; Steinert, M.; Stadler, M.; Surup, F. Viriditins from Byssochlamys spectabilis, their stereochemistry and biosynthesis. *Tetrahedron Lett.* **2020**, *61*, 151446. [CrossRef]

6. Al-Qasim, M.; Abu-Gharbieh, W.; Assas, K. Nematophagal ability of Jordanian isolates of Paecilomyces variotii on the root-knot nematode Meloidogyne javanica. *Nematol. Mediterr.* **2009**, *37*, 53–57.

7. Perveen, Z.; Shahzad, S. A comparative study of the efficacy of Paecilomyces species against root-knot nematode Meloidogyne incognita. *Pak. J. Nematol.* **2013**, *31*, 125–135.

8. Ahmad, R.Z.; Sidi, B.B.; Endrawati, D.; Ekawasti, F. Paecilomyces lilacinus and P. variotii as a predator of nematode and trematode eggs. *IOP Conf. Ser. Earth Environ. Sci.* **2019**, *299*, 012056. [CrossRef]

9. Rodrigo, S.; Santamaria, O.; Halecker, S.; Lledó, S.; Stadler, M. Antagonism betweenByssochlamys spectabilis(anamorphPaecilomyces variotii) and plant pathogens: Involvement of the bioactive compounds produced by the endophyte. *Ann. Appl. Boil.* **2017**, *171*, 464–476. [CrossRef]

10. Mohammadi, S.; Soltani, J.; Piri, K. Soilborne and invertebrate pathogenic Paecilomyces species show activity against pathogenic fungi and bacteria. *J. Crop. Prot.* **2016**, *5*, 377–387. [CrossRef]

11. Zhang, P.; Li, X.M.; Wang, J.N.; Wang, B.G. Oxepine-containing diketopiperazine alkaloids from the algal-derived endophytic fungus Paecilomyces variotii EN-291. *Helv. Chim. Acta* **2015**, *98*, 800–804. [CrossRef]

12. Hajano, J.U.D.; Lodhi, M.; Pathan, M.A.; Khanzada, A.; Shah, G.S. In-vitro evaluation of fungicides, plant extracts and bio-control agents against rice blast pathogen Magnaporthe oryzae couch. *Pak. J. Bot.* **2012**, *44*, 1775–1778.

13. Nakajima, M.; Itoi, K.; Takamatsu, Y.; Sato, S.; Furukawa, Y.; Furuya, K.; Honma, T.; Kadotani, J.; Kozasa, M.; Haneishi, T. Cornexistin: A new fungal metabolite with herbicidal activity. *J. Antibiot.* **1991**, *44*, 1065–1072. [CrossRef]

14. Song, X.; Zhang, L.H.; Peng, A.T.; Cheng, B.P.; Ling, J.F. First Report of Paecilomyces variotii Isolated From Citrus Psyllid (Diaphorina citri), the Vector of Huanglongbing of Citrus, in China. *Plant Dis.* **2016**, *100*, 2526. [CrossRef]

15. Hyung, J.J.; Kang, H.; Jong, J.J.; Soo, K.Y. Paecilomyces Variotii Extracts for Preventing and Treating Infections Caused by Fish Pathogenic Microorganisms. KR Patent 2013051523, 2013.

16.	García-Peña, E.; Hernández, S.; Auria, R.; Revah, S. Correlation of Biological Activity and Reactor Performance in Biofiltration of Toluene with the Fungus Paecilomyces variotii CBS115145. *Appl. Environ. Microbiol.* **2005**, *71*, 4280–4285. [CrossRef]

17.	García-Peña, E.; Ortiz, I.; Hernandez, S.; Revah, S. Biofiltration of BTEX by the fungus Paecilomyces variotii. *Int. Biodeterior. Biodegrad.* **2008**, *62*, 442–447. [CrossRef]

18.	Liu, Z.; Liu, G.; Cai, H.; Shi, P.; Chang, W.; Zhang, S.; Zheng, A.; Xie, Q.; Ma, J. Paecilomyces variotii: A Fungus Capable of Removing Ammonia Nitrogen and Inhibiting Ammonia Emission from Manure. *PLoS ONE* **2016**, *11*, e0158089. [CrossRef]

19.	Steiner, B.; Aquino, V.R.; Paz, A.A.; Silla, L.M.D.R.; Zavascki, A.; Goldani, L. Paecilomyces variotii as an Emergent Pathogenic Agent of Pneumonia. *Case Rep. Infect. Dis.* **2013**, *2013*, 1–3. [CrossRef]

20.	Santamaria, O.; Lledó, S.; Rodrigo, S.; Poblaciones, M.J. Effect of Fungal Endophytes on Biomass Yield, Nutritive Value and Accumulation of Minerals in Ornithopus compressus. *Microb. Ecol.* **2017**, *74*, 841–852. [CrossRef]

21.	Lu, C.; Liu, H.; Jiang, D.; Wang, L.; Jiang, Y.; Tang, S.; Hou, X.; Han, X.; Liu, Z.; Zhang, M.; et al. Paecilomyces variotii extracts (ZNC) enhance plant immunity and promote plant growth. *Plant Soil* **2019**, *441*, 383–397. [CrossRef]

22.	Sánchez-Montesinos, B.; Diánez, F.; Moreno-Gavíra, A.; Gea, F.J.; Santos, M. Plant Growth Promotion and Biocontrol of Pythium ultimum by Saline Tolerant Trichoderma Isolates under Salinity Stress. *Int. J. Environ. Res. Public Health* **2019**, *16*, 2053. [CrossRef]

23.	Raja, N. Biopesticides and Biofertilizers: Ecofriendly Sources for Sustainable Agriculture. *J. Fertil. Pestic.* **2013**, *4*, e112. [CrossRef]

24.	Adesemoye, A.; Kloepper, J.W. Plant–microbes interactions in enhanced fertilizer-use efficiency. *Appl. Microbiol. Biotechnol.* **2009**, *85*, 1–12. [CrossRef] [PubMed]

25.	Vandenberghe, L.; Garcia, L.M.B.; Rodrigues, C.; Camara, M.C.; Pereira, G.; De Oliveira, J.; Soccol, C.R. Potential applications of plant probiotic microorganisms in agriculture and forestry. *AIMS Microbiol.* **2017**, *3*, 629–648. [CrossRef]

26.	Martínez, F.D.; Santos, M.; Carretero, F.; Marin, F. Trichoderma saturnisporum,a new biological control agent. *J. Sci. Food Agric.* **2015**, *96*, 1934–1944. [CrossRef]

27.	Fernando, D.; Santos, M.; Francisco, C.; Francisco, M. Biostimulant Activity of Trichoderma saturnisporum in Melon (Cucumis melo). *HortScience* **2018**, *53*, 810–815. [CrossRef]

28.	Smith, S.E.; Manjarrez, M.; Stonor, R.; McNeill, A.; Smith, F.A. Indigenous arbuscular mycorrhizal (AM) fungi contribute to wheat phosphate uptake in a semi-arid field environment, shown by tracking with radioactive phosphorus. *Appl. Soil Ecol.* **2015**, *96*, 68–74. [CrossRef]

29.	Bhardwaj, D.; Ansari, M.; Sahoo, R.K.; Tuteja, N. Biofertilizers function as key player in sustainable agriculture by improving soil fertility, plant tolerance and crop productivity. *Microb. Cell Factories* **2014**, *13*, 66. [CrossRef]

30.	Begum, N.; Qin, C.; Ahanger, M.A.; Raza, S.; Khan, M.I.; Ashraf, M.; Ahmed, N.; Zhang, L. Role of Arbuscular Mycorrhizal Fungi in Plant Growth Regulation: Implications in Abiotic Stress Tolerance. *Front. Plant Sci.* **2019**, *10*, 1068. [CrossRef]

31.	Khan, M.R.; Mohiddin, F.A. Trichoderma: Its Multifarious Utility in Crop Improvement. In *Crop Improvement through Microbial Biotechnology*; Elsevier BV: Amsterdam, The Netherlands, 2018; pp. 263–291.

32.	Murali, M.; Amruthesh, K.N. Plant Growth-promoting Fungus Penicillium oxalicum Enhances Plant Growth and Induces Resistance in Pearl Millet Against Downy Mildew Disease. *J. Phytopathol.* **2015**, *163*, 743–754. [CrossRef]

33.	Shimizu, K.; Hossain, M.M.; Kato, K.; Kubota, M.; Hyakumachi, M. Induction of Defense Responses in Cucumber Plants by Using the Cell-free Filtrate of the Plant Growth-Promoting Fungus Penicillium simplicissimum GP17-2. *J. Oleo Sci.* **2013**, *62*, 613–621. [CrossRef] [PubMed]

34.	Bitas, V.; McCartney, N.; Li, N.; Demers, J.; Kim, J.-E.; Kim, H.-S.; Brown, K.M.; Kang, S. Fusarium Oxysporum Volatiles Enhance Plant Growth Via Affecting Auxin Transport and Signaling. *Front. Microbiol.* **2015**, *6*, 351. [CrossRef] [PubMed]

35.	Kojima, H.; Hossain, M.; Kubota, M.; Hyakumachi, M. Involvement of the salicylic acid signaling pathway in the systemic resistance induced in Arabidopsis by plant growth-promoting fungus Fusarium equiseti GF19-1. *J. Oleo Sci.* **2013**, *62*, 415–426. [CrossRef] [PubMed]

36. Zhou, L.S.; Tang, K.; Guo, S.-X. The Plant Growth-Promoting Fungus (PGPF) Alternaria sp. A13 Markedly Enhances Salvia miltiorrhiza Root Growth and Active Ingredient Accumulation under Greenhouse and Field Conditions. *Int. J. Mol. Sci.* **2018**, *19*, 270. [CrossRef]

37. Islam, S.; Akanda, A.M.; Sultana, F.; Hossain, M. Chilli rhizosphere fungus Aspergillus spp. PPA1 promotes vegetative growth of cucumber (Cucumis sativus) plants upon root colonisation. *Arch. Phytopathol. Plant Prot.* **2013**, *47*, 1231–1238. [CrossRef]

38. Naznin, H.A.; Kimura, M.; Miyazawa, M.; Hyakumachi, M. Analysis of Volatile Organic Compounds Emitted by Plant Growth-Promoting Fungus Phoma sp. GS8-3 for Growth Promotion Effects on Tobacco. *Microbes Environ.* **2012**, *28*, 42–49. [CrossRef]

39. Schwyn, B.; Neilands, J. Universal chemical assay for the detection and determination of siderophores. *Anal. Biochem.* **1987**, *160*, 47–56. [CrossRef]

40. Louden, B.C.; Haarmann, D.; Lynne, A.M. Use of Blue Agar CAS Assay for Siderophore Detection. *J. Microbiol. Boil. Educ.* **2011**, *12*, 51–53. [CrossRef]

41. Nautiyal, C.S. An efficient microbiological growth medium for screening phosphate solubilizing microorganisms. *FEMS Microbiol. Lett.* **1999**, *170*, 265–270. [CrossRef]

42. Lima-Rivera, D.L.; Lopez-Lima, D.; Desgarennes, D.; Velázquez-Rodríguez, A.S.; Carrión, G. Phosphate solubilization by fungi with nematicidal potential. *J. Soil Sci. Plant Nutr.* **2016**, *16*, 507–524. [CrossRef]

43. Fiske, C.H.; Subbarow, Y. The colorimetry determination of phosphorous. *J. Biol. Chem.* **1925**, *66*, 375–400.

44. Murali, M.; Amruthesh, K.N.; Sudisha, J.S.; Niranjana, R.; Shetty, H.S. Screening for plant growth promoting fungi and their ability for growth promotion and induction of resistance in pearl millet against downy mildew disease. *J. Phytol.* **2012**, *4*, 30–36.

45. Dickson, A.; Leaf, A.L.; Hosner, J.F. Quality Appraisal of White Spruce and White Pine Seedling Stock in Nurseries. *For. Chron.* **1960**, *36*, 10–13. [CrossRef]

46. Peterson, R.L.; Wagg, C.; Pautler, M. Associations between microfungal endophytes and roots: Do structural features indicate function? *Botany* **2008**, *86*, 445–456. [CrossRef]

47. Della Mónica, I.F.; Saparrat, M.C.; Godeas, A.M.; Scervino, J.M. The co-existence between DSE and AMF symbionts affects plant P pools through P mineralization and solubilization processes. *Fungal Ecol.* **2015**, *17*, 10–17. [CrossRef]

48. Huang, W.-K.; Cui, J.-K.; Liu, S.-M.; Kong, L.-A.; Wu, Q.-S.; Peng, H.; He, W.-T.; Sun, J.-H.; Peng, D.-L. Testing various biocontrol agents against the root-knot nematode (Meloidogyne incognita) in cucumber plants identifies a combination of Syncephalastrum racemosum and Paecilomyces lilacinus as being most effective. *Boil. Control.* **2016**, *92*, 31–37. [CrossRef]

49. Goettel, M.S.; Roberts, D.W. Mass production, formulation and field application of entomopathogenic fungi. In *Biological Control of Locusts and Grasshopper*; Lomer, C.J., Prior, C., Eds.; CAB International: Wallingford, UK, 1992.

50. Lane, B.S.; Trinci, A.P.; Gillespie, A.T. Influence of cultural conditions on the virulence of conidia and blastospores of Beauveria bassiana to the green leafhopper, *Nephotettix virescens. Mycol. Res.* **1991**, *95*, 829–833. [CrossRef]

51. Vessey, J.K. Plant growth promoting rhizobacteria as biofertilizers. *Plant Soil* **2003**, *255*, 571–586. [CrossRef]

52. Schuster, A.; Schmoll, M. Biology and biotechnology of Trichoderma. *Appl. Microbiol. Biotechnol.* **2010**, *87*, 787–799. [CrossRef]

53. Saha, M.; Sarkar, S.; Sarkar, B.; Sharma, B.K.; Bhattacharjee, S.; Tribedi, P. Microbial siderophores and their potential applications: A review. *Environ. Sci. Pollut. Res.* **2015**, *23*, 3984–3999. [CrossRef] [PubMed]

54. Vala, A.K.; Vaidya, S.Y.; Dube, H.C. Siderophore production by facultative marine fungi. *Indian J. Mar. Sci.* **2000**, *29*, 339–340.

55. Renshaw, J.C.; Robson, G.; Trinci, A.P.; Wiebe, M.; Livens, F.R.; Collison, D.; Taylor, R. Fungal siderophores. Structures, functions and applications. *Mycol. Res.* **2002**, *106*, 1123–1142. [CrossRef]

56. Ivanova, E.G.; Doronina, N.V.; Trotsenko, Y.A. Aerobic Methylobacteria Are Capable of Synthesizing Auxins. *Microbiology* **2001**, *70*, 392–397. [CrossRef]

57. Khalid, A.; Arshad, M.; Zahir, Z. Screening plant growth-promoting rhizobacteria for improving growth and yield of wheat. *J. Appl. Microbiol.* **2004**, *96*, 473–480. [CrossRef] [PubMed]

58. Egamberdieva, D.; Kucharova, Z. Selection for root colonizing bacteria stimulating wheat growth in saline soils. *Biol. Fertil. Soil* **2009**, *45*, 563–571. [CrossRef]

59. Ali, S. Endophytic fungi from Caralluma acutangula can secrete plant growth promoting enzymes. *Fresenius Environ. Bull.* **2019**, *28*, 2688–2696.

60. Waqas, M.; Khan, A.L.; Shahzad, R.; Ullah, I.; Khan, A.R.; Lee, I.-J. Mutualistic fungal endophytes produce phytohormones and organic acids that promote japonica rice plant growth under prolonged heat stress*. *J. Zhejiang Univ. Sci. B* **2015**, *16*, 1011–1018. [CrossRef]

61. Waqas, M.; Khan, A.L.; Lee, I.-J. Bioactive chemical constituents produced by endophytes and effects on rice plant growth. *J. Plant Interact.* **2013**, *9*, 478–487. [CrossRef]

62. Kucey, R.M.N. Effect of Penicillium bilaji on the Solubility and Uptake of P and Micronutrients from Soil by Wheat. *Can. J. Soil Sci.* **1988**, *68*, 261–270. [CrossRef]

63. Khan, M.S.; Zaidi, A.; Ahemad, M.; Oves, M.; Wani, P.A. Plant growth promotion by phosphate solubilizing fungi–current perspective. *Arch. Agron. Soil Sci.* **2010**, *56*, 73–98. [CrossRef]

64. Bilal, S.; Shahzad, R.; Khan, A.L.; Kang, S.-M.; Imran, Q.M.; Al-Harrasi, A.; Yun, B.-W.; Lee, I.-J. Endophytic Microbial Consortia of Phytohormones-Producing Fungus Paecilomyces formosus LHL10 and Bacteria Sphingomonas sp. LK11 to Glycine max L. Regulates Physio-hormonal Changes to Attenuate Aluminum and Zinc Stresses. *Front. Plant Sci.* **2018**, *9*. [CrossRef] [PubMed]

65. Etesami, H.; Alikhani, H.A.; Hosseini, H.M. Indole-3-acetic acid (IAA) production trait, a useful screening to select endophytic and rhizosphere competent bacteria for rice growth promoting agents. *MethodsX* **2015**, *2*, 72–78. [CrossRef] [PubMed]

66. Mastouri, F.; Björkman, T.; Harman, G.E. Seed Treatment withTrichoderma harzianumAlleviates Biotic, Abiotic, and Physiological Stresses in Germinating Seeds and Seedlings. *Phytopathology* **2010**, *100*, 1213–1221. [CrossRef]

67. Cochran, H.C. Effect of seed size on uniformity of pimento transplants (*Capsicum annuum* L.) at harvest time. *J. Am. Sot. Hort. Sci.* **1974**, *99*, 234–235.

68. Waqas, M.; Khan, A.L.; Hamayun, M.; Kamran, M.; Kang, S.M.; Kim, Y.H.; Lee, I.J. Assessment of endophytic fungi cultural filtrate on soybean seed germination. *Afr. J. Biotechnol.* **2012**, *11*, 15135–15143.

69. Jia, C.; Liu, Z.; Zhang, M.; Jia, J.; Zheng, L.; Wang, Q.; Wang, H. Effects of extracts from Paecilomyces variotii on the yield and quality of cherry radish (*Raphanus sativus* L. var. radculus pers). *J. Agric. Res. Environ.* **2019**, *36*, 176–183.

70. Anis, M.; Abbasi, W.; Zaki, M.J. Bioefficacy of microbial antagonists against Macrophomina phaseolina on sunflower. *Pak. J. Bot.* **2010**, *42*, 2935–2940.

71. Maitlo, S.A.; Rajput, N.A.; Syed, R.N.; Khanzada, M.A.; Rajput, A.Q.; Lodhi, A.M. Microbial control of Fusarium wilt of chickpea caused by *Fusarium oxysporum* f. sp. ciceris. *Pak. J. Bot.* **2019**, *51*, 2261–2268. [CrossRef]

Article

Nematicidal Evaluation and Active Compounds Isolation of *Aspergillus japonicus* ZW1 against Root-Knot Nematodes *Meloidogyne incognita*

Qiong He [1], Dongya Wang [1], Bingxue Li [1], Ambreen Maqsood [1,2]  and Haiyan Wu [1,*]

[1] Guangxi Key Laboratory of Agric-Environment and Agric-Products Safety, Agricultural College of Guangxi University, Nanning 530004, China; heqiong3344@163.com (Q.H.); wdy15677131171@163.com (D.W.); 18437958381@163.com (B.L.); ambreenagrarian@gmail.com (A.M.)

[2] Department of Plant Pathology, Faculty of Agriculture and Environmental Sciences, The Islamia University of Bahawalpur, Bahawalpur 63100, Pakistan

* Correspondence: whyzxb@gmail.com

Received: 23 July 2020; Accepted: 16 August 2020; Published: 19 August 2020

Abstract: The root-knot nematode is one of the most damaging plant-parasitic nematodes worldwide, and the ecofriendly alternative approach of biological control has been used to suppress nematode populations. Here the nematicidal activity of *Aspergillus japonicus* ZW1 fermentation filtrate against *Meloidogyne incognita* was evaluated in vitro and in greenhouse, and the effects of *A. japonicus* ZW1 fermentation filtrate on seed germination and the active compound of *A. japonicus* ZW1 fermentation filtrate were determined. The 2-week fermentation filtrate (2-WF) of *A. japonicus* ZW1 exhibited markedly inhibitory effects on egg hatching, and 5% 2-WF showed potential nematicidal activities on second-stage juveniles (J2s); the mortality of J2s was 100% after 24 h exposure. The internal contents of nematodes were degraded and remarkable protruded wrinkles were present on the body surface of J2s. The nematicidal activity of the fermentation was stable after boiling and was not affected by storage time. A germination assay revealed that 2-WF did not have a negative effect on the viability and germination of corn, wheat, rice, cowpeas, cucumbers, soybeans, or tomato seeds. The pot-grown study confirmed that a 20% fermentation broth solution significantly reduced root galls and egg numbers on tomatoes, and decreased galls and eggs by 47.3% and 51.8% respectively, over Czapek medium and water controls. The active compound from the *A. japonicus* ZW1 fermentation filtrate was isolated and identified as 1,5-Dimethyl Citrate hydrochloride ester on the basis of nuclear magnetic resonance (NMR) and LC-MS (liquid chromatograph-mass spectrometer) techniques. Thus, fermentation of *A. japonicus* ZW1 could be considered a potential new biological nematicide for the control of *M. incognita*.

Keywords: biocontrol *Aspergillus japonicus*; root-knot nematode; fermentation filtrate; biological control; seed germination

1. Introduction

Root-knot nematodes (*Meloidogyne* spp.) are economically important worldwide pathogens causing considerable damage to many crops, including cucumbers, tomatoes, rice [1–4], and even cotton [5,6]. *Meloidogyne incognita* is an important species of root-knot nematodes worldwide due to its direct impact on crop yields [7–9]. Specifically, it is capable of causing an estimated yield loss of 5–43% within vegetable crops cultivated in tropical and subtropical areas [10] and estimated $100 billion loss per year worldwide [11].

Due to their short life cycle and high reproduction rates, these root-knot nematodes have been particularly challenging to control. Previously, chemical nematicides are efficiently used to suppress nematode populations, such as fenamiphos, sebufos, dazomet, and carbofuran [12]; however, these have been found to be harmful to both the eco-environment and human health due to their toxic effects. Thus, as a result of these negative impacts and the significant economic losses which can result from nematodes, new and alternative biological control options are urgently needed [13]. Therefore, the use of biological agents to suppress the population of plant-parasitic nematodes could provide an alternative strategy to sustainably manage plant-parasitic nematodes. Using biofumigation instead of harmful fumigants (like synthetic nematicide methyl bromide) to control nematodes is an increasingly feasible method of parasitic nematode management [14]. Plants such as *Melia azedarach* have been found to be potential sources of biofumigation plant material to control *Meloidogyne* spp. on tomato [15]. Moreover, microbial agents for the control of plant-parasitic nematodes is also a potential method; such as bacteria [16,17], fungi [18,19] and actinomycetes [20], which are nematophagous or antagonistic for root-knot nematodes. Specifically, *Arthrobotrys irregularis*, *Pochonia chlamydosporium*, *Paecilomyces lilacinus*, *Myrothecium verrucaria*, bacteria *Pasteuria usgae*, *Bacillus firmus*, *Burkholderia cepacia*, *Pseudomonas fluorescens*, and *Streptomyces avermitilis* [21,22] have been commercially used in many countries for the control of plant-parasitic nematodes. Some potential microbial sources were constantly obtained, volatiles from beneficial bacteria (*Bacillus* sp., *Paenibacillus* sp. and *Xanthomonas* sp.) can control *M. graminicola* second-stage juveniles (J2s) on rice and significantly reduced infection of susceptible rice [23]. Co-inoculation of *Streptomyces* spp. strains KPS-E004 and KPS-A032 showed success in suppressing root-knot nematode [24].

In our previous study, *A. japonicus* ZW1 culture filtrate was shown to have marked nematicidal activity against *M. incognita*. As a result, the main objective of this work was to evaluate the potential biological control of *A. japonicus* ZW1 against root-knot nematodes including: (1) the nematicidal activity of *A. japonicus* ZW1 fermentation filtrate on eggs and J2s within pot and in vitro experiments; (2) electron microscopic evaluation of J2 bodies after treatment with 2-week fermentation filtrate (2-WF); (3) effect of boiling and storage time on nematicidal activity stability of the fermentation filtrate; and (4) evaluation for the effect of *A. japonicus* ZW1 fermentation filtrate on the germination of various crop seeds.

2. Materials and Methods

2.1. Nematode Preparation

Tomato seeds (cv. Xin Bite 2 F1) were sourced from Yashu Garden Seeds Co., Ltd., (Guangzhou, China) and were used to generate seedlings for culturing the *M. incognita*. For the nematodes culture, one-month-old tomato seedlings were transplanted into pots ($7 \times 7 \times 8$ cm) with second stage juveniles of root-knot nematode-infected peat moss (Gui Yu Xin Nong Technology Co., Ltd., Nanning, China) and maintained at 25 °C with a 14 h light (22000 Lux) and 10 h dark photoperiod treatment within a GXZ-280C incubator (Jiangnan Instrument Factory, Ningbo, China). Tomato roots were collected 35 days after inoculation and were gently rinsed with tap water. Eggs were then extracted with 1% NaOCl [25] and hatched at 25 °C using the modified Baermann funnel method [26]. Eggs were put in 30 ìm pore sieves, nested in petri dishes (6 cm-diameter) containing 3 mL distilled water, and the fresh J2s in water were then collected on the day of experiment and used for subsequent experimentation.

2.2. Fermentation Filtrate Preparation

A. japonicus ZW1 from soil was deposited in the China Center for Type Culture Collection (accession number CCTCC No. M 2014641) and GenBank (accession number KR708636.1). One cm^2 potato dextrose agar (PDA) with a fresh culture of *A. japonicus* ZW1 (cultured 3-5 days at 25 °C) was inoculated in triangular flasks with 100 mL Czapek medium (NaNO$_3$ 0.2 g, KCl 0.05 g, FeSO$_4$ 0.001 g, K$_2$HPO$_4$ 0.1 g, MgSO$_4$ 0.05 g, Sucrose 3.0 g, H$_2$O 100 mL) and incubated in a MQD-S2R shaker (Minquan Instrument Co., Ltd., Shanghai, China) at 150 rpm and 25 °C [27] for 3 consecutive weeks, with 10 triangular flasks replicates per week. Czapek medium without inoculation was used as a negative control. At the end of the 3-week period, fermentation broth from a total of 30 conical flasks was then filtered using 0.45 μm Millipore filters (Whatman, Clifton, NJ, USA) and 1-week fermentation filtrate (1-WF), 2-WF, and 3-week fermentation filtrate (3-WF) were prepared. The concentration of 2.5% (i.e., fermentation filtrate volume: sterilized water volume = 1:39), 5% (1:19), 10% (1:9), 20% (1:4) and 50% (1:1) of 1-week fermentation filtrate(1-WF), 2-week fermentation filtrate (2-WF), and 3-week fermentation filtrate (3-WF) were used and 20% Czapek medium and sterilized water were used as control.

2.3. Effect of Fermentation Filtrate on Meloidogyne Incognita Egg Hatching

Fresh eggs were treated with 2.5%, 5%, 10%, 20%, and 50% 1-WF, 2-WF, and 3-WF; and also 20% Czapek medium and sterilized water as controls. The specific experimental conditions were as follows: approximately 100 eggs and 200 μL of different concentrations of fermentation filtrate were dispensed into each well of 96-well plate, with 4 replicates for each treatment. Additionally, all experiments were performed in triplicate. The initial number of eggs was counted, and the hatched J2s were recorded using an inverted microscope (Ti-S, Nikon Instruments Inc., Tokyo, Japan) at 0, 3, 6, 9, 12, 15 d after exposure in the dark at 25 °C. The cumulative hatching rate was calculated using the following formula: cumulative hatching rate = (the number of hatched J2s)/(the initial number of eggs) × 100%.

2.4. Nematicidal Activity of Fermentation Filtrate on Meloidogyne Incognita J2s

Approximately 60 fresh J2s were contained in each well of a 96-well plate and treated with 200 μL of 2.5%, 5%, 10%, 20%, and 50% 1-WF, 2-WF, and 3-WF, 20% Czapek medium and sterilized water. The number of dead nematodes were counted using a Ti-S Nikon microscope (Nikon Instruments Inc., Tokyo, Japan) at 6, 12, 24, 48 h after treatment with the solutions and pictures were taken at each time point except for 48 h. It wax determined whether he bodies of dead J2s were straight and lacking movement even after mechanical prodding [28,29]. The test was conducted at 25 °C in the dark and the experiment was replicated 4 times. J2 mortality was calculated for each well as follows: mortality = (the number of dead J2s/total J2s) × 100%. This experiment was performed a total of three times.

2.5. Scanning Electron Microscopy Observations

J2s were treated with 10% 2-WF for 10 h and subsequently analyzed with scanning electron microscopy (SEM) using the approach as described below [30,31]. In preparation for the microscopic evaluations, J2 specimens were fixed in 2.5% glutaraldehyde with 0.1 M phosphate buffer (pH 7.2) at 4 °C overnight and subsequently washed 3 times in 0.1 M phosphate buffer. Afterwards, they were then fixed in 1% osmium tetroxide for 2 h, washed 3 times in 0.1 M phosphate buffer again, dehydrated in a graded series of ethanol, critical point dried with Quorum K850 critical dryers (Emitech, East Sussex, England, UK) and finally sputter coated with MSP-2S gold-palladium (IXRF, Austin, TX, USA). Prepared J2 specimens were observed using a SU8100 scanning electron microscope (Hitachi, Tokyo, Japan) operating at 3.0 kV accelerating voltage.

2.6. Transmission Electron Microscopy Observations

The technical approach was very similar to the aforementioned method described for 'scanning electron microscopy observations', however, after J2s were dehydrated with ethanol, they were subsequently embedded in Araldite (Sigma-Aldrich, Sigma-Aldrich LLC., Darmstadt, Germany). To enable evaluation of the specimens, ultrathin sections (70 nm) were obtained using an EM UC7 ultramicrotome (Leica, Wetzlar, Germany) with a Diatome Ultra 45° diamond knife (Diatome Ltd., Helmstrasse Nidau, Switzerland). Sectioned samples were then stained with uranyl acetate and lead citrate using carbon film copper 500 mesh [30,32]. Sections of the J2 bodies were then observed using an HT7700 transmission electron microscope (Hitachi, Tokyo, Japan) operating at an 80.0 kV accelerating voltage.

2.7. Greenhouse Experiment

Thirty day old (3–4 leaf stage) healthy tomato seedlings (cv. Xin Bite 2 F1) were transplanted in a pot (785 cm^3) containing 250 g autoclaved and dried peat moss. A total of 2000 fresh J2s were inoculated in each pot at 3 days after transplanting. Subsequently, 130 mL of 20% and 50% 2-WF were used in this experiment and applied in pots. 20% of Czapek medium and tap water were utilized as controls. A randomized design with 6 replicates for each treatment group was used for the pot experiment and all materials were maintained after inoculation at 25 °C in a greenhouse with a 14 h light and 10 h dark photoperiod. Thirty-five days after transplantation, tomato roots were collected and gently washed with tap water to remove residual materials. Plant height, root fresh weight, and the total number of galls and eggs per plant root system were determined. The eggs were extracted separately from plants with a 1% NaOCl method as previously described [25] and were subsequently collected in beakers with water. Afterwards, 50 µL of a well-mixed egg suspension solution were transferred to a counting dish to enable egg count determination. Eggs were counted three times and the total number of eggs in the entire suspension was calculated. This experiment was repeated twice.

2.8. Effect of Boiling and Storage Time on Nematicidal Activity Stability of Fermentation Filtrate

Two-hundred mL of fresh 2-WF was dispensed into two 100 mL beakers respectively. One of the beakers was boiled in a microwave oven at 100 °C, whereas the second beaker was maintained at room temperature. The fermentation filtrate from two beakers were diluted to 10% and sterilized water was used as a control. Nematicidal activity was then conducted as described above and the experiment was triplicated.

For the analysis of storage time, the experiment was set up for 1-, 2-, and 3-week old 2-WF at 4 °C and 25 °C in dark, respectively; with 4 replicates for each treatment. After storage, the 2-WF solution was filtered through a sterile 0.45 µm polyethersulfone filter (Whatman, Clifton, NJ, USA) and subsequently diluted to a 10% solution in sterilized water. Sterilized water alone was used as a negative control. The nematicidal activity was measured as described above and this experiment was repeated 3 times.

2.9. Evaluation of the Strain Fermentation Filtrate on The Germination of Crop Seeds

In this study, the effect of 2-WF of *A. japonicus* ZW1 was evaluated on seed germination of various crops, e.g., from corn (Qingnong 13), wheat (Mianmai 41), cowpeas (Shanlv), cabbage (Green column), cucumbers (Liaoning 8), rice (Teyou 09103), tomatoes (Hongyingguo 808), and soybeans (Ludou 4). First, healthy seeds were surface sterilized with 2% NaOCl for 3 min and subsequently rinsed 5 times with sterilized water [33]. Seeds were treated with 10% and 20% 2-WF in triplicates across 3 independent experiments, with sterilized water used as a negative control. The sterilized crop seeds were then exposed to the fermentation filtrate in a moist chamber and incubated for several days in the dark at room temperature (25 °C). Sprouted seeds were counted every day until the seed germination rate no longer changed. The seed germination rate was calculated as: (number of germinated seed/total tested seeds) × 100%.

2.10. Isolation and Structural Determination of Aspergillus Japonicus ZW-1 Nematicidal Metabolites

Eight litre of *A. japonicus* ZW-1 2-week fermentation broth was filtered through 8 layers of muslin gauze, then concentrated to 500 mL using rotary evaporation (Hei-VAP Core ML G3, Instruments GmbH & Co. Heidolph, KG, Schwabach, Germany) at 55 °C. The crude extract (15.6 g) from *A. japonicus* ZW-1 fermentation broth was extracted with 1-butanol and evaporated at 40 °C until dry, dissolved in methanol (MeOH) and chromatographed on methylated sephadex LH20 (Beijing Solarbio Science & Technology Co., Ltd., Beijing, China) using MeOH as eluent to give two fractions, the two fractions were dissolved in distilled water to make 2.0 mg mL^{-1} aqueous solution for activity assay. One fraction showed activity against J2. This active fraction was dissolved in the chloroform, at which point white crystals formed. The solution was filtered through cotton which was then washed 20 times using chloroform and dried at room temperature to get the purified active compound.

The chemical structures of the active compound were determined using nuclear magnetic resonance (NMR) analysis and high-resolution electrospray ionization mass spectrometry (HR-ESI-MS) analysis. ^{1}H nuclear magnetic resonance (NMR) and ^{13}C NMR spectra were acquired in MeOH with a Bruker AVANCE III HD600 spectrometer (Bruker Corporation, Faellanden, Switzerland) at 600 MHz for ^{1}H NMR spectra and 125 MHz for ^{13}C NMR spectra using tetramethylsilane as the internal standard. HR-ESI-MS analysis was performed using a Waters E2695 model ion trap mass spectrometer (Waters, Milford, MA, USA) [34]. The nematicidal activity of active compounds at different concentrations (1.25, 1.00, 0.75, 0.50, 0.25 mg mL^{-1}) was measured as described above and this experimental approach was repeated 3 times. Sterilized water was used as a control.

2.11. Statistical Analysis

Data were analyzed using SPSS 19.0. software (SPSS Inc. Chicago, IL, USA) and statistical significance was calculated using a one-way analysis of variance (ANOVA). The means of different parameters for each treatment group were compared among each other using a Fisher's protected least significant difference (LSD) test at $p < 0.05$. All figures for statistical analyses were made using Sigma Plot 10.0 (SPSS Inc., Chicago, IL, USA).

3. Results

3.1. Effect of Fermentation Filtrates on Hatching of Meloidogyne Incognita Eggs

The fermentation filtrate of *A. japonicus* ZW1 at various concentrations and different time points showed significant nematicidal activity against cumulative hatching rate of eggs. The cumulative hatching rate of eggs increased over time in the 1-WF, 2-WF, and 3-WF treatments (Figure 1). In relative comparison to 1-WF, *M. incognita* eggs exhibited higher sensitivity to 2-WF and 3 WF. Fifteen days after incubation, the cumulative hatching rates in 20% and 50% 1-WF were 71.1% and 30.1%, respectively, and were significantly lower in comparison to 2.5%, 5%, and 10% 1-WF and controls ($p < 0.05$). For the 2-WF treated samples, cumulative hatching rates in 5%, 10%, 20%, and 50% 2-WF were 42.5%, 36.0%, 24.3%, and 6.4%, respectively, 15 d after incubation. These values were significantly lower than that of the 2.5% 2-WF and control treatments ($p < 0.05$). Cumulative hatching rates in 5%, 10%, 20%, and 50% 3-WF treatments were 53.0%, 42.2%, 34.6%, and 21.2%, respectively, 15 d after incubation. These results were significantly lower than that of the 2.5% 2-WF and control treatments ($p < 0.05$).

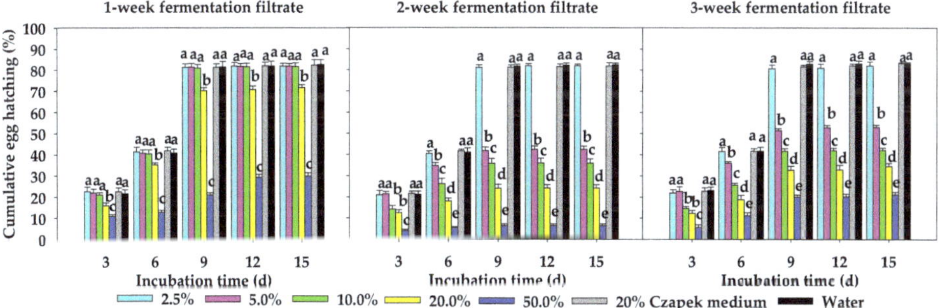

Figure 1. Cumulative *Meloidogyne incognita* eggs hatching rates in *Aspergillus japonicus* ZW1 fermentation filtrate. The bars represent the standard error. The same letter is not significantly different ($p < 0.05$) according to a Fisher's protected least significant difference (LSD) test.

3.2. Nematicidal Activity of Fermentation Filtrates on Meloidogyne Incognita J2s

The time of culturing influenced the nematicidal activity of the fermentation filtrate on J2s (Figure 2). In comparison to the 1-WF and control treatments, the mortality of J2s was higher in 2-WF and 3-WF treatments at different time points post incubation. In the 1-WF treatment, the mortality of J2s was less than 3.3% and no significant difference was observed after treatment for a 6 to 48 h period. Conversely, application of 2-WF and 3-WF resulted in a significantly higher mortality of J2s at different concentrations of the fermentation filtrates as compared to the controls ($p < 0.05$). When investigating 50% 2-WF and 3-WF, the mortality of J2s reached 100% after a 6 h incubation period. After the 48 h incubation period, the mortality of 2.5% 2-WF and 3-WF treatments reached 56.1% and 56.8%, respectively, and were all significantly higher than the controls ($p < 0.05$). From a morphological perspective, treatment with 2-WF resulted in differences in the J2 when compared to the controls (Figure 3). Specifically, microscopic observations revealed that the bodies of J2s in the 2-WF treatment were either straight or arched without movements at 6 h post-incubation (Figure 3, A2). However, bubbles (Figure 3, Bu) appeared in the body of J2s over time and protruded wrinkles (Figure 4, Wr) on the body surface and areas of intensive cytoplasmic vacuolization were observed (such as damaged areas; Figure 5, Da) at 10 h post-exposure to treatment with 2-WF.

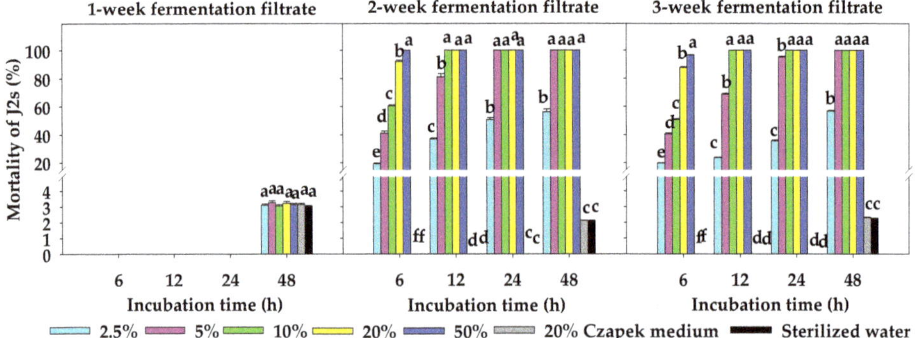

Figure 2. The mortality of *Meloidogyne incognita* J2s in *Aspergillus japonicus* ZW1 fermentation filtrate. Means with the same letter in each group designate no significant differences ($p < 0.05$) based on analysis with a Fisher's protected LSD test.

3.3. Greenhouse Experiment

Treatment with fermentation broth of *A. japonicus* ZW1 resulted in a significant reduction in the number of root galls and eggs per plant as compared to controls (Table 1). The number of root galls and eggs were 8.2 and 3488.9 per plant in the 50% fermentation broth treatment, respectively; whereas 16.8 and 6020 were observed per plant in the 20% fermentation broth treatment, respectively. In both treatments, the number of root galls and eggs was significantly lower than what was observed in controls ($p < 0.05$). The 50% fermentation broth decreased root galls by 78.6% and eggs by 69.4% per plant in comparison to treatment with the Czapek medium control (38.4 root galls and 11413.3 eggs) and 79.9% root galls and 72.0% eggs per plant compared with the tap water control (40.8 root galls and 12480.0 eggs, respectively), and root galls and eggs from the 20% fermentation broth treatment decreased by 56.3% and 47.3% per plant compared with the Czapek medium control (38.4 root galls and 11413.3 eggs, respectively), and 58.8% and 51.8% compared with the tap water control (40.8 root galls and 12480.0 eggs, respectively).

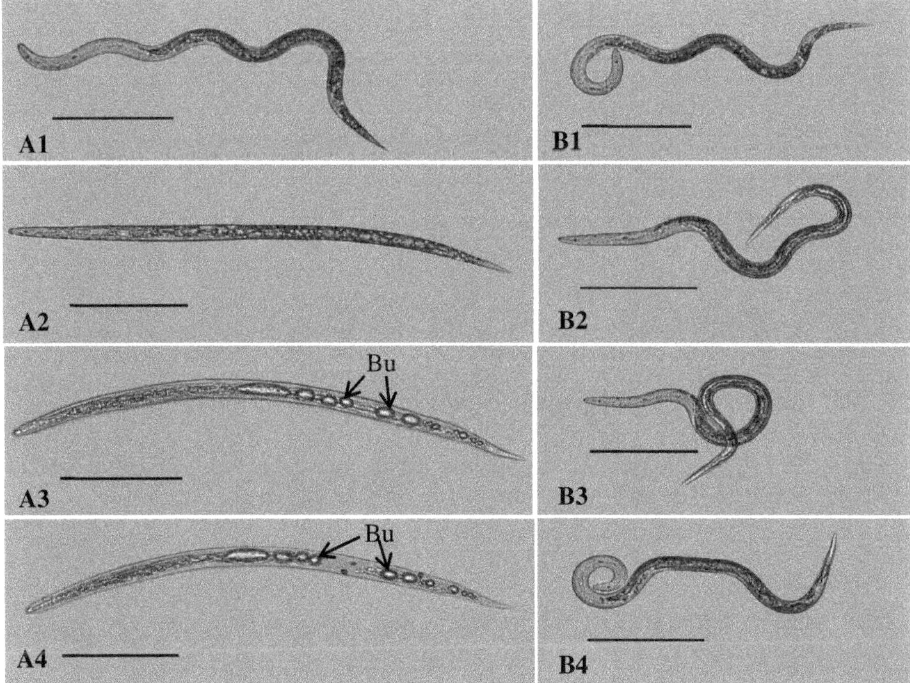

Figure 3. Morphology of second-stage juveniles of *Meloidogyne incognita* treated with 10% 2-week fermentation filtrate (2-WF) of *Aspergillus japonicus* ZW1. **A1–A4** were treated with 10% 2-WF; **B1–B4** were treated with sterilized water; **A1** and **B1** were treated at 0 h; **A2** and **B2** were treated at 6 h; **A3** and **B3** were treated at 12 h; and **A4** and **B4** were treated at 24 h. **Bu:** bubbles. Scale bars of **A1–A4** and **B1–B4** were 100 μm.

Table 1. Effect of *Aspergillus japonicus* ZW1 fermentation broth on the formation of galls and eggs on roots and the growth of tomato plants infected with *Meloidogyne incognita*.

Treatments	Plant Height (cm)	Fresh Root Weight (g)	Root Galls per Plant	Egg Number per Plant
50% Fermentation Broth	26.6 ± 0.6 a	0.6 ± 0.3 a	8.2 ± 1.7 c	3488.9 ± 155.6 d
20% Fermentation Broth	26.5 ± 0.6 a	0.9 ± 0.2 a	16.8 ± 1.4 b	6020.0 ± 214.9 c
Czapek Medium Control	26.9 ± 0.5 a	0.7 ± 0.1 a	38.4 ± 4.3 a	11413.3 ± 338.9 b
Tap Water Control	26.4 ± 0.6 a	0.8 ± 0.2 a	40.8 ± 3.8 a	12480.0 ± 200.4 a

Values represent means ± standard error of six replicate plants per treatment using the combination of two different experiments. Means with the same letter were not significantly different ($p < 0.05$) according to a Fisher's protected LSD test.

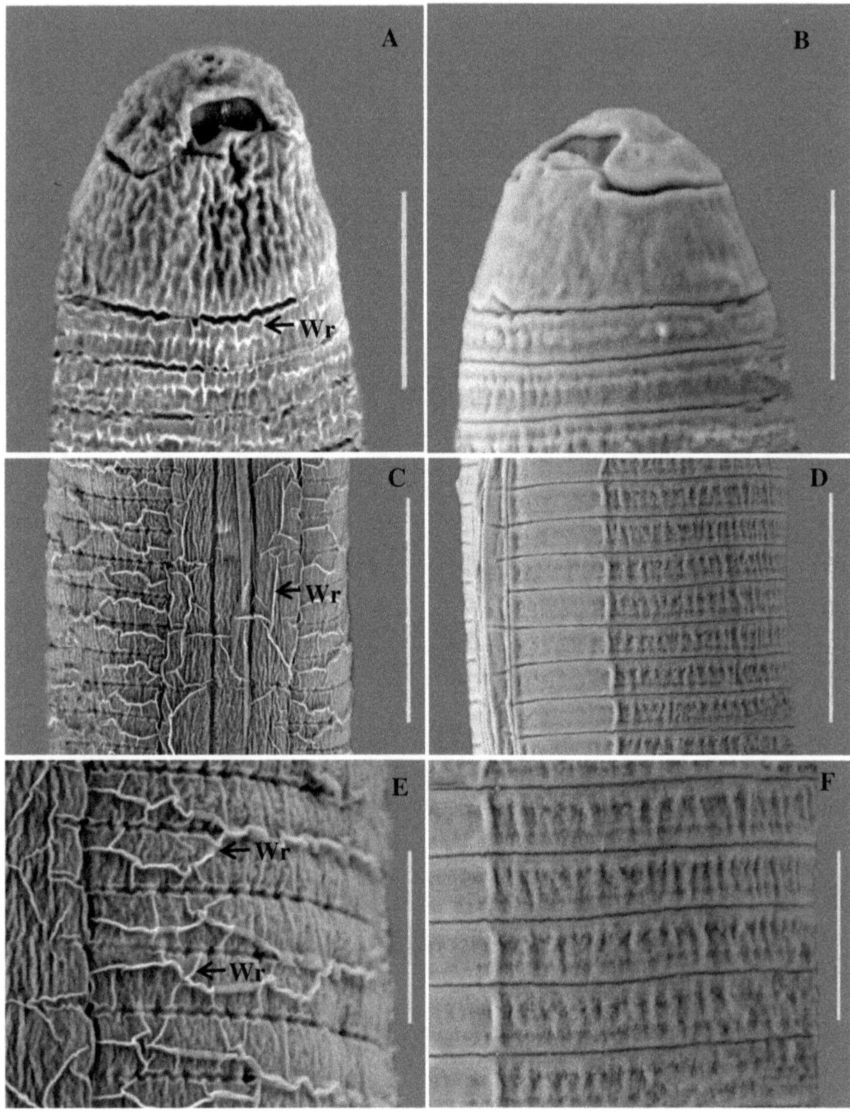

Figure 4. Visualization of the effect of 10% 2-WF of *Aspergillus japonicus* ZW1 on the morphology of *Meloidogyne incognita* J2s with scanning electron microscopy. (**A,C,E**) J2s treated with 10% 2-WF. (**B,D,F**) J2s treated with sterilized water. (**A,B**) Head region of J2. (**C–F**) The lateral field of J2. Scale bars of (**A,B,E,F**) and (**C,D**) were 2 and 5 μm, respectively. **Wr**: protruded wrinkles (black arrow).

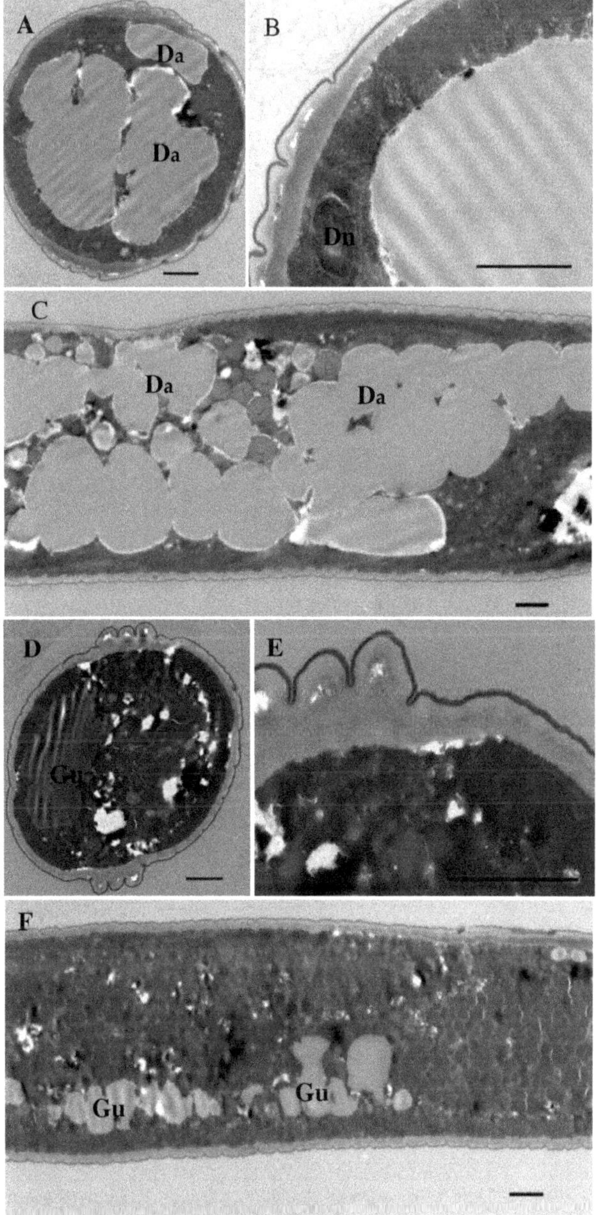

Figure 5. Cross-sections of *Meloidogyne incognita* J2 treated with 10% 2-WT of *Aspergillus japonicus* ZW1. (**A–C**) J2s treated with *A. japonicus* ZW1 fermentation filtrate. (**D–F**) J2s treated with sterilized water. Scale bars of **A**, **B**, **C**, **D**, **E**, and **F** were 2 μm. **Da**: damaged and area. **Gu**: gut. **Dn**: destructed nuclei.

3.4. Effect of Boiling and Storage Time on the Nematicidal Activity of Fermentation Filtrate

The mortality of J2s in fresh and boiled 10% 2-WF did not display any significant differences (Table 2). After a 48 h incubation period, the mortality of J2 reached 100.0% in both fermentation filtrates and was significantly higher than what was observed in the sterilized water treatment ($p < 0.05$).

Table 2. Nematicidal activity of the boiled fermentation filtrate of *Aspergillus japonicus* ZW1 on *Meloidogyne incognita* J2s.

Treatment with 10% 2-WF	Incubation Time (h)			
	6	12	24	48
Untreated	44.9 ± 5.6 a	91.2 ± 3.3 a	91.9 ± 3.4 a	99.2 ± 0.8 a
Boiled	40.5 ± 4.7 a	93.6 ± 2.4 a	96.6 ± 2.0 a	99.0 ± 1.0 a
Sterilized Water	0.0 ± 0.0 b	0.0 ± 0.0 b	0.0 ± 0.0 b	0.1 ± 0.1 b

Values represent means ± standard deviation of three replicates. Means with the same letter are not significantly different ($p < 0.05$) according to a Fisher's protected LSD test.

No significant difference was observed in the mortality of J2s exposed to different storage conditions of 10% 2-WF (Table 3). Specifically, they all reached 100% mortality after a 48 h incubation period, which was higher than the sterilized water treatment ($p < 0.05$).

Table 3. Mortality of *Meloidogyne incognita* J2s in *Aspergillus japonicus* ZW1 fermentation filtrate under different storage conditions.

Treatments with 10% 2-WF	Storage Time	Incubation Time (h)			
		6	12	24	48
	1-week	58.9 ± 5.3 a	99.4 ± 0.6 a	100.0 ± 0.0 a	100.0 ± 0.0 a
4 °C	2-week	58.3 ± 2.2 a	98.8 ± 0.7 a	100.0 ± 0.0 a	100.0 ± 0.0 a
	3-week	60.8 ± 2.2 a	100.0 ± 0.0 a	100.0 ± 0.0 a	100.0 ± 0.0 a
	1-week	62.1 ± 1.8 a	100.0 ± 0.0 a	100.0 ± 0.0 a	100.0 ± 0.0 a
25 °C	2-week	55.6 ± 3.3 a	99.0 ± 0.6 a	100.0 ± 0.0 a	100.0 ± 0.0 a
	3-week	58.4 ± 4.4 a	98.7 ± 0.8 a	99.6 ± 0.4 a	100.0 ± 0.0 a
Sterilized Water	–	0.0 ± 0.0 b	0.0 ± 0.0 b	0.0 ± 0.0 b	1.4 ± 0.8 b

Values represent the means ± standard error of four replicates; means with the same letter are not significantly different ($p < 0.05$) according to a Fisher's protected LSD test.

3.5. Effect of Fermentation Filtrate on Seed Germination

The 20% and 10% 2-WF did not influenced the germination of corn, rice, tomato, cowpea, and cucumber seeds (Table 4). Two days after incubation with 10% 2-WF, the wheat seed germination rate was 85.4% and was significantly higher than what was observed in the control ($p < 0.05$). After an extended period of time beyond the 48-h time period, this value did not increase any further. For soybean seeds treated with 10% 2-WF, germination was significantly lower than what was observed in sterilized water ($p < 0.05$) at day 1; however, there were no statistically significant differences 2–5 days post-incubation across 20% and 10% 2-WF and sterilized water treatments. For cabbage seeds, germination in 20% 2-WF was significantly lower than what was observed in 10% 2-WF and control treatments ($p < 0.05$).

Table 4. Seed germination (%) in different concentrations of 2-week *Aspergillus japonicus* ZW1 fermentation filtrate.

Seeds	Treatments	Incubation Time (d)					
		1	2	3	4	5	6
Wheat	20%	64.6 ± 4.5 a	78.1 ± 1.8 ab	78.1 ± 1.8 ab	78.1 ± 1.8 ab	78.1 ± 1.8 ab	–
	10%	63.5 ± 5.5 a	85.4 ± 3.8 a	85.4 ± 3.8 a	85.4 ± 3.8 a	85.4 ± 3.8 a	–
	Sterilized Water	60.4 ± 4.6 a	72.5 ± 1.3 b	73.5 ± 0.9 b	73.5 ± 0.9 b	73.5 ± 0.9 b	–
Corn	20%	10.4 ± 1.0 a	83.3 ± 1.0 a	88.5 ± 2.8 a	88.5 ± 2.8 a	88.5 ± 2.8 a	–
	10%	11.5 ± 4.5 a	81.3 ± 6.5 a	90.6 ± 4.8 a	90.6 ± 4.8 a	90.6 ± 4.8 a	–
	Sterilized Water	9.4 ± 4.8 a	78.1 ± 4.8 a	89.6 ± 2.1 a	89.6 ± 2.1 a	89.6 ± 2.1 a	–
Rice	20%	3.0 ± 1.8 a	94.0 ± 1.8 a	97.0 ± 1.8 a	97.0 ± 1.8 a	97.0 ± 1.8 a	–
	10%	4.0 ± 2.7 a	96.0 ± 4.0 a	97.0 ± 3.0 a	97.0 ± 3.0 a	97.0 ± 3.0 a	–
	Sterilized Water	1.0 ± 1.0 a	90.9 ± 3.0 a	93.9 ± 1.8 a	93.9 ± 1.7 a	93.9 ± 1.8 a	–
Tomato	20%	0.0	48.1 ± 2.9 a	73.1 ± 3.2 a	81.8 ± 3.3 a	87.5 ± 1.8 a	87.5 ± 1.8 a
	10%	0.0	51.6 ± 2.7 a	69.5 ± 3.6 a	79.5 ± 4.6 a	89.3 ± 3.3 a	91.3 ± 3.2 a
	Sterilized Water	0.0	46.6 ± 8.2 a	72.8 ± 4.2 a	85.0 ± 4.3 a	90.0 ± 2.5 a	91.0 ± 1.6 a
Soybean	20%	33.3 ± 2.8 ab	90.6 ± 4.8 a	100.0 ± 0.0 a	100.0 ± 0.0 a	100.0 ± 0.0 a	–
	10%	26.0 ± 5.2 b	86.5 ± 1.0 a	96.8 ± 1.8 a	99.0 ± 1.0 a	99.0 ± 1.0 a	–
	Sterilized Water	41.8 ± 3.8 a	90.6 ± 3.6 a	100.0 ± 0.0 a	100.0 ± 0.0 a	100.0 ± 0.0 a	–
Cowpea	20%	56.3 ± 3.6 a	99.0 ± 1.0 a	100.0 ± 0.0 a	100.0 ± 0.0 a	100.0 ± 0.0 a	–
	10%	64.3 ± 7.3 a	100.0 ± 0.0 a	100.0 ± 0.0 a	100.0 ± 0.0 a	100.0 ± 0.0 a	–
	Sterilized Water	65.6 ± 3.6 a	99.0 ± 1.0 a	99.0 ± 1.0 a	99.0 ± 1.0 a	99.0 ± 1.0 a	–
Cucumber	20%	100.0	100.0	100.0	100.0	100.0	–
	10%	100.0	100.0	100.0	100.0	100.0	–
	Sterilized Water	100.0	100.0	100.0	100.0	100.0	–
Cabbage	20%	65.1 ± 4.6 b	72.1 ± 6.9 b	72.1 ± 6.9 b	72.1 ± 6.9 b	72.1 ± 6.9 b	–
	10%	89.0 ± 1.2 a	95.1 ± 1.8 a	96.1 ± 2.5 a	96.1 ± 2.5 a	96.1 ± 2.5 a	–
	Sterilized Water	77.9 ± 4.0 a	95.0 ± 3.7 a	95.0 ± 3.7 a	95.0 ± 3.7 a	95.0 ± 3.7 a	–

Values represent the means ± standard error of four replicates; means with the same letter were not significantly different ($p < 0.05$) according to a Fisher's protected LSD test.

3.6. Structural Confirmation of Nematicidal Substance from 2-WF

The active compound was a pale-yellow crystal, which can dissolve easily in water. The ^{1}H NMR spectrum in MeOH exhibited signals due to two methyl groups at δ 3.68 (each 3H, s, 7, 8-CH$_3$), 2.95, 2.85 (each 2H, AB system, d, J = 12.0 Hz, 2, 4-CH$_2$). The ^{13}C NMR and heteronuclear multiple-quantum correlation spectra revealed two carbonyl carbons at δ$_C$ 175.00 (s, C-6), 170.46 (s, C-1, C-5), two methoxy groups 72.84 (s, C-3), 50.78 (q, C-7, C-8), 42.63 (t, C-2, C-4). The electrospray ionization mass spectrometry (ESI-MS) data of active compound was identified the molecular formula of C$_8$H$_{12}$O$_7$ by the [M]$^-$ ion signal at m/z 219 [M]$^-$. The structure of the active compound was determined to be 1,5-Dimethyl Citrate hydrochloride ester (C$_8$H$_{12}$O$_7$ HCl, Figure 6) by the analysis of its spectroscopic data and comparison with the values in the literature [35].

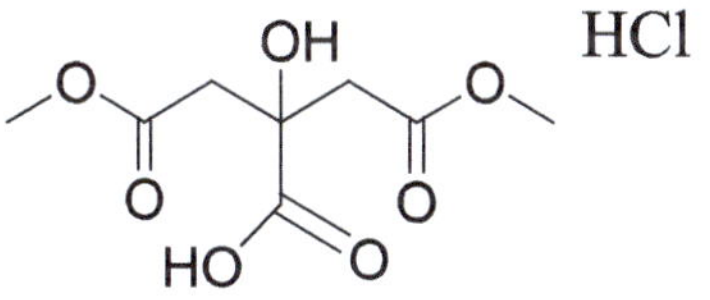

Figure 6. Chemical structures of active compound from *Aspergillus japonicus* ZW1 fermentation filtrate.

3.7. Effect of 1,5-Dimethyl Citrate Hydrochloride Ester on Meloidogyne Incognita J2s

1,5-Dimethyl Citrate hydrochloride ester had a strong toxic activity against J2s at low concentrations, and J2s mortality increased with the duration of exposure in different concentration of 1,5-Dimethyl Citrate hydrochloride ester (Table 5). There were significant differences in mortality between concentrations and control after exposure ($p < 0.05$). The mortality of J2s in concentrations of 1.25, 1.00, 0.75, 0.50, and 0.25 mg mL^{-1} of 1,5-Dimethyl Citrate hydrochloride ester were 91.7%, 57.7%,

36.9%, 20.8%, and 3.3% respectively at 48 h after exposure, which were significantly higher than that of sterilized water ($p < 0.05$).

Table 5. Mortality (%) of *Meloidogyne incognita* J2s in different concentrations of active compound from *Aspergillus japonicus* ZW-1 fermentation filtrate.

Concentration mg/mL	Incubation Time (h)			
	6	12	24	48
1.25	63.4 ± 0.9 a	72.9 ± 0.5 a	78.8 ± 0.6 a	91.7 ± 0.5 a
1.00	39.9 ± 0.7 b	44.4 ± 0.6 b	47.1 ± 0.4 b	57.7 ± 0.5 b
0.75	23.3 ± 0.8 c	31.4 ± 0.3 c	34.1 ± 0.7 c	36.9 ± 0.7 c
0.50	2.0 ± 0.3 d	4.8 ± 0.1 d	7.9 ± 0.2 d	20.8 ± 0.7 d
0.25	0.0 ± 0.0 e	0.0 ± 0.0 e	1.6 ± 0.1 e	3.3 ± 0.1 e
Sterilized Water	0.0 ± 0.0 e	0.0 ± 0.0 e	0.0 ± 0.0 f	0.0 ± 0.0 f

Values represent the means ± standard error of four replicates; means with the same letter each column were not significantly different ($p < 0.05$) according to a Fisher's protected LSD test.

Nematicidal activity of 1,5-Dimethyl Citrate hydrochloride ester was evaluated by comparing the median lethal concentrations (LC50) for different concentrations on *M. incognita* J2s under different exposure times. The concentrations at which 50% of the dead *M. incognita* J2s (LC50) were 1.0373, 0.9646, 0.9397, and 0.7614 mg mL^{-1} 1,5-Dimethyl Citrate hydrochloride ester for 6, 12, 24, and 48 h respectively. The LC50 values were decreasing with the enhanced of exposure time (Table 6).

Table 6. Toxicity of active compound to *Meloidogyne incognita* J2s at different treatment durations.

Exposure Time (h)	Slope (±SE)	Correlation Coefficient	LC50 (95%CI)	LC90 (95%CI)
6	4.8790(±0.2118)	0.9881	1.0373 (0.9112–1.1808)	1.5283 (1.2756–1.8312)
12	5.1225(±0.2843)	0.9800	0.9646 (0.8229–1.1308)	1.4059 (1.1282–1.7520)
24	5.1099(±0.1618)	0.9760	0.9397 (0.7922–1.1145)	1.9421 (1.4234–2.6498)
48	5.4928(±0.2180)	0.9596	0.7614 (0.6261–0.9260)	1.5469 (1.0971–2.1811)

LC-lethal concentration expressed in mg/mL active compound with 95% confidence intervals (CI). SE, standard error.

4. Discussion

In general, the management of parasitic nematodes is a challenging process and current control strategies are mostly dependent upon the application of nematicides [36]. However, many effective nematicides have been restricted for usage and have been banned from the market in recent years due to environmental concerns [37]. Biological options are gaining attention as promising new tools due to their environmentally-friendly and non-toxic characteristics. The potential for using microbes in controlling plant-parasitic nematodes has been documented [38] and effective microbes have been obtained from soil, plants, and the surface of nematodes [39–41]. *Aspergillus* spp. are very common in soil and are lethal to the nematode population; *A. niger* and *A. candidus* were the potential fungal agents to be used against plant-parasitic nematodes [35,42,43]. The results of this study indicated that fermentation of the *A. japonicus* ZW1 from soil was found to not only inhibit egg hatching but was also toxic to nematodes in vitro. The 2-WF was shown to be more toxic to J2s than 1-WF and 3-WF; this effect showed the presence of more active compounds in 2-WF, worth previous characterization. The similar behavior of several fungi and bacteria were also studied against plant parasitic nematodes. Among them a culture filtrate of the rhizosphere bacterium *Pseudoxanthomonas japonensis* isolated from soil exhibited strong nematicidal activity against the *M. incognita* [30]; a metabolite of *Xylaria grammica* KCTC 13121BP isolated from lichen showed strong J2 killing and egg-hatching inhibitory effects [44];

and a culture medium of *Stenotrophomonas maltophilia* and *Rhizobium nepotum* isolated from the surface of nematodes reduced the pathogenicity of wild pine wood nematodes [39].

Natural products have many limitations, such as natural laccases, which have poor stability of enzymatic activity [45]. As a result, it was important to determine and assess if the novel environmentally-friendly nematicides could be stable for practical and durable application opportunities. Consequently, in our present study, we were interested to determine the durability of the novel biological filtrates. Importantly, the toxic activity of the *A. japonicus* ZW1 fermentation filtrate was not effected by boiling, storage time (1-, 2-week, and 3-week) and warm/cold conditions (25 °C and 4 °C). Usually, the surface coating of nematodes was considered to play an important role in the external protection of nematode bodies, sensing, and communication [46,47]. The microbes and plant produced several acidic metabolites or proteinases that specifically degraded the outer membrane of host cells during primary infection [42,48,49]. In our study, wrinkles on the surface of the body of J2s in 2-WF were observed with scanning electron microscopy, and internal bubbles appeared in their body over time. Additionally, other prominent changes such as intensive cytoplasmic vacuolization areas were observed using transmission electron microscopy; suggesting that the activity of compounds produced by *A. japonicus* ZW1 targeted the skin of nematodes and changed its permeability [50]. Previous research showed acidoid (acetic acid) damage the nuclei of cells and led to intensive cytoplasmic vacuolization areas in the body of J2 *M. incognita* [28]. Nematicidal metabolites from the endophytic fungus *Chaetomium globosum* YSC5 significantly reduced the reproduction of *M. javanica* as well [51]. In our present study, nematicidal compound 1,5-Dimethyl Citrate hydrochloride ester from *A. japonicus* ZW1, first isolated and identified on the basis of NMR, LC-MS techniques, was different with the nematicidal compounds produced by *A. niger* (oxalic acid) and *A. candidus* (Citric acid and 3-hydroxy-5-methoxy-3-(methoxycarbonyl)-5-oxopentanoic acid). *M. incognita* J2 mortality reached 100% at 1 day, and egg hatching was suppressed by 95.6% at 7 days after treated with 2 mmol L^{-1} (180 µg mL^{-1}) oxalic acid [42]. 3-hydroxy-5-methoxy-3-(methoxycarbonyl)- 5-oxopentanoic acid was an isomer of 1,5-Dimethyl Citrate, which increased the mean percentage of immobile *Ditylenchus destructor* by 50% at a concentration of 50 mg mL^{-1} after exposure for 72 h [35]. In our study, *M. incognita* J2 treated with 1,5-Dimethyl Citrate hydrochloride ester, mortality reached 91.7% at 48 h after exposure to 1.25 mg mL^{-1} concentration, the LC50 was 0.7614 mg mL^{-1}, which exhibited the most potent toxic activity against the J2 of *M. incognita*. However, the interesting thing was that in in vitro bioassay, fermentation of the strain exhibited better nematicidal effects, and the mortality of J2s reached 100% after exposed to 5% concentration (approximately 100 µg mL^{-1} 1,5-Dimethyl Citrate hydrochloride ester) *A. japonicus* ZW1 fermentation filtrate at 24 h. Our speculation is that the nematicidal effect originated 1,5-Dimethyl Citrate hydrochloride ester combined with some other compounds produced by *A. japonicus* ZW1. Thus, we still need further study to find and proved other nematicidal activity compounds by metabonomics analysis.

No effect on the seed germination of corn, wheat, rice, cowpeas, cucumbers, soybeans, and tomatoes was observed for the 10% and 20% 2-WF treatments. In whole pot experiments, treatment with the fermentation broth of the strain suppressed root galls and egg populations for tomatoes. As a result, these results suggested that *A. japonicus* ZW1 produced and excreted metabolites that were toxic to root-knot nematodes but did not exert negative effects on seed germination. Thus, *A. japonicus* showed desirable, effective, and safe biocontrol properties against *M. incognita* for both in vitro and greenhouse conditions. Taken together, these observations suggest that the fermentation filtrate of *A. japonicus* ZW1 is safe for use as a biological control fungus against root-knot nematodes. However, further studies are warranted and necessary to evaluate the in vivo efficacy of the strain against root-knot nematodes or other plant-parasitic nematodes.

5. Conclusions

A. japonicus ZW1 fermentation filtrate exhibited a potential biocidal activity on *M. incognita* in vitro and in vivo. The *A. japonicus* ZW1 2-week fermentation filtrate exhibited markedly inhibitory effects on egg hatching and nematicidal activities on J2s followed by 3-week fermentation filtrate. The *A. japonicus* ZW1 filtrate penetrated the body wall of *M. incognita* and caused intensive cytoplasmic vacuolization with remarkable protruded wrinkles appearing on the body surface of the J2s. Moreover, the nematicidal activity of the fermentation was stable after a boiling treatment and was not affected by storage time. *A. japonicus* ZW1 fermentation filtrate had no negative effect on the viability and germination of corn, wheat, rice, cowpeas, cucumbers, soybeans, and tomato seeds. The main active compound of 1,5-Dimethyl Citrate hydrochloride ester was first isolated and identified from the *A. japonicus* ZW1 fermentation filtrate. Finally, this work highlights the relevance of *A. japonicus* ZW1 fermentation filtrate as a potential new biological nematicide resource for the control of *M. incognita*.

Author Contributions: Conceptualization, Q.H., A.M. and H.W.; methodology, Q.H., A.M.; software, D.W.; B.L. and A.M.; validation, Q.H.; formal analysis, Q.H., A.M., D.W. and B.L.; investigation, A.M. and H.W.; resources, H.W.; data curation, Q.H., A.M. and B.L.; writing—original draft preparation, Q.H.; writing—review and editing, Q.H., A.M., D.W., and H.W.; and supervision, H.W. All authors have read and agreed to the published version of the manuscript.

Funding: This study was supported by the National Natural Science Foundation of China (31660511, 31460464), Guangxi Innovation Team of National Modern Agricultural Technology System (nycytxgxcxtd-10-04), Construction Project of Characteristic Specialty and Experimental Training Teaching Base (Center)-Characteristic Specialty-Plant Protection (2018–2020).

Conflicts of Interest: All authors declare they have no conflict of interest.

References

1. Fanelli, E.; Cotroneo, A.; Carisio, L.; Troccoli, A.; Grosso, S.; Boero, C.; Boero, C.; Capriglia, F.; Luca, F.D. Detection and molecular characterization of the rice root-knot nematode *Meloidogyne graminicola* in Italy. *Eur. J. Plant Pathol.* **2017**, *149*, 467–476. [CrossRef]

2. Kayani, M.Z.; Mukhtar, T.; Hussain, M.A. Effects of southern root knot nematode population densities and plant age on growth and yield parameters of cucumber. *Crop Prot.* **2017**, *92*, 207–212. [CrossRef]

3. Besnard, G.; Thi-Phan, N.; Ho-Bich, H.; Dereeper, A.; Nguyen, H.T.; Quénéhervé, P.; Aribi, J.; Bellafiore, S. On the close relatedness of two rice-parasitic root-knot nematode species and the recent expansion of *Meloidogyne graminicola* in Southeast Asia. *Genes* **2019**, *10*, 175. [CrossRef]

4. Bozbuga, R.; Dasgan, H.Y.; Akhoundnejad, Y.; Imren, M.; Günay, O.C.; Toktay, H. Effect of *Mi* gene and nematode resistance on tomato genotypes using molecular and screening assay. *Cytol. Genet.* **2020**, *54*, 154–164. [CrossRef]

5. Alves, G.C.S.; Ferri, P.H.; Seraphin, J.C.; Fortes, G.A.C.; Rocha, M.R.; Santos, S.C. Principal Response Curves analysis of polyphenol variation in resistant and susceptible cotton after infection by a root-knot nematode (RKN). *Physiol. Mol. Plant Pathol.* **2016**, *96*, 19–28. [CrossRef]

6. Lopes, C.M.L.; Cares, J.E.; Perina, F.J.; Nascimento, G.F.; Mendona, J.S.F.; Moita, A.W.; Castagnone-Sereno, P.; Carneiro, R.M.D.G. Diversity of *Meloidogyne incognita* populations from cotton and aggressiveness to *Gossypium* spp. accessions. *Plant Pathol.* **2019**, *68*, 816–824. [CrossRef]

7. Trudgill, D.L.; Blok, V.C. Apomictic, polyphagous root-knot nematodes: Exceptionally successful and damaging biotrophic root pathogens. *Annu. Rev. Phytopathol.* **2001**, *39*, 53–77. [CrossRef]

8. Onkendi, E.M.; Kariuki, G.M.; Marais, M.; Moleleki, L.N. The threat of root-knot nematodes (*Meloidogyne* spp.) in Africa: A review. *Plant Pathol.* **2014**, *63*, 727–737. [CrossRef]

9. Janati, S.; Houari, A.; Wifaya, A.; Essarioui, A.; Mimouni, A.; Hormatallah, A.; Sbaghi, M.; Dababat, A.A.; Mokrini, F. Occurrence of the root-knot nematode species in vegetable crops in Souss region of Morocco. *Plant Pathol. J.* **2018**, *34*, 308–315.

10. Seid, A.; Fininsa, C.; Mekete, T.; Decraemer, W.; Wesemael, W.M.L. Tomato (*Solanum lycopersicum*) and root-knot nematodes (*Meloidogyne* spp.)–a century-old battle. *Nematology* **2015**, *17*, 995–1009. [CrossRef]

11. Mukhtar, T.; Hussain, M.A.; Kayani, M.Z.; Aslam, M.N. Evaluation of resistance to root-knot nematode (*Meloidogyne incognita*) in okra cultivars. *Crop Prot.* **2014**, *56*, 25–30. [CrossRef]

12. Patel, B.K.; Patel, H.R. Effect of physical, cultural and chemical methods of management on population dynamics of phytonematodes in bidi tobacco nursery. *Tob. Res.* **1999**, *25*, 51–60.

13. Sikora, R.A. Management of the antagonistic potential in agricultural ecosystems for the biological control of plant parasitic nematodes. *Annu. Rev. Phytopathol.* **1992**, *30*, 245–270. [CrossRef]

14. Brennan, R.J.B.; Glaze-Corcoran, S.; Wick, R.; Hashemi, M. Biofumigation: An alternative strategy for the control of plant parasitic nematodes. *J. Integr. Agric.* **2020**, *19*, 1680–1690. [CrossRef]

15. Ntalli, N.; Monokrousos, N.; Rumbos, C.; Kontea, D.; Zioga, D.; Argyropoulou, M.D.; Menkissoglu-Spiroudi, U.; Tsiropolos, N.G. Greenhouse biofumigation with *Melia azedarch* controls *Meloidogyne* spp. and enhances soil biological activity. *J. Pest Sci.* **2017**, *91*, 29–40. [CrossRef]

16. Stirling, G.R.; Wong, E.; Bhuiyan, S. *Pasteuria*, a bacterial parasite of plant-parasitic nematodes: Its occurrence in Australian sugarcane soils and its role as a biological control agent in naturally-infested soil. *Australas. Plant Pathol.* **2017**, *46*, 563–569. [CrossRef]

17. Viljoen, J.J.F.; Labuschagne, N.; Fourie, H.; Sikora, R.A. Biological control of the root-knot nematode *Meloidogyne incognita* on tomatoes and carrots by plant growth-promoting rhizobacteria. *Trop. Plant Pathol.* **2019**, *44*, 284–291. [CrossRef]

18. Hussain, M.; Zouhar, M.; Rysanek, P. Suppression of *Meloidogyne incognita* by the entomopathogenic fungus *Lecanicillium muscarium*. *Plant Dis.* **2018**, *102*, 977–982. [CrossRef]

19. Hussain, M.; Maòasová, M.; Zouhar, M.; Rysanek, P. Comparative virulence assessment of different nematophagous fungi and chemicals against northern root-knot nematodes, *Meloidogyne hapla*, on carrots. *Pak. J. Zool.* **2020**, *52*, 199–206. [CrossRef]

20. Nimnoi, P.; Pongsilp, N.; Ruanpanun, P. Monitoring the efficiency of *Streptomyces galilaeus* strain KPS-C004 against root knot disease and the promotion of plant growth in the plant-parasitic nematode infested soils. *Biol. Control* **2017**, *114*, 158–166. [CrossRef]

21. Dong, L.Q.; Zhang, K.Q. Microbial control of plant-parasitic nematodes: A five-party interaction. *Plant Soil* **2006**, *288*, 31–45. [CrossRef]

22. Li, J.; Zou, C.G.; Xu, J.P.; Ji, X.L.; Niu, X.M., Yang, J.K.; Huang, X.W.; Zhang, K.Q. Molecular mechanisms of nematode-nematophagous microbe interactions: Basis for biological control of plant-parasitic nematodes. *Annu. Rev. Phytopathol.* **2015**, *53*, 67–95. [CrossRef]

23. Bui, H.X.; Hadi, B.A.R.; Oliva, R.; Schroeder, N.E. Beneficial bacterial volatile compounds for the control of root-knot nematode and bacterial leaf blight on rice. *Crop Prot.* **2020**, *135*, 104792. [CrossRef]

24. Nimnoi, P.; Ruanpanun, P. Suppression of root-knot nematode and plant growth promotion of chili (*Capsicum flutescens* L.) using co-inoculation of *Streptomyces* spp. *Biol. Control* **2020**, *145*, 104244. [CrossRef]

25. Hussey, R.S.; Barker, K.R. A comparison of methods of collecting inocula of *Meloidogyne* ssp. including a new technique. *Plant Dis. Rep.* **1973**, *57*, 1025–1028.

26. Wu, H.Y.; Silva, J.O.; Becker, J.S.; Becker, J.O. Fluazaindolizine mitigates plant-parasitic nematode activity at sublethal dosages. *J. Pest Sci.* **2020**. [CrossRef]

27. Hahn, M.H.; Mio, L.L.M.D.; Kuhn, O.J.; Duarte, H.D.S.S. Nematophagous mushrooms can be an alternative to control *Meloidogyne javanica*. *Biol. Control* **2019**, *138*, 104024. [CrossRef]

28. Choi, I.H.; Kim, J.; Shin, S.C.; Park, I.K. Nematicidal activity of monoterpenoids against the pine wood nematode (*Bursaphelencus xylophilus*). *Russ. J. Nematol.* **2007**, *15*, 35–40.

29. Hajji-Hedfi, L.; Larayedh, A.; Hammas, N.C.; Regaieg, H.; Horrigue-Raouani, N. Biological activities and chemical composition of *Pistacia lentiscus* in controlling *Fusarium* wilt and root-knot nematode disease complex on tomato. *Eur. J. Plant Pathol.* **2019**, *155*, 281–291. [CrossRef]

30. Holland, R.J.; Williams, K.L.; Khan, A. Infection of *Meloidogyne javanica* by *Paecilomyces lilacinus*. *Nematology* **1999**, *1*, 131–139. [CrossRef]

31. Janssen, T.; Karssen, G.; Topalović, O.; Coyne, D.; Bert, W. Integrative taxonomy of root-knot nematodes reveals multiple independent origins of mitotic parthenogenesis. *PLoS ONE* **2017**, *12*. [CrossRef] [PubMed]

32. Ntalli, N.; Ratajczak, M.; Oplos, C.; Menkissoglu-Spiroudi, U.; Adamaki, Z. Acetic acid, 2-undecanone, and (e)-2-decenal ultrastructural malformations on *Meloidogyne incognita*. *J. Nematol.* **2016**, *48*, 248–260. [CrossRef] [PubMed]

33. Zhao, J.; Liu, W.; Liu, D.; Lu, C.; Zhang, D.; Wu, H.; Dong, D.; Meng, L. Identification and evaluation of *Aspergillus tubingensis* as a potential biocontrol agent against grey mould on tomato. *J. Gen. Plant Pathol.* **2018**, *84*, 148–159. [CrossRef]

34. Hu, Y.; Li, J.; Li, J.; Zhang, F.; Wang, J.; Mo, M.; Liu, Y. Biocontrol efficacy of *Pseudoxanthomonas japonensis* against *Meloidogyne incognita* and its nematostatic metabolites. *FEMS Microbiol. Lett.* **2019**, *366*, fny287. [CrossRef]

35. Shemshura, O.N.; Bekmakhanova, N.E.; Mazunina, M.N.; Meyer, S.L.F.; Rice, C.P.; Masler, E.P. Isolation and identification of nematode-antagonistic compounds from the fungus *Aspergillus candidus*. *FEMS Microbiol. Lett.* **2016**, *363*, fnw26. [CrossRef]

36. Desaeger, J.A.; Watson, T.T. Evaluation of new chemical and biological nematicides for managing *Meloidogyne javanica* in tomato production and associated double-crops in Florida. *Pest Manag. Sci.* **2019**, *75*, 3363–3370. [CrossRef]

37. Sissell, K. EPA bans carbofuran residues; sued over endosulfan. *Chem. Week* **2008**, *170*, 29.

38. Liang, L.M.; Zou, C.G.; Xu, J.Q.; Zhang, K.Q. Signal pathways involved in microbe-nematode interactions provide new insights into the biocontrol of plant-parasitic nematodes. *Philos. Trans. R. Soc. B* **2019**, *374*, 20180317. [CrossRef]

39. Liu, K.C.; Zeng, F.L.; Ben, A.L.; Han, Z.M. Pathogenicity and repulsion for toxin-producing bacteria of dominant bacteria on the surface of American pine wood nematodes. *J. Phytopathol.* **2017**, *165*, 580–588. [CrossRef]

40. Liu, M.J.; Hwang, B.S.; Zhi, J.C.; Li, W.J.; Park, D.J.; Seo, S.T.; Seo, S.T.; Kim, C.J. Screening, isolation and evaluation of a nematicidal compound from actinomycetes against the pine wood nematode, *Bursaphelenchus xylophilus*. *Pest Manag. Sci.* **2019**, *75*, 1585–1593. [CrossRef]

41. Ponpandian, L.N.; Rim, S.O.; Shanmugam, G.; Jeon, J.; Park, Y.H.; Lee, S.K.; Bae, H. Phylogenetic characterization of bacterial endophytes from four *Pinus* species and their nematicidal activity against the pine wood nematode. *Sci. Rep.* **2019**, *9*, 12457. [CrossRef] [PubMed]

42. Jang, J.Y.; Choi, Y.H.; Shin, T.S.; Kim, T.H.; Shin, K.S.; Park, H.W.; Kim, Y.; Kim, H.; Choi, G.J.; Jang, K.S.; et al. Biological control of *Meloidogyne incognita* by *Aspergillus niger* F22 producing oxalic acid. *PLoS ONE* **2016**, *11*. [CrossRef] [PubMed]

43. Jin, N.; Liu, S.M.; Peng, H.; Huang, W.K.; Kong, L.A.; Wu, Y.H.; Chen, Y.P.; Ge, F.Y.; Jian, H.; Peng, D.L. Isolation and characterization of *Aspergillus niger* NBC001 underlying suppression against *Heterodera glycines*. *Sci. Rep.* **2019**, *9*, 591. [CrossRef] [PubMed]

44. Kim, T.Y.; Jang, J.Y.; Yu, N.H.; Chi, W.J.; Bae, C.H.; Yeo, J.H.; Park, A.R.; Hur, J.S.; Park, H.W.; Park, J.Y.; et al. Nematicidal activity of grammicin produced by *Xylaria grammica* KCTC 13121BP against *Meloidogyne incognita*. *Pest Manag. Sci.* **2018**, *74*, 384–391. [CrossRef] [PubMed]

45. Kang, C.; Ren, D.; Zhang, S.; Zhang, X.; He, X.; Deng, Z.; Huang, C.; Guo, H. Effect of polyhydroxyl compounds on the thermal stability and structure of laccase. *Pol. J. Environ. Stud.* **2019**, *28*, 3253–3259. [CrossRef]

46. Spiegel, Y.; McClure, M.A. The surface coat of plant-parasitic nematodes: Chemical composition, origin, and biological role—A review. *J. Nematol.* **1995**, *27*, 127–134.

47. Curtis, R.H.C. Plant-nematode interactions: Environmental signals detected by the nematode's chemosensory organs control changes in the surface cuticle and behaviour. *Parasite* **2008**, *15*, 310–316. [CrossRef]

48. Djian, C.; Pijarowski, L.; Ponchet, M.; Arpin, N.; Favre-Bonvin, J. Acetic acid: A selective nematicidal metabolite from culture filtrates of *Paecilomyces Lilacinus* (Thom) Samson and *Trichoderma Longibrachiatum* Rifai. *Nematologica* **1991**, *37*, 101–112.

49. Phiri, A.M.; Pomerai, D.D.; Buttle, D.J.; Behnke, J.M.B. Developing a rapid throughput screen for detection of nematicidal activity of plant cysteine proteinases: The role of *Caenorhabditis elegans* cystatins. *Parasitology* **2014**, *141*, 164–180. [CrossRef]

50. Jatala, P. Biological control of plant-parasitic nematodes. *Annu. Rev. Phytopathol.* **1986**, *24*, 453–489. [CrossRef]

51. Khan, B.; Yan, W.; Wei, S.; Wang, Z.Y.; Zhao, S.S.; Cao, L.L.; Rajput, N.A.; Ye, Y.H. Nematicidal metabolites from endophytic fungus *Chaetomium globosum* YSC5. *FEMS Microbiol. Lett.* **2019**, *366*, fnz169. [CrossRef] [PubMed]

 **agronomy**

Article

Influence of Metal-Resistant *Staphylococcus aureus* Strain K1 on the Alleviation of Chromium Stress in Wheat

Fanrong Zeng [1], Munazza Zahoor [2], Muhammad Waseem [3], Alia Anayat [4], Muhammad Rizwan [2], Awais Ahmad [5], Tahira Yasmeen [2,*], Shafaqat Ali [2,6,*], Mohamed A. El-Sheikh [7], Mohammed Nasser Alyemeni [7] and Leonard Wijaya [7]

1   School of Agriculture, Yangtze University, 88 Jingmi Road, Jingzhou 434025, China; fanrong.zeng@yangtzeu.edu.cn
2   Department of Environmental Sciences and Engineering, Government College University, Faisalabad 38000, Pakistan; anglicmind5@gmail.com (M.Z.); mrazi1532@yahoo.com (M.R.)
3   Department of Microbiology, Government College University, Faisalabad 38000, Pakistan; muhammad.waseem@gcuf.edu.pk
4   Soil & Water Testing Lab., Ayub Agricultural Research Institute, Jhang Road, Faisalabad 38000, Pakistan; aliaanayat@gmail.com
5   Department of Chemistry, the University of Lahore, Lahore 54590, Pakistan; awaisahmed@gcuf.edu.pk
6   Department of Biological Sciences and Technology, China Medical University, Taichung 40402, Taiwan
7   Department of Botany and Microbiology, College of Science, King Saud University, Riyadh 11451, Saudi Arabia; melsheikh@ksu.edu.sa (M.A.E.-S.); mnyemeni@ksu.edu (M.N.A.); leon077@gmail.com (L.W.)
*   Correspondence: tahirayasmeen@gcuf.edu.pk (T.Y.); shafaqataligill@yahoo.com or shafaqataligill@gcuf.edu.pk (S.A.)

Received: 20 July 2020; Accepted: 26 August 2020; Published: 9 September 2020

Abstract: Chromium (Cr) is recognized as a toxic metal that has detrimental effects on living organisms; notably, it is discharged into soil by various industries as a result of anthropogenic activities. Microbe-assisted phytoremediation is one of the most emergent and environmentally friendly methods used for the detoxification of pollutants. In this study, the alleviative role of *Staphylococcus aureus* strain K1 was evaluated in wheat (*Triticum aestivum* L.) under Cr stress. For this, various Cr concentrations (0, 25, 50 and 100 mg·kg^{-1}) with and without peat-moss-based bacterial inoculum were applied in the soil. Results depicted that Cr stress reduced the plants' growth by causing oxidative stress in the absence of *S. aureus* K1 inoculation. However, the application of *S. aureus* K1 regulated the plants' growth and antioxidant enzymatic activities by reducing oxidative stress and Cr toxicity through conversion of Cr^{6+} to Cr^{3+}. The Cr^{6+} uptake by wheat was significantly reduced in the *S. aureus* K1 inoculated plants. It can be concluded that the application of *S. aureus* K1 could be an effective approach to alleviate the Cr toxicity in wheat and probably in other cereals grown under Cr stress.

Keywords: chromium; *Staphylococcus aureus*; wheat; oxidative stress; antioxidants

---

1. Introduction

Environmental pollution by toxic metals has dramatically increased because of various man-made actions taken while revolutionizing industries and urban life. Although these activities have substantially improved the living standards of humans, they have, at same time, deteriorated the environment [1]. Direct or indirect discharge of sewage and industrial effluent into surface water bodies has resulted in augmentation of chromium (Cr) and other toxic metals in soils [2], causing toxicity to plants [3], animals and humans [4]. In agricultural systems, Cr can easily move to different parts of crops and accumulate there to be later consumed by the animals and humans [5]. Soil contamination by Cr and other heavy metals impacts biodiversity negatively and badly disturbs the living entities in the soil [6].

Major origins of Cr contamination are the leather industry [7,8], mining [1], steel industry, paint industry, wood preservatives, volcanic eruption and weathering [9]. Chromium exerts negative effects on plants by reducing the plant height and root growth, interrupting the germination process, causing disproportion in nutrient levels, exerting harmful effects on photosynthesis, retarding soil microbial activities, inhibiting enzyme activity and stimulating the formation of reactive oxygen species (ROS) which result in induction of oxidative stress in plants [1,10,11]. Chromium can cause different malfunctions in human biological systems that may lead to the death of affected persons [12–14]. Wastewater effluents from the industries are discharged directly into water bodies that are utilized mostly for irrigation purposes. Farmers have to rely on this untreated contaminated water due to limited resources and inadequate sanitation facilities [15].

A major staple food across the world is wheat (*Triticum aestivum* L.), which fulfils food requirements of about 50% of the worldwide population [16]. Amongst wheat-producing countries, Pakistan comes ninth in the world. Wheat subsidizes necessary amino acids, vitamins and minerals, dietary fibers and phytochemicals in our diet [17]. Wheat can accumulate higher Cr concentration in stems followed by leaves and grains [18]. According to the literature, increased heavy metal accumulation in wheat tissues has become a potential source of food chain contamination that can cause serious abnormalities to human biological systems [19,20]. Crops may also have the ability to reduce the Cr from Cr^{6+} to Cr^3. This reduction process is likely to happen in roots as a detoxification mechanism [21]. There are a number of remediation methods used to treat sites contaminated with toxic metals. Presently, scientists have made rampant use of biologically centered techniques to deal with such toxic contaminants in order to remove them from environmental entities, including water, air and soil, or at least make them less damaging to the ecosystem [22]. The phytoremediation technique is a modernized method with a lower budget and environmentally sustainable system [23]; it destroys contaminants by using plants along with their rhizospheric microorganisms. Microbial-assisted phytoremediation helps to deal with toxic heavy metals by stabilizing or transforming them to less toxic forms in carrier materials such as soil, shallow water, sediments or groundwater [24]. Microbes have the capability to modify their genetic sequences in response to variation in environmental factors [25]. In soil polluted with heavy metals, microbes assist the plants by producing various growth-regulating substances, such as organic acids, hormones, siderophores and enzymes, that help in plant growth promotion by involving diverse mechanisms, namely acidification, precipitation, redox reactions and chelation [26]. Likewise, roots excrete beneficial nutrients to support the successful colonization and growth of microbes [26]. Chromium-reducing bacteria have the capability to remediate Cr toxicity by reducing Cr^{6+} into Cr^{3+} in the rhizosphere through bioaccumulation and biosorption mechanisms [27]. *Staphylococcus aureus* is a Gram-positive, ubiquitous and round-shaped facultative anaerobe that grows in clusters, forming a biofilm on surfaces. It can grow in a range of growth temperature from 7 to 48 °C, with 37 °C as the optimal temperature for growth [28]. It was isolated from tannery effluent and characterized as a chromium-reducing bacterium. The application of phytoremediation, along with Cr-resistant bacteria for detoxification of Cr^{6+}, has been considered a safe, effective and economical technique over customary techniques [29,30]. In this study, the alleviative role of *Staphylococcus aureus* strain K1 under Cr stress was evaluated in wheat plants. It was hypothesized that microbes (such as *Staphylococcus aureus* strain K1) may alleviate Cr toxicity in wheat by enhancing antioxidant enzymatic activities of wheat while reducing oxidative stress through biotransformation (Cr^{6+} into Cr^{3+}) and biosorption of Cr.

2. Materials and Methods

2.1. Soil Preparation

Sandy clay loam soil was brought from nursery and was air-dried without direct sunlight. After air-drying, soil sieving was done by a mesh with a pore size of 2 mm. Soil was then sterilized at a temperature of 121 °C for 20–30 min for the purpose of removing any kind of contaminant or bacteria that can cause hindrance in further findings [31]. Chromium solutions of different concentrations were prepared from stock solution of $K_2Cr_2O_7$ in the laboratory, and soil was spiked with final Cr concentrations of 0, 25, 50 and 100 mg·kg^{-1} of soil.

These different concentrations of Cr were taken to determine the maximum concentration of hexavalent Cr tolerable by strain K1. However, in case of Cr reduction, the lower concentration of Cr was used due to the fact that Cr is found in lower concentrations in the natural environment, especially in industrial effluents [31]. The concentrations of Cr used were similar to those used in the literature and were chosen considering the fact that, in field conditions, we had to establish the reduction ability of this particular strain rather than its maximum potential to survive in response to metal stress [15]. The soil was added in the pots (5 kg soil per pot) with proper mixing following the treatment plan. Electrical conductivity and pH from saturated soil were determined by making a soil-to-water ratio of 1:25. Soil was extracted with ammonium bicarbonate diethylenetriaminepentaacetic acid (AB-DTPA) solution for the measurement of bioavailable trace elements in the soil [32]. Soil organic matter was determined following the prescribed method [33]. Soil physicochemical characteristics are given in Table 1.

Table 1. Soil characterization of pot experiment.

Soil Properties	Unit	Values
Texture Properties		**Sandy Clay Loam**
pH	-	7.71
Sand	%	63.7
Clay	%	21.9
Silt	%	14.4
Electrical Conductivity (EC)	dSm^{-1}	4.77
Soluble Ion Values		
Cl$^-$	mmol$_c$L^{-1}	7.15
Ca^{2+} + Mg^{2+}	mmol$_c$L^{-1}	14.92
CO$_3{}^{2-}$	mmol$_c$L^{-1}	0.85
OM	%	0.90
CEC	cmol$_c$kg^{-1}	13.2
HCO$_3{}^-$	mmol$_c$L^{-1}	3.84
Metal Concentration		
Available Cr^{6+}	mg·kg^{-1}	0.04
Available Zn^{2+}	mg·kg^{-1}	0.72
Available Cu^{2+}	mg·kg^{-1}	0.23

2.2. Segregation of Cr-Resistant Bacteria

A modified method of serial dilution was adopted to isolate the Cr-tolerant bacteria from metal-contaminated industrial effluent [34]. For this, ten-fold serial dilutions (10^{-1}, 10^{-2}, 10^{-3} and 10^{-4}) were prepared from samples of collected wastewater using sterilized distilled water [34]. Then, 0.1 mL from each dilution was added to petri plates having Tryptic Soy Agar complemented with 0.5 mM Cr^{6+}. Morphologically different colonies were picked and transferred to petri plates supplemented with gradually elevated levels (0.0, 0.5, 2.5, 5.0, 10.0, 15.0, 20, 22 and 23 mM) of Cr^{6+} [35]. The bacteria that showed maximum resistance to the highest concentration of hexavalent Cr were selected for use in further studies.

2.3. Bacterial Identification

Molecular characterization was carried out through the amplification of 16S rDNA gene via polymerase chain reaction (PCR) using the following universal primers: 27F (5′-AAACTCAAATGAATTGACGG-3′) and 1492R (5′-ACGGGCGGTGTGTAC-3′) [36]. For genomic DNA extraction, Favorgen DNA extraction kit was used following the manufacturer's guideline. The initial denaturation temperature was set at 94 °C for a period of 5 min, and this was followed by 40 recurring cycles of denaturing DNA at 94 °C for 45 s, annealing at 53 °C for 45 s and elongation at 72 °C for 60 s. Final extension was set at 72 °C for 10 min, and this was followed by temperature being held at 4 °C [37]. PCR product (5 μL) was loaded in gel wells, and the reaction was allowed to complete; the product was then visualized using Gel Documentation System (Slite 200 W) under ultraviolet light [37]. After validation, 30 μL PCR product was delivered to Macrogen (Seoul, Korea) for the purpose of sequencing. ChormasPro (v1.7.1) software was used for correction of sequences that were submitted to GenBank for accession number. A phylogenetic tree was constructed by downloading similar partial 16S rDNA gene sequences from the NCBI BLAST database with the help of computer software MEGA (v7.0.) [38].

2.4. Bacterial Inoculum Preparation

In order to obtain pure inoculum of *S. aureus* strain K1, an individual isolated colony was inoculated in 250 mL sterilized nutrient broth and incubated at 150 rpm on orbital rotary shaker for 48 h (at 37 °C). The pure culture was harvested via centrifugation at $6000 \times g$ for 10 min, and the supernatant was discarded. The pellet was washed with sterilized distilled water and resuspended in 100 mL of normal saline (0.85% NaCl) solution. Overall, cell density for the inoculum was maintained at 1×10^8 CFU mL^{-1} [39].

2.5. Seed Coating and Pot Experiment

For this study, seeds of wheat variety Sehar were taken from Ayub Agriculture Research Institute, Faisalabad, Pakistan. Seeds were first washed thoroughly with distilled water, and this was followed by surface sterilization using 10% hydrogen peroxide (H_2O_2) for 30 min [40]. The sterilized seeds were immersed in double volume of bacterial suspension (1×10^8 CFU mL^{-1}) and kept at 37 ± 2 °C on a rotary shaker (90 rpm) for 2 h. To facilitate the attachment of bacterial inoculum to the seeds, carboxymethyl cellulose (CMC) (2%) was added to the suspension as a sticking agent. Seeds were dried under shade after 2 h of inoculation for further experimental use. Uninoculated sterilized seeds were used as control. Clay and peat moss in equal parts (1:1) were mixed and the seeds were added to this mixture, which was shaken well for proper coating and incubated overnight in the dark. The completely randomized design had a total of eight treatments, with three replicates for each treatment. A total of eight seeds per pot were sown, and thinning was performed to result in four seedlings per pot after 3 weeks of seed germination.

2.6. Treatments

The experiment was conducted in plastic pots using different concentrations of Cr (0, 25, 50 and 100 mg·kg^{-1}) in the presence and absence of bacterial inoculation. Different treatments were as follows: T1 (Control), 0 mg·kg^{-1} Cr; T2, 25 mg·kg^{-1} Cr; T3, 50 mg·kg^{-1} Cr; T4, 100 mg·kg^{-1} Cr; T5, 0 mg·kg^{-1} Cr + *S. aureus* K1; T6, 25 mg·kg^{-1} Cr + *S. aureus* K1; T7, 50 mg·kg^{-1} Cr + *S. aureus* K1; T8, 100 mg·kg^{-1} Cr + *S. aureus* K1.

2.7. Plant Harvesting

At 135 days after seed sowing, plants were harvested at maturity. The height and spike lengths of plants were measured with a meter rod. Shoots, roots, spikes and grains were separated properly.

Then, 0.1 M HCl was used to remove the metals from the root surface, and the roots were washed with distilled water. Samples of roots and shoots were kept in a hot air oven (70 °C) for a period of 72 h. Afterwards, dry weight was recorded and samples were crushed to small pieces and processed for further analyses.

2.8. Determination of Chlorophyll Contents and Gas Exchange Parameters

At 8 weeks after seed germination, fresh leaf samples were taken to determine chlorophyll contents using acetone (85% *v/v*) for pigment extraction. These leaf samples were kept in the dark at 4 °C for 24 h. Centrifugation of samples was done to get the supernatant. Absorbance was recorded by spectrophotometer at three different wavelengths (470, 647 and 664.5 nm), and final chlorophyll contents were calculated by following the prescribed method [41]. Photosynthetic rate, transpiration rate and stomatal conductance of samples were recorded 8 weeks after seed germination on a fully sunny day using an infrared gas analyzer (IRGA, LCA-4, Analytical Development Company, Hoddesdon, UK).

2.9. Determination of Reactive Oxygen Species and Antioxidant Enzyme Activities

At 2 months after seed sowing, fresh leaves of plants were sampled for the estimation of reactive oxygen species (ROS) through the assessment of electrolyte leakage (EL) and the contents of malondialdehyde (MDA) and hydrogen peroxide (H_2O_2). Additionally, the activities of enzymes such as superoxide dismutase (SOD), peroxidase (POD), catalase (CAT) and ascorbate peroxidase (APX) were assessed. For the EL estimation, distilled water tubes were used to place leaf samples. Samples were autoclaved at 32 °C for period of 2 h, and the observed EC of the solution was termed as EC_1. Afterwards, this solution was autoclaved at 121 °C for 20 min to measure EC_2, and finally EL was calculated using the following equation as described by Dionisio-Sese and Tobita [42]:

$$EL = (EC_1/EC_2) \times 100$$

The concentration of MDA was measured using the method of Heath and Packer (1968) as modified by Dhindsa et al. [43] and Zhang and Kirham [44]. Hydrogen peroxide was recorded through homogenization of samples in phosphate buffer 50 mM (pH 6.5) and centrifugation followed by addition of 20% H_2SO_4 (*v/v*). Samples were centrifuged once more for 15 min, and readings were taken by spectrophotometer at 410 nm absorbance [45]. A spectrophotometer was utilized to record the activities of antioxidant enzymes such as SOD, POD, CAT and APX. Fresh leaf samples were crushed in liquid nitrogen (N_2), and 0.05 M phosphate buffer (pH 7.8) was utilized for the purpose of standardization. This was followed by centrifugation at 4 °C on 12,000× *g* for a period of 10 min. Supernatant was collected for the sake of antioxidant enzyme activity measurements. The method of Zhang [46] was employed to measure SOD and POD activities, while the Aebi method [47] was used for CAT activity. APX contents were estimated using the method of Nakano and Asada [48].

2.10. Estimation of Cr Contents in Plants

Digestion of dry shoot and root samples was performed for 1 g of each sample in 4:1 (*v/v*) ratio of HNO_3:$HClO_4$ as described by Rehman et al. [49]. Finally, the digested samples were run on an atomic absorption spectrophotometer for the estimation of Cr concentrations in the processed samples.

2.11. Statistical Analysis

The IBM SPSS Statistics for Windows, Version 21.0, was used for the data analyses, using the analysis of variance (ANOVA) tool at a 5% probability level. Tukey's HSD post hoc test was performed for multiple comparison of triplicates.

3. Results

The current study was envisaged to assess the capability of metal-resistant *Staphylococcus aureus* strain K1 to ameliorate the Cr stress in wheat plants.

3.1. Growth Characteristics of Isolate K1

The bacterial strain K1, capable of tolerating a Cr concentration of up to 22 mM, was selected for further studies. Morphologically, it is characterized by Gram-positive cocci ($\approx$1 μm) with yellowish golden color. Chemically, it is oxidase- and coagulase-negative and catalase-positive (Table 2). The BLASTn investigation showed that it has a close resemblance (99%) to *Staphylococcus aureus* strain ATCC 12600 (NR_115606.1) and *Staphylococcus aureus* strain NBRC 100910 (MG971399.1). The similar 16S rDNA gene sequences from GenBank were used to carry out phylogenetic analysis, which also confirmed that the isolate K1 belongs to *Staphylococcus aureus*; therefore, it was named *Staphylococcus aureus* strain K1 (KX685332). This was done in order to remain confident that the bacterial strain used in this study was *Staphylococcus aureus* strain K1, as culture media can sometimes be contaminated with other bacteria.

Table 2. Biochemical and morphological characteristics of *S. aureus* strain K1.

Sr. No.	Characteristic	*Staphylococcus aureus* **K1**
1	Morphology	Convex, round
2	Color	Yellowish, golden
3	Gram-reaction	+ve
4	Catalase	+ve
5	Coagulase plasma reaction	−ve

3.2. Effect of S. aureus K1 Contact Time on Chromium (Cr^{6+}) Reduction

Staphylococcus aureus K1 exhibited optimum growth at pH 8 and 35 °C. Under optimum growth conditions, the effect of contact time on bacterial ability to reduce the hexavalent Cr in the medium was observed. It was observed that the Cr reduction of *S. aureus* K1 increased with increasing contact time (Figure 1). It was found that 26%, 45%, 71%, 80% and 99% Cr^{6+} (initial metal concentration = 1 mM) was removed from the medium by *Staphylococcus aureus* K1 after 2, 4, 8, 16 and 24 h of incubation, respectively (Figure 1).

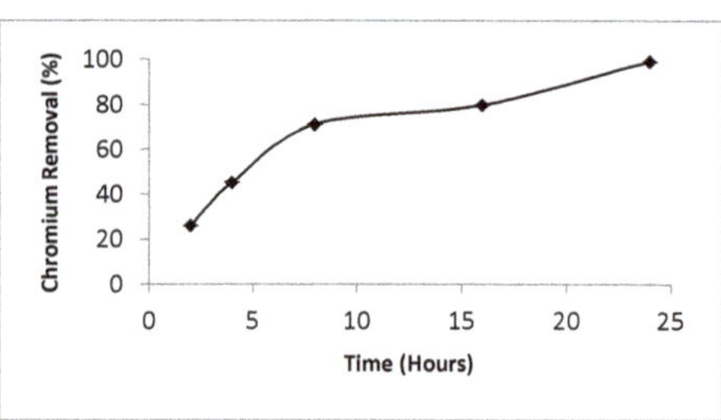

Figure 1. The impact of contact time (hours) impact on the Cr removal ability of *Staphylococcus aureus* K1.

3.3. Effect of S. aureus K1 on Plant Growth Promotion

Chromium stress substantially decreased the growth of wheat plants. A significant reduction in the length of shoots (31.18%), roots (32.02%) and spikes (40.70%) and the dry weight of shoots (34.29), roots (44.17) and grains (31.06%) of the plant was observed at 100 mg·kg^{-1} Cr concentration alone as compared to *S. aureus* K1 inoculated seeds + 100 mg·kg^{-1} Cr concentration (Figure 2). A significant change in shoot and root length was observed in inoculated plants as compared to uninoculated plants at all levels of Cr. Wheat plants stressed with 50 mg·kg^{-1} of Cr showed an observable reduction in growth attributes; however, this decrease was minimized in inoculated plants compared to uninoculated plants, as shown in Figure 2. The growth was gradually decreased when the Cr concentration in the growth medium increased from 25 to 100 mg·kg^{-1} (Figure 2A–D). Moreover, the maximum growth reduction was noticed with 100 mg·kg^{-1} of Cr stress. The data regarding plant growth attributes indicated that inoculation with *S. aureus* K1 significantly improved the wheat growth and dry biomass under Cr stress conditions.

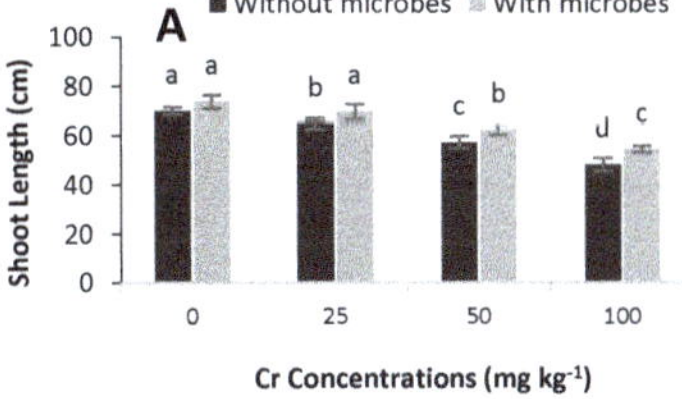
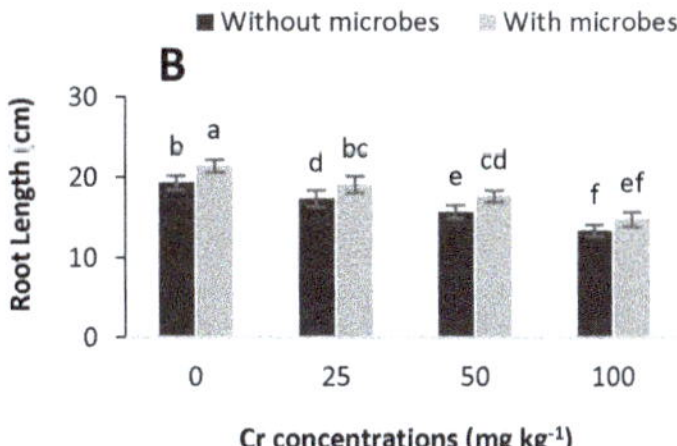
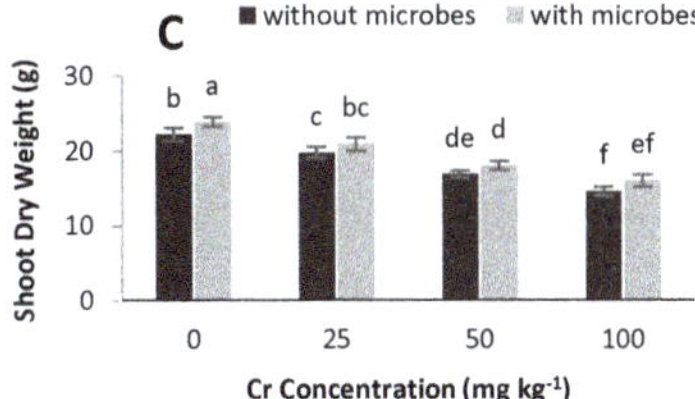
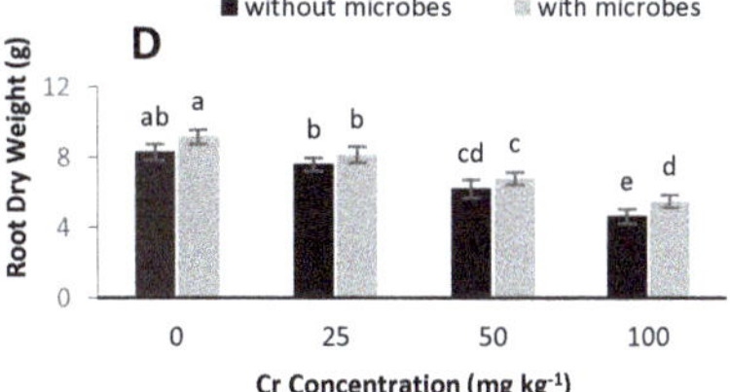
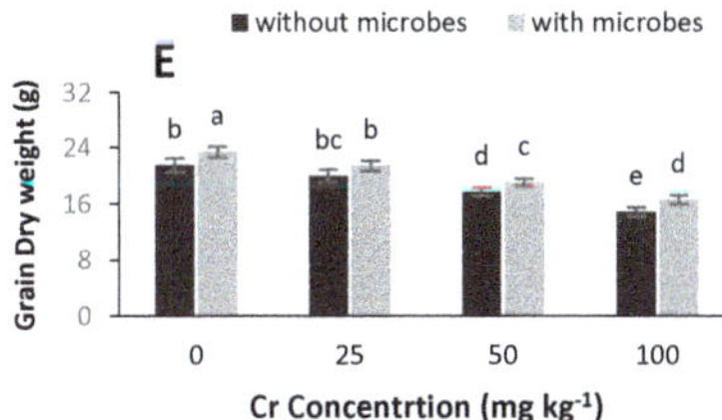

Figure 2. Influence of the different Cr levels (0, 25, 50, 100 mg·kg^{-1}), with and without peat-moss-based microbial inoculation, on length of shoot (**A**) and root (**B**), dry weight of shoot (**C**) and root (**D**) and grain dry weight (**E**) of wheat. Bars indicate the mean of three replicates with standard deviation (SD). Different bars with lowercase letters show noteworthy changes among various treatments at $p < 0.05$.

3.4. IRGA Parameters and Chlorophyll Contents

IRGA parameters such as transpiration rate, stomatal conductance and photosynthetic rate gradually reduced under increased Cr concentrations alone. The transpiration rate was greater at 25 mg·kg^{-1} of Cr stress and decreased with increasing Cr stress levels at concentrations from 50 to 100 mg·kg^{-1}. Without microbial inoculation, transpiration rate decreased by 12%, 21% and 32% under 25, 50 and 100 mg·kg^{-1} Cr stress, respectively, as compared to control (Figure 3A). Similarly, stomatal conductance and photosynthetic rate in uninoculated plants also reduced with increasing Cr concentrations. Stomal conductance decreased by 9%, 25%, 45% and photosynthetic rate decreased by 12%, 25% and 46% under 25, 50 and 100 mg·kg^{-1} Cr stress, respectively, as shown in Figure 3B,C. These results explain the effective role of bacterial inoculation in improving gas exchange attributes in Cr-stressed wheat plants by comparing uninoculated plants. Similarly, chlorophyll a contents decreased by 9.40%, 26.21% and 40.08% in inoculated plants and by 10.66%, 28.02% and 41.87% in uninoculated plants under 25, 50 and 100 mg·kg^{-1} Cr stress, respectively, as shown in Figure 3D.

On the other hand, as compared to untreated control, chlorophyll b was reduced by 15.36%, 27.27% and 40.80% in uninoculated wheat plants and by 14.44%, 27.24% and 40.63% in inoculated wheat plants under 25, 50 and 100 mg·kg^{-1} Cr stress, respectively, as shown in Figure 3E. A gradual decrease in carotenoid contents was also observed in inoculated and uninoculated plants with increasing level of Cr stress, where inoculated plants showed 6%, 19% and 27% reduction in carotenoid contents while uninoculated plants showed 9%, 19% and 28% reduction under 25, 50 and 100 mg·kg^{-1} Cr stress, respectively (Figure 3F)

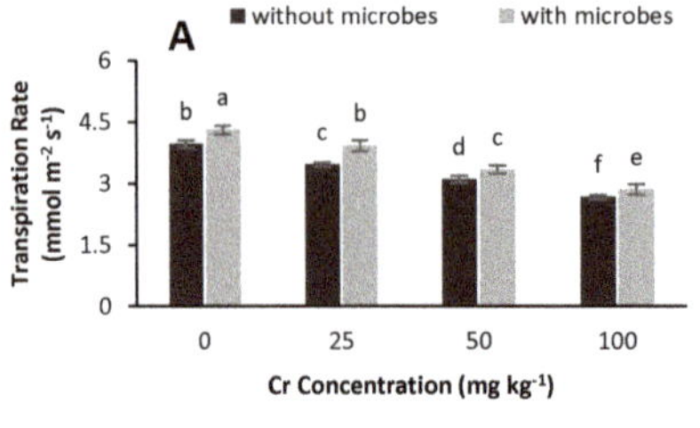
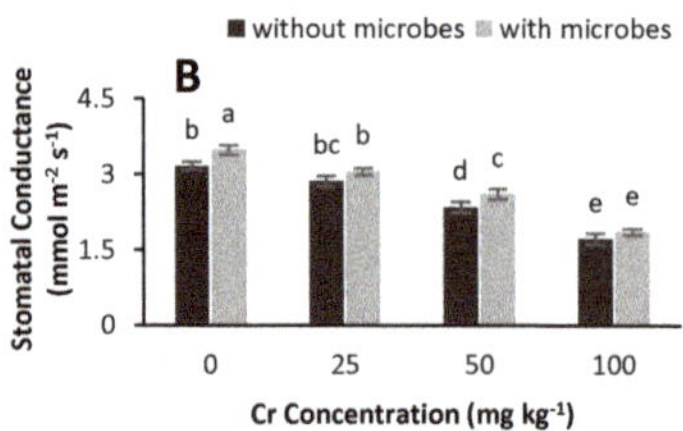
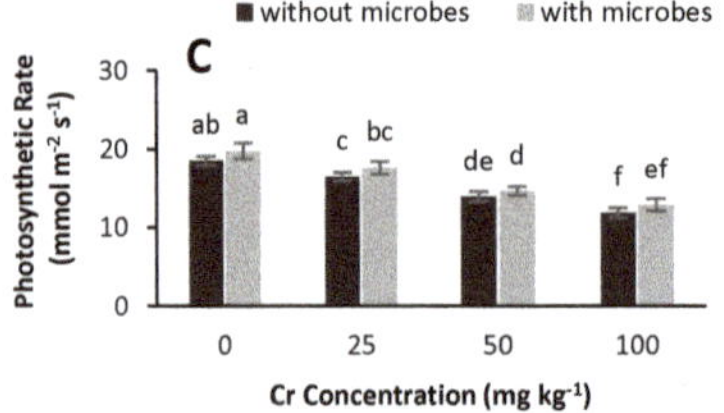
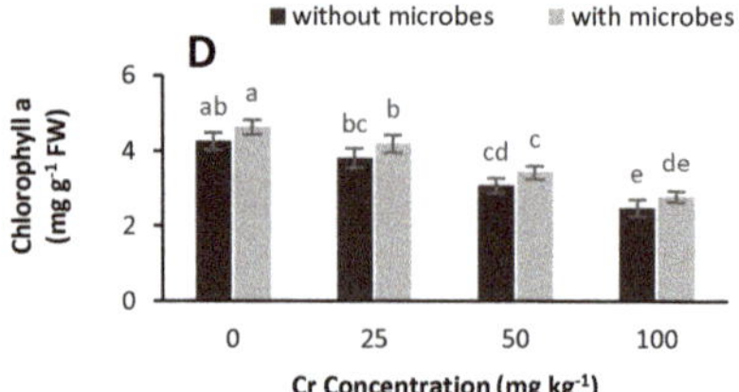
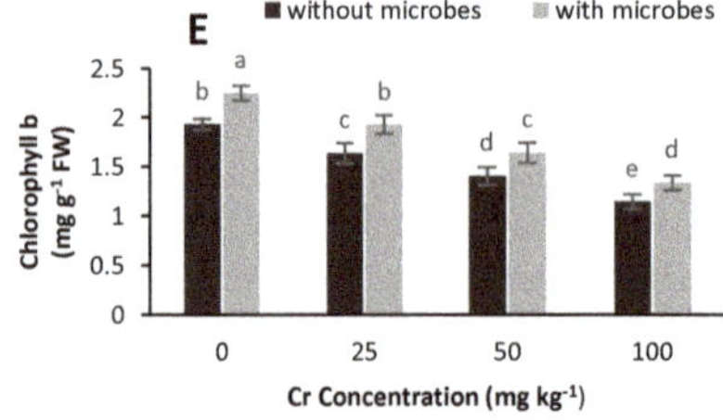
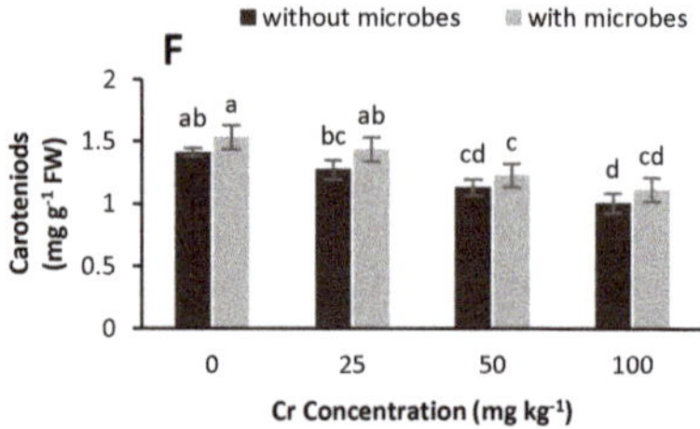

Figure 3. Influence of the different Cr levels (0, 25, 50, 100 mg·kg^{-1}), with and without peat-moss-based microbial inoculation, on transpiration rate (**A**), stomatal conductance (**B**), photosynthetic rate (**C**), chlorophyll a (**D**), chlorophyll b (**E**) and carotenoids (**F**) of wheat plants. Bars indicate the mean values and standard deviation of three replicates. Different bar letters show significant changes among various treatments at $p < 0.05$.

3.5. Estimation of EL, MDA and H_2O_2

A substantial increase in EL was noted in both roots and shoots of wheat plants under Cr stress, as shown in Figure 4A,B. Uninoculated wheat plants showed more EL in leaves and roots under all Cr levels (0, 25, 50 and 100 mg·kg^{-1}) as compared to inoculated plants. EL in uninoculated leaves was increased by 17.98%, 36.40% and 56.52% and EL in uninoculated roots increased by 9%, 32% and 53% under 25, 50 and 100 mg·kg^{-1} Cr, respectively. On the other hand, inoculation with *S. aureus* K1 increased EL in leaves by 15.83%, 33.26% and 55.90% and in roots by 13%, 33% and 56%, under 25, 50 and 100 mg·kg^{-1} Cr, respectively (Figure 4A,B).

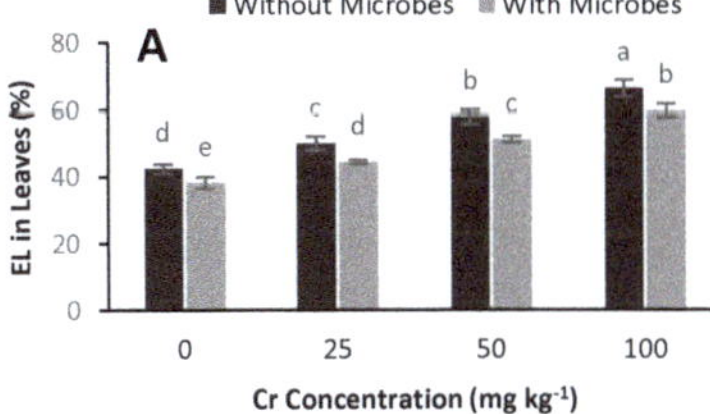
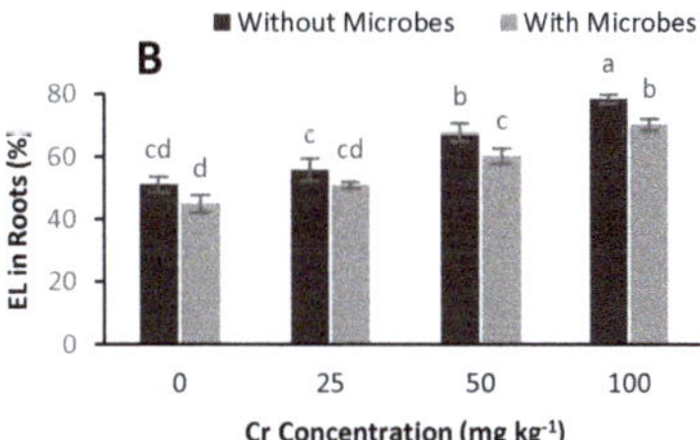

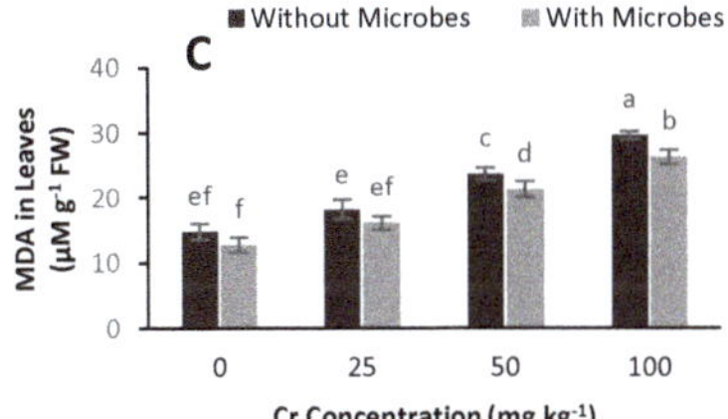
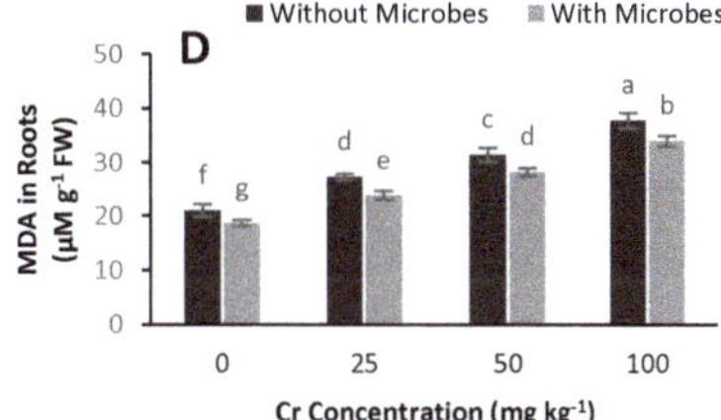

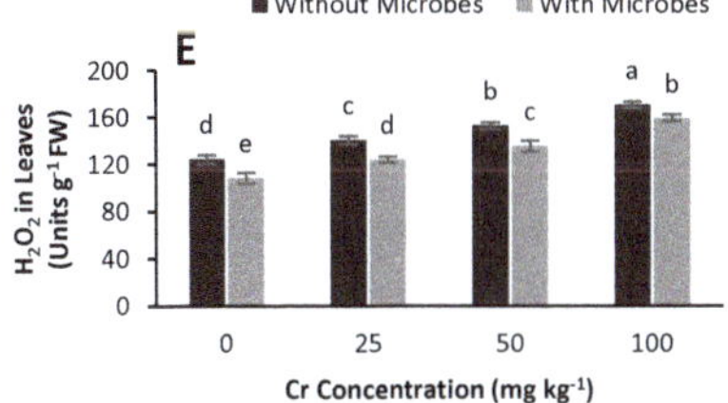
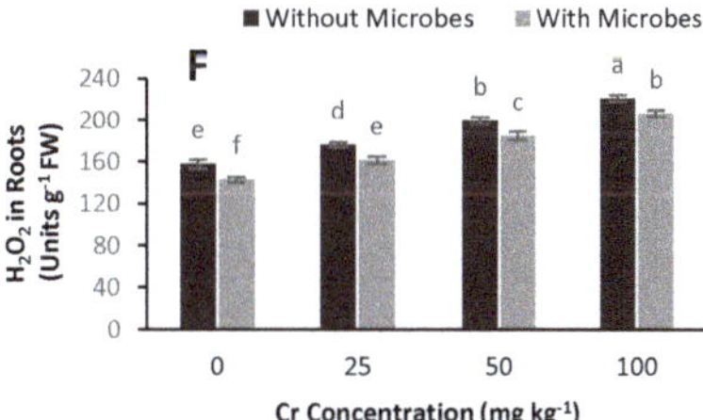

Figure 4. Influence of the different Cr levels (0, 25, 50, 100 mg·kg^{-1}), with and without peat-moss-based microbial inoculation, on EL in leaves (**A**), EL in roots (**B**), MDA in leaves (**C**), MDA in roots (**D**), H_2O_2 in leaves (**E**) and H_2O_2 in roots (**F**) of wheat plants. Bars indicate the mean values and standard deviation of three replicates. Different bar letters show significant changes among various treatments at $p < 0.05$.

There was a noticeable increase in MDA content of leaves, showing lipid peroxidation due to high level of Cr stress, as shown in Figure 4C,D. Maximum MDA contents were observed in leaves and roots of uninoculated plants under 100 mg·kg^{-1} Cr stress as compared to their respective controls. However, inoculation with *S. aureus* K1 reduced MDA content in all the plants of varying level of Cr stress compared to uninoculated plants. Likewise, a gradual rise in H_2O_2 of wheat leaves was observed with increasing levels of Cr (Figure 4E,F). Furthermore, a noteworthy decrease in H_2O_2 content was observed in *S. aureus* K1 inoculated plants, both Cr-stressed and control.

3.6. Effect of S. aureus on Antioxidant Enzyme Activities

The findings revealed that SOD activity in leaves and roots was significantly higher at the 25 mg·kg^{-1} Cr level but gradually decreased with increasing Cr levels, both in uninoculated and inoculated plants. SOD activity increased by 19.59%, 5.22% and 6.98% in uninoculated plant leaves and by 17.58%, 5.22% and 3.08% in uninoculated plant roots under 25, 50 and 100 mg·kg^{-1} Cr treatments, respectively. However, inoculation with *S. aureus* K1 enhanced the SOD activity by 24.71%, 9.64% and 3.51% in leaves and 20.83%, 9.49%, and 4.34% in roots under 25, 50 and 100 mg·kg^{-1} Cr, respectively (Figure 5A,B). As compared to noncontaminated treatments (control), a decline in the CAT activity was observed under Cr contamination (Figure 5C,D). Inoculation with *S. aureus* K1 provoked a substantial increase in the activity of the CAT enzyme in wheat leaves (Figure 5C). CAT activity in roots also improved (114.31 Units·g^{-1} FW) under bacterial inoculation as compared to uninoculated plants (102.66 Units g^{-1} FW) at 25 mg·kg^{-1} Cr (Figure 5D). Moreover, abridged CAT activity was noticed at the highest level of Cr stress (100 mg·kg^{-1}); activity at this level was increased by 5.52% in leaves and 3.63% in roots for uninoculated plants, while inoculated plants showed increase of 5.06% in leaves and 1.37% in roots, as shown in Figure 5C,D. The POD activity substantially ($p < 0.05$) increased due to addition of Cr as compared to control (Figure 5E,F). There was a noticeable reduction in POD activity in leaves under bacterial inoculation with *S. aureus* strain K1 (22.27%, 11.99% and 0.21%) as compared to uninoculated treatments (21.63%, 10.12% and 2.92%) (Figure 5E). There was a substantial increase in the activity of the APX enzyme observed under Cr stress in wheat plants, as shown in Figure 5G,H. There was increase in APX activity in plant shoots and roots, with the maximum production occurring at the Cr concentration of 25 mg·kg^{-1}, and the APX activity decreased at the highest Cr level in the growth medium (Figure 5G,H). Furthermore, the maximum APX activity was observed in roots without inoculation at Cr concentration of 25 mg·kg^{-1}, as shown in Figure 5H.

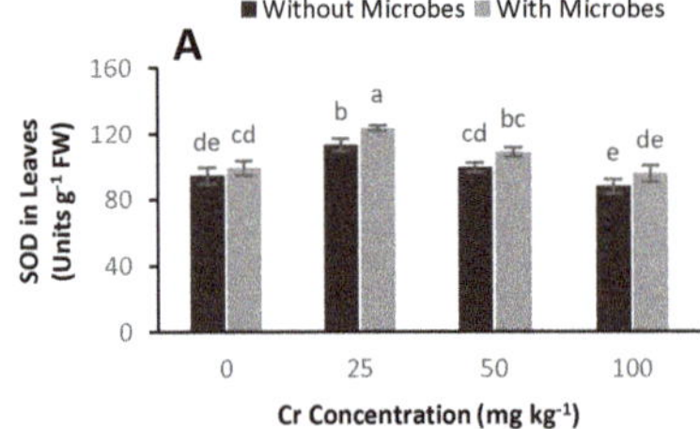
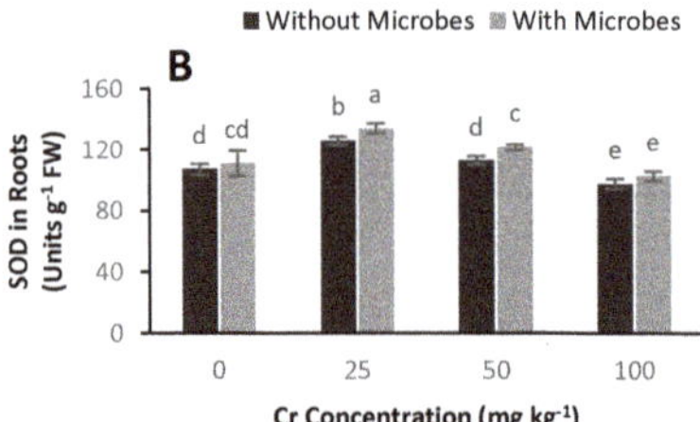

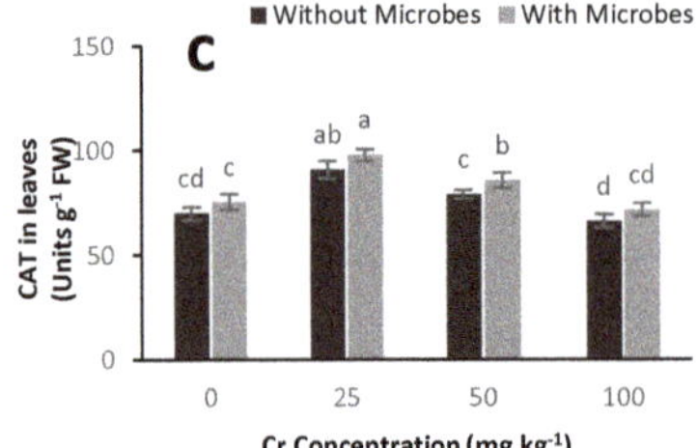
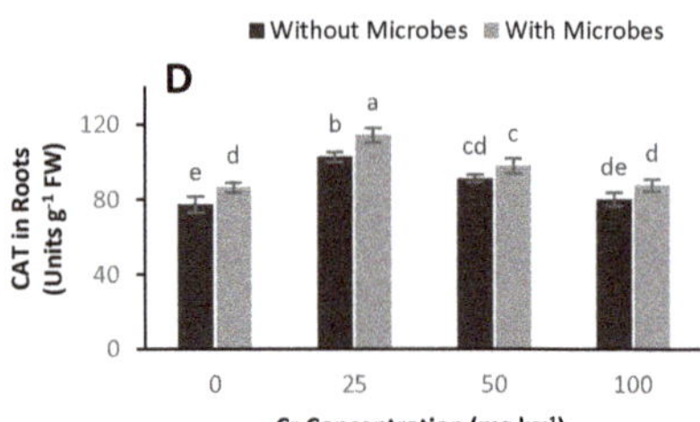

Figure 5. *Cont.*

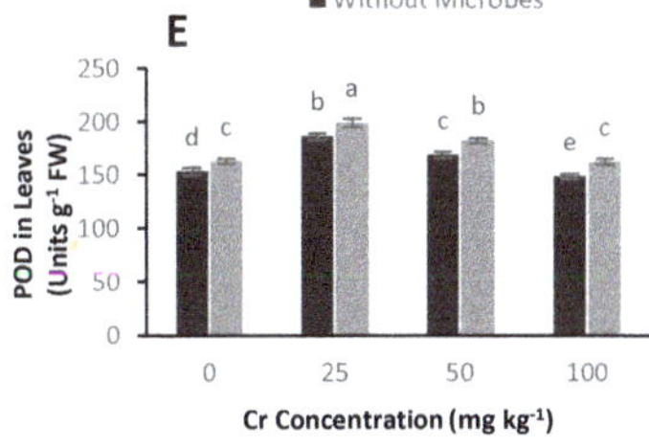

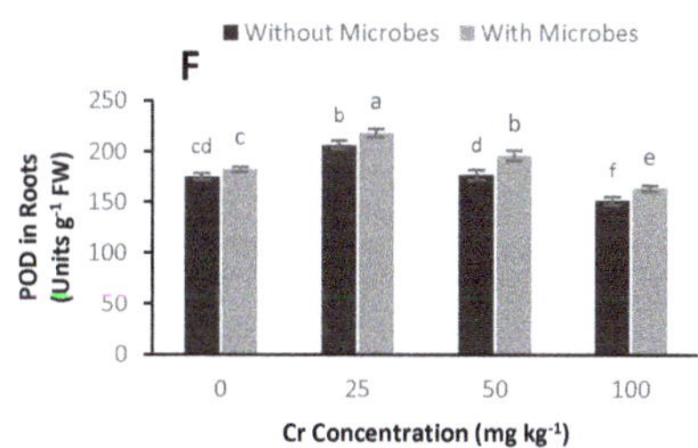

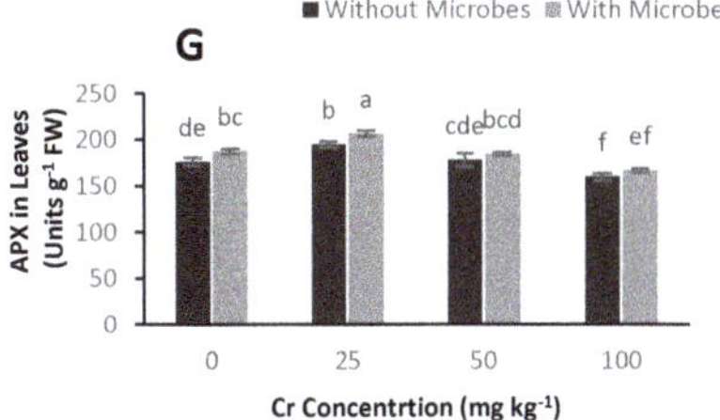

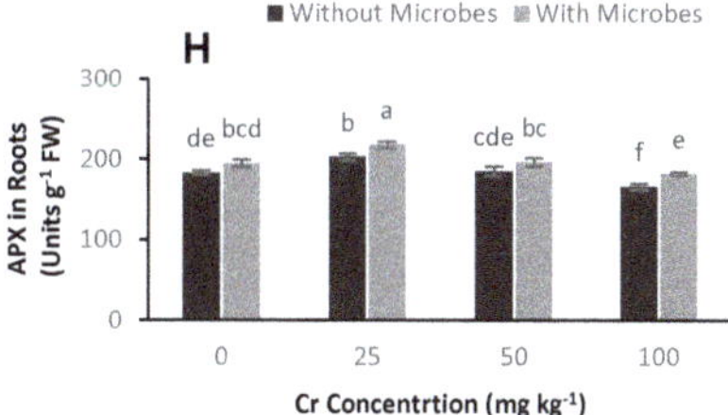

Figure 5. Influence of the different Cr levels (0, 25, 50, 100 mg·kg^{-1}), with and without peat-moss-based microbial inoculation, on SOD in leaves (**A**), SOD in roots (**B**), CAT in leaves (**C**), CAT in roots (**D**), POD in leaves (**E**), POD in roots (**F**), APX in leaves (**G**) and APX in roots (**H**) of wheat plants. Bars indicate the mean values with standard SD of three replicates. Different bar letters show significant changes among various treatments at $p < 0.05$.

3.7. Cr Accumulation in Plants

The data regarding Cr accumulation in shoots and roots of the wheat plants are shown in Figure 6A,B. With increasing concentration of applied Cr, a gradual increase in Cr concentrations was observed in roots and shoots in a dose-additive manner. In addition, inoculation of *S. aureus* K1 significantly decreased the Cr concentrations both in shoots and roots as compared to uninoculated plants.

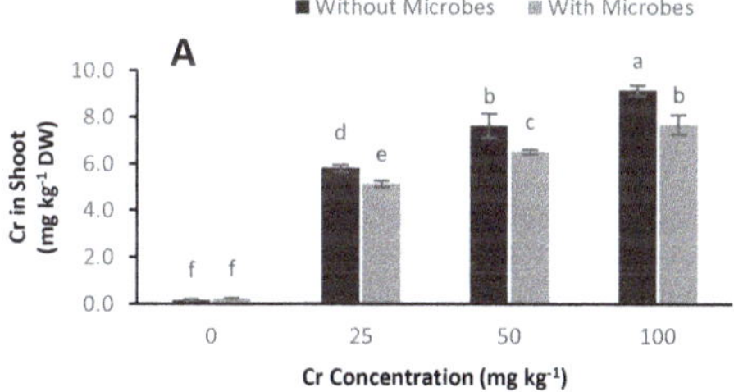

Figure 6. *Cont.*

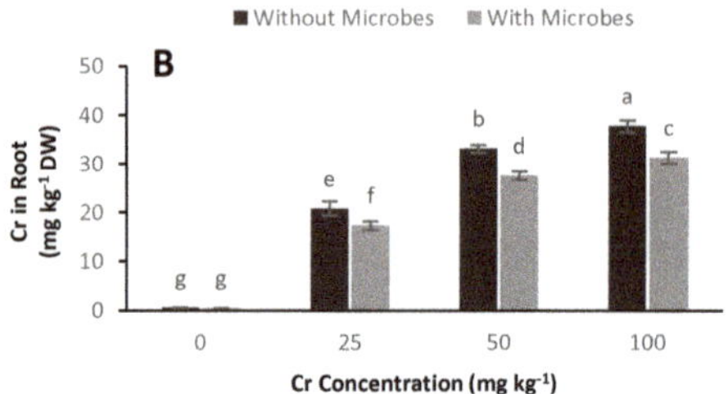

Figure 6. Influence of the different Cr levels (0, 25, 50, 100 mg·kg^{-1}), with and without peat-moss-based microbial inoculation, on Cr concentrations in shoots (**A**) and roots (**B**) of wheat plants. Bars indicate the mean values with standard SD of three replicates. Different bar letters show significant changes among various treatments at $p < 0.05$.

4. Discussion

The major objective of our research was to appraise the effectiveness of *Staphylococcus aureus* K1 treatment in reducing the toxic effects of Cr stress in wheat plants. An indigenous bacterial strain, *Staphylococcus aureus* K1 (GenBank accession no. KX685332), capable of tolerating up to 22 mM of Cr^{6+} was isolated from a metal-polluted environment. Numerous research studies with similar metal-tolerant bacterial isolations from metal-contaminated sites have been reported [35,50,51]. Our results also supported the findings of Mustapha and Halimoon [52], who isolated a total of 21 isolates from electroplating industries and reported that merely 5 of them were Cr-tolerant (up to 50 mg·L^{-1}). The results of the current study show that *S. aureus* K1 increased plant growth parameters under Cr metal stress (Figure 2).

4.1. Detoxification of Metals by S. aureus K1

Microbes have a number of metal resistance mechanisms involving chromosomes, transposon-encoded genes or plasmids. These mechanisms are mostly plasmid-facilitated and show resistance to some particular anion or cation [53]. Metals can have different impacts inside cells depending upon their concentration [53]; once a certain level is exceeded, bacteria respond with the initiation of a number of resistance mechanisms, including metallothioneins, P-type ATPases, CDF transporters and RND efflux pumps [54]. The genes located on plasmids, chromosomes or transposons that are responsible for resistance can easily be transferred to new community members from their point of location [53,55].

The genotype of bacteria, the nature and type of the metal and the pH of the culturing media are among the factors responsible for showing the degree of tolerance of microbes to various metals (Hg, Co, Pb, Ag, Zn, Mn, Cu, Cr) [56]. This kind of resistance against toxic heavy metals might be recognized by employing a number of potential methods like bioaccumulation of heavy metals by microbes, ion exclusion and low-molecular-weight binding protein production [57,58]. Elevated levels of metal resistance systems in bacterial cells are an indication of environmental heavy metal bioavailability [59]. The results of Chudobova et al. [60] showed a maximum resistance and capability of *S. aureus* strains under Cd^{2+} and Zn^{2+} ions. This resistance observed in *S. aureus* might be due to the efflux system containing a P-type ATPase transport system acting against Cd^{2+} ions [53,61].

4.2. Effect of S. aureus on Plant Growth Promotion under Cr Metal Stress

Different wheat varieties may differ in their response to different concentration of Cr in the soil. This could be attributed to various biological aspects of wheat varieties, as different wheat varieties show differences in growth parameters (e.g., leaf size). A heavy metal like Cr can easily make its way to aerial portions of plants, where it will affect their shoot metabolism at the cellular level and cause severe damage to minerals, water and nutrients, consequently retarding plant growth [10,62].

However, bacterial inoculation may improve the nutritional requirements of both micro- (Mn, Zn, Cu and Fe) and macronutrients (N, P and K) by modifying host physiology, which results in changed uptake pattern of roots. Similarly, a recent investigation done by Islam et al. [63] showed an increase in Fe and K concentrations in maize plants under Cr stress due to bacterial inoculations. According to an observation, plants with bacterial inoculation showed a reduction in metal accumulation in their aerial parts, which might be due to delayed translocation of metals from roots to upper parts [64]. Similar observations were recorded in this current research. Moreover, we isolated *S. aureus* K1 from wastewater that was contaminated with Cr, so the microbes may have the capability of performing metal detoxification as a part of their metabolic system. There was substantial improvement in plant growth and leaf pigments due to inoculation of specific microbes [63].

4.3. Chlorophyll Contents

Higher chlorophyll contents were observed in plants with bacterial inoculation compared to uninoculated plants (Figure 3). However, with further increasing metal concentrations, a reduction in chlorophyll contents was noted. This is in agreement with the findings of another research study, where chlorophyll a and chlorophyll b in wheat plants decreased with increasing concentrations of Pb in the growth medium [65].

4.4. ROS Species and Antioxidant Enzyme Production

Reactive oxygen species can be produced in plants when exposed to Cr^{6+}, which may damage the photosynthetic apparatus and protein complex of thylakoid membranes and result in inhibition of chlorophyll production [66]. In adverse conditions, plants release MDA contents; this reveals the level of lipid peroxidation, as MDA is the last decomposition product of membrane lipid peroxidation [67]. The increase in MDA contents found in the present study is indicative of imbalance between the generation and removal of free radicals in the cells [68]. The decreased lipid peroxidation with *S. aureus* K1 inoculation under Cr stress could be due to the increase in ROS-scavenging enzyme production in plants. This may be supported by a previously published study which revealed that the gene profile of metal detoxifying enzymes was activated by bacterial inoculation to deal with metal stress [69]. Reactive oxygen species are generated in response to stress caused by heavy metals like hexavalent Cr, and plants have a detoxifying antioxidant enzyme system for their maintenance. These enzymes are POD, SOD, APX and CAT, and they work alongside other non-enzymatic antioxidants. The activities performed by antioxidant enzymes in plants under metal stress are extremely variable and dependent on plant species, metal concentration, metal ions and exposure time period [70]. At low metal concentration, SOD activity may increase, but it becomes constant with increased metal concentration [71]. The enhancement in CAT activity was also noted in a number of plants under metal stress [72]. An increase in CAT activity was also observed as an adaptive trait of isolate CPSB21 [73]. Increased antioxidant enzyme activities in plants with inoculation of CPSB21 may be due to increases in mRNA/gene expression of antioxidant enzymes as compared to uninoculated plants [74].

4.5. Reduction of Cr Concentration in Plants by Bacterial Inoculation

A significant difference was found between uninoculated and *S. aureus* K1 inoculated plants in terms of Cr concentration. In contaminated soil, the results showed that the level of Cr was higher in the roots of wheat plants than it was in the shoots, which may be due to decreased translocation of Cr from roots to shoots of plants [75,76]. Immobilization of Cr in root cell vacuoles may lead to higher Cr accumulation in roots, which can cause toxicity in plants [77]. In the present study, inoculation of wheat plants with Cr-resistant microbes decreased the Cr concentration and its translocation from soil to roots and upper parts of wheat plants. The reduction of hexavalent Cr (Cr^{6+}) to trivalent Cr (Cr^{3+}) by bacterial isolates may be the reason for the improved growth of wheat plants [78] and hence the decreased level of the Cr contents in soil. Hasnain and Sabri [79] also reported a pattern of decreased Cr uptake and accumulation in roots and shoots of wheat plants inoculated with *Pseudomonas* sp.

A decrease in Cr concentration in soil was observed after wheat plant harvesting. This decrease was recorded in uninoculated Cr-contaminated wheat plants as a result of increased accumulation and uptake of Cr in roots and shoots [80]. Such decrease may also be due to Cr^{6+} reduction into Cr^{3+} under the influence of bacterial inoculation [78,81]. Scientists are also considering the use genetically engineered microorganisms (GEM), which may be well adjusted to their local environment (both climatic and soil) for effective elimination of heavy metals from contaminated soils [58,82,83].

5. Conclusions

The outcomes of this study indicate that the application of the peat-moss-based microbial inoculum improved plant growth and yield parameters and comparatively decreased metal accumulation by the plants. Overall, gas exchange attributes and chlorophyll contents increased with *S. aureus* K1 inoculation. This research study concluded that *S. aureus* K1 reduced the toxicity of Cr in wheat plants. The Cr-resistant *S. aureus* K1 supported the plant growth, decreased and detoxified Cr in plants and allowed better production of wheat in a Cr-contaminated environment. However, in-depth exploration (i.e., at the molecular level) of the alleviative mechanisms in plants should be conducted in future studies.

Author Contributions: F.Z., M.R., T.Y., S.A., M.A.E.-S., M.N.A. and L.W.; Data curation, M.Z., M.W., A.A. (Alia Anayat) and T.Y.; Formal analysis, M.Z., M.W., A.A. (Alia Anayat), M.R. and T.Y.; Funding acquisition, S.A., M.N.A. and L.W.; Investigation, M.R., A.A. (Awais Ahmad), T.Y. and M.A.E.-S.; Methodology, M.A., M.W., M.R., A.A. (Awais Ahmad), T.Y. and S.A.; Project administration, F.Z., T.Y. and S.A.; Resources, F.Z., M.W., S.A., M.A.E.-S., M.N.A. and L.W.; Software, F.Z., M.Z., M.A.E.-S. and L.W.; Supervision, M.R., T.Y. and S.A.; Validation, M.Z., M.W., A.A. (Alia Anayat) and A.A. (Awais Ahmad); Visualization, M.Z., M.W., A.A. (Alia Anayat) and A.A. (Awais Ahmad); Writing—original draft, F.Z., M.Z., S.A. and L.W.; Writing—review & editing, F.Z., T.Y. and M.N.A. All authors have read and agreed to the published version of the manuscript.

Funding: This research was supported by the Higher Education Commission, Pakistan (grant number 20-3653/NRPU/R&D/HEC/14/437). The authors would like to extend their sincere appreciation to the Researchers Supporting Project Number (RSP-2020/182), King Saud University, Riyadh, Saudi Arabia.

Acknowledgments: We are grateful to Government College University, Faisalabad, Pakistan, for its support. The authors would like to extend their sincere appreciation to the Researchers Supporting Project Number (RSP-2020/182), King Saud University, Riyadh, Saudi Arabia.

Conflicts of Interest: The authors declare that they have no conflict of interest.

References

1. Singh, H.P.; Mahajan, P.; Kaur, S.; Batish, D.R.; Kohli, R.K. Chromium toxicity and tolerance in plants. *Environ. Chem. Lett.* **2013**, *11*, 229–254. [CrossRef]

2. Kunjam, M. Studies on selected heavy metals on seed germination and plant growth in pea plant (*Pisum sativum*) grown in solid medium. *J. Pharmacogn. Phytochem.* **2015**, *3*, 85–87.

3. Habiba, U.; Ali, S.; Farid, M.; Shakoor, M.B.; Rizwan, M.; Ibrahim, M.; Abbasi, G.H.; Hayat, T.; Ali, B. EDTA enhanced plant growth, antioxidant defense system, and phytoextraction of copper by *Brassica napus* L. *Environ. Sci. Pollut. Res.* **2014**, *22*, 1534–1544. [CrossRef] [PubMed]

4. Adrees, M.; Ali, S.; Iqbal, M.; Bharwana, S.A.; Siddiqi, Z.; Farid, M.; Ali, Q.; Saeed, R.; Rizwan, M. Mannitol alleviates chromium toxicity in wheat plants in relation to growth, yield, stimulation of anti-oxidative enzymes, oxidative stress and Cr uptake in sand and soil media. *Ecotoxicol. Environ. Saf.* **2015**, *122*, 1–8. [CrossRef] [PubMed]

5. Mathur, S.; Kalaji, H.M.; Kalaji, H.M. Investigation of deleterious effects of chromium phytotoxicity and photosynthesis in wheat plant. *Photosynthetica* **2016**, *54*, 185–192. [CrossRef]

6. Kamran, M.A.; Mufti, R.; Mubariz, N.; Syed, J.H.; Bano, A.; Javed, M.T.; Munis, M.F.H.; Tan, Z.; Chaudhary, H.J. The potential of the flora from different regions of Pakistan in phytoremediation: A review. *Environ. Sci. Pollut. Res.* **2013**, *21*, 801–812. [CrossRef]

7. Daud, M.K.; Mei, L.; Variath, M.T.; Ali, S.; Li, C.; Rafiq, M.T.; Zhu, S.J. Chromium (VI) Uptake and Tolerance Potential in Cotton Cultivars: Effect on Their Root Physiology, Ultramorphology, and Oxidative Metabolism. *BioMed Res. Int.* **2014**, *2014*, 1–12. [CrossRef]

8. Afshan, S.; Ali, S.; Bharwana, S.A.; Rizwan, M.; Farid, M.; Abbas, F.; Ibrahim, M.; Mehmood, M.A.; Abbasi, G.H. Citric acid enhances the phytoextraction of chromium, plant growth, and photosynthesis by alleviating the oxidative damages in *Brassica napus* L. *Environ. Sci. Pollut. Res.* **2015**, *22*, 11679–11689. [CrossRef]

9. Nagajyoti, P.C.; Lee, K.D.; Sreekanth, T.V.M. Heavy metals, occurrence and toxicity for plants: A review. *Environ. Chem. Lett.* **2010**, *8*, 199–216. [CrossRef]

10. Shanker, A. Chromium toxicity in plants. *Environ. Int.* **2005**, *31*, 739–753. [CrossRef]

11. Kamran, M.A.; Eqani, S.A.M.A.S.; Bibi, S.; Xu, R.-K.; Monis, M.F.H.; Katsoyiannis, A.; Bokhari, H.; Chaudhary, H.J. Bioaccumulation of nickel by E. sativa and role of plant growth promoting rhizobacteria (PGPRs) under nickel stress. *Ecotoxicol. Environ. Saf.* **2016**, *126*, 256–263. [CrossRef] [PubMed]

12. Dotaniya, M.L.; Das, H.; Meena, V.D. Assessment of chromium efficacy on germination, root elongation, and coleoptile growth of wheat (*Triticum aestivum* L.) at different growth periods. *Environ. Monit. Assess.* **2014**, *186*, 2957–2963. [CrossRef] [PubMed]

13. Lenka, S. Impacts of fertilizers use on environmental quality. In Proceedings of the National Seminar on Environmental Concern for Fertilizer Use in Future, Bidhan Chandra Krishi Viswavidyalaya, Kalyani, India, 26 February 2016.

14. Saha, J.; Selladurai, R.; Coumar, M.V.; Dotaniya, M.L.; Kundu, S.; Patra, A.K. *Industrial Activities in India and Their Impact on Agroecosystem*; Springer Science and Business Media LLC: Berlin/Heidelberg, Germany, 2017; Volume 10, pp. 229–249.

15. Khan, M.Y. Effect of microbial inoculation on wheat growth and phytostabilization of chromium contaminated soil. *Pak. J. Bot.* **2013**, *45*, 27–34.

16. Rizwan, M.; Ali, S.; Abbas, T.; Zia-Ur-Rehman, M.; Hannan, F.; Keller, C.; Al-Wabel, M.I.; Ok, Y.S. Cadmium minimization in wheat: A critical review. *Ecotoxicol. Environ. Saf.* **2016**, *130*, 43–53. [CrossRef] [PubMed]

17. Shewry, P.R.; Hey, S.J. The contribution of wheat to human diet and health. *Food Energy Secur.* **2015**, *4*, 178–202. [CrossRef]

18. Awan, N.; Fatima, A.; Farhan, M.; Awan, N. Comparative Analysis of Chromium and Cadmium in Various Parts of Wheat and Maize. *Pol. J. Environ. Stud.* **2019**, *28*, 1561–1566. [CrossRef]

19. Carlin, D.J.; Naujokas, M.F.; Bradham, K.D.; Cowden, J.; Heacock, M.; Henry, H.F.; Lee, J.S.; Thomas, D.J.; Thompson, C.; Tokar, E.J.; et al. Arsenic and Environmental Health: State of the Science and Future Research Opportunities. *Environ. Health Perspect.* **2016**, *124*, 890–899. [CrossRef]

20. Abbas, Q.; Yousaf, B.; Liu, G.; Zia-Ur-Rehman, M.; Ali, M.U.; Munir, M.A.M.; Hussain, S.A. Evaluating the health risks of potentially toxic elements through wheat consumption in multi-industrial metropolis of Faisalabad, Pakistan. *Environ. Sci. Pollut. Res.* **2017**, *24*, 26646–26657. [CrossRef]

21. Lytle, C.M.; Lytle, F.W.; Yang, N.; Qian, J.-H.; Hansen, D.; Zayed, A.; Terry, N. Reduction of Cr(VI) to Cr(III) by Wetland Plants: Potential for In Situ Heavy Metal Detoxification. *Environ. Sci. Technol.* **1998**, *32*, 3087–3093. [CrossRef]

22. Mishra, J.; Singh, R.; Arora, N.K. Alleviation of Heavy Metal Stress in Plants and Remediation of Soil by Rhizosphere Microorganisms. *Front. Microbiol.* **2017**, *8*, 1706. [CrossRef]

23. Ashraf, S.; Ali, Q.; Zahir, Z.A.; Ashraf, S.; Asghar, H.N. Phytoremediation: Environmentally sustainable way for reclamation of heavy metal polluted soils. *Ecotoxicol. Environ. Saf.* **2019**, *174*, 714–727. [CrossRef] [PubMed]

24. Kong, Z.; Glick, B.R. *The Role of Plant Growth-Promoting Bacteria in Metal Phytoremediation*; Elsevier: Amsterdam, The Netherlands, 2017; pp. 97–132.

25. Ryan, R.P.; Monchy, S.; Cardinale, M.; Taghavi, S.; Crossman, L.; Avison, M.B.; Berg, G.; Van Der Lelie, D.; Dow, J.M. The versatility and adaptation of bacteria from the genus Stenotrophomonas. *Nat. Rev. Genet.* **2009**, *7*, 514–525. [CrossRef] [PubMed]

26. Taj, Z.Z.; Rajkumar, M. *Perspectives of Plant Growth-Promoting Actinomycetes in Heavy Metal Phytoremediation*; Springer Science and Business Media LLC: Berlin/Heidelberg, Germany, 2016; pp. 213–231.

27. Mishra, S.; Bharagava, R.N. Toxic and genotoxic effects of hexavalent chromium in environment and its bioremediation strategies. *J. Environ. Sci. Health Part C* **2015**, *34*, 1–32. [CrossRef] [PubMed]

28. Wirtanen, G.; Salo, S. Biofilm risks. In *Handbook of Hygiene Control in the Food Industry*; Elsevier: Amsterdam, The Netherlands, 2016; pp. 55–79.

29. Ahemad, M. Enhancing phytoremediation of chromium-stressed soils through plant-growth-promoting bacteria. *J. Genet. Eng. Biotechnol.* **2015**, *13*, 51–58. [CrossRef]

30. Jing, Y.X.; Yan, J.L.; He, H.D.; Yang, D.J.; Xiao, L.; Zhong, T.; Yuan, M.; De Cai, X.; Bin Li, S. Characterization of Bacteria in the Rhizosphere Soils ofPolygonum Pubescensand Their Potential in Promoting Growth and Cd, Pb, Zn Uptake byBrassica napus. *Int. J. Phytoremediation* **2013**, *16*, 321–333. [CrossRef]

31. Mukhtar, S. Potential of sunflower (*Helianthus annuus* L.) for phytoremediation of nickel (Ni) and lead (Pb) contaminated water. *Pak. J. Bot.* **2010**, *42*, 4017–4026.

32. Soltanpour, P.N. Use of ammonium bicarbonate DTPA soil test to evaluate elemental availability and toxicity 1. *Commun. Soil Sci. Plant Anal.* **1985**, *16*, 323–338. [CrossRef]

33. Moodie, C.D.; Smith, H.W.; McCreery, R.A. Laboratory Manual for Soil Fertility. *Soil Sci.* **1951**, *71*, 400. [CrossRef]

34. Lucious, S. Heavy metal tolerance and antibiotic sensitivity of bacterial strains isolated from tannery effluent. *Asian J. Exp. Biol. Sci.* **2013**, *4*, 597–606.

35. Zhang, X.; Krumholz, L.R.; Yu, Z.; Chen, Y.; Liu, P.; Li, X. A Novel Subspecies of *Staphylococcus aureus* from Sediments of Lanzhou Reach of the Yellow River Aerobically Reduces Hexavalent Chromium. *J. Bioremediation Biodegrad.* **2013**, *4*. [CrossRef]

36. Turner, S.; Pryer, K.M.; Miao, V.P.W.; Palmer, J.D. Investigating deep phylogenetic relationships among cyanobacteria and plastids by small subunit rRNA sequence analysis. *J. Eukaryot. Microbiol.* **1999**, *46*, 327–338. [CrossRef] [PubMed]

37. Moyersoen, B.; E Beever, R.; Martin, F. Genetic diversity of Pisolithus in New Zealand indicates multiple long-distance dispersal from Australia. *New Phytol.* **2003**, *160*, 569–579. [CrossRef]

38. Tamura, K.; Peterson, D.; Stecher, G.; Nei, M.; Kumar, S.; Peterson, N. MEGA5: Molecular Evolutionary Genetics Analysis Using Maximum Likelihood, Evolutionary Distance, and Maximum Parsimony Methods. *Mol. Boil. Evol.* **2011**, *28*, 2731–2739. [CrossRef] [PubMed]

39. Hadi, F.; Bano, A. Effect of diazotrophs (Rhizobium and Azatebactor) on growth of maize (*Zea mays* L.) and accumulation of lead (Pb) in different plant parts. *Pak. J. Bot.* **2010**, *42*, 4363–4370.

40. Wu, S.; Cheung, K.; Luo, Y.; Wong, M.H. Effects of inoculation of plant growth-promoting rhizobacteria on metal uptake by Brassica juncea. *Environ. Pollut.* **2006**, *140*, 124–135. [CrossRef] [PubMed]

41. Lichtenthaler, H.K. Chlorophylls and carotenoids: Pigments of photosynthetic biomembranes. In *Methods in Enzymology*; Elsevier: Amsterdam, The Netherlands, 1987; pp. 350–382.

42. Dionisio-Sese, M.L.; Tobita, S. Antioxidant responses of rice seedlings to salinity stress. *Plant Sci.* **1998**, *135*, 1–9. [CrossRef]

43. Plumb-Dhindsa, P.; Dhindsa, R.S.; Thorpe, T.A. Leaf Senescence: Correlated with Increased Levels of Membrane Permeability and Lipid Peroxidation, and Decreased Levels of Superoxide Dismutase and Catalase. *J. Exp. Bot.* **1981**, *32*, 93–101. [CrossRef]

44. Zhang, J.; Kirkham, M. Drought-Stress-Induced Changes in Activities of Superoxide Dismutase, Catalase, and Peroxidase in Wheat Species. *Plant Cell Physiol.* **1994**, *35*, 785–791. [CrossRef]

45. Jana, S.; Choudhuri, M.A. Glycolate metabolism of three submersed aquatic angiosperms: Effect of heavy metals. *Aquat. Bot.* **1981**, *11*, 67–77. [CrossRef]

46. Zhang, X. The measurement and mechanism of lipid peroxidation and SOD, POD and CAT activities in biological system. In *Research Methodology of Crop Physiology*; Agriculture Press: Beijing, China, 1992; pp. 208–211.

47. Aebi, H. Catalase in vitro. *Methods Enzym.* **1984**, *105*, 121–126.

48. Nakano, Y.; Asada, K. Hydrogen Peroxide is Scavenged by Ascorbate-specific Peroxidase in Spinach Chloroplasts. *Plant Cell Physiol.* **1981**, *22*, 867–880. [CrossRef]

49. Rehman, M.Z.-U.; Rizwan, M.; Ghafoor, A.; Naeem, A.; Ali, S.; Sabir, M.; Qayyum, M.F. Effect of inorganic amendments for in situ stabilization of cadmium in contaminated soils and its phyto-availability to wheat and rice under rotation. *Environ. Sci. Pollut. Res.* **2015**, *22*, 16897–16906. [CrossRef] [PubMed]

50. Khan, Z.; Nisar, M.A.; Hussain, S.Z.; Arshad, M.N.; Rehman, A. Cadmium resistance mechanism in Escherichia coli P4 and its potential use to bioremediate environmental cadmium. *Appl. Microbiol. Biotechnol.* **2015**, *99*, 10745–10757. [CrossRef] [PubMed]

51. Oaikhena, E.E.; Makaije, D.B.; Denwe, S.D.; Namadi, M.M.; Haroun, A.A. Bioremediation Potentials of Heavy Metal Tolerant Bacteria Isolated from Petroleum Refinery Effluent. *Am. J. Environ. Prot.* **2016**, *5*, 29. [CrossRef]

52. Mustapha, M.U.; Halimoon, N. Screening and Isolation of Heavy Metal Tolerant Bacteria in Industrial Effluent. *Procedia Environ. Sci.* **2015**, *30*, 33–37. [CrossRef]

53. Bruins, M.R.; Kapil, S.; Oehme, F.W. Microbial Resistance to Metals in the Environment. *Ecotoxicol. Environ. Saf.* **2000**, *45*, 198–207. [CrossRef]

54. Nies, D.H. Efflux-mediated heavy metal resistance in prokaryotes. *FEMS Microbiol. Rev.* **2003**, *27*, 313–339. [CrossRef]

55. Sørensen, S.J.; Bailey, M.; Hansen, L.H.; Kroer, N.; Wuertz, S. Studying plasmid horizontal transfer in situ: A critical review. *Nat. Rev. Genet.* **2005**, *3*, 700–710. [CrossRef]

56. Kashif, M.; Ngaini, Z.; Harry, A.V.; Vekariya, R.L.; Ahmad, A.; Zuo, Z.; Alarifi, A. An experimental and DFT study on novel dyes incorporated with natural dyes on titanium dioxide (TiO_2) towards solar cell application. *Appl. Phys. A* **2020**, *126*, 1–13. [CrossRef]

57. Nies, D.H. Microbial heavy-metal resistance. *Appl. Microbiol. Biotechnol.* **1999**, *51*, 730–750. [CrossRef]

58. Das, S.; Dash, H.R.; Chakraborty, J. Genetic basis and importance of metal resistant genes in bacteria for bioremediation of contaminated environments with toxic metal pollutants. *Appl. Microbiol. Biotechnol.* **2016**, *100*, 2967–2984. [CrossRef] [PubMed]

59. Roosa, S.; Wattiez, R.; Prygiel, E.; Lesven, L.; Billon, G.; Gillan, D.C. Bacterial metal resistance genes and metal bioavailability in contaminated sediments. *Environ. Pollut.* **2014**, *189*, 143–151. [CrossRef] [PubMed]

60. Chudobova, D.; Dostalova, S.; Ruttkay-Nedecký, B.; Guran, R.; Rodrigo, M.A.M.; Tmejová, K.; Krizkova, S.; Zitka, O.; Adam, V.; Kizek, R. The effect of metal ions on *Staphylococcus aureus* revealed by biochemical and mass spectrometric analyses. *Microbiol. Res.* **2015**, *170*, 147–156. [CrossRef]

61. Nies, D.H.; Silver, S.D. Ion efflux systems involved in bacterial metal resistances. *J. Ind. Microbiol. Biotechnol.* **1995**, *14*, 186–199. [CrossRef] [PubMed]

62. Chatterjee, J.; Kumar, P.; Sharma, P.N.; Tewari, R.K. Chromium toxicity induces oxidative stress in turnip. *Indian J. Plant Physiol.* **2015**, *20*, 220–226. [CrossRef]

63. Islam, F.; Yasmeen, T.; Arif, M.S.; Riaz, M.; Shahzad, S.M.; Imran, Q.; Ali, I. Combined ability of chromium (Cr) tolerant plant growth promoting bacteria (PGPB) and salicylic acid (SA) in attenuation of chromium stress in maize plants. *Plant Physiol. Biochem.* **2016**, *108*, 456–467. [CrossRef]

64. Islam, F.; Yasmeen, T.; Ali, Q.; Ali, S.; Arif, M.S.; Hussain, S.; Rizvi, H. Influence of Pseudomonas aeruginosa as PGPR on oxidative stress tolerance in wheat under Zn stress. *Ecotoxicol. Environ. Saf.* **2014**, *104*, 285–293. [CrossRef]

65. Lamhamdi, M.; El Galiou, O.; Bakrim, A.; Nóvoa-Muñoz, J.C.; Arias-Estévez, M.; Aarab, A.; Lafont, R. Effect of lead stress on mineral content and growth of wheat (*Triticum aestivum*) and spinach (*Spinacia oleracea*) seedlings. *Saudi J. Boil. Sci.* **2013**, *20*, 29–36. [CrossRef]

66. Oliveira, H. Chromium as an Environmental Pollutant: Insights on Induced Plant Toxicity. *J. Bot.* **2012**, *2012*, 1–8. [CrossRef]

67. Arif, M.S.; Yasmeen, T.; Shahzad, S.M.; Riaz, M.; Rizwan, M.; Iqbal, S.; Asif, M.; Soliman, M.H.; Ali, S. Lead toxicity induced phytotoxic effects on mung bean can be relegated by lead tolerant Bacillus subtilis (PbRB3). *Chemosphere* **2019**, *234*, 70–80. [CrossRef]

68. Sytar, O.; Kumar, A.; Latowski, D.; Kuczyńska, P.; Strzałka, K.; Prasad, M.N.V. Heavy metal-induced oxidative damage, defense reactions, and detoxification mechanisms in plants. *Acta Physiol. Plant.* **2012**, *35*, 985–999. [CrossRef]

69. Islam, F.; Yasmeen, T.; Ali, Q.; Mubin, M.; Ali, S.; Arif, M.S.; Hussain, S.; Riaz, M.; Abbas, F. Copper-resistant bacteria reduces oxidative stress and uptake of copper in lentil plants: Potential for bacterial bioremediation. *Environ. Sci. Pollut. Res.* **2015**, *23*, 220–233. [CrossRef]

70. Sharma, S.S.; Dietz, K.-J. The relationship between metal toxicity and cellular redox imbalance. *Trends Plant Sci.* **2009**, *14*, 43–50. [CrossRef] [PubMed]

71. Štolfa, I.; Pfeiffer, T.Ž.; Špoljarić, D.; Teklić, T.; Lončarić, Z. *Heavy Metal-Induced Oxidative Stress in Plants: Response of the Antioxidative System*; Springer Science and Business Media LLC: Berlin/Heidelberg, Germany, 2015; pp. 127–163.

72. Fayiga, A.O.; Ma, L.Q.; Cao, X.; Rathinasabapathi, B. Effects of heavy metals on growth and arsenic accumulation in the arsenic hyperaccumulator *Pteris vittata* L. *Environ. Pollut.* **2004**, *132*, 289–296. [CrossRef] [PubMed]

73. Vital, S.A.; Fowler, R.W.; Virgen, A.; Gossett, D.R.; Banks, S.W.; Rodriguez, J. Opposing roles for superoxide and nitric oxide in the NaCl stress-induced upregulation of antioxidant enzyme activity in cotton callus tissue. *Environ. Exp. Bot.* **2008**, *62*, 60–68. [CrossRef]

74. Gururani, M.A.; Upadhyaya, C.P.; Baskar, V.; Venkatesh, J.; Nookaraju, A.; Park, S.W. Plant Growth-Promoting Rhizobacteria Enhance Abiotic Stress Tolerance in Solanum tuberosum Through Inducing Changes in the Expression of ROS-Scavenging Enzymes and Improved Photosynthetic Performance. *J. Plant Growth Regul.* **2012**, *32*, 245–258. [CrossRef]

75. Huffman, E.W.D.; Allaway, W.H. Chromium in plants. Distribution in tissues, organelles, and extracts and availability of bean leaf chromium to animals. *J. Agric. Food Chem.* **1973**, *21*, 982–986. [CrossRef]

76. Zayed, A.; Lytle, C.M.; Qian, J.-H.; Terry, N. Chromium accumulation, translocation and chemical speciation in vegetable crops. *Planta* **1998**, *206*, 293–299. [CrossRef]

77. Shanker, A.; Djanaguiraman, M.; Sudhagar, R.; Chandrashekar, C.; Pathmanabhan, G. Differential antioxidative response of ascorbate glutathione pathway enzymes and metabolites to chromium speciation stress in green gram ((L.) R.Wilczek. cv CO 4) roots. *Plant Sci.* **2004**, *166*, 1035–1043. [CrossRef]

78. Salunkhe, P.; Dhakephalkar, P.; Paknikar, K.M. Bioremediation of hexavalent chromium in soil microcosms. *Biotechnol. Lett.* **1998**, *20*, 749–751. [CrossRef]

79. Hasnain, S.; Sabri, A.N. Growth stimulation of *Triticum aestivum* seedlings under Cr-stresses by non-rhizospheric pseudomonad strains. *Environ. Pollut.* **1997**, *97*, 265–273. [CrossRef]

80. Mishra, S.; Shanker, K.; Srivastava, M.; Srivastava, S.; Shrivastav, R.; Dass, S.; Prakash, S. A study on the uptake of trivalent and hexavalent chromium by paddy (*Oryza sativa*): Possible chemical modifications in rhizosphere. *Agric. Ecosyst. Environ.* **1997**, *62*, 53–58. [CrossRef]

81. Cheung, K.H.; Gu, J.-D. Reduction of chromate ($CrO42-$) by an enrichment consortium and an isolate of marine sulfate-reducing bacteria. *Chemosphere* **2003**, *52*, 1523–1529. [CrossRef]

82. Gupta, S.; Singh, D. Role of Genetically Modified Microorganisms in Heavy Metal Bioremediation. In *Advances in Environmental Biotechnology*; Springer Science and Business Media LLC: Berlin/Heidelberg, Germany, 2017; Volume 3, pp. 197–214.

83. Ali, B.; Wang, B.; Ali, S.; Ghani, M.A.; Hayat, M.T.; Yang, C.; Xu, L.; Zhou, W. 5-Aminolevulinic Acid Ameliorates the Growth, Photosynthetic Gas Exchange Capacity, and Ultrastructural Changes Under Cadmium Stress in *Brassica napus* L. *J. Plant Growth Regul.* **2013**, *32*, 604–614. [CrossRef]

 *agronomy*

Article

Rock Phosphate-Enriched Compost in Combination with Rhizobacteria; A Cost-Effective Source for Better Soil Health and Wheat (*Triticum aestivum*) Productivity

Motsim Billah [1], Matiullah Khan [2], Asghari Bano [3,*], Sobia Nisa [4], Ahmad Hussain [5], Khadim Muhammad Dawar [6], Asia Munir [7] and Naeem Khan [8,*] 

[1] Department of Soil Science, University of Haripur, Haripur 21120, Pakistan; motsimbillah@gmail.com
[2] Pakistan Agriculture Research Council, NARC, Islamabad 44000, Pakistan; mukhan65pk@yahoo.co.uk
[3] Department of Biosciences, University of Wah, Wah Cantt. 47000, Pakistan
[4] Department of Microbiology, University of Haripur, Haripur 21120, Pakistan; sobia@uoh.edu.pk
[5] Department of Forestry, University of Haripur, Haripur 21120, Pakistan; ahmad.hussain@uoh.edu.pk
[6] Department of Soil and Environmental Science, The University of Agriculture, Peshawar 25000, Pakistan; khadimdawar@yahoo.com
[7] Soil and Water Testing Laboratory, Rawalpindi 46000, Pakistan; asiamunir800@gmail.com
[8] Department of Agronomy, Institute of Food and Agricultural Sciences, University of Florida, Gainesville, FL 32611, USA
* Correspondence: bano.asghari@gmail.com (A.B.); naeemkhan@ufl.edu (N.K.)

Received: 20 July 2020; Accepted: 9 September 2020; Published: 14 September 2020

Abstract: Organic materials from various sources have been commonly adopted as soil amendments to improve crop productivity. Phosphorus deficiency and fixation in alkaline calcareous soils drives a reduction in crop production. A two-year field experiment was conducted to evaluate the impact of rock phosphate enriched composts and chemical fertilizers both individually and in combination with plant growth promoting rhizobacteria (PGPR) on wheat productivity and soil chemical and biological and biochemical properties. The present study demonstrates significant increments in crop agronomic and physiological parameters with *Pseudomonas* sp. inoculated $RPEC_1$ (rock phosphate + poultry litter + *Pseudomonas* sp.) over the un-inoculated untreated control. However, among all other treatments i.e., $RPEC_2$ (rock phosphate + poultry litter solubilized with *Proteus* sp.), RPC (rock phosphate + poultry litter), HDP (half dose inorganic P from Single Super Phosphate-SSP 18% P_2O_5) and SPLC (poultry litter only); $RPEC_1$ remained the best by showing increases in soil chemical properties (available phosphorus, nitrate nitrogen, extractable potassium), biochemical properties (alkaline phosphatase activity) and biological properties (microbial biomass carbon and microbial biomass phosphorus). Economic analysis in terms of Value Cost Ratio (VCR) showed that the seed inoculation with *Pseudomonas* sp. in combination with $RPEC_1$ gave maximum VCR (3.23:1) followed by $RPEC_2$ (2.61:1), FDP (2.37:1), HDP (2.05:1) and SPLC (2.03:1). It is concluded that inoculated rock phosphate (RP) enriched compost ($RPEC_1$) can be a substitute to costly chemical fertilizers and seed inoculation with *Pseudomonas* sp. may further increase the efficiency of composts.

Keywords: available phosphorus; enriched compost; PGPR; poultry litter; rock phosphate; wheat

1. Introduction

The function of fertilizers for maximum crop production in under-developed countries is customary and well recognized. Nevertheless, the increasing prices of inorganic phosphate fertilizers and the extensive use of chemical fertilizers in agriculture, is also under debate due to environmental concerns and for consumer health reasons [1]. Reduction of agrochemicals for crop production is of great concern for sustainable agriculture [2]. Moreover, inorganic phosphate fertilizers are not totally soluble in soil matrix due to precipitation reactions with ions of Al and Fe in acidic, and Ca in alkaline calcareous soils [3]. Moreover, high dose application of chemical fertilizers creates negative impacts such as changes in soil pH through alkalization and acidification, pollution of water resources through runoff, suppression of microorganisms and friendly insects, fixation of nutrients, degradation of soil structure due to increased decomposition of organic matter [4]. The research workers are required to look for substitutes to inorganic fertilizers [5], which are cost-effective and environmentally friendly. The use of rock phosphate (RP) as an alternative for P fertilizer is gaining attention in sustainable agriculture through microbial solubilization [6] and preparation of RP-enriched compost [7]. The mixing of RP with organic materials such as animal feces, plant residues and inoculation with acid-producing microbes may enhance P solubility from RP because when organic materials decompose, more soluble P is released due to the action of organic acids produced by the microbes [8]. The incorporation of organic residues either singly or in conjunction with a cheap source of mining element as rock phosphate may help to improve soil quality and productivity [9]. Rock phosphate enriched compost which was solubilized by phosphate solubilizing fungi and applied on a mung-bean crop, significantly enhanced yield and P-uptake [10].

Various RP enriched composts and inorganic fertilizers such as diammonium phosphate (DAP) were applied on wheat in a pot experiment. The data revealed that RP enriched composts showed no significant performance in the earlier stages of wheat growth but at maturity, it gave higher grain yield, nutrient uptake and increased fertility status of P and K in the soils [11]. Isolated phosphate-solubilizing fungi from phosphate mines of China were reported to have efficient biofertilizers and P solubilizers with the capacity to enhance the growth of wheat [12]. Colonization of soil by nonindigenous phosphate-solubilizing microorganisms depend both on their interactions with indigenous microorganisms associated with plants and their ability to utilize diverse substrates in soil [13]. The role of phosphate-solubilizing microorganisms in phosphate solubilization has been attributed mainly to their abilities to reduce the pH of the surroundings by the production of organic acids [13]. Preparing the RP-enriched compost with phosphate solubilizing microbes may not only compensate for the higher cost of manufacturing fertilizers, but also provide a sustainable source of available phosphorus to growing plants in alkaline soils [14].

Plant growth promoting rhizobacteria (PGPR) are important inoculants for integrated nutrient management [15] which help in dissolving inorganic P by excreting organic acids and chelation of P cations to release P in soil solution [16]. It was reported that there are several PGPR inoculants currently commercialized that promote growth either by suppression of plant disease, improved nutrient acquisition, or phytohormone production [17]. Generally, phytohormone in plants plays a vital role in cell division, proliferation, and differentiation, vascular tissue alteration, responses to light and gravity, general root and shoot architecture, seed and tuber germination, ethylene synthesis, vegetative growth processes, fruit development [18–20], initiation of lateral and floral organ and organogenesis [21], initiation of rooting, foliation and flowering [22], formation of lateral and adventitious roots [23], and increasing the growth of cambium and size of xylem cells [24]. Bacterial phytohormone production is widely distributed among plant-associated bacteria and is still considered the primary mechanism that enhances the growth and yield of plants [25].

PGPRs influence direct growth promotion of plants by fixing atmospheric nitrogen, solubilizing insoluble phosphates, secreting hormones such as IAA, GAs, and Kinetins besides ACC (1-aminocycloprapane-1-carboxylic acid) deaminase production [26], that helps in the regulation of ethylene. Amongst the majority of influential P solubilizers, bacterial strains from the genera *Pseudomonas*, *Bacillus*, *Rhizobium* and *Enterobacter* are of great importance. Application of phosphate solubilizing microbes in the production of compost can help to increase the interest of farmers to use organic phosphatic fertilizers in alkaline soils [14]. Therefore, this study aimed to evaluate the availability of phosphorus from RP enriched compost with the application of PGPRs and its comparative effectiveness with inorganic fertilizers (Single Super Phosphate) on soil nutrient status, wheat growth and production.

2. Material and Methods

2.1. Experimental Site and Treatments

Two-year field experiments at National Agricultural Research Centre, Islamabad (73°70′ E and 33°39′ N with an altitude 610 masl during growing months Nov, 2010-May, 2011 and Nov, 2011-May, 2012), were conducted on wheat (var. GA-2002). Soil textural class of the experimental site was silty loam. The meteorological data during the growing season (2010–2012) of wheat is given in Table 1. Different composts being prepared during the previous experiments [27] were used in the study for their effectiveness to get better crop production. The treatments included; Control (Untreated un-inoculated); SPLC (Simple poultry litter compost); RP (rock phosphate 18.5% P_2O_5); $RPEC_1$ (rock phosphate + poultry litter solubilized with *Pseudomonas* sp. during composting process); $RPEC_2$ (rock phosphate + poultry litter solubilized with *Proteus* sp. during the composting process); FDP (Full dose inorganic P from Single Super Phosphate-SSP18% P_2O_5); HDP (Half dose inorganic P from Single Super Phosphate-SSP 18% P_2O_5). Treatments were applied at a rate of 100 kg P ha^{-1}, respectively from composts as well as from inorganic fertilizers on a total P basis during seed bed preparation. The nutrient status of different composts, is given in Table 2. The recommended dose of nitrogen at the rate of 100 kg ha^{-1} was equally applied to each plot (4 m × 3 m) either from inorganic fertilizer (Urea-46% N) or compost on a nutrient basis. However, SPLC was applied at the rate of 4.5t ha^{-1}. There were three replications for each treatment. All the fertilizer treatments were applied to respective plots at the same time of sowing.

Table 1. Meteorological data during the growing seasons of wheat crop (2010–2012).

Months	2010–2011			2011–2012		
	Av. Temp (°C)	Rainfall (mm)	R.H (%)	Av. Temp. (°C)	Rainfall (mm)	R.H (%)
Nov.	18.03	4.13	64.2	16.32	7.09	65.52
Dec.	18.18	26.97	64.89	19.23	0	62.74
Jan.	15.5	8.32	71.29	15.55	59.06	68.48
Feb.	11.91	78.73	75.84	14.21	44.12	70.07
March	15.55	53.19	63.02	16.03	15.95	58.65
April	15.27	53.96	61.08	14.87	40.93	54.68
May	17.71	17.29	44.75	17.84	9.47	38.44
Mean	16.02	34.66	63.58	16.29	25.23	59.8

Adopted from CAEWRI, National Agricultural Research Centre, Islamabad. Av. Temp—Average temperature; R.H—relative humidity, mm = millimeter.

Table 2. Nutrient composition of different composts applied as treatments in the experiments.

Compost	Av. P (%)	Total N (%)	TOC (%)	C:N
SPLC	0.35	1.35	19.36	14.34
$RPEC_1$	1.72	1.29	16.3	12.66
$RPEC_2$	1.24	1.28	17.7	13.83

SPLC—Simple poultry litter; $RPEC_1$—Poultry litter + rock phosphate + *Pseudomonas* sp.; $RPEC_2$—Poultry litter + rock phosphate + *Proteus* sp.; Av. P—Available phosphorus; N—Nitrogen; TOC—Total organic carbon; C:N; Carbon–nitrogen ratio.

2.2. Seed Inoculation

The PGPR strains; *Pseudomonas* sp. (Accession no. KF307201) and *Proteus* sp. (Accession no. KF307202) were used at 6 × 10^8 CFU/mL for seed inoculation. Wheat seeds were inoculated with cultures for 4 h and then the seeds were shade dried before sowing. The inoculants were applied individually as well as in combination with organic and inorganic fertilizer treatments.

2.3. Yield, Physiology and Plant Nutrient Analysis

Growth and yield parameters; the number of tillers, grain yield and total dry matter yield were recorded at the time of harvesting. However, for the determination of dry matter yield, the aerial part of the plant from each plot was harvested. Then the spikes were separated from harvested plants of respective treatments and the grains of each pot were weighed to calculate grain yield [kg ha^{-1}].

Chlorophyll and phytohormones (IAA, GA) were analyzed in flag leaves of the wheat plants. Chlorophyll was recorded by using SPAD chlorophyll meter [Konica Minolta, Langenhagen, Germany], while leaf IAA and GA were extracted through the method of Kettner and Doerffling [28] and analyzed on HPLC (Agilent 1100, Waldbronn, Germany) using UV detector and C-18 column (39 × 300 mm). Methanol, acetic acid, and water (30:1:70) were used as mobile phase. The wavelength used for the detection of IAA was 280 nm [29] whereas for GA, it was adjusted at 254 nm. These hormones were identified on the basis of retention time and peak area of the standards. Pure IAA and GA_3 (Sigma Chemicals Co. Ltd. St. Louis, Missouri, USA) were used as standard for identification and quantification of plant hormones. The above ground plants were harvested from each plot, dried at 70 °C for 48 h, ground at the grinding mill and samples were stored in Ziploc polyethylene bags at room temperature till nutrient analysis. Total phosphorus in plant samples and in seeds was analyzed through Olsen and Sommers [30]. However, phosphorus concentration in shoot was used for the calculation of plant P uptake (kg ha^{-1}).

2.4. Soil Analysis

Initial soil samples were taken for physicochemical properties (Table 3). Soil samples (0–30 cm) were analyzed for the texture [31], organic matter [32], total P [33]. However, soil samples were extracted through Ammonium Bicarbonate Diphenyl Triamine Penta Acetic Acid (AB-DTPA) solution for determination of available P, NO_3-N and extractable K following the method of Soltanpour and Schwab [34] and soil pH (1:5 soil–water) using the method of Mclean [35]. Undisturbed soil samples were collected for soil bulk density (g cm^{-3}) using stainless steel cylinders [36]. Soil phosphatase activity was determined by the method of Tabatabai and Bremner [37], whereas, microbial biomass carbon and microbial biomass phosphorus was determined following the method adopted by Steel and Torriej [38]. For determining post-harvest soil properties, soil samples were collected after 6 days of wheat crop harvesting.

Table 3. Physicochemical, biological and biochemical properties of soil.

Properties	2010–2011	2011–2012
Texture	Silty Loam	Silty Loam
Sand	18%	18%
Silt	52%	50%
Clay	30%	32%
pH	7.48	7.5
Ec (dSm^{-1})	0.45	0.46
Bulk density (g cm^{-3})	1.43	1.42
NO_3-N (mg kg^{-1})	3.23	3.4
Total Phosphorus (mg kg^{-1})	500	482
Available Phosphorus (mg kg^{-1})	2.7	2.8
Extractable potassium (mg kg^{-1})	96	92

Table 3. *Cont.*

Properties	2010–2011	2011–2012
Organic matter (%)	0.86	0.84
Cu (μg g^{-1})	1.1	0.78
Fe (μg g^{-1})	54.38	58.98
Zn (μg g^{-1})	1.7	1.64
Mn (μg g^{-1})	1	1.32
Microbial Biomass carbon (mg kg^{-1})	83	84
Microbial Biomass phosphorus (mg kg^{-1})	7	8
Alkaline phosphatase activity (μg g^{-1})	110	112

2.5. Statistical Analysis

The experiment was laid down following the randomized complete block design (RCBD) with split plot design. Different soil amendments (composts and inorganic fertilizers) were assigned with the main plot while PGPRs were placed in sub-plots of the field. Analysis of variance (ANOVA) was conducted with the General Linear Models and means were compared according to the Tukey HSD test with Statistix 8.1 [39]. Two years of data were pooled because there were not interactions between the two years and year was included as a random effect in statistical model.

3. Results

Means of two-year data (2010–2011 and 2011–2012) are provided here due to the result similarity trend from both the years.

3.1. Yield and Yield Components

The data for the number of tillers showed 36%, 34%, 30%, 24% and 21% increases with un-inoculated RPEC$_1$, FDP, RPEC$_2$, SPLC and HDP, respectively, over un-inoculated untreated control (Table 4). The treatment RP did not show any significant increase over un-inoculated untreated control. Seed inoculation with PGPRs without any fertilizer treatment did not show any difference with un-inoculated control. However, seed inoculation with *Pseudomonas* sp. in combination with RPEC$_1$ treatment showed a maximum 5% increase in the number of tillers over un-inoculated RPEC$_1$ and FDP.

C—Control (un-inoculated untreated), SPLC—Simple poultry compost, RPEC$_1$—Rock phosphate enriched compost inoculated with *Pseudomonas* species, RPEC$_2$—Rock phosphate enriched compost inoculated with *Proteus* species, RP—Rock phosphate, HDP—Half dose inorganic P fertilizer, FDP—Full dose inorganic P fertilizer.

The data presented in Table 4, showed that there was a significant ($p \leq 0.05$) effect of PGPR on grain yield of wheat crop. A maximum (18%) increase in grain yield was recorded in plants inoculated with *Pseudomonas* sp. which was 4% higher than inoculation with *Proteus* sp. Without inoculation, maximum (67%) increase in grain yield was recorded with the application of RPEC$_1$ which was 4%, 9% and 16% higher than FDP, RPEC$_2$ and SPLC, respectively over un-inoculated untreated control. However, RPEC$_2$ showed a 52% increase in grain yield over control. The interactive effect of fertilizers × PGPR, was highly significant ($p \leq 0.05$) for grain yield. *Pseudomonas* sp. inoculated RPEC$_1$ and FDP gave maximum (10%) increase over un-inoculated RPEC$_1$ and FDP treatments, respectively. The *Proteus* sp. in combination with RPEC$_1$ also showed 3% increase over un-inoculated RPEC$_1$ treatment, whereas the treatment RP produced minimum grain yield showing 14% increase over un-inoculated RP treatment.

The data in Table 4 show that the treatment RPEC$_1$ resulted in a maximum increase in dry matter yield which was 3.8%, 16%, 27% higher than FDP, RPEC$_2$ and SPLC respectively, over un-inoculated untreated control. The stimulatory effect of PGPR was recorded on dry matter yield. However, the interactive effect of PGPR and fertilizer treatments was significant for dry matter yield of wheat crop. The combination of *Pseudomonas* sp. with RPEC$_1$ gave the maximum increase (62%) similar to

FDP (60%) while with *Proteus* sp. in combination with RPEC$_1$ showed 56% increase over un-inoculated untreated control. RP inoculation with *Proteus* sp. showed nonsignificant difference with un-inoculated untreated control.

Table 4. Effects of plant growth promoting rhizobacteria (PGPR), P-enriched compost and inorganic fertilizers on yield and yield components on wheat.

Treatments	C	SPLC	RPEC$_1$	RPEC$_2$	RP	HDP	FDP
Number of Tillers (m^{-2})							
Without inoculation	260 [e]	323 [c]	354 [a]	339 [b]	261 [e]	315 [c,d]	347 [ab]
	(±2.22)	(±1.42)	(±2.83)	(±3.53)	(±3.12)	(±3.9)	(±2.88)
Proteus sp.	263 [e]	331 [c]	361 [a]	347 [b]	264 [e]	321 [d]	352 [ab]
	(±2.98)	(±2.17)	(±2.67)	(±4.94)	(± 4.24)	(±3.87)	(±4.72)
Pseudomonas sp.	267 [f]	343 [d]	372 [a]	356 [b,c]	269 [f]	326 [e]	364 [ab]
	(±3.61)	(±3.14)	(±4.88)	(±5.01)	(±3.9)	(±2.84)	(±2.92)
Grain yield (kg ha^{-1})							
Without inoculation	2177 [f,g]	3120 [d]	3629 [a]	3317 [c]	2195 [f]	3020 [d,e]	3495 [b]
	(±29.22)	(±36.09)	(±210.74)	(±24.36)	(±29.032)	(±25.78)	(±31.16)
Proteus sp.	2474 [f,g]	3319 [e]	3731 [a]	3524 [c]	2498 [f]	3511 [c,d]	3620 [b]
	(±20)	(±25.45)	(±33.22)	(±24.67)	(±36.103)	(±23.24)	(±29.2)
Pseudomonas sp.	2571 [f,j]	3433 [d]	3987 [a]	3639 [c]	2592 [f]	3364 [d,e]	3848 [b]
	(±32.89)	(±34.36)	(±39.84)	(±23.48)	(±22.86)	(±38.96)	(±29.07)
Dry matter yield (kg ha^{-1})							
Without inoculation	8955 [f,g]	10,804 [d]	13,714 [a]	11,773 [c]	8969 [f]	10,746 [d,e]	13,208 [b]
	(±38.43)	(±31.9)	(±26.7)	(±32.33)	(±24.41)	(±35.9)	(±24.47)
Proteus sp.	8978 [e]	10,930 [c]	13,926 [a]	12,077 [c]	8987 [e]	10,806 [c,d]	13,358 [b]
	(±28.97)	(±39.93)	(±35.55)	(±26.23)	(±34.48)	(±37.13)	(±27.2)
Pseudomonas sp.	9005 [f,g]	11,138 [d]	14,503 [a]	12,651 [c]	9050 [f]	10,891 [e]	14,382 [a,b]
	(±34.02)	(±0.67)	(±37.14)	(±36.29)	(±45)	(±34.65)	(±25.37)

All the treatments sharing common letter are similar otherwise they differ significantly at $p \leq 0.05$.

3.2. Leaf Chlorophyll, IAA and GA Contents

Mean data recorded for chlorophyll contents in flag leaves of wheat crop showed that there was a significant ($p \leq 0.05$) difference for the treatments (Figure 1). Among the un-inoculated treatments, RPEC$_1$ showed the highest (28%) increase which was 2%, 6%, 12% and 25% higher than FDP, RPEC$_2$, SPLC and HDP, respectively. Seed inoculation with *Pseudomonas* sp. resulted in an increase (4%) in chlorophyll content over un-inoculated treatments. However, the interactive effect of treatments (PGPRs × fertilizer) showed 29% increase followed by FDP (27%) over un-inoculated untreated control. While the treatment RP showed a nonsignificant difference when applied in combination with *Pseudomonas* as well as *Proteus* sp.

Data in Figure 2 show that RPEC$_1$ resulted maximum (12%) increase in IAA content, having a similar effect as with FDP, followed by RPEC$_2$ showing a 9% increase, while HDP and SPLC showed a similar effect (i.e., 7% increase) in IAA content over un-inoculated untreated control. The inoculation of seeds with *Pseudomonas* sp. gave higher values of IAA by showing 6% increase over un-inoculated treatments. The interactive effect of PGPR × Fertilizer was nonsignificant except for *Pseudomonas* sp. which showed a 20% increase, when used in combination with RPEC$_1$ and FDP treatments.

The treatment RPEC$_1$ resulted in a 13% increase in GA content, followed by FDP (11%), RPEC$_2$ (9%), SPLC (6%), while HDP resulted only a 4% increase over un-inoculated untreated control (Figure 3). However, seed inoculation with *Pseudomonas* sp. showed a maximum (5%) increase over un-inoculated RPEC$_2$ treatment. The data showed that the treatments RPEC$_1$ and FDP in combination with *Pseudomonas* sp. showed a maximum (16%) increase in GA contents of flag leaves. PGPR inoculation with RP and HDP showed a nonsignificant difference among respective un-inoculated treatments.

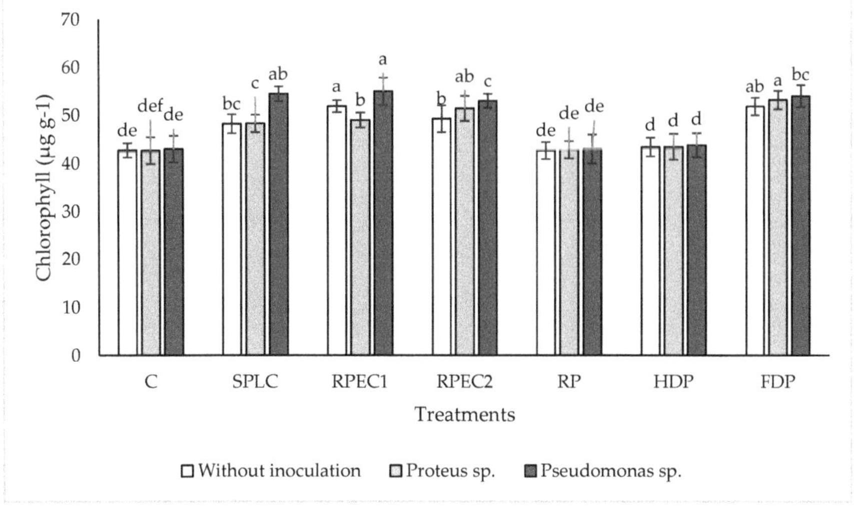

Figure 1. Effects of PGPR, P-enriched compost and inorganic fertilizers on leaf chlorophyll concentration ($\mu g\ g^{-1}$). C—Control; SPLC—Poultry litter only; RPEC1—Rock phosphate + poultry litter solubilized with *Pseudomonas* sp. during the composting process; $RPEC_2$—Rock phosphate + poultry litter solubilized with *Proteus* sp. during composting process), RP—Rock phosphate + poultry litter; HDP—Half dose inorganic P from Single Super Phosphate-SSP 18% P_2O_5; FDP—Chemical fertilizer (Single Super Phosphate). All the treatments sharing a common letter are similar, otherwise they differ significantly at $p \leq 0.05$.

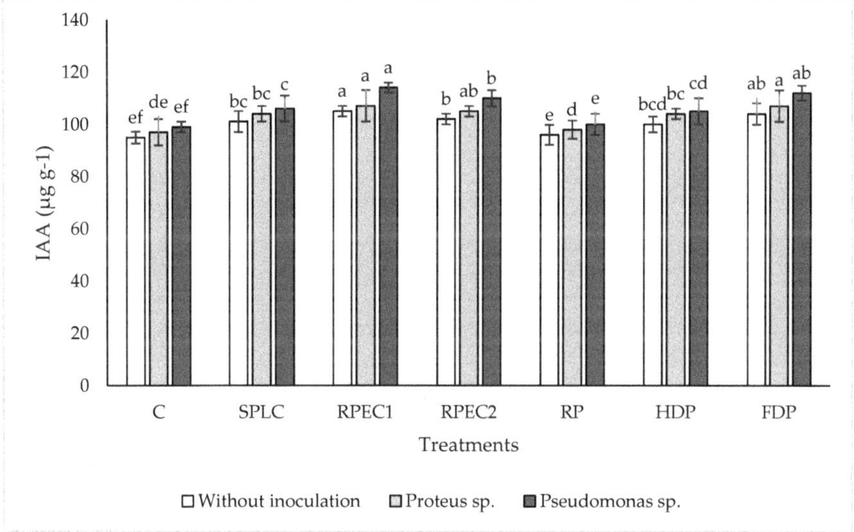

Figure 2. Effects of PGPR, P-enriched compost and inorganic fertilizers on leaf IAA concentration ($\mu g\ g^{-1}$) in wheat. C—Control; SPLC—Poultry litter only; RPEC1—Rock phosphate + poultry litter solubilized with *Pseudomonas* sp. during the composting process; $RPEC_2$—Rock phosphate + poultry litter solubilized with *Proteus* sp. during composting process), RP—Rock phosphate + poultry litter; HDP—Half dose inorganic P from Single Super Phosphate—SSP 18% P_2O_5; FDP- Chemical fertilizer (Single Super Phosphate). All the treatments sharing common letter are similar otherwise they differ significantly at $p \leq 0.05$.

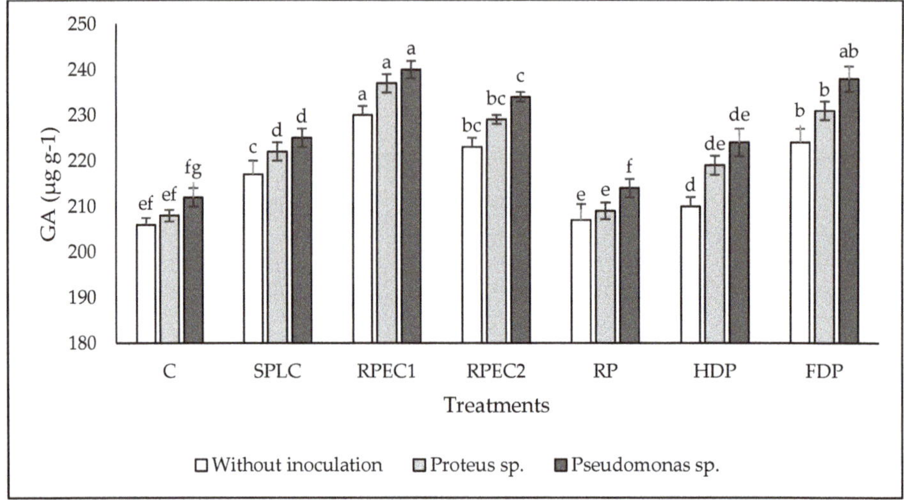

Figure 3. Effects of PGPR, P-enriched compost and inorganic fertilizers on leaf GA concentration ($\mu g\ g^{-1}$) in wheat. C—Control; SPLC—Poultry litter only; RPEC1—Rock phosphate + poultry litter solubilized with *Pseudomonas* sp. during the composting process; RPEC$_2$—Rock phosphate + poultry litter solubilized with *Proteus* sp. during composting process), RP—Rock phosphate + poultry litter; HDP—Half dose inorganic P from Single Super Phosphate—SSP 18% P$_2$O$_5$; FDP—Chemical fertilizer (Single Super Phosphate). All the treatments sharing common letter are similar otherwise they differ significantly at $p \leq 0.05$.

3.3. Plant Phosphorus Uptake and Seed Phosphorus

The data presented in Figure 4 show that the phosphorus uptake was maximum (70%) due to the application of RPEC$_1$ followed by RPEC$_2$ (63%) and FDP (60%), while RP treatment showed no significant difference compared to un-inoculated untreated control. Seed inoculation with *Pseudomonas* sp. resulted in a maximum (7%) increase in P-uptake over un-inoculated treatments. The interaction of fertilizer treatments and PGPRs showed that RPEC$_1$ in combination with *Pseudomonas* sp. showed maximum increase (88%) in P-uptake followed by *Proteus* sp. inoculated RPEC$_1$ (79%) over untreated un-inoculated control.

The seed phosphorus content showed a 61% increase following application of RPEC$_1$ over un-inoculated untreated control, which was 12%, 17%, 33% and 41% higher than FDP, RPEC$_2$, SPLC and HDP, respectively (Figure 5). The application of *Pseudomonas* sp. alone also resulted in an increase (3.5%) in seed P contents over un-inoculated untreated control whereas the interactive effect of PGPR × Fertilizer was nonsignificant.

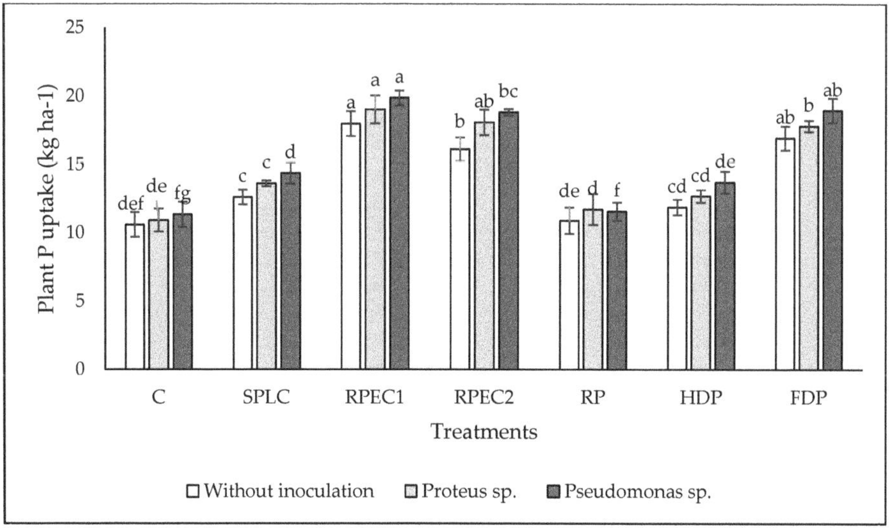

Figure 4. Effects of PGPR, P-enriched compost and inorganic fertilizers on plant P uptake (kg ha^{-1}) in wheat. C—Control; SPLC—Poultry litter only; RPEC1—Rock phosphate + poultry litter solubilized with *Pseudomonas* sp. during the composting process; RPEC$_2$—Rock phosphate + poultry litter solubilized with *Proteus* sp. during composting process), RP—Rock phosphate + poultry litter; HDP— Half dose inorganic P from Single Super Phosphate—SSP 18% P$_2$O$_5$; FDP—Chemical fertilizer (Single Super Phosphate). All the treatments sharing common letter are similar otherwise they differ significantly at $p \leq 0.05$.

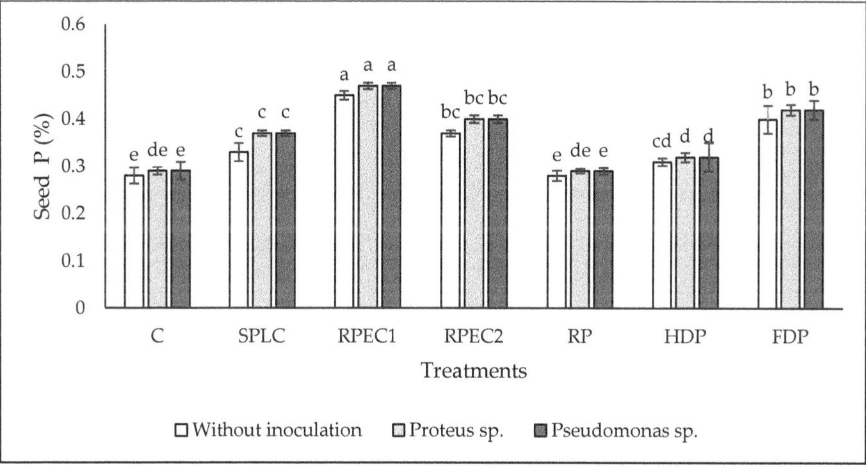

Figure 5. Effects of PGPR, P-enriched compost and inorganic fertilizers on plant P uptake (kg ha^{-1}) in wheat. C—Control; SPLC—Poultry litter only; RPEC1—Rock phosphate + poultry litter solubilized with *Pseudomonas* sp. during the composting process; RPEC$_2$—Rock phosphate + poultry litter solubilized with *Proteus* sp. during composting process), RP—Rock phosphate + poultry litter; HDP— Half dose inorganic P from Single Super Phosphate—SSP 18% P$_2$O$_5$; FDP—Chemical fertilizer (Single Super Phosphate). All the treatments sharing common letter are similar otherwise they differ significantly at $p \leq 0.05$.

3.4. Soil Properties

3.4.1. Available P, Nitrate Nitrogen and Extractable Potassium

The post-harvest soil analysis for phosphorus availability showed that the treatments significantly increased the P availability (Table 5). The treatment $RPEC_1$ resulted in a significant increase over un-inoculated untreated control, the value of which was 37%, 82%, and 130% higher than SPLC, HDP and RP, respectively. The PGPR seed inoculation effect was significant ($p \leq 0.05$) for post-harvest available soil P contents. The phosphorus content was increased (20% and 9%) in the rhizosphere of plants treated with *Pseudomonas* sp. and *Proteus* sp. respectively. In combination with *Pseudomonas* sp., the treatment $RPEC_1$ gave maximum increase (3.43-fold) over un-inoculated untreated control, while $RPEC_1$ in combination with *Proteus* sp. and *Pseudomonas* sp. in combination with FDP showed similar results by giving a 3.17-fold increase over the un-inoculated untreated control. The PGPRs (*Pseudomonas* sp. and *Proteus* sp.) in combination with $RPEC_2$ showed a similar effect for increase [8%] in available P over un-inoculated $RPEC_2$. The treatment RP in combination with *Pseudomonas* sp. resulted in 30% increase over un-inoculated RP, which was 16.5% higher than *Proteus* sp. inoculated RP.

C—Control (un-inoculated untreated), SPLC—Simple poultry compost, $RPEC_1$—Rock phosphate enriched compost inoculated with *Pseudomonas* species, $RPEC_2$—Rock phosphate enriched compost inoculated with *Proteus* species, RP—Rock phosphate, HDP—Half dose inorganic P fertilizer, FDP—Full dose inorganic P fertilizer. Soil samples for nutrient and biological analyses were collected two days after wheat harvesting.

Table 5. Effects of PGPR, P-enriched compost and inorganic fertilizers on post-harvest soil of wheat in field experiments.

Treatments	C	SPLC	$RPEC_1$	$RPEC_2$	RP	HDP	FDP
Available P (mg kg^{-1})							
Without inoculation	2.8 [g]	6.0 [e]	10.9 [a]	9.7 [b]	4.4 [f]	7.8 [d]	8.76 [c]
Proteus sp.	2.9 [g]	6.3 [e]	11.6 [a]	10.1 [b]	4.9 [f]	8.4 [c,d]	8.90 [c]
Pseudomonas sp.	3.2 [g]	6.9 [e]	14.4 [a]	10.48 [b]	5.7 [f]	9.2 [c,d]	9.74 [b,c]
Nitrate nitrogen (mg kg^{-1})							
Without inoculation	3.07 [f]	3.87 [a,b,c]	3.98 [a]	3.92 [a,b]	3.07 [f]	3.39 [e]	3.78 [b,c,d]
Proteus sp.	3.08 [e]	3.88 [a,b]	4.04 [a]	4.07 [a]	3.07 [e]	3.42 [d]	3.81 [b,c]
Pseudomonas sp.	3.0 [f]	3.89 [b,c]	4.20 [a]	4.14 [a,b]	3.08 [f]	3.44 [e]	3.85 [c,d]
Extractable potassium (mg kg^{-1})							
Without inoculation	95.50 [c,d]	106 [a,b]	109.5 [a]	106 [a,b]	96.5 [c,d]	101 [a,b,c]	102 [a,b,c]
Proteus sp.	96.33 [d]	108 [a,b]	110 [a]	107 [a,b]	97 [c,d]	102 [a,b,c]	103 [a,b,c]
Pseudomonas sp.	97.2 [d]	108 [a,b]	112 [a]	109 [a,b]	98 [c,d]	104 [b,c]	104 [b,c]
Alkaline phosphatase (μg PNP g^{-1} hr^{-1})							
Without inoculation	117 [d]	136 [b]	151 [a]	143 [a,b]	118 [d]	127 [c]	133 [b,c]
Proteus sp.	118 [f,g]	142 [b,c]	157 [a]	147 [b]	121 [f]	132 [d,e]	137.5 [c,d]
Pseudomonas sp.	120 [e]	155.5 [b]	167 [a]	156 [b]	122 [e]	136 [c,d]	143.5 [c]
Microbial biomass carbon (μg g^{-1})							
Without inoculation	84 [f,g]	112 [c]	136 [a]	123 [b]	87 [e,f]	93 [e]	106 [c,d]
Proteus sp.	89 [f,g]	119 [c]	144 [a]	137 [a,b]	92 [f]	103 [e]	114 [c,d]
Pseudomonas sp.	93 [f,g]	127 [c]	159 [a]	146 [b]	96 [f]	114 [d,e]	120 [c,d]
Microbial biomass phosphorus (μg g^{-1})							
Without inoculation	7 [d]	14 [b]	19 [a]	18 [a]	7 [d]	10 [b,c]	11 [b,c]
Proteus sp.	8 [d]	16 [b,c]	21 [a]	19 [a,b]	8 [d]	11 [c,d]	12 [c,d]
Pseudomonas sp.	8 [e]	18 [c]	26 [a]	24 [a,b]	8 [e]	13 [d]	13 [d]

All the treatments sharing a common letter are similar otherwise they differ significantly at $p \leq 0.05$.

Mean data for post-harvest soil nitrate-nitrogen showed significant differences with the application of fertilizer treatments (Table 5). The treatments RPEC$_1$ and RPEC$_2$ resulted a 36% increase followed by SPLC, FDP and HDP showing 29%, 26% and 14% increase over un-inoculated untreated control, respectively. The treatment RP showed nonsignificant difference with the control. There was a nonsignificant effect of seed inoculation on NO$_3$-N over un-inoculated control. However, the maximum increase was recorded by the application of *Pseudomonas* sp. which was 4% higher over un-inoculated control treatments. The interactive effect of PGPR and fertilizer treatments was nonsignificant with SPLC, RP, HDP and FDP, while the treatments RPEC$_1$ and RPEC$_2$ showed 36% and 35% increases over un-inoculated untreated control. The mean data showed that all the treatments increased extractable potassium except RP (Table 5). A maximum increase (15%) in the content of extractable K was recorded following the treatment RPEC$_1$ followed by SPLC and RPEC$_2$ which were significantly similar in their effect by showing 11% and 12% increase over un-inoculated untreated control, respectively. The treatments FDP and HDP also showed similar effects and increased extractable K by 7% over control. Seed inoculation with *Pseudomonas* sp. showed 2% increase in seed phosphorus contents over un-inoculated untreated control. The interactive effect of PGPR × Fertilizers, was nonsignificant, whereas *Pseudomonas* sp. inoculation in combination with RPEC$_1$ showed maximum (17%) phosphorus contents over un-inoculated untreated control.

3.4.2. Alkaline Phosphatase and Microbial Biomass

Alkaline phosphatase activity was significantly ($p \leq 0.05$) increased as a result of different treatments (Table 5). The treatment RPEC$_1$ resulted a maximum increase (29%) over un-inoculated untreated control, which was 5.6%, 11%, 13.5% and 19% higher than RPEC$_2$, SPLC, FDP and HDP respectively. *Pseudomonas* sp. inoculation showed 8% increase over un-inoculated treatments, which was 2.5% higher than *Proteus* sp. inoculation. Significant ($p \leq 0.05$) increase in alkaline phosphatase activity was recorded due to the combine effects of PGPRs with different fertilizer treatments. The inoculation of *Pseudomonas* sp. in combination with the treatment RPEC$_1$ showed maximum (43%) increase over untreated un-inoculated control. The treatments RPEC$_2$ and SPLC in combination with *Pseudomonas* sp. and RPEC$_1$ in combination with *Proteus* sp. showed similar effect and increased the alkaline phosphatase activity by 34% over un-inoculated untreated control. The treatment FDP in combination with *Pseudomonas* sp. showed a 23% increase over untreated un-inoculated control; the effect of which was significantly similar to un-inoculated RPEC$_2$ treatment. The PGPRs (*Pseudomonas* sp. and *Proteus* sp.) in combination with HDP showed a similar effect but significantly lower percentage increase than un-inoculated HDP treatment.

Mean data showed that the fertilizer treatments significantly ($p \leq 0.05$) improved the microbial biomass carbon contents (Table 5). Significant increase (65%) was recorded in microbial biomass carbon contents in RPEC$_1$ treatment over control; the increase for RPEC$_1$ was also 9%, 22%, 30%, 43%, and 60% higher than that of RPEC$_2$, SPLC, FDP, HDP and RP, respectively. Inoculation of seeds with *Pseudomonas* sp. showed maximum (16%) increase in microbial biomass carbon over un-inoculated treatments, the values of which were 8% higher than *Proteus* sp. inoculated treatments. The interactive effect of PGPR inoculation to seeds and fertilizer treatments was also significant with microbial biomass carbon contents. Among the *Pseudomonas* sp. inoculated treatments, RPEC$_1$ showed 89% increase in microbial biomass carbon over un-inoculated untreated control, while the treatment RPEC$_2$ showed 74% increase, which was significantly similar with *Proteus* sp. inoculated RPEC$_1$ treatment. However, *Proteus* sp. inoculated RPEC$_2$ increased microbial biomass carbon by 63% and showed a nonsignificant difference with un-inoculated RPEC$_1$ treatment. *Pseudomonas* sp. in combination with FDP showed a nonsignificant difference with un-inoculated RPEC$_2$ but was 13% higher than the un-inoculated FDP treatment. The treatment RP in combination with *Pseudomonas* sp. increased microbial biomass by 10% over untreated un-inoculated control.

The data in Table 5 show that microbial biomass phosphorus (MBP) increased significantly ($p \leq 0.05$) with the application of fertilizer treatments than the un-inoculated untreated control. The maximum increase (1.75-fold) in MBP was recorded from the treatment RPEC$_1$ which was 7%, 37% and 83%

higher than $RPEC_2$, SPLC and FDP, respectively while FDP showed nonsignificant difference with HDP. The treatment RP showed a nonsignificant difference with un-inoculated untreated control. There was also a significant effect of PGPR on microbial biomass phosphorus. *Pseudomonas* sp. inoculation increased microbial biomass P by 33% over un-inoculated treatments, the values of which were 14% higher than *Proteus* sp. inoculated treatments. *Pseudomonas* sp. inoculation with $RPEC_1$ showed a maximum increase (2.7-fold) in MBP followed by $RPEC_2$ (2.43-fold) over un-inoculated untreated control. The treatment $RPEC_1$ in combination with *Proteus* sp. showed an (61%) increase over un-inoculated untreated control which was at par with *Pseudomonas* sp. inoculated SPLC. The treatments FDP, HDP increased (46%) microbial biomass P showing nonsignificant difference with each other and RP showed 12% increase in microbial biomass P with PGPR inoculation over un-inoculated untreated control.

3.5. Economic Analysis

The economic analysis of applied treatments (Table 6) in terms of value cost ratio (VCR) showed that $RPEC_1$ performed best with and without seed inoculation. Among the un-inoculated treatments, $RPEC_1$ showed maximum VCR (2.72) followed by $RPEC_2$ (2.14), FDP (1.94) while the minimum (0.06) VCR was received from RP. Seed inoculation with *Pseudomonas* sp. in combination with $RPEC_1$ superseded all of the treatments resulting in maximum VCR (3.23). Hence, rock phosphate enriched compost alone or more so in combination with phosphate solubilizing bacteria (PSB), can perform better than chemical fertilizers. The economic analysis revealed that RPEC could be an economically feasible substitute to costly chemical fertilizers for sustainable crop production.

Table 6. Economic Analysis of the applied products presented as value cost ratio (VCR).

Treatments	Grain Yield	Increase in Yield	Increased Yield Value	Cost of Inputs	Net Return	VCR
	kg ha^{-1}			Rs. ha^{-1}		
Control	2177	-	-	-	-	-
SPLC	3120	943	28,290	17,760	10,530	1.59:1
$RPEC_1$	3629	1452	43,560	16,010	27,550	2.72:1
$RPEC_2$	3317	1140	34,200	16,010	18,190	2.14:1
RP	2195	18	540	9285	−8745	0.06:1
HDP	3020	843	25,290	13,435	11,855	1.88:1
FDP	3495	1318	39,540	20,385	19,155	1.94:1
Proteus sp. (1S)	2474	297	8910	7310	1600	1.22:1
SPLC + 1S	3319	1142	34,260	18,560	15,700	1.85:1
$RPEC_1$+1S	3731	1554	46,620	16,810	29,810	2.77:1
$RPEC_2$+1S	3524	1347	40,410	16,810	23,600	2.40:1
RP+1S	2498	321	9630	10,085	−455	0.95:1
HDP+1S	3511	1334	40,020	14,235	25,785	2.01:1
FDP+1S	3620	1443	43,290	21,185	22,105	2.04:1
Pseudomonas (2S)	2571	394	11,820	7310	4510	1.62:1
SPLC + 2S	3433	1256	37,680	18,560	19,120	2.03:1
$RPEC_1$+ 2S	3987	1810	54,300	16,810	37,490	3.23:1
$RPEC_2$+ 2S	3639	1462	43,860	16,810	27,050	2.61:1
RP+ 2S	2592	415	12,450	10,085	2365	1.23:1
HDP+ 2S	3364	1187	35,610	14,235	21,375	2.05:1
FDP+ 2S	3848	1671	50,130	21,185	28,945	2.37:1

Increase in yield = Yield of treatment − Yield of control, Increased yield value = Grain price × increase in yield, Net return = Increased yield value − cost of inputs, Value cost ratio (VCR) = Increased yield value/cost of inputs, Poultry litter = Rs. 1.5 kg^{-1}, Rock phosphate = Rs. 5 kg^{-1}, Single super phosphate (SSP) = Rs. 25 kg^{-1}, Urea = Rs. 30 kg^{-1}, Labor charges for compost preparation = Rs. 5250, Seed inoculant = Rs. 200 L^{-1}, Wheat grain price = Rs. 30 kg^{-1}, SPLC—Simple Poultry litter, $RPEC_1$—Rock Phosphate Enriched Compost solubilized with *Pseudomonas* sp., $RPEC_2$—Rock phosphate enriched compost solubilized with *Proteus* sp. 1S-Seed inoculation with *Proteus* sp., 2S—Seed inoculation with *Pseudomonas* sp., RP—Rock phosphate, HDP—Half dose of inorganic fertilizer, FDP—Full dose of inorganic fertilizer. Rs—Refer to national currency (Rupees).

4. Discussion

The use of plant growth promoting rhizobacteria (PGPR) in combination with organic (composts, rock phosphate) and inorganic (chemical fertilizers) phosphorus sources significantly increased the number of tillers per plant and yield components of wheat crop. The results are in conformity to the findings of Akhtar et al. [40] who recorded increase in plant height, the number of tillers, grain yield and 1000 grain weight of wheat with the use of compost and PGPR inoculation.

Maximum grain yield was obtained by the application of RPEC$_1$ which was higher than the full dose of inorganic P fertilizers (FDP), irrespective of the PGPR seed inoculants. The observed yield increase from RPEC$_1$ was indicative of the high P availability and greater photosynthesis as observed by an increase in chlorophyll content and dry matter production, which was maximum in RPEC$_1$ over other treatments. Plant P availability as the key factor for maximum plant growth and higher crop production [41]. Although a full dose of P (FDP) as inorganic fertilizer (SSP) is a source of readily available phosphorus necessary for early growth of the plants, at the site of SSP application, production of the least soluble Ca-P compounds due to surface adsorption and precipitation, reduce P availability [10]. The organic acids produced due to compost might have reduced P exchange sites through chelation and released more soluble forms of plant available P [42] compared to SSP which could help increase growth and yield of wheat. Seed inoculation with *Pseudomonas* sp. increased the grain yield with fertilizer treatments, however the maximum increase was recorded from *Pseudomonas* sp. inoculation with RPEC$_1$ followed by the inoculated FDP treatment. Microbial community in the root rhizosphere might have taken part to release fixed phosphorus through organic acids production which ultimately increased the yield of wheat. Afzal and Bano [43] reported that seed inoculation with PGPR in combination with P fertilizer increased the grain yield of wheat which was 30–40% higher than the un-inoculated P fertilizer. It was reported that organic manures and bio-fertilizers have a high impact on nutrient uptake, physiological process of wheat, and also on water holding capacity of the soil which ultimately increase grain yield of the crop [44]. Amujoyegbe et al. [45] recorded higher grain yield of maize due to the application of chicken manure in combination with microbes compared to chemical fertilizer and chicken manure alone. An association of agronomic traits with grain yield and a positive correlation of 1000 grain weight with grain yield was previously demonstrated in PGPR + manure treated plants of wheat [44].

Increase in the dry matter yield due to the application of RPEC$_1$ compared to FDP may be due to higher vegetative growth, chlorophyll content and the maximum number of tillers during the crop growth, while mobilization of phosphorus due to dissolution of rock phosphate from RPEC$_1$ might have taken part in the physiological processes leading to maximum biomass yield. Higher yields of mung-bean were recorded due to bio-inoculated RP enriched compost having higher citrate soluble, water soluble P and organic P, maximum microbial biomass carbon and acid phosphatase activity compared to un-inoculated composts [7]. Similarly, Nishanth and Biswas [11] prepared enriched composts with *Aspergillus awamori* inoculation and tested these on the wheat crop, which gave maximum biomass production in comparison to composts prepared without inoculants. Hossain et al. [46] reported an increase in grain and straw yield of wheat crop with the application of phosphate solubilizing bacteria (PSB) along with different levels of phosphorus. In concurrence with the present results, an increase in dry matter and grain yield of agronomic crops due to phosphate solubilizing microorganisms in combination with different P fertilizers were reported earlier by different workers [47–49].

Phosphorus plays an important role in chlorophyll production and regulation. It has been reported that the partitioning of photosynthates between leaves and reproductive organs is regulated by the availability of phosphorus to the plants [50]. Maximum increase in chlorophyll contents in flag leaves were recorded due to the application of RPEC$_1$ followed by FDP and SPLC. Zafar et al. [51] reported an increase in chlorophyll contents by 10–89% over control in leaves of maize crop following application of P fertilizers in the form of compost and inorganic fertilizers. The PGPR in combination with compost was recorded to be stimulatory for chlorophyll production; this was confirmed for *Pseudomonas* sp. in combination with RPEC$_1$. Seed inoculation with *Pseudomonas* sp. alone or in combination with P fertilizers, was more efficient for improving chlorophyll contents in flag leaves of wheat plants. Naseem and Bano [52] reported that the seed inoculation with *Pseudomonas* sp.

and *Bacillus cereus* increased chlorophyll contents by 8–13% in leaves of wheat crop. An increase in chlorophyll contents with the application of organic manure was also recorded [53].

PGPR alone or in combination with fertilizers showed a significant effect on IAA and GA contents of wheat flag leaves, however, maximum increase was recorded as a result of $RPEC_1$ application followed by FDP and $RPEC_2$. Among the PGPRs, *Pseudomonas* sp. performed better than the *Proteus* sp. Indole Acetic Acid synthesis by bacteria may have various regulatory effects in plant–bacterial interactions and significant effect on plant growth promotion [54]. Generally, phytohormones in plants plays an important role in cell division, proliferation, and differentiation, vascular tissue alteration, responses to light and gravity, general root and shoot architecture, seed and tuber germination, organ differentiation, peak predominance, ethylene synthesis, vegetative growth processes, fruit development and aging. These results are in accordance with the findings of Saharan and Nehra [55], who reported that the phytohormone production through PGPR (*Pseudomonas, Azotobacter, Azospirillum*) may contribute to growth and yield of the crop. IAA acts as a signal molecule for cell expansion, division and differentiation. Higher counts of genus *Pseudomonas* were recorded [56] in winter wheat cultivars and described the developmental phase of wheat crops as a key factor in higher population of the microbes. The GA and IAA were reported to be produced by bacterial strains such as *Bacillus* and *Pseudomonas* [57] and inoculation of wheat with *Pseudomonas* sp. gave maximum increase in growth and yield [58]. Khan et al. [59] found an increase in IAA and GA contents in leaves of wheat inoculated with *Pseudomonas* and *Bacillus* strains. Sivasankari et al. [60] isolated bacterial strains from black gram (*Vigna mungo*) rhizosphere soil and reported maximum IAA production from *Pseudomonas* sp. than *Proteus* sp.

Phosphors uptake increased with the application of RP enriched compost ($RPEC_1$) which would be due to phosphorus in the soluble form. Higher concentration of macronutrients due to the decomposition of organic materials in the soil were recorded [61]. Incorporation of organic materials can enhance phosphorus availability in the soil solution by decreasing P sorption/fixation through chelation [62]. Phosphorus also plays an efficient role in plant photosynthesis, respiration, formation of cell membrane, glycolysis and enzymes activities [63] showing that the growth and development of all crops are dependent upon P availability [64]. The presence of P as an integral part of nucleotides, phospholipids, phosphoproteins, and coenzymes shows its importance for life [65]. An increase in P-uptake due to enriched compost in the present study was due to the maximum available P as well as total organic and readily available carbon. Sharma et al. [66] reported increased N uptake (18–38 kg ha^{-1}), P uptake (2.7–6.6 kg ha^{-1}), and K uptake by (16–41 kg ha^{-1}) in the rice–wheat system when inoculated with *Pseudomonas striata*. It was reported by Nishanth and Biswas [11] that RP enriched compost inoculated with *Aspergillus awamori* can significantly enhance P uptake in wheat crop, which was recorded as 78% more efficient compared to DAP. Ghaderi et al. [67] reported 51%, 29% and 62% release of phosphorus from iron hydroxides by the application of *Pseudomonas putida, Pseudomonas fluorescens, and Pseudomonas fluorescens*, respectively. Shrivastava [10] reported that inoculation of microbes with P enriched manure show maximum P uptake in mung-bean crop compared to SSP fertilizer. The P-enriched compost in combination with effective microbes (EM) can enhance N and P uptakes of the cowpea crop [14].

Crop growth is regulated by the nutrient supply from organic or chemical fertilizer sources. Organic materials are considered to be the best source for nutrient supply to plants but with slow release until the crop maturity, which may create a delay in crop maturity or cause high nutrients content in the produce [68]. Maximum P concentration in the wheat seeds with the application of $RPEC_1$ might be due to a slow release process resulting in P accumulation in the seeds due to mobility of the phosphorus from soil to plant process. The integrated management of P fertilizers at the root zone can increase the mobility of P from plant roots through physiological adaptive mechanisms [69]. Seed inoculation with *Pseudomonas* sp. showed an increase in seed phosphorus. According to Son et al. [70] soybean seed P content increased with inoculation of phosphate solubilizing microorganisms.

Organic and inorganic amendments have a great impact on soil properties [71]; however, while the application of fertilizer increases P availability at all crop growth stages compared to control treatment, the RP compost showed maximum P availability at later stages of wheat crop growth [11]. The increase in post-harvest soil P availability with the application of RP enriched compost may be due to mineralization of both RPEC and soil organic P, and chelation of P through ligand exchange reactions to reduce P fixation throughout the crop growth stages. The ligand exchange reactions can increase P mobilization through organic and phosphate anions adsorption with Fe and Al sites [72]. Slow release of P through mineralization of organic P fraction from enriched compost was reported previously [73]. Organic acids produced by phosphate solubilizing microorganisms are sources of H^+ ions which help mineralize tri-calcium phosphate of RP to mono-calcium phosphate; the available form of phosphorus for better plant growth [74].

The application of compost treatments showed significantly higher nitrate-nitrogen contents in post-harvest soil compared to control. Higher nitrate nitrogen content from compost treated plots would be due to reduced nitrate leaching from the soil [75]. Sommers and Giordano [76] stated that all inorganic N in soil amended with Municipal Solid Waste (MSW) compost was available for plant uptake, but 5 to 75% of the organic N will be mineralized within 1 year after application. The findings are in accordance with the results of Baziramakenga et al. [77] who reported an increase in inorganic nitrogen (NO_3-N) contents of snap-bean post-harvest soil with the application of compost of de-inking paper residues and poultry manure. The reason for higher NO_3-N contents due to the application of composts is attributed to the formation of phospho-protein due to the interaction with rock phosphate from enriched compost, which is less susceptible to volatilization. The proteins are decomposed by soil bacteria and change into ammonium that is further nitrified by nitrifying bacteria. This form of nitrogen from compost is slowly available to plants having fewer chances of loss through volatilization. The escape of ammonia from soil decreases if the nitrogen source is compost, organic manure or green manure [78]. The presence of phosphate preserves the nitrogen resulting in a decrease in the number of denitrifying bacteria [79]. The slow release process of nutrients from enriched compost might be another reason for higher nitrate-nitrogen contents than inorganic fertilizers (FDP) in post-harvest soil. Adeli et al. [80] reported higher residual soil NO_3-N contents after cotton crops with the application of poultry manure compared to inorganic fertilizers. Seed inoculation with PGPR (*Pseudomonas* sp. and *Proteus* sp.) showed an increase in post-harvest nitrate-nitrogen contents. Canbolat et al. [81] also reported increase in soil post-harvest nitrogen contents with application of *Pseudomonas putida* compared to the control on barley crop.

Extractable potassium contents increased in post-harvest soil with the application of compost compared to inorganic P fertilizer (FDP). A significantly higher concentration of K with the application of enriched compost compared to FDP and control might be due to the higher water-soluble potassium present in the enriched composts. Stratoon et al. [82] reported that K in composts remains in water soluble forms and thus does not need to be mineralized before becoming available to plants. The increase in soil extractable K by rock phosphate enriched compost (RPEC) may be related to the direct addition to the available K pool of the soils, and to the reduction of K fixation and increase the release of K from the soil solid phase due to the interaction of organic matter and/or soil microorganisms with K-bearing minerals [77]. It was revealed that potassium in manure and compost is highly plant-available and can be used similar to K fertilizer application [83].

Soil enzymes (alkaline phosphatase and acid phosphatase) play a vital role in conversion of fixed soil phosphorus to plant available form [7]. The increase in alkaline phosphatase activities with the application of RP enriched compost in the present study may be due to the availability of organic C which consequently increased the soil phosphatase activity [84] and the compost might have provided considerable carbon and nitrogen for maximum growth of microbes. It has been emphasized that C and N are interlinked with P mineralization by microbes [85] and Shrivastava et al. [10] concluded that the availability of metabolizable C plays a significant role to increase soil phosphatase activity with the application of P enriched manure on mungbean crop. Soil enzymes such as acid and alkaline phosphatases help to increase mineralization of P_0 to P_i by creating a strong relation between bio-available and unavailable P in the soil [86]. Some researchers [87,88] believe that there is an inverse relationship between available P and phosphatases due to negative feedback of phosphate ions on PHO genes, that suppress phosphatase synthesis by microbes [89]. However, phosphatase activity was not affected by the use of rock phosphate as a phosphate source in RP enriched compost [9] showing long persistence and least biodegradation of enzymes with the application of compost. But Pascual et al. [90] endorsed the decrease in phosphatase activities with time span due to exhaustion of biodegradable substrates by microbial activity.

Microbial biomass is an important factor assessing soil quality and its ability to provide energy for nutrient recycling and transformation in the soils [91]. Kiani et al. [92] documented the microbial biomass responses to different land management systems including fertilizer addition and organic amendment application and identified suitable soil quality indicators. The microbial biomass carbon (MBC) acts as substrate supplying entity for microbial communities in soil [93]. In the present study, maximum MBC with the application of RP enriched compost compared to poultry litter compost is due to higher percentage of existing microbial biomass carbon, mineralizable nitrogen and water-soluble carbon in the former compost. The results are in conformity to the findings of Meena et al. [94] who reported an increase in soil microbial biomass carbon with the application of enriched compost compared to ordinary compost as well as inorganic fertilizer. Previously Ayed et al. [95] found an increase in microbial biomass carbon with the application of compost compared to inorganic chemical fertilizer and control in wheat crop.

The maximum microbial biomass phosphorus (MBP) with the application of RPEC and inorganic P fertilizer could be due to the transformation of labile and nonlabile inorganic phosphorus to the organic pool through microbial activity of compost. As Leytem et al. [96] reported the assimilation of various fractions of P into microbial biomass which ultimately provides available P for plants since most organic phosphorus in microbial cells is hydrolysable. Microbial biomass phosphorus may also help in calcareous soils by providing plant available P with application of manure [97] by the mechanism in which P is immobilized and transformed to labile P, which is safe from fixation and transmitted to available P [98]. The results in the present study showed less microbial biomass P with the addition of a full recommended dose of inorganic P fertilizer, indicating that soil existing organic carbon was limited to support microbial growth and activity. Minimum microbial biomass was recorded with the addition of high P inorganic fertilizer to the soil in an incubation study.

5. Conclusions

The present research revealed that enrichment of rock phosphate with poultry litter and PGPR during the process of composting improves nutrient availability and biological properties of the compost. Application of RP enriched compost in field experiment increased yield and yield components of the wheat crop compared to the full recommended dose of inorganic fertilizer and control. Moreover, seed inoculation with PGPR showed significant results to improve the agronomic effectiveness of RP enriched compost. Chemical (availability of phosphorus) and biological (microbial biomass C & P, alkaline and acid phosphatase activities) properties of post-harvest soil improved with the application of RP enriched compost. It can be concluded that RP enriched compost may be an alternative to chemical fertilizer to improve the growth and yield of the crop.

Author Contributions: Conceptualization, M.B., A.B. and N.K.; Methodology, M.B., M.K., K.M.D., and N.K.; Formal analysis, M.B., M.K., S.N. and N.K.; Software, M.B., K.M.D. and N.K.; Data curation, S.N., K.M.D. and A.M.; Validation, M.B., A.B., A.H., A.M. and N.K., Investigation, M.B., A.B., S.B., A.H. and N.K., Resources, M.B., M.K., A.B., S.N., A.M. and N.K.; writing—original draft preparation, M.B., M.K., A.B. and N.K.; writing—review and editing, M.B., M.K., A.B., A.H., A.S. and N.K.; supervision, M.B., A.B. and N.K. All authors have read and agreed to the published version of the manuscript.

Funding: This research received no external funding.

Acknowledgments: The author is thankful to Higher Education Commission Islamabad Pakistan and Pakistan Agriculture Research Council, Islamabad, Pakistan for the assistance to conduct this research.

Conflicts of Interest: The authors declare no conflict of interest.

References

1. Elkoca, E.; Turan, M.; Donmez, M.F. Effects of single, dual and triple inoculations with Bacillus subtilis, Bacillus megaterium and Rhizobium leguminosarum bv. Phaseoli on nodulation, nutrient uptake, yield and yield parameters of common bean [Phaseolus vulgaris l. cv.'elkoca-05']. *J. Plant Nutr.* **2010**, *33*, 2104–2119. [CrossRef]
2. Aydin, E.; Šimanský, V.; Horák, J.; Igaz, D. Potential of Biochar to Alternate Soil Properties and Crop Yields 3 and 4 Years after the Application. *Agronomy* **2020**, *10*, 889. [CrossRef]
3. Gyaneshwar, P.; Kumar, G.N.; Parekh, L.J.; Poole, P.S. Role of soil microorganisms in improving P nutrition of plants. *Plant Soil* **2002**, *245*, 83–93. [CrossRef]
4. Ali, S.; Xie, L. Plant Growth Promoting and Stress mitigating abilities of Soil Born Microorganisms. *Recent Pat. Food Nutr. Agric.* **2019**, *15*. [CrossRef] [PubMed]
5. Khan, N.; Bano, A. Exopolysaccharide producing rhizobacteria and their impact on growth and drought tolerance of wheat grown under rainfed conditions. *PLoS ONE* **2019**, *14*, e0222302. [CrossRef] [PubMed]
6. Reddy, M.S.; Kumar, S.; Babita, K. Bio-solubilization of poorly soluble rock phosphates by Aspergillus tubingensis and Aspergillus niger. *Bioresour. Technol.* **2002**, *84*, 187–189. [CrossRef]
7. Achal, V.; Savant, V.V.; Reddy, M.S. Phosphate solubilization by a wild type strain and UV-induced mutants of Aspergillus tubingensis. *Soil Biol. Biochem.* **2007**, *39*, 695–699. [CrossRef]
8. Puente, M.E.; Bashan, Y.; Li, C.Y.; Lebsky, V.K. Microbial populations and activities in the rhizoplane of rock-weathering desert plants. I. Root colonization and weathering of igneous rocks. *Plant Biol.* **2004**, *6*, 629–642. [CrossRef]
9. Gaind, S.; Pandey, A.K.; Lata. Microbial biomass, P-nutrition, and enzymatic activities of wheat soil in response to phosphorus enriched organic and inorganic manures. *J. Environ. Sci. Health Part B* **2006**, *41*, 177–187. [CrossRef]
10. Shrivastava, M.; Kale, S.P.; D'Souza, S.F. Rock phosphate enriched post-methanation bio-sludge from kitchen waste based biogas plant as P source for mungbean and its effect on rhizosphere phosphatase activity. *Eur. J. Soil Biol.* **2011**, *47*, 205–212. [CrossRef]
11. Nishanth, D.; Biswas, D.R. Kinetics of phosphorus and potassium release from rock phosphate and waste mica enriched compost and their effect on yield and nutrient uptake by wheat (Triticum aestivum). *Bioresour. Technol.* **2007**, *99*, 3342–3353. [CrossRef] [PubMed]
12. Naseem, H.; Ahsan, M.; Shahid, M.A.; Khan, N. Exopolysaccharides producing rhizobacteria and their role in plant growth and drought tolerance. *J. Basic Microbiol.* **2018**, *58*, 1009–1022. [CrossRef] [PubMed]
13. Reyes, I.; Valery, A.; Valduz, Z. Phosphate-solubilizing microorganisms isolated from rhizospheric and bulk soils of colonizer plants at an abandoned rock phosphate mine. In *First International Meeting on Microbial Phosphate Solubilization*; Springer: Dordrecht, Germany, 2007; pp. 69–75.
14. Chen, Y.P.; Rekha, P.D.; Arun, A.B.; Shen, F.T.; Lai, W.A.; Young, C.C. Phosphate solubilizing bacteria from subtropical soil and their tricalcium phosphate solubilizing abilities. *Appl. Soil Ecol.* **2006**, *34*, 33–41. [CrossRef]
15. Zayed, G.; Abdel-Motaal, H. Bio-active composts from rice straw enriched with rock phosphate and their effect on the phosphorous nutrition and microbial community in rhizosphere of cowpea. *Bioresour. Technol.* **2005**, *96*, 929–935. [CrossRef]
16. Adesemoye, A.O.; Obini, M.; Ugoji, E. Comparison of plant growth-promotion with Pseudomonas aeruginosa and Bacillus subtilis in three vegetables. *Braz. J. Microbiol.* **2008**, *39*, 423–426. [CrossRef] [PubMed]
17. Khan, N.; Bano, A. Modulation of phytoremediation and plant growth by the treatment with PGPR, Ag nanoparticle and untreated municipal wastewater. *Int. J. Phytoremediat.* **2016**, *18*, 1258–1269. [CrossRef]
18. Duca, D.; Lorv, J.; Patten, C.L.; Rose, D.; Glick, B.R. Indole-3-acetic acid in plant–microbe interactions. *Antonie Van Leeuwenhoek* **2014**, *106*, 85–125. [CrossRef]
19. Davies, P.J. The plant hormones: Their nature, occurrence and functions. In *Plant Hormones*; Davies, P.J., Ed.; Springer: Dordrecht, Germany, 2010; pp. 1–15.
20. Ahemad, M.; Kibret, M. Mechanisms and applications of plant growth promoting rhizobacteria: Current perspective. *J. King Saud Univ. Sci.* **2014**, *26*, 1–20. [CrossRef]

21. Chandler, J.W. The hormonal regulation of flower development. *J. Plant. Growth Regul.* **2011**, *30*, 242–254. [CrossRef]

22. Phillips, K.A.; Skirpan, A.L.; Liu, X.; Christensen, A.; Slewinski, T.L.; Hudson, C.; Barazesh, S.; Cohen, J.D.; Malcomber, S.; McSteen, P. Vanishing tassel2 encodes a grass-specifc tryptophan aminotransferase required for vegetative and reproductive development in maize. *Plant Cell* **2011**, *23*, 550–566. [CrossRef]

23. Bellini, C.; Pacurar, D.; Perrone, L. Adventitious roots and lateral roots: Similarities and differences. *Annu. Rev. Plant Biol.* **2014**, *65*, 639–666. [CrossRef] [PubMed]

24. Uggla, C.; Moritz, T.; Sandberg, G.; Sundberg, B. Auxin as a positional signal in pattern formation in plants. *Proc. Natl. Acad. Sci. USA* **1996**, *93*, 9282–9286. [CrossRef]

25. Chen, T.; Chen, Z.; Ma, G.H.; Du, B.H.; Shen, B.; Ding, Y.Q.; Xu, K. Diversity and potential application of endophytic bacteria in ginger. *Genet. Mol. Res.* **2014**, *13*, 4918–4931. [CrossRef] [PubMed]

26. Glick, B.R. The enhancement of plant growth by free-living bacteria. *Can. J. Microbiol.* **1995**, *41*, 109–117. [CrossRef]

27. Saharan, B.S.; Nehra, V. Plant growth promoting rhizobacteria: A critical review. *Life Sci. Med. Res.* **2011**, *21*, 30.

28. Billah, M.; Bano, A. Role of plant growth promoting rhizobacteria in modulating the efficiency of poultry litter composting with rock phosphate and its effect on growth and yield of wheat. *Waste Manag. Res.* **2015**, *33*, 63–72. [CrossRef]

29. Whitelaw, M.A. Growth promotion of plants inoculated with phosphate-solubilizing fungi. In *Advances in Agronomy*; Academic Press: Cambridge, MA, USA, 1999; Volume 69, pp. 99–151.

30. Kettner, J.; Dörffling, K. Biosynthesis and metabolism of abscisic acid in tomato leaves infected with Botrytis cinerea. *Planta* **1995**, *196*, 627–634. [CrossRef]

31. Olsen, S.; Sommers, L.E. Phosphorus. In *Methods of Soil Analysis: Part 2 Chemical and Microbiological Properties*; Page, A.L., Ed.; American Society of Agronomy: Madison, WI, USA; Soil Science Society of America: Madison, WI, USA, 1982; pp. 403–430.

32. Sarwar, M.; Arshad, M.; Martens, D.A.; Frankenberger, W.T. Tryptophan-dependent biosynthesis of auxins in soil. *Plant Soil* **1992**, *147*, 207–215. [CrossRef]

33. Blake, G.R.; Hartge, K.H. Bulk density. In *Methods of Soil Analysis: Part 1 Physical and Mineralogical Methods*; American Society of Agronomy: Madison, WI, USA; Soil Science Society of America: Madison, WI, USA, 1986; Volume 5.1, pp. 363–375.

34. Soltanpour, P.N.; Schwab, A.P. A new soil test for simultaneous extraction of macro-and micro-nutrients in alkaline soils. *Commun. Soil Sci. Plant Anal.* **1977**, *8*, 195–207. [CrossRef]

35. McLean, E.O. Soil pH and lime requirement. In *Methods of Soil Analysis: Part 2 Chemical and Microbiological Properties*; American Society of Agronomy: Madison, WI, USA; Soil Science Society of America: Madison, WI, USA, 1983; Volume 9.2.2, pp. 199–224.

36. Brookes, P.C.; Powlson, D.S.; Jenkinson, D.S. Measurement of microbial biomass phosphorus in soil. *Soil Biol. Biochem.* **1981**, *14*, 319–329. [CrossRef]

37. Tabatabai, M.A.; Bremner, J.M. Use of p-nitrophenyl phosphate for assay of soil phosphatase activity. *Soil Biol. Biochem.* **1969**, *1*, 301–307. [CrossRef]

38. Steel, R.G.D.; Torrie, J.H. *Principles and Procedures of Statistics*; HcGraw-Hill: New York, NY, USA, 1960.

39. Khan, N.; Bano, A. Effects of exogenously applied salicylic acid and putrescine alone and in combination with rhizobacteria on the phytoremediation of heavy metals and chickpea growth in sandy soil. *Int. J. Phytoremediat.* **2018**, *20*, 405–414. [CrossRef] [PubMed]

40. Akhtar, M.J.; Asghar, H.N.; Shahzad, K.; Arshad, M. Role of plant growth promoting rhizobacteria applied in combination with compost and mineral fertilizers to improve growth and yield of wheat (Triticum aestivum L.). *Pak. J. Bot.* **2009**, *41*, 381–390.

41. Grant, C.A.; Flaten, D.N.; Tomasiewicz, D.J.; Sheppard, S.C. The importance of early season phosphorus nutrition. *Can. J. Plant Sci.* **2001**, *81*, 211–224. [CrossRef]

42. Illmer, P.; Schinner, F. Solubilization of inorganic calcium phosphates—Solubilization mechanisms. *Soil Biol. Biochem.* **1995**, *27*, 257–263. [CrossRef]

43. Afzal, A.; Bano, A. Rhizobium and phosphate solubilizing bacteria improve the yield and phosphorus uptake in wheat (Triticum aestivum). *Int. J. Agric. Biol.* **2008**, *10*, 85–88.

44. Gangwar, K.S.; Singh, K.K.; Sharma, S.K.; Tomar, O.K. Alternative tillage and crop residue management in wheat after rice in sandy loam soils of Indo-Gangetic plains. *Soil Tillage Res.* **2006**, *88*, 242–252. [CrossRef]

45. Amujoyegbe, B.J.; Opabode, J.T.; Olayinka, A. Effect of organic and inorganic fertilizer on yield and chlorophyll content of maize (*Zea mays* L.) and sorghum *Sorghum bicolour* (L.) Moench. *Afr. J. Biotechnol.* **2007**, *6*. [CrossRef]

46. Hossain, M.B.; Sattar, M.A.; Islam, M.Z. Characterization of P and characterization of phosphate solubilizing bacteria on wheat. *J. Agric. Res.* **2004**, *42*, 295–303.

47. Iftikhar, R.; Khaliq, I.; Ijaz, M.; Rashid, M.A.R. Association analysis of grain yield and its components in spring wheat (Triticum aestivum L.). *Am. Eurasian J. Agric. Environ. Sci.* **2015**, *12*, 389–392.

48. Khan, M.S.; Zaidi, A. Synergistic effects of the inoculation with plant growth-promoting rhizobacteria and an arbuscular mycorrhizal fungus on the performance of wheat. *Turk. J. Agric. For.* **2007**, *31*, 355–362.

49. Sharif, M.; Khan, M.; Khan, M.A.; Wahid, F.; Marwat, K.B.; Khattak, A.M.; Naseer, M. Effect of rock phosphate and farmyard manure applied with effective microorganisms on the yield and nutrient uptake of wheat and sunflower crops. *Pak. J. Bot.* **2015**, *47*, 219–226.

50. Dell, B.; Huang, L. Physiological response of plants to low boron. *Plant Soil* **1997**, *193*, 103–120. [CrossRef]

51. Zafar, M.; Abbasi, M.K.; Arjumend, T.; Jabran, K. Impact of compost, inorganic phosphorus fertilizers and their combination on maize growth, yield, nutrient uptake and soil properties. *J. Anim. Plant Sci.* **2012**, *22*, 1036–1041.

52. Naseem, H.; Bano, A. Role of plant growth-promoting rhizobacteria and their exopolysaccharide in drought tolerance of maize. *J. Plant Interact.* **2014**, *9*, 689–701. [CrossRef]

53. Jala-Abadi, A.L.; Siadat, S.A.; Bakhsandeh, A.M.; Fathi, G.; Saied, K.A. Effect of organic and inorganic fertilizers on yield and yield components in wheat (T. aestivum and T. durum) genotypes. *Adv. Environ. Biol.* **2012**, *6*, 756–763.

54. Khan, N.; Zandi, P.; Ali, S.; Mehmood, A.; Adnan Shahid, M.; Yang, J. Impact of salicylic acid and PGPR on the drought tolerance and phytoremediation potential of Helianthus annus. *Front. Microbiol.* **2018**, *9*, 2507. [CrossRef]

55. Khan, N.; Bano, A. Growth and Yield of Field Crops Grown Under Drought Stress Condition Is Influenced by the Application of PGPR. In *Field Crops: Sustainable Management by PGPR*; Springer: Cham, Switzerland, 2019; pp. 337–349.

56. Osborne, D.J.; Jackson, M.B. *Cell Separation in Plants: Physiology, Biochemistry and Molecular Biology*; Springer Science & Business Media: Berlin/Heidelberg, Germany, 2013; Volume 35.

57. Wachowska, U.; Okorski, A.; Głowacka, K. Population structure of microorganisms colonizing the soil environment of winter wheat. *Plant Soil Environ.* **2006**, *52*, 39–44.

58. Gutiérrez-Mañero, F.J.; Ramos-Solano, B.; Probanza, A.N.; Mehouachi, J.R.; Tadeo, F.; Talon, M. The plant-growth-promoting rhizobacteria Bacillus pumilus and Bacillus licheniformis produce high amounts of physiologically active gibberellins. *Physiol. Plant.* **2012**, *111*, 206–211. [CrossRef]

59. Khan, N.; Bano, A.; Curá, J.A. Role of Beneficial Microorganisms and Salicylic Acid in Improving Rainfed Agriculture and Future Food Safety. *Microorganisms* **2020**, *8*, 1018. [CrossRef]

60. Sivasankari, B.; Kavikumar, N.; Anandharaj, M. Indole-3-acetic acid production and enhanced plant growth promotion by indigenous bacterial species. *J. Curr. Res. Sci.* **2013**, *1*, 331–335.

61. Walley, F.L.; Germida, J.J. Response of spring wheat [Triticum aestivum] to interactions between Pseudomonas species and Glomus clarum NT4. *Biol. Fertil. Soils* **1997**, *24*, 365–371. [CrossRef]

62. Ghosh, P.K.; Bandyopadhyay, K.K.; Manna, M.C.; Mandal, K.G.; Misra, A.K.; Hati, K.M. Comparative effectiveness of cattle manure, poultry manure, phosphocompost and fertilizer-NPK on three cropping systems in vertisols of semi-arid tropics. II. Dry matter yield, nodulation, chlorophyll content and enzyme activity. *Bioresour. Technol.* **2004**, *95*, 85–93. [CrossRef] [PubMed]

63. Brady, N.C.; Weil, R.R.; Weil, R.R. *The Nature and Properties of Soils*; Prentice Hall: Upper Saddle River, NJ, USA, 2008; Volume 13, pp. 662–710.

64. Wu, C.; Wei, X.; Sun, H.L.; Wang, Z.Q. Phosphate availability alters lateral root anatomy and root architecture of Fraxinus mandshurica Rupr. seedlings. *J. Integr. Plant Biol.* **2005**, *47*, 292–301. [CrossRef]

65. Wyngaard, N.; Cabrera, M.L.; Jarosch, K.A.; Bünemann, E.K. Phosphorus in the coarse soil fraction is related to soil organic phosphorus mineralization measured by isotopic dilution. *Soil Biol. Biochem.* **2007**, *96*, 107–118. [CrossRef]

66. Sharma, S.N.; Prasad, R.; Davari, M.; Ram, M.; Dwivedi, M.K. Effect of phosphorus management on production and phosphorus balance in a rice (*Oriza sativa* L.)-wheat (*Triticum aestivum*) cropping system. *Arch. Agron. Soil Sci.* **2011**, *57*, 655–667. [CrossRef]

67. Ghaderi, A.; Aliasgharzad, N.; Oustan, S.; Olsson, P.A. Efficiency of three Pseudomonas isolates in releasing phosphate from an artificial variable-charge mineral [iron III hydroxide]. *Soil Environ.* **2008**, *27*, 71–76.

68. Halvin, J.L.; Beaton, J.D.; Tisdale, S.L.; Nelson, W.L. *Soil Fertility and Fertilizers: An Introduction to Nutrient Management*; Prentice Hall: Upper Saddle River, NJ, USA, 2005.

69. Muhammad, A.; Khan, M.A.; Awan, I.U.; Khan, E.A.; Khan, A.A.; Ghulam, J. Effect of single and combined use of various organic amendments on wheat grown over green manured soil: I. Growth and yield attributes. *Pak. J. Nutr.* **2011**, *10*, 640–646.

70. Son, T.T.N.; Diep, C.N.; Giang, T.T.M. Effect of Bradyrhizobia and phosphate solubilizing bacteria application on soybean in rotational system in the Mekong delta. *Omonrice* **2006**, *14*, 48–57.

71. Belay, A.; Claassens, A.; Wehner, F.C. Effect of direct nitrogen and potassium and residual phosphorus fertilizers on soil chemical properties, microbial components and maize yield under long-term crop rotation. *Biol. Fertil. Soils* **2002**, *35*, 420–427.

72. Trolove, S.N.; Hedley, M.J.; Kirk, G.J.D.; Bolan, N.S.; Loganathan, P. Progress in selected areas of rhizosphere research on P acquisition. *Soil Res.* **2003**, *41*, 471–499. [CrossRef]

73. Biswas, D.R.; Narayanasamy, G. Research note: Mobilization of phosphorus from rock phosphate through composting using crop residue. *Fertil. News* **2002**, *47*, 53–56.

74. Khan, N.; Bano, A.; Zandi, P. Effects of exogenously applied plant growth regulators in combination with PGPR on the physiology and root growth of chickpea (Cicer arietinum) and their role in drought tolerance. *J. Plant Interact.* **2018**, *1*, 239–247. [CrossRef]

75. Shiralipour, A.; McConnell, D.B.; Smith, W.H. Physical and chemical properties of soils as affected by municipal solid waste compost application. *Biomass Bioenergy* **1992**, *3*, 261–266. [CrossRef]

76. Sommers, L.E.; Giordano, P.M. Use of nitrogen from agricultural, industrial, and municipal wastes. In *Nitrogen in Crop Production*; American Society of Agronomy, Inc.: Madison, WI, USA; Crop Science Society of America, Inc.: Madison, WI, USA; Soil Science Society of America, Inc.: Madison, WI, USA, 1984; pp. 207–220.

77. Baziramakenga, R.; Lalande, R.R.; Lalande, R. Effect of de-inking paper sludge compost application on soil chemical and biological properties. *Can. J. Soil Sci.* **2001**, *81*, 561–575. [CrossRef]

78. Shen, Y.; Lin, H.; Gao, W.; Li, M. The effects of humic acid urea and polyaspartic acid urea on reducing nitrogen loss compared with urea. *J. Sci. Food Agric.* **2020**. [CrossRef]

79. Iqbal, M.K.; Shafiq, T.; Hussain, A.; Ahmed, K. Effect of enrichment on chemical properties of MSW compost. *Bioresour. Technol.* **2010**, *10*, 5969–5977. [CrossRef]

80. Adeli, A.; Tewolde, H.; Sistani, K.; Rowe, D. Comparison of broiler litter and commercial fertilizer at equivalent N rates on soil properties. *Commun. Soil Sci. Plant Anal.* **2010**, *41*, 2432–2447. [CrossRef]

81. Canbolat, M.Y.; Barik, K.; Çakmakçi, R.; Sahin, F. Effects of mineral and biofertilizers on barley growth on compacted soil. *Acta Agric. Scand. Sec. B Soil Plant Sci.* **2006**, *56*, 324–332. [CrossRef]

82. Stratton, M.L.; Rechcigl, J.E. Agronomic benefits of agricultural, municipal, and industrial by-products and their co-utilization: An overview. In *Beneficial Co-Utilization of Agricultural, Municipal and Industrial By-Products*; Springer: Dordrecht, Germany, 1998; pp. 9–34.

83. Khan, N.; Bano, A. Rhizobacteria and Abiotic Stress Management. In *Plant Growth Promoting Rhizobacteria for Sustainable Stress Management*; Springer: Singapore, 2019; pp. 65–80.

84. Medina, A.; Vassileva, M.; Barea, J.M.; Azcón, R. The growth-enhancement of clover by Aspergillus-treated sugar beet waste and Glomus mosseae inoculation in Zn contaminated soil. *Appl. Soil. Ecol.* **2006**, *33*, 87–98. [CrossRef]

85. Allison, S.D.; Vitousek, P.M. Responses of extracellular enzymes to simple and complex nutrient inputs. *Soil Biol. Biochem.* **2005**, *37*, 937–944. [CrossRef]

86. Olander, L.P.; Vitousek, P.M. Regulation of soil phosphatase and chitinase activityby N and P availability. *Biogeochemistry* **2000**, *49*, 175–191. [CrossRef]

87. Moscatelli, M.C.; Lagomarsino, A.; De Angelis, P.; Grego, S. Seasonality of soil biological properties in a poplar plantation growing under elevated atmospheric CO_2. *Appl. Soil Ecol.* **2005**, *30*, 162–173. [CrossRef]

88. Criquet, S.; Braud, A.; Nèble, S. Short-term effects of sewage sludge application on phosphatase activities and available P fractions in Mediterranean soils. *Soil Biol. Biochem.* **2007**, *39*, 921–929. [CrossRef]

89. Khan, N.; Bano, A. Role of plant growth promoting rhizobacteria and Ag-nano particle in the bioremediation of heavy metals and maize growth under municipal wastewater irrigation. *Int. J. Phytoremediat.* **2016**, *18*, 211–221. [CrossRef]

90. Pascual, J.A.; Moreno, J.L.; Hernández, T.; García, C. Persistence of immobilised and total urease and phosphatase activities in a soil amended with organic wastes. *Bioresour. Technol.* **2002**, *82*, 73–78. [CrossRef]

91. Wang, F.E.; Chen, Y.X.; Tian, G.M.; Kumar, S.; He, Y.F.; Fu, Q.L.; Lin, Q. Microbial biomass carbon, nitrogen and phosphorus in the soil profiles of different vegetation covers established for soil rehabilitation in a red soil region of southeastern China. *Nutr. Cycl. Agroecosyst.* **2004**, *68*, 181–189. [CrossRef]

92. Kiani, M.; Hernandez-Ramirez, G.; Quideau, S.; Smith, E.; Janzen, H.; Larney, F.J.; Puurveen, D. Quantifying sensitive soil quality indicators across contrasting long-term land management systems: Crop rotations and nutrient regimes. *Agric. Ecosyst. Environ.* **2017**, *248*, 123–135. [CrossRef]

93. Mohammadi, K. Soil microbial activity and biomass as influenced by tillage and fertilization in wheat production. *Am. Eurasian J. Agric. Environ. Sci.* **2011**, *10*, 330–337.

94. Meena, M.D.; Biswas, D.R. Phosphorus and potassium transformations in soil amended with enriched compost and chemical fertilizers in a wheat–soybean cropping system. *Commun. Soil Sci. Plant. Anal.* **2014**, *45*, 624–652. [CrossRef]

95. Ayed, L.B.; Hassen, A.; Jedidi, N.; Saidi, N.; Bouzaiane, O.; Murano, F. Microbial C and N dynamics during composting process of urban solid waste. *Waste Manag. Res.* **2007**, *25*, 24–29. [CrossRef] [PubMed]

96. Leytem, A.B.; Turner, B.L.; Raboy, V.; Peterson, K.L. Linking manure properties to phosphorus solubility in calcareous soils: Importance of the manure carbon to phosphorus ratio. *Soil Sci. Soc. Am. J.* **2005**, *69*, 1516–1524. [CrossRef]

97. Khan, N.; Bano, A. Role of PGPR in the Phytoremediation of Heavy Metals and Crop Growth under Municipal Wastewater Irrigation. In *Phytoremediation*; Springer: Cham, Switzerland, 2018; pp. 135–149.

98. Ayaga, G.; Todd, A.; Brookes, P.C. Enhanced biological cycling of phosphorus increases its availability to crops in low-input sub-Saharan farming systems. *Soil. Biol. Biochem.* **2006**, *38*, 81–90. [CrossRef]

Article

Co-Inoculation of Rhizobacteria and Biochar Application Improves Growth and Nutrientsin Soybean and Enriches Soil Nutrients and Enzymes

Dilfuza Jabborova [1,2,3,]*, Stephan Wirth [3], Annapurna Kannepalli [2],
Abdujalil Narimanov [1], Said Desouky [4], Kakhramon Davranov [5], Riyaz Z. Sayyed [6],
Hesham El Enshasy [7,8,9], Roslinda Abd Malek [7], Asad Syed [10] and Ali H. Bahkali [10]

1   Laboratory of Medicinal Plants Genetics and Biotechnology, Institute of Genetics and Plant Experimental Biology, Uzbekistan Academy of Sciences, Tashkent Region, Kibray 111208, Uzbekistan; narimanov63@list.ru
2   Division of Microbiology, ICAR-Indian Agricultural Research Institute, Pusa, New Delhi 110012, India; annapurna96@gmail.com
3   Leibniz Centre for Agricultural Landscape Research (ZALF), D-15374 Müncheberg, Germany; swirth@zalf.de
4   Botany and Microbiology Department, Faculty of Science, Al-Azhar University, Cairo 11651, Egypt; dr_saidesouky@yahoo.com
5   Institute of Microbiology, Academy of Sciences of Uzbekistan, Tashkent 100128, Uzbekistan; k-davranov@mail.ru
6   Department of Microbiology, PSGVP Mandal's, Arts, Science & Commerce College, Shahada 425409, Maharashtra, India; sayyedrz@gmail.com
7   Institute of Bioproduct Development (IBD), Universiti Teknologi, Malaysia (UTM), Skudai, Johor Bahru 81310, Malaysia; henshasy@ibd.utm.my (H.E.E.); roslinda@ibd.utm.my (R.A.M.)
8   School of Chemical and Energy Engineering, Faculty of Engineering, Universiti Teknologi, Malaysia (UTM), Skudai, Johor Bahru 81310, Malaysia
9   City of Scientific Research and Technology Application, New Burg Al Arab, Alexandria 21934, Egypt
10 Department of Botany and Microbiology, College of Science, King Saud University, P.O. Box 2455, Riyadh 11451, Saudi Arabia; asadsayyed@gmail.com (A.S.); abahkali@ksu.edu.sa (A.H.B.)
*   Correspondence: dilfuzajabborova@yahoo.com

Received: 11 July 2020; Accepted: 27 July 2020; Published: 6 August 2020

Abstract: Gradual depletion in soil nutrients has affected soil fertility, soil nutrients, and the activities of soil enzymes. The applications of multifarious rhizobacteria can help to overcome these issues, however, the effect of co-inoculation of plant-growth promoting rhizobacteria (PGPR) and biochar on growth andnutrient levelsin soybean and on the level of soil nutrients and enzymes needs in-depth study. The present study aimed to evaluate the effect of co-inoculation of multifarious *Bradyrhizobium japonicum* USDA 110 and *Pseudomonas putida* TSAU1 and different levels (1 and 3%) of biochar on growth parameters and nutrient levelsin soybean and on the level of soil nutrients and enzymes. Effect of co-inoculation of rhizobacteria and biochar (1 and 3%) on the plant growth parameters and soil biochemicals were studied in pot assay experiments under greenhouse conditions. Both produced good amounts of indole-acetic acid; (22 and 16 µg mL^{-1}), siderophores (79 and 87%SU), and phosphate solubilization (0.89 and 1.02 99 g mL^{-1}). Co-inoculation of *B. japonicum* with *P. putida* and 3% biochar significantly improved the growth and nutrient content ofsoybean and the level of nutrients and enzymes in the soil, thus making the soil more fertile to support crop yield. The results of this research provide the basis of sustainable and chemical-free farming for improved yields and nutrients in soybean and improvement in soil biochemical properties.

Keywords: biochar; *Bradyrhizobium japonicum*; *Pseudomonas putida*; plant growth; plant nutrients; soil enzymes; soil nutrients; soybean

---

1. Introduction

The global climate scenario is experiencing a drastic depletion of soil nutrients due to various anthropogenic activities, burning of fossil fuel, and excess use of agrochemicals [1]. Applications of plant-growth promoting rhizobacteria (PGPR) and biochar have been advocated as an effective, cheap, and sustainable approach for the replenishment of crop health, crop nutrients, and soil nutrients and enzymes and for improving and sustaining soil fertility [2]. Furthermore, these amendments have a positive impact on the growth [3], development, and yield of several crops [4,5]. Various reports claimed that the application of plant growth-promoting rhizobacteria (PGPR) and biochar improves plant growth, plant nutrients, and physicochemical properties of soil [6–8]. Moreover, such applications of biochar also keep a check onatmospheric CO_2 levels [9] and, thus, contribute todecrease global warming effects [10], while the use of PGPR to increase soil fertility and plant nutrients will help to reduce the doses of agrochemicals in the field [11].

A wide variety of symbiotic bacteria, such as *Rhizobium* sp. and *B. japonicum*, etc., have been reported to promote seed germination, the growth of root and shoot, andthe level of nutrients in soybean and also improve soil biochemical properties [4,5].Rhizobia-legumes symbiosis plays a vital role in increasing crop yields, reducing the use of inorganic nitrogen fertilizers and improving soil fertility [12]. Rhizobial species are commonly used as inoculants in various parts of the world for improving the yield of legumes. Co-inoculation with multifarious *Bradyrhizobium* sp. and *Pseudomonas* sp. improves plant growth, plant, and soil nutrients and enzymes through the production of siderophores [13], phytohormones [14], enzymes [15], exopolysaccharide [16], stress tolerance [17], and phosphate solubilization [18–23], etc. Thus, several studies reported increases in nodules number, nodule weight, nitrogen fixed, plant growth, and yield of legumes due to co-inoculation with plant growth promoting *Bradyrhizobium* sp. and *Pseudomonas* sp. [12–14], while the combination of biochar with PGPR further increases root length, shoot length, nodule per plant, seed number, and yield of crops [5].

The activity of PGPR bioinoculants helps in improving the level of extracellular soil enzymes that facilitates the decomposition of soil organic matter and ensures the availability of nutrients in the soil [15]. Among the soil enzymes, proteases and acid and alkaline phosphomonoesterase are the major enzymes that mediate the hydrolysis of the protein and phosphate (P) into bioavailable amino acids, organic nitrogen, and soluble P [16]. However, the activities of these enzymes are governed by many factors, such as soil properties, soil organic matter level, and the presence of organic compounds [24]. We hypothesized that co-inoculation with *B. japonicum*+*P. putida* and biochar would facilitate the beneficial effects on soybean plant growth, plant nutrients, and soil nutrients and enzymes.

The present study was aimed at evaluating the effects of co-inoculation of multiple plant growth-promoting traits positive in *Bradyrhizobium japonicum* USDA 110 and *Pseudomonas putida* TSAU1 and different levels (1 and 3%) of biochar on seed germination, growth parameters, and nutrient levels in soybean and the level of nutrients and enzymes in soil. The outcome of this study may provide a better way of increasing soil fertility and increasing the growth and yield of soybean. This approach has multiple dimensions; as utilization of biochar is not only a cheaper option but will also help in solving the management issues of biochar, it is expected to minimize the doses of agrochemicals and produce chemical-free food. The consortium effect of PGPR and application of biochar provide excellent benefits to the farmers as theyincur less investment and yield more crop productivity, and this organically grown crop has more demand with a good selling price.

2. Materials and Methods

2.1. Bacterial Culture, Soybean, and Biochar

B. japonicum USDA 110 and *P. putida* TSAU1 strains were collected from the culture collection of the Department of Microbiology and Biotechnology, National University of Uzbekistan, Tashkent, Uzbekistan. Soybean (*Glycine max* L. Merr.) seeds were obtained from Leibniz Centre for Agricultural Landscape Research (ZALF), Müncheberg, Germany.

The maize biochar (MBC) was collected from the Leibniz-Institute for Agriculture Engineering and Bioeconomy (ATB), Potsdam, Germany. Pyrolysis of MBC was carried out at 600 °C for 30 min and the chemical compositions of MBC were analyzed according to the method of Reibe et al. [25].

2.2. Screening for the Production of PGP Metabolites

B. japonicum USDA 110 and *P. putida* TSAU1 strains were screened for phosphate (P) solubilization on Pikovoskaya's agar and in Pikovoskaya's broth [26] for the production of indole-3-acetic acid (IAA)according to the method of Brick et al. [27], for production and estimation of siderophore according to the method of Patel et al. [28] and Payne [29], and the production and estimation of aminocyclopropane-1-carboxylate deaminase (ACCD) activity according to the method of Penrose and Glick [30]. The ACCD activity was measured as the amount of α-keto-butyrate produced per mg protein per h.

2.3. Surface Sterilization, Germination, and Bacterization of Seeds

Soybean seeds were sorted to eliminate broken, small, infected seeds and sterilized with 10% sodium hypochlorite solution for 5 min and washed three times with sterile, distilled water. Seeds were germinated in 85 mm × 15 mm tight-fitting plastic Petri dishes with 5 mL of water. *B. japonicum* USDA 110 and *P. putida* TSAU 1 broth rich in PGP metabolites were used for the inoculation of germinated seeds. Germinated seeds were first placed with sterile forceps into bacterial suspension (5×10^6 CFU g^{-1}) for 10 min before planting, were air-dried, and then planted in plastic pots containing 400 g sandy loamy soil.

2.4. Experimental Design

The effect of rhizobacteria on the growth of soybean was studied in pot experiments in a greenhouse at ZALF, Müncheberg, Germany during July 2015. All the experiments were carried out in a randomized block design (RBD) with three replications. Experimental treatments included un-inoculated control (soil without biochar and soil with two levels of biochar (1 and 3%)), inoculation with *B. japonicum* USDA 110 (soil without biochar and soil with two levels of biochar (1 and 3%)), and co-inoculation with *B. japonicum* USDA 110 and *P. putida* TSAU 1 strains (soil without biochar and soil with two levels of biochar (1 and 3%)). The plants were grown in greenhouse conditions at 24 °C during the day and 16 °C at night for 30 days.

2.5. Measurement of Plant Growth Parameters and Plant Nutrients

Plants harvested after 30 days were subjected to the measurement of seed germination rate, root length, shoot length, root dry weight, shoot dry weight, and the number of nodules per plant of soybean. Plant nutrients, such as nitrogen (N), phosphorus (P), potassium (K), magnesium (Mg), sodium (Na), and calcium (Ca) were estimated from crushed plant tissue with an inductively coupled plasma optical emission spectrometer (ICP-OES; iCAP 6300 Duo, Thermo Fischer Scientific Inc., Waltham, MA, USA) via Mehlich-3 extraction [30]. The nitrogen and phosphorus contents of root and shoot were determined from dried powdered biomass. For nitrogen estimation, 1 g of plant biomass was digested with 10 mL concentrated H_2SO_4 and 5 g catalyst mixture in the digestion tube. The mixture was allowed to cool and then processed for distillation. The distillate was collected and titrated with H_2SO_4 blank (without leaf) Total nitrogen was calculated from the blank and sample titer reading [31]. For the estimation of P content, plant P was extracted with 0.5 N NaHCO$_3$ (pH8.5)and treated with ascorbic acid in an acidic medium [32]. The intensity of blue color produced was measured and the amount of P was calculated from the standard curve of P. For the estimation of potassium content of plant biomass, 25 mL of ammonium acetate solution was added in 5 g of the biomass sample, the content was shaken for 5 min and filtered, and the amount of K from the filtrate was measured [33]. For the estimation of Na, Mg, and Ca, 1 g of plant extract was mixed with 80 mL of 0.5 N HCl for 5 min at 25 °C followed by measurement of concentrations of these elements in the filtrate [34].

2.6. Analysis of Soil Nutrient and Soil Enzymes

The rootsoil (10 g) of experimental pots was air-dried soil, shaken with 100 mL ammonium acetate (0.5 M) for 30 min to effectively displace the available nutrients, and adhered to soil minerals. The soil organic carbon (SOC), nitrogen (N), phosphate (P), and potassium (K) content of soil were determined by the dry combustion method according to the method of Sims [35] and Nelson and Sommers [36] using a CNS analyzer (TruSpec, Leco Corp., St. Joseph, MI, USA). For this purpose, 10 mL of 1 N $K_2Cr_2O_7$ and 20 mL of concentrated $H_2SO_4$was added in 1g soil, mixed thoroughly and diluted with 200 mL of distilled water followed by the addition of 10 mL each of H_3PO_4 and sodium fluoride. The resulting solution was used for the elemental analysis. Blank (without soil) served as control. Soil Organic Carbon (SOC) of soil sample was calculated with the help of blank and sample titer reading.

The acid and alkaline phosphomonoesterase activities were assayed according to the method of Tabatabai and Bremner [37].Moist soil (0.5 g) was placed in a 15 mL vial, and 2 mL of modified universal buffer (MUB) (pH 6.5 for the acid phosphatase assay or pH 11 for the alkaline phosphatase assay) and 0.5 mL of p-nitrophenyl phosphate substrate solution (0.05 M) were added to the vial, sequentially. The assay and control batches were replicated 3 times. The concentration of p-nitrophenol (p-NP) produced in the assays of acid and alkaline phosphomonoesterase activities were calculated from a p-NP calibration curve after subtracting the absorbance of the control at 400 nm. Protease activity was assayed according to the method of Ladd and Butler [38]. For this, 0.5 g of soil was weighed into a glass vial, and 2.5 mL of phosphate buffer (0.2 M, pH of 7.0) and 0.5 mL of N-benzoyl-L-arginine amide (BAA) substrate solution (0.03 M) were added. The ammonium released was calculated by relating the measured absorbance at 690 nm.

2.7. Statistical Analyses

All the experiments were performed in three replicates and the average of triplicate was considered. Experimental data were analyzed with the StatView Software (SAS Institute, Cary, NC, USA, 1998) using ANOVA. The significance of the effect of treatment was determined by the magnitude of the *p*-value ($p < 0.05 < 0.001$).

3. Results

3.1. Analysis of Maize Biochar

Analysis of pyrolyzed maize biochar contained (g%) dry weight: 92.85, ash: 18.42, total C: 75.16, N: 1.65, P: 5.26, and K: 31.12 with a pH of 9.89 and electrical conductivity of 3.08.

3.2. Screening for the Production of PGP Metabolites

Both the cultures under study produced a wide variety of PGP traits. *B. japonicum* USDA 110 and *P. putida* TSAU1 produced 22 and 16 µg mL^{-1} of IAA, 79 and 87% siderophore, and 0.89 and 1.02 99 g mL^{-1} phosphate solubilization, respectively.

3.3. Measurement of Plant Growth Parameters and Plant Nutrients

The effect of rhizobacteria and biochar levels indicated a significant improvement in the seed germination rate and growth of the soybean plant treated with biochar and rhizobacteria over the control plant (without biochar treatment). The addition of different levels of biochar, inoculation of *B. japonicum* USDA 110, and *P. putida* strain TSAU 1 with biochar and without biocharshowed variable increases in the growth parameters. Addition of 3% biochar alone enhanced the seed germination by 15%, root length by 20% (Figure 1a), shoot length by 41% (Figure 1a), root dry weight by 22% (Figure 1b), and shoot dry weight by 13% (Figure 1b), as compared to the control plant (without biochar). Individual addition of *B. japonicum* USDA 110 and *P. putida* strains TSAU 1 with varying levels of biochar (1–3%) and without biochar also promoted the growth of the plant. However, a co-inoculation with *B. japonicum* USDA 110 and *P. putida* strains TSAU 1 with 3% biochar resulted in significant increasesin seed germination and plant growth attributes. Increases in seed germination by 20%, root length by 76% (Figure 1a), shoot length by 41% (Figure 1a), root dry weight by 56% (Figure 1b), shoot dry weight by 59% (Figure 1b), and number of nodules per plant by 57% (Figure 1c) were recorded over the control plant treated with 3% biochar alone.

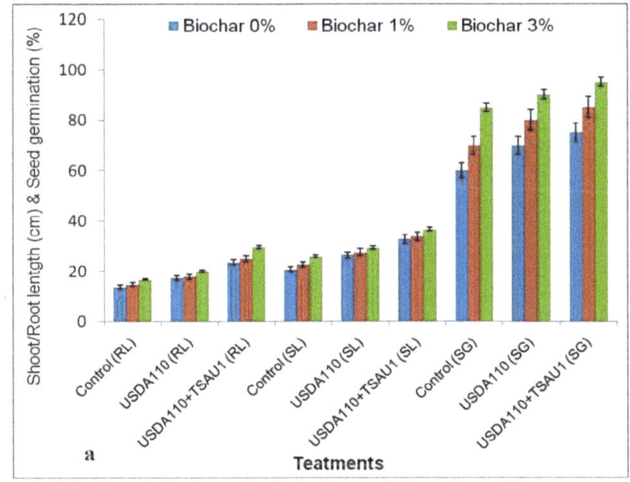

RL= Root length SL= Shoot length, SG= Seed germination *Significant at *P* (0.01)

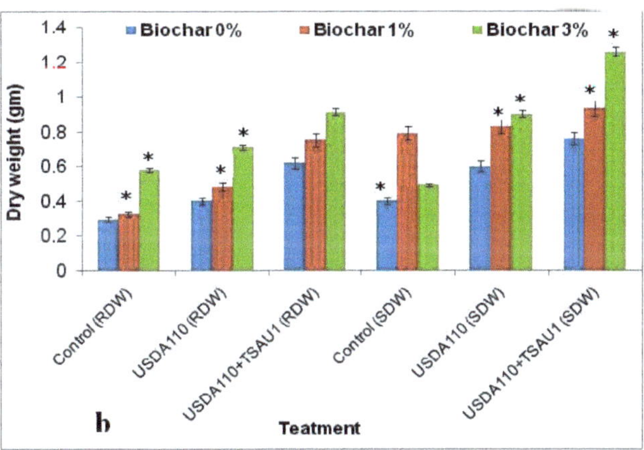

RDW = Root dry weight, SDW = Shoot dry weight, *Significant at *P* (0.01)

Figure 1. *Cont.*

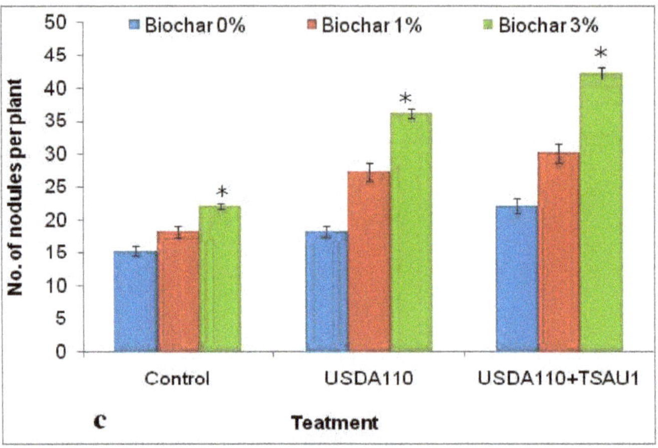

*Significant at *P* (0.01)

Figure 1. Effect of rhizobacteria and biochar concentrations on (**a**) root length [cm] and shoot length [cm], (**b**) dry weight of the root [g] and dry weight of the shoot [g], and (**c**) number of nodules. Plant growth parameters were measured after 30 days of growth of plant growth under greenhouse conditions.* = values significant at *p* 0.01.

Analysis of nutrients in a soybean plant (before sowing and after harvesting) revealedthat treatments with 1 and 3% biochar improved the content of total N, P, K, Mg, Na, and Ca in the plant. The inoculation of *B. japonicum* USDA 110 alone (0% biochar) increased N content by 36%, P content by 8.3%, K content by 5.6%, Mg content by 4.8%, Na content by 30%, and Ca content by 2.88%. However, the co-inoculation of *B. japonicum* USDA 110 and *P.putida* TSAU1 with 3% biochar showed a significant improvement in N content by 62.85%, P content by 7.42, K content by 76.85%, Mg content by 5.14%, Na content by 20%, and Ca content by 28%, as compared to the control (without biochar) (Table 1).

Table 1. Effect of rhizobacteria and biochar levels on plant nutrients.

Biochar Application	Treatments	N (%)	P (%)	K (%)	Mg (%)	Na (%)	Ca (%)
0%	Control	1.75 + 0.01	0.24 + 0.01	1.40 + 0.04	0.39 + 0.10	0.02 + 0.00	0.82 + 0.03
	TSAU1	2.00 + 0.02 *	0.25 + 0.04	1.41 + 0.02	0.43 + 0.02	0.06 + 0.01 *	0.51 + 0.03
	USDA 110	2.39 + 0.02 *	0.26 + 0.04	1.49 + 0.02	0.47 + 0.02	0.08 + 0.01 *	1.07 + 0.03
	USDA110+TSAU1	2.60 + 0.02 *	0.27 + 0.02	2.09 + 0.15 *	0.62 + 0.01 *	0.09 + 0.01 *	1.17 + 0.01 *
1%	Control	1.77 + 0.02	0.27 + 0.03	2.33 + 0.02	0.66 + 0.02	0.03 + 0.01	0.95 + 0.03
	USDA 110	2.51 + 0.02 *	0.28 + 0.02	2.52 + 0.04	0.52 + 0.02	0.07 + 0.01 *	1.25 + 0.03
	USDA+TSAU 1	2.64 + 0.02 *	0.32 − 0.02	2.33 + 0.03	0.68 + 0.02	0.13 + 0.04 *	1.21 + 0.02
3%	Control	1.91 + 0.02	0.28 + 0.01	2.41 + 0.02	0.64 + 0.02	0.03 + 0.03	1.09 + 0.02
	USDA 110	2.27 + 0.01	0.37 + 0.01	3.64 * + 0.01	0.48 + 0.01	0.02 + 0.01	0.99 + 0.01
	USDA+TSAU1	2.85 * + 0.01	0.35 + 0.01 *	3.72 * + 0.01	0.39 + 0.01	0.03 + 0.01	1.04 + 0.01

Values are the average of three replicates ± values are standard deviations. Plant nutrient contents were measured after 30 days of growth of plant under greenhouse conditions. * = values significant at p 0.01.

3.4. Estimation of Soil Nutrient Content and Soil Enzymes

Analysis of soil nutrient content revealed that the inoculation of soybean with *B. japonicum* USDA 110 alone (3% biochar) increased N content by 73%, P content by 173%, and K content by 17%, as compared to the control of 3% biochar. *B. japonicum* USDA 110 alone (3% biochar) significantly enhanced the N content by 98% and K content by 117%, as compared to the control without biochar (Table 2).

Table 2. Effect of rhizobacteria and biochar levels on soil nutrients.

Biochar Application	Treatments	SOC (%)	Total N (%)	P (mg)	K (mg)
0%	Control	21.09 ± 0.01	0.080 ± 0.01	4.29 ± 0.03	2.95 ± 0.02
	TSAU1	23.06 ± 0.01	0.082 ± 0.01	4.43 ± 0.03	3.05 ± 0.02
	USDA 110	27.08 ± 0.01	0.083 ± 0.01	4.60 ± 0.02 *	3.27 ± 0.03 *
	USDA+TSAU1	29.04 ± 0.02 *	0.094 ± 0.8 *	4.88 ± 0.02 *	5.58 ± 0.03 *
1%	Control	25.09 ± 0.01	0.091 ± 0.01	4.22 ± 0.03	4.83 ± 0.02
	USDA 110	29.06 ± 0.01	0.101 ± 0.02 *	6.14 ± 0.01 *	5.44 ± 0.01 *
	USDA+TSAU1	32.07 ± 0.8 *	0.164 ± 0.03 *	16.67 ± 0.05 *	5.68 ± 0.02 *
3%	Control	25.09 ± 0.01	0.094 ± 0.01	6.02 ± 0.01	5.35 ± 0.03
	USDA 110	33.05 ± 0.01	0.163 ± 0.01 *	16.47 ± 0.01 *	6.30 ± 0.01 *
	USDA+TSAU1	41.08 ± 0.01 *	0.170 ± 0.01 *	18.33 ± 0.01 *	8.49 ± 0.01 *

Values are the average of three replicates. ± values are standard deviations. * = values significant at *p* 0.01. Soil nutrient contents were measured after 30 days of growth of plant under greenhouse conditions.

The lowest level of these elements was evident in the soil without biochar treatment. The highest values of SOC, N, P, and K were observed in soil amended with 3% biochar and co-inoculation with *B. japonicum* USDA 110 and *P. putida* TSAU1 vis-à-vis the lowest value found in soil with *B. japonicum* USDA 110 and *P. putida* TSAU1 alone or in combination but without biochar and soil with no bioinoculants and no biochar treatments (Table 2).

Co-inoculation of soybean with of *B. japonicum* USDA 110 and *P. Putida* TSAU 1 strains enhanced nutrient contents of soil compared to all other treatments. The combination with *B. japonicum* USDA 110 and *P. putida* TSAU 1 (3% biochar) significantly increased N content by 80%, P content by 204%, and K content by 58% compared to the control of 3% biochar. When co-inoculated with *B. japonicum* USDA 110 and *P. putida* TSAU 1 (3% biochar)the N content rose by 11% and K content by 35% compared to variants inoculated with *B. japonicum* USDA 110 alone.

The addition of biochar to soil increased the activity of soil protease and acid and alkaline phosphomonoesterase. Substantial increases of 25.05%, 21.02%, and 23.02% in the activities of protease and acid and alkaline phosphomonoesterase, respectively, were evident due to the co-inoculation of *B. japonicum* USDA 110 and *P. Putida* TSAU1 (0% biochar). A combination of this treatment with 1% biochar further improved the activities of these enzymes. However, the activities of these enzymes were significantly improved due to the co-inoculation of *B. japonicum* USDA 110 and *P. Putida* TSAU 1 with 3% biochar. 2-fold, 1.52-fold, and 1.25-fold increases in the activities of protease and acid and alkaline phosphomonoesterase, respectively, were evident due to co-inoculation with two bioinoculants and 3% biochar (Table 3).

Table 3. Effect of rhizobacteria and biochar levels on soil enzymes.

Biochar Application	Treatments	Protease Activity (μg NH$_4^+$-N g^{-1}h^{-1})	Acid Phosphomonoesterase Activity (μg pNPg^{-1}h^{-1})	Alkaline Phosphomonoesterase Activity (μg pNPg^{-1}r^{-1})
0%	Control	19.2 ± 0.05	650.3 ± 30.1	300.1 ± 16.3
	TSUA1	20.1 ± 0.05	697.1 ± 20.1	317.1 ± 12.3
	USDA 110	23.5 ± 0.10	703.3 ± 34.5	365.6 ± 18.1
	USDA+TSAU 1	25.8 ± 0.19 *	780.6 ± 38.8 *	380.2 ± 20.4 *
1%	Control	21.4 ± 0.07	766.3 ± 35.7	370.5 ± 19.5
	USDA 110	25.8 ± 0.20 *	820.9 ± 45.3 *	425.3 ± 21.6 *
	USDA+TSAU 1	27.7 ± 0.18 *	940.6 ± 43.2 *	482.2 ± 20.8 *
3%	Control	24.3 ± 0.09	810.3 ± 37.6	420.6 ± 19.5
	USDA 110	28.5 ± 0.11 *	911.8 ± 46.3 *	483.5 ± 21.2 *
	USDA+TSAU 1	30.8 ± 0.15 *	1020.4 ± 48.6 *	535.7 ± 25.2 *

Values are the average of three replicates. ± values are standard deviations. * = values significant at p 0.01. Soil enzyme levels were measured after 30 days of inoculation of PGPR and biochar.

4. Discussion

4.1. Screening for the Production of PGP Metabolites

PGPR is known to produce a wide variety of plant-beneficial metabolites that help in plant nutrition and the overall vigor of the plant [39–42]. Production of IAA, siderophore, and P solubilization have been reported in various species of *Bradyrhizobium*, including *B. japonicum* [42–45] and *P. putida* [46,47]. Sayyed et al. [48] reported the production of siderophores from *P. fluorescence* NCIM5096 isolated from the groundnut field rhizosphere. Shaikh et al. [49] reported the production of siderophore from *P. aeruginosa* isolated from the banana field rhizosphere. Pandya et al. [50] reported the production of siderophore and phytohormones, such as IAA and gibberellins in *Pseudomonas* sp. *Rhizobium* sp., and *Azotobacter* sp. isolated from the sugarcane field rhizosphere. Theyobserved higher yields of phytohormones in *Pseudomonas* sp., as compared to the other isolates. Wani et al. [40] reported the production of siderophore in soil bacterium *P. aeruginosa* RZS9. They claimed a further increase in siderophore yield following the optimization of the process by a statistical approach. Jabborova et al. [14] reported the production of siderophore, IAA, and enzymes, such as protease, cellulose, lipase, P solubilization, and antifungal activity in nine endophytic PGPR strains. Sayyed et al. [51] reported the production of copious amounts of siderophore in *P. fluorescence* NCIM 5096 and *P. putida* NCIM2847.

4.2. Measurement of Plant Growth Parameters and Plant Nutrients

An increase in seed germination is due to the phytohormone production, while plant growth promotion during the symbiotic association is due to the nitrogen and other nutrients supplied by the bacterial symbiont. Sayyed et al. [48] reported plant growth-promoting effects of siderophore producing *P. fluorescence* NCIM5096 in wheat and groundnut. Wani et al. [40] reported the plant growth-promoting effects and antifungal-activities production of siderophore producing *P. aeruginosa*. Pandya et al. [50] reported that the inoculation of siderophore and phytohormone producing *Pseudomonas* sp., *Rhizobium* sp., and *Azotobacter* sp. promoted growth in wheat. Jabborova et al. [14] found that inoculation of siderophore, IAA, and enzymes producing P-solubilizing endophytic PGPR strains promoted the growth of medicinal plants. Sayyed et al. [13] observed growth promotion in wheat due to the inoculation of siderophore-producing *P. fluorescence* NCIM 5096 and *P.putida* NCIM2847.

Masciarelli et al. [45] reported a significant increase in the number of root nodules in soybean due to inoculation with *B. japonicum*. Egamberdieva et al. [23] reported the synergistic effect of co-inoculation of *B. japonicum* and *P. putida* to be more effective in increasing nodulation in soybean.Several researchers reported that biochar increased plant growth, nodule number, and yield in different crops [3,5,47]. Pandit et al. [7] claimed that the application of 3% biochar promoted the growth of maize. Uzoma et al. [52] recorded a significant increase in the productivity of biocharrized maize, as compared to a control under sandy soil conditions. Increased growth, more nodulation, and improved yield of soybean after the application of biochar were also reported by Iijima et al. [53].

The addition of organically rich biochar and inoculation with PGPR plays a vital role in increasing the soil microbial activity that provides more nutrition to the plant [54]. Egamberdieva et al. [55] reported significant ($p < 0.05$) increases in N, P, K, and Mg contents in chickpea plants treated with *Mesorhizobium ciceri* and biochar. It has been reported that the biochar amendment improves the water-holding capacity of soil [56], which increases the availability of minerals and nutrients [55]. Shen et al. [57] reported the positive effect of biochar amendment on the plant uptake of plant nutrients. Prendergast et al. [58] claimed that the addition of biochar can induce changes in nutrient availability and may provide additional N, P, K, Mg, Na, Ca. Shen et al. [57] observed an increase in P uptake in plants due to the application of biochar. Egamberdieva et al. [55] observed a significant increase in K content in chickpea roots and shoots treated with *M. ciceri* and biochar. Wang et al. [59] observed similar results and claimed an increasing level of K and Mg uptake in soybean due to the addition of bamboo biochar. Ma et al. [60] reported a positive effect of co-inoculation of *B. japonicum* and biochar on N and other nutrient contents in soybean root and shoot biomass. An increase in N content may be due to the positive impact of biochar on the nodule number that contributes more N to the shoot and root biomass.

4.3. Estimation of Soil Nutrients and Soil Enzymes

Since biochar is an organically rich amendment, its addition is expected to increase the level of soil nutrients. Egamberdievaet al. [55] reported a two-fold rise in SOC, N, P, K, and Mg concentrations in soil amended with biochar, and a three-fold increase in these nutrients in the soil treated with biochar and inoculation with *M. ciceri*. Similar results were reported by Wang et al. [61]. An increase in the soil's organic carbon and other nutrients can also be correlated with increased mineralization due to increased enzyme activity. A linear relationship between soil nutrients and the activities of soil enzymes involved in mineralization has been proposed by Ouyang et al. [62]. Fall et al. [63] reported significant ($p < 0.05$) increases in SOC, available N, soluble P, and total nitrogen upon the application of biochar at a higher rate (12 t ha^{-1}). They also recorded an increase in rice rhizospheric carboxylate secretions. Głodowska et al. [6] suggested a combination of biochar and *B. japonicum* strain 532 C, which significantly increased the number of nodules and the growth of soybean. The combination with biochar and *B. japonicum* resulted in enhanced nodulation, nodule biomass, and shoot biomass of soybean [63]. Numerous studies have shown that biocharapplication increases the nutrient contents of plants and soil and improves soil fertility [7,62–64]. Egamberdieva et al. [55] found that inoculation of *B. japonicum* USDA 110 halophilic *P. putida* TSAU1 promoted growth, protein content, nitrogen, and phosphorus uptake and improved the root-system architecture of soybean. Their results indicated that the synergistic effect of co-inoculation of these two strains significantly improved plant growth, nitrogen, phosphorus contents, and contents of soluble leaf proteins as compared with the inoculation with *B. japonicum* USDA 110 alone or the control.

Masciarelli et al. [45] found that co-inoculation of soybean plants with *B. Amyloliquefaciens* subsp Plantarum and *B. japonicum* showed significant improvement in plant growth parameters and nodulation. They found that inoculation of *B. amyloliquefaciens* subsp. Plantarum with *B. japonicum* enhanced the ability of *B.japonicum* to colonize host plant roots and increase the number of nodules. Phosphomonoesterase (E.C. 3.1.3.2) in the soil is either of plant-root or microbial origin. It plays a major role in P solubilization in soils and in making P available to plants [40]. Acid phosphomonoesterase is dominant in acidic soil, while alkaline phosphomonoesterase occurs in the alkaline soil. The presence of these enzymes and their level in the soil is directly related to the extent of P solubilization and, hence, the amount of soluble P in the soil. Non-nitrogen fixers, such as *Pseudomonas* sp. assimilate nitrogen through the decomposition of protein–nitrogen to low molecular nitrogenous compounds and increase the soil nitrogen and, thus, soil fertility. Extracellular proteases enter the soil via microbial production.

Co-inoculation of *B. japonicum* and *P.putida* along with the application of biochar has been reported to enhance the activities of a wide variety of enzymes in soil [60]. The increase in activities of soil enzyme may be due to increased microbial activity as a result of the addition of consortium of organisms and the addition of biochar that contains good amounts of carbon, nitrogen, and minerals to support cell proliferation and, therefore, enzyme activities [60]. Egamberdieva et al. [55] demonstrated a 2-fold increase in protease and a 40% increase in acid phosphomonoesterase activity due to the addition of biochar. The positive effect on the activities of the soil enzymes can be attributed to the stimulating effect of biochar on microbial activity [60]. The enhancement in the soil enzyme activities due to rhizobial inoculation was also observed by Fall et al. [63]. Ouyang et al. [62] reported that the addition of biochar increases the activities of soil enzymes and attributed this increased enzyme activity to the availability of nutrients and increased microbial activities brought by the addition of biochar to the soil Egamberdieva et al. [55] and Ma et al. [60] also reported the positive effect of increasing the level of biochar on protease activity. Oladele [64] reported a significant ($p < 0.05$) increase in soil enzymes, such as invertase, alkaline phosphatase, urease, and catalase as a result of the higher application of biochar. It has been reported that with the amendment of more biochar, more soil proteins adhere to the surfaces of biochar pores, make the protein (substrate) unavailable in the soil, and cause a decrease in protease activity [22]. However, we report increased protease activity with an increase in the biochar amendment to the soil.

5. Conclusions

The application of biochar positively affects the growth and nodulation of soybean by increasing nutrient contents, such as N, P, and K in soil. Inoculation with *B. japonicum* USDA 110 alone increased the number of nodules, the length and dry weight of roots, and the length of shoots of soybean, as compared to the control. *B. japonicum* enhanced the total N content, P content, and K content of the soil, as compared to controls with biochar and without biochar, respectively. Co-inoculation with *B. japonicum* USDA 110 and *P. putida* TSAU 1 significantly increased the growth of soybean, nutrient contents in soybeanand soil, and activities of soil protease and acid and alkaline monophosphoeserase, as compared to the control. However, the combined application of *B. japonicum* USDA 110 and *P. putida* TSAU 1 and biochar (3%) showed pronounced positive effects on growth and vigor of soybean, nutrient levels in plant biomass and soil, and activities of soil enzymes. Thus, the co-inoculation with rhizobia and application of biochar offers the best eco-friendly and chemical-freestrategy for the sustainable increase in the yield and replenishment of nutrients in soybean and soil and increase in soil biochemical properties. In general, consortia of PGPR and biochar application improves plant growth, contents of plant and soil nutrients, and soil enzyme activities, which influence soil nutrient retention, nutrient availability, and improve crop growth.The present study demonstrates that application of *B.japonicum* alone has the capacity to improve soybean growth, nutrient contents, and improve soil biochemical properties, however, the co-inoculation of this symbiont along with *P.putida* has a more positive effect on plant growth and soil biochemicals, and co-inoculation of these rhizobia in combination with biochar possesses the capacity to significantly improve the growth and nutrient contents in soybean as well as nutrients and enzyme activities in soil. However, to claim the bio-efficacy potential of the co-inoculation of rhizobacteria and application of biochar needs multiple field studies over the season and in different agro-climatic zones.

Author Contributions: D.J. designed the experiments and wrote the manuscript; S.W. and K.D. supervised the work; A.N. and S.D. performed the methodology; R.Z.S., S.W., H.E.E., and A.S. interpreted the data and edited the paper; A.K. and R.A.M. performed analysis, and A.H.B. conceived the funding. All authors have read and approved the paper. All authors have read and agreed to the published version of the manuscript.

Funding: This work was funded by The Researchers Supporting Project Number RSP-2020/15, King Saud University, Riyadh, Saudi Arabia, Allcosmos Industries Sdn. Bhd research project No.R.J130000.7344.4B200, and German Academic Exchange Service DAAD 2019, 57440916 for funding this research and UTM-TNCPI research fund for the payment of APC.

Acknowledgments: The authors extend their appreciation to The Researchers Supporting Project Number (RSP-2020/15), King Saud University, Riyadh, Saudi Arabia. We thank colleagues at Leibniz Centre for Agricultural Landscape Research (ZALF), Müncheberg, Germany for providing necessary support laboratory and greenhouse facilities, namely the Experimental Field Station and the Central Laboratory.

Conflicts of Interest: The authors declare no conflict of interest.

References

1. Sarfraz, R.; Hussain, A.; Sabir, A.; Fekih, I.B.; Ditta, A.; Xing, S. Role of biochar and plant growth promoting rhizobacteria to enhance soil carbon sequestration—A review. *Environ. Monit. Assess.* **2019**, *191*, 251. [CrossRef] [PubMed]

2. Zhang, A.; Bian, R.; Pan, G.; Cui, L.; Hussain, Q.; Li, L.; Zheng, J.; Zheng, J.; Zhang, X.; Han, X. Effects of biochar amendment on soil quality, crop yield and greenhouse gas emission in a Chinese rice paddy: A field study of 2 consecutive rice growing cycles. *Field Crops Res.* **2012**, *127*, 153–160. [CrossRef]

3. Asai, H.; Samson, B.K.; Stephan, H.M.; Songyikhangsuthor, K.; Homma, K.; Kiyono, Y.; Inoue, Y.; Shiraiwa, T.; Horie, T. Biochar amendment techniques for upland rice production in Northern Laos: 1. Soil physical properties, leaf SPAD and grain yield. *Field Crops Res.* **2009**, *111*, 81–84. [CrossRef]

4.   Saxena, J.; Rana, G.; Pandey, M. Impact of addition of biochar along with *Bacillus* sp. on growth and yield of French beans. *Sci. Hortic.* **2013**, *162*, 351–356. [CrossRef]

5.   Agboola, K.; Moses, S. Effect of biochar and cowdung on nodulation, growth and yield of soybean (*Glycine max* L. Merrill). *Int. J. Agric. Biosci.* **2015**, *4*, 154–160.

6.   Głodowska, M.; Schwinghamer, T.; Husk, B.; Smith, D. Biochar based inoculants improve soybean growth and nodulation. *Agric. Sci.* **2017**, *8*, 1048–1064. [CrossRef]

7.   Pandit, N.; Mulder, J.; Hale, S.; Martinsen, V.; Schmidt, H.; Cornelissen, G. Biochar improves maize growth by alleviation of nutrient stress in a moderately acidic low-input Nepalese soil. *Sci. Total Environ.* **2018**, *625*, 1380–1389. [CrossRef]

8.   Chen, H.; Ma, J.; Wei, J.; Gong, X.; Yu, X.; Guo, H.; Zhao, Y. Biochar increases plant growth and alters microbial communities via regulating the moisture and temperature of green roof substrates. *Sci. Total Environ.* **2018**, *635*, 333–342. [CrossRef]

9.   Laird, D. The charcoal vision: A win–win–win scenario for simultaneously producing bioenergy, permanently sequestering carbon, while improving soil and water quality. *Agron. J.* **2008**, *100*, 178–181. [CrossRef]

10.   Hossain, M.; Strezov, V.; Chan, K.; Nelson, P. Agronomic properties of wastewater sludge biochar and bioavailability of metals in production of cherry tomato (*Lycopersicone sculentum*). *Chemosphere* **2010**, *78*, 1167–1171. [CrossRef]

11.   Zhang, H.; Prithiviraj, B.; Charles, T.; Driscoll, B.; Smith, D. Low-temperature tolerant *Bradyrhizobium japonicum* strains allowing improved nodulation and nitrogen fixation of soybean in a short season (cool spring) area. *Eur. J. Agron.* **2003**, *19*, 205–213. [CrossRef]

12.   Egamberdieva, D.; Jabborova, D.; Wirth, S. Alleviation of salt stress in legumes by co-inoculation with *Pseudomonas* and *Rhizobium*. In *Plant-Microbe Symbiosis: Fundamentals and Advances*; Springer: Singapore, 2013; pp. 291–303. [CrossRef]

13.   Sayyed, R.; Gangurde, N.; Patel, P.; Josh, S.; Chincholkar, S. Siderophore production by *Alcaligenes faecalis* and its application for growth promotion in *Arachis hypogaea*. *Indian J. Biotechnol.* **2010**, *9*, 302–307.

14.   Jabborova, D.; Annapurna, K.; Fayzullaeva, M.; Sulaymonov, K.; Kadirova, D.; Jabbarov, Z.; Sayyed, R. Isolation and characterization of endophytic bacteria from ginger (*Zingiberofficinale* Rosc.). *Ann. Phytomed.* **2020**, *9*, 116–121. [CrossRef]

15.   Buckley, S.; Allen, D.; Brackin, R.; Jämtgård, S.; Näsholm, T.; Schmidt, S. Microdialysis as an in situ technique for sampling soil enzymes. *Soil Biol. Biochem.* **2019**, *135*, 20–27. [CrossRef]

16.   Sayyed, R.; Patel, P.; Shaikh, S. Plant growth promotion and root colonization by EPS producing *Enterobacter* sp. RZS5 under heavy metal contaminated soil. *Indian J. Exp. Biol.* **2015**, *53*, 116–123.

17.   Sagar, A.; Riyazuddin, R.; Shukla, P.; Ramteke, P.; Sayyed, R. Heavy metal stress tolerance in *Enterobacter* sp. PR14 is mediated by plasmid. *Indian J. Exp. Biol.* **2020**, *58*, 115–121.

18.   Elkoca, E.; Kantar, F.; Sahin, F. Influence of nitrogen-fixing and phosphorus solubilizing bacteria on the nodulation, plant growth, and yield of chickpea. *J. Plant Nutr.* **2007**, *31*, 157–171. [CrossRef]

19.   Jabborova, D.; Davranov, K. Effect of phosphorus and nitrogen concentrations on root colonization of soybean (*Glycine max* L.) by *Bradyrhizobium japonicum* and *Pseudomonas putida*. *Int. J. Adv. Biotechnol. Res.* **2015**, *6*, 418–424.

20.   Egamberdieva, D.; Jabborova, D.; Berg, G. Synergistic interactions between *Bradyrhizobium japonicum* and the endophyte *Stenotrophomonas rhizophila* and their effects on growth, and nodulation of soybean under salt stress. *Plant Soil* **2016**, *405*, 35–45. [CrossRef]

21.   Egamberdieva, D.; Wirth, S.; Jabborova, D.; Räsänen, L.; Liao, H. Coordination between *Bradyrhizobium* and *Pseudomonas* alleviates salt stress in soybean through altering root system architecture. *J. Plant Interact.* **2017**, *12*, 100–107. [CrossRef]

22.   Jabborova, D.; Enakiev, Y.; Davranov, K.; Begmatov, S. Effect of co inoculation with *Bradyrhizobiumjaponicum* and *Pseudomonas putida* on root morph-architecture traits, nodulation and growth of soybean in response to phosphorus supply under hydroponic conditions. *Bulg. J. Agric. Sci.* **2018**, *24*, 1004–1011.

23.   Egamberdieva, D.; Jabborova, D.; Wirth, S.; Alam, P.; Alyemeni, M.; Ahmad, P. Interaction of magnesium with nitrogen and phosphorus modulates symbiotic performance of soybean with *Bradyrhizobiumjaponicum* and its root architecture. *Front. Microbiol.* **2018**, *9*, 1.

24.   Holik, L.; Vranová, V. Proteolytic activity in meadow soil after the Application of phytohormones. *Biomolecules* **2019**, *9*, 507. [CrossRef] [PubMed]

25. Reibe, K.; Götz, K.; Roß, C.; Döring, T.; Ellmer, F.; Ruess, L. Impact of quality and quantity of biochar and hydrochar on soil Collembola and growth of spring wheat. *Soil Biol. Biochem.* **2015**, *83*, 84–87. [CrossRef]

26. Nautiyal, C. An efficient microbiological growth medium for screening phosphate solubilizing microorganisms. *FEMS Microbiol. Lett.* **1999**, *170*, 265–270. [CrossRef] [PubMed]

27. Bric, J.; Bostock, R.; Silverstone, S. Rapid in situ assay for indoleacetic acid production by bacteria immobilized on a nitrocellulose membrane. *Appl. Environ. Microbiol.* **1991**, *57*, 535–538. [CrossRef]

28. Patel, P.; Shaikh, S.; Sayyed, R. Modified chrome azurol S method for detection and estimation of siderophores having affinity for metal ions other than iron. *Environ. Sustain.* **2018**, *1*, 81–87. [CrossRef]

29. Payne, S. Detection, isolation, and characterization of siderophores. In *Methods in Enzymol*; Elsevier: Amsterdam, The Netherlands, 1994; pp. 329–344. [CrossRef]

30. Penrose, D.; Glick, B. Methods for isolating and characterizing ACC deaminase–containing plant growth–promoting rhizobacteria. *Physiol. Plant.* **2003**, *118*, 10–15. [CrossRef]

31. Labconco, C. *A Guide to Kjeldahl Nitrogen Determination Methods and Apparatus*; Labconco Corporation: Houston, TX, USA, 1998.

32. Upadhyay, A.; Sahu, R. Determination of total nitrogen in soil and plant. In *Laboratory Manual on Advances in Agro-Technologies for Improving Soil, Plant and Atmosphere Systems*; CAFT: Jabalpur, India, 2012; pp. 18–19.

33. Upadhyay, A.; Sahu, R. Determination of potassium in soil and plant. In *Laboratory Manual on Advances in Agro-Technologies for Improving Soil, Plant and Atmosphere Systems*; CAFT: Jabalpur, India, 2012; pp. 23–35.

34. Sahrawat, K. Determination of calcium, magnesium, zinc and manganese in plant tissue using a dilute HCl extraction method. *Commun. Soil Sci. Plant Anal.* **1987**, *18*, 947–962. [CrossRef]

35. Sims, J. Soil test phosphorus: Principles and methods. Methods of Phosphorus Analysis for Soils, Sediments, Residuals and Waters. *Southern Coop. Ser. Bull.* **2009**, *408*, 9–19.

36. Nelson, D.; Sommers, L. Total carbon, organic carbon, and organic matter. *Methods Soil Anal. Part 2 Chem. Microbiol. Prop.* **1983**, *9*, 539–579. [CrossRef]

37. Tabatabai, M.; Bremner, J. Use of p-nitrophenyl phosphate for assay of soil phosphatase activity. *Soil Biol. Biochem.* **1969**, *1*, 301–307. [CrossRef]

38. Ladd, J.; Butler, J. Short-term assays of soil proteolytic enzyme activities using proteins and dipeptide derivatives as substrates. *Soil Biol. Biochem.* **1972**, *4*, 19–30. [CrossRef]

39. Patel, P.; Shaikh, S.; Sayyed, R. Dynamism of PGPR in bioremediation and plant growth promotion in heavy metal contaminated soil. *Indian J. Exp. Biol.* **2016**, *54*, 286–290.

40. Wani, S.; Shaikh, S.; Sayyed, R. Statistical-based optimization and scale-up of siderophore production process on laboratory bioreactor. *3Biotech* **2016**, *6*, 69.

41. Saxena, B.; Rani, A.; Sayyed, R.; El-Enshasy, H.A. Analysis of Nutrients, Heavy Metals and Microbial Content In Organic and Non-Organic Agriculture Fields of Bareilly Region-Western Uttar Pradesh, India. *Biosci. Biotechnol. Res. Asia* **2020**, *17*, 399–406. [CrossRef]

42. Sagar, A.; Sayyed, R.; Ramteke, P.; Sharma, S.; Najat Marraiki Elgorban, A.; Syed, A. ACC deaminase and antioxidant enzymes producing halophilic *Enterobacter* sp. PR14 promotes the growth of rice and millets under salinity stress. *Plant Physiol. Mol. Biol.* **2020**. [CrossRef]

43. Seneviratne, M.; Gunaratne, S.; Bandara, T.; Weerasundara, L.; Rajakaruna, N.; Seneviratne, G.; Vithanage, M. Plant growth promotion by *Bradyrhizobium japonicum* under heavy metal stress. *S. Afr. J. Bot.* **2016**, *105*, 19–24. [CrossRef]

44. Mubarik, N.; Mahagiani, I.; Wahyudi, A. Production of IAA by Bradyrhizobium sp. *World Acad. Sci. Eng. Technol.* **2013**, 152–155. [CrossRef]

45. Masciarelli, O.; Llanes, A.; Luna, V. A new PGPR co-inoculated with *Bradyrhizobium japonicum* enhances soybean nodulation. *Microbiol. Res.* **2014**, *169*, 609–615. [CrossRef]

46. Meliani, A.; Bensoltane, A.; Benidire, L.; Oufdou, K. Plant growth-promotion and IAA secretion with *Pseudomonas fluorescens* and *Pseudomonas putida*. *Res. Rev. J. Bot. Sci.* **2017**, *6*, 16–24.

47. Genesio, L.; Miglietta, F.; Baronti, S.; Vaccari, F.P. Biochar increases vineyard productivity without affecting grape quality: Results from a four years field experiment in Tuscany. *Agric. Ecosyst. Environ.* **2015**, *201*, 20–25. [CrossRef]

48. Sayyed, R.; Naphade, B.; Joshi, S.; Gangurde, N.; Bhamare, H.; Chincholkar, S. Consortium of *A. feacalis* and *P. fluorescens* promoted the growth of *Arachis hypogea* (Groundnut). *Asian J. Microbiol. Biotechnol. Environ. Sci.* **2009**, *48*, 83–86.

Agronomy **2020**, *10*, 1142

49. Shaikh, S.; Patel, P.; Patel, S.; Nikam, S.; Rane, T.; Sayyed, R. Production of biocontrol traits by banana field *fluorescent Pseudomonads* and comparison with chemical fungicide. *Indian J. Exp. Biol.* **2014**, *52*, 917–920. [PubMed]

50. Pandya, N.; Desai, P.; Sayyed, R. Antifungal, and phytohormone production ability of plant growth-promoting rhizobacteria associated with the rhizosphere of sugarcane. *World J. Microbiol. Biotechnol.* **2011**, *13*, 112–116.

51. Sayyed, R.; Badgujar, M.; Sonawane, H., Mhaske, M.; Chincholkar, S. Production of microbial iron chelators (siderophores) by *fluorescent Pseudomonads*. *Indian J. Biotechnol.* **2005**, *4*, 484–490.

52. Uzoma, K.; Inoue, M.; Andry, H.; Fujimaki, H.; Zahoor, A.; Nishihara, E. Effect of cow manure biochar on maize productivity under sandy soil condition. *Soil Use Manag.* **2011**, *27*, 205–212. [CrossRef]

53. Iijima, M.; Yamane, K.; Izumi, Y.; Daimon, H.; Motonaga, T. Continuous application of biochar inoculated with root nodule bacteria to subsoil enhances yield of soybean by the nodulation control using crack fertilization technique. *Plant Prod. Sci.* **2015**, *18*, 197–208. [CrossRef]

54. Lehmann, J.; Rillig, M.C.; Thies, J.; Masiello, C.A.; Hockaday, W.C.; Crowley, D. Biochar effects on soil biota–a review. *Soil Biol. Biochem.* **2011**, *43*, 1812–1836. [CrossRef]

55. Egamberdieva, D.; Li, L.; Ma, H.; Wirth, S.; Bellingrath-Kimura, S. Soil amendment with different maize biochars improves chickpea growth under different moisture levels by improving symbiotic performance with *Mesorhizobium ciceri* and soil biochemical properties to varying degrees. *Front. Microbiol.* **2019**, *10*, 2423. [CrossRef]

56. Bruun, E.; Petersen, C.; Hansen, E.; Holm, J.; Hauggaard-Nielsen, H. Biochar amendment to coarse sandy subsoil improves root growth and increases water retention. *Soil Use Manag.* **2014**, *30*, 109–118. [CrossRef]

57. Shen, Q.; Hedley, M.; Camps Arbestain, M.; Kirschbaum, M. Can biochar increase the bioavailability of phosphorus? *J. Soil Sci. Plant Nutr.* **2016**, *16*, 268–286. [CrossRef]

58. Prendergast-Miller, M.T.; Duvall, M.; Sohi, S. Localisation of nitrate in the rhizosphere of biochar-amended soils. *Soil Biol. Biochem.* **2011**, *43*, 2243–2246. [CrossRef]

59. Wang, C.; Alidoust, D.; Yang, X.; Isoda, A. Effects of bamboo biochar on soybean root nodulation in multi-elements contaminated soils. *Ecotoxicol. Environ. Saf.* **2018**, *150*, 62–69. [CrossRef] [PubMed]

60. Ma, H.; Egamberdieva, D.; Wirth, S.; Bellingrath-Kimura, S.D. Effect of biochar and irrigation on soybean-*Rhizobium* symbiotic performance and soil enzymatic activity in field rhizosphere. *Agronomy* **2019**, *9*, 626. [CrossRef]

61. Wang, Y.; Yin, R.; Liu, R. Characterization of biochar from fast pyrolysis and its effect on chemical properties of the tea garden soil. *J. Anal. Appl. Pyrolysis* **2014**, *110*, 375–381. [CrossRef]

62. Ouyang, L.; Tang, Q.; Yu, L.; Zhang, R. Effects of amendment of different biochars on soil enzyme activities related to carbon mineralisation. *Soil Res.* **2014**, *52*, 706–716. [CrossRef]

63. Fall, D.; Bakhoum, N.; NourouSall, S.; Zoubeirou, A.; Sylla, S.; Diouf, D. Rhizobial inoculation increases soil microbial functioning and gum arabic production of 13-year-old *senegaliasenegal* (L.) britton, trees in the north part of Senegal. *Front. Plant Sci.* **2016**, *7*, 1355. [CrossRef]

64. Oladele, S. Effect of biochar amendment on soil enzymatic activities, carboxylate secretions and upland rice performance in a sandy clay loam Alfisol of Southwest Nigeria. *Sci. Afr.* **2019**, *4*, e00107. [CrossRef]

Article

Evaluating Biochar-Microbe Synergies for Improved Growth, Yield of Maize, and Post-Harvest Soil Characteristics in a Semi-Arid Climate

Maqshoof Ahmad [1,*], Xiukang Wang [2,*], Thomas H. Hilger [3], Muhammad Luqman [1], Farheen Nazli [4], Azhar Hussain [1], Zahir Ahmad Zahir [5], Muhammad Latif [6], Qudsia Saeed [7], Hina Ahmed Malik [8] and Adnan Mustafa [9]

[1] Department of Soil Science, The Islamia University of Bahawalpur, Bahawalpur 63100, Pakistan; luqmansidhu229@gmail.com (M.L.); azharhaseen@gmail.com (A.H.)

[2] College of Life Sciences, Yan'an University, Yan'an 716000, China

[3] Hans-Ruthenberg Institute, University of Hohenheim, 70593 Stuttgart, Germany; thomas.hilger@uni-hohenheim.de

[4] Pesticide Quality Control Laboratory, Punjab Agriculture Department, Government of Punjab, Bahawalpur 63100, Pakistan; farheenmaqshoof@gmail.com

[5] Institute of Soil and Environmental Sciences, University of Agriculture, Faisalabad 38000, Pakistan; zazahir@yahoo.com

[6] Department of Agronomy, The Islamia University of Bahawalpur, Bahawalpur 63100, Pakistan; mlatifiub@gmail.com

[7] College of Natural Resources and Environment, Northwest Agriculture and Forestry University, Xianyang 712100, China; syedaqudsia.saeed@yahoo.com

[8] Institute of Food and Agriculture Sciences, University of Florida, Gainesville, FL 110690, USA; Hmalik@ufl.edu

[9] National Engineering Laboratory for Improving Quality of Arable Land, Institute of Agricultural Resources and Regional Planning, Chinese Academy of Agricultural Sciences, Beijing 100081, China; adnanmustafa780@gmail.com

* Correspondence: maqshoof_ahmad@yahoo.com (M.A.); wangxiukang@yau.edu.cn (X.W.)

Received: 4 June 2020; Accepted: 16 July 2020; Published: 21 July 2020

Abstract: Arid and semi-arid regions are characterized by high temperature and low rainfall, leading to degraded agricultural soils of alkaline calcareous nature with low organic matter contents. Less availability of indigenous nutrients and efficacy of applied fertilizers are the major issues of crop production in these soils. Biochar application, in combination with plant growth promoting rhizobacteria with the ability to solubilize nutrients, can be an effective strategy for improving soil health and nutrient availability to crops under these conditions. Experiments were planned to evaluate the impact of biochar obtained from different sources in combination with acid-producing, nutrient-solubilizing *Bacillus* sp. ZM20 on soil biological properties and growth of maize (*Zea mays* L.) crops under natural conditions. Various biochar treatments, viz. wheat (*Triticum aestivum* L.) straw biochar, Egyptian acacia (*Vachellia nilotica* L.) biochar, and farm-yard manure biochar with and without *Bacillus* sp. ZM20, were used along with control. Soil used for pot and field trials was sandy loam in texture with poor water holding capacity and deficient in nutrients. Results of the pot trial showed that fresh and dry biomass, 1000 grain weight, and grain yield was significantly improved by application of biochar of different sources with and without *Bacillus* sp. ZM20. Application of biochar along with *Bacillus* sp. ZM20 also improved soil biological properties, i.e., soil organic matter, microbial biomass carbon, ammonium, and nitrate nitrogen. It was also observed that a combined application of biochar with *Bacillus* sp. ZM20 was more effective than a separate application of biochar. The results of wheat straw biochar along with *Bacillus* sp. ZM20 were better as compared to farm-yard manure biochar and Egyptian acacia biochar. Maximum increase (25.77%) in grain yield was observed in the treatment where wheat straw biochar (0.2%) was applied in combination with

Bacillus sp. ZM20. In conclusion, combined application of wheat straw biochar (0.2%) inoculated with *Bacillus* sp. ZM20 was the most effective treatment in improving the biological soil properties, plant growth, yield, and quality of maize crop as compared to all other treatments.

Keywords: aridity; *Bacillus* sp.; biochar; nutrient availability; organic matter; soil health

---

1. Introduction

The current world population is about 7.6 billion, which is increasing at an exponential rate and will be about 9.8 billion in 2050 and is further expected to rise to 11.2 billion in 2100, as reported by the United Nations [1]. About half of the added population will be concentrated in less developed countries. Due to this reason, there will be a marked decrease in agricultural lands, as most of the productive lands will be used for constructing new housing societies and infrastructure [2]. To feed the world population, utilization of less productive soils, and bringing such soils into the agricultural system by fighting desertification, salinization, and soil pollution is the major challenge for the scientific community [3]. Moreover, increasing per-hectare yield of the major crops along with exploring the unutilized arable lands can be helpful to meet the challenge of food requirements.

Maize, being the staple food of most of the world population, is an important cereal crop [4]. Its total production is even more than rice and wheat crops [5]. Maize has gained its popularity to meet the world food requirements due to higher yield per unit area as compared to other staple crops [6]. Although the per-acre yield of maize is adequate, it is an exhaustive crop that needs more nutrients and that is why it depletes more nutrients from the soil [7]. It has high demand for phosphatic- and zinc-containing fertilizers as compared to other major crops; therefore, nutrient deficiency is experienced more in the maize crop [8].

Biochar can be effective to rehabilitate degraded lands by improving the soil physical properties, nutrient-holding capacity, and soil carbon contents, leading to improvement in soil productivity [9]. It is a carbon-rich compound that is produced through a process known as pyrolysis and has beneficial implications as a potential soil amendment [10]. Use of biochar has gained popularity as a carbon negative material which resists environmental change as it draws carbon from the atmosphere into the soil and persists for hundreds to thousands of years [11]. Recent interest has been developed to use biochar as a soil amendment for improving soil quality through mitigation of soil salinization, soil acidity, and metal contamination, along with improvement in soil productivity [12–15]. Biochar application to soil positively affects the properties of soil, including soil structure, water retention capacity, fertility, and carbon sequestration of degraded soil [16,17]. It also improves soil microbial activity due to presence of micropores in biochar which allow the sorption of dissolved organic matter, thus, helping speed up the soil rehabilitation process [18]. However, the success highly depends upon the types and rates of biochar application, the nature of feedstock, and soil and climate variations. In this regard, utilizing biochar with other soil amendments such as plant growth-promoting rhizobacteria (PGPR) has proved to be a better approach to conserving the environment, resulting in increased efficacy and cost-effectiveness [3,9].

The use of microorganisms with the aim of improving nutrient availability for plants is an important practice and is considered necessary for agriculture these days [19]. The PGPR are the bacteria that inhabit either the rhizosphere, the soil in the immediate vicinity of plant roots, or inside the plant tissues, helping the plants with better growth through some direct and indirect mechanisms [20,21]. There are certain PGPR species which can solubilize insoluble mineral compounds in soil through the production of organic acids along with some other growth-promoting mechanisms [22,23]. Among these, phosphate solubilizing rhizobacteria [24], zinc solubilizing rhizobacteria [25], and potassium solubilizing rhizobacteria [26] are well documented. These nutrient-solubilizing bacterial species also have multiple plant growth-promoting traits such as siderophores production, chitin decomposition, hydrogen cyanide production, ammonia production, etc. [24]. They effectively colonize plant roots, thus helping the improvement of plant growth and nutrient acquisition [27,28]. These bacteria can also induce tolerance against different biotic and abiotic stresses in plants through several indirect mechanisms [27,29]. Moreover, bacterial inoculation improves soil health by fixing atmospheric nitrogen [30,31], production of plant hormones, siderophores and exopolysaccharides [32], and phytoremediation of heavy metals and other organic pollutants [33,34].

The integrated use of biochar and PGPR can reduce the use of chemical fertilizers for crop production in addition to improving soil health through increased soil organic matter contents, enhanced soil aggregation, better microbial activity, and increased soil fertility [35,36]. The improvement in soil health and maize growth has also been reported by the combined use of biochar and PGPR under water-stressed conditions [37]. Work on the use of biochar for increasing soil fertility and remediating the polluted soil has been carried out, but the use of biochar as soil amendment for improving the soil health, growth, and yield of maize in the degraded soils of arid and semi-arid regions has been least explored. It has been hypothesized that the use of biochar and PGPR can help improve barren desert soils to productive farmlands, and release the pressure off the ever-decreasing cultivated areas. Keeping with this view, current study was conducted to investigate the potential of biochar obtained from different sources along with acid-producing, nutrient-solubilizing *Bacillus* sp. for improving soil biological properties, growth, and yield of maize crop in desert regions.

2. Materials and Methods

2.1. Biochar Preparation and Characterization

Biochar was prepared from Egyptian acacia (*Vachellia nilotica* L.) stem, wheat straw, and dairy manure pyrolyzed at 450 °C following the method of Naeem et al. [38]. The dried branches of Egyptian acacia were chopped in small pieces of 2–3 inches and further dried at 105 °C for one hour. The oven-dried biomass was pyrolyzed at 450 °C. Finally, the prepared biochar was crushed into smaller particles for even distribution in soil. Before charring, the dairy manure was air-dried and sieved ($\leq$2 mm), then pyrolyzed in a muffle furnace at 450 °C. Similarly, the air-dried, chopped wheat straw was also pyrolyzed at 450 °C. The weight of biomass used for each type of biochar was recorded prior and after pyrolysis. After cooling, the biochar was passed through a 250-μm sieve and stored in a refrigerator (at 4 °C) before use.

The biochar produced from different sources was analyzed for chemical characteristics (Table 1). Prepared biochar was analyzed for its turnover rate made from the pyrolysis of feedstock. The biochar production rate was calculated by using the total weight of raw material used to prepare that biochar. Biochar yield was estimated by following the method of Al-Wabel et al. [39]. The pH and electrical conductivity (EC) of the biochar was measured using a 1:20 solid/solution ratio after shaking for ninety minutes in deionized water in a mechanical shaker. The carbon contents of biochar were assessed using the loss-on-ignition approach [40,41]. Total nitrogen (N) contents were measured using Kjeldahl distillation equipment [42].

Table 1. Rate and physicochemical properties of biochar; values are the mean of three replications ± SE.

Parameters	Egyptian Acacia Biochar	Farmyard Manure Biochar	Wheat Straw Biochar
Turnover rate (%)	28.45 ± 1.27	23.31 ± 0.94	38.76 ± 1.78
pH	9.3 ± 0.25	8.67 ± 0.12	8.43 ± 0.21
EC (dS cm^{-1})	1.85 ± 0.03	2.1 ± 0.02	1.9 ± 0.03
Bulk density (g cm^{-3})	0.36 ± 0.01	0.43 ± 0.01	0.46 ± 0.03
Nitrogen (%)	0.31 ± 0.02	0.28 ± 0.01	0.47 ± 0.02
Carbon (%)	68.45 ± 3.24	44.21 ± 2.16	52.67 ± 1.89

2.2. Collection of Rhizobacterial Strains

Pre-selected and characterized rhizobacterial strain *Bacillus* sp. *ZM20*, accession number KX086260, with strong ability to produce organic acids [23] under zinc deficient conditions was collected from the Soil Microbiology and Biotechnology Laboratory, Department of Soil Science, the Islamia University of Bahawalpur.

2.3. Soil Sampling and Analysis

A bulk soil sample (0–15 cm) was taken from the experimental field and the soil that was used for pot trial. The soil samples were air-dried and sieved through 2-mm sieve followed by analysis for basic soil characteristics (Table 2) as per standard protocols. The pH, EC, and organic matter were measured according to the method of Nelson and Sommers [43]. The available N was analyzed by the Kjeldhal method [42], while for available phosphorus (P), Olsen's method [44] was used. The extractable potassium (K) was measured using a flame photometer (Model; Model BWB-XP, BWB Technologies, UK). The saturation percentage referring to the field capacity of soil was estimated by oven-drying the soil sample at 105 °C to a constant weight, followed by calculations according to the method as described by Sarfraz et al. [45]. All the chemicals were of analytical grade (Sigma-Aldrich, Unichem, Merck) supplied by Wahid Scientific Store, Lahore, Pakistan.

Table 2. Characteristics of the soils; values are the mean of three replications ± SE.

Parameter	Pot Trial	Field Trial
EC_e (dS m^{-1})	1.6 ± 0.01	1.8 ± 0.01
pH	8.1 ± 0.04	7.9 ± 0.02
Organic matter (%)	0.39 ± 0.02	0.47 ± 0.02
Available N (%)	0.024 ± 0.001	0.059 ± 0.003
Available P (mg kg^{-1})	3.7 ± 0.01	4.5 ± 0.03
Extractable K (mg kg^{-1})	53 ± 1.68	77 ± 3.21
Saturation percentage (%)	33 ± 0.76	36 ± 0.71
Water-holding capacity (Inches ft^{-1})	1.27 ± 0.05	1.29 ± 0.03
Textural class	Sandy loam	Sandy loam

2.4. Pot Trial

A pot trial was conducted in the wire house to evaluate the impact of biochar obtained from different sources in combination with acid-producing, nutrient-solubilizing *Bacillus* sp. ZM20 on soil biological properties, and the growth and yield of maize crops (Pioneer-30Y80) under natural environmental conditions in the February–March sowing season. The experiment was conducted in a wire house with natural growth conditions, protecting the experimental units from animals and birds with wire only. Various biochar treatments viz. Egyptian acacia biochar (0.1%), Egyptian acacia biochar (0.2%), farmyard manure (FYM) biochar (0.1%), FYM biochar (0.2%), wheat straw biochar (0.1%), and wheat straw biochar (0.2%) with and without *Bacillus* sp. ZM20 were used along with controls. Soil (8 kg per pot^{-1}) used to fill the pots (height 12″, diameter 12″) was sandy loam in texture with poor water-holding capacity (1.27 inches ft^{-1}) and deficient in nutrients (Table 2), as analyzed by following the standard protocols as defined by Ryan et al. [46]. The pots were arranged in the wire house following a completely randomized design (CRD) in factorial arrangement with three replications. Maize seeds were inoculated with a slurry of *Bacillus* sp. ZM20 prepared by mixing the inoculum, sugar solution, and peat in the ratio (04:01:05). The inoculated seeds were used to sow in one set of treatments while in the other set un-inoculated maize seeds were sown. The recommended doses of P and K at the rate of 90 kg ha^{-1} and 60 kg ha^{-1} while half of the recommended dose of N (120 kg ha^{-1}) were applied as basal doses in the form of diammonium phosphate, sulfate of potash, and urea. The remaining dose of N was applied in two splits. Good quality tap water meeting the irrigation quality criteria [47] was used to irrigate pots, and all other agronomic practices were carried out according to requirements. Growth and yield parameters were recorded at the time of harvesting and grain samples were collected to analyze for N, P, and K.

2.5. Field Trial

A field trial was conducted in the February–March (Spring 2019) sowing season to verify the results of the pot trial and further recommendation to farming community. The same treatment plan was followed as observed in the pot trial. The field trial was conducted in the field area of the Department of Soil Science, The Islamia University of Bahawalpur, Pakistan. The soil of the experimental field was sandy loam in texture with poor water-holding capacity and deficient in nutrients (Table 2). The randomized complete block design was used for the field trial with a factorial arrangement and three replications. The size of the plots was 22′ × 16′ with a row-to-row distance of 2.5′. Maize seeds were inoculated before sowing by following the same procedure as described above in the pot trial. The recommended doses of P and K at the rate of 90 and 60 kg ha^{-1} while half of the recommended dose of N (120 kg ha^{-1}) were applied as basal doses in the form of diammonium phosphate, sulfate of potash, and urea. The remaining dose of N was applied in two splits. Canal water was used for irrigation purposes and all other agronomic practices were carried out according to requirements. Growth and yield parameters were recorded at the time of harvesting and grain samples were collected to analyze for N, P, and K.

2.6. Nutrient Analyses in Grains

Grains were digested according to the protocol as described by Wolf [48]. The P in grain samples was analyzed using a UV-visible spectrophotometer (Agilent Carry 60, USA), while K in grains was determined on a flame photometer (Model; Model BWB-XP, BWB Technologies, Newbury, UK) by following the standard methods [46]. For the analysis of N in grains, an automatic digestion unit (DK 6), semi automatic distillation unit (UDK 126) of Kjeldahl apparatus (VELP Sci., Italy) was used, followed by standard titration as described in the Kjeldahl method [49]. All the chemicals were of analytical grade (Sigma-Aldrich, Unichem, Merck) supplied by Wahid Scientific Store, Lahore, Pakistan.

2.7. Post-Harvest Soil Sample Collection and Analysis

The post-harvest soils samples were collected from the pot (harvested in July) and field trials (harvested in July), and analyzed for organic matter, microbial biomass carbon (MBC), ammonium N, and nitrate N under pot and field conditions. The composite soil sampling method was used, and the samples were air-dried and sieved through a 2-mm sieve before analysis. The prepared soil samples were stored in a refrigerator at 4 °C and analyzed within seven days. The organic matter contents were measured according to the method of Nelson and Sommers [43]. For the analysis of microbial biomass carbon (MBC), chloroform fumigation and extraction methods were used [50,51]. For the analyses of ammoniacal N and nitrate N in soil, the methods of Kamphake et al. [52] and Sims and Jackson [53], respectively, were used. All the chemicals were of analytical grade (Sigma-Aldrich, Unichem, Merck) supplied by Wahid Scientific Store, Lahore, Pakistan. Replicated measurements were always performed to ensure the accuracy of the data.

2.8. Statistical Analysis

All data reported here are means of three replicates which were analyzed using one-way analysis of variance (ANOVA) in Statistix 8.1. The mean values were compared through a least significant difference (LSD) test as described by Steel et al. [54].

3. Results

3.1. Pot Trial

Integrated use of biochar and *Bacillus* sp. ZM20 improved soil properties in the pot trial. Results (Figure 1A) showed that FYM (Farm yard manure) biochar treatments increased the organic matter in the pot trial. The application of biochar without inoculation increased the soil organic matter contents, but the results of Egyptian acacia biochar at both levels were non-significant when compared with the control under un-inoculated and inoculated sets of treatments. Under inoculated treatments, the maximum organic matter (0.449%) was observed in the treatment where wheat straw biochar (0.2%) was applied in combination with *Bacillus* sp. ZM20; this treatment was, however, non-significant with FYM biochar application (0.2%), and wheat straw biochar application (0.1%) under inoculated treatments. Combined inoculation of biochar and *Bacillus* sp. ZM20 showed better results than separate application of biochar in all treatments.

The application of biochar from different sources significantly improved the MBC in the soil (Figure 1B). Maximum improvement in MBC under an un-inoculated set of treatments was observed by the application of wheat straw biochar (0.2%), which was statistically at par with the application of wheat straw biochar (0.1%) and FYM biochar (0.2%). These treatments, however, were significantly better than all other treatments under un-inoculated conditions. The application of biochar from all sources in the presence of *Bacillus* sp. ZM20 was significantly better than separate use, except for Egyptian acacia biochar (0.1%), where the increase was non-significant with that of the respective un-inoculated treatment. Maximum MBC (342 mg kg^{-1}) was observed in the treatment where wheat straw biochar was applied (0.2%), and it was statistically similar with that of the wheat straw biochar (0.1%) treatment.

The sole and combined application of biochar and inoculated with *Bacillus* sp. ZM20 to improve the ammonium N and nitrate N in the pot trial was observed (Figure 1C,D). The application of biochar separately, and in combination with *Bacillus* sp. ZM20, significantly enhanced the ammonium N and nitrate N, except for Egyptian acacia biochar treatments, which gave non-significant improvement in both cases as compared to the control. A maximum increase in ammonium N and nitrate N was recorded due to the combined application of wheat straw biochar (0.2%) and *Bacillus* sp. ZM20 as compared to the inoculated control. Overall, inoculated treatments showed better results regarding ammonium N and nitrate N than un-inoculated treatments.

The results of the impact of the integrated use of biochar and *Bacillus* sp. ZM20 on plant height (Figure 2A) revealed that the separate as well as combined use of biochar with *Bacillus* sp. ZM20 significantly improved plant height in the pot trial, except for Egyptian acacia biochar, at both levels, which was statistically non-significant compared to the control. In the inoculated treatment, the combined use of wheat straw biochar (0.2%) and *Bacillus* sp. ZM20 was carried out and showed the maximum plant height. In the case of root length, the results of separate applications of wheat straw biochar at both levels and FYM biochar (0.2%) were significantly better than those of the control; however, other treatments gave non-significant improvement in root length when compared with the control. The combined use of biochar and *Bacillus* sp. ZM20 was better than the separate use of biochar in improving the root length, but the results were non-significant with un-inoculated treatments in all cases. Maximum improvement in root length as compared to control was observed with the combined use of wheat straw biochar (0.2%) and *Bacillus* sp. ZM20; however, it was statistically non-significant with the treatments of wheat straw biochar (0.1%) and FYM biochar (0.2%) in combination with *Bacillus* sp. ZM20 (Figure 2B). Application of biochar significantly improved the shoot fresh and dry biomass separately and in combination with *Bacillus* sp. ZM20 as compared to respective controls. Maximum improvement in shoot fresh biomass and shoot dry biomass were observed with the application of wheat straw biochar in combination with *Bacillus* sp. ZM20 (Figure 2C,D). The improvement due to the inoculation of *Bacillus* sp. ZM20 in both the parameters over the respective un-inoculated treatment was non-significant in all the cases.

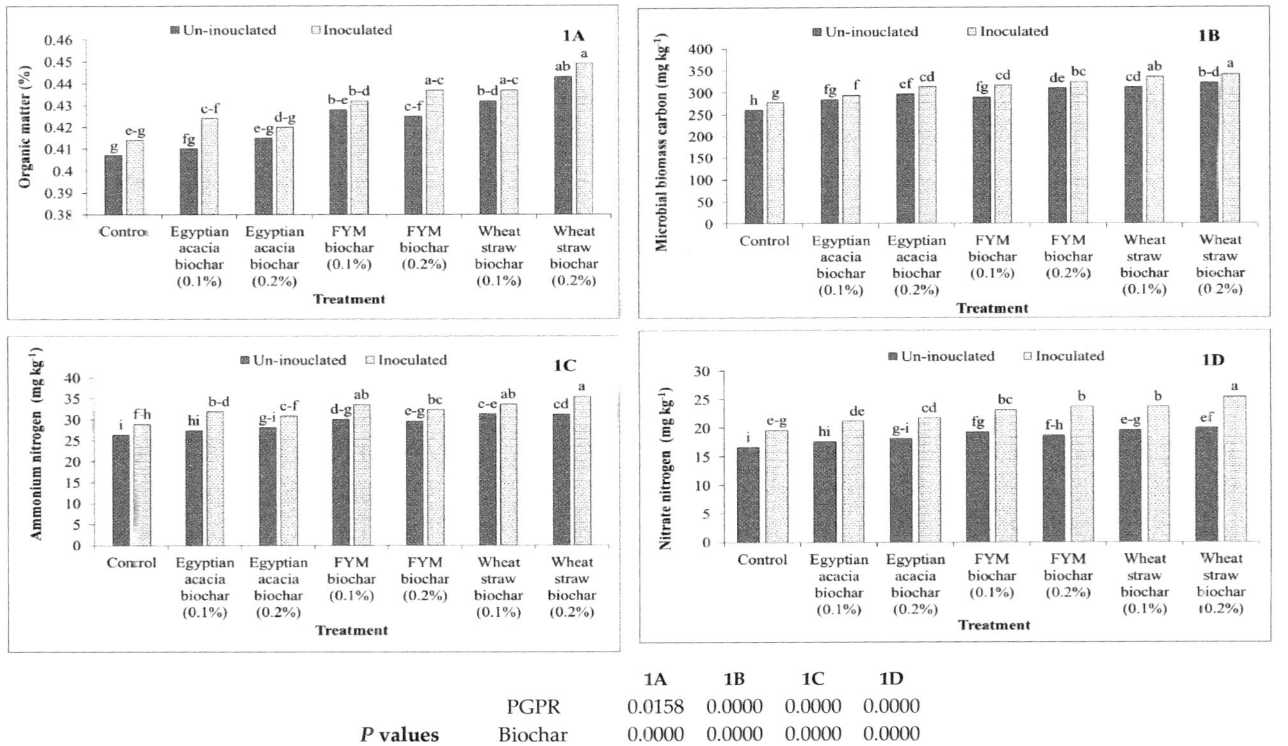

		1A	1B	1C	1D
	PGPR	0.0158	0.0000	0.0000	0.0000
P values	Biochar	0.0000	0.0000	0.0000	0.0000
	PGPR + Biochar	0.9401	0.6795	0.6394	0.3057

Figure 1. Effects of biochar with and without *Bacillus* sp. ZM20 on organic matter (**A**): Least significant difference (LSD) 0.0156, microbial biomass carbon (**B**): LSD 14.290, ammonium N (**C**): LSD 2.096, and nitrate N (**D**): LSD 1.6386 under pot trial; ($n = 3$); bars sharing same letters are statistically not different from each other at $p \leq 0.05$.

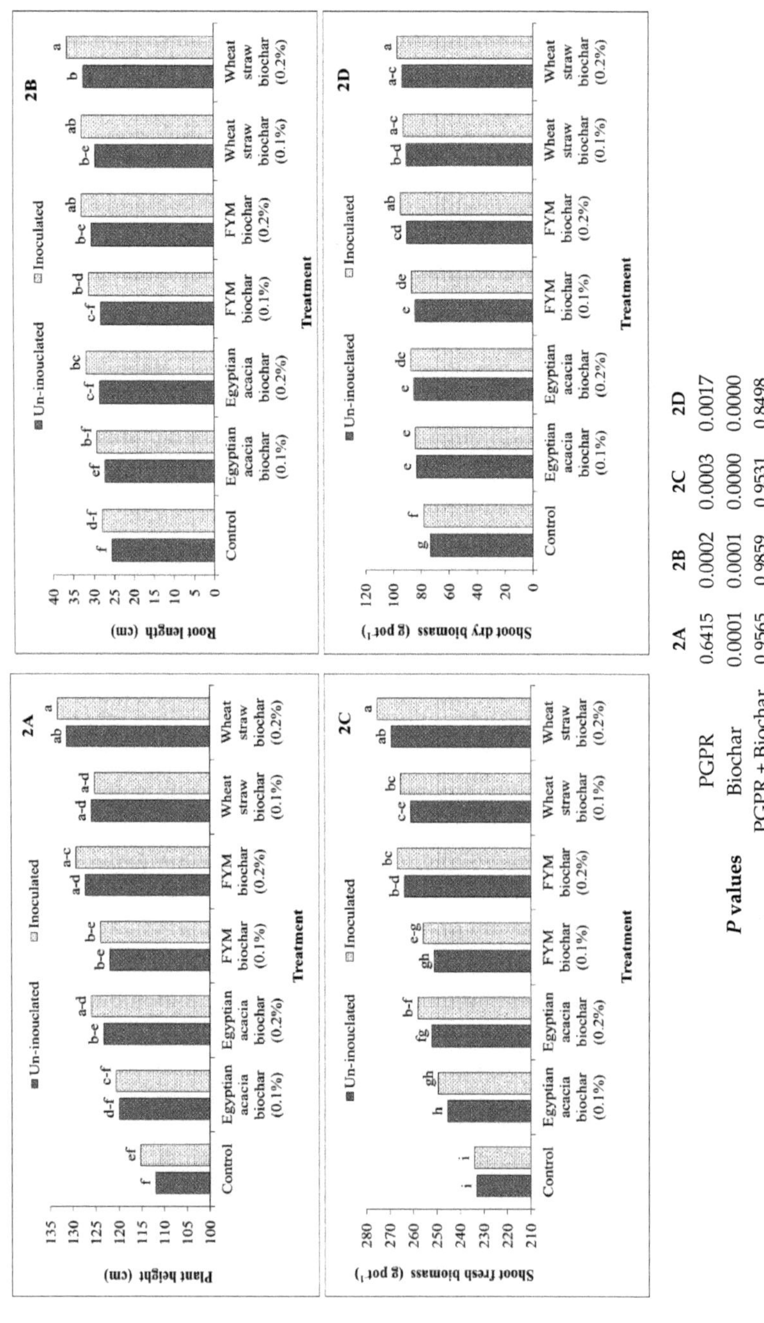

Figure 2. Effects of biochar with and without *Bacillus* sp. ZM20 on plant height (**A**): LSD 3.6996, root length (**B**): LSD 8.7945, shoot fresh biomass (**C**): LSD 6.8333 and shoot dry biomass (**D**): 4.6543 of maize under pot trial; (*n* = 3); bars sharing same letters are statistically not different from each other at *p* ≤ 0.05.

Results of the effects of separate as well as combined applications of biochar inoculated with *Bacillus* sp. ZM20 significantly improved the root fresh biomass in the pot trial (Table 3). A maximum increase (29.63%) in root fresh biomass was observed due to the combined use of wheat straw biochar (0.2%) with *Bacillus* sp. ZM20. The results of the separate use of biochar (all treatments) without inoculum were, however, non-significant with the control in most of the cases, except wheat straw biochar (0.2%), which gave significantly better results than the control. The results of the improvement in root dry biomass were non-significant with the control due to the biochar application with and without in most of the cases, except for FYM biochar (0.2%) and wheat straw biochar (0.2%) in both sets of treatments, i.e., inoculated and un-inoculated. A maximum increase (23.36%) in root dry biomass was observed due to the combined use of wheat straw biochar (0.2%) with *Bacillus* sp. ZM20 (Table 3).

Table 3. Effects of biochar with and without *Bacillus* sp. ZM20 on root fresh biomass, root dry biomass, 100 grain weight, and grain yield of maize in pot trial.

Treatment	Un-Inoculated	Inoculated	Un-Inoculated	Inoculated
	Root Fresh Biomass (g pot^{-1})		Root Dry Biomass (g pot^{-1})	
Control	28.00 [f]	36.00 [c–e]	14.33 [d]	15.67 [c,d]
Egyptian acacia biochar (0.1%)	29.33 [f]	38.33 [b–d]	15.33 [c,d]	17.33 [a–d]
Egyptian acacia biochar (0.2%)	31.67 [e,f]	42.00 [a,b]	16.33 [a–d]	17.67 [a–c]
FYM biochar (0.1%)	30.67 [e,f]	41.33 [a–c]	16.00 [b–d]	17.67 [a–c]
FYM biochar (0.2%)	33.00 [d–f]	43.67 [a,b]	17.67 [a–c]	19.00 [a,b]
Wheat straw biochar (0.1%)	31.00 [e,f]	43.00 [a,b]	17.33 [a–d]	18.33 [a–c]
Wheat straw biochar (0.2%)	35.67 [d,e]	46.67 [a]	18.00 [a–c]	19.33 [a]
LSD ($p \leq 0.05$)	5.5636		3.1110	
p value — PGPR	0.0000		0.0192	
p value — Biochar	0.0020		0.0314	
p value — PGPR + Biochar	0.9586		0.9996	
	100 Grain Weight (g)		Grain Yield (g pot^{-1})	
Control	18.67 [c]	19.67 [b,c]	103.3 [f]	106.3 [e,f]
Egyptian acacia biochar (0.1%)	20.33 [b,c]	21.33 [a–c]	114.3 [d,e]	116.0 [d]
Egyptian acacia biochar (0.2%)	21.00 [a–c]	21.67 [a–c]	116.3 [d]	121.0 [b–d]
FYM biochar (0.1%)	21.00 [a–c]	21.67 [a–c]	116.0 [d]	121.7 [b–d]
FYM biochar (0.2%)	21.67 [a–c]	23.67 [a,b]	122.0 [b–d]	127.0 [a–c]
Wheat straw biochar (0.1%)	21.33 [a–c]	23.00 [a,b]	120.0 [c,d]	126.0 [a–c]
Wheat straw biochar (0.2%)	22.33 [a–c]	24.67 [a]	129.0 [a,b]	133.7 [a]
LSD ($p \leq 0.05$)	4.3030		8.9777	
p value — PGPR	0.1039		0.0134	
p value — Biochar	0.1421		0.0000	
p value — PGPR + Biochar	0.9956		0.9755	

Values sharing same letter(s) within a parameter are statistically non-significant with each other at 5% level of probability; values are the mean of three replications ± SE. Lower case words show difference in treatment means.

Results regarding the effects of separate and combined applications of biochar with *Bacillus* sp. ZM20 on 100-grain weight are presented in (Table 3). In most of the cases, regarding the use of all types of biochar, individually as well as in combination with *Bacillus* sp. ZM20, the results were nonsignificant with the control, except for the wheat straw biochar (0.2%) application in combination with *Bacillus* sp. ZM20. Statistical analyses showed that all treatments of the separate and combined uses of biochar showed significantly better results than the control in improving the grain yield of maize in the pot trial. A maximum increase (25.77%) in grain yield was observed in the treatment where wheat straw biochar (0.2%) was applied in combination with *Bacillus* sp. ZM20 (Table 3). Data (Table 4) showed that the use of all types of biochar, individually as well as in combination with *Bacillus* sp. ZM20, gave non-significant results in improving the stover yield in all the cases when compared to the control.

Table 4. Effects of biochar with and without *Bacillus* sp. ZM20 on stover yield, and nitrogen, phosphorus and potassium concentration in grains of maize in pot trial.

Treatment	Un-Inoculated	Inoculated	Un-Inoculated	Inoculated
	Stover Yield (g pot^{-1})		Nitrogen Conc. in Grains (%)	
Control	20.33 [c]	23.00 [a–c]	2.17 [g]	2.19 [f,g]
Egyptian acacia biochar (0.1%)	22.00 [b,c]	24.33 [a–c]	2.21 [f,g]	2.21 [e–g]
Egyptian acacia biochar (0.2%)	23.00 [a–c]	26.33 [a,b]	2.24 [c–g]	2.25 [c–f]
FYM biochar (0.1%)	26.00 [a–c]	25.67 [a–c]	2.22 [d–g]	2.25 [c–f]
FYM biochar (0.2%)	25.00 [a–c]	27.00 [a,b]	2.28 [a–e]	2.30 [a–c]
Wheat straw biochar (0.1%)	24.33 [a–c]	26.33 [a,b]	2.28 [b–e]	2.29 [a–d]
Wheat straw biochar (0.2%)	25.67 [a–c]	28.00 [a]	2.32 [a,b]	2.35 [a]
LSD (*p* ≤ 0.05)	5.9763		0.0705	
p value — PGPR	0.0737		0.1857	
p value — Biochar	0.2071		0.0000	
p value — PGPR + Biochar	0.9862		0.9971	
	Phosphorus Conc. in Grains (%)		Potassium Conc. in Grains (%)	
Control	0.373 [e]	0.390 [c–e]	2.57 [h]	2.60 [g,h]
Egyptian acacia biochar (0.1%)	0.380 [d,e]	0.403 [b–d]	2.61 [g,h]	2.63 [f,g]
Egyptian acacia biochar (0.2%)	0.390 [c–e]	0.410 [a–c]	2.64 [e–g]	2.69 [c–e]
FYM biochar (0.1%)	0.387 [c–e]	0.407 [a–c]	2.62 [g,h]	2.67 [d–f]
FYM biochar (0.2%)	0.397 [b–e]	0.417 [a,b]	2.70 [b–d]	2.74 [a,b]
Wheat straw biochar (0.1%)	0.400 [b–d]	0.410 [a–c]	2.67 [d–f]	2.72 [b,c]
Wheat straw biochar (0.2%)	0.407 [a–c]	0.430 [a]	2.74 [a,b]	2.77 [a]
LSD (*p* ≤ 0.05)	0.0250		0.0499	
p value — PGPR	0.0000		0.0003	
p value — Biochar	0.0000		0.0000	
p value — PGPR + Biochar	0.0000		0.9546	

Values sharing same letter(s) within a parameter are statistically non-significant with each other at 5% level of probability; values are the mean of three replications ± SE. Lower case words show difference in treatment means.

Results (Table 4) showed that N concentration in grains of maize was significantly improved due to the separate as well as combined application of all types of biochar and *Bacillus* sp. ZM20. The results of Egyptian acacia biochar (both levels) and FYM biochar (0.1%) gave non-significant results in both sets of treatments. A maximum increase (7.30%) in N concentration in maize grains was observed due to the combined use of wheat straw biochar (0.2%) with *Bacillus* sp. ZM20. The results of the impact of biochar (all treatments), with and without inoculum, on P concentration in maize grains were non-significant with the control in most of the cases, except for the wheat straw biochar (at both levels) application in un-inoculated set of treatments, and the FYM biochar (0.1%) and wheat straw biochar (0.2%) application in combination with *Bacillus* sp. ZM20. Similar results were observed in case of K concentration in maize grains, where the maximum improvement (6.5%) over control was observed due to combined use of wheat straw biochar (0.2%) and *Bacillus* sp. ZM20 (Table 4).

3.2. Field Trial

Results (Figure 3) showed that all treatments significantly increased the organic matter and MBC under field conditions, except the application of Egyptian acacia biochar (0.1%), which gave non-significant improvement in organic matter when compared with the control under the un-inoculated treatment (Figure 3A). Under inoculated treatments, maximum improvement (5.78%) in organic matter contents over the control was observed in treatment where wheat straw biochar (0.2%) was applied in combination with *Bacillus* sp. ZM20; this treatment was, however, non-significant with the use of wheat straw biochar (0.1%) under inoculated treatments. The combined inoculation of biochar and *Bacillus* sp. ZM20 showed better results than the separate application of biochar in all treatments.

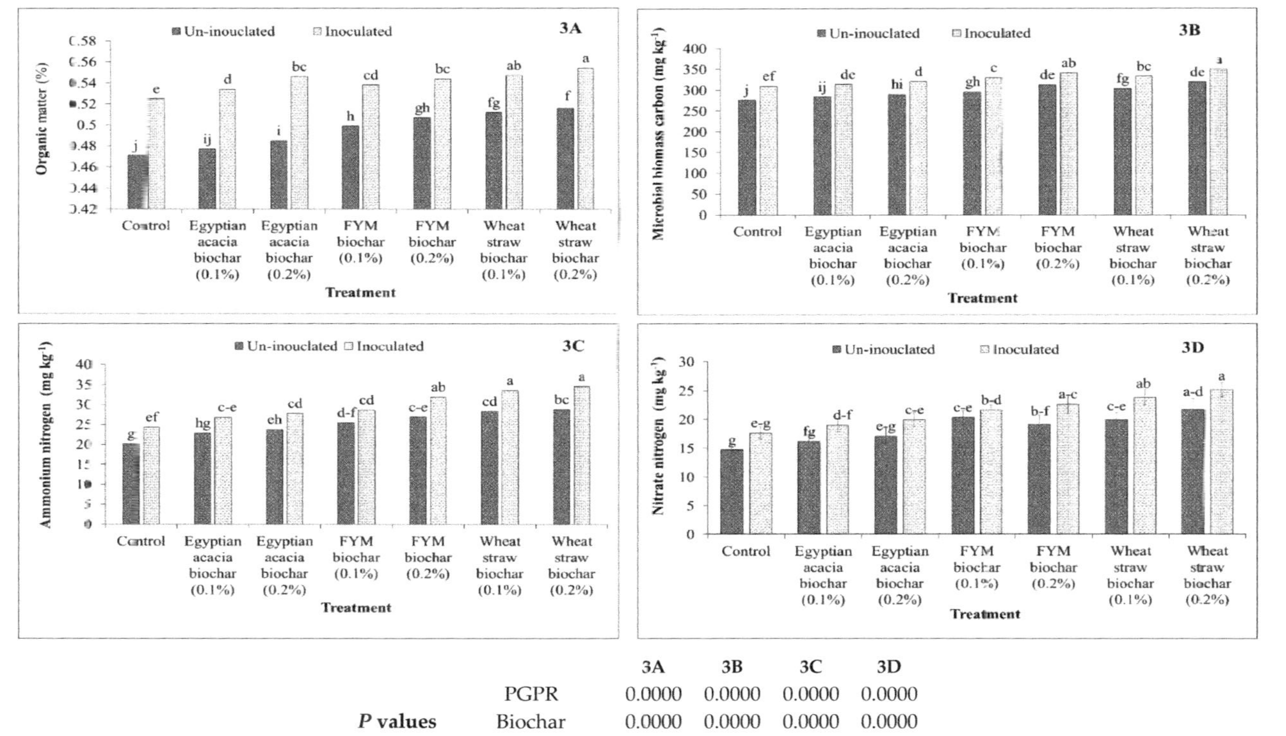

		3A	3B	3C	3D
	PGPR	0.0000	0.0000	0.0000	0.0000
P values	Biochar	0.0000	0.0000	0.0000	0.0000
	PGPR + Biochar	0.0002	0.9811	0.9286	0.9537

Figure 3. Effects of biochar with and without *Bacillus* sp. ZM20 on organic matter (**A**), microbial biomass carbon (**B**), ammonium nitrogen (**C**), and nitrate nitrogen (**D**) under field trial ($n = 3$); bars sharing same letters are statistically not different from each other at $p \leq 0.05$.

The application of biochar from different sources also significantly improved the MBC under field conditions in a semi-arid climate (Figure 3B). A maximum improvement (23.39%) in MBC under an un-inoculated set of treatments was observed by the application of wheat straw biochar (0.2%), which was statistically at par with application of FYM biochar (0.2%). These treatments, however, were significantly better than all other treatments under un-inoculated conditions. The application of biochar from all sources in the presence of *Bacillus* sp. ZM20 was significantly better than the separate use, except for Egyptian acacia biochar (0.1%), where the increase was non-significant with that of respective un-inoculated treatment. A maximum MBC (22.89%) was observed in the treatment where the wheat straw biochar (0.2%) was applied in combination with *Bacillus* sp. ZM20, and it was statistically similar with that of the FYM biochar (0.2%) treatment.

The application of biochar separately and in combination with *Bacillus* sp. ZM20 significantly enhanced the ammonium N and nitrate N, except for the Egyptian acacia biochar treatments (both levels), which gave non-significant improvement in both cases as compared to the control (Figure 3C,D). A maximum increase in ammonium N (22.61%) and nitrate N (29.59%) as compared to the inoculated control was recorded due to the combined application of wheat straw biochar (0.2%) and *Bacillus* sp. ZM20. Overall, inoculated treatments showed better results regarding ammonium N and nitrate N than un-inoculated treatments.

The results of the impact of integrated use of biochar and *Bacillus* sp. ZM20 on plant height (Table 5) under field conditions revealed that the separate as well as combined use of biochar with *Bacillus* sp. ZM20 improved plant height, but that this improvement was statistically non-significant with the control in most of the cases. In the inoculated treatment, the combined use of wheat straw biochar (0.2%) and *Bacillus* sp. ZM20 was carried out and showed the maximum improvement (12.7%) in plant height. The application of biochar separately and in combination with *Bacillus* sp. ZM20 significantly improved the shoot fresh and dry biomass as compared to respective controls. Maximum improvement in shoot fresh biomass (16.6%) and shoot dry biomass (20.75%) was observed with the application of wheat straw biochar (0.2%) in combination with *Bacillus* sp. ZM20 (Table 5). Results regarding the effects of separate and combined applications of biochar with *Bacillus* sp. ZM20 on 1000-grain weight and grain yield showed a significant improvement in most of the cases, except for Egyptian acacia biochar (0.1%), which gave non-significant improvement when compared with the control. A maximum increase (21.9%) in grain yield was observed in the treatment where wheat straw biochar (0.2%) was applied in combination with *Bacillus* sp. ZM20 (Table 6).

Table 5. Effects of biochar with and without *Bacillus* sp. ZM20 on plant height, shoot fresh biomass, shoot dry biomass, and 1000-grain weight of maize in field trial.

Treatment	Un-Inoculated	Inoculated	Un-Inoculated	Inoculated
	Plant Height (cm)		Shoot Fresh Biomass (g pot^{-1})	
Control	136.3 [f]	139.3 [e,f]	243.6 [g]	245.3 [g]
Egyptian acacia biochar (0.1%)	141.0 [d–f]	143.3 [c–f]	255.0 [f]	260.3 [e,f]
Egyptian acacia biochar (0.2%)	144.0 [e,f]	150.3 [a–c]	262.3 [e]	266.0 [d,e]
FYM biochar (0.1%)	143.7 [c–f]	146.7 [b–e]	261.7 [e,f]	265.7 [d,e]
FYM biochar (0.2%)	147.0 [b–e]	154.3 [a,b]	273.3 [b,c]	278.0 [b,c]
Wheat straw biochar (0.1%)	146.3 [b–e]	149.7 [a–d]	271.0 [c,d]	276.3 [b,c]
Wheat straw biochar (0.2%)	153.0 [a,b]	157.0 [a]	279.0 [a,b]	286.0 [a]
LSD (*p* ≤ 0.05)	8.9702		7.2668	
p value — PGPR	0.0174		0.0022	
p value — Biochar	0.0003		0.0000	
p value — PGPR + Biochar	0.9774		0.9678	

Table 5. *Cont.*

Treatment	Un-Inoculated	Inoculated	Un-Inoculated	Inoculated
	Plant Height (cm)		**Shoot Fresh Biomass (g pot^{-1})**	
	Shoot Dry Biomass (g pot^{-1})		**1000-Grain Weight (g)**	
Control	79.67 ^{g}	80.33 ^{f,g}	222.33 ^{g}	232.67 ^{e,f}
Egyptian acacia biochar (0.1%)	84.33 $^{d-f}$	83.67 $^{e-g}$	229.00 ^{f,g}	237.67 ^{d,e}
Egyptian acacia biochar (0.2%)	85.00 ^{d,e}	87.67 $^{c-e}$	236.00 $^{b-f}$	243.33 $^{b-d}$
FYM biochar (0.1%)	84.33 $^{d-f}$	87.00 $^{c-e}$	234.00 ^{e,f}	242.67 $^{b-d}$
FYM biochar (0.2%)	88.67 ^{c,d}	95.00 ^{a,b}	243.00 $^{b-d}$	248.33 ^{a,b}
Wheat straw biochar (0.1%)	88.67 ^{c,d}	93.67 ^{a,b}	239.33 $^{c-e}$	246.33 $^{a-c}$
Wheat straw biochar (0.2%)	91.33 ^{b,c}	97.00 ^{a}	247.67 ^{a,b}	251.33 ^{a}
LSD ($p \leq 0.05$)	4.5478		7.6490	
p value — PGPR	0.0008		0.0000	
p value — Biochar	0.0000		0.0000	
p value — PGPR + Biochar	0.2527		0.8961	

Values sharing same letter(s) with in a parameter are statistically non-significant with each other at 5% level of probability; values are the mean of three replications ± SE.

Table 6. Effects of biochar with and without *Bacillus* sp. ZM20 on grain yield, and nitrogen, phosphorus, and potassium concentrations in grain of maize in field trial.

Treatment	Un-Inoculated	Inoculated	Un-Inoculated	Inoculated
	Grain Yield (t ha^{-1})		**Nitrogen Conc. in Grains (%)**	
Control	7.40 ^{f}	7.90 ^{d,e}	2.20 ^{i}	2.24 $^{g-i}$
Egyptian acacia biochar (0.1%)	7.57 ^{e,f}	8.23 ^{c,d}	2.23 ^{h,i}	2.31 ^{e,f}
Egyptian acacia biochar (0.2%)	8.07 ^{d}	8.60 ^{c}	2.28 ^{f,g}	2.34 $^{b-e}$
FYM biochar (0.1%)	7.93 ^{d,e}	8.57 ^{c}	2.27 $^{f-h}$	2.34 $^{b-e}$
FYM biochar (0.2%)	8.33 ^{c,d}	9.10 ^{b}	2.33 $^{c-e}$	2.38 ^{a,b}
Wheat straw biochar (0.1%)	8.27 ^{c,d}	9.10 ^{b}	2.31 $^{d-f}$	2.35 $^{b-d}$
Wheat straw biochar (0.2%)	8.63 ^{c}	9.63 ^{a}	2.37 $^{a-c}$	2.41 ^{a}
LSD ($p \leq 0.05$)	0.4485		0.0434	
p value — PGPR	0.0000		0.0000	
p value — Biochar	0.0000		0.0000	
p value — PGPR + Biochar	0.6905		0.7951	
	Phosphorus Conc. in Grain (%)		**Potassium Conc. in Grain (%)**	
Control	0.393 ^{g}	0.403 $^{e-g}$	2.64 ^{i}	2.69 $^{g-i}$
Egyptian acacia biochar (0.1%)	0.400 ^{fg}	0.417 $^{c-f}$	2.68 ^{h,i}	2.74 $^{d-g}$
Egyptian acacia biochar (0.2%)	0.410 $^{d-g}$	0.423 $^{b-d}$	2.71 $^{f-h}$	2.80 $^{b-d}$
FYM biochar (0.1%)	0.407 $^{d-g}$	0.420 $^{b-e}$	2.71 $^{f-h}$	2.78 $^{c-e}$
FYM biochar (0.2%)	0.420 $^{b-e}$	0.437 ^{a,b}	2.76 $^{c-f}$	2.85 ^{a,b}
Wheat straw biochar (0.1%)	0.413 $^{c-f}$	0.430 $^{a-c}$	2.74 $^{e-h}$	2.82 ^{b,c}
Wheat straw biochar (0.2%)	0.430 $^{a-c}$	0.447 ^{a}	2.82 ^{b,c}	2.89 ^{a}
LSD ($p \leq 0.05$)	0.0197		0.0622	
p value — PGPR	0.0004		0.0000	
p value — Biochar	0.0001		0.0000	
p value — PGPR + Biochar	0.9981		0.9359	

Values sharing same letter(s) with in a parameter are statistically non-significant with each other at 5% level of probability; values are the mean of three replications ± SE. Lower case words show difference in treatment means.

The results in Table 6 show that the N concentration in grains of maize was significantly improved due to the separate as well as combined application of all types of biochar and *Bacillus* sp. ZM20, except for Egyptian acacia biochar (0.1%), which gave non-significant improvement in the

N concentration in grains under an un-inoculated set of treatments. A maximum increase (7.6%) in the N concentration in maize grains was observed due to the combined use of wheat straw biochar (0.2%) with *Bacillus* sp. ZM20. The results of the impact of biochar (all treatments), with and without inoculum, on P and K concentrations in maize grains were significantly better than the control in most of the cases. A maximum improvement in both the parameters over the control was observed due to combined use of wheat straw biochar (0.2%) and *Bacillus* sp. ZM20 (Table 6).

4. Discussion

The present study was conducted in arid and semi-arid regions on sandy loam soil characterized by low rainfall and high temperature, associated with low organic matter content. Due to low organic matter, biochar in combination with bacterial inoculation can have the ability to improve the soil health and crop yield under such a scenario. The application of biochar can be effective at rehabilitating degraded lands by improving the soil structure, nutrient- and water-holding capacity, and soil carbon contents, leading to improvement in soil productivity [9,55]. A carbon-rich compound called charcoal is produced through a process known as pyrolysis and has beneficial implications such as soil amendment for improving soil health and crop yield [10,56]. The physicochemical properties of biochar are crucial in determining its functionality and impact on plant growth and soil health [57]. It was observed that biochar contains a high carbon-to-nitrogen ratio (Table 1), which makes it stable against decomposition. The carbon contents of Egyptian acacia biochar were higher compared to the other two sources, but wheat straw biochar had a higher turnover rate as compared to the other sources. In previous studies, scientists have also reported that biochar is rich in carbon contents along with other nutrients like C, N, and S [58,59] which have shown promising results in improving crop growth and yield characteristics similar to the findings of the current study.

In this study, the application of biochar improved the soil biological properties (soil organic matter contents, MBC), along with improvement in ammonium and nitrate N contents in soil (Figures 1 and 3). The increase in the levels of biochar increased the content of organic matter and MBC in studied soil. The presence of high carbon and other nutrients might have helped in the improvement of soil fertility as reported by Oni et al. [17], suggesting that biochar application positively affects the soil structure, water retention capacity, fertility, and soil carbon sequestration, leading to improvement in crop growth and productivity. Similarly, biochar application increased the ratio of below-ground biomass to above-ground biomass due to an increase in water-holding capacity, as reported previously [60], and a reduction in soil strength [61]. The integration of biochar and PGPR is a win-win strategy as biochar provides a niche for microbes due to its microporous structure, which in turn increases microbial activity and hence the sorption of dissolved organic matter [18]. The increase in carbon and organic matter contents in the present study due to the addition of different biochar types is in good agreement with Shenbagavalli and Mahimairaja [62]. The integrated use of biochar and PGPR can improve soil health through increasing soil organic matter contents, enhancing soil aggregation, promoting better microbial activity, and increasing soil fertility [35,36]. In our study, the integrated use of biochar inoculated with *Bacillus* sp. ZM20 was significantly better in improving soil organic matter and MBC, which might have supported crop growth. Our results are in good agreement with previous reports by Ullah et al. [37], in which they reported the increased growth, physiology, and production of crops under the combined application of biochar and PGPR. This increase in growth and yield of wheat in present study might be attributed to enhanced supply of nutrients that are scarcely available in the soil including nitrogen, phosphorus, zinc, and iron. This may also be due to the positive effects of applied PGPR which are well recognized candidates equipped with plenty of mechanisms, i.e., the production of siderophores that helps in iron acquisition, synthesis of plant growth regulators, and exopolysaccharides [15,32,36].

Biochar application as a soil amendment increases the growth parameters of plants (plant root and shoot growth), and their nutrient uptake by improving the water status of plants and water-use efficiency [63,64], thus leading to improved yield of crop plants. In the present study, the application of biochar from different sources improved the maize root and shoot growth and nutrient uptake, along with the yield and yield contributing factors (Tables 3 and 5). A maximum increase (25.77%) in grain yield was observed in the treatment where wheat straw biochar (0.2%) was applied in combination with *Bacillus* sp. ZM20. This might be due to the enhanced water-holding capacity of the soil [65] that resulted in enhanced nutrient availability [65], thus improving the growth of crop plants under the applied biochar [38] and PGPR [28]. As stated by Hussain et al. [31], the combined use of PGPR and biochar at the rate of 0.5 tons/ha have shown enhanced water-holding capacity of the soil, and hence the growth and yield of maize (*Zea mays* L.). Recently, Shen et al. [66] reported that biochar application improved plant growth; however, willow woodchip biochar was significantly better than pine-based biochar in improving plant growth and nutrient uptake of *Lotus pedunculatus*. The improved growth and yield characteristics of maize under the applied biochar are in good agreement with previous studies [14,67,68]. The enhanced soil characteristics and crop growth responses in the present study under the application of biochar and PGPR might be attributed to the differences in soil characteristics and the alkaline nature of biochar in the soil studied here.

The PGPR inhabits either the rhizosphere, the soil in the immediate vicinity of plant roots, or inside the plant tissues, and helps the plants exhibit better growth through some direct and indirect mechanisms [20,21]. The phosphorus-solubilizing bacteria (PSB), zinc-solubilizing bacteria (ZSB), and potassium-solubilizing bacteria (KSB) can increase plant nutrient availability along multiple plant growth promoting traits, such as siderophores production, chitin decomposition, hydrogen cyanide production, and ammonia production [25,26,69]. In the current study, the combined use of biochar and *Bacillus* sp. ZM20 improved maize growth, the uptake of N, P, and K, and the yield, which might be due to solubilization of nutrients through acid production, along with other growth-promoting characteristics such as siderophore production, exopolysaccharides production, and HCN production exhibited by this strain, as reported in previous studies [14,25]. The application of biochar improves the quality of soil and makes it conducive for better microbial activity [70]. Previous studies have reported that the integrated use of biochar and *Pseudomonas fluorescens* enhanced the growth of cucumber by improving plant–water relations under water deficit conditions. It has been reported that PGPR effectively colonize plant rhizosphere, thus helping in improving the growth, yield, and nutrient acquisition [29]. One possible reason behind increased uptake of N, P, and K in the present study (Tables 4 and 6) might be due to the promoting effects of PGPR and the applied biochar, which resulted in enhanced nutrient use efficiency, as has been reported previously [19,28,36]. Moreover, the presence of biochar in addition to PGPR might have helped to increase the sorption capacity of soil, resulting in higher mineral (NPK) concentration in wheat grains (Tables 4 and 6). These results are substantiated with those reported previously [37].

5. Conclusions

Low organic matter and depleted nutrients are the major issues of agricultural soils in arid and semi-arid regions. In the present study, the application of biochar from different sources significantly improved soil biological properties, growth, yield, and quality of maize grains. The integrated use of biochar and *Bacillus* sp. ZM20 was more effective as compared to the separate application. Biochar application along with *Bacillus* sp. ZM20 also improved soil biological properties, i.e., soil organic matter, MBC. Moreover, the biochar source and rate also influenced the soil properties and plant growth with different degrees of efficacy. The use of wheat straw biochar along with inoculated with the *Bacillus* sp. ZM20 bacterial strain was better as compared to farm-yard manure biochar and Egyptian acacia biochar. It is concluded that the combined application of wheat straw biochar (0.2%) and *Bacillus* sp. ZM20 was the most effective treatment in improving the soil properties, plant growth, yield, and quality of maize crops as compared to all other treatments in the pot and field trials.

Author Contributions: Conceptualization, M.A. and A.M.; methodology, M.A., A.H. and Z.A.Z.; software, H.A.M. and X.W.; validation, Z.A., Q.S., X.W. and A.M.; formal analysis, A.H., Q.S. and F.N.; investigation,

M.L. (Muhammad Latif), A.M., H.A.M. and M.A.; resources, A.M.; data curation, M.L. (Muhammad Luqman), M.A. and F.N.; writing—original draft preparation, M.A.; writing—review and editing, A.M., X.W., T.H.H. and Z.A.Z.; visualization, Z.A.Z., A.H., T.H.H. and A.M.; supervision, Z.A.Z. All authors have read and agreed to the published version of the manuscript.

Funding: This research received no external funding.

Acknowledgments: The authors acknowledge the Soil Microbiology and Biotechnology Laboratory, Department of Soil Science, and the Islamia University of Bahawalpur, Pakistan for providing research facilities. This work was supported by the Higher Education Commission of Pakistan (grant number SRGP-785).

Conflicts of Interest: The authors report no competing interests either financially or otherwise.

References

1. United Nations. The World Population Prospects: The 2017 Revision, Published by the UN Department of Economic and Social Affairs. 2017. Available online: https://www.un.org/development/desa/en/news/population/world-population-prospects-2017.html (accessed on 2 April 2020).
2. Peerzado, M.B.; Magsi, H.; Sheikh, M.J. Land use conflicts and urban sprawl: Conversion of agriculture lands into urbanization in Hyderabad, Pakistan. *J. Saudi Soc. Agric. Sci.* **2019**, *18*, 423–428. [CrossRef]
3. Shrivastava, P.; Kumar, R. Soil salinity: A serious environmental issue and plant growth promoting bacteria as one of the tools for its alleviation. *Saudi J. Biol. Sci.* **2015**, *22*, 123–131. [CrossRef] [PubMed]
4. Nuss, E.T.; Tanumihardjo, S.A. Maize: A paramount staple crop in the context of global nutrition. *Compr. Rev. Food Sci. Food Saf.* **2010**, *9*, 417–436. [CrossRef]
5. Wang, Y.; Gao, F.; Gao, G.; Zhao, J.; Wang, X.; Zhang, R. Production and cultivated area variation in cereal, rice, wheat and maize in China (1998–2016). *Agronomy* **2019**, *9*, 222. [CrossRef]
6. Tandzi, L.N.; Mutengwa, C.S. Estimation of maize (*Zea mays* L.) yield per harvest area: Appropriate methods. *Agronomy* **2020**, *10*, 29. [CrossRef]
7. Das, A.; Patel, D.; Munda, G.C.; Ghosh, P.K. Effect of organic and inorganic sources of nutrients on yield, nutrient uptake and soil fertility of maize (*Zea mays*)—Mustard (*Brassica campestris*) cropping system. *Indian J. Agric. Sci.* **2010**, *80*, 85–88.
8. Thilakarathna, M.S.; Raizada, M.N. A review of nutrient management studies involving finger millet in the semi-arid tropics of Asia and Africa. *Agronomy* **2015**, *5*, 262–290. [CrossRef]
9. Cybulak, M.; Sokołowska, Z.; Boguta, P. Impact of biochar on physicochemical properties of haplic luvisol soil under different land use: A plot experiment. *Agronomy* **2019**, *9*, 531. [CrossRef]
10. Glaser, B.; Wiedne, K.; Seeling, S.; Schmidt, H.P.; Gerber, H. Biochar organic fertilizers from natural resources as substitute for mineral fertilizers. *Agron. Sustain. Dev.* **2015**, *35*, 667–678. [CrossRef]
11. Duku, H.M.; Gu, S.; Hagan, E.B. Biochar production potentials in Ghana-a review. *Renew. Sustain. Energy Rev.* **2011**, *15*, 3539–3551. [CrossRef]
12. Ding, Y.; Liu, Y.; Liu, S.; Huang, X.; Li, Z.; Tan, X.; Zeng, G.; Zhou, L. Potential benefits of biochar in agricultural soils: A review. *Pedosphere* **2017**, *27*, 645–661. [CrossRef]
13. Palansooriya, K.N.; Ok, Y.S.; Award, Y.M.; Lee, S.S.; Sung, J.K.; Kautsospyros, A.; Moon, D.H. Impact of biochar application on upland agriculture: A review. *J. Environ. Manag.* **2019**, *234*, 52–64. [CrossRef] [PubMed]
14. Naveed, M.; Ramzan, N.; Mustafa, A.; Samad, A.; Niamat, B.; Yaseen, M.; Ahmad, Z.; Hasanuzzaman, M.; Sun, N.; Shi, W.; et al. Alleviation of salinity induced oxidative stress in *Chenopodium quinoa* by Fe biofortification and biochar-endophyte interaction. *Agronomy* **2020**, *10*, 168. [CrossRef]
15. Sabir, A.; Naveed, M.; Bashir, M.A.; Hussain, A.; Mustafa, A.; Zahir, Z.A.; Kamran, M.; Ditta, A.; Núñez-Delgado, A.; Saeed, Q.; et al. Cadmium mediated phytotoxic impacts in *Brassica napus*: Managing growth, physiological and oxidative disturbances through combined use of biochar and *Enterobacter* sp. MN17. *J. Environ. Manag.* **2020**, *265*, 110522. [CrossRef] [PubMed]
16. Awad, Y.M.; Lee, S.S.; Kim, K.H.; Ok, Y.S.; Kuzyakov, Y. Carbon and nitrogen mineralization and enzyme activities in soil aggregate-size classes: Effects of biochar, oyster shells, and polymers. *Chemosphere* **2018**, *198*, 40–48. [CrossRef] [PubMed]
17. Oni, B.A.; Oziegbe, O.; Olawole, O.O. Significance of biochar application to the environment and economy. *Ann. Agric. Sci.* **2019**, *64*, 222–236. [CrossRef]

18. Hameed, A.; Hussain, S.A.; Yang, J.; Ijaz, M.U.; Liu, Q.; Suleria, H.A.R.; Song, Y. Antioxidants potential of the filamentous fungi (*Mucor circinelloides*). *Nutrients* **2017**, *9*, 1101. [CrossRef]

19. Bargaz, A.; Karim, L.; Chtouki, M.; Zeroual, Y.; Driss, D. Soil microbial resources for improving fertilizers efficiency in an integrated plant nutrient management system. *Front. Microbiol.* **2018**, *9*, 1606. [CrossRef]

20. Santoyo, G.; Moreno-Hagelsieb, G.; Orozco-Mosqueda, M.C.; Glick, B.R. Plant growth-promoting bacterial endophytes. *Microbiol. Res.* **2016**, *183*, 92–99. [CrossRef]

21. Ahmad, M.; Nadeem, S.M.; Zahir, Z.A. Plant-microbiome interactions in agroecosystem: An application. In *Microbiome in Plant Health and Disease*; Kumar, V., Ed.; Springer Nature: Singapore, 2019; pp. 251–291.

22. Hussain, A.; Zahir, Z.A.; Asghar, H.N.; Ahmad, M.; Jamil, M.; Naveed, M.; Akhtar, M.F.Z. Zinc solubilizing bacteria for zinc biofortification in cereals: A step towards sustainable nutritional security. In *Role of Rhizospheric Microbes in Soil. Volume 2: Nutrient Management and Crop Improvement*; Meena, V.S., Ed.; Springer: New Delhi, India, 2018; pp. 203–227.

23. Mumtaz, M.Z.; Barry, K.M.; Baker, A.L.; Nichols, D.S.; Ahmad, M.; Zahir, Z.A.; Britz, M.L. Production of lactic and acetic acids by *Bacillus* sp. ZM20 and *Bacillus cereus* following exposure to zinc oxide: A possible mechanism for Zn solubilization. *Rhizosphere* **2019**, *12*, 100170. [CrossRef]

24. Ahmad, M.; Ahmad, I.; Hilger, T.H.; Nadeem, S.M.; Akhtar, M.F.; Jamil, M.; Hussain, A.; Zahir, Z.A. Preliminary study on phosphate solubilizing *Bacillus subtilis* strain Q3 and *Paenibacillus* sp. strain Q6 for improving cotton growth under alkaline conditions. *PeerJ* **2018**, *6*, e5122. [CrossRef] [PubMed]

25. Mumtaz, M.Z.; Ahmad, M.; Jamil, M.; Asad, S.A.; Hafeez, F. *Bacillus* strains as potential alternate for zinc biofortification of maize grains. *Int. J. Agric. Biol.* **2018**, *20*, 1779–1786.

26. Saha, M.; Maurya, B.R.; Meena, V.S.; Bahadur, I.; Kumar, A. Identification and characterization of potassium solubilizing bacteria (KSB) from Indo-Gangetic Plains of India. *Biocatal. Agric. Biotechnol.* **2016**, *7*, 202–209. [CrossRef]

27. Ahmad, M.; Naseer, I.; Hussain, A.; Mumtaz, M.Z.; Mustafa, A.; Hilger, T.H.; Zahir, Z.A.; Minggang, X. Appraising endophyte—Plant symbiosis for improved growth, nodulation, nitrogen fixation and abiotic stress tolerance: An experimental investigation with chickpea (*Cicer arietinum* L.). *Agronomy* **2019**, *9*, 621. [CrossRef]

28. Ali, M.A.; Naveed, M.; Mustafa, A.; Abbas, A. The good, the bad, and the ugly of rhizosphere microbiome. In *Probiotics and Plant Health*; Springer: Singapore, 2017; pp. 253–290.

29. Nazli, F.; Najm-ul-Seher; Khan, M.Y.; Jamil, M.; Nadeem, S.M.; Ahmad, M. Soil microbes and plant health. In *Plant Disease Management Strategies for Sustainable Agriculture through Traditional and Modern Approaches, Sustainability in Plant and Crop Protection*; IUl Haq, I., Ijaz, S., Eds.; Springer Nature: Basel, Switzerland, 2020; pp. 111–135.

30. Naseer, I.; Ahmad, M.; Nadeem, S.M.; Ahmad, I.; Najm-ul-Seher; Zahir, Z.A. Rhizobial inoculants for sustainable agriculture: Prospects and applications. In *Biofertilizers for Sustainable Agriculture and Environment, Soil Biology*; Giri, B., Ed.; Springer Nature: Basel, Switzerland, 2019; pp. 245–284.

31. Hussain, A.; Ahmad, M.; Mumtaz, M.Z.; Ali, S.; Sarfraz, R.; Naveed, M.; Jamil, M.; Damalas, C.A. Integrated application of organic amendments with *Alcaligenes* sp. AZ9 improves nutrient uptake and yield of maize (*Zea mays*). *J. Plant Growth Regul.* **2020**, *9*, 1–16. [CrossRef]

32. Khan, N.; Bano, A. Exopolysaccharide producing rhizobacteria and their impact on growth and drought tolerance of wheat grown under rainfed conditions. *PLoS ONE* **2019**, *14*, e0222302. [CrossRef] [PubMed]

33. Pramanik, K.; Mitra, S.; Sarkar, A.; Soren, T.; Maiti, T.K. Characterization of cadmium resistant *Klebsiella pneumoniae* MCC 3091 promoted rice seedling growth by alleviating phytotoxicity of cadmium. *Environ. Sci. Pollut. Res.* **2017**, *24*, 24419–24437. [CrossRef]

34. Saeed, Z.; Naveed, M.; Imran, M.; Bashir, M.A.; Sattar, A.; Mustafa, A.; Hussain, A.; Xu, M. Combined use of *Enterobacter* sp. MN17 and zeolite reverts the adverse effects of cadmium on growth, physiology and antioxidant activity of *Brassica napus*. *PLoS ONE* **2019**, *14*, e0213016. [CrossRef]

35. Ijaz, M.; Tahir, M.; Shahid, M.; Ul-Allah, S.; Sattar, A.; Sher, A.; Mahmood, K.; Hussain, M. Combined application of biochar and PGPR consortia for sustainable production of wheat under semiarid conditions with a reduced dose of synthetic fertilizer. *Braz. J. Microbiol.* **2019**, *50*, 449–458. [CrossRef]

36. Hussain, A.; Ahmad, M.; Mumtaz, M.Z.; Nazli, F.; Farooqi, M.A.; Khalid, I.; Iqbal, Z.; Arshad, H. Impact of integrated use of enriched compost, biochar, humic acid and *Alcaligenes* sp. AZ9 on maize productivity and soil biological attributes in natural field conditions. *Ital. J. Agron.* **2019**, *14*, 101–107. [CrossRef]

37. Ullah, N.; Ditta, A.; Khalid, A.; Mehmood, S.; Rizwan, M.S.; Ashraf, M.; Mubeen, F.; Imtiaz, M.; Iqbal, M.M. Integrated effect of algal biochar and plant growth promoting rhizobacteria on physiology and growth of maize under deficit irrigations. *J. Soil Sci. Plant Nutr.* **2019**, *20*, 346–356. [CrossRef]

38. Naeem, M.A.; Khalid, M.; Ahmad, Z.; Naveed, M. Low pyrolysis temperature biochar improve growth and nutrient availability of maize on typic Calciargid. *Commun. Soil Sci. Plant Anal.* **2015**, *47*, 41–51. [CrossRef]

39. Al-Wabel, M.I.; Al-Omran, A.; El-Naggar, A.H.; Nadeem, M.; Usman, A.R.A. Pyrolysis temperature induced changes in characteristics and chemical composition of biochar produced from conocarpus wastes. *Bioresour. Technol.* **2013**, *131*, 374–379. [CrossRef] [PubMed]

40. Nelson, D.W.; Sommers, L.E. Total carbon, organic carbon, and organic matter. In *Methods of Soil Analysis. Part 3. Chemical Methods. Soil Science of America and American Society of Agronomy*; Black, C.A., Ed.; ACSESS: Madison, WI, USA, 1996; pp. 961–1010.

41. Ryan, J.; Estefan, G.; Rashid, A. *Soil and Plant Analysis Laboratory Manual*, 2nd ed.; International Center for Agriculture in Dry Areas (ICARDA): Aleppo, Syria, 2001; p. 172.

42. Jackson, M.L. *Soil Chemical Analysis*; Prentice Hall Inc.: New York, NY, USA, 1962.

43. Nelson, D.W.; Sommers, L.E. Total carbon, organic carbon and organic matter. In *Methods of Soil Analysis. Part 2: Chemical and Microbiological Properties*; Agronomy Monographs; SSSA: Madison, WI, USA, 1982; pp. 570–571.

44. Watanabe, F.S.; Olsen, S.R. Test of an ascorbic acid method for determining phosphorus in water and $NaHCO_3$ extracts from soil. *Soil Sci. Soc. Am. Proc.* **1965**, *29*, 677–678. [CrossRef]

45. Sarfraz, M.; Ashraf, Y.; Ashraf, S. A Review: Prevalence and antimicrobial susceptibility profile of listeria species in milk products. *Matrix Sci. Media* **2017**, *1*, 3–9. [CrossRef]

46. Ryan, J. *Methods of Soil, Plant, and Water Analysis: A Manual for the West Asia and North Africa Region*; International Center for Agricultural Research in the Dry Areas (ICARDA): Beirut, Lebanon, 2017.

47. Ayers, R.S.; Westcot, D.W. *Water Quality for Agriculture, FAO Irrigation and Drainage Papers 29 (Rev.1)*; Food and Agriculture Organization of the United Nations: Rome, Italy, 1994; pp. 1–11.

48. Wolf, B. The comprehensive system of leaf analysis and its use for diagnosing crop nutrient status. *Commun. Soil Sci. Plant Anal.* **1982**, *13*, 1035–1059. [CrossRef]

49. Bullock, D.; Moore, K. Protein and Fat Determination in Corn. In *Seed Analysis. Modern Methods of Plant Analysis*; Linskens, H.F., Jackson, J.F., Eds.; Springer: Berlin/Heidelberg, Germany, 1992; Volume 14, pp. 181–197.

50. Jenkinson, D.S.; Ladd, J.N. *Microbial Biomass in Soil, Measurement and Turn Over*; Marcel Dekker: New York, NY, USA, 1981.

51. Bremner, E.; Kessel, V. Extractability of microbial ^{14}C and ^{15}N following addition of variable rates of labeled glucose and ammonium sulphate to soil. *Soil Biol. Biochem.* **1990**, *22*, 707–713. [CrossRef]

52. Kamphake, L.J.; Hannah, S.A.; Cohen, J.M. Automated analysis for nitrate by hydrazine reduction. *Water Res.* **1967**, *1*, 205–216. [CrossRef]

53. Sims, J.R.; Jackson, D.G. Rapid analysis of soil nitrate with chromotropic acid. *Soil Sci. Soc. Am. J.* **1971**, *35*, 603–606. [CrossRef]

54. Steel, R.G.D.; Torrie, J.H.; Dickey, D.A. Principles and Procedures of Statistics. In *A Biometrical Approach*, 3rd ed.; McGraw Hill Book Co.: New York, NY, USA, 2007.

55. Bashir, M.A.; Naveed, M.; Ahmad, Z.; Gao, B.; Mustafa, A.; Núñez-Delgado, A. Combined application of biochar and sulfur regulated growth, physiological, antioxidant responses and Cr removal capacity of maize (*Zea mays* L.) in tannery polluted soils. *J. Environ. Manag.* **2020**, *259*, 110051. [CrossRef]

56. Fidel, R.B.; Laird, D.A.; Thompson, M.L.; Lawrinenko, M. Characterization and quantification of biochar alkalinity. *Chemosphere* **2017**, *167*, 367–373. [CrossRef] [PubMed]

57. Zimmerman, R.A.; Gao, B.; Ahn, M.Y. Positive and negative carbon mineralization priming effects among a variety of biochar-amended soils. *Soil Biol. Biochem.* **2011**, *43*, 1169–1179. [CrossRef]

58. Novak, J.M.; Cantrell, K.B.; Watts, D.W. Compositional and thermal evaluation of lignocellulosic and poultry litter chars via high and low temperature pyrolysis. *Bioenergy Res.* **2013**, *6*, 114–130. [CrossRef]

59. Mierzwa-Hersztek, M.; Gondek, K.; Limkowicz-Pawlas, A.; Baran, A.; Bajda, T. Sewage sludge biochars management-ecotoxicity, mobility of heavy metals, and soil microbial biomass. *Environ. Toxicol. Chem.* **2017**, *37*, 1197–1207. [CrossRef] [PubMed]

60. Karhu, K.; Mattila, T.; Irina, B.; Regina, K. Biochar addition to agricultural soil increased CH$_4$ uptake and water holding capacity—Results from a short-term pilot field study. *Agric. Ecosyst. Environ.* **2011**, *140*, 309–313. [CrossRef]

61. Steinbeiss, S.; Gleixner, G.; Antonietti, M. Effect of biochar amendment on soil carbon balance and soil microbial activity. *Soil Biol. Biochem.* **2009**, *41*, 1301–1310. [CrossRef]

62. Shenbagavalli, S.; Mahimairaja, S. Characterization and effect of biochar on nitrogen and carbon dynamics in soil. *Int. J. Adv. Biol. Res.* **2012**, *2*, 249–255.

63. Haider, G.; Koyro, H.W.; Azam, F.; Steffens, D.; Müller, C.; Kam-mann, C. Biochar but not humic acid product amendment affected maize yields via improving plant-soil moisture relations. *Plant Soil* **2015**, *395*, 141–157. [CrossRef]

64. Bruun, E.W.; Petersen, C.T.; Hansen, E.; Holm, J.K.; Hauggaard-Nielsen, H. Biochar amendment to coarse sandy subsoil improves root growth and increases water retention. *Soil Use Manag.* **2014**, *30*, 109–118. [CrossRef]

65. Rogovska, N.; Laird, D.A.; Rathke, S.J.; Karlen, D.L. Biochar impact on midwestern mollisols and maize nutrient availability. *Geoderma* **2014**, *23*, 340–347. [CrossRef]

66. Shen, Q.; Hedley, M.; Arbestain, M.C.; Kirschbaum, M.U.F. Can biochar increase the bioavailability of phosphorus? *J. Soil Sci. Plant Nutr.* **2016**, *16*, 268–286. [CrossRef]

67. Kamran, M.; Malik, Z.; Parveen, A.; Zong, Y.; Abbasi, G.H.; Rafiq, M.T.; Shaaban, M.; Mustafa, A.; Bashir, S.; Rafay, M.; et al. Biochar alleviates Cd phytotoxicity by minimizing bioavailability and oxidative stress in pak choi (*Brassica chinensis* L.) cultivated in Cd-polluted soil. *J. Environ. Manag.* **2019**, *250*, 109500. [CrossRef]

68. Naveed, M.; Mustafa, A.; Azhar, A.Q.; Kamran, M.; Zahir, Z.A.; Núñez-Delgado, A. *Burkholderia phytofirmans* PsJN and tree twigs derived biochar together retrieved Pb-induced growth, physiological and biochemical disturbances by minimizing its uptake and translocation in mung bean (*Vigna radiata* L.). *J. Environ. Manag.* **2020**, *257*, 109974. [CrossRef]

69. Mustafa, A.; Naveed, M.; Saeed, Q.; Ashraf, M.N.; Hussain, A.; Abbas, T.; Kamran, M.; Minggang, X. Application potentials of plant growth promoting rhizobacteria and fungi as an alternative to conventional weed control methods. In *Crop Production*; IntechOpen: London, UK, 2019.

70. Yang, S.; Chen, X.; Jiang, Z.; Ding, J.; Sun, X.; Xu, J. Effects of biochar application on soil organic carbon composition and enzyme activity in paddy soil under water-saving irrigation. *Int. J. Environ. Res. Public Health* **2020**, *17*, 333. [CrossRef]

Article

Chlorophyll Fluorescence Parameters and Antioxidant Defense System Can Display Salt Tolerance of Salt Acclimated Sweet Pepper Plants Treated with Chitosan and Plant Growth Promoting Rhizobacteria

Muneera D. F. ALKahtani [1], Kotb A. Attia [2], Yaser M. Hafez [3], Naeem Khan [4], Ahmed M. Eid [5], Mohamed A. M. Ali [6] and Khaled A. A. Abdelaal [7,*]

1   Biology Department, College of Science, Princess Nourah bint Abdulrahman University, Riyadh POX 102275-11675, Saudi Arabia; mdfkahtani@gmail.com

2   Center of Excellence in Biotechnology Research, King Saud University, Riyadh POX 2455-11451, Saudi Arabia; kattia1.c@ksu.edu.sa

3   Plant Pathology and Biotechnology Lab., Faculty of Agriculture, Kafrelsheikh University, Kafr El Sheikh 33516, Egypt; hafezyasser@gmail.com

4   Department of Agronomy, Institute of Food and Agricultural Sciences, University of Florida, Gainesville, FL 32611, USA; naeemkhan@ufl.edu

5   Botany and Microbiology Department, Faculty of Science, Al-Azhar University, Nasr City 11884, Egypt; aeidmicrobiology@azhar.edu.eg

6   Department of Horticulture, Faculty of Agriculture, New Valley University, El-Kharga, New Valley 72511, Egypt; Mohamedahmedali77@aun.edu.eg

7   Excellence Center (EPCRS), Plant Pathology and Biotechnology Lab., Faculty of Agriculture, Kafrelsheikh University, Kafr El Sheikh 33516, Egypt

*   Correspondence: khaled.elhaies@gmail.com

Received: 15 July 2020; Accepted: 1 August 2020; Published: 12 August 2020

Abstract: Salinity stress deleteriously affects the growth and yield of many plants. Plant growth promoting rhizobacteria (PGPR) and chitosan both play an important role in combating salinity stress and improving plant growth under adverse environmental conditions. The present study aimed to evaluate the impacts of PGPR and chitosan on the growth of sweet pepper plant grown under different salinity regimes. For this purpose, two pot experiments were conducted in 2019 and 2020 to evaluate the role of PGPR (*Bacillus thuringiensis* MH161336 10^{6-8} CFU/cm^3) applied as seed treatment and foliar application of chitosan (30 mg dm^{-3}) on sweet pepper plants (cv. Yolo Wonder) under two salinity concentrations (34 and 68 mM). Our findings revealed that, the chlorophyll fluorescence parameter (*Fv/Fm* ratio), chlorophyll *a* and *b* concentrations, relative water content (RWC), and fruit yield characters were negatively affected and significantly reduced under salinity conditions. The higher concentration was more harmful. Nevertheless, electrolyte leakage, lipid peroxidation, hydrogen peroxide (H_2O_2), and superoxide (O_2^-) significantly increased in stressed plants. However, the application of *B. thuringiensis* and chitosan led to improved plant growth and resulted in a significant increase in RWC, chlorophyll content, chlorophyll fluorescence parameter (*Fv/Fm* ratio), and fruit yield. Conversely, lipid peroxidation, electrolyte leakage, O_2^-, and H_2O_2 were significantly reduced in stressed plants. Also, *B. thuringiensis* and chitosan application regulated the proline accumulation and enzyme activity, as well as increased the number of fruit plant^{-1}, fruit fresh weight plant^{-1}, and total fruit yield of sweet pepper grown under saline conditions.

Keywords: sweet pepper; salinity; *Bacillus*; chitosan; chlorophyll fluorescence; fruit yield

1. Introduction

Sweet pepper belongs to Solanacease family. It is an annual plant in the cultivated lands in many countries, however it is grown as a perennial plant in tropical areas. It is one of the most widespread and popular vegetables, and has a greatest economic importance worldwide [1]. It is the richest source of different antioxidants and vitamins and has several health benefits [2]. However, salinity is a very significant factor that threatens the production of economic plants such as sweet pepper [1], strawberry plants [3], and cucumber plants [4]. Salinity damages plant growth and proliferation by creating water stress and cytotoxicity due to the excess in uptake of ions, such as sodium and chloride. Furthermore, salinity is usually accompanied by oxidative stress due to the generation of reactive oxygen species [5,6]. Salinity stress adversely affects morpho-physiological characters of sweet pepper such as plant height and leaf area which are significantly reduced [7]. Likewise, chlorophyll *a* and *b* as well as RWC were reduced under salinity in cucumber [4]. Photosynthesis is harmfully affected by salinity through the reduction in stomatal conductance. Also, salinity led to increased ion toxicity and negatively affected nutrients uptake, especially potassium uptake, so the salt stressed plants showed low membrane stability [8]. The chlorophyll fluorescence parameters were adversely affected with salinity and the content of chlorophyll pigments significantly decreased in cucumber [9]. Also, the study of Misra et al. [10] pointed out that salt stress causes photoinhibition in PSII and decreases its activity. Salt stress led to decreased chlorophyll concentrations, leaf area and mungbean yield [11] and led to an increase in the accumulation of Na^+, decreasing the uptake of mineral nutrients such as nitrogen and potassium [12]. The high level of Na^+ was associated with the ROS accumulation such as H_2O_2 and O^{2-}. The excessive formation of ROS causes protein oxidation and lipid peroxidation under several stresses mainly under salinity stress [1,13]. Previous studies have shown that the adverse effects of salinity stress on leaf number, plant length, fresh and dry weights of shoots, and plant yield also increases with the increase in NaCl concentration [14–16].

According to salinity concentrations, the plants are classified to euhalophytes or glycophytes. Euhalophytes have the salinity thresholds of 250 mM NaCl, i.e., euhalophytes are able to complete their life cycle upon salinities exceeding 250 mM NaCl. Glycophytes cannot grow under high salinity concentrations and their response to salinity differs in terms of osmotic regulation, photosynthetic electron transport, chlorophyll content, and reactive oxygen species (ROS) formation as well as antioxidant defense system [1,7]. The excessive accumulation of ROS under stress, such as salinity [1,17], drought [18,19], and biotic stress factors [20–23], results in the activation of the enzymatic and non-enzymatic antioxidant system to enhance stress tolerance in plants to cope with increased accumulation of ROS [24]. The antioxidative system also consists of some of the non-enzymatic systems, such as salicylic acid and carotenoids. Nonetheless, the enzymatic defense system contains ascorbate peroxidases (APX), glutathione reductases (GR), superoxide dismutases (SOD), catalases (CAT), and peroxidases (POD), which protect the plant tissues against stress factors [25]. Also, the plants have adaptive mechanisms to salinity stress through morphological, anatomical, and biochemical changes. Euhalophytes can cope with salinity stress through different mechanisms, such as salt exclusion, salt elimination, salt succulence and salt redistribution [7]. Furthermore, EL%, lipid peroxidation, and ROS were increased significantly under salinity, as these parameters are signals to various stresses, such as salinity, drought, and heat [26–29], that enable plants to respond to a particular stress. Some plants protect themselves from salinity stress by maintaining ion homeostasis and transportation of the excess salt to the vacuole or sequestering in the older tissues which ultimately are sacrificed, thereby defending itself from salinity stress [30]. Meanwhile, other plants keep the ion concentration in the cytoplasm at a low level. Membranes along with their linked components play an essential role in retaining ion concentration within the cytosol during the period of stress by regulating ion uptake and transport [31,32]. Chlorophyll fluorescence is a fast method for photosynthetic processes measurements [33] and provides a lot of information about the plant status under abiotic and biotic stresses to understand the mechanisms of photosynthesis and how plants respond to various stresses [34]. Chlorophyll fluorescence parameters are important indicators used to measure the quantum yield of photosystem II (PSII), display the plant response to stress and the harmful effects, particularly on photosynthesis and chlorophyll concentrations [35].

Chitosan or chitin is a natural polysaccharide consisting of two molecules of D-glucosamine and naturally present in the cell walls of many organisms such as crabs, shrimp, fungi, and the exoskeleton of insects [36]. In the agricultural field, it improves the morpho-physiological parameters and alleviates the injurious effect of abiotic stresses through stress transduction pathway [37]. Application of chitosan led to increased plant tolerance to many stresses in various plants [38,39], enhance growth characters and improve germination rate of many plants [38,40]. The fruit yield of tomato plants was improved with chitosan treatments [41]. Under drought, barley plants treated with chitosan showed a significant increase in chlorophyll, RWC, total soluble sugar, and grain yield [42]. Plant growth-promoting rhizobacteria (PGPR) can prompt plant tolerance to stress through some chemical and physical changes which are identified as induced systemic tolerance [43]. The application of PGPR led to improved growth and yield production [44]. Under stress conditions, PGPR can improve the injurious impacts and enhance the yield production under salt conditions [45], as a bio-fertilizer in sugar beet and sweet sorghum plants [20,46,47] and as a bio-control agent [48–50]. There are many PGPR strains, such as *Bacillus, Azotobacter, Azospirillum, Pseudomonas, Rhizobium,* and *Serratia,* which can be used in improving plant growth even under various stress factors [51,52] by the production of antioxidants, phytohormones and vitamins [53]. There is a lot of information about the effect of PGPR, nevertheless studies about chitosan and its effects on plants under salinity stress are still scarce and have not yet been fully understood. Hence, in this research, we focus on the effect of chitosan and *Bacillus thuringiensis* MH161336 in alleviating the harmful effect of salinity to improve chlorophyll fluorescence parameters, chlorophyll concentration, enzymes activity, and fruit yield of sweet pepper.

2. Materials and Methods

2.1. Experiments Preparation and Plant Materials

Two pot experiments were conducted at Kafrelsheikh University, Agricultural Botany Department during two summer seasons 2019 and 2020, to evaluate the effect of seed treatment with plant growth promoting rhizobacteria (*B. thuringiensis* MH161336 10^{6-8} CFU/cm^3) and foliar spray with chitosan 30 mg·dm^{-3} on sweet pepper plants under salinity (sodium chloride at 34 and 68 mM). The physio-biochemical characters were done at Plant Pathology & Biotechnology Lab., and EPECRS Excellence Center, Kafrelsheikh University. The seeds of sweet pepper (*Capsicum annuum* L.) cv. Yolo Wonder (obtained from a private agricultural company) were divided into three groups (the first group was treated with *B. thuringiensis* and the others without treatments). Seed treatment was done with *B. thuringiensis.* Thereby, the seeds underwent surface sterilization by sodium hypochlorite 2.5% for 5 min, 70% ethanol for 1 min, and were then washed 5 times by sterile distilled water. *B. thuringiensis* MH161336 which was isolated from the halophytic plant *Spergularia marina* (obtained from Dr. Ahmed Eid), *B. thuringiensis* pure cultures were grown in nutrient broth at 35 ± 2 °C on a shaker at 180× *g*. Bacterial cultures were diluted in sterilized distilled water to reach a final concentration of 10^{6-8} CFU/cm^3 [54]. Sterilized seeds were incubated with bacterial suspensions at room temperature for 6 h and sown in the nursery in foam trays on 7th and 3rd January in the two seasons, respectively. After forty-five days from the sowing, the transplantation was done in pots 50 cm^3 in diameter, each one containing two seedlings and the pots were divided into three groups (control, *B. thuringiensis* treatment and chitosan treatment 30 mg·dm^{-3}). The plants irrigated with two concentrations (34 and 68 mM) of saline water (was prepared from NaCl) and the group of chitosan treatment was treated with chitosan 30 mg·dm^{-3} twice after 20 and 40 days from transplanting. The compound fertilizer containing nitrogen, phosphorus, and potassium (NPK) (135:40:35 kg·ha^{-1}) was used as recommended in two doses, the first dose after 12 days from transplanting and the second at the flowering stage initiation. The experiments were in a completely randomized design with 4 replicates, the physiological and biochemical studies were done at 80 days from transplanting. The chemical and physical characters of experimental soil were determined [55] and are presented in Table 1.

Table 1. Chemical and physical characters of the experimental soil before conducting the experiments in 2019 and 2020 seasons.

Seasons	PH	* EC Ds/m	Mechanical Analysis			Soil Texture	Organic Matter (%)	Total N (%)	Total P (ppm)
			Sand%	Silt%	Clay%				
2019	8.11	0.464	21.96	23.98	47.4	Clay	1.79	0.158	8.8
2020	8.16	0.483	22.17	24.29	46.8	Clay	1.82	0.149	8.2

Seasons	Soluble Cations				Soluble Anions		
	Na^+	K^+	Ca^{++}	Mg^{++}	HCO_3^-	SO_4^{--}	Cl^-
2019	2.14	0.18	2.02	2.4	4.2	2.07	0.6
2020	2.19	0.17	2.04	2.3	4.3	1.93	0.5

* EC = Electrical conductivity.

2.2. Biochemical and Physiological Characters

The physiological and biochemical characters were recorded at 80 days from transplanting.

2.2.1. Relative Water Content (RWC%)

According to Sanchez et al. [56], twenty leaf discs were used to determine RWC, the fresh weight (FW) for the discs was determined, the same discs were saved in petri dishes containing distilled water for 1 h to determine the turgid weight (TW), after that the discs were dried for 24 h at 80 °C to determine the dry weight (DW). Relative water content (RWC%) was calculated as follows: RWC = (FW − DW)/(TW − DW) × 100.

2.2.2. Determination of Chlorophyll *a* and *b* Concentrations

The extraction was done using *N-N* Dimethyl formamide, whereby 5 cm^3 *N-N* Dimethyl formamide was added to 1 g fresh leaves and kept in the refrigerator overnight, and the absorbance was measured at 647 and 664 nm according to Moran [57].

2.2.3. Electrolyte Leakage Assay (EL%)

Ten discs (1 cm^2) of sweet pepper leaves were placed into flasks containing deionized water (25 cm^3). Flasks were shaken for 20 h, initial electrical conductivity was recorded for each vial and then flasks were immersed in a water bath at 80 °C for 1 h. The vials were shaken for 20 h at 21 °C. Final conductivity was measured for each flask. Electrolyte leakage % was calculated according to Szalai et al. [58] with the following formula: initial conductivity/final conductivity × 100.

2.2.4. Chlorophyll Fluorescence Parameter

Chlorophyll *a* fluorescence parameter (*Fv/Fm* ratio) was measured at 80 days from the sowing using a chlorophyll fluorometer (PEA, Hansatech Instrument Ltd., version 1.21, Norfolk, UK).

According to Schreiber [59], middle-aged sweet pepper leaves were placed in the dark for 30 min to stimulate the reaction of photosystem II. The minimum chlorophyll fluorescence (*Fo*) was measured using a measuring beam of <0.1 µmol m^{-2}·s^{-1}. The maximum fluorescence (*Fm*) was determined after a 1 s saturating pulse (>3500 µmol·m^{-2}·s^{-1}). Variable fluorescence (*Fv*) was determined by the difference between the maximum fluorescence and the minimum fluorescence (*Fm* − *Fo*). The maximum efficiency of PSII was determined as the ratio of (*Fv*) to (*Fm*) as follows: *Fv/Fm* = (*Fm* − *Fo*)/*Fm*.

2.2.5. Proline Determination

According to Bates et al. [60], proline was assayed in sweet pepper plants, 0.5 g fresh leaf in 3% sulphosalicylic acid and centrifuged for 20 min at 3000× g. Then, 2 cm^3 of glacial acetic acid and 2 cm^3 ninhydrin reagent was boiled with 2 cm^3 supernatant for 1 h, the reaction was completed in an ice bath, and proline was separated using toluene. Proline was determined as $\mu g \cdot g^{-1}$ FW using a spectrophotometer at 520 nm.

2.2.6. Determination of Lipid Peroxidation

According to Davenport et al. [61], lipid peroxidation was measured as malondialdehyde (MDA) using 100 mg fresh leaves in 1% trichloro acetic acid and centrifuged at 10,000× g for 5 min. 0.5% thiobarbituric acid was then added, and mixture was boiled at 95 °C for half an hour. The samples were placed on an ice bath and centrifuged for 5 min at 5000× g, the measurements were done using spectrophotometer at 532 and 600 nm. MDA (nmol·g^{-1} FW) = [6.45 × (A532 − A600) − (0.56 × A450)] × V^{-1}W, where V = volume (cm^3); W = weight (g).

2.2.7. Determination of Superoxide (O_2^-) and Hydrogen Peroxide (H_2O_2)

Sweet pepper leaves were vacuum infiltrated with 10 mM potassium salicylate buffer containing 0.1 *w/v*% nitro blue tetrazolium (NBT) or 0.1 *w/v*% 3,3-diaminobenzidine (DAB). The leaves were incubated in the light for 140 min and two hours, respectively. The samples were cleared with trichloroacetic acid in ethanol: chloroform 4:1 *v/v* for 1 day, the samples were washed and placed in 50% glycerol. O_2^- and H_2O_2 were determined as nmol·g^{-1} FW according to Huckelhoven et al. [62] using a ChemiImager 4000 digital imaging system (Alpha Innotech Corp., San Leandro, CA, USA).

2.2.8. Assay of Enzymes Activity

For the determination of enzymes, 0.5 g fresh leaves were homogenized in 3 cm^3 of 50 mM Tris buffer at 0–4 °C, containing 1 mM EDTA-Na$_2$ and 7.5% polyvinyl pyrrolidone. The samples were centrifuged 12,000× g for 20 min at 4 °C and the total soluble enzyme activities were measured using spectrophotometer in the supernatant [63]. Catalase activity (CAT) was determined through the decomposition of H_2O_2 by catalase results in the decrease of the ultraviolet absorption of H_2O_2 at 240 nm, catalase activity can be calculated from this decrease. The reaction mixture contained 2.15 cm^3, 2 cm^3 0.1 M Na-phosphate buffer, 100 μL H_2O_2, and 50 μL leaves extract. The solution is mixed, and the absorptions were recorded at 240 nm according to Aebi [64]. Peroxidase activity (POX) was calculated according to Hammerschmidt et al. [65]. The reaction mixture consisted of 2.9 cm^3 of a 100 mM sodium phosphate buffer containing 0.25% (*v/v*) guaiacol and 100 mM H_2O_2. The reaction was done with adding 100 μL of crude enzyme extract, the changes in absorbance were recorded every 30 s intervals for 3 min at 470 nm, the activity was determined for min^{-1}·g^{-1} fresh weight. Activity of superoxide dismutase (SOD) was measured according to Mishra et al. [66]. Then, we add 290 μL of a mixture containing 100 mM potassium phosphate buffer, 0.1 mM EDTA, 11 mm^3 xanthine, cytochrome-c, and 0.002 units of xanthine oxidase to 20 μg of protein extracts was prepared. Xanthine oxidase regulation produced an increase in the absorbance due to the reduction of cytochrome-c (0.025 ± 0.005 min^{-1}). Activity of SOD was stated by McCord and Fridovich [67]. According to Goldberg and Spooner [68], GR activity was measured, approximately 0.05 cm^3 enzyme extract was mixed with 1 cm^3 phosphate buffer combined with EDTA, 0.1 cm^3 glutathione, and 0.1 cm^3 NADPH, the absorbance was determined at 340 nm.

2.2.9. Fruit Yield

The harvest date starts at 120 days from transplanting to determine number of fruit plant^{-1}, fruit fresh weight plant^{-1} (g), and total fruit yield (ton hectare^{-1}).

2.3. Statistical Analysis

Statistical analysis was done using analysis of variance (ANOVA) procedures according to the method of Gomez and Gomez [69] using the MSTAT-C statistical software package. The means between treatments were compared by Duncan [70] when the difference was significant ($P \leq 0.05$). The correlation analysis was done using XLSTAT 2014.5.03 statistical software.

3. Results

3.1. Effect on Relative Water Content (RWC%)

The presented results in Figure 1A showed a significant decrease in RWC in sweet pepper under two salinity concentrations (57.6% at the low concentration (34 mM) (S1) and 52% at the high concentration (68 mM) (S2) comparing with control plants (74.6%) as the mean of the two seasons. Likewise, the results in Figure 1 revealed that seed treatment with *B. thuringiensis* showed a significant increase in RWC in stressed plants (65.7% compared with 57.6% at the low concentration and 60.8% compared with 52% at the high concentration). Furthermore, chitosan application at 30 mg dm^{-3} caused a significant increase in RWC (71.5% compared with 57.6% at the low concentration of salinity) and (67.1% compared with 52% at the high concentration) as a mean of both seasons in the stressed plants. The best treatment under salinity conditions was chitosan at 30 mg·dm^{-3} which achieved 71.5% when compared with control plants 74.6% without any significant difference.

3.2. Effect on Chlorophyll a and b Concentrations

It is obvious from the achieved results in Figure 1B–C that chlorophyll was significantly reduced in stressed plants; chlorophyll *a* significantly decreased at low concentration of salinity (2 mg·g^{-1} FW^{-1}) compared with control (2.85 mg·g^{-1} FW^{-1}) as the mean of both seasons. Furthermore, the high salinity concentration caused a significant reduction in chlorophyll *a* (1.25 mg·g^{-1} FW^{-1}) in stressed plants compared to control (2.85 mg·g^{-1} FW^{-1}). Similarly, salinity stress led to a significant decrease in chlorophyll *b* concentration, the two concentrations caused significant decreases (0.84 and 0.55 mg·g^{-1} FW^{-1} respectively) compared with control (2.85 mg·g^{-1} FW^{-1}). Nonetheless, seed treatment with *B. thuringiensis* and chitosan application led to significant increases in chlorophyll *a* and *b*. The greatest result was obtained with chitosan (S1 + Chitosan) treatment (2.85 mg·g^{-1} FW^{-1}) in the stressed plants with the low salinity concentration compared to the stressed plants (S1) without treatments (2 mg·g^{-1} FW^{-1}).

3.3. Effect on Electrolyte Leakage (EL%)

The presented data in Figure 1D exhibited that EL% significantly increased in the stressed plants, the low salinity concentration caused significant increase (42.3%) comparing with control (13.8%) as the mean of two seasons. Furthermore, the high salinity concentration was more harmfully effective and caused a significant increase in EL% (52.6%) compared with control (13.8%). Nevertheless, chitosan application 30 mg dm^{-3} and seed treatment with *B. thuringiensis* led to significant decrease in EL% in the stressed plants under the two concentrations. Seed treatment with *B. thuringiensis* caused a positive effect and significant decrease in EL% (30.2% and 37.6%) in the stressed plants at the two concentrations compared with untreated plants (42.3% and 52.6%), respectively. Furthermore, EL% was reduced significantly in the stressed treated plants with chitosan 30 mg dm^{-3} (21.7% and 27.2%) that compared with the stressed untreated plants (42.3% and 52.6%).

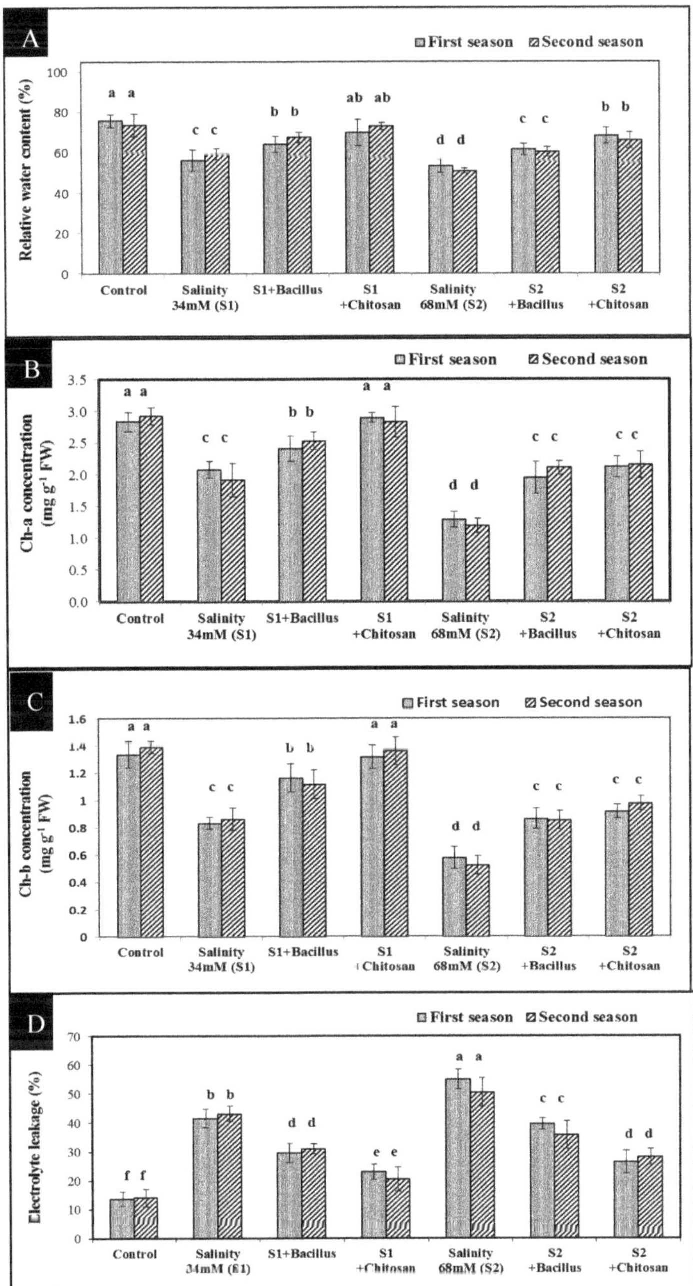

Figure 1. Effect of *B. thuringiensis* and chitosan on relative water content (**A**) chlorophyll *a*, (**B**) chlorophyll *b*, (**C**) and electrolyte leakage (**D**) under two salinity concentrations in sweet pepper plants during two seasons [first season (2019) and second season (2020)]. Data is the mean (±SE) of four replicates. Different letters above the data columns indicate significant differences between the samples determined by ANOVA, Duncan's multiple range test at 0.05 level.

3.4. Effect on Proline Concentration

It could be noted from Figure 2A that the exposed plants to salinity at (S1) and (S2) caused a significant increase in proline concentration, the high concentration of salinity (S2) achieved the high concentration of proline (24 µg·g^{-1} FW) comparing to control (9.1 µg·g^{-1} FW) as the mean of both seasons in sweet pepper. Application of seed treatment with *B. thuringiensis* and chitosan application in stressed plants led to regulate proline accumulation when compared with the control and the stressed untreated plants. *B. thuringiensis* seed treatment led to the regulation of proline accumulation in the stressed plants (12.7 µg·g^{-1} FW at the low concentration of salinity and 13.6 µg·g^{-1} FW at the high concentration comparing to the stressed untreated plants 17.4 and 24 µg·g^{-1} FW) at the two concentrations, respectively. Chitosan application had a significant effect on proline content (9.8 and 12.2 µg·g^{-1} FW) compared with stressed untreated plants (17.4 and 24 µg·g^{-1} FW) at the two concentrations, respectively. The difference was not significant between the both seasons.

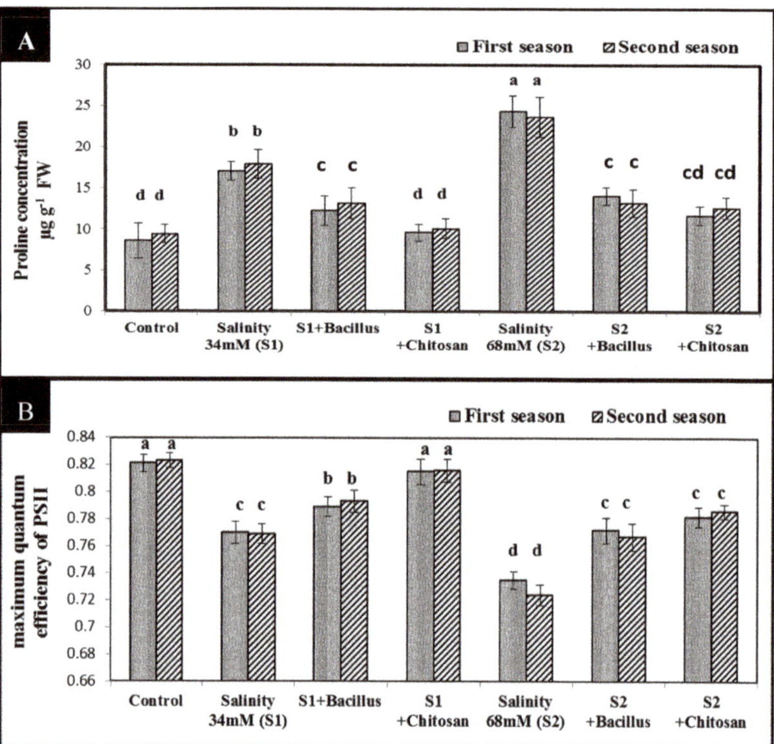

Figure 2. Effect of *B. thuringiensis* and chitosan on proline content (**A**) and maximum efficiency of PSII (*Fv/Fm*) (**B**) under two salinity concentrations in sweet pepper during two seasons. Data is the mean (±SE) of four replicates. Different letters above the data columns indicate significant differences between the samples determined by ANOVA, Duncan´s multiple range test at 0.05 level.

3.5. Effect on Chlorophyll Fluorescence Parameter (Fv/Fm)

Our results in Figure 2B indicated that chlorophyll fluorescence parameters were adversely affected under salinity conditions. The maximum efficiency of PSII (*Fv/Fm*) significantly reduced in sweet pepper (0.790) at the low salinity concentration and (0.729) at the high salinity concentration, respectively comparing to the control (0.822). However, seed treatment with *B. thuringiensis* caused significant increase in *Fv/Fm* ratio in the stressed plants (0.791) at the low concentration of salinity and (0.769) at the high salinity concentration when compared with the stressed untreated plants (0.790) at the low concentration and (0.729) at the high salinity concentration. Likewise, under the two concentrations, chitosan caused a significant increase in *Fv/Fm* ratio. The best treatment was chitosan at the low salinity concentration (0.815) compared with control (0.822).

3.6. Effect on Lipid Peroxidation as Malondialdehyde

According to the findings in Figure 3, lipid peroxidation (MDA) significantly increased in sweet pepper (11.35 and 13.8 μmol·g^{-1} FW) at the two salinity concentrations, respectively as the mean of both seasons when compared with control plants (6.75 μmol·g^{-1} FW). Nevertheless, MDA significantly decreased in the stressed plants according to seed treatment with *B. thuringiensis* and chitosan treatment. *B. thuringiensis* treatment had a positive effect on MDA and led to significant reduction in the MDA content at the two salinity concentrations (8.8 and 10.5 μmol·g^{-1} FW) when compared with the stressed untreated plants (11.35 and 13.8 μmol·g^{-1} FW). The application of chitosan significantly reduced MDA content in sweet pepper under the two salinity concentrations (7 and 7.85 μmol·g^{-1} FW) when compared with stressed untreated plants (11.35 and 13.8).

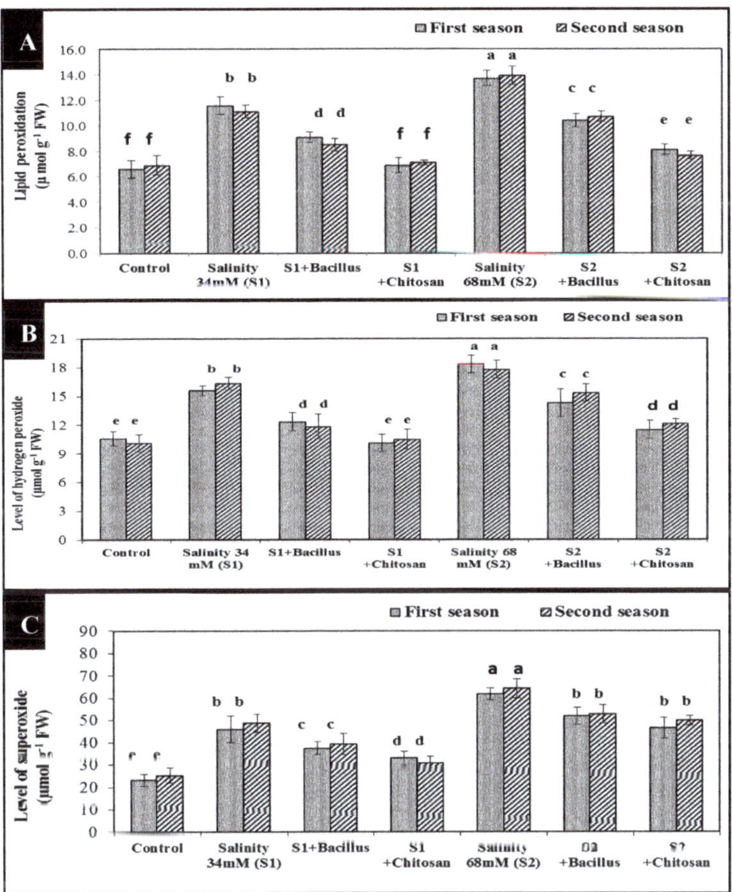

Figure 3. Effect of *B. thuringiensis* and chitosan on lipid peroxidation (**A**), H_2O_2 (**B**) and O_2^- (**C**) under two salinity concentrations in sweet pepper during two seasons. Data is the mean (±SE) of four replicates. Different letters above the data columns indicate significant differences between the samples determined by ANOVA, Duncan's multiple range test at 0.05 level.

3.7. Effect on O_2^- and H_2O_2

ROS, mainly O_2^- and H_2O_2 significantly increased under the both salinity concentrations (Figure 3). O_2^- significantly increased (47.4 and 63 units) at the two salinity concentrations compared with control (24.16 units). Conversely, *B. thuringiensis* treatment caused a significant decrease in O_2^- in the salt stressed plants (38.3 and 52.3 units) in comparison with stressed untreated plants (47.4 and 63 units). Also, chitosan treatment caused a significant decrease in O_2^- (31.7 and 48.3 units) when compared with the stressed untreated plants (47.4 and 63 units).

Salinity stress caused a significant increase in H_2O_2 in sweet pepper (16 and 18.1 units) at the two concentrations, respectively as compared to control (10.3 units). However, the levels of H_2O_2 were decreased significantly according to *B. thuringiensis* seed treatment and chitosan application in the stressed plants at the two salinity concentrations. Chitosan application gave the best and most significant results (10.3 and 11.8 units) compared to stressed untreated plants (16 and 18.1 units) at the two salinity concentrations, respectively.

3.8. Effect on the Activity of Catalase (CAT), Peroxidase Activity (POX), Superoxide Dismutase (SOD) and Glutathione Reductase (GR) Enzymes

Salinity stress at both concentrations caused significant increases in CAT, POX, SOD and GR enzyme (Figure 4). CAT activity significantly increased in the stressed plants (124.8 and 149.3 mM H_2O_2 g^{-1} FW min^{-1}) at the two salinity concentrations, respectively, when compared with control (78.6 mM H_2O_2 g^{-1} FW min^{-1}).

However, chitosan treatment and *B. thuringiensis* seed treatment caused significant reduction in CAT activity at both salinity concentrations. Chitosan with the low salinity concentration (S1 + Chitosan) gave the best result (85.8 mM H_2O_2 g^{-1} FW min^{-1}) as compared to stressed untreated plants (124.8 mM H_2O_2 g^{-1} FW min^{-1}) and control plants (78.6 mM H_2O_2 g^{-1} FW min^{-1}). Moreover, POX, SOD and GR activities significantly increased in the stressed plants at the low salinity concentration (0.6 μmol tetra-gualacol g^{-1} FW min^{-1}, 74.5 and 0.59 unit/cm^3) as compared to control plants (0.24, 38.3 and 0.36), also, the enzymes activity significantly increased in the stressed plants at the high concentration (0.76, 98.7 unit mg^{-1} FW min^{-1} and 0.59 unit/cm^3) respectively. Nevertheless, chitosan application and seed treatment with *B. thuringiensis* caused a significant reduction in POX, SOD, and GR activity in the stressed plants at the two salinity concentrations compared to the stressed untreated plants.

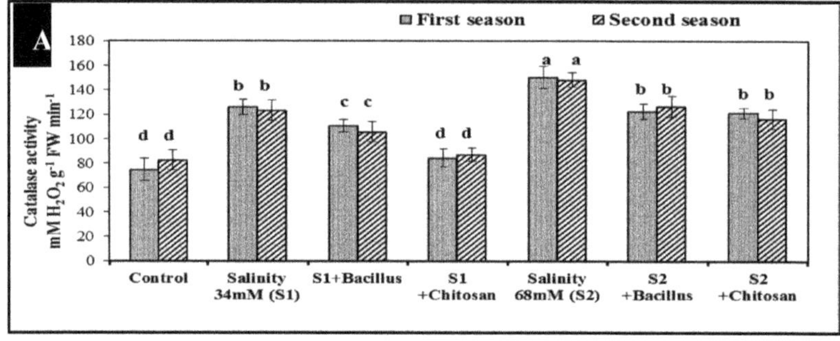

Figure 4. *Cont.*

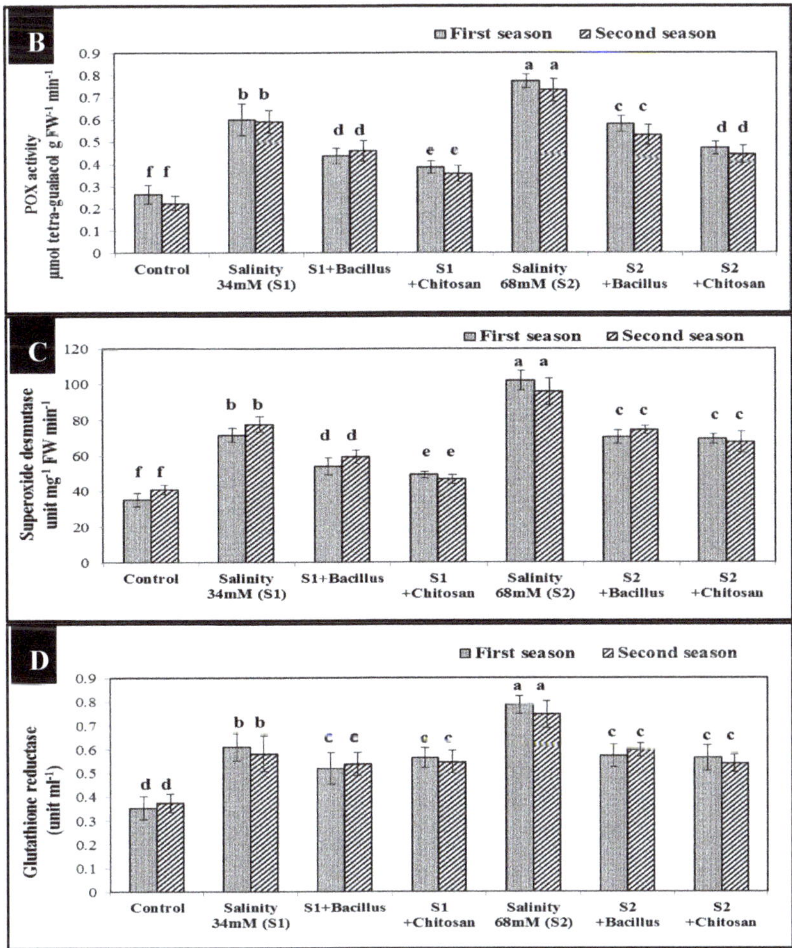

Figure 4. Effect of *B. thuringiensis* and chitosan on the activity of CAT (**A**), POX (**B**), SOD (**C**) and GR (**D**) under two salinity concentrations in sweet pepper during two seasons. Data is the mean (±SE) of four replicates. Different letters above the data columns indicate significant differences between the samples determined by ANOVA, Duncan´s multiple range test at 0.05 level.

3.9. Effect on Number of Fruit Plant⁻¹, Fruit Fresh Weight Plant⁻¹ and Total Fruit Yield (Ton Hectare⁻¹).

In the present study, the results in Figure 5 point out that salinity at the both concentrations caused a significant decrease in number of fruit plant⁻¹ (7.7 and 4.8 fruit), fresh weight of fruit plant⁻¹ (524.5 and 356.4 g) and total fruit yield hectare⁻¹ (7.05 and 5 ton) as the mean of the both seasons when compared to control plants (15.7 fruit plant⁻¹, 974 g plant⁻¹ and 14.9 ton hectare⁻¹). However, *B. thuringiensis* and chitosan significantly increased the number of fruit plant⁻¹, fruit fresh weight (g plant⁻¹) and total fruit yield (ton hectare⁻¹) in the stressed plants compared with untreated plants. Interestingly enough, under the both salinity concentrations, chitosan application gave the best results and significantly increased the number of fruit plant⁻¹ (14.9 and 12.7), fruit fresh weight plant⁻¹ (911 and 527 g plant⁻¹), and total fruit yield (14 and 10.8 ton hectare⁻¹) as the mean of the both seasons.

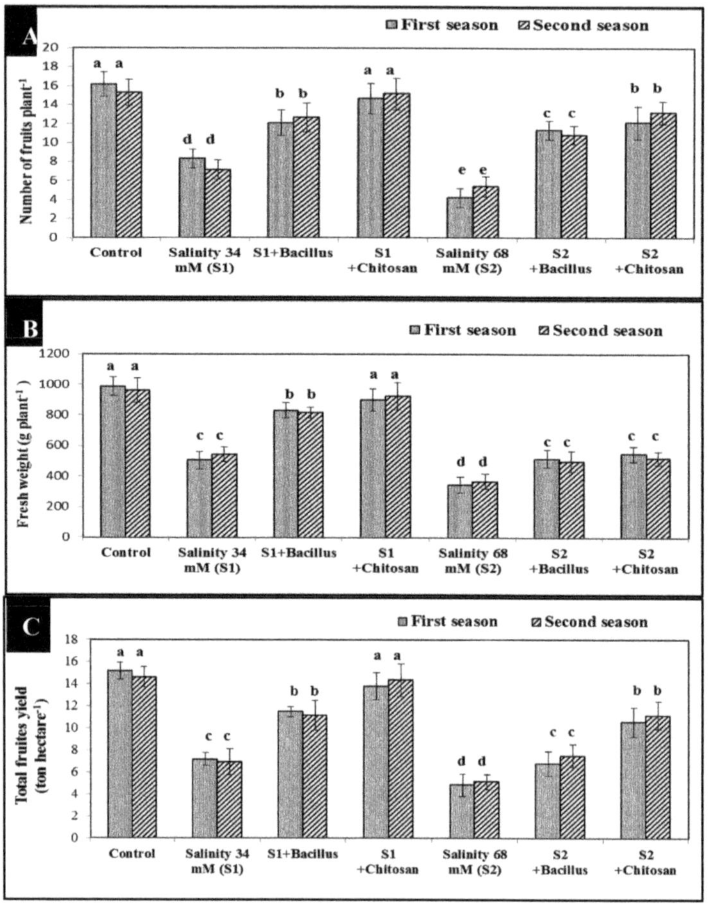

Figure 5. Effect of *B. thuringiensis* and chitosan on number of fruit plant^{-1}(**A**), fruit fresh weight plant^{-1} (**B**) and total fruit yield (ton hectare^{-1}) (**C**) under two salinity concentrations in sweet pepper during two seasons. Data is the mean (±SE) of four replicates. Different letters above the data columns indicate significant differences between the samples determined by ANOVA, Duncan's multiple range test at 0.05 level.

3.10. Correlation Studies

In the present study chlorophyll *a* was positively and significantly correlated with chlorophyll *b* (r = 0.99), number of fruits (r = 0.98), RWC (r = 0.97), GR (r = 0.80) and MDA (r = 0.75). Among the treatment it has a negative correlation with salinity stress @ 34 mM (r = −0.05), salinity stress @ 68 mM (r = −0.02), however, a positive correlation was noted among the chlorophyll *a* and treatments of Bacillus sp. and chitosan (Figure 6 and Supplementary Table S1). A similar trend of relationship was shown by chlorophyll *b*. Proline showed highly positive correlation with MDA (r = 0.98), H_2O_2 (r = 0.97), SOD (r = 0.96) and GR (r = 0.96) but was negatively correlated with the treatments Bacillus sp. (r = −0.04) and chitosan (r = −0.05). A very similar correlation was observed among all the studies. Antioxidant enzymes and H_2O_2 concentration that were highly correlated with each other also showed a negative correlation with the treatments of *Bacillus* sp. and chitosan. The number of fruits showed a highly significant correlation with chlorophyll *a* and *b* (r = 0.98) and with RWC (r = 0.95). However, this trait was inversely related to the treatments of salinity @ 34 mM (r = − 0.15) and @ 68 mM (r = − 0.25).

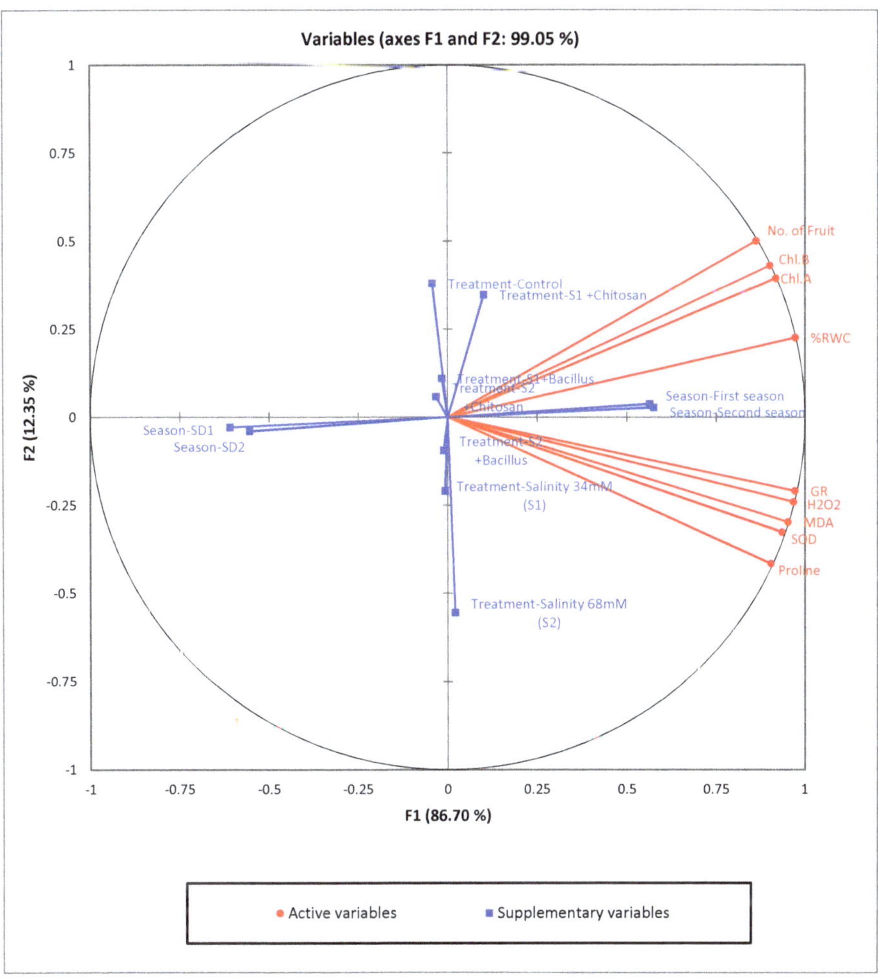

Figure 6. Circle of correlation between variables and factors for sweet pepper.

4. Discussion

Salinity stress adversely affects plant growth, inhibiting plant development and reducing fruit yield of sweet pepper. The present data revealed the deleterious effects of salinity at the two different concentrations (34 and 68 mM) on RWC. This might be due to the injurious influence of salinity on the cell wall structure [71], thereby increasing ethylene concentration, which reduces the growth of roots [44]. This effect causes changes in cell wall properties, the reduction in osmotic potential, and the decrease in water balance [72], consequently reducing RWC in sweet pepper [1]. These deleterious impacts of salinity were overcome by seed treatment with *B. thuringiensis* and treating stressed sweet pepper with chitosan. The pivotal role of *B. thuringiensis* under salinity stress could be due to the formation of Indole-acetic acid which causes enhancement of root growth and increased water uptake [73]. Likewise, PGPR can produce exopolysaccharides (EPSs) which aggregate with soil particles and improve soil structure as well as water uptake [74]. Further, the application of PGPR causes a decay in the soil bulk density and enhances the availability of soil water. Chitosan application positively affects RWC in stressed plants, this progressive effect of chitosan could be due to the positive role of chitosan on water availability in stressed plants. These valuable effects were documented in barley under drought [19].

Chlorophyll *a* and *b* are very important pigments in the process of photosynthesis, in this process, two reactions take place. One such reaction is the light reaction, in which NADPH and ATP are produced, and the second is the dark reaction, in which carbon dioxide is fixed [75]. Demonstrated data revealed a significant decrease in chlorophyll content under the two salinity concentrations, this decrease in chlorophyll was more considerable at the high concentration (68 mM) than at lower concentration (34.mM) and this might be due to the damaging effect of salinity on the chloroplast structure [3,76], that decrease energy transport from PSII to PSI [77] and, consequently, reduce the chlorophyll formation in stressed sweet pepper plants. The harmful effect of salinity on the content of chlorophyll was also due to reduction in stomatal conductance and destruction of biochemical processes [78]. These findings are in accordance with those reported by Abdelaal et al. [1] in sweet pepper under salinity stress. Also, Asrar et al. [79] indicated that a high salinity concentration caused harmful effects on PSII and decreased chloroplast proteins as well as chlorophyll concentrations. This decrease in chlorophyll concentrations is related to the reduction in RWC under high salt concentration.

Conversely, inoculation of seeds with *B. thuringiensis* mitigates the adverse effects of salinity on the content of chlorophyll that improve the overall growth and proliferation of plants under stressful environments [80]. Beside this, the application of chitosan had also synergistic effects on the contents of chlorophyll *a* and *b*. This increase in the content of chlorophyll with the application of chitosan may be attributed to the fact that chitosan is a rich source for amino acids which increase the chloroplast number and chlorophyll formation. These results are in harmony with the findings of Possingham [81]. During the present study, a significant increase was found in EL% under two different salt concentrations mainly. The higher salt concentration was more effective and significantly increased the EL%. This negative influence of salinity on EL% may be due to its damaging impacts on the cytoplasmic membrane and permeability process. Previously, a similar result was reported by Abdelaal et al. [1] in sweet pepper. Contrariwise, EL% significantly reduced in stressed plants as a result of seed treatment with *B. thuringiensis* and chitosan, these valuable effects of *B. thuringiensis* treatment and chitosan application is attributed to the positive roles of *B. thuringiensis* and chitosan on membrane stability and an improvement in the selective permeability of cell plasma membrane.

In the present study, the chlorophyll fluorescence parameter was adversely affected under two salinity concentrations. Salinity stress causes a significant decrease to maximum efficiency of PSII (*Fv/Fm*). This adverse effect of salinity on (*Fv/Fm*) might be due to its role in the inhibition of electron transport and the reaction centers at the PSII sites as well as destroys the oxygen-evolving complex [82–84]. Also, salinity stress has a negative effect on enzymes activity and decreases the activity of water splitting enzyme complexes and electron transport chains resulting in decrease *Fv/Fm* [85]. However, seed treatment with *B. thuringiensis* and the application of chitosan caused a significant increase *Fv/Fm* ratio in the stressed plants. These results are credited to the helpful role of *B. thuringiensis* and chitosan in increasing the production of protective metabolites, increasing N and K content as well as the number of chloroplasts under stress [81,86], and consequently, improving the chlorophyll fluorescence parameter. The obtained results indicated that proline significantly increased in the stressed plants under both the salinity concentrations (34 and 68 mM). This impact of salinity may be due to its role in reducing the proline oxidation to glutamate, consequently increasing the proline content [87]. Proline is one of the most important osmoprotectants, plays a key role in osmotic regulation, and protects the plants under stress [1,8]. Chitosan application and seed treatment with *B. thuringiensis* regulated proline content under salinity conditions. Seed inoculation with *B. thuringiensis* positively regulated proline content under stress because this species regulates the osmotic balance under saline conditions. Similar results for proline production under saline conditions were also reported by Egamberdieva et al. [88].

Salinity could hamper plant growth and increase lipid peroxidation, O_2^-, and H_2O_2. A significant increase was noted in the mentioned parameters during the present study. These reactive compounds can damage lipids and proteins, essential for the process of photosynthesis and electron transport chain. Islam et al. [18] noted similar results in two wheat cultivars grown under saline conditions. However, in the present study, a significant decrease was noted in the lipid peroxidation upon treatment with chitosan. This may be due to the involvement of chitosan in cell protection from oxidative stress under salinity conditions. Similarly, O_2^- and H_2O_2 were significantly reduced with chitosan due to the presence of hydroxyl and amino groups which react with ROS, thus chitosan can scavenge superoxide radicals [89]. Chitosan derived from the pathogen is recognized by a specific cellular receptor resulting in enhancing the defense response to abiotic and biotic stresses [90]. The positive effect of chitosan in the plant cell protection was also noted in plants under drought stress [20]. Interestingly, seed treatment with *B. thuringiensis* led to improved cell membrane stability and decreased the formation of MDA in the stressed sweet pepper, this effect of *B. thuringiensis* is due to its improved phenol content and defense enzyme system [91]. Also, *B. thuringiensis* causes decreases in O_2^- and H_2O_2 by increasing reactive oxygen scavenging enzyme activity [92].

Enzymes up-regulation (CAT, POX, SOD, and GR) is involved in the mitigation of salinity stress in sweet peppers compared with control plants. The significant increase in these enzymes is a natural defense system, which helps to cope with salinity stress and reduces the osmotic and toxic effects by scavenging ROS. Our results are in agreement with those reported by Abdelaal et al. [17] and Foyer et al. [93]. Nevertheless, it was clear from our results that the application of seed treatment with *B. thuringiensis* led to improved and regulated up-regulation of CAT, POX, SOD, and GR in the stressed sweet pepper. The induction of these enzymes is involved in the mitigation of salt stress in sweet pepper treated with *Bacillus*. A similar trend of enzyme activity was recorded in the findings of Kohler et al. [94]. Likewise, chitosan application causes an increase in enzymes activity to protect the plant from oxidative damage and reduce lipid peroxidation as well as scavenge O_2^- due to its structure and protective role in sweet pepper plants subjected to salinity stress. These results are in agreement with those reported by Hafez et al. [19]. The presented study showed that two salinity concentrations caused a significant reduction in the number of fruit plant^{-1}, fruit fresh weight plant^{-1}, and total fruit yield. This harmful impact of salinity may be due to the decrease in reproductive organs, such as pollen grains in stressed plants [95], and also due to the decrease in water absorption, nutrients uptake, and chlorophyll content [1,4], resulting in a significant decrease in fruit yield [96]. The vital role of *B. thuringiensis* might be due to the formation of growth regulators such as gibberellins, auxin, and cytokinins, as well as an increase in proline content [87], up-regulation of essential enzymes and solubilization of nutrients [89], and an increase in the number of fruits and fruit yield hectar^{-1} in sweet pepper. These findings are in agreement with the previous results reported by Hafez et al. [19], Katiyar et al. [36], and Hidangmayum et al. [37].

5. Conclusions

The present research concluded that seeds treated with *B. thuringiensis* and foliar application of chitosan 30 mg dm^{-3} on sweet pepper plants under two salinity concentrations (34 and 68 mM) led to an improvement of the adverse effects of salinity and enhanced the growth and yield of sweet pepper. RWC, chlorophyll *a* and *b* concentrations, chlorophyll fluorescence parameters, and fruit yield characters significantly increased with *B. thuringiensis* and chitosan treatments in sweet pepper under two salinity concentrations. Conversely, lipid peroxidation, electrolyte leakage, and reactive oxygen species (O_2^- and H_2O_2) were decreased significantly as a result of *B. thuringiensis* and chitosan treatments. Overall, seed treatment with *B. thuringiensis* and chitosan foliar application was an effective and cheaper approach to cope with the deleterious effects of salinity on sweet pepper by improving the chlorophyll fluorescence parameters, proline accumulation, and up-regulation of enzymes activity as well as the enhancement of fruit yield characters.

Supplementary Materials: The following are available online at http://www.mdpi.com/2073-4395/10/8/1180/s1, Table S1: Correlation matrix among different treatments and quantitative traits of sweet pepper.

Author Contributions: Conceptualization, K.A.A.A., Y.M.H., K.A.A., and M.D.F.A.; methodology, K.A.A.A., Y.M.H., K.A.A., M.D.F.A., A.M.E., and M.A.M.A.; software, K.A.A.A., Y.M.H., K.A.A., M.D.F.A., and M.A.M.A.; formal analysis, K.A.A.A., Y.M.H., K.A.A., A.M.E., M.A., and N.K.; investigation, K.A.A.A., Y.M.H., K.A.A., A.M.E., M.A.M.A., and N.K.; resources, K.A.A.A., Y.M.H., K.A.A., A.M.E., M.A.M.A., and N.K.; data curation, K.A.A.A., Y.M.H., K.A.A., A.M.E., M.A.M.A., and N.K.; writing-original draft preparation, K.A.A.A., Y.M.H., K.A.A., A.M.E., M.A.M.A., M.D.F.A., and N.K.; writing-review and editing, K.A.A.A., Y.M.H., A.M.E., M.A., N.K., and M.D.F.A.; visualization, K.A.A.A., Y.M.H., and K.A.A.; funding acquisition, M.D.F.A., K.A.A.A., M.A., A.M.E., and K.A.A. All authors have read and agreed to the published version of the manuscript.

Funding: This research was funded by the Deanship of Scientific Research at Princess Nourah bint Abdulrahman University through the Fast-track Research Funding Program.

Acknowledgments: The authors extend their appreciation to all members of PPBL and EPCRS excellence center, Faculty of Agriculture, Kafrelsheikh University, and the Deanship of Scientific Research at Princess Nourah bint Abdulrahman University Fast-track Research Funding Program.

Conflicts of Interest: The authors declare no conflict of interest

References

1. Abdelaal, K.A.A.; EL-Maghraby, L.M.; Elansary, H.; Hafez, Y.M.; Ibrahim, E.I.; El-Banna, M.; El-Esawi, M.; Elkelish, A. Treatment of Sweet Pepper with Stress Tolerance-Inducing Compounds Alleviates Salinity Stress Oxidative Damage by Mediating the Physio-Biochemical Activities and Antioxidant Systems. *Agronomy* **2020**, *10*, 26. [CrossRef]

2. Chavez-Mendoza, C.; Sanchez, E.; Muñoz-Marquez, E.; Sida-Arreola, J.P.; Flores-Cordova, M.A. Bioactive compounds and antioxidant activity in different grafted varieties of bell pepper. *Antioxidants* **2015**, *4*, 427–446. [CrossRef] [PubMed]

3. El-Banna, M.F.; Abdelaal, K.A.A. Response of Strawberry Plants Grown in the Hydroponic System to Pretreatment with H_2O_2 before Exposure to Salinity Stress. *J. Plant Prot. Mansoura Univ.* **2018**, *9*, 989–1001. [CrossRef]

4. Helaly, M.N.; Mohammed, Z.; El-Shaeery, N.I.; Abdelaal, K.A.A.; Nofal, I.E. Cucumber grafting onto pumpkin can represent an interesting tool to minimize salinity stress. Physiological and anatomical studies. *Middle East J. Agric. Res.* **2017**, *6*, 953–975.

5. Isayenkov, S.V.; Maathuis, F.J. Plant salinity stress: Many unanswered questions remain. *Front. Plant Sci.* **2019**, *10*, 80. [CrossRef] [PubMed]

6. Hernández, J.A.; Almansa, M.S. Short-term effects of salt stress on antioxidant systems and leaf water relations of pea leaves. *Physiol. Plant.* **2002**, *115*, 251–257. [CrossRef]

7. Koyro, H.W. Effect of salinity of growth, photosynthesis, water relations and solute composition of the potential cash crop halophyte *Plantago coronopus* (L.). *Environ. Exp. Bot.* **2006**, *56*, 136–146. [CrossRef]

8. Gorhman, J.; Hardy, C.; Wyn, R.G.; Jones, L.R.; Law, C.N. Chromosomal location of a K/Na discrimination character in the D genome of wheat. *Theor. Appl. Genet.* **1987**, *74*, 584–588.

9. Harizanova, A.; Koleva-valkova, L. Effect of silicon on photosynthetic rate and the chlorophyll fluorescence parameters at hydroponically grown cucumber plants under salinity stress. *J. Cent. Eur. Agric.* **2019**, *20*, 953–960. [CrossRef]

10. Misra, N.; Ansari, M.; Gupta, A. Differential response of scavenging of reactive oxygen species in green gram genotype grown under salinity stress. *Am. J. Plant Physiol.* **2006**, *1*, 41–53. [CrossRef]

11. Hasan, M.K.; El Sabagh, A.; Sikdar, M.S.; Alam, M.J.; Ratnasekera, D.; Barutcular, C.; Abdelaal, K.A.A.; Islam, M.S. Comparative adaptable agronomic traits of blackgram and mungbean for saline lands. *Plant Arch.* **2017**, *17*, 589–593.

12. Carmen, B.; Roberto, D. Soil bacteria support and protect plants against abiotic stresses. In *Abiotic Stress in Plants Mechanisms and Adaptations*; Shanker, A., Ed.; Pub. In Tech: London, UK, 2011; pp. 143–170.

13. Islam, F.; Yasmeen, T.; Ali, S.; Ali, B.; Farooq, M.A.; Gill, R.A. Priming induced antioxidative responses in two wheat cultivars under saline stress. *Acta Physiol. Plant.* **2015**, *37*, 153. [CrossRef]

14. Memon, S.A.; Hou, X.; Wang, L.J. Morphological analysis of salt stress response of pak Choi. *Electron. J. Environ. Agric. Food Chem.* **2010**, *9*, 248–254.

15. Khan, N.; Bano, A. Role of plant growth promoting rhizobacteria and Ag-nano particle in the bioremediation of heavy metals and maize growth under municipal wastewater irrigation. *Int. J. Phytorem.* **2016**, *18*, 211–221. [CrossRef]

16. Gama, P.B.S.; Inanaga, S.; Tanaka, K.; Nakazawa, R. Physiological response of common bean (*Phaseolus vulgaris* L.) seedlings to salinity stress. *Afr. J. Biotechnol.* **2007**, *6*, 79–88.

17. Abdelaal, K.A.; Mazrou, Y.S.; Hafez, Y.M. Silicon Foliar Application Mitigates Salt Stress in Sweet Pepper Plants by Enhancing Water Status, Photosynthesis, Antioxidant Enzyme Activity and Fruit Yield. *Plants* **2020**, *9*, 733. [CrossRef]

18. Abdelaal, K.A.; Hafez, Y.M.; Sabagh, A.E.L.; Saneoka, H. Ameliorative effects of Abscisic acid and yeast on morpho-physiological and yield characteristics of maize plant (*Zea mays* L.) under drought conditions. *Fresenius Environ. Bull.* **2017**, *26*, 7372–7383.

19. Hafez, Y.; Attia, K.; Alamery, S.; Ghazy, A.; Al-Doss, A.; Ibrahim, E.; Rashwan, E.; El-Maghraby, L.; Awad, A.; Abdelaal, K.A. Beneficial Effects of Biochar and Chitosan on Antioxidative Capacity, Osmolytes Accumulation, and Anatomical Characters of Water-Stressed Barley Plants. *Agronomy* **2020**, *10*, 630. [CrossRef]

20. Abdelaal, K.A.; Hafez, Y.M.; Badr, M.M.; Youseef, W.A.; Esmaeil, S.M. Biochemical, histological and molecular changes in susceptible and resistant wheat cultivars inoculated with stripe rust fungus *Puccinia striiformis* f.sp. *tritici*. *Egypt. J. Biol. Pest Control* **2014**, *24*, 421–429.

21. Abdelaal, K.A.; Omara, I.R.; Hafez, Y.M.; Esmail, S.M.; EL Sabagh, A. Anatomical, biochemical and physiological changes in some Egyptian wheat cultivars inoculated with *Puccinia graminis* f.sp. *tritici*. *Fresenius Environ. Bull.* **2018**, *27*, 296–305.

22. Esmail, S.M.; Omara, R.I.; Abdelaal, K.A.; Hafez, M. Histological and biochemical aspects of compatible and incompatible wheat-Puccinia striiformis interactions. *Physiol. Mol. Plant Pathol.* **2019**, *106*, 120–128. [CrossRef]

23. Omara, R.I.; El-Kot, G.A.; Fadel, F.M.; Abdelaal, K.A.; Saleh, E.M. Efficacy of Certain Bioagents on Patho-Physiological Characters of Wheat Plants under Wheat Leaf Rust Stress. *Physiol. Mol. Plant Pathol.* **2019**, *106*, 102–108. [CrossRef]

24. Hafez, Y.M.; Abdelaal, K.A. Investigation of susceptibility and resistance mechanisms of some Egyptian wheat cultivars (*Triticum aestivum* L.) inoculated with *Blumeria graminis* f.sp. *tritici* using certain biochemical, molecular characterization and SEM. *J. Plant Prot. Path. Mansoura Univ.* **2015**, *6*, 431–454.

25. Khan, N.; Bano, A.; Zandi, P. Effects of exogenously applied plant growth regulators in combination with PGPR on the physiology and root growth of chickpea (*Cicer arietinum*) and their role in drought tolerance. *J. Plant Interact.* **2018**, *13*, 239–247. [CrossRef]

26. Abdelaal, K.A.; Attia, K.A.; Alamery, S.F.; El-Afry, M.M.; Ghazy, A.I.; Tantawy, D.S.; Al-Doss, A.A.; El-Shawy, E.E.; Abu-Elsaoud, A.M.; Hafez Y., M. Exogenous Application of Proline and Salicylic Acid can Mitigate the Injurious Impacts of Drought Stress on Barley Plants Associated with Physiological and Histological Characters. *Sustainability* **2020**, *12*, 1736. [CrossRef]

27. Abdelaal, K.A. Effect of Salicylic acid and Abscisic acid on morpho-physiological and anatomical characters of faba bean plants (*Vicia faba* L.) under drought stress. *J. Plant Prod. Mansoura Univ.* **2015**, *6*, 1771–1788. [CrossRef]

28. Abdelaal, K.A.; Hafez, Y.M.; El-Afry, M.; Tantawy, D.S.; Alshaal, T. Effect of some osmoregulators on photosynthesis, lipid peroxidation, antioxidative capacity, and productivity of barley (*Hordeum vulgare* L.) under water deficit stress. *Environ. Sci. Pollut. Res.* **2018**, *25*, 30199–30211. [CrossRef]

29. Elkelish, A.; Qari, S.H.; Mazrou, Y.M.; Abdelaal, K.A.; Hafez, Y.M.; Abu-Elsaoud, A.M.; Batiha, G.; El-Esawi, M.; El Nahhas, N. Exogenous Ascorbic Acid Induced Chilling Tolerance in Tomato Plants Through Modulating Metabolism, Osmolytes, Antioxidants, and Transcriptional Regulation of Catalase and Heat Shock Proteins. *Plants* **2020**, *10*, 431. [CrossRef]

30. Zhu, J.-K. Regulation of ion homeostasis under salt stress. *Curr. Opin. Plant Biol.* **2003**, *6*, 441–445. [CrossRef]

31. Sairam, R.K.; Tyagi, A. Physiology and molecular biology of salinity stress tolerance in plants. *Curr. Sci.* **2004**, *86*, 407–421.

32. Gupta, B.; Huang, B. Mechanism of salinity tolerance in plants: Physiological, biochemical, and molecular characterization. *Int. J. Genom.* **2014**, *18*. [CrossRef] [PubMed]

33. Maxwell, K.; Johnson, G.N. Chlorophyll fluorescence—A practical guide. *J. Exp. Bot.* **2000**, *51*, 659–668. [CrossRef] [PubMed]

34. Murchie, E.H.; Lawson, T. Chlorophyll fluorescence analysis: A guide to good practice and understanding some new applications. *J. Exp. Bot.* **2013**, *64*, 3983–3998. [CrossRef] [PubMed]

35. Kalaji, M.H.; Goltsev, V.; Zuk-Golaszewska, B.; Zivcak, M.; Brestic, M. *Chlorophyll Fluorescence: Understanding Crop Performance-Basics and Applications*; Taylor and Francis: New York City, NY, USA, USA 2017; p. 222. ISBN 9781498764490.

36. Katiyar, D.; Hemantaranjan, A.; Singh, B. Chitosan as a promising natural compound to enhance potential physiological responses in plant: A review. *Indian J. Plant Physiol.* **2015**, *20*, 1–9. [CrossRef]

37. Hidangmayum, A.; Dwivedi, P.; Katiyar, D.; Hemantaranjan, A. Application of chitosan on plant responses with special reference to abiotic stress. *Physiol. Mol. Biol. Plants* **2019**, *25*, 313–326. [CrossRef] [PubMed]

38. Al-Tawaha, A.M.; Seguin, P.; Smith, D.L.; Beaulieu, C. Foliar application of elicitors alters isofl avone concentrations and other seed characteristics of field-grown soybean. *Can. J. Plant Sci.* **2006**, *86*, 677–684. [CrossRef]

39. Kim, H.J.; Chen, F.; Wang, X.; Rajapakse, N.C. Effect of chitosan on the biological properties of sweet basil (*Ocimum basilicum* L.). *J. Agric. Food Chem.* **2005**, *53*, 3696–3701. [CrossRef]

40. Balal, R.M.; Shahid, M.A.; Javaid, M.M.; Iqbal, Z.; Liu, G.D.; Zotarelli, L.; Khan, N. Chitosan alleviates phytotoxicity caused by boron through augmented polyamine metabolism and antioxidant activities and reduced boron concentration in *Cucumis sativus* L. *Acta Physiol. Plant.* **2017**, *39*, 31. [CrossRef]

41. Monirul, I.M.; Humayun, K.M.; Mamun, A.N.K.; Monirul, I.; Pronabananda, D. Studies on yield and yield attributes in tomato and chilli using foliar application of oligo-chitosan. *GSC Biol. Pharm. Sci.* **2018**, *3*, 020–028.

42. Yang, J.; Kloepper, J.W.; Ryu, C.M. Rhizosphere bacteria help plants tolerate abiotic stress. *Trends Plant Sci.* **2009**, *14*, 1–4. [CrossRef]

43. Noel, T.C.; Sheng, C.; Yost, C.K.; Pharis, R.P.; Hynes, M.F. *Rhizobium leguminosarum* as a plant growth-promoting rhizobacterium: Direct growth promotion of canola and lettuce. *Can. J. Microbiol.* **1996**, *42*, 279–283. [CrossRef] [PubMed]

44. Mayak, S.; Tirosh, T.; Glick, B.R. Plant growth-promoting bacteria that confer resistance in tomato plants to salt stress. *Plant Physiol. Biochem.* **2004**, *42*, 565–572. [CrossRef]

45. Abdelaal, K.A.; Tawfik, S.F. Response of Sugar Beet Plant (*Beta vulgaris* L.) to Mineral Nitrogen Fertilization and Bio-Fertilizers. *Int. J. Curr. Microbiol. App. Sci.* **2015**, *4*, 677–688.

46. Abdelaal, K.A.; Badawy, S.A.; Abdel Aziz, R.M.; Neana, S.M.M. Effect of mineral nitrogen levels and biofertilizer on morphophysiological characters of three sweet sorghum varieties (*Sorghum bicolor* L. Moench). *J. Plant Prod. Mansoura Univ.* **2015**, *6*, 189–203. [CrossRef]

47. Abdelaal, K.A. Pivotal Role of Bio and Mineral Fertilizer Combinations on Morphological, Anatomical and Yield Characters of Sugar Beet Plant (*Beta vulgaris* L.). *Middle East J. Agric. Res.* **2015**, *4*, 717–734.

48. Hafez, Y.M.; Attia, K.; Kamel, S.; Alamery, S.; El-Gendyd, S.; Al-Doss, A.; Mehiar, F.; Ghazy, A.; Abdelaal, K.A. *Bacillus subtilis* as a bio-agent combined with nano molecules can control powdery mildew disease through histochemical and physiobiochemical changes in cucumber plants. *Physiol. Mol. Plant Pathol.* **2020**, *111*, 101489. [CrossRef]

49. Khan, N.; Bano, A.; Ali, S.; Babar, M.A. Crosstalk amongst phytohormones from planta and PGPR under biotic and abiotic stresses. *Plant Growth Regul.* **2020**, *90*, 189–203. [CrossRef]

50. Kousar, B.; Bano, A.; Khan, N. PGPR Modulation of Secondary Metabolites in Tomato Infested with *Spodoptera litura*. *Agronomy* **2020**, *10*, 778. [CrossRef]

51. Abou-Attia, F.A.M.; Abdelaal, K.A. Effect of Bio and Mineral fertilization on the main insect pests and some characters of sugar beet plants. *J. Agric. Sci. Mansoura Univ.* **2007**, *32*, 1471–1485.

52. Egamberdieva, D. Pseudomonas chlororaphis: A salt-tolerant bacterial inoculant for plant growth stimulation under saline soil conditions. *Acta Physiol. Plant.* **2012**, *34*, 751–756. [CrossRef]

53. Khan, N.; Bano, A.M.; Babar, A. Impacts of plant growth promoters and plant growth regulators on rainfed agriculture. *PLoS ONE* **2020**, *15*, e0231426. [CrossRef] [PubMed]

54. Zhao, S.; Zhou, N.; Zhao, Z.-Y.; Zhang, K.; Wu, G.-H.; Tian, C.-Y. Isolation of endophytic plant growth-promoting bacteria associated with the halophyte *Salicornia europaea* and evaluation of their promoting activity under salt stress. *Curr. Microbiol.* **2016**, *73*, 574–581. [CrossRef] [PubMed]

55. Page, A.L. *Methods of Soil Analysis. Part 2, Chemical and Microbiological Properties*, 2nd ed.; Soil Science Society of America Inc.: Madison, WI, USA, 1982.

56. Sanchez, F.J.; de Andrés, E.F.; Tenorio, J.L.; Ayerbe, L. Growth of epicotyls, turgor maintenance and osmotic adjustment in pea plants (*Pisum sativum* L.) subjected to water stress. *Field Crop. Res.* **2004**, *86*, 81–90. [CrossRef]

57. Moran, R. Formulae for Determination of Chlorophyllous Pigments Extracted with N,N-Dimethylformamide 1. *Plant Physiol.* **1982**, *69*, 1376–1381. [CrossRef] [PubMed]

58. Szalai, G.; Janda, T.; Padi, E.; Szigeti, Z. Role of light in post-chilling symptoms in maize. *J. Plant Physiol.* **1996**, *148*, 378–383. [CrossRef]

59. Schreiber, U. *Pulse-Amplitude (RAM) Fluorometry and Saturation Pulse Method in Chlorophyll Fluorescence: A Signature of Photosynthesis*; Papageorgiou, G., Govindjee, Eds.; Springer: Dordrecht, The Netherlands, 2004; pp. 279–319.

60. Bates, L.S.; Waldren, R.P.; Teare, I.D. Rapid determination of free proline for water-stress studies. *Plant Soil* **1973**, *39*, 205–207. [CrossRef]

61. Davenport, S.B.; Gallego, S M.; Benavides, M.P.; Tomaro, M.L. Behaviour of antioxidant defense system in the adaptive response to salt stress in *Helianthus annuus* L. cells. *Plant Growth Regul.* **2003**, *40*, 81 88. [CrossRef]

62. Huckelhoven, R.; Fodor, J.; Preis, C.; Kogel, K.H. Hypersensitive cell death and papilla formation in barley attacked by the powdery mildew fungus are associated with hydrogen peroxide but not with salicylic acid accumulation. *Plant Physiol.* **1999**, *1119*, 1251–1260. [CrossRef]

63. Hafez, Y.M.; Abdelaal, K.A.; Eid, M.E.; Mehiar, F.F. Morpho-physiological and biochemical responses of barley plants (*Hordeum vulgare* L.) against barley net blotch disease with application of non-traditional compounds and fungicides. *Egypt. J. Biol. Pest Cont.* **2016**, *26*, 261–268.

64. Aebi, H. Catalase in vitro. In *Methods in Enzymology; Oxygen Radicals in Biological Systems*; Academic Press: Cambridge, MA, USA, 1984; Volume 105, pp. 121–126.

65. Hammerschmidt, R.; Nuckles, E.M.; Kuć, J. Association of enhanced peroxidase activity with induced systemic resistance of cucumber to *Colletotrichum lagenarium*. *Physiol. Plant Pathol.* **1982**, *20*, 73–82. [CrossRef]

66. Mishra, N.P.; Mishra, R.K.; Singhal, G.S. Changes in the activities of anti-oxidant enzymes during exposure of intact what leaves to strong visible light at different temperatures in the presence of protein synthesis inhibitors. *Plant Physiol.* **1993**, *102*, 903–910. [CrossRef] [PubMed]

67. McCord, J.M.; Fridovich, I. Superoxide Dismutase: An enzymatic reaction for erythrocuprein (hemocuprein). *J. Biol. Chem.* **1969**, *244*, 6049–6055. [PubMed]

68. Goldberg, D.M.; Spooner, R.J. *Methods of Enzymatic Analysis*, 3rd ed.; Verlag Chemie: Weinheim, Germany, 1983; pp. 258–265.

69. Gomez, K.A.; Gomez, A.A. *Statistical Procedures for Agricultural Research*, 2nd ed.; Wiley Inter Science: New York City, NY, USA, 1984; pp. 1–690.

70. Duncan, B.D. Multiple ranges and multiple F-test. *Biometrics* **1955**, *11*, 1–42. [CrossRef]

71. Wang, Y.; Nil, N. Changes in chlorophyll, ribulose biphosphate carboxylase-oxygenase, glycine betaine content, photosynthesis and transpiration in *Amaranthus tricolor* leaves during salt stress. *J. Hortic. Sci. Biotechnol.* **2000**, *75*, 623–627. [CrossRef]

72. Parvin, K.; Hasanuzzaman, M.; Bhuyan, M.H.M.B.; Nahar, K.; Mohsin, S.M.; Fujita, M. Comparative Physiological and Biochemical Changes in Tomato (*Solanum lycopersicum* L.) under Salt Stress and Recovery: Role of Antioxidant Defense and Glyoxalase Systems. *Antioxidants* **2019**, *8*, 350. [CrossRef]

73. Wang, Y.; Li, K.; Li, X. Auxin redistribution modulates plastic development of root system architecture under salt stress in Arabidopsis thaliana. *J. Plant Physiol.* **2009**, *166*, 1637–1645. [CrossRef]

74. Naseem, H.; Ahsan, M.; Shahid, M.A.; Khan, N. Exopolysaccharides producing rhizobacteria and their role in plant growth and drought tolerance. *J. Basic. Microbiol.* **2018**, *58*, 1009–1022. [CrossRef]

75. Allakhverdiev, S.I.; Nishiyama, Y.; Miyairi, S.; Yamamoto, H.; Inagaki, N.; Kanesaki, Y.; Murata, N. Salt stress inhibits the repair of photodamaged photosystem II by suppressing the transcription and translation of psbA genes in Synechocystis. *Plant Physiol.* **2002**, *130*, 1443–1453. [CrossRef]

76. Khan, N.; Bano, A.; Curá, J.A. Role of Beneficial Microorganisms and Salicylic Acid in Improving Rainfed Agriculture and Future Food Safety. *Microorganisms* **2020**, *8*, 1018. [CrossRef]

77. Wang, W.; Wang, C.; Pan, D.; Zhang, Y.; Luo, B.; Ji, J. Effects of drought stress on photosynthesis and chlorophyll fluorescence images of soybean (*Glycine max*) seedlings. *Int. J. Agric. Biol. Eng.* **2018**, *11*, 196–201. [CrossRef]

78. Tavakkoli, E.; Rengasamy, P.; McDonald, G. High concentrations of Na^+ and Cl^- ions in soil solution have simultaneous detrimental effects on growth of faba bean under salinity stress. *J. Exp. Bot.* **2010**, *61*, 4449–4459. [CrossRef] [PubMed]

79. Asrar, H.; Hussain, T.; Midhat, S.; Hadi, S.; Gul, B.; Nielsen, B.L.; Khan, M.A. Salinity induced changes in light harvesting and carbon assimilating complexes of *Desmostachya bipinnata* (L.) staph. *Environ. Exp. Bot.* **2017**, *135*, 86–95. [CrossRef]

80. Shah, G.; Jan, M.; Afreen, M.; Anees, M.; Rehman, S.; Daud, M.K.; Malook, I.; Jamil, M. Halophilic bacteria mediated phytoremediation of salt-affected soils cultivated with rice. *J. Geochem. Explor.* **2017**, *174*, 59–65. [CrossRef]

81. Possingham, J.V. Plastid replication and development in the life cycle of higher plants. *Annu. Rev. Plant Physiol.* **1980**, *31*, 113–129. [CrossRef]

82. Plazek, A.; Rapacz, M.; Hura, K. Relationship between quantum efficiency of PSII and cold-induced plant resistance to fungal pathogens. *Acta Physiol. Plant.* **2004**, *26*, 141–148.

83. Mehta, P.; Jajoo, A.; Mathur, M.; Bharti, S. Chlorophyll a fluorescence study revealing effects of high salt stress on photosystem II in wheat leaves. *Plant Physiol. Biochem.* **2010**, *48*, 16–20. [CrossRef]

84. Kalaji, H.; Rastogi, A.; Živčgk, M.; Brestic, M. Prompt chlorophyll fluorescence as a tool for crop phenotyping: An example of barley landraces exposed to various abiotic stress factors. *Photosynthetica* **2018**. [CrossRef]

85. Yang, D.L.; Lian, J.R.; Ping, C.X.; Wei, L. Quantitative Trait Loci Mapping for Chlorophyll Fluorescence and Associated Traits in Wheat (*Triticum aestivum*). *J. Integr. Plant Biol.* **2007**, *49*, 646–654. [CrossRef]

86. Li, Z.; Zhang, Y.; Zhang, X.; Merewitz, E.; Peng, Y.; Ma, X.; Yan, Y. Metabolic pathways regulated by chitosan contributing to drought resistance in white clover. *J. Proteome Res.* **2017**, *16*, 3039–3052. [CrossRef]

87. Khan, N.; Bano, A.; Rahman, M.A.; Rathinasabapathi, B.; Babar, M.A. UPLC-HRMS-based untargeted metabolic profiling reveals changes in chickpea (*Cicer arietinum*) metabolome following long-term drought stress. *Plant Cell Environ.* **2019**, *42*, 115–132. [CrossRef]

88. Egamberdieva, D.; Davranov, K.; Wirth, S.; Hashem, A.; Abd_Allah, E.F. Impact of soil salinity on the plant-growth–promoting and biological control abilities of root associated bacteria. *Saudi J. Biol. Sci.* **2017**, *24*, 1601–1608. [CrossRef] [PubMed]

89. Prashanth, H.K.V.; Dharmesh, S.M.; Rao, K.S.; Tharanathan, R.N. Free radical-induced chitosan depolymerized products protect calf thymus DNA from oxidative damage. *Carbohydr. Res.* **2007**, *342*, 190–195. [CrossRef] [PubMed]

90. Malerba, M.; Cerana, R. Chitosan Effects on Plant Systems. *Int. J. Mol. Sci.* **2016**, *23*, 996. [CrossRef] [PubMed]

91. Sharma, I.P.; Sharma, A.K. Physiological and biochemical changes in tomato cultivar PT-3 with dual inoculation of mycorrhiza and PGPR against root-knot nematode. *Symbiosis* **2017**, *71*, 175–183. [CrossRef]

92. Bianco, C.; Defez, R. Medicago truncatula improves salt tolerance when nodulated by an indole-3-acetic acid-overproducing *Sinorhizobium meliloti* strain. *J. Exp. Bot.* **2009**, *60*, 3097–3107. [CrossRef]

93. Foyer, C.H.; Descourvieres, P.; Kunert, K.J. Protection against oxygen radicals: An important defense mechanism studied in transgenic plants. *Plant Cell Environ.* **1994**, *17*, 507–523. [CrossRef]

94. Kohler, J.; Caravaca, F.; Roldàn, A. An AM fungus and a PGPR intensify the adverse effects of salinity on the stability of rhizosphere soil aggregates of *Lactuca sativa*. *Soil Biol. Biochem.* **2010**, *42*, 429–434. [CrossRef]

95. Hu, Y.; Yu, W.; Liu, T.; Shafi, M.; Song, L.; Du, X.; Huang, X.; Yue, Y.; Wu, J. Effects of paclobutrazol on cultivars of Chinese bayberry (*Myrica rubra*) under salinity stress. *Photosynthetica* **2017**, *55*, 443–453. [CrossRef]

96. Hayat, R.; Ali, S.; Amara, U.; Khalid, R.; Ahmed, I. Soil beneficial bacteria and their role in plant growth promotion: A review. *Ann. Microbiol.* **2010**, *60*, 579–598. [CrossRef]

 *agronomy*

Article

Rhizobacteria Isolated from Saline Soil Induce Systemic Tolerance in Wheat (*Triticum aestivum* L.) against Salinity Stress

Noshin Ilyas [1,*], Roomina Mazhar [1], Humaira Yasmin [2], Wajiha Khan [3], Sumera Iqbal [4], Hesham El Enshasy [5,6,7,*] and Daniel Joe Dailin [5,6]

[1] Department of Botany, PMAS-Arid Agriculture University, Rawalpindi 46300, Pakistan; roominamazhar83@gmail.com
[2] Department of Bio-Sciences, COMSATS University, Islamabad 45550, Pakistan; humaira.yasmin@comsat.edu.pk
[3] Department of Biotechnology, COMSATS University Islamabad, Abbottabad Campus, Abbottabad 22010, Pakistan; wajihak@cuiatd.edu.pk
[4] Department of Botany, Lahore College for Women University, Lahore 54000, Pakistan; sumeraiqbal2@yahoo.com
[5] Institute of Bioproduct Development (IBD), Universiti Teknologi Malaysia (UTM), Skudai, Johor 81310, Malaysia; jddaniel@utm.my
[6] School of Chemical and Energy Engineering, Faculty of Engineering, Universiti Teknologi Malaysia (UTM), Skudai, Johor 81310, Malaysia
[7] City of Scientific Research and Technology Applications (SRTA), New Burg Al Arab, Alexandria 21934, Egypt
* Correspondence: noshinilyas@yahoo.com (N.I.); henshasy@ibd.utm.my (H.E.E.)

Received: 8 June 2020; Accepted: 23 June 2020; Published: 10 July 2020

Abstract: Halo-tolerant plant growth-promoting rhizobacteria (PGPR) have the inherent potential to cope up with salinity. Thus, they can be used as an effective strategy in enhancing the productivity of saline agro-systems. In this study, a total of 50 isolates were screened from the rhizospheric soil of plants growing in the salt range of Pakistan. Out of these, four isolates were selected based on their salinity tolerance and plant growth promotion characters. These isolates (SR_1. SR_2, SR_3, and SR_4) were identified as *Bacillus* sp. (KF719179), *Azospirillum brasilense* (KJ194586), *Azospirillum lipoferum* (KJ434039), and *Pseudomonas stutzeri* (KJ685889) by 16S rDNA gene sequence analysis. In vitro, these strains, in alone and in a consortium, showed better production of compatible solute and phytohormones, including indole acetic acid (IAA), gibberellic acid (GA), cytokinin (CK), and abscisic acid (ABA), in culture conditions under salt stress. When tested for inoculation, the consortium of all four strains showed the best results in terms of improved plant biomass and relative water content. Consortium-inoculated wheat plants showed tolerance by reduced electrolyte leakage and increased production of chlorophyll a, b, and total chlorophyll, and osmolytes, including soluble sugar, proline, amino acids, and antioxidant enzymes (superoxide dismutase, catalase, peroxidase), upon exposure to salinity stress (150 mM NaCl). In conclusion, plant growth-promoting bacteria, isolated from salt-affected regions, have strong potential to mitigate the deleterious effects of salt stress in wheat crop, when inoculated. Therefore, this consortium can be used as potent inoculants for wheat crop under prevailing stress conditions.

Keywords: salinity; PGPR; wheat; compatible solutes; antioxidant enzymes

1. Introduction

Globally, the production rate of agriculture is far less than the estimated food requirement of the ever-increasing population and the gap will be widened over time [1] (GAP Report, 2018).

Agro-ecosystems are influenced by environmental and climatic conditions, farming techniques, and management practices. It is estimated that internationally, salinity affects 22% of the total cultivated and 33% of the total irrigated agricultural area, which is increasing at an alarming rate of 10% annually. Pakistan is also facing severe salinity issues and a total area of 6.30 million hectares is salt affected, out of which 1.89 million hectares is marked as saline [2].

Due to a higher concentration of sodium chloride (NaCl), plants growing in salt-affected soils suffer from both hyperosmotic and hyperionic effects. These stresses result in reduced water uptake; altered ion and mineral absorption rates; increased production of reactive oxygen species, causing disorganization of the cell membrane; and reduction of metabolic activities [3]. Halophytes adapt themselves to saline conditions by adjusting their physiological activities, maintaining their water balance by osmotic adjustments, producing compatible solutes, and modifying the antioxidant system [4]. Some plants overcome salinity stress through the production of osmolytes, particularly glycine betaine, proline, soluble sugars, and proteins [5].

Improvement in the crop yield of saline soils requires a multidimensional approach consisting of salt-tolerant varieties or amelioration by chemical neutralizers, but there is a dire need for eco-friendly sustainable approaches. Rhizobacteria, showing potential to improve plant growth, are termed as plant growth-promoting rhizobacteria (PGPR) [6]. PGPR have the potential to improve plant growth through various mechanisms, including better plant growth, the production of phytohormones, and amelioration of stresses [7]. Due to the natural coping mechanisms of PGPR, their inoculation can help the amelioration of various abiotic stresses in plants. PGPR inoculation can help to improve the growth and yield of crops, particularly in regions prone to drought and salt stress [8,9]. Natural halotolerant PGPR strains have better potential for the amelioration of salt stress in regional crops for sustainable yields. These native PGPR strains are well acclimated to indigenous conditions and the plant–microbe interactions can help the plants to tolerate stress [10].

In this study, native halotolerant PGPR strains were isolated from local saline soils, and their ability to promote plant growth when inoculated under salt stress was investigated. The objective of the present research was to focus on the evaluation of isolated bacterial strains to stimulate salinity tolerance and the promotion of wheat growth, as well as the identification and characterization of the candidate strain both bio-physiochemically and genetically. This study provides a basis to identify and characterize PGPR from natural saline conditions and testing their potential for improving salinity tolerance in wheat, the major staple crop across the world.

2. Materials and Methods

2.1. Soil Sampling and Physicochemical Analysis

The rhizospheric soil of four halophytes namely, *Abutilon bidentatum*, *Maytenus royleanus*, *Leptochloa fusca* (Kallar grass), and *Dedonia viscose*, was collected from a salt range of Pakistan (313–360 m.a.s.l; 32°23–33°00 north latitude and 71°30–73°30 east longitude). The rhizospheric soil was sieved and stored at 4 °C for future analysis. Rhizospheric soil was analyzed for pH and electrical conductivity (EC) [11], soil texture, macro and micronutrients [12], and available nutrients [13].

2.2. Strain Isolation and HaloTolerance Assay

Rhizobacteria were isolated from rhizospheric soil of *Abutilon bidentatum*, *Maytenus royleanus*, *Leptochloa fusca*, and *Dedonia viscose* by using the serial dilution and spread plate techniques [14]. The soil suspension was made by adding 1 g of soil in 9 mL of Milli-Q distilled water. An aliquot of soil suspension was inoculated on Luria-Bertani (LB) agar plates and incubated at 28 ± 2 °C for 48 h. The obtained colonies were purified by sub-culturing. The colony-forming unit (CFU) was calculated according to the formula given by [15]:

$$CFU/g = (\text{colonies number} \times \text{dilution factor/volume of inoculum}).$$

Distinct bacterial colonies were examined for colony characteristics (shape, size, margin, elevation, appearance, texture, pigmentation, and optical properties) as well as for cellular characteristics (cell shape, gram testing) [16]. QTS-24 kits were used to determine the carbon/nitrogen (C/N) source utilization pattern of bacterial isolates. Isolated bacterial strains were tested for their halotolerance abilities by growing them in LB media supplemented with NaCl (2%, 4%, 6%, 8%, 10%, 15%) [16].

2.3. Plant Growth-Promoting (PGP) Traits

All the bacterial isolates were evaluated for their PGP characteristics. Phosphorous (P) solubilization was done by spot inoculating overnight grown cultures onto pikovaskaya's agar (Sigma) containing tri-calcium phosphate as an insoluble P source [17]. The colonies, which produced clearing zones in the pikovaskaya's agar plates, were considered positive for phosphorous solubilization. Total solubilized phosphate was measured by using the phosphomolybdate blue color method [18]. Modified pikovaskaya's broth medium was inoculated with each strain and incubated at 30 °C for 5 days. The cultures were centrifuged at 6000 rpm for 15 min. The supernatant (500 µL) was mixed with 40 µL of 2,4-dinitrophenol, after which 20 µL of dilute sulfuric acid were added, followed by 5 mL of chromogenic reagent, and the volume was diluted to 50 mL using sterilized water and absorbance was recorded at 680 nm. Siderophore production was done by spot inoculation on chrome azurol S (CAS) media as described by Schwyn and Neilands [19]. Bacterial strains were spot inoculated on petri plates containing CAS media. An uninoculated plate was taken as the control. After inoculation, plates were incubated at 28 °C for 5–7 days and observed for the formation of an orange zone around the bacterial colonies. Bacterial isolates were tested for hydrogen cyanide production through the method of Lorck [20]. Bacterial strains were streaked on nutrient agar medium (pre-soaked in 0.5% picric acid and 2% sodium carbonate *w/v*), supplemented with glycine (4.4 g/L). Plates were sealed with parafilm paper and incubated at 30 °C for 4 days. The appearance of an orange or red color indicates the production of hydrogen cyanide.

2.4. Germination Experiment

This experiment was carried in the Plant Physiology Laboratory of PMAS-Arid Agriculture University. Seeds of the wheat variety (Galaxy 2013) obtained from the National Agricultural Research Centre, Islamabad were surface sterilized by treatment with sodium hypochlorite (1%) solution for 5 min. After, seeds were successively washed with distilled water. All the isolated strains were tested for germination attributes. Sterilized seeds of wheat were placed in pre-soaked filter paper in Petri dishes. NaCl solution (50 mM, 100 mM, 150 mM, 200 mM) was given instead of normal water. The germination experiment was carried out under laboratory conditions with an average photoperiod of 10 h day/14 h night at 24 °C. The germination percentage, seedling vigor index, and promptness index were measured for each treatment [21]. Four strains were selected for further analysis, based upon their efficacy in the germination experiment and were labeled as SR_1, SR_2, SR_3, and SR_4.

2.5. Production of Osmolytes

To analyze proline and total soluble sugars, the supernatant of PGPR grown in LB broth supplemented with NaCl concentrations (0%, 2%, 4%, 6%, 8%, and 10%) were analyzed as described by Upadhyay et al. [22]. For the estimation of the proline contents, centrifugation of the culture broth was done at $1000\times g$ for 10 min and the supernatant was used for estimation. Total soluble sugar (TSS) was estimated by mixing 1 mL of supernatant with 4 mL of anthrone reagent, the mixture was later boiled in a water bath for 8 min. After rapid cooling, the optical density was measured at 630 nm, and the amount of TSS was calculated from a standard curve.

2.6. Phytohormone Production

The ability of four selected halotolerant strains to produce phytohormones (IAA, GA, CK, ABA) in the culture media was measured by the method of Tien et al. [23]. The extraction of hormones was done by centrifugation of bacterial cultures at 10,000 rpm for 15 min. For adjustment of the pH (2.8), 1 N HCl was used. In the next step, an equal volume of ethyl acetate was used for hormone extraction. The resulting solution was evaporated at 35 °C and the end residue was mixed in 1500 µL of methanol. Finally, the samples were run on High Performance Liquid Chromatography (HPLC) (Agilent 1100), which had a C18 column (39 × 300 mm) and a UV detector. For standardization of HPLC, pure grade chemicals of the hormones IAA, CK, GA, and ABA (Sigma Chemical Co., St. Louis, MO, USA) were dissolved in HPLC-grade methanol and were used. The wavelength used for the detection was as follows: IAA at 280 nm; and GA, CK, and ABA at 254 nm. The phytohormone content of LB media, without inoculum, was used to normalize the data.

2.7. 16S rRNA Gene Sequence and Phylogenetic Analysis

DNA was extracted from pure LB broth cultures as described by Chen and Kuo [24]. Amplification of genomic DNA of isolated strains was done as described by Weisburg et al. [25]. The PCR was carried out for amplification of the 16S rRNA gene with universal nucleotide sequence forward primer (fd1) AGAGTTTGATCCTGGCTCAG, and reverse primer (rd1) (AAGGAGGTGATCCAGCC). DNA was purified and sequenced on an automated sequencer by gel purification kits (JET quick, Gel Extraction Spin Kit, GENOMED). The strains were identified by using a nearly complete sequence of the 16s rRNA gene on (BLAST) NCBI by comparing sequence homology with other strains. The maximum parsimony method was used for the analysis of evolutionary linkages [26].

2.8. Plant Inoculation

A pot experiment was conducted in the greenhouse of the Botany Department, PMAS-AAUR, Rawalpindi. A complete randomized design was applied with three replications. Each selected halotolerant strain was grown overnight in LB media. To obtain a cell pellet, the supernatant was discarded after centrifugation at 3000 rpm for 3 min. The cell pellet was washed three times with autoclaved water and the absorbance was recorded with a spectrophotometer at 600 nm to obtain the desired concentration, i.e., 10^7 CFU. Ten sterilized seeds were sown in each pot (containing 10 kg of soil) in the greenhouse with the day 10 h/14 h night at a temperature of 21/15 °C. Soil moisture was maintained at 15 ± 1%. Four strains and their consortium were evaluated under two treatment controls and 150 mM NaCl stress. The salt level was maintained with EC of 4.0 dS m^{-1} (first irrigation) or 8.5 dS m^{-1} (second irrigation). Plants were harvested after 45 days of sowing. Fresh and dry biomass was recorded. Leaf area was measured with the help of a leaf area meter. All the samples were collected in zipper bags and stored at −20 °C freezer for further biochemical assays. The percent of water content was determined by measuring the ratio between the fresh and dry weight of the upper fully developed leaf by using the following formula [27]:

$$RWC = [FW - DW]/[TW - DW] \times 100 \tag{1}$$

2.9. Electrolyte Leakage (%)

Electrolyte leakage was determined by the method of Srairam [28]. Leaf discs weighing 0.1 g were heated in 10 mL of distilled water for 30 min at 40 °C and the electrical conductivity (C1) was recorded. The same discs were then heated at 100 °C and again electrical conductivity (C2) was recorded. Whereas, calculations were done by the following formula:

$$MSI = [1 - (C1/C2)] \times 100 \tag{2}$$

2.10. Chlorophyll and Carotenoid Content

Leaf chlorophyll a, b, total chlorophyll, and carotenoid contents were estimated by the method of Arnon [29]. Fresh leaves (0.5 g) were ground in 10 mL of 80% acetone. The readings of the filtrate were measured at 470 nm, 663 nm, and 645 nm. Calculations were done by the following equations:

$$\text{Chla (mg/g)} = [12.7A_{663} - 2.69A_{645}]\ (v/w) \tag{3}$$

$$\text{Chlb (mg/g)} = [22.9A_{645} - 4.68A_{663}]\ (v/w) \tag{4}$$

$$\text{Total chlorophyll (mg/g)} = [(20.2A_{645} + 8.02A_{663})\ v/w] \tag{5}$$

$$\text{Carotenoids content(mg/g)} = (1000\ A_{470} - 1.8\ \text{Chl}_a - 85.02\ \text{Chl}_b)/198 \tag{6}$$

where A is the optical density at a specific wavelength.

2.11. Proline Content

Proline contents were determined by following the protocols of Bates [30]. Fresh leaves (0.5 g) were homogenized with 10 mL of sulfosalicylic acid (3.0%). The solution was filtered, and the filtrate was mixed with equal amounts of glacial acetic acid and ninhydrin reagent. The mixture was heated for 1 h in a water bath at 90 °C and the reaction was stopped by transferring the mixture to ice. Toluene (1 mL) was added to the mixture and the solution was mixed and the solution separated into two layers. The upper layer was isolated in separate test tubes and the reading was measured at 520 nm. Proline was determined as follows:

$$\text{Proline} = (\text{Reading of sample} \times \text{Diluted concentration} \times \text{K value})/\text{material weight} \tag{7}$$

2.12. Total Soluble Sugar and Amino Acid

Soluble sugars were estimated after the method of Dubois et al. [31]. Ground plant tissue (0.1 g) was mixed with 3 mL of 80% methanol. The solution was heated in a water bath for 30 min at 70 °C. An equal volume of extract (0.5 mL) and 5% phenol was mixed with concentrated sulphuric acid (1.5 mL) and was again incubated in the dark for 30 min. The absorbance of the sample was checked at 490 nm and the calculations were done by applying the following formula:

$$\text{Sugar (µg/mL)} = \text{Absorbance of sample} \times \text{Dilution factor} \times \text{K value} \tag{8}$$

Fresh tissue in grams.

The standard curve was prepared for glucose solution, which was used for the determination of the amount of sugar, expressed in mg g^{-1} fw^{-1}.

The Ninhydrin method was used for the determination of free amino acids [32]. Leaf extract (1 mL) was mixed with the same volume of 0.2 M citrate buffer (pH-5) and 80% ethanol, and 2 mL of the ninhydrin reagent. The absorbance of the reaction mixture was taken to 570 nm. Amino acids were computed with the equation:

Amino acids = Absorption × volume × Diluted concentration/Sample weight × 1000.

The amino acid, leucine, was used for preparing the standard curve, and results were expressed in mg of amino acid per g of dry tissue.

2.13. Total Protein Content

The concentration of protein was quantified by the Bradford assay [33]. Bovine serum albumin was used as a standard. Proteins were extracted by dissolving 0.2 g of leaf samples in 4 mL of sodium phosphate buffer (pH 7), and 0.5 mL of the extract was mixed with 3 mL of Comassive bio red dye. The optical density of the solution was measured at 595 nm. Protein was determined by:

$$\text{Protein} = \text{Reading of extract} \times \text{Diluted concentration} \times \text{value of K/sample weight} \qquad (9)$$

2.14. Antioxidant Enzyme Assay

Enzyme extract was prepared by grinding one gram of leaf in liquid nitrogen. The obtained powder was added in 10 mL of 50 mM phosphate buffer (pH 7.0) and was mixed with 1 mM Ethylene Diamine Tetra Acetic acid (EDTA) and 1% polyvinylpyrrolidone (PVP). The whole mixture was centrifuged at $13,000 \times g$ for 20 min at 4 °C. The supernatant was used for the enzyme assay.

The catalase (CAT) content was estimated by observing the degradation of H_2O_2 at 240 nm [34]. Catalase activity (U mg protein^{-1}) was calculated from the molar absorption coefficient of 40 mm^{-1}cm^{-1} for H_2O_2. Peroxidase dismutase (POD) was determined by following the procedure of Rao [35]. The reaction mixture consisted of 10 μL of crude enzyme extract, 20 μL of 100 mM guaiacol, 10 μL of 100 mM H_2O_2, and 160 μL of 50 mM sodium acetate (pH 5.0). Absorbance was recorded at 450 nm.

Superoxide dismutase (SOD) activity was done by using the procedure of Giannopolitis and Ries [36]. The composition of the reaction mixture was 50 mM sodium phosphate buffer (pH 7.8), 0.1 M tris-HCL, 14 mM methionine, 1.05 mM riboflavin, 0.03% TritonX-100, 50 mM nitroblue tetrazolium chloride (NBT), 100 mM EDTA, and 20 μL enzyme extracts. After adding riboflavin, the glass tubes were illuminated for 5 min, and reactions were stopped by turning off lights. The absorbance was recorded at 560 nm.

2.15. Statistical Analysis

Three replicates were used for the mean and standard deviation values of the data. The obtained data were further analyzed by Duncan's multiple range tests using MSTAT-C version 1.4.2. The correlation coefficient of the data was done using the software Statistix version 8.1. Mean values were compared by the least significant difference (LSD) at $p \leq 0.05$ [37]. The heatmap for the correlation coefficient was prepared by using web tool clustvis (https://biit.cs.ut.ee/clustvis/).

3. Results

3.1. Soil Analysis

Analysis of the rhizospheric soil samples of all four plants showed the soil was sandy clay loam with an EC range of 0.76–0.85 dSm^{-1}, pH in the range of 7.99–8.12, high Na/K ratio, and a low concentration of nutrients (Table 1).

Table 1. Physiochemical properties of the rhizosphere soil and rhizobacterial population.

Host Plant Species	pH	EC (cSm^{-1})	Soil Texture	SAR (mmol/L)	OC (%)	Macronutrient (meq/L)						Available Nutrients (kg/ha)			PGPR Population (cfu $\times 10^5$ g^{-1} of Soil)
						CO_3	HCO_3	Cl	Ca^+	Na^+	K^+	N	P	K	
Abutilom bicicntura	7.99	0.82	Sandy clay loam	39.6	0.72	3.9	15	50	3	82.95	0.3	240	200	320	69
Mavtenus royleanus	8.12	0.85	Sandy clay loam	40.3	0.54	4	14	48	2.5	80.6	0.4	238	196	325	64
Kallar grass	7.80	0.76	Sandy clay loam	42.5	0.62	4.2	17	47	2.8	78.9	0.5	243	190	315	62
Dedonia viscoce	8.01	0.79	Sandy clay loam	38.1	0.64	4.5	17.5	48	3.1	81.2	0.4	237	205	330	65

3.2. Isolation and Screening of Salt-Tolerant PGPR Strains

A total of 50 isolates were obtained from the rhizospheric soil of four halophytic plants. Among all isolates, 90% of colonies were round, creamy, and had entire margins (Supplementary Materials Table S1). Further, 78% of isolates were Gram-negative and rod-shaped (Supplementary Materials Table S2).

In the halotolerant assay, 70% of strains were able to grow up to 6%, 20% strains showed tolerance at 10%, while four strains SR_1, SR_2, SR_3, and SR_4 were able to grow at 15% NaCl (Supplementary Materials Table S2). These four strains also showed positive results for phosphorous solubilization, hydrogen cyanide, and siderophore production (Supplementary Materials Table S3).

3.3. Effect of Bacterial Isolates on Germination of Wheat

Salt stress resulted in a considerable reduction in the germination parameters of the wheat seeds. Under salt-stressed conditions, the seedling vigor index and germination index showed a 12.5% and 31% decrease compared to the control. Though most of the strains showed a significant increase in seed germination, four strains SR_1, SR_2, SR_3, and SR_4 showed prominent results (14.28%, 35%, 42%, and 55%), respectively, as compared to the non-inoculated control under the salt stress condition (Supplementary Materials Table S4).

3.4. Identification of Isolates

Initially, the four strains were identified based on the C/N source utilization pattern (Supplementary Materials Table S5). Molecular identification of the screened halotolerant strains was done based on 16S rRNA sequences and on the comparison of the 1500-bp sequence of 16S rRNA gene subjected to BLAST to confirm the relatedness with other bacterial strains. The isolate SR_1 (1485 base pair) was closely related (98% nucleotide identity) to sequences of bacteria annotated as *Bacillus* strain JQ 926435 in the GenBank database. The sequence of SR_2 (1480 base pairs) was 99% identical to *Azospirillum brasilense* DQ 288686.1, SR_3 (1482 base pairs), and 96% identical to strain *Azospirillum lipoferum* accession no. M. 5906.1. Furthermore, the isolated strain SR_4 showed a 99% homology with *Pseudomonas stutzeri* JQ 926435. The accession numbers of the identified strains were obtained from NCBI and are given in Table 2.

Table 2. Molecular identification of the isolates based on partial 16S rDNA analysis.

No	Isolates	Base Pair Length	Similarity (%)	Strain Identification	Accession No.
1	SR_1	1485	98%	*Bacillus* sp.	KF719179
2	SR_2	1480	99%	*Azospirillum brasilense*	KJ194586
3	SR_3	1482	96%	*Azospirillum lipoferum*	KJ434039
4	SR_4	1263	99%	*Pseudomonas stutzeri*	KJ685889

Further phylogenetic analysis of the identified bacteria was conducted in MEGA4 software to determine their affiliation [38]. The evolutionary history was inferred using the maximum parsimony method [26]. The results are shown in Supplementary Materials Figures S1–S4.

3.5. Production of Phytohormones

Based on the halotolerance assays, PGP traits, and germination assay results, four isolates were selected for further analysis. All the halotolerant PGPR strains showed the production of phytohormones in liquid culture (Figure 1). Halotolerant PGPR strains were able to produce IAA (0.5–2.1 µg mL^{-1}), gibberellic acid (1.5–2.5 µg mL^{-1}), CK (0.39–0.64 µg mL^{-1}), and ABA (1.9–3.4 µg mL^{-1}). The PGPR strains SR_2 and SR_3 produced higher concentrations of phytohormones than those of SR_1 and SR_4; however, the bacterial consortium produced maximum concentrations of IAA (2.1 µg mL^{-1}), gibberellic acid (2.5 µg mL^{-1}), CK (0.64 µg mL^{-1}), and ABA (3.4 µg mL^{-1}).

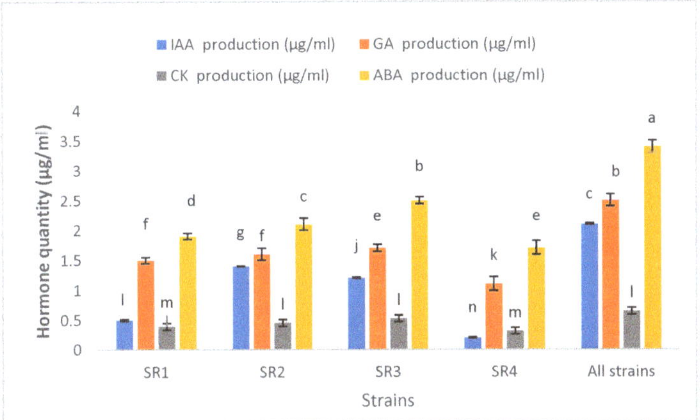

Figure 1. Production of phytohormones (Indole Acetic Acid (IAA), Gibberellic Acid (GA), Cytokinin (CK), and Abscisic Acid (ABA) by PGPR strains and their consortium in culture media. (SR$_1$: Inocualted with *Bacillus* sp; SR$_2$: Inocualted with *Azospirillum brasilense*; SR$_3$: Inocualted with *Azospirillum lipoferum*; SR$_4$: Inocualted with *Pseudomonas stutzeri*; Consortium is a combination of all four strains *Bacillus* sp, *Azospirillum brasilense, Azospirillum lipoferum, Pseudomonas stutzeri*). This data displays the means and standard deviation (*n* = 3). Different letters show significant differences between treatments (*p* < 0.05).

3.6. Production of Compatible Solutes

A considerable amount of proline was produced by all the screened halotolerant strains when subjected to different salinity levels. Production of proline by SR$_2$ and SR$_3$ was the highest in the 10% saline condition than the control. The maximum amount of proline (12.1 µg mg^{-1}) was produced by the bacterial consortium, which was 23% greater than SR$_2$ and SR$_3$. For the carbohydrate contents, a significant amount of soluble sugars was recorded by all the strains (Figure 2). The production of soluble sugars was more pronounced at different salinity levels than the control. The bacterial strains SR$_2$ and SR$_3$ produced a greater amount of (89–111 µg mg^{-1}) soluble sugar as compared to the control, but the consortium of bacterial isolates recorded the maximum values at 10% NaCl (222 µg mg^{-1}) (Figure 3).

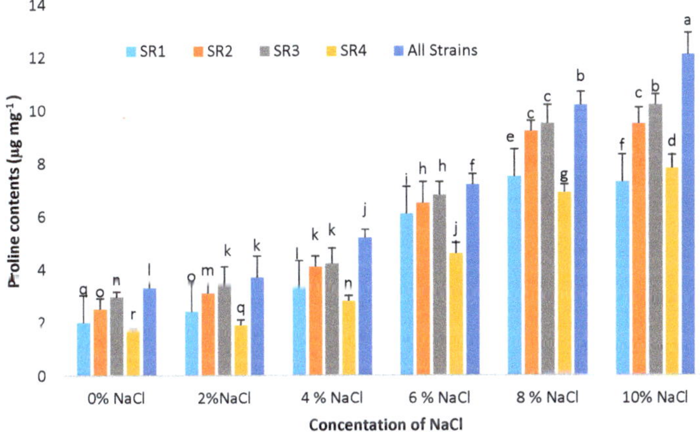

Figure 2. Production of proline by PGPR strains and their consortium in culture media supplemented with different concentrations of NaCl (2%, 4%, 6%, 8%, and 10%). The treatment details are the same as in Table 3. This data displays the means and standard deviation (*n* = 3). Different letters show significant differences between treatments (*p* < 0.05).

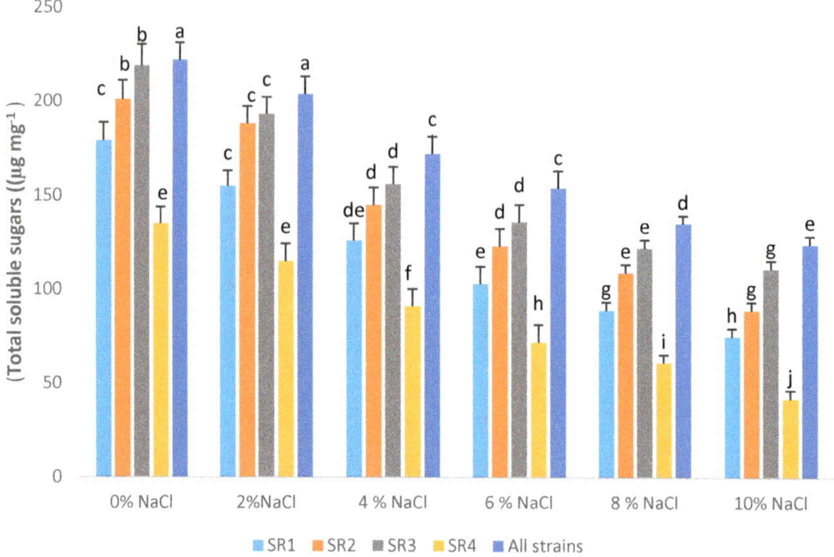

Figure 3. Production of total soluble sugar by PGPR strains and their consortium in culture media supplemented with different concentrations of NaCl (2%, 4%, 6%, 8%, and 10%). The treatment details are the same as in Table 3. This data displays the means and standard deviation ($n = 3$). Different letters show significant differences between treatments ($p < 0.05$).

3.7. Effect of PGPR Inoculation on the Biomass of Wheat (Triticum aestivum L.) Plants Grown under Salinity Stress

The overall decrease of 30% in the plant biomass of wheat plants was observed due to salt stress. However, the bacterial isolates exerted a significant positive influence on wheat growth and resulted in an increase in the biomass of plants in the control and stressed conditions, respectively. The relative increase in the fresh and dry biomass due to bacterial isolates ranged between 39% and 67% as compared to the uninoculated plants under saline conditions.

The best results were obtained when plants were inoculated with a consortium of all four isolated strains, which caused an increase of 93% in stress and 60% in controlled conditions. Moreover, pronounced results were also encountered for dry biomass, when plants were inoculated with a consortium, which resulted in an increase of 65.4% in salt stress and 78.7% in control conditions (Table 3).

Table 3. Effect of inoculation of halotolerant PGPR on the fresh and dry biomass and leaf area of wheat plants grown under salinity stress.

Treatments	Fresh Biomass (g)		Dry Biomass (g)		Leaf Area (cm^2)	
	0 mM	150 mM	0 mM	150 mM	0 mM	150 mM
Control	10 ± 1g	7.3 ± 0.4i	3.3 ± 0.1c	2.2 ± 0.04d	140 ± 12e	120 ± 17f
SR$_1$	11 ± 0.5f	8.2 ± 0.9h	3.5 ± 0.4c	2.8 ± 0.06d	150 ± 15d	130 ± 14f
SR$_2$	13.2 ± 0.9e	10.3 ± 0.7g	4.1 ± 0.1b	3.5 ± 0.09c	167 ± 12c	140 ± 24e
SR$_3$	16.9 ± 1.2 b	13.9 ± 1.4d	4.7 ± 0.5b	3.8 ± 0.03 c	177 ± 17b	147 ± 17e
SR$_4$	11.50.98f	7.8 ± 0.54i	3.6 ± 0.5c	3.0 ± 0.08d	160 ± 14c	135 ± 12f
Consortium	20.3 ± 1.8a	14.1 ± 1.9c	5.9 ± 0.2a	4.3 ± 0.03b	186 ± 19a	152 ± 13d

This data displays the means and standard deviation ($n = 3$). Different letters show significant differences ($p < 0.05$). (SR$_1$: Inocualted with *Bacillus* sp; SR$_2$: Inocualted with *Azospirillum brasilense*; SR$_3$: Inocualted with *Azospirillum lipoferum*; SR$_4$: Inocualted with *Pseudomonas stutzeri*; Consortium is a combination of all four strains *Bacillus* sp, *Azospirillum brasilense*, *Azospirillum lipoferum*, *Pseudomonas stutzeri*).

3.8. Effect on the Membrane Stability Index and Water Content

Results of the percent electrolytic leakage showed that the inoculation remains significant under stress as well as normal conditions However, co-inoculation with bacterial consortium successfully decreased (34%) the ionic discharge at the 150 mM NaCl level compared to the control (Table 4). Furthermore, the percent of water content showed a significant reduction of 33% in wheat plants under salt stress as compared to the uninoculated control plants. More pronounced results were obtained with SR_2 and SR_3, causing an increase of 10.5% and 17.54% in the stress condition. The consortium-inoculated plants recorded the maximum amount of water of 21% and 17.64% in the stress and control conditions. A similar trend was observed by SR_1 and SR_4 (Table 4).

Table 4. Effect of inoculation of halotolerant PGPR strains on the leaf water content and electrolyte leakage of wheat plants grown under salinity stress.

Treatments	Percent Water Content		Electrolyte Leakage (%)	
	0 mM	150 mM	0 mM	150 mM
Control	85 ± 1.5b	57 ± 0.9e	33 ± 0.4d	55 ± 0.5a
SR_1	86.3 ± 1.6b	60 ± 1d	30 ± 0.3d	50 ± 0.45a
SR_2	89 ± 1.9b	63 ± 1.4d	26.2 ± 0.25e	41.3 ± 0.33c
SR_3	95 ± 2.1a	67 ± 1.7d	25.7 ± 0.4e	42.08 ± 0.11c
SR_4	88.7 ± 1.9b	61 ± 1.15d	31.2 ± 0.22e	47.8 ± 0.44b
Consortium	97 ± 2.0a	70 ± 1.75c	22.1 ± 0.22f	35.2 ± 0.23d

This data displays the means and standard deviation ($n = 3$). Different letters show significant differences ($p < 0.05$). Treatment details are the same as in Table 3.

3.9. Chlorophyll Contents

Salinity stress negatively affected the photosynthetic pigments of wheat plants. A considerable decrease of 30.4%, 22%, and 25% was observed in chlorophyll a, b, and total chlorophyll. The response to the consortium was effective ($p \leq 0.05$) and resulted in a 13.23%, 12.49%, 12.9%, and 11.76% increase as compared to the control under salt-stress conditions (Table 5).

Table 5. Effect of halotolerant PGPR on the chlorophyll a, chlorophyll b, total chlorophyll, and carotenoid contents of wheat plants grown under salinity stress.

	Chlorophyll a (mg/g Fresh Weight)		Chlorophyll b (mg/g Fresh Weight)		Total Chlorophyll (mg/g Fresh Weight)		Carotenoid (mg/g Fresh Weight)	
Treatments	0 mM	150 mM	0 mM	150 mM	0 mM	150 mM	0 mM	150 mM
Control	1.06 ± 0.01d	0.59 ± 0.01h	0.27 ± 0.02d	0.12 ± 0.01h	1.18 ± 0.10e	0.86 ± 0.05f	46.9 ± 0.1f	65.8 ± 0.15k
SR_1	1.13 ± 0.03b	0.75 ± 0.03g	0.29 ± 0.04b	0.13 ± 0.02g	1.26 ± 0.09d	1.01 ± 0.03k	47.3 ± 0.3e	67.6 ± 0.5i
SR_2	1.18 ± 0.04c	0.81 ± 0.02f	0.32 ± 0.03c	0.15 ± 0.02f	1.33 ± 0.7c	1.13 ± 0.02j	50.5 ± 0.4c	69.8 ± 0.4h
SR_3	1.2 ± 0.05c	0.85 ± 0.04g	0.33 ± 0.05c	0.17 ± 0.03f	1.37 ± 0.8b	1.18 ± 0.04k	51.1 ± 0.2b	69.2 ± 0.5i
SR_4	1.12 ± 0.02b	0.77 ± 0.03b	0.28 ± 0.01b	0.13 ± 0.02g	1.13 ± 0.6f	1.05 ± 0.01d	48.2 ± 0.4d	68.1 ± 0.6j
Consortium	1.4 ± 0.04a	0.9 ± 0.02e	0.35 ± 0.05a	0.19 ± 0.04e	1.59 ± 0.5a	1.25 ± 0.03h	52.8 ± 0.6a	70.4 ± 0.8g

This data displays the means and standard deviation ($n = 3$). Different letters show significant differences ($p < 0.05$). Treatment details are the same as in Table 3.

3.10. Proline Contents

Salinity stress increased proline accumulation in wheat plants. A considerable increase of 50% in the proline content of wheat plants was recorded in saline stress conditions as compared to their respective control. Inoculation with halotolerant PGPR increased the levels of proline in the leaves. All four inoculants increased the proline contents in the range of 18–36%, respectively. The accumulation of proline was maximum in consortium-treated plants, with an increase of 46.67% under stress conditions (Table 6).

Table 6. Effects of halotolerant PGPR on the total soluble sugar, amino acid, protein, and proline contents of wheat plants grown under salinity stress.

Treatments	Total Soluble Sugar (μg g^{-1} FW)		Total Amino Acid (μg g^{-1} FW)		Proline (μg g^{-1} FW)	
	0 mM	150 mM	0 mM	150 mM	0 mM	150 mM
Control	27 ± 2d	33 ± 5i	330 ± 10g	368 ± 20e	40 ± 03d	120 ± 5j
SR$_1$	29 ± 3d	35 ± 7h	345 ± 12 f	379 ± 17e	44 ± 05d	128 ± 6i
SR$_2$	31 ± 5c	39 ± 6g	360 ± 15e	401 ± 27c	51 ± 3c	130 ± 7g
SR$_3$	33 ± 3c	33 ± 1.0f	370 ± 24e	420 ± 25b	54 ± 3b	135 ± 9h
SR$_4$	29 ± 4 d	36 ± 8h	350 ± 12f	387 ± 24.4d	43 + 3 c	125 ± 5i
Consortium	39 ± 5 c	43 ± 5e	381 ± 10d	439 ± 15a	57 ± 4b	145 ± 7f

This data displays the means and standard deviation ($n = 3$). Different letters show significant differences ($p < 0.05$). Treatment details are the same as in Table 3.

3.11. Amino Acid Content

The amino acid content was highest in the consortium of halotolerant PGPR strains, with an increase of 19.29% and 15.54% under salt stress and control conditions. Moreover, plants inoculated with SR$_2$ and SR$_3$ contained 10% and 14.1% greater concentrations of amino acids as compared to the uninoculated stressed plants (Table 6).

3.12. Total Soluble Sugar

Salinity stress produced a significant increase of 12.5% for the soluble sugar contents of wheat plants as compared to the control. The best outcomes were obtained when plants were inoculated with SR$_2$ and SR$_3$, which resulted in an increase of 9.52% and 15.87%, respectively, under stress conditions. However, a more prominent effect was revealed with the inoculation of a consortium of strains, with an increase of 28.57% and 23.2%, respectively, under the stress and control condition (Table 6).

3.13. Antioxidants Enzyme Assay

The antioxidant enzymes of the wheat plants showed a significant increase under salinity stress. Inoculation with all four halotolerant PGPR improved the production of antioxidant enzymes in plants. However, the best results were shown by the consortium of all strains. The consortium increased the superoxide dismutase activity by 21.4% as compared to stressed plants. Similarly, a significant increase of 16% in the catalase activity was recorded by the inoculation with the consortium. A significant increase of 34.4% in the peroxidase content of plants was recorded as compared to the control (Table 7).

Table 7. Effects of halotolerant PGPR on the antioxidant enzymes activity of wheat plants grown under salinity stress.

Treatments	Superoxide Dismutase (EU mg^{-1} Protein)		Catalase (EU mg^{-1} Protein)		Peroxidase (EU mg^{-1} Protein)	
	0 mM	150 mM	0 mM	150 mM	0 mM	150 mM
Control	0.74 ± 0.06k	1.83 ± 0.02f	2.5 ± 0.03h	4.13 ± 0.02f	144 ± 3f	255 ± 5f
SR$_1$	0.76 ± 0.04j	1.85 ± 0.01e	2.7 ± 0.02k	4.3 ± 0.04e	148 ± 7j	260 ± 4d
SR$_2$	0.8 ± 0.03i	1.9 ± 0.04c	3.01 ± 0.04i	4.7 ± 0.09c	153 ± 4i	263 ± 6c
SR$_3$	0.82 ± 0.05h	1.91 ± 0.03b	3.12 ± 0.05h	4.8 ± 0.10b	155 ± 6h	267 ± 7b
SR$_4$	0.78 ± 0.3j	1.85 ± 0.4d	2.6 ± 0.04j	4.5 ± 0.05d	150 ± 4j	257 ± 5d
Consortium	0.86 ± 0.07g	1.96 ± 0.05a	3.25 ± 0.05g	5.05 ± 0.04a	162 ± 3g	270 ± 6a

This data displays the means and standard deviation ($n = 3$). Different letters show significant differences ($p < 0.05$). Treatment details are the same as in Table 3.

3.14. Heatmap Responses of Pearson's Correlation Coefficient (r)

From the heat map analysis, the data of the osmolyte production, electrolyte leakage, chlorophyll contents, antioxidant enzymes, and halotolerant PGPR showed positive correlations (Figure 4). A comparative analysis of the parameters related to salinity tolerance (presented by green boxes) showed that salinity tolerance had a positive correlation with amino acid, osmotic potential, soluble sugars, proline, SOD, POD, and CAT activities (Figure 5).

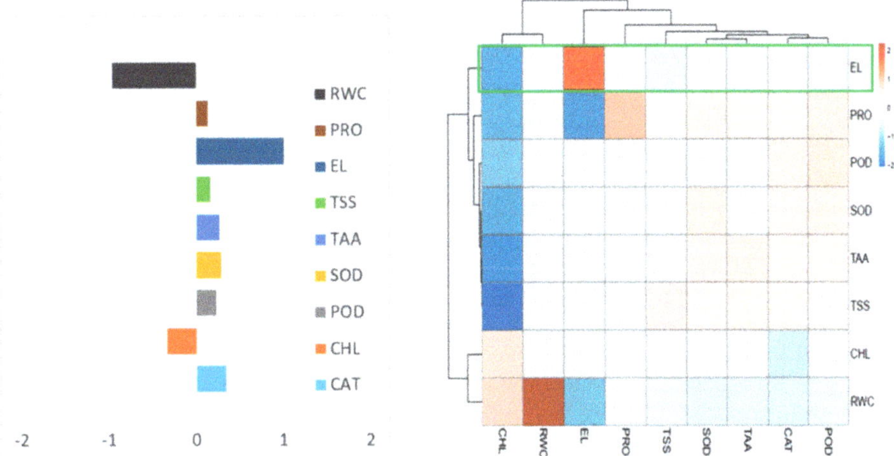

Figure 4. Heatmap of the correlation coefficient (r) for the antioxidant enzymes, stress determinants, and relative water content of wheat leaves treated with bacterial isolates and their consortium. Whereas, EL = Electrolyte leakage, Pro = Proline, POD = Peroxidase, SOD = Superoxide dismutase, CHL = Total chlorophyll, TAA = Total amino acids, TSS = Total soluble sugars, RWC = relative water content.

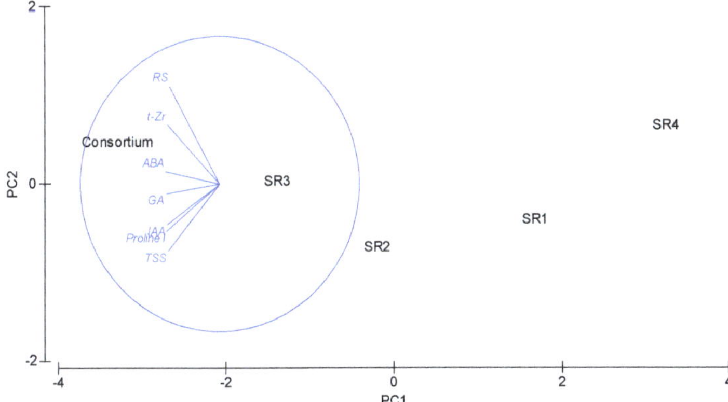

Figure 5. Principle component analysis (PCA) of phytohormones, proline, total soluble sugars, and reducing sugars of halotolerant bacterial isolates and their consortium grown under salt stress in culture conditions. Whereas, IAA = Indole acetic acid, GA = gibberellic acid, CK = Trans zeatin riboside, RS = Reducing sugars, TSS = Total soluble sugars.

4. Discussion

Soil bacteria associated with rhizosphere have been known as growth promotors as well as biotic and abiotic stress alleviators [8]. Bacteria associated with the roots of halophytes and saline soil, capable of tolerating higher levels of salts, are termed as halotolerant [39]. In the current study, bacterial isolates SR_1, SR_2, SR_3, and SR_4 showed the best salt tolerance abilities among all 50 bacterial isolates from the roots–soil interface of plants growing in the saline area. Phenotypic and molecular genotyping (16S RNA sequencing) of four potent isolates proved that SR_2 and SR_3 strains belong to the *Azospirillum* genus (*Azospirillum brasilense* and *Azospirillum lipoferum*) and the other two (SR$_I$ and SR$_4$) belong to the genus *Bacillus* (*Bacillus sturtezi*) and *Pseudomonas* (*Paeudomonas stutzeri*) (Table 2). These beneficial PGPR belonged to different genera, which indicate that plant growth promotion has been distributed across different taxons Halotolerant strains from the genera of *Pseudomonas*, *Bacillus*, *Azospirillum*, *Klebsiella*, and *Ochromobacter* have shown remarkable performance in the amelioration of salt stress in a wide range of crops [40].

Halotolerant PGPR has been reported to promote plant growth as well as mitigate salinity stress [41]. In the current study, we attempted to identify the key mechanisms used by halotolerant strains to alleviate the salinity stress in wheat plants by regulating plant defense mechanisms. The ability of halotolerant PGPR to produce phytohormones is associated with improved growth of plants under saline conditions [42]. The halotolerant PGPR produced IAA, GA, CK, and ABA. The results showed that *Azospirillum* strains produced higher amounts of GA, IAA, and CK than those of *Bacillus* and *Pseudomonas* strains in liquid media (Figure 1). The production of hormones by halotolerant PGPR is thoroughly supported by previous literature and many halotolerant strains of *Azotobacter*, *Bacillus*, *Arthrobacter*, *Azospirillum*, and *Pseudomonas* have been shown to produce IAA, GA, CK, and ABA [43]. These phytohormones regulate the stress defense responses in plants. They influence all aspects of plant growth, like cell wall elongation (IAA), cell division (CK), germination (gibberellin), and stress tolerance (ABA) [44–46]. Various reports suggest that these phytohormones produced under salinity stress help plants to survive and impart tolerance in them under abiotic stresses [46].

Here, the results proved that rhizobacteria secrete more compatible solutes (soluble sugars and proline) in culture media supplemented with a higher NaCl (10%) content. Various studies documented that bacterial cells can accumulate a considerable amount of compatible solutes inside their cells, acting as osmolytes and helping them to survive under severe osmotic stress [47].

Salinity is one of the common factors that can limit agricultural productivity due to its effects on seed germination, plant growth, and crop yield. Wheat is an important staple crop, but as it is a moderately salt-tolerant crop, high salt stress strictly limits its growth and development. Salt stress ultimately reduces the crop yield and nutritive value of wheat. The regulation of physiological, enzymatic, and biochemical changes in plants after inoculation with PGPR helps to alleviate salt or drought stress [40,48].

We demonstrated that salinity reduced the growth and development and relative water content of wheat plants. It also caused curling and wilting of leaves, early leaf senescence, and ultimately a reduction in the growth of plants. This is consistent with what was found in a previous study that salinity restricts cell differentiation and the cell cycle due to osmotic and ionic stress, deficiency of nutrients, oxidative damage, and limited water uptake, which affects plant germination, growth development, and physiological processes, ultimately leading to growth inhibition [49].

In this study, a consortium of four strains produced a prominent result for the dry biomass and leaf area than the control and individual inoculants. These results are in line with Walker et al. [50], who reported that inoculation with a consortium of *Azospirillum-Pseudomonas-Glomus* improved the root architecture in maize under salinity. A better adaptability of PGPR to stress conditions is correlated with efficient root colonization, phosphate solubilization, and nitrogen fixation abilities [51]. From the results, it is clear that salinized plants inoculated with halotolerant strains and their consortium exhibited a higher relative water content of leaves. Rakshapal et al. [52] also observed that PGPR-treated plants not only cope with stress but also that these microbes help to maintain higher water levels in comparison to control plants.

Salinity decreases the photosynthetic efficiency of plants and results in the production of reactive oxygen species (ROS), which cause damage to DNA, proteins, and membranes [53]. We described the results of photosynthetic pigments of wheat plants, which showed that treatment with a consortium showed a pronounced effect of reducing the damage caused by salinity on the photosynthetic apparatus. A similar pattern of results was reported by El-Esawi et al. [54], who observed an increase in the photosynthetic efficiency of plants by PGPR inoculation under salinity.

Salt stress can develop more discharge of electrolytes through the misplacement of Ca associated with membranes. As a result, the permeability of the membrane is destroyed and accumulates a higher efflux of electrolytes inside plant cells/tissue [55]. In the current study, the successive increase in the electrolyte leakage of wheat plants was observed at 150 mM salt stress than the control. These results are inconsistent with the Bojórquez-Quintal et al. [56], who found salt stress enhances electrolyte leakage and the generation of reactive oxygen species (ROS), having a detrimental effect on plant growth. Our results showed that inoculation with halotolerant PGPR tends to decrease the injurious effect of saline stress and decrease the potential electrolytic leakage of ions in stress-treated plants. This is consistent with what was found in previous studies [57,58].

In the present study, the concentration of compatible solutes was also increased in inoculated wheat plants under salt stress (Table 7). The accumulation of compatible solutes, particularly proline, free amino acid, and soluble sugar, is correlated with the adaptability of the plant to stress conditions. We reported that halotolerant PGPR produces compatible osmolytes, which help the plants to maintain their ionic balance. PGPR also induce osmolyte accumulation [59] and phytohormone signaling [40], which facilitates plants in overcoming the initial osmotic shock after salinization. In a previous study, it was found that rice inoculation with salt-tolerant *Bacillus amyloliquefaciens* under salinity increased the plant's salt tolerance and affected the expression of genes involved in osmotic and ionic stress response mechanisms [60].

Proline is the most important osmolyte, which is produced in plants by the hydrolysis of proteins under osmotic stress [61]. From the results, it is clear that a consortium of halotolerant PGPR plants improved proline levels under salt stress. These results are in line with Wang et al. [62]. The production of osmolytes helps the plant to maintain a high turgor potential, prevent oxidative damage by scavenging reactive oxygen species, and protect the membrane structure [63].

We also reported a pronounced increase in the production of soluble sugars with a consortium of halotolerant strains in wheat under salinity stress. PGPR can stimulate carbohydrate metabolism and transport, which results in changes in the source–sink relations, photosynthesis, and growth rate. In previous reports, seeds inoculated with *B. aquimaris* strains showed an increased production of total soluble sugars in wheat under salinity conditions, which resulted in higher biomass and plant growth [64].

An increase in the antioxidant enzyme activity of wheat plants grown under salinity stress was observed by a consortium of halotolerant PGPR strains. This indicates that these bacteria can help the plant to combat the deleterious effects of ROS generated during salinity stress. These results tie well with the previous studies, where an increase in antioxidant enzyme activity under salinity stress was proven to be associated with salt tolerance [65]. Moreover, Wang et al. [66] reported that the application of PGPR strains alleviates the oxidative damage induced by abiotic stresses, including salinity, by augmenting the activity of antioxidant enzymes.

5. Conclusions

In summary, crop inoculations with halotolerant PGPR consortium can serve as a potential tool for alleviating salinity stress. Halotolerant PGPR strains have developed several mechanisms to cope with salinity, particularly the potential to produce phytohormones and compatible solutes. Halotolerant PGPR strains can induce salinity tolerance in plants by activating key defense mechanisms like the production of osmoregulators as well as activating ROS scavenging enzymes. Natural microflora adapted to saline conditions can be used for the development of microbial consortia for crop inoculation, ultimately leading to the formulation of biofertilizer for salt-stressed areas. However, further investigation is needed to observe their performance in field conditions.

Supplementary Materials: The following are available online at http://www.mdpi.com/2073-4395/10/7/989/s1, Table S1: Morphology of isolates from rhizosphere of plants from saline soil, Table S2: Preliminary screening data of isolated strains (+ indicates groeth, − indicates no growth), Table S3: Growth characters of isolated strains, Table S4: Effect of Isolates on germination attributes of wheat, Table S2: Carbon/Nitrogen source utilization pattern determined by QTS -24 kits, Figure S1: Phylogenetic analysis of strain SR1, Figure S2: Phylogenetic analysis of strain SR2, Figure S3: Phylogenetic analysis of strain SR3, Figure S4: Phylogenetic analysis of strain SR4.

Author Contributions: Conceptualization, N.I. Writing—Original Draft, N.I. and R.M. Formal Analysis, H.Y. and W.K. Investigation, R.M. Proofreading, H.E.E. and D.J.D. Editing, H.Y., W.K. and S.I. Formatting, D.J.D. Writing—Review, H.Y., S.I. and W.K. Supervision, N.I. Facilitation, H.E.E. Review, H.E.E. and D.J.D. All authors have read and agreed to the published version of the manuscript.

Funding: This research was funded in part by Allcosmos Industries Sdn. Bhd. through research project No. R.J130000.7344.4B200. The APC was supported by UTM-TNCPI research fund.

Conflicts of Interest: Authors have no conflict of interest.

Ethical Statement: Not applicable.

References

1. GAP Report. Global Agricultural Productivity Report (GAP Report) Global Harvest Initiative, Washington. 2018. Available online: https://globalagriculturalproductivity.org/wp-content/uploads/2019/01/GHI_2018-GAP-Report_FINAL-10.03.pdf (accessed on 5 May 2020).

2. Abbas, R.; Rasul, S.; Aslam, K.; Baber, M.; Shahid, M.; Mobeen, F.; Naqqash, T. Halo-tolerant PGPR: A hope for cultivation of saline soils. *J. King Saud Univ. Sci.* **2019**, *31*, 1195–1201. [CrossRef]

3. Mishra, J.; Fatima, T.; Arora, N.K. Role of secondary metabolites from plant growth-promoting rhizobacteria in combating salinity stress. In *Plant Microbiome: Stress Response*; Ahmad, P., Egamberdieva, D., Eds.; Springer: Singapore, 2018; pp. 127–163.

4. Yasin, N.A.; Khan, W.; Ahmad, S.R.; Ali, A.; Ahmad, A.; Akram, W. Imperative roles of halo-tolerant plant growth-promoting rhizobacteria and kinetin in improving salt tolerance and growth of black gram (*Phaseolusmungo*). *Environ. Sci. Pollut. Res.* **2017**, *25*, 4491–4505. [CrossRef]

5. Abdel-Latef, A.A.H.; Chaoxing, H. Does Inoculation with *Glomus mosseae* Improve Salt Tolerance in Pepper Plants? *J. Plant Growth Regul.* **2014**, *33*, 644–653. [CrossRef]

6. Liu, K.; McInroy, J.A.; Hu, C.-H.; Kloepper, J.W. Mixtures of Plant-Growth-Promoting Rhizobacteria Enhance Biological Control of Multiple Plant Diseases and Plant-Growth Promotion in the Presence of Pathogens. *Plant Dis.* **2018**, *102*, 67–72. [CrossRef]

7. Khan, A.; Zhao, X.Q.; Javed, M.T.; Khan, K.S.; Bano, A.; Shen, R.F. *Bacillus pumilus* enhances tolerance in rice (*Oryza sativa* L.) to combined stresses of NaCl and high boron due to limited uptake of Na+. *Environ. Exp. Bot.* **2016**, *124*, 120–129. [CrossRef]

8. Yasmin, H.; Nosheen, A.; Naz, R.; Keyani, R.; Anjum, S. Regulatory role of rhizobacteria to induce drought and salt stress tolerance in plants. In *Field Crops: Sustainable Management by PGPR. Sustainable Development and Biodiversity*; Maheshwari, D., Dheeman, S., Eds.; Springer: Cham, Switzerland, 2019; Volume 23, p. 23.

9. Niu, X.; Song, L.; Xiao, Y.; Ge, W. Drought-tolerant plant growth-promoting rhizobacteria associated with foxtail millet in a semi-arid agroecosystem and their potential in alleviating drought stress. *Front. Microbiol.* **2018**, *8*, 2580. [CrossRef] [PubMed]

10. Chu, T.N.; Tran, B.T.H.; Van-Bui, L.; Hoang, M.T.T. Plant growth-promoting rhizobacterium *Pseudomonas* PS01 induces salt tolerance in *Arabidopsis thaliana*. *BMC Res. Notes* **2019**, *12*, 11. [CrossRef] [PubMed]

11. Radojevic, M.; Bashkin, V.N. *Practical Environmental Analysis*; The Royal Society of Chemistry: Cambridge, UK, 1999; p. 466.

12. Soltanpour, P.N.; Schwab, A.P. A new soil test for simultaneous extraction of macro- and micro-nutrients in alkaline soils. *Commun. Soil Sci. Plant Anal.* **1977**, *8*, 195–207. [CrossRef]

13. Kingston, H.M. *Microwave Assisted Acid Digestion of Sediments, Sludges, Soils and Oils*; Duquesne University: Pittsburgh, PA, USA, 1994; p. 4.

14. Somasegaran, P.; Hoben, H.J. Counting rhizobia by a plant infection method. In *Handbook for Rhizobia*; Springer: New York, NY, USA, 1994; pp. 58–64.

15. James, G.C. *Native Sherman Rockland Community College, State University of New York*; The Benjamin/Coming Publishing Company Inc.: San Francisco, CA, USA, 1987; pp. 75–80.

16. Robinson, R.J.; Fraaije, B.A.; Clark, I.M.; Jackson, R.W.; Hirsch, P.R.; Mauchline, T.H. Endophytic bacterial community composition in wheat (*Triticum aestivum*) is determined by plant tissue type, developmental stage and soil nutrient availability. *Plant Soil* **2016**, *405*, 381–396. [CrossRef]

17. Pikovskaya, R. Mobilization of phosphorus in soil in connection with vital activity of some microbial species. *Mikrobiologiya* **1948**, *17*, 362–370.

18. Murphy, J.A.M.E.S.; Riley, J.P. A modified single solution method for the determination of phosphate in natural waters. *Anal. Chim. Acta* **1962**, *27*, 31–36. [CrossRef]

19. Schwyn, B.; Neilands, J.B. Universal chemical assay for the detection and determination of siderophores. *Anal. Biochem.* **1987**, *160*, 47–56. [CrossRef]

20. Lorck, H. Production of hydrocyanic acid by bacteria. *Physiol. Plant.* **1948**, *1*, 142–146. [CrossRef]

21. Noreen, Z.; Ashraf, M.; Hassan, M.U. Inter-accessional variation for salt tolerance in pea (*Pisum sativum* L.) at germination and screening stage. *Pak. J. Bot.* **2007**, *39*, 2075–2085.

22. Upadhyay, S.K.; Singh, J.S.; Singh, D.P. Exopolysaccharide-producing plant growth-promoting rhizobacteria under salinity condition. *Pedosphere* **2011**, *21*, 214–222. [CrossRef]

23. Tien, T.M.; Gaskins, M.H.; Hubbell, D.H. Plant growth substances produced by *Azospirillum brasilense* and their effect on the growth of pearl millet (*Pennisetum americanum* L.). *Appl. Environ. Microbiol.* **1979**, *37*, 1016–1024. [CrossRef] [PubMed]

24. Chen, W.P.; Kuo, T.T. A simple and rapid method for the preparation of gram-negative bacterial genomic DNA. *Nucleic Acids Res.* **1993**, *21*, 2260. [CrossRef]

25. Weisburg, W.G.; Barns, S.M.; Pelletier, D.A.; Lane, D.J. 16S ribosomal DNA amplification for phylogenetic study. *J. Bacteriol.* **1991**, *173*, 697–703. [CrossRef]

26. Eck, R.V.; Dayhoff, M.O. *Atlas of Protein Sequence and Structure*; National Biomedical Research Foundation: Silverspring, MD, USA, 1966; p. 12.

27. Unyayer, S.; Keles, Y.; Cekic, F.O. The antioxidative response of two tomato species with different tolerances as a result of drought and cadmium stress combination. *Plant Soil Environ.* **2005**, *51*, 57–64. [CrossRef]

28. Sriram, S.; Raguchander, T.; Babu, S.; Nandakumar, R.; Shanmugam, V.; Vidhyasekaran, P. Inactivation of phytotoxin produced by the rice sheath blight pathogen *Rhizoctonia solani*. *Can. J. Microbiol.* **2000**, *46*, 520–524. [CrossRef]

29. Arnon, D.I. Copper enzymes in isolated chloroplasts. Polyphenol oxidase in *Beta vulgaris*. *Plant Physiol.* **1949**, *24*, 1–15. [CrossRef] [PubMed]

30. Bates, L.S.; Waldren, R.P.; Teare, I.D. Rapid determination of free proline for water stress studies. *Plant Soil* **1973**, *39*, 205–207. [CrossRef]

31. Dubois, M.; Gilles, K.A.; Hamilton, J.K.; Rebers, P.A.; Smith, F. A colorimetric method for the determination of sugars. *Nature* **1951**, *168*, 167. [CrossRef] [PubMed]

32. Hamilton, P.B.; VanSlyke, D.D. Amino acid determination with ninhydrin. *J. Biol. Chem.* **1943**, *150*, 231–233.

33. Bradford, M. A rapid and sensitive method for the quantitation of microgram, quantitation of proteins utilizing the principle of protein dye-binding. *Anal. Biochem.* **1976**, *72*, 248–254. [CrossRef]

34. Aebi, H. Catalase in Vitro. *Methods Enzymol.* **1984**, *105*, 121–126. [CrossRef]

35. Rao, M.V.; Paliyath, G.; Ormrod, D.P. Ultraviolet-B- and ozone induced biochemical changes in antioxidant enzymes of *Arabidopsis thaliana*. *Plant Physiol.* **1996**, *110*, 125–136. [CrossRef]

36. Giannopolitis, C.N.; Ries, S.K. Superoxide dismutase in higher plants. *Plant Physiol.* **1977**, *59*, 309–314. [CrossRef]

37. Steel, R.G.D.; Torrie, G.H. *Principles and Procedures of Statistics Singapore*, 2nd ed.; McGraw Hill Book Co Inc.: New York, NY, USA, 1980.

38. Tamura, K.; Dudley, J.; Nei, M.; Kumar, S. MEGA4: Molecular evolutionary genetics analysis (MEGA) software version 4.0. *Mol. Boil. Evol.* **2007**, *24*, 1596–1599. [CrossRef]

39. Ramadoss, D.; Lakkineni, V.K.; Bose, P.; Ali, S.; Annapurna, K. Mitigation of salt stress in wheat seedlings by halotolerant bacteria isolated from saline habitats. *SpringerPlus* **2013**, *2*, 6. [CrossRef]

40. Egamberdieva, D.; Wirth, S.; Bellingrath-Kimura, S.D.; Mishra, J.; Arora, N.K. Salt-Tolerant Plant Growth Promoting Rhizobacteria for Enhancing Crop Productivity of Saline Soils. *Front. Microbiol.* **2019**, *10*, 2791. [CrossRef] [PubMed]

41. Sarkar, A.; Ghosh, P.K.; Pramanik, K.; Mitra, S.; Soren, T.; Pandey, S. A halo-tolerant *Enterobacter* sp. displaying ACC deaminase activity promotes rice seedling growth under salt stress. *Microbiol. Res.* **2018**, *169*, 20–32. [CrossRef] [PubMed]

42. Patel, D.; Saraf, M. Influence of soil ameliorants and microflora on induction of antioxidant enzymes and growth promotion of (*Jatropha curcas* L.) under saline condition. *Eur. J. Soil Biol.* **2013**, *55*, 47–54. [CrossRef]

43. Abd-Allah, E.F.; Alqarawi, A.A.; Hashem, A.; Radhakrishnan, R.; Al-Huqail, A.A.; Al-Otibi, F.A. Endophytic bacterium *Bacillus subtilis* (BERA 71) improves salt tolerance in chickpea plants by regulating the plant defense mechanisms. *J. Plant Interact.* **2017**, *3*, 37–44. [CrossRef]

44. TrParray, A.P.; Jan, S.; Kamili, A.N.; Qadri, R.A.; Egamberdieva, D.; Ahmad, P. Current perspectives on plant growth promoting rhizobacteria. *J. Plant Growth Regul.* **2016**, *35*, 877–902. [CrossRef]

45. Salomon, M.; Bottini, R.; de Souza Filho, G.A.; Cohen, A.C.; Moreno, D.; Gil, M. Bacteria isolated from roots and rhizosphere of *Vitis vinifera* retard water losses, induce abscisic acid accumulation and synthesis of defense-related terpenes in in vitro cultured grapevine. *Physiol. Plant.* **2014**, *151*, 359–374. [CrossRef] [PubMed]

46. Bottini, R.; Cassán, F.; Piccoli, P. Gibberellin production by bacteria and its involvement in plant growth promotion and yield increase. *Appl. Microbiol. Biotechnol.* **2004**, *65*, 497–503. [CrossRef]

47. Parida, A.K.; Das, A.B. Salt tolerance and salinity effect on plants: A review. *Ecotoxicol. Environ. Saf.* **2005**, *60*, 324–349. [CrossRef]

48. Kumar, A.; Verma, J.P. Does plant-microbe interaction confer stress tolerance in plants: A review? *Microbiol. Res.* **2018**, *207*, 41–52. [CrossRef]

49. Isayenkov, S.V.; Maathuis, F.J.M. Plant Salinity Stress: Many Unanswered Questions Remain. *Front. Plant Sci.* **2019**, *10*, 80. [CrossRef]

50. Walker, V.; Couillerot, O.; Felten, A.V.; Bellvert, F.; Jansa, J.; Maurhofer, M.; Bally, R.; Moënne-Loccoz, Y.; Comte, G. Variation of secondary metabolite levels in maize seedling roots induced by inoculation with *Azospirillum, Pseudomonas* and *Glomus* consortium under field conditions. *Plant Soil* **2012**, *356*, 151–163. [CrossRef]

51. Nadeem, S.M.; Zahir, Z.A.; Naveed, M.; Arshad, M. Preliminary investigations on inducing salt tolerance in maize through inoculation with rhizobacteria containing ACC deaminase activity. *Can. J. Microbiol.* **2007**, *53*, 1141–1149. [CrossRef] [PubMed]

52. Rakshapal, S.; Sumit, K.S.; Rajendra, P.P.; Alok, K. Technology for improving essential oil yield of *Ocimum basilicum* L. (sweet basil) by application of bioinoculant colonized seeds under organic field conditions. *Ind. Crop. Prod.* **2013**, *45*, 335–342. [CrossRef]

53. Islam, F.; Yasmeen, T.; Arif, M.S.; Ali, S.; Ali, B.; Hameed, S. Plant growth promoting bacteria confer salt tolerance in *Vigna radiata* by up-regulating antioxidant defense and biological soil fertility. *Plant Growth Regul.* **2016**, *80*, 23–36. [CrossRef]

54. El-Esawi, M.A.; Alaraidh, I.A.; Alsahli, A.A.; Alamri, S.A.; Ali, H.M.; Alayafi, A.A. *Bacillus firmus* (SW5) augments salt tolerance in soybean (*Glycine max* L.) by modulating root system architecture, antioxidant defense systems and stress-responsive genes expression. *Plant Physiol. Biochem.* **2018**, *132*, 375–384. [CrossRef]

55. Garg, N.; Manchanda, G. Role of arbuscular mycorrhizae in the alleviation of ionic, osmotic and oxidative stresses induced by salinity in *Cajanus cajan* (L.) Millsp. (pigeonpea). *J. Agron. Crop Sci.* **2009**, *195*, 110–123. [CrossRef]

56. Bojórquez-Quintal, E.; Velarde-Buendía, A.; Ku-González, Á.; Carillo-Pech, M.; Ortega-Camacho, D.; Echevarría-Machado, I.; Pottosin, I.; Martínez-Estévez, M. Mechanisms of salt tolerance in habanero pepper plants (*Capsicum chinense* Jacq.): Proline accumulation, ions dynamics and sodium root-shoot partition and compartmentation. *Front. Plant Sci.* **2014**, *5*, 605. [CrossRef]

57. Kaya, C.; Tuna, A.L.; Okant, A.M. Effect of foliar applied kinetin and indole acetic acid on maize plants grown under saline conditions. *Turk. J. Agric. For.* **2010**, *34*, 529–538.

58. Habib, S.H.; Kausar, H.; Saud, H.M. Plant Growth-Promoting Rhizobacteria Enhance Salinity Stress Tolerance in Okra through ROS-Scavenging Enzymes. *BioMed Res. Int.* **2016**, *2016*, 6284547. [CrossRef]

59. Bremer, E.; Kramer, R. Responses of microorganisms to osmotic stress. *Annu. Rev. Microbiol.* **2019**, *73*, 313–334. [CrossRef]

60. Nautiyal, C.S.; Srivastava, S.; Chauhan, P.S.; Seem, K.; Mishra, A.; Sopory, S.K. Plant growth-promoting bacteria *Bacillus amyloliquefaciens* NBRISN13 modulates gene expression profile of leaf and rhizosphere community in rice during salt stress. *Plant Physiol. Biochem.* **2013**, *66*, 1–9. [CrossRef] [PubMed]

61. Krasensky, J.; Jonak, C. Drought, salt, and temperature stress-induced metabolic rearrangements and regulatory networks. *J. Exp. Bot.* **2012**, *63*, 1593–1608. [CrossRef] [PubMed]

62. Wang, Q.; Dodd, I.C.; Belimov, A.A.; Jiang, F. Rhizosphere bacteria containing 1-aminocyclopropane-1-carboxylate deaminase increase growth and photosynthesis of pea plants under salt stress by limiting Na+ accumulation. *Funct. Plant Biol.* **2016**, *43*, 161–172. [CrossRef] [PubMed]

63. Prado, F.E.; Boero, C.; Gallardo, M.; González, J.A. Effect of NaCl on germination, growth, and soluble sugar content in *Chenopodium quinoa* wild seeds. *Bot. Bull. Acad. Sin.* **2000**, *41*, 27–343.

64. Upadhyay, S.K.; Singh, D.P. Effect of salt-tolerant plant growth-promoting rhizobacteria on wheat plants and soil health in a saline environment. *Plant Biol. (Stuttg.)* **2015**, *17*, 288–293. [CrossRef] [PubMed]

65. Moghaddam, M.; Farhadi, N.; Panjtandoust, M.; Ghanati, F. Seed germination, antioxidant enzymes activity and proline content in medicinal plant *Tagetes minuta* under salinity stress. *Plant Biosyst.* **2019**. [CrossRef]

66. Wang, C.J.; Yang, W.; Wang, C.; Gu, C.; Niu, D.D.; Liu, H.X.; Wang, Y.P.; Guo, J. Induction of drought tolerance in cucumber plants by a consortium of three plant growth-promoting rhizobacterium strains. *PLoS ONE* **2012**, *7*, e52565. [CrossRef]

Article

Volatile Organic Compounds from Rhizobacteria Increase the Biosynthesis of Secondary Metabolites and Improve the Antioxidant Status in *Mentha piperita* L. Grown under Salt Stress

Lorena del Rosario Cappellari, Julieta Chiappero, Tamara Belén Palermo, Walter Giordano and Erika Banchio *

INBIAS Instituto de Biotecnología Ambiental y Salud (CONICET—Universidad Nacional de Río Cuarto), Campus Universitario, 5800 Río Cuarto, Argentina; lcappellari@exa.unrc.edu.ar (L.d.R.C.); jchiappero@exa.unrc.edu.ar (J.C.); tpalermo@exa.unrc.edu.ar (T.B.P.); wgiordano@exa.unrc.edu.ar (W.G.)
* Correspondence: ebanchio@exa.unrc.edu.ar; Fax: +54-358-4676232

Received: 4 July 2020; Accepted: 27 July 2020; Published: 29 July 2020

Abstract: Salinity is a major abiotic stress factor that affects crops and has an adverse effect on plant growth. In recent years, there has been increasing evidence that microbial volatile organic compounds (mVOC) play a significant role in microorganism–plant interactions. In the present study, we evaluated the impact of microbial volatile organic compounds (mVOC) emitted by *Bacillus amyloliquefaciens* GB03 on the biosynthesis of secondary metabolites and the antioxidant status in *Mentha piperita* L. grown under 0, 75 and 100 mM NaCl. Seedlings were exposed to mVOCs, avoiding physical contact with the bacteria, and an increase in NaCl levels produced a reduction in essential oil (EO) yield. Nevertheless, these undesirable effects were mitigated in seedlings treated with mVOCs, resulting in an approximately a six-fold increase with respect to plants not exposed to mVOCs, regardless of the severity of the salt stress. The main components of the EOs, menthone, menthol, and pulegone, showed the same tendency. Total phenolic compound (TPC) levels increased in salt-stressed plants but were higher in those exposed to mVOCs than in stressed plants without mVOC exposure. To evaluate the effect of mVOCs on the antioxidant status from salt-stressed plants, the membrane lipid peroxidation was analyzed. Peppermint seedlings cultivated under salt stress and treated with mVOC showed a reduction in malondialdehyde (MDA) levels, which is considered to be an indicator of lipid peroxidation and membrane damage, and had an increased antioxidant capacity in terms of DPPH (2,2-diphenyl–1-picrylhydrazyl) radical scavenging activity in relation to plants cultivated under salt stress but not treated with mVOCs. These results are important as they demonstrate the potential of mVOCs to diminish the adverse effects of salt stress.

Keywords: mVOCs; Plant growth promoting rhizobacteria; PGPR; *Mentha piperita*; *Bacillus amyloliquefaciens* GB03; salt stress; secondary metabolites; MDA; DPPH

---

1. Introduction

Many aromatic plants, such as *Mentha piperita* L. (peppermint), are important sources of essential oil (EO) production. The EOs are generated and stored in glandular trichomes, where they form complex mixtures of secondary metabolites (SM) mainly composed of the volatile mono- and sesquiterpenes responsible for the characteristic aromas of various plant species [1,2]. Therefore, the quality of aromatic plants is recognized by the composition and concentration of these components for each species. Furthermore, the quantity and quality of SM is determined by environmental factors including temperature, soil quality, light intensity, and/or water availability [3].

Biotic and abiotic stresses are major constraints on crop yield, with environmental stress representing a strong restriction on increasing crop productivity as well as affecting the use of natural resources. A soil is considered to be saline when the ion concentration reaches an electrical conductivity of >4 dS m^{-1}, measured on a saturated soil at 25 °C, and consequently interferes with the growth of species of agricultural interest [4]. Salinity impacts agricultural production in most crops by affecting the physical-chemical properties of the soil and the ecological balance of the cultivated area [5]. As salinity affects many aspects of the physiology and metabolism of the plants, the presence of soluble salts in general has a negative consequence for the plant's growth by decreasing the water potential and thus restricting the absorption of water by the roots (osmotic effect). In addition, the absorption of specific saline ions leads to their accumulation in tissues in concentrations at which they can become toxic and induce physiological disorders (ionic toxicity) in the plant, with high concentrations of saline ions being able to modify the absorption of essential nutrients and leading to nutritional imbalances (nutritional effect) [6]. These effects are reflected by a decrease in germination, vegetative growth, and reproductive development [4,7].

Plant tolerance to salt stress is linked to the use of different strategies, including osmotic adjustment, the exclusion of toxic ions from the aerial part, translocation of photoassimilates to underground organs, an increased growth of the root system, and ensuring the availability of water and nutrients, among others. Furthermore, salinity can produce an accumulation of reactive oxygen species (ROS) [6], which may lead to a deterioration of photosynthetic pigments, lipid peroxidation, alterations in the selective permeability of the cell membranes, protein denaturation, and DNA mutations [8–10]. Damage of the cell membrane produces small hydrocarbons such as malondialdehyde (MDA), which is a sign of membrane cellular damage. Plants have well-described protection and repair systems that mitigate ROS damage. In addition, certain species have developed protective mechanisms that include enzymatic and non-enzymatic components [11,12].

Plant growth promoting rhizobacteria (PGPR) are beneficial microorganisms capable of colonizing the rhizosphere of plants and benefiting them both directly and indirectly [13]. It is well known that PGPR functions in different ways: synthesizing specific compounds for the plants, helping the uptake of nutrients, and protecting the plants from diseases [14–16]. In general, it has been observed that the negative effects that salinity produces in plant development can be mitigated by the use of microorganisms as inoculants, which is an alternative technology to improve the abiotic stress tolerance capacity of plants [17–21]. In this regard, considerable attention has been focused on understanding the molecular, physiological, and morphological mechanisms underlying rhizobacterial-mediated stress tolerance. In fact, the mechanisms by which these bacteria mediate abiotic stress tolerance continue to be widely studied, largely because they are difficult to elucidate [22,23].

Advances in research have revealed that certain PGPR strains are capable of emitting microbial volatile organic compounds (mVOCs) [24–28]. These compounds mainly consist of an abundant and very complex mixture of compounds, including alcohols, alkanes, alkenes, esters, ketones, sulfur, and terpenoids, characterized by their low molecular weight and high vapor pressure under normal conditions, which can vaporize significantly and enter the atmosphere. The analysis of mVOCs is a developing research area that has an effect on the applied agricultural, medical, and biotechnical applications, with a related interesting mVOC database containing available information regarding microbial volatiles having been published [29]. Recent studies have also provided new insights into the participation of mVOCs in inter- and intra-specific communication [30]. These compounds have been observed to have the ability to promote plant growth and induce systemic resistance (ISR) against pathogenic organisms, thereby improving the well-being of crops [24,27,28,31,32]. VOCs from *Paraburkholderia phytofirmans* have been shown to increase plant growth rate and tolerance to salinity, reproducing the effects of direct bacterial inoculation of roots [32]. Thus, the emission of mVOCs is currently recognized as being a very relevant aspect in microorganism–plant interactions [17,21,28,33,34].

We have previously demonstrated that both the direct inoculation of PGPR and exposure to VOCs emitted by these rhizobacteria stimulate the biosynthesis of SM and increase the biomass production in different aromatic plants [25,26,35–39]. Although there are few reports about the effects of mVOCs emitted by rhizobacteria on the SM yield of aromatic plants under conditions of abiotic stress, studies related to the emission of volatile organic compounds with biological activity by rhizobacteria is a novel area attracting increasing interest.

It should also be noted that it is necessary to examine the use of fertilizers and chemical synthesis pesticides related to the concentration of salts in the soil in order to develop sustainable agriculture, as this is key to assessing the proposal of alternative and complementary strategies. Taking this into consideration, among the possible alternatives, the use of microbial inoculants, considered to be a clean technology aligned with the principles of sustainable agriculture, becomes more relevant. Thus, the present study was founded on the hypothesis that the investigation of mVOCs with respect to the description of their biological functions and ecological roles is crucial for elucidating the mechanisms related to the control of critical biological processes in plant health and that this could also offer useful benefits to confront agronomic and environmental complications. In this present study, the aim was to explore the potential of mVOCs in ameliorating salinity effects in *M. piperita,* with an important objective of the study being to evaluate the role of mVOCs in EOs and the phenolic compound levels, as well as their function in the antioxidant status of plants grown under salt stress conditions.

2. Materials and Methods

2.1. Bacterial Strains and In Vitro Plant Treatments

2.1.1. Bacterial Cultures

Bacillus amyloliquefaciens GB03 (originally described as *Bacillus subtilis* GB03) [40] strain was grown on LB (Luria-Bertani) medium for routine use and maintained in nutrient broth with 15% glycerol at −80 °C for storage. The bacterial culture was grown overnight at 30 °C and centrifuged at 120, washed twice in 0.9% NaCl by Eppendorf centrifugation (4300× g, 10 min, 4 °C), re-suspended in sterile water, and adjusted to a final concentration of ~10^9 CFU/ mL for use as an inoculum.

2.1.2. Plant Micropropagation

The *M. piperita* plant is a commercially cultivated crop grown in the Traslasierra valley (Córdoba province, Argentina). Young shoots from peppermint were surface-disinfected and micropropagated, as previously described by Santoro et al. [26].

2.1.3. In Vitro Exposure to mVOCs

Single nodes from aseptically cultured plantlets were planted in sterilized glass jars (250 mL) containing 50 mL MS (Murashige and Skoog) solid media with 0.8% (*w/v*) agar and 3% (*w/v*) sucrose. Then, a small (10 mL) glass vial containing ca. 3 mL of Hoagland media with 0.8% (*w/v*) agar and 3% (*w/v*) sucrose was introduced into each jar. The small vial was inoculated with GB03 (50 μL), which served as the source of bacterial volatiles, with sterile water used in the control. Plants were exposed to mVOCs without having any physical contact with the rhizobacteria. Jars containing plants and bacteria were covered with aluminum foil, sealed with parafilm to avoid contamination, and placed in a growth chamber under controlled conditions (16/8-h light/dark cycle), temperature (22 ± 2 °C) and relative humidity (~70%). After 30 days, all plants were collected [38].

2.1.4. Treatments

MS media (plant growth media) and Hoagland media (bacterial growth media) were supplemented with different salt concentrations: 0, 75, and 100 mM NaCl. For each experimental set, both the plant and bacteria were grown under the same concentration of NaCl but without contact with each other. Salt level concentrations were selected based on previous observations: at lower concentrations (25 and

50 mM), plant growth was not affected, and at higher levels (125 and 150 mM), the rooting capacity decreased significantly. Experiments were repeated three times (10 jars per treatment; 1 plant/jar).

2.2. Essential Oil Extraction and Analysis

Shoot samples were individually weighed and subjected to hydrodistillation in a Clevenger-like apparatus for 40 min. The volatile fraction was collected in dichloromethane, and β-pinene (1 μL in 50 μL ethanol) was added as an internal standard (as it was previously reported, β-pinene is not present in peppermint plants [37]). The major *M. piperita* EO components, which comprise ~60% of the total oil volume, are limonene, linalool, (−) menthone, (−) menthol, and (+) pulegone. These compounds were quantified in relation to the standard added during the distillation procedure described above. The flame ionization detector (FID) response factors for each compound generated essentially equivalent areas (differences $p < 0.05$).

Chemical analyses were performed using a Perkin-Elmer Q-700 gas chromatograph (GC), equipped with a CBP−1 capillary column (30 m × 0.25 mm, film thickness 0.25 μm) and a mass selective detector. Analytical conditions were as follows: injector temperature 250 °C; detector temperature 270 °C; oven temperature programmed from 60 °C (3 min) to 240 °C at 4°/min; carrier gas = helium at a constant flow rate of 0.9 mL/min; source 70 eV. The oil components ((−) menthone, (−) menthol, and (+) pulegone) were established by comparison of the diagnostic ions (NIST 2014 library) and GC retention times with those of the respective authentic standard compounds purchased from Sigma-Aldrich [34]. GC analysis was performed using a Shimadzu GC-RIA gas chromatograph fitted with a 30 m × 0.25 mm fused silica capillary column coated with Supelcowax 10 (film thickness 0.25 μm). The GC operating conditions were as follows: injector and detector temperatures 250 °C; oven temperature programmed from 60 °C (3 min) to 240 °C at 4°/min; detector = FID; carrier gas = nitrogen at a constant flow rate of 0.9 mL/min.

2.3. Total Phenolic Content (TPC) Determination

The total phenolic content of the extract was determined by the Folin–Ciocalteu method, as previously described by Cappellari et al. [41]. The TPC were expressed in terms of μg gallic acid (a common reference compound) equivalent per g plant fresh weight using the standard curve.

2.4. Antioxidant Activity

The capacity of radical scavenging in extracts against stable DPPH• (2,2-diphenyl−1-picrylhydrazyl) was determined by the Brand-Williams et al. method [42] with minor modifications, as previously described by Chiappero et al. [43]. A calibration curve was obtained using ascorbic acid, and the scavenging capacity of the plant extracts was expressed as mM ascorbic acid equivalents (AAE) per g fresh weight (mM AEE/g FW). All experiments were performed in triplicate for each experimental unit.

2.5. Lipid Peroxidation

Lipid peroxidation was measured by quantifying the malondialdehyde (MDA) production using the thiobarbituric acid reaction. The MDA content was measured following the method of Heath and Packer [44], with some modifications, as reported by Chiappero et al. [43]. The amount of MDA was determined by its molar extinction coefficient (155 mM^{-1} cm^{-1}), which was expressed as μmol MDA/g FW (grams of fresh weight). The experiments were performed in triplicate for each experimental unit.

2.6. Statistical Analysis

Data were subjected to a two-way analysis of variance (ANOVA) (mVOcs × salt stress), followed by a comparison of multiple treatment levels with those of the control, using the post hoc Fisher LSD test. Infostat software version 2018 (Group Infostat, Universidad Nacional de Córdoba, Argentina) was used for the statistical analysis. Principal component analysis (PCA) using Infostat statistical package was conducted. The analysis of extracts shows the relationships among the treatments (mVOCs exposure and salt stress conditions) and the different variables measured (EO, TPC, lipid peroxidation (MDA), and antioxidant capacity (AAE)). At least 15 observations were used for each treatment in the multivariate dataset.

3. Results

3.1. Essential Oil

Peppermint plants subjected to salt stress showed a reduction in EO content. Plants grown under 75 or 100-mM salt concentrations and those not treated with mVOCs revealed a 50% decrease in EO yield ($p < 0.05$) (Figure 1). When plants were treated with mVOCs under control conditions, the EO content rose approximately 3.3 times compared to plants not exposed to mVOCs (Figure 1). When plants were grown under salt stress conditions and treated with mVOC, positive effects of mVOCs on EO yields were detected. The levels of EOs increased approximately 5.6 and 6.5-fold in plants grown under 75 or 100 mM and treated with mVOCs, respectively, in relation to plants subjected to salt conditions but not treated with mVOCs, with a statistically significant interaction effect between salt stress and mVOCs being found ($p < 0.05$).

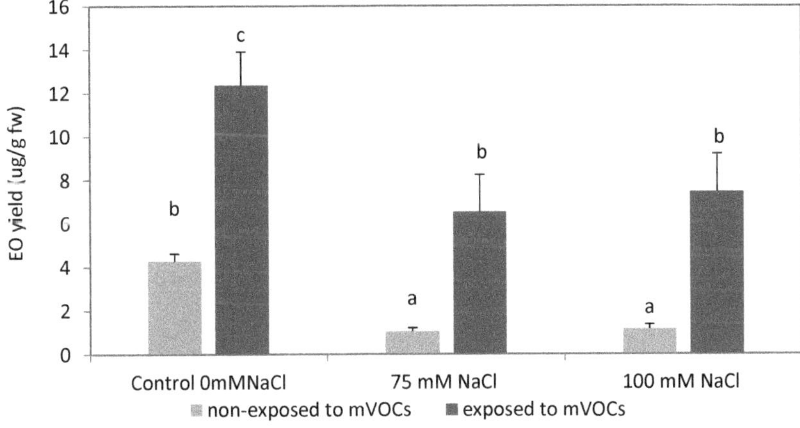

Figure 1. Essential oil yield in *Mentha piperita* plants grown under different salt concentrations (0, 75, and 100 mM NaCl) and exposed to *B. amyloliquefaciens* GB03 mVOCs (mean ± SE). Values followed by the same letter in a column are not significantly different according to Fisher's LSD test ($p < 0.05$).

Regarding the main compounds of the EOs, growing under salt stressed conditions resulted in a decrease in menthone and menthol (Table 1); although menthol content was approximately 3.5 times lower in plants grown under 75 or 100 mM concentrations and not treated with mVOCs ($p < 0.05$), the effect on menthol concentration was not statistically significant but followed the same trend as for menthone, which was significant. However, the pulegone concentration was not significantly different for control plants exposed to salt. For plants treated with mVOCs, the levels of menthone and pulegone increased approximately 2 and 3-fold, respectively, compared to those of the corresponding controls at each salinity level. However, the menthol concentration was not modified by mVOC exposure. In plants submitted to 75 mM NaCl and treated with GB03 mVOCs, the concentrations of menthone, menthol, and pulegone were approximately 6.7, 5.8, and 3.4-fold higher, respectively, in relation to plants subjected to salt conditions but not treated to mVOCs and similar to plants treated to mVOCs and not salt stressed. At 100 mM NaCl, the menthone and pulegone contents revealed the same tendency, with an increase observed in plants treated with mVOCs ($p < 0.05$), but the menthol concentration was not modified by the mVOCs (Table 1).

Table 1. Concentrations of main essential oil (EO) compounds in *Mentha piperita* grown under salt stress media (0, 75, and 100 mM NaCl) and exposed to *B. amyloliquefaciens* GB03 mVOCs emission (mean ± SE). Values are mean ± standard error (SE).

NaCl Concentration	(−)-Menthone (µg/g fw)	(−)-Menthol (µg/g fw)	(+)-Pulegone (µg/g fw)
0 mM			
control	0.99± 0.28 *b*	1.07± 0.15 *a*	1.18± 0.14 *a*
B. amyloliquefaciens GB03	2.27± 0.42 *c*	1.14± 0.23 *a*	5.29± 0.54 *c*
75 mM			
control	0.25± 0.05 *a*	0.10± 0.05 *a*	0.55± 0.12 *a*
B. amyloliquefaciens GB03	1.55± 0.17 *bc*	0.81± 0.03 *a*	2.73± 0.41 *b*
100 mM			
control	0.26± 0.05 *a*	0.22± 0.08 *a*	0.56± 0.13 *a*
B. amyloliquefaciens GB03	1.35± 0.49 *b*	0.63± 0.03 *a*	2.87± 0.79 *b*

Means followed by the same letter in a given column are not significantly different according to Fisher's LSD test ($p < 0.05$).

3.2. Total Phenolic Content

The level of TPC in plants subjected to salt stress conditions increased with the severity of the NaCl concentration ($p < 0.05$), both in plants exposed and not exposed to mVOCs. In plants grown under salt conditions (75 or 100 mM), the TPC levels rose by 15 and 50%, respectively, in relation to control plants (Figure 2). In addition, the plants subjected to both concentrations of NaC and treated with GB03 VOCs registered an increase in TPC compared to non-exposed plants ($p < 0.05$), but no statistically significant interaction effect was found ($p > 0.05$). The highest TPC concentrations were detected in plants treated with salt 100 mM and mVOCs.

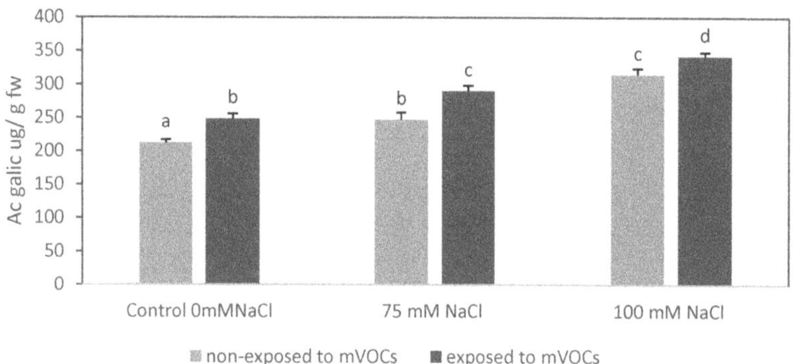

Figure 2. Total phenolic content of *Mentha piperita* plants grown under salt stress media (0, 75, and 100 mM NaCl) and exposed to *B. amyloliquefaciens* GB03 mVOCs emission (mean ± SE). Values followed by the same letter in a column are not significantly different according to Fisher's LSD test ($p < 0.05$).

3.3. Radical Scavenging Capacity

The antioxidant capacity of the DPPH• radical scavenger increased 2.6 and 3.6-fold in peppermint leaves grown under 75 and 100 mM NaCl conditions, respectively ($p < 0.05$) (Figure 3). Moreover, when plants were subjected to salt conditions and treated with mVOCs, the antioxidant capacity increased ($p < 0.05$) by 50% and 30% for 75 and 100 mM NaCl, respectively, in relation to salt stressed plants not exposed to mVOCs. The highest levels of antioxidant activity were observed when plants were exposed to VOCs and grown under 100 mM NaCl conditions, with the ascorbic acid equivalents (AAE) increasing 4.75-fold with respect to control plants (not exposed to mVOCs).

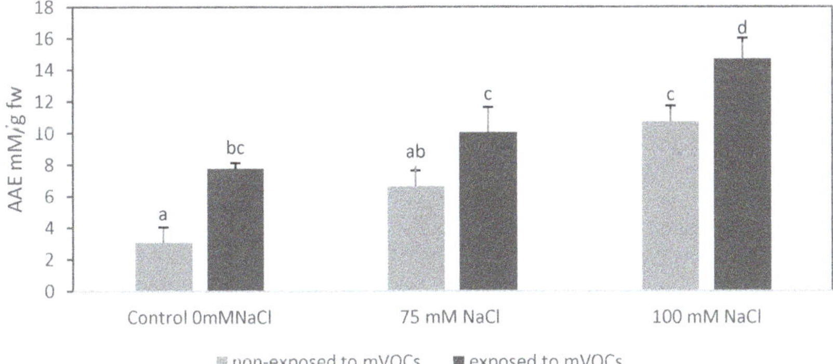

Figure 3. Antioxidant activity expressed as ascorbic acid equivalents (AAE) in *Mentha piperita* grown under salt stress media (0, 75, and 100 mM NaCl) and exposed to *B. amyloliquefaciens* GB03 mVOCs emission (mean ± SE). Values followed by the same letter in a column are not significantly different according to Fisher's LSD test ($p < 0.05$).

3.4. Lipid Peroxidation

Oxidative damage to the membrane lipids was observed due to salt stress, as shown by the MDA levels (Figure 4), with the highest MDA levels being observed ($p < 0.05$) at the higher salt concentration. The lipid peroxidation increased 1.4 and 2-fold in 75 and 100 mM NaCl treated plants, respectively, in relation to control plants. For plants treated with mVOCs and subjected to salt stress, the MDA content was approximately 25% lower than for plants stressed and not treated with mVOCs (75 and 100 mM NaCl plants).

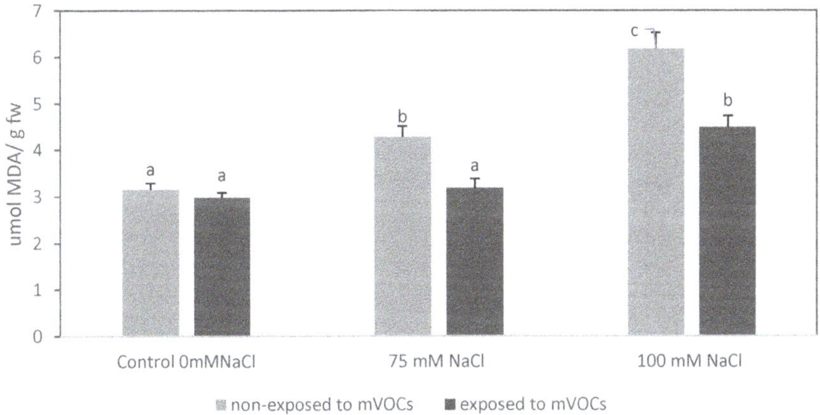

Figure 4. Malondialdehyde (MDA) content in *Mentha piperita* grown under salt stress media (0, 75, and 100 mM NaCl) and exposed to *B. amyloliquefaciens* GB03 mVOCs emission (mean ± SE). Values followed by the same letter in a column are not significantly different according to Fisher's LSD test ($p < 0.05$).

3.5. Principal Component Analysis

PCA represents a graphic image that simplifies the visualization and perception of the dataset and the variables. We used the PCA to extract and reveal the relationships among the factors (growth conditions and exposure to mVOCs) and different variables as EO, TPC, lipid peroxidation (MDA), and antioxidant capacity (AAE) in the multivariate analysis (Figure 5). The plot defined by the first two principal components was enough to explain most of the variations in the data (96.8%) and give a cophenetic correlation coefficient of 0.997. The PCA (Figure 5) showed that 100 mM NaCl (high salt concentrations) combined with exposure to mVOCs was strongly associated with TPC content and antioxidant capacity (AAE), as revealed by the circle in Figure 5. Considering the relationships among variables, a strong positive correlation (acute angle) was observed between TPC levels and AAE. There were also positive correlations found among MDA levels with no mVOC exposure and 100 mM NaCl. In addition, in PC2, positive relationships were observed between AAE, EO, and TPC with mVOC exposure.

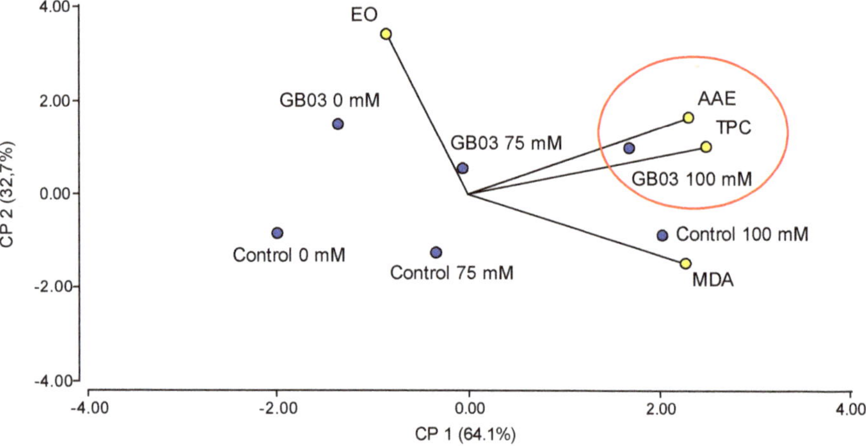

Figure 5. Principal component analysis for the physiological response of *Mentha piperita* grown under different salt stress concentrations (0, 75, and 100 mM NaCl) and *B. amyloliquefaciens* GB03 mVOCs emission. PRO: proline, TPC: total phenolic content, and MDA: lipid peroxidation were determined by estimating the amount of malondialdehyde (MDA); AEE: DPPH radical scavenging capacity.

4. Discussion

Salinity is one of the most important environmental factors diminishing plant yield, mainly in arid and semi-arid environments. The responses of plants to salt stress are intricate and affect several components, with plants having the ability to respond via signal transduction pathways by adjusting their metabolism [45,46]. These responses can differ in relation to toxic ion uptake, ion compartmentation and/or exclusion, osmotic regulation, CO_2 assimilation, photosynthetic electron transport, chlorophyll content and fluorescence, ROS generation, and antioxidant defenses [45–48].

PGPR make a significant contribution to the protection against abiotic stress through their biological activities at the rhizosphere, as exopolysaccharides production (EPS), phytohormones and 1-aminocyclopropane- 1-carboxylate (ACC) deaminase synthesis, induction of the accumulation of osmolytes and antioxidants, upregulating or downregulating the stress responsive genes, and by changes in the root morphology and volatile compounds [17–21,49,50]. In addition, in recent years, an increasing number of PGPR VOC studies have demonstrated an effect against abiotic stresses [7,38,51].

In the present study, we found that when peppermint plants were subjected to salt stress, the EO yield decreased by 50% for both concentrations evaluated (75 and 100 mM NaCl). Additionally, there was a corresponding decrease in the main compounds menthone, menthol, and pulegone. Comparable effects were reported in *M. arvensis* grown under 100, 300, and 500 mM NaCl, with a reduction of 31%, 54%, and 67%, respectively [52]. In contrast, Karray-Bouraoui et al. [53] noted an enhanced *M. pulegium* EO yield of about 2.75-fold under 50-mM salt stress conditions, with a higher density of glandular trichomes on the leaves. Furthermore, Neffati and Marzouk [54] showed that the compounds of *Coriandrum sativum* L. oil were modified by salinity and were revealed to be dependent on salt level treatment. There are contradictory reports concerning changes in EO yield in relation to salt stress. An increase in EOs and in their composition in response to low levels of salinity was reported in *Satureja hortensis* [55], in sage [56] and in thyme [57]. In contrast, other studies reported a decrease in EOs in lemon balm and in sweet marjoram [58]. Additionally, Ben Taarit et al. [59] reported that the compositions of EOs of *Salvia officinalis* were altered in moderate or high salt stress, in controls and in plants grown under 25 mM NaCl, with the major compound of the EOs being viridiflorol, whereas at higher levels (50 and 75 mM NaCl), 1, 8-cineole was predominant, and at 100 mM NaCl, manool was the principal compound.

The EO yield variations reported under abiotic stress could have resulted from the fact that their production is affected by different physiological, biochemical, metabolic, and genetic factors, which are complex to isolate from one another. In addition, the geographical, seasonal, developmental, and organ variations all contribute to EO yield, as do anatomical and hormonal factors [60–63]. The impact of salt stress on the EO levels probably was due to acclimation processes in stressed plants. Whereas in the initial stage of stress, the metabolism is severely affected, later, the acclimatization processes may reduce the secondary metabolite biosynthesis [64,65].

In the present study, the EO content in salt stressed plants treated with mVOCs showed a 5.6 and 6.5-fold increase with respect to their respective controls (plants grown under 75 or 100 mM NaCl and not treated with mVOCs, respectively), demonstrating that GB03 mVOCs have the capacity to reverse the negative effects of salinity on the EO yield. In fact, mVOCs induced salt tolerance in plants in a previous study of ours, with peppermint plants subjected to salt stress conditions and treated with GB03 VOCs having a higher shoot fresh weight, root dry weight, and total chlorophyll content compared to controls [38]. In this sense, the biosynthesis of terpenoids is affected by the primary metabolism—for example, the photosynthesis for carbon and energy supply. Factors that increase biomass production may have an impact on the relationships among the primary and secondary metabolisms, causing an increased biosynthesis of secondary metabolites [66]. Related to this, augmented plant biomass seems to lead to a larger availability of substrate for monoterpene biosynthesis [35,67].

We have also observed that abscisic acid (ABA) was not connected to salt tolerance generated in plants subjected to salt stress and treated with VOCs [38]. This observation suggests that GB03 VOCs protection against osmosis is ABA independent [68]. The jasmonic acid (JA) levels were similar in salt treated plants, when treated with mVOCs or not. In contrast, the salicylic acid (SA) levels were higher in plants subjected to salt and treated with mVOCs compared to plants subjected to salt conditions and not treated with mVOCs. SA is an important signal molecule for modulating plant responses to stress [38]. Chemical analysis using Solid Phase Microextraction (SPME) fibers of the VOC emissions from GB03 grown under salt conditions revealed the release of a total of seven components, belonging to the following four classes: hydrocarbons (cyclohexane, dodecane, undecane and hexadecane), ketones (acetoin), aldehydes (benzaldehyde), and ethers (2-butanone-3metioxy-3 methyl). The relative quantity of acetoin, the major VOC compound emitted by GB03, enhanced with salt concentration [38]. Concerning the complex profile of compounds, VOC emission is strongly affected by the collection methodology employed, the growth medium, and the density of the bacterium [50,69,70]. For instance, Farag et al. [71] identified a higher number of compounds from GB03 VOCs than Cappellari and Banchio [38], probably due to the different collection methodology used.

It has also been reported that plants treated with GB03 mVOCs and grown in a saline media accumulated less Na + through the regulation of the Na transporter. The GB03 VOCs decreased the Na level in *Arabidopsis* by decreasing Na uptake and/or increasing Na exudation [49]. Furthermore, they led to an acidification of the rhizosphere [72]. Certain bacterial VOCs activate closure of the stomata, reducing the water evaporation [73], and are also involved in biofilm formation, which maintains soil moisture content and increases drought tolerance in plants [51,74,75]. In addition, mVOCs emitted by PGPR also act as a biocontrol against several phytopathogens and trigger plant defense responses through the induction of systemic resistance (ISR) [24,71,76]. For example, the production of EOs is related to the defense response system [63], since numerous terpenes have antimicrobial activity [77]. Similarly, monoterpene synthesis is induced by herbivore feeding in *Minthostachys mollis* [78] and several plant species, suggesting that these compounds protect leaves from future attacks [67,79–81]. Consequently, as mentioned above, endogenous SA levels increased in plants cultivated under salt conditions and treated with GB03, with previous observations suggesting that the biosynthesis of *M. piperita* monoterpenes is SA and JA dependent [82].

A rise in TPC levels in different tissues under salt conditions has also been described in different plant species [83–85]. A consequence of abiotic stress is superoxide production, which leads to a detoxification mechanism. Related to this, phenolics are synthesized by many plant species for protection against abiotic stress conditions, and their levels are correlated with antioxidant activity [63,86]. Salinity stress induces metabolic and physiological reactions, as well as drastically decreasing the CO_2 uptake due to stomatal restrictions. As a consequence, the consumption of reduction equivalents (NADPH 2+) for CO_2 fixation via the Calvin cycle decreases significantly, leading to oxidative stress and an oversupply of reduction equivalents, with the metabolic processes being moved to biosynthetic activities that consume reduction equivalents. Hence, the biosynthesis of reduced compounds, such as phenols, is increased [63,85,87]. Among the SM found in *M. piperita* are phenolic compounds such as caffeic acid, rosmarinic acid, eriocitrin, and luteolin- 7-O-glucoside [88,89], with their proportion in leaves being approximately 19–23% of dry weight [90–92]. Here, we found that peppermint plants either subjected to salt conditions and/or treated with GB03 VOCs produced a positive effect on the TPC content compared to the respective control plants. Plants grown under 100 mM NaCl and treated with VOCs revealed a higher TPC content. In fact, phenolic compounds are important and powerful agents in scavenging free radicals [93–96]. The antioxidant capacity of phenolic compounds is due to their high reactivity as hydrogen or electron donors, to the particularity of the polyphenol-derived radical to stabilize and delocalize the unpaired electron, and to their capacity to chelate transition metal ions [92,97].

In a previous study, we observed that direct inoculation as well as drought stress in *M. piperita* increased TPC and phenylalanine ammonia lyase (PAL) activity, with the latter being responsible for the synthesis of phenolic compounds [41,43]. In agreement, the TPC was observed to increase in different plant species submitted to abiotic stress [86]—for example, in *T. vulgaris* subjected to drought stress [96] and in *M. pulegium* under salt stress [98]. Conversely, Rahimi et al. [99] and Alhaithloul et al. [100] described a reduction in TPC in *M piperita* plants subjected to drought stress. However, in *Tagetes minuta* plants inoculated with *P. fluorescens* WCS417r and *Azospirillum brasilense*, and in chickpea inoculated with *P. fluorescens* [101], TPC levels increased significantly [36]. Jayapala et al. [102] reported the induction of resistance against pathogens through enhancement of the activities of defense-related enzymes and a higher accumulation of TPC in chili plants inoculated with *Bacillus* sp. Furthermore, Tahir et al. [27] revealed that *Bacillus* sp. mVOCs negatively influence the development of the pathogen *R. solanacearum* by activating ISR in tobacco plants. Molecular studies have shown that resistance is the consequence of an increase in the SM levels and defense-related enzymes, including PAL.

Phenolic compounds are antioxidants that may be required for scavenging ROS and protecting the lipid membrane from oxidative stress [12]. For example, Fagopyrum *esculentum* plants grown under media with increasing salt concentrations revealed a concentration-dependent increase in the accumulation of phenolic compounds, resulting in a higher DPPH free radical scavenging potential [103]. This effect was corroborated in the present study in plants subjected to salinity environments and treated with mVOCs, which showed a heightened antioxidant capacity, as revealed by the high levels of AAE detected in the DPPH• scavenging assay and by the low amounts of MDA. The highest levels of antioxidant activity were observed when plants were grown under 100 mM NaCl and mVOC. The GB03 mVOCs decreased the MDA levels in plants subjected to salt stress, to similar levels as those in control plants. In contrast, after water deficit treatment in peppermint plants, heightened amounts of MDA, as a cell membrane damage index, were detected [99]. Additionally, peppermint growing under control conditions was revealed to be more effective in scavenging DPPH free radicals and had a higher reducing power than when exposed to drought and heat stress. This observation provides signals that tissues of peppermint subjected to heat and/or drought stress contain fewer antioxidants and reducing compounds [100].

The PCA analysis showed that plants subjected to high salt concentrations combined with exposure to mVOCs strongly affected the TPC content and antioxidant capacity (AAE). This relationship was also detected in drought-stressed peppermint plants inoculated with GB03 [43].

In plants that were inoculated and subjected to osmotic stress, similar results in MDA reduction were observed to those reported for cucumber plants inoculated with a consortium of PGPR under drought stress conditions [104], as well as those in white clover and *M. arvensis* inoculated under saline conditions [51,105]. The decrease in the leaf MDA content resulting from mVOC treatment suggests its ability to reduce the peroxidation of cell membrane lipids under salt stress and to protect the leaf cell from damage. Moreover, Gopinath et al. [106] reported in *Nicotiana tabacum* that when callus was exposed to volatile compounds from *Bacillus badius* M12 and the volatile, 2,3- butanediol, this led to increased antioxidant activity by the expression of SOD, a key antioxidant enzyme. In addition, treatment with mVOCs from GB03 and *Pseudomonas simiae* increased choline and glycine betaine biosynthesis in *Arabidopsis* [51,68]. These osmolytes have positive effects on enzyme and membrane integrity, along with adaptive roles in mediating osmotic adjustment in plants subjected to stress conditions [107]. In another investigation, 2,3-butanediol was found to induce plant production of nitric oxide (NO) and hydrogen peroxide [108], and it was reported that NO regulates antioxidant enzymes at the level of activity and gene expression [109]. At the same time, the plant hormone SA is required for plant growth under abiotic stress [7,17,73]. Finally, an increase in the SA levels was shown in peppermint plants subjected to salt stress and treated with GB03 VOCs [38].

5. Conclusions

Salt stresses affect the growth and productivity of crop plants and are detrimental to the plants, thereby reducing their yield. Thus, it is necessary to improve the technologies of abiotic stress management. In recent decades, several studies have shown that PGPR has the ability to ameliorate the negative effects of salt or water. However, only a few reports have been published on PGPR VOCs as elicitors of tolerance to abiotic stress in aromatic and medicinal plants. The GB03 VOCs have been shown to increase plant growth and chlorophyll content and lead to better morphological characteristics in *M. piperita* plants subjected to salt stress. The results shown in the present study establish that for peppermint plants grown in the laboratory under salt media, the volatiles emitted by GB03 significantly increased SM production and improved the antioxidant status. This suggests that the accumulation of SMs is a plant strategy to avoid oxidative damage caused by ROS, a direct result of salt stress. Bacterial volatiles are promising candidates for a rapid non-invasive technique to increase SM production in aromatic and medicinal crops growing under abiotic stress conditions. In addition, this is a potentially useful system for the production of SMs, which have remarkable biological activities and are often exploited as medicinal and food ingredients for therapeutic, aromatic, and culinary purposes. However, future studies are still necessary to elucidate how plants modulate and perceive PGPR VOC-elicited abiotic tolerance.

Author Contributions: L.d.R.C., J.C., and T.B.P. performed the experiments; E.B. designed the research and analyzed the data, L.d.R.C., E.B., and W.G. wrote the manuscript. All authors read, revised, and approved the final manuscript. All authors have read and agreed to the published version of the manuscript.

Funding: This study was supported by grants from the Secretaría de Ciencia y Técnica de la Universidad Nacional de Río Cuarto, the Agencia Nacional de Promoción Científica y Tecnológica (ANPCyT) PICT 0636/14, Argentina, and financial support was given to E.B. by the Georg Forster Research Fellowship of the Alexander von Humboldt Foundation. E.B. and W.G. are Career Members of CONICET. L.C., J.C., and T.B.P. received fellowships from CONICET.

Acknowledgments: This research was supported by grants from the Secretaría de Ciencia y Técnica de la Universidad Nacional de Río Cuarto, Consejo Nacional de Investigaciones Científicas y Técnicas (CONICET), and Agencia Nacional de Promoción Científica y Tecnológica (ANPCyT). E.B. and W.G. are Career Members of CONICET. E.B. obtained financial support from a Georg Forster Research Fellowship of the Alexander von

Humboldt Foundation. L.C., J.C., and T.B.P. have fellowships from CONICET. The authors are grateful to Paul Hobson, native speaker, for editorial assistance.

Conflicts of Interest: The authors declare no conflict of interest.

References

1. McConkey, M.E.; Gershenzon, J.; Croteau, R.B. Developmental regulation of monoterpene biosynthesis in the glandular trichomes of peppermint. *Plant Physiol.* **2000**, *122*, 215–224. [CrossRef] [PubMed]
2. Lange, B.M.; Mahmoud, S.S.; Wildung, M.R.; Turner, G.W.; Davis, E.M.; Lange, I.; Baker, R.C.; Boydston, R.A.; Croteau, R.B. Improving peppermint essential oil yield and composition by metabolic engineering. *Proc. Natl. Acad. Sci. USA* **2011**, *108*, 16944–16949. [CrossRef] [PubMed]
3. Ramakrishna, A.; Ravishankar, G.A. Influence of abiotic stress signals on secondary metabolites in plants. *Plant Signal. Behav.* **2011**, *6*, 1720–1731. [PubMed]
4. Khan, N.; Bano, A.; Curá, J.A. Role of Beneficial Microorganisms and Salicylic Acid in Improving Rainfed Agriculture and Future Food Safety. *Microorganisms* **2020**, *8*, 1018. [CrossRef] [PubMed]
5. Upadhyay, S.K.; Singh, J.S.; Singh, D.P. Exopolysaccharide-producing plant growth-promoting rhizobacteria under salinity condition. *Pedosphere* **2011**, *21*, 214–222. [CrossRef]
6. Zhu, J.K. Plant salt tolerance. *Trends Plant Sci.* **2001**, *6*, 66–71. [CrossRef]
7. Liu, X.M.; Zhang, H. The effects of bacterial volatile emissions on plant abiotic stress tolerance. *Front. Plant Sci.* **2015**, *6*, 774. [CrossRef]
8. Mittler, R. Oxidative stress, antioxidants and stress tolerance. *Trends Plant Sci.* **2002**, *7*, 405–410. [CrossRef]
9. Mithofer, A.; Schulze, B.; Boland, W. Biotic and heavy metal stress response in plants: Evidence for common signals. *FEBS Lett.* **2004**, *566*, 1–5. [CrossRef]
10. Noctor, G.; Mhamdi, A.; Foyer, C.H. The roles of reactive oxygen metabolism in drought: Not so cut and dried. *Plant Physiol.* **2014**, *164*, 1636–1648. [CrossRef]
11. Reddy, A.R.; Chaitanya, K.V.; Vivekanandan, M. Drought-induced responses of photosynthesis and antioxidant metabolism in higher plants. *J. Plant Physiol.* **2004**, *161*, 1189–1202. [CrossRef] [PubMed]
12. Khan, N.; Bano, A.; Ali, S.; Babar, M.A. Crosstalk amongst phytohormones from planta and PGPR under biotic and abiotic stresses. *Plant Growth Reg.* **2020**, *90*, 189–203. [CrossRef]
13. Kloepper, J.W.; Lifshitz, R.; Zablotowicz, R.M. Free-living bacterial inoculation for enhancing crop productivity. *Trends Biotecnol.* **1989**, *7*, 39–49. [CrossRef]
14. Vessey, J.K. Plant growth promoting rhizobacteria as biofertilizers. *Plant Soil.* **2003**, *255*, 571–586. [CrossRef]
15. Van Loon, L.C. Plant responses to plant growth-promoting rhizobacteria. *Eur. J. Plant Pathol.* **2007**, *119*, 243–254. [CrossRef]
16. Hassan, M.; Boersma, M.; Lawaju, B.R.; Lawrence, K.S.; Liles, M.; Kloepper, J.W. Effects of secondary metabolites produced by PGPR amended with orange peel on the mortality of second-stage juveniles of *Meloidogyne incognita. Plant Health* **2019**, *108*, Abstracts of Presentations, subsection S2.
17. Farag, M.A.; Zhang, H.; Ryu, C.M. Dynamic chemical communication between plants and bacteria through airborne signals: Induced resistance by bacterial volatiles. *J. Chem. Ecol.* **2013**, *39*, 1007–1018. [CrossRef]
18. Timmusk, S.; Islam, A.; Abd El, D.; Lucian, C.; Tanilas, T.; Ka nnaste, A.; Behers, L.; Nevo, E.; Seisenbaeva, G.; Stenström, E.; et al. Drought-tolerance of wheat improved by rhizosphere bacteria from harsh environments: Enhanced biomass production and reduced emissions of stress volatiles. *PLoS ONE* **2014**, *9*, e96086. [CrossRef]
19. Vurukonda, S.S.K.P.; Vardharajula, S.; Shrivastava, M.; SkZ, A. Enhancement of drought stress tolerance in crops by plant growth promoting rhizobacteria. *Microbiol. Res.* **2016**, *184*, 13–24. [CrossRef]
20. Enebe, M.C.; Babalola, O.O. The influence of plant growth-promoting rhizobacteria in plant tolerance to abiotic stress: A survival strategy. *Appl. Microbiol. Biotechnol.* **2018**, *102*, 7821–7835. [CrossRef]
21. Khan, N.; Zandi, P.; Ali, S.; Mehmood, A.; Adnan Shahid, M.; Yang, J. Impact of salicylic acid and PGPR on the drought tolerance and phytoremediation potential of Helianthus annus. *Front. Microbiol.* **2018**, *9*, 2507. [CrossRef] [PubMed]
22. Kumar, A.; Verma, J.P. Does plant-Microbe interaction confer stress tolerance in plants: A review? *Microbiol. Res.* **2018**, *207*, 41–52. [CrossRef] [PubMed]
23. Liu, H.; Brettell, L.E.; Qiu, Z.; Singh, B.K. Microbiome-Mediated Stress Resistance in Plants. *Trends Plant Sci.* **2020**, *25*, 733–743. [CrossRef] [PubMed]

24. Ryu, C.M.; Farag, M.A.; Hu, C.H.; Reddy, M.S.; Kloepper, J.W.; Pare, P.W. Bacterial volatiles induce systemic resistance in Arabidopsis. *Plant Physiol.* **2004**, *134*, 1017–1026. [CrossRef]

25. Banchio, E.; Xie, X.; Zhang, H.; Paré, P.W. Soil bacteria elevate essential oil accumulation and emissions in sweet basil. *J. Agric. Food Chem.* **2009**, *57*, 653–657. [CrossRef]

26. Santoro, M.; Zygadlo, J.; Giordano, W.; Banchio, E. Volatile organic compounds from rhizobacteria increase biosynthesis of essential oils and growth parameters in peppermint (*Mentha piperita*). *Plant Physiol. Biochem.* **2011**, *49*, 1077–1082. [CrossRef]

27. Tahir, H.A.; Gu, Q.; Wu, H.; Raza, W.; Hanif, A.; Wu, L.; Colman, M.V.; Gao, X. Plant Growth Promotion by Volatile Organic Compounds Produced by *Bacillus subtilis* SYST2. *Front. Microbiol.* **2017**, *8*, 171. [CrossRef]

28. Naseem, H.; Ahsan, M.; Shahid, M.A.; Khan, N. Exopolysaccharides producing rhizobacteria and their role in plant growth and drought tolerance. *J. Basic Microbiol.* **2018**, *58*, 1009–1022. [CrossRef]

29. Lemfack, M.C.; Gohlke, B.-O.; Toguem, S.M.T.; Preissner, S.; Piechulla, B.; Preissner, R. mVOC 2.0: A database of microbial volatiles. *Nucleic Acids Res.* **2018**, *46*, 1261–1265. [CrossRef]

30. Kanchiswamy, C.N.; Malnoy, M.; Maffei, M.E. Chemical diversity of microbial volatiles and their potential for plant growth and productivity. *Front. Plant Sci.* **2015**, *6*, 151. [CrossRef]

31. Lee, B.; Farag, M.A.; Park, H.B.; Kloepper, J.W.; Lee, S.H.; Ryu, C.M. Induced Resistance by a long-chain bacterial volatile: Elicitation of plant systemic defense by a CVolatile Produced by *Paenibacillus polymyxa*. *PLoS ONE* **2012**, *7*, e48744. [CrossRef] [PubMed]

32. Khan, N.; Bano, A. Rhizobacteria and Abiotic Stress Management. In *Plant Growth Promoting Rhizobacteria for Sustainable Stress Management*; Springer: Singapore, 2019; pp. 65–80.

33. Hossain, M.J.; Ran, C.; Liu, K.; Ryu, C.M.; Rasmussen-Ivey, C.R.; Williams, M.A.; Hassan, M.K.; Choi, S.-K.; Jeong, H.; Newman, M.; et al. Deciphering the conserved genetic loci implicated in plant disease control through comparative genomics of *Bacillus amyloliquefaciens* subsp. plantarum. *Front. Plant Sci.* **2015**, *6*, 631. [CrossRef] [PubMed]

34. Ledger, T.; Rojas, S.; Timmermann, T.; Pinedo, I.; Poupin, M.J.; Garrido, T.; Richter, P.; Tamayo, J.; Donoso, R. Volatile-mediated effects predominate in *Paraburkholderia phytofirmans* growth promotion and salt stress tolerance of *Arabidopsis thaliana*. *Front. Microbiol.* **2016**, *7*, 1838. [CrossRef] [PubMed]

35. Banchio, E.; Bogino, P.; Santoro, M.V.; Torres, L.; Zygadlo, J.; Giordano, W. Systemic induction of monoterpene biosynthesis in *Origanum x majoricum* by soil bacteria. *J. Agric. Food Chem.* **2010**, *58*, 650–654. [CrossRef] [PubMed]

36. Cappellari, L.; Santoro, M.V.; Nievas, F.; Giordano, W.; Banchio, E. Increase of secondary metabolite content in marigold by inoculation with plant growth-promoting rhizobacteria. *Appl. Soil Ecol.* **2013**, *70*, 16–22. [CrossRef]

37. Cappellari, L.R.; Santoro, M.V.; Reinoso, H.; Travaglia, C.; Giordano, W.; Banchio, E. Anatomical, morphological, and phytochemical effects of inoculation with plant growth-promoting rhizobacteria on peppermint (*Mentha piperita*). *J. Chem. Ecol.* **2015**, *41*, 149–158. [CrossRef] [PubMed]

38. Cappellari, L.; Banchio, E. Microbial Volatile Organic Compounds Produced by *Bacillus amyloliquefaciens* GBAmeliorate the Effects of Salt Stress in *Mentha piperita* Principally Through Acetoin Emission. *J. Plant Growth Regul.* **2020**, *39*, 764–775. [CrossRef]

39. Santoro, M.V.; Bogino, P.C.; Nocelli, N.; Cappellari, L.; Giordano, W.; Banchio, E. Analysis of plant growth-promoting effects of fluorescent pseudomonas strains isolated from *Mentha piperita* rhizosphere and effects of their volatile organic compounds on essential oil composition. *Front. Microbiol.* **2016**, *7*, 1085. [CrossRef]

40. Choi, S.K.; Jeong, H.; Kloepper, J.W.; Ryu, C.M. Genome sequence of *Bacillus amyloliquefaciens* GB03, an active ingredient of the first commercial biological control product. *Gen Announc.* **2014**, *2*, 01092–01098. [CrossRef]

41. Cappellari, L.R.; Chiappero, J.; Santoro, M.; Giordano, W.; Banchio, E. Inducing phenolic production and volatile organic compounds emission by inoculating *Mentha piperita* with plant growth-promoting rhizobacteria. *Sci. Hortic.* **2017**, *220*, 193–198. [CrossRef]

42. Brand-Williams, W.; Cuvelier, M.E.; Berset, C.L.W.T. Use of a free radical method to evaluate antioxidant activity. *LWT Food Sci. Technol.* **1995**, *28*, 25–30. [CrossRef]

43. Chiappero, J.; Cappellari, L.; Sosa Alderete, L.G.; Palermo, T.B.; Banchio, E. Plant growth promoting rhizobacteria improve the antioxidant status in *Mentha piperita* grown under drought stress leading to an enhancement of plant growth and total phenolic content. *Ind. Crops Prod.* **2019**, *139*, 111553. [CrossRef]

44. Heath, R.L.; Packer, L. Photoperoxidation in isolated chloroplasts. Kinetics and stoichiometry of fatty acid peroxidation. *Arch. Biochem. Biophys.* **1968**, *125*, 189–198. [CrossRef]

45. Huang, G.-T.; Ma, S.-L.; Bai, L.-P.; Zhang, L.; Ma, H.; Jia, P.; Liu, J.; Zhong, M.; Guo, Z.-F. Signal transduction during cold, salt, and drought stresses in plants. *Mol. Biol. Rep.* **2012**, *39*, 969–987. [CrossRef] [PubMed]

46. Forni, C.; Duca, D.; Glick, B.R. Mechanisms of plant response to salt and drought stress and their alteration by rhizobacteria. *Plant Soil.* **2017**, *410*, 335–356. [CrossRef]

47. Khan, N.; Bano, A. Effects of exogenously applied salicylic acid and putrescine alone and in combination with rhizobacteria on the phytoremediation of heavy metals and chickpea growth in sandy soil. *Int. J. Phytoremed.* **2018**, *20*, 405–414. [CrossRef]

48. Acosta-Motos, J.R.; Ortuño, M.F.; Bernal-Vicente, A.; Diaz-Vivancos, P.; Sanchez-Blanco, M.J.; Hernandez, J.A. Plant Responses to Salt Stress: Adaptive Mechanisms. *Agronomy* **2017**, *7*, 18. [CrossRef]

49. Zhang, H.; Kim, M.S.; Sun, Y.; Dowd, S.E.; Shi, H.; Paré, P.W. Soil bacteria confer plant salt tolerance by tissue-specific regulation of the sodium transporter HKT1. *Mol. Plant-Microbe Interact.* **2008**, *21*, 737–744. [CrossRef]

50. Yang, J.; Kloepper, J.W.; Ryu, C.M. Rhizosphere bacteria help plants tolerate abiotic stress. *Trends Plant Sci.* **2009**, *14*, 1–4. [CrossRef]

51. Vaishnav, A.; Kumari, S.; Jain, S.; Varma, A.; Choudhary, D.K. Putative bacterial volatile-mediated growth in soybean (*Glycine max* L. Merrill) and expression of induced proteins under salt stress. *J. Appl Microbiol.* **2015**, *119*, 539–551. [CrossRef]

52. Bharti, N.; Barnawal, D.; Awasthi, A.; Yadav, A.; Kalra, A. Plant growth promoting rhizobacteria alleviate salinity induced negative effects on growth, oil content and physiological status in *Mentha arvensis*. *Acta Physiol. Plant.* **2014**, *36*, 45–60. [CrossRef]

53. Karray-Bouraoui, N.; Ksouri, R.; Falleh, H.; Rabhi, M.; Grignon, C.; Lachaal, M. Effects of environment and development stage on phenolic content and antioxidant activities of Tunisian *Mentha pulegium* L. *J. Food Biochem.* **2009**, *34*, 79–89. [CrossRef]

54. Neffati, M.; Marzouk, B. Changes in essential oil and fatty acid composition in coriander (*Coriandrum sativum* L.) leaves under saline conditions. *Ind. Crops Prod.* **2008**, *28*, 137–142. [CrossRef]

55. Baher, Z.F.; Mirza, M.; Ghorbanli, M.; Rezaii, M.B. The influence of water stress on plant height, herbal and essential oil yield and composition in *Satureja hortensis* L. *Flavour Frag. J.* **2002**, *17*, 275–277. [CrossRef]

56. Hendawy, S.F.; Khalid, K.A. Response of sage (*Salvia officinalis* L.) plants to zinc application under different salinity levels. *J. Appl. Sci. Res.* **2005**, *1*, 147–155.

57. Ezz El-Din, A.A.; Aziz, E.E.; Hendawy, S.F.; Omer, E.A. Response of *Thymus vulgaris* L. to salt stress and Alar (B9) in newly reclaimed soil. *J. Appl. Sci. Res.* **2009**, *5*, 2165–2170.

58. Shalan, M.N.; Abdel-Latif, T.A.T.; Ghadban, E.A. Effect of water salinity and some nutritional compounds of the growth and production of sweet marjoram plants (*Marjorana hortensis* L.). *Egypt. J. Agric. Res.* **2006**, *84*, 959.

59. Ben Taarit, M.K.; Msaada, K.; Hosni, K.; Marzouk, B. Changes in fatty acid and essential oil composition of sage (*Salvia officinalis* L.) leaves under NaCl stress. *Food Chem.* **2010**, *9*, 951–956. [CrossRef]

60. Mehmood, A.; Hussain, A.; Irshad, M.; Hamayun, M.; Iqbal, A.; Khan, N. In vitro production of IAA by endophytic fungus Aspergillus awamori and its growth promoting activities in Zea mays. *Symbiosis* **2019**, *77*, 225–235. [CrossRef]

61. Gleadow, R.M.; Woodrow, I.E. Defense chemistry of cyanogenic *Eucalyptus cladocalyx* seedlings is affected by water supply. *Tree Physiol.* **2002**, *22*, 939–945. [CrossRef]

62. Falk, K.L.; Tokuhisa, J.G.; Gershenzon, J. The effect of sulfur nutrition on plant glucosinolate content: Physiology and molecular mechanisms. *Plant Biol.* **2007**, *9*, 573–581. [CrossRef] [PubMed]

63. Khan, N.; Bano, A.; Rahman, M.A.; Rathinasabapathi, B.; Babar, M.A. UPLC-HRMS-based untargeted metabolic profiling reveals changes in chickpea (*Cicer arietinum*) metabolome following long-term drought stress. *Plant Cell Environ.* **2019**, *42*, 115–132. [CrossRef] [PubMed]

64. Harb, A.; Awad, D.; Samarah, N. Gene expression and activity of antioxidant enzymes in barley (*Hordeum vulgare* L.) under controlled severe drought. *J. Plant Interact.* **2015**, *10*, 109–116. [CrossRef]

65. Kleinwächter, M.; Paulsen, J.; Bloem, E.; Schnug, E.; Selmar, D. Moderate drought and signal transducer induced biosynthesis of relevant secondary metabolites in thyme (*Thymus vulgaris*), greater celandine (*Chelidonium majus*) and parsley (*Petroselinum crispum*). *Ind. Crops Prod.* **2015**, *64*, 158–166. [CrossRef]

66. Pott, D.M.; Osorio, S.; Vallarino, J.G. From Central to Specialized Metabolism: An Overview of Some Secondary Compounds Derived from the Primary Metabolism for Their Role in Conferring Nutritional and Organoleptic Characteristics to Fruit. *Front. Plant Sci.* **2019**, *10*, 835. [CrossRef] [PubMed]

67. Harrewijn, P.; Van Oosten, A.M.; Piron, P.G.M. *Natural Terpenoids as Messengers: A Multidisciplinary Study of Their Production, Biological Functions and Practical Applications*; Kluwer Academic Publishers: London, UK, 2001.

68. Zhang, H.; Murzello, C.; Sun, Y.; Kim, M.S.; Xie, X.; Jeter, R.M.; Zak, J.C.; Dowd, S.E.; Paré, P.W. Choline and osmotic-stress tolerance induced in Arabidopsis by the soil microbe *Bacillus subtilis* (GB03). *Mol. Plant Microbe Interact.* **2010**, *23*, 1097–1104. [CrossRef] [PubMed]

69. Blom, D.; Fabbri, C.; Connor, E.C.; Schiestl, F.P.; Klauser, D.R.; Boller, T.; Eberl, L.; Weisskopf, L. Production of plant growth modulating volatiles is widespread among rhizosphere bacteria and strongly depends on culture conditions. *Environ. Microbiol.* **2011**, *13*, 3047–3058. [CrossRef]

70. Rath, M.; Mitchell, T.R.; Gold, S.E. Volatiles produced by *Bacillus mojavensis* RRC101 act as plant growth modulators and are strongly culture-dependent. *Microbiol. Res.* **2018**, *208*, 76–84. [CrossRef]

71. Farag, M.A.; Ryu, C.M.; Sumner, L.W.; Pare, P.W. GC-MS SPME profiling of rhizobacterial volatiles reveals prospective inducers of growth promotion and induced systemic resistance in plants. *Phytochemistry* **2006**, *67*, 2262–2268. [CrossRef]

72. Khan, N.; Ali, S.; Shahid, M.A.; Kharabian-Masouleh, A. Advances in detection of stress tolerance in plants through metabolomics approaches. *Plant Omics* **2017**, *10*, 153. [CrossRef]

73. Cho, S.M.; Kang, B.R.; Han, S.H.; Anderson, A.J.; Park, J.Y.; Lee, Y.H.; Cho, B.H.; Yang, K.Y.; Ryu, C.M.; Kim, Y.C. 2R, 3R-butanediol, a bacterial volatile produced by *Pseudomonas chlororaphis* O6, is involved in induction of systemic tolerance to drought in *Arabdopsis thaliana*. *Mol. Plant-Microbe Interact.* **2008**, *21*, 1067–1075. [CrossRef]

74. Naseem, H.; Bano, A. Role of plant growth-promoting rhizobacteria and their exopolysaccharide in drought tolerance of maize. *J. Plant Interact.* **2014**, *9*, 689–701. [CrossRef]

75. Chen, L.; Liu, Y.; Wu, G.; Njeri, K.V.; Shen, Q.; Zhang, N.; Zhang, R. Induced maize salt tolerance by rhizosphere inoculation of *Bacillus amyloliquefaciens* SQR. *Physiol. Plant.* **2016**, *158*, 34–44. [CrossRef] [PubMed]

76. Ikram, M.; Ali, N.; Jan, G.; Guljan, F.; Khan, N. Endophytic fungal diversity and their interaction with plants for agriculture sustainability under stressful condition. *Recent Pat. Food Nutr. Agric.* **2019**, *12*. [CrossRef] [PubMed]

77. Sangwan, N.S.; Farooqi, A.H.A.; Shabih, F.; Sangwan, R.S. Regulation of essential oil production in plants. *Plant Growth Regul.* **2001**, *24*, 3–21. [CrossRef]

78. Banchio, E.; Zygadlo, J.; Valladares, G. Quantitative variations in the essential oil of *Minthostachys mollis* (Kunth.) Griseb. in response to insects with different feeding habits. *J. Agric. Food Chem.* **2005**, *53*, 6903–6906. [CrossRef]

79. Hartmann, T. Plant-derived secondary metabolites as defensive chemicals in herbivorous insects: A case study in chemical ecology. *Planta* **2004**, *219*, 1–4. [CrossRef] [PubMed]

80. Freeman, B.C.; Beattie, G.A. An overview of plant defenses against pathogens and herbivores. *Plant Health Instr.* **2008**. [CrossRef]

81. Kim, Y.S.; Choi, Y.E.; Sano, H. Plant vaccination: Stimulation of defense system by caffeine production in planta. *Plant Signal Behav.* **2010**, *5*, 489–493. [CrossRef]

82. Cappellari, L.; Santoro, V.M.; Schmidt, A.; Gershenzon, J.; Banchio, E. Induction of essential oil production in *Mentha x piperita* by plant growth promoting bacteria was correlated with an increase in jasmonate and salicylate levels and a higher density of glandular trichomes. *Plant Physiol. Biochem.* **2019**, *141*, 142–153. [CrossRef]

83. Parida, A.K.; Das, A.B. Salt tolerance and salinity effects on plants: A review. *Ecotoxicol. Environ. Saf.* **2005**, *60*, 324–349. [CrossRef] [PubMed]

84. Kousar, B.; Bano, A.; Khan, N. PGPR Modulation of Secondary Metabolites in Tomato Infested with *Spodoptera litura*. *Agronomy* **2020**, *10*, 778. [CrossRef]

85. Ellenberger, J.; Siefen, N.; Krefting, P.; Schulze Lutum, J.-B.; Pfarr, D.; Remmel, M.; Schröder, L.; Röhlen-Schmittgen, S. Effect of UV Radiation and Salt Stress on the Accumulation of Economically Relevant Secondary Metabolites in Bell Pepper Plants. *Agronomy* **2020**, *10*, 142. [CrossRef]

86. Selmar, D.; Kleinwächter, M. Influencing the product quality by deliberately applying drought stress during the cultivation of medicinal plants. *Ind. Crops Prod.* **2013**, *42*, 558–566. [CrossRef]

87. Singh, D.; Prabha, R.; Meena, K. Induced accumulation of polyphenolics and flavonoids in cyanobacteria under salt stress protects organisms through enhanced antioxidant activity. *Am. J. Plant Sci.* **2014**, *5*, 726–735. [CrossRef]

88. Dorman, H.J.; Koşar, M.; Başer, K.H.; Hiltunen, R. Phenolic profile and antioxidant evaluation of *Mentha × piperita* L. (peppermint) extracts. *Nat. Prod. Commun.* **2009**, *4*, 535–542. [CrossRef]

89. Farnad, N.; Heidari, R.; Aslanipour, B. Phenolic composition and comparison of antioxidant activity of alcoholic extracts of Peppermint (*Mentha piperita*). *J. Food Meas. Charact.* **2014**, *8*, 113–121. [CrossRef]

90. McKay, D.L.; Blumberg, J.B. A review of the bioactivity and potential health benefits of peppermint tea (*Mentha piperita* L.). *Phytother. Res.* **2006**, *20*, 619–633. [CrossRef]

91. Khan, N.; Bano, A. Modulation of phytoremediation and plant growth by the treatment with PGPR, Ag nanoparticle and untreated municipal wastewater. *Int. J. Phytoremed.* **2016**, *18*, 1258–1269. [CrossRef]

92. Riachi, L.G.; De Maria, C.A.B. Peppermint antioxidants revisited. *Food Chem.* **2015**, *176*, 72–81. [CrossRef]

93. Bagues, M.; Hafsi, C.; Yahia, Y.; Souli, I.; Boussora, F.; Nagaz, K. Modulation of Photosynthesis, Phenolic Contents, Antioxidant Activities, and Grain Yield of Two Barley Accessions Grown under Deficit Irrigation with Saline Water in an Arid Area of Tunisia. *Pol. J. Environ. Stud.* **2019**, *28*, 3071–3080. [CrossRef]

94. Awika, J.M.; Rooney, L.W.; Wu, X.; Prior, R.L.; Cisneros-Zevallos, L. Screening Methods to Measure Antioxidant Activity of Sorghum (*Sorghum bicolor*) and Sorghum Products. *J. Agric. Food Chem.* **2003**, *51*, 6657–6662. [CrossRef] [PubMed]

95. Agati, G.; Tattini, M. Multiple functional roles of flavonoids in photoprotection. *New Phytol.* **2010**, *156*, 786–793. [CrossRef] [PubMed]

96. Khalil, N.; Fekry, M.; Bishr, M.; El-Zalabani, S.; Salama, O. Foliar spraying of salicylic acid induced accumulation of phenolics, increased radical scavenging activity and modified the composition of the essential oil of water stressed *Thymus vulgaris* L. *Plant Physiol. Biochem.* **2018**, *123*, 65–74. [CrossRef] [PubMed]

97. Oh, J.; Jo, H.; Cho, A.R.; Kim, S.J.; Han, J. Antioxidant and antimicrobial activities of various leafy herbal teas. *Food Control* **2013**, *31*, 403–409. [CrossRef]

98. Oueslati, S.; Karray-Bouraoui, N.; Attia, H.; Rabhi, M.; Ksouri, R.; Lachaal, M. Physiological and antioxidant responses of *Mentha pulegium* (Pennyroyal) to salt stress. *Acta Physiol. Plant.* **2010**, *32*, 289–296. [CrossRef]

99. Rahimi, Y.; Taleei, A.; Ranjbar, M. Long-term water deficit modulates antioxidant capacity of peppermint (*Mentha piperita* L.). *Sci. Hortic.* **2018**, *237*, 36–43. [CrossRef]

100. Alhaithloul, H.A.; Soliman, M.H.; Ameta, K.L.; El-Esawi, M.A.; Elkelish, A. Changes in Ecophysiology, Osmolytes, and Secondary Metabolites of the Medicinal Plants of *Mentha piperita* and *Catharanthus roseus* Subjected to Drought and Heat Stress. *Biomolecules* **2020**, *10*, 43. [CrossRef]

101. Khan, N.; Bano, A. Role of PGPR in the Phytoremediation of Heavy Metals and Crop Growth under Municipal Wastewater Irrigation. In *Phytoremediation*; Springer: Cham, Switzerland, 2018; pp. 135–149.

102. Jayapala, N.; Mallikarjunaiah, N.; Puttaswamy, H.; Gavirangappa, H.; Ramachandrappa, N.S. Rhizobacteria Bacillus spp. induce resistance against anthracnose disease in chili (*Capsicum annuum* L.) through activating host defense response. *Egypt. J. Biol. Pest Control.* **2019**, *29*, 45. [CrossRef]

103. Lim, J.H.; Park, K.J.; Kim, B.K.; Jeong, J.W.; Kim, H.J. Effect of salinity stress on phenolic compounds and carotenoids in buckwheat (*Fagopyrum esculentum* M.) sprout. *Food Chem.* **2012**, *135*, 1065–1070. [CrossRef]

104. Wang, C.J.; Yang, W.; Wang, C.; Gu, C.; Niu, D.D.; Liu, H.X.; Wang, Y.P.; Guo, J.H. Induction of drought tolerance in cucumber plants by a consortium of three plant growth-promoting rhizobacterium strains. *PLoS ONE* **2012**, *7*, e52565. [CrossRef] [PubMed]

105. Han, Q.Q.; Lü, X.P.; Bai, J.P.; Qiao, Y.; Paré, P.W.; Wang, S.M.; Zhang, J.L.; Wu, Y.N.; Pang, X.P.; Xu, W.B.; et al. Beneficial soil bacterium *Bacillus subtilis* (GB03) augments salt tolerance of white clover. *Front. Plant Sci.* **2014**, *5*, 525. [CrossRef] [PubMed]

106. Gopinath, S.; Kumaran, K.S.; Sundararaman, M.A. New initiative in micropropagation: Airborne bacterial volatiles modulate organogenesis and antioxidant activity in tobacco (*Nicotiana tabacum* L.) callus. *In Vitro Cell. Dev. Biol. Plant* **2015**, *51*, 514–523. [CrossRef]

107. Giri, J. Glycinebetaine and abiotic stress tolerance in plants. *Plant Signal. Behav.* **2011**, *6*, 1746–1751. [CrossRef]

108. Mehmood, A.; Hussain, A.; Irshad, M.; Khan, N.; Hamayun, M.; Ismail; Afridi, S.G.; Lee, I.J. IAA and flavonoids modulates the association between maize roots and phytostimulant endophytic Aspergillus fumigatus greenish. *J. Plant Interact.* **2018**, *1*, 532–542. [CrossRef]

109. Groß, F.; Durner, J.; Gaupels, F. Nitric oxide, antioxidants and prooxidants in plant defence responses. *Front. Plant Sci.* **2013**, *4*, 419. [CrossRef]

Review

Insights into the Physiological and Biochemical Impacts of Salt Stress on Plant Growth and Development

Muhammad Adnan Shahid [1,*], Ali Sarkhosh [2,*], Naeem Khan [3], Rashad Mukhtar Balal [4], Shahid Ali [5], Lorenzo Rossi [6], Celina Gómez [7], Neil Mattson [8], Wajid Nasim [9] and Francisco Garcia-Sanchez [10]

[1] Department of Agriculture, Nutrition and Food Systems, University of New Hampshire, Durham, NH 03824, USA
[2] Horticultural Sciences Department, IFAS-University of Florida, Gainesville, FL 32611, USA
[3] Department of Agronomy, IFAS-University of Florida, Gainesville, FL 32611, USA; naeemkhan@ufl.edu
[4] Department of Horticulture, College of Agriculture, University of Sargodha, Sargodha 40100, Pakistan; uaf_rashad@yahoo.com
[5] Plant Epigenetic and Development, Northeast Forestry University Harbin, Heilongjiang 150040, China; shahidcafi926@gmail.com
[6] Horticultural Sciences Department, University of Florida, IFAS-Indian River Research and Education Center, Fort Pierce, FL 34945, USA, l.rossi@ufl.edu
[7] Environmental Horticulture Department, IFAS-University of Florida, Gainesville, FL 32611, USA; cgomezv@ufl.edu
[8] Horticulture Section, School of Integrative Plant Science, Cornell University, Ithaca, NY 14853, USA; nsm47@cornell.edu
[9] Department of Agronomy, Faculty of Agriculture and Environmental Sciences, The Islamia University of Bahawalpur (IUB), Punjab 63100, Pakistan; wajid.nasim@iub.edu.pk
[10] CEBAS-CSIC, Departamento De Nutrición Vegetal, Campus Universitario de Espinardo, Espinardo, 30100 Murcia, Spain; fgs@cebas.csic.es
* Correspondence: muhammad.shahid@unh.edu (M.A.S.); sarkhosha@ufl.edu (A.S.)

Received: 11 April 2020; Accepted: 27 June 2020; Published: 30 June 2020

Abstract: Climate change is causing soil salinization, resulting in crop losses throughout the world. The ability of plants to tolerate salt stress is determined by multiple biochemical and molecular pathways. Here we discuss physiological, biochemical, and cellular modulations in plants in response to salt stress. Knowledge of these modulations can assist in assessing salt tolerance potential and the mechanisms underlying salinity tolerance in plants. Salinity-induced cellular damage is highly correlated with generation of reactive oxygen species, ionic imbalance, osmotic damage, and reduced relative water content. Accelerated antioxidant activities and osmotic adjustment by the formation of organic and inorganic osmolytes are significant and effective salinity tolerance mechanisms for crop plants. In addition, polyamines improve salt tolerance by regulating various physiological mechanisms, including rhizogenesis, somatic embryogenesis, maintenance of cell pH, and ionic homeostasis. This research project focuses on three strategies to augment salinity tolerance capacity in agricultural crops: salinity-induced alterations in signaling pathways; signaling of phytohormones, ion channels, and biosensors; and expression of ion transporter genes in crop plants (especially in comparison to halophytes).

Keywords: abiotic stresses; cell membrane stability; climate change; osmolytes; polyamines

---

1. Introduction

1.1. Overview of Salinity

Abiotic stresses like salinity, drought, and high temperature have undesirable effects on crop productivity and quality, and negative trends in sustainable agriculture [1]. Salinity in particular is an important limiting factor, causing low yield with inferior quality. Climate change is considered one of the major contributing factors to soil salinization, leading to land degradation and desertification [2]. According to Flowers et al. [3], high salt concentration is responsible for negative impacts on 7% of total land surface, and 5% of cultivated land. Poor irrigation water quality is another important factor contributing to soil salinization [4]. For these reasons, soil salinization is a reported major cause of reductions in the productivity of irrigated and rainfed lands of the world [5,6].

The adverse effects of salinization on plants are evident from negative growth trends from alteration or inhibition of biochemical and physiological processes. Plants can be classified as glycophytes or halophytes by their ability to survive under high salt concentrations [7]. Glycophytes are plants that are severely affected by saline conditions both at the cellular and whole-plant level. Under saline conditions, these plants exhibit greater accumulation of solutes, and ionic and osmotic stresses confer nutritional imbalances, which limit the productivity of these plants. The majority of terrestrial plants are glycophytes, including crop plants [1,8].

Conversely, halophytes regulate their biochemical and physiological processes through ionic compartmentalization, production of osmolytes and compatible solutes, enzymatic changes, and absorption of selective ions. These adaptations promote seed germination, succulence, and salt exclusion for these plants in a saline environment [9,10]. Halophyte succulence keeps the proportion of ions to water in balance under high salt conditions by maintaining high water contents. Succulence is expressed as large cell size, reduced growth and surface area per tissue volume, and increased water constituents. Interestingly, halophytes also have a greater number of mitochondria, indicating that more energy is required to survive under saline conditions [11,12]. Halophytes also have less sodium and chloride ion accumulation in their cytoplasms, allowing their chloroplasts to survive even while the plant experiences salinity shocks [13,14]. Halophytes also have a specialized system for salt excretion from the plant tissues via specific glands. These glands are characteristic of halophytic leaves. The leaves will remove the salts onto the leaf surface before the salts can reach the shoots of the plant. The presence of halophytes is limited to habitats with plentiful water (e.g., salt marshes, etc.). "Salt hairs," which regulate water loss, replace "secretary glands", if a plant is adapted to a relatively drier climate as compared to marshes [15,16]. Hydathodes are another adaptation by plants to remove excessive salts, with less stomatal conductance and transpiration water loss [17,18].

Various undesirable effects appear because of high salt concentration. Ion imbalance is one of the major consequences. A high concentration of Na and Cl ions, as an example, can lead to biochemical processes which can prove to be fatal for the plants [19–21]. Sodium and chloride toxicity not only induce nutritional disorders but also cause physiological drought by lowering the osmotic potential of the soil solutions [22]. Soil salinity prevents the plant from taking up water from the soil, resulting in a decline in cellular water, thus affecting cell turgor. Soil salinity also adversely affects photosynthetic activity in the plant and encourages the production of reactive oxygen species (ROS), thus reducing plant growth [23,24].

The identification of salt stress by plant species and their subsequent response is controlled by signals—signals which are generated by ions, osmotic differential, hormones, or ROS [25]. These signals bind to their respective receptors and initiate the physiological mechanisms which enable a plant to adapt to stress conditions (Figure 1). Under abiotic stress conditions, three types of signal transduction have been categorized, i.e., the ionic signaling pathway, the osmolyte regulation pathway, and the gene regulation pathway [26]. For signal transduction under salinity stress, the ionic stress signaling pathway has been elucidated. Calcium (Ca) occupies a central position in this regard. It induces signal transduction in plants to adapt to stress conditions [27]. High cytosolic Ca concentration initiates many processes involving enzymatic activity regulation, ion channel performance, and gene expression [28]. Exogenously applied calcium regulates K^+/Na^+ selectivity, and thus confers salt adaptation by improving signal transduction. Glycinebetaine is reported to maintain signal transduction and ion homeostasis under salt-stressed conditions [29,30].

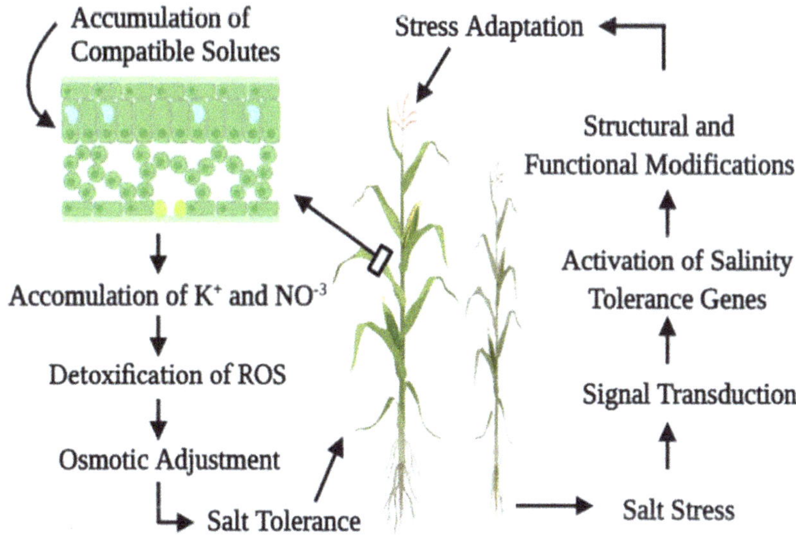

Figure 1. Salt stress signals that bind to their respective receptors and initiate the physiological and molecular mechanisms to enable a plant to survive under stressed conditions.

Glycophytes and halophytes mediate serious effects of salinity at the cellular level by inducing changes in the plasma membrane and cytoplasm. As a tolerance mechanism, the plant alters the structure and composition of their plasma membrane, especially lipid and protein contents. The cell membrane is usually the foremost target of any stress [31]. Salt stress also alters the cell cytoplasmic viscosity and composition [32,33].

It is essential to understand each tolerance mechanism at the cellular level in order to understand each tolerance mechanism at the plant level. The protoplasmic features studied at the cellular level include plasma membrane permeability, cytoplasmic viscosity, cytoplasmic streaming, and cell solute potential. Cytoplasmic viscosity describes the water contents of the cytoplasm in conjunction with its inter-macromolecular interactions. Cell solute potential represents the solute contents of the cell. Cell membrane permeability significantly increases with the increase in salinity [34,35]. The ability of the plasma membrane to repair, regenerate, and maintain its integrity stabilizes the cell structure and function under stress conditions. It is mainly dependent on the composition of the plasma membrane (i.e., mainly lipid contents). Saline conditions result in enhanced lipid peroxidation [36].

The cell membrane stability technique is widely utilized to judge the behavior of various plant genotypes in response to salt stress [37]. Thus, it can contribute to assessing the salt tolerance of plant genotypes. Cell membrane stability is also reported to correlate with potassium (K) ions, osmotic potential, osmotic adjustment, and relative water contents [38,39]. The differing patterns of cell membrane permeability help in characterizing genotypes as tolerant or sensitive. In a saline environment, salt-sensitive genotypes show marked alterations, whereas salt-tolerant genotypes show minor changes. Salinity also alters the degree of saturation of membrane fatty acids and membrane fluidity [40].

The salt-tolerant genotypes have high cytoplasmic viscosity due to augmentation in hydrophilic cytoplasmic proteins and other macromolecules [41]. In more sensitive genotypes, a saline environment results in a high concentration of salts in plant cells, lowering the solute potential [42]. Salinity inflicts serious irregularities during cell division, one of the various metabolic processes which face severe alterations in a saline environment. A saline environment especially alters the leaf anatomy by affecting mitochondria and vacuoles [43,44], plant leaf area, and stomatal thickness [45]. One way plants exhibit tolerance to saline environments is portioning or compartmentalization of toxic ions. This mechanism enables salt-tolerant plant species to retain toxic levels of harmful ions in vacuoles and inhibit their interference with cytoplasmic metabolic activities [46]. The Na^+ and Cl^- partitioning in the vacuole stimulates higher concentrations of K^+ and organic osmolytes in the cytoplasm in order to adjust osmotic pressure of the ions in the vacuole [47].

1.2. Salinity and Morphological Attributes

Halophytes have the unique feature of succulence, a feature which keeps the ionic uptake in proper proportion with water, by maintaining high water contents (Figure 2). Succulence results in large cell size, reduced growth [48,49], reduced surface area per tissue volume, and increased water constituents. The salt tolerance is evident as maintenance of vegetative growth and yield and lower necrotic percentage [50,51]. Moreover, halophytes also have a greater number of mitochondria, indicating that more energy is required to survive under saline conditions [11,21]. Sodium (Na) accumulation causes necrosis in old leaves, initiating from tips and then extending towards the leaf base. It also decreases leaf life span, net productivity, and crop yield [52]. Biomass reduction and foliar damage become more prominent with time and at higher salinity levels. High salt concentration caused a reduction in fresh fruit yield in various vegetables [31,53–55]. However, salinity treatments did not prove harmful for vegetative growth and the number of flowers [53,55]. In contrast to the above report, Chartzoulakis and Klapaki, [56], Kaya et al. [57], and Giuffrida et al. [58], noted a reduction in fruit numbers and fruit weight under salt stress. Saline conditions were reported as producing non-significant results on some growth attributes, water status, and tissue concentration of major nutrients [59]. Salt treatments caused a significant reduction in plant height, root length, and dry weight [60–62]. Salt stress caused reduction in fresh and dry weight of cotton seedlings [63,64] and seed germination percentage in wheat varieties [35,65]. Tomato (*Solanum lycopersicum*) and pepper plants grown under salt stress experienced a reduction in dry weight, plant height [57,66], fruit weight, and relative water contents [67]. Broad bean, which is a green vegetable, experienced a significant reduction in plant height, leaf area, pod weight, number of pods per plant, seed yield, number of seeds, and product quality due to salt stress. However, a significant positive trend was observed between dry leaf matter, specific leaf weight, and salinity [68]. Suppression in seedling growth and dry matter accumulation was observed in Indian mustard (*Brassica juncea* L.) because of salt stress, which was ameliorated by putrescine application [69]. Exogenously applied sugar beet extract and shikimic acid on salt stressed eggplant and tomato crops respectively displayed a marked influence on fresh fruit weight, number of fruits, and shoot and root fresh and dry weight [54,70].

According to Caines and Shennan [71], root growth is more susceptible to saline conditions than shoot growth, but both are affected, making them reasonable indicators of salinity damage. A similar suppression in the shoot and root growth was also observed by Evers et al. [72] and Gao et al. [73] in the case of *Solanum tuberosum* L. Under high salt concentrations, potato root and shoot development was hindered [74,75]. High salt concentrations reduced the leaf area and increased the root:shoot ratio in wild type (Ailsa Craig) and ABA-deficit mutant (*notabilis*) tomato genotypes [76]. Suppression of fresh and dry weight of tomato plants because of salt stress can be alleviated by promoting the growth of *Achromobacter piechaudii* in the tomato growth media. *A. piechaudii* also caused a reduction in ethylene production by tomato seedlings, the opposite effect of salt stress on tomato seedlings [77]. High ethylene levels proved to be harmful to the growth of the plant [78]. Interestingly, Hu et al. [79] reported that potato root growth could be improved through brassinosteroid application. Salt stress caused a reduction in marketable yield of pepper plants grown hydroponically. Saline conditions also imparted negative features to the fruit quality in terms of fruit pulp thickness and firmness. It also resulted in increased fructose, glucose, and *myo*-inositol fruit concentrations [4,42]. The salt-sensitive pepper genotypes showed maximum damage and experienced severe chlorosis and necrosis, whereas tolerant genotypes were slightly less affected. Sodium exclusion can be regarded as a criterion to allocate salt stress tolerance status to pepper genotypes [80]. Korkmaz et al. [81] have reported that the exogenous application of glycine betaine can reduce the effect of salinity in pepper plants.

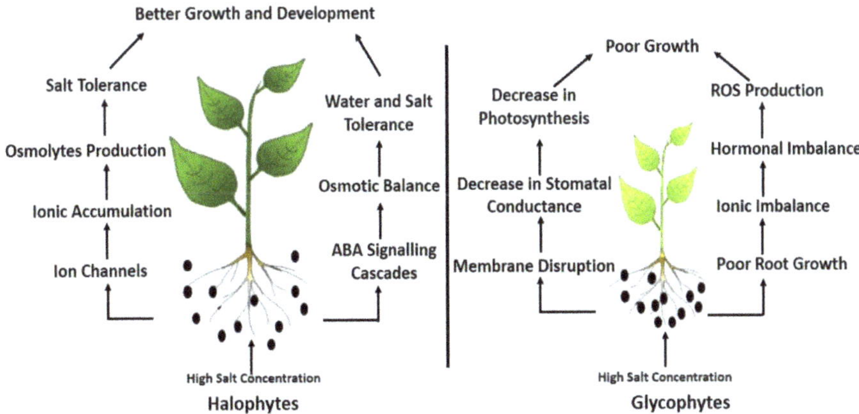

Figure 2. Regulation of physiological and biochemical process in halophytes through ionic compartmentalization, osmotic adjustment, enzymatic activities, polyamines, and stress signaling regulation.

High salt concentration has been observed to cause detrimental effects on leguminous crops. In fact, saline conditions induced smaller sized nodules, reduced nodule volume per plant, less nodulation, and inferior plant growth. At the cellular level, saline conditions caused drastic alteration in the mechanism of nodule formation. It reduced the turgor of the peripheral cells of the nodule, altered its zonation, enhanced the infection thread enlargement, reduced the release of bacteria from infection threads, and stimulated the electron-dense material (phenolics) and its accumulation in vacuoles. Salinity caused a reduction in nitrogen-fixing ability of the nodules, which has the outcome of reduced respiration rate and protein synthesis [82–84]. Salinity tolerance patterns vary considerably among leguminous crops. It is distinctly apparent in *Vicia faba*, *Glycine max*, *Pisum sativum*, and *Casuarina glauca*, which can be categorized as salt-sensitive [85,86]. In addition, germination of seeds faces serious limitations upon exposure to salt stress [87,88]. Halophytes and glycophytes differ significantly in their germination behavior. Halophytes have the ability to maintain their germination mechanism to an extent with the advent of salinity. However, a sharp decline in germination occurs for glycophytes under salt stress. Imbibition is affected by the lower solute potential of the soil solution. It results in enzymatic deregulations and imbalances in source-sink relationships and ratios of different plant growth regulators present in the seed and required for efficient seed germination [89,90].

1.3. Salinity and Physiological Attributes

Photosynthesis is of prime importance in the production of adenosine triphosphate (ATP), which provides the energy required for CO_2 fixation to sugars. Various abiotic stresses alter photosynthetic mechanisms [91] by disrupting thylakoid membranes, modifying the electron transport chain, altering enzymatic activity and protein synthesis, and changing Calvin cycle patterns. All these abnormalities cause deregulation in ATP synthesis [92], which can lead to the deficiency of certain ions due to ion degradation and synthesis inhibition [93–96].

Stepien and Klobus [97] reported a decline in the relative water content of cucumber (*Cucumis sativus*) leaves after exposure to saline solutions. They attributed their results to higher Na and reduced K content, a situation which reduces photosynthesis because of the antagonistic competition of Na for ion uptake. Gas exchange is inversely related to the concentration of Na and chloride ions [98,99], and parameters like photosynthesis, stomatal conductance, and transpiration tend to be negatively influenced by saline treatments [100,101]. Photosynthesis of pepper plants has been reported to be lower when the entire root system is affected by salt stress compared to partial root exposure, and factors like stomatal conductance and transpiration are similarly affected by complete or partial salt stress [99,102]. A low stomatal conductance was reported in wild type (Ailsa Craig) and ABA-deficit mutant (*notabilis*) tomato genotypes, which was negatively correlated with increasing xylem ABA for both genotypes [76]. Stomatal conductance is often correlated with photosynthetic efficiency, which is a prerequisite for higher biomass production and yield [10]. In addition, salinity is reported to reduce the maximum quantum efficiency of photosystem II (PSII) [48,94,103–106]. A distinct correlation has been found between Na ion contents and chlorophyll fluorescence, which is often used to estimate salt tolerance of plants [48]. Higher salt levels are also known to alter photosynthesis via non-stomatal limitations, including variations in photosynthetic enzyme activity and changes in the concentration of chlorophyll and carotenoids [107,108]. Pepper leaves have exhibited a significant reduction in chlorophyll pigment under saline conditions [109]. A similar reduction in chlorophyll *a* and *b* contents has been reported in melon by Kaya et al. [57]. This chlorophyll degradation under salt stress can be attributed to an enzyme called chlorophyllase [110–113] and to the absolute concentration of chloride and Na in the leaves [98,114–118]. Although carotenoid content has been reported to decline in response to salinity, anthocyanin pigments typically increase as a result of salinity.

Plant antioxidants include secondary metabolites (phenolic compounds), which are generated in response to stress conditions. These secondary metabolites may include tocopherol, which serves to stabilize membrane integrity [119,120], ascorbic acid, carotenoids, flavonoids, and glutathione [90,121]. Tocopherol plays a key role as a signaling molecule between cells [93]. Ascorbic acid is a significant antioxidant involved in plant adaptation [122] and occurs abundantly in cell organelles and apoplasts [99]. It has the ability to scavenge superoxide, hydroxyl, and singlet oxygen. Carotenoids are found in chloroplasts and reported to aid in light reception for photosynthesis. Moreover, they are also protective compounds which scavenge ROS [95,123]. Putrescine is reported to increase the level of carotenoids and glutathione in Indian mustard (*B. juncea* L.) against salt stress. Putrescine supplementation inhibits ROS generation by accelerating antioxygenic enzymes and therefore aiding in the maintenance of chloroplastic membranes and the NADP$^+$/NADPH ratio [124–126].

C$_4$ plants are reported to be more resistant against salinity stress than C$_3$ plants by having a better capacity to preserve the photosynthetic apparatus against oxidative stress [127,128]. Other physiological responses that are indirectly related to salinity stress include changes in water use efficiency and evapotranspiration, which can benefit from the use of beneficial bacteria [77,129–131]. In addition, higher salt concentration can stimulate the accumulation of spermine and spermidine, which contribute to the induction of salt tolerance in plants and lead to maintenance of fruit quality (Figure 3) [132–135].

Figure 3. Bunches, grapevine cultivar Shiraz. Control bunch (**A**); Salt-stressed bunch (**B**); showing symptoms of coulure and millerandage. Salt treatment was applied from budburst until veraison via fertigation with 35 mM NaCl added to control nutrient solution [136].

2. Salinity and Water Relations

Water is an essential constituent of plant cells, supporting almost all physiological and biochemical processes contributing to plant growth and development [8]. High salt concentration hinders the movement of water from the soil to the plants by reducing the water conductivity of roots [9] and affecting relative water content at the cellular level [31,99]. Alterations in rootzone water status have profound effects in the way plants respond to higher salt concentration [1,93], but adverse effects of salinity stress can be mitigated to an extent with proper irrigation and nutrient management and by controlling the evaporation rate of plants. A recent review by van Zelm et al. [137] listed root hydraulic conductivity, osmotic potential, relative water content, leaf water potential, water-use efficiency, stomatal conductance, and transpiration, among others, as typical parameters that are commonly used to indicate water-relation responses to salinity stress, which are determinant for plant growth.

Others have shown a relationship between water potential and photosynthetic parameters. For example, a decline in CO_2 assimilation has been reported with the depression of water potential [126,138]. To achieve active growth, plants maintain positive turgor pressure and regulate osmotic potential. Under saline conditions plants face osmotic stress due to alterations in water potential [48,97,139]. Osmotic stress triggers cellular and whole plant responses [44,140–143], and plants cope with that osmotic stress through regulating ionic homeostasis, which induces tolerance against toxic ions and growth under unfavorable conditions [8,144]. In order to maintain osmotic adjustment, plants lower the cellular osmotic potential to help improve water uptake and adjust ionic concentration in cells [42,145]. For that purpose, osmolytes appear to be of major consideration. These may be sugars, polyols, or amino acids. Osmotic adjustment appears to be a significant and effective salinity resistant mechanism in crop plants, which can be exploited efficiently through selection and breeding efforts that target salinity tolerance in different plant species.

3. Salinity and Biochemical Attributes

Plants have the ability to sustain their life under a saline environment through synthesis and accumulation of compatible solutes in the cytosol. These are soluble compounds with low molecular mass. These chemical compounds can maintain physiological and biochemical processes, without having interference in these processes. The chief components of compatible solutes are sugar alcohols (mannitol, sorbitol, ononitol), quaternary ammonia compounds (glycine betaine, proline betaine), proline, and tertiary sulfonium compounds. They act to scavenge the reactive oxygen species (ROS) and inhibit lipid peroxidation, hence preventing damage at the cellular level. These compatible solutes act in favor of osmotic adjustment and prevent ROS damage at the cellular level [146,147]. These maintain macromolecular conformation in the cytosol, which may be changed due to the accumulation of charged ions, under saline conditions [8,97]. These organic compounds are termed as compatible due to their consistency with the cell's metabolism [148], and their lowering of the water potential without altering cell water contents. These organic compounds are hydrophilic in nature, having the ability to replace water present on protein surfaces [149,150], without interfering with their structure and function. These solutes play a key role in preventing the drastic effects of high ion concentration on enzymatic activities [24,151–153]. Such important roles of compatible solutes lead to osmoregulation of plant cells under osmotic stress. In addition to osmoregulation, these organic compounds have a distinct role in protein stabilization, maintenance of membrane integrity, protection of OEC of PSII from dissociation [154], and scavenging of reactive oxygen species (ROS). Mannitol, sorbitol, glycerol, proline, ononitol, and pinitol have been reported to scavenge ROS species [155].

Mannitol (sugar alcohol) metabolism in higher plants is a superior attribute, contributing to salt and osmotic stress tolerance while playing a significant role as a compatible solute. It also improves plant responses under biotic stress as well, like under pathogen infestation [156]. Mannitol is reported to be synthesized at the same time with either sucrose or raffinose saccharide. In salt-tolerant species, mannitol accumulation increases, indicating that high mannitol levels contribute to salt tolerance. Mannitol acts to scavenge the reactive oxygen species, thus protecting protein molecules [157,158]. Pinitol and ononitol have been reported to accumulate under various stresses, predominantly drought and salt stress [159,160]. Interestingly, polyols can be used as a potential biochemical marker for genetically engineered stress resistance plant genotypes [161,162].

3.1. Salinity and Proline

Proline is an osmolyte, an amino acid, which is thought to play a significant role in inducing tolerance in plants against stressed conditions [163]. Salt stress can result in elevation in proline levels [74]. Ethephon, when used with sodium chloride in spinach, also increased proline levels [164]. The importance of proline is highlighted by its existence in bacteria with a relationship to plants experiencing water or salinity stress. High proline levels can serve as a nitrogen source for plants during recovery [165]. The precursor of proline synthesis is glutamate, involving pyrroline carboxylic acid synthetase and pyrroline carboxylic reductase [166]. An increase was noted in the activity of pyrroline-5-carboxylate synthetase (P5CS) and a decline was recorded in proline dehydrogenase activity in potato seedlings under salt stress. These changes of enzymatic activity were more pronounced in salt-sensitive cultivars [74,167]. However, an increase in proline contents of potato clones was recorded upon salt exposure [50,168]. It serves to stabilize ultra-structural changes in cells, scavenge ROS (reactive oxygen species), and maintain cellular redox potential. Under stress conditions, a higher accumulation of proline is reported in cell cytosol, strengthening the ability of the cell to make ionic adjustments. Its accumulation is linearly related to stress tolerance in plants [169]. Proline biosynthesis is reported to be mediated by Ca [170–172] and abscisic acid [8]. Previously, contrasting views about proline accumulation were also reported in plants under stress [19,158], where it appeared as a salt stress injury symptom, e.g., rice [173] and sorghum [174].

Some plant genotypes do not respond to proline accumulation, but their salt tolerance potential can be enhanced through the exogenous application of proline [31,175]. It may be helpful in counteracting the harmful effects through osmo-protection, resulting in a higher growth rate. Proline also increases the activities of antioxidant enzymes like SOD (superoxide dismutase) and POD (peroxidase) [176]. Proline is not reported to scavenge ROS directly, but through enhanced antioxidant enzyme activity. It is reported to be more effective in mitigating the drastic effects of salinity than glycine betaine [177]. Proline used at higher concentrations may prove to be lethal for the plant, causing ultra-structural damages leading to ROS generation [178]. The effective dose of proline varies with genotype and plant developmental stage [179–182]. Proline accumulation has been reported for drought sensitive and tolerant barley genotypes grown under saline conditions. Under salt stress, a considerable amount of proline was present, with relatively lower quantities in root tissues. Proline accumulation is reported to be more prominent in tolerant genotypes [183].

3.2. Salinity and Polyamines

Polyamines are multivalent compounds consisting of two or more amino groups. In higher plants, these are identified as putrescine, spermidine, and spermine [147,184]. These are involved in various physiological mechanisms including rhizogenesis, somatic embryogenesis, maintenance of cell pH and ionic balance [29], pollen and flower formation, abscission, senescence, and dormancy. Endogenous polyamine synthesis can be stimulated by cytokinin [185,186]. These compounds act to stabilize macromolecules like DNA and RNA. Moreover, polyamines have a significant role in numerous abiotic and biotic stresses [151,187]. At the cellular level, polyamines contribute to regulating the plasma membrane potential, ionic homeostasis, and tolerance against salinity [188,189]. Exogenously applied polyamine or ornithine caused a reduction in proline accumulation in plant tissues under salt stress. However, an alternate trend was observed in the case of non-stressed beans [99,190]. Putrescine is characterized as a de-stressor agent and a nitrogen source under stressed conditions [191]. Putrescine has been reported to reverse the biomass reduction in Indian mustard [68,192]. Its production in plant cells follows two alternative pathways: conversion from ornithine or arginine. Putrescine is then converted to spermidine and subsequently to spermine by addition of an aminopropyl group. Spermine deficiency caused Ca ion imbalance in *Arabidopsis thaliana*, thus indicating spermine as a maintainer of plant cell ionic homeostasis under salt stress [193].

3.3. Salinity and Glycine-Betaine

Glycine-betaine (GB) is present in a wide range of organisms, from bacteria to higher plants and animals. In addition to being involved in osmoregulation, it maintains and regulates the performance of PSII protein complexes by protecting extrinsic regulatory protein against denaturation. It also stabilizes macromolecules, due to its ability to form strong bonds with water [136]. It protects these macromolecules during drought and thermal stress, which is why it is sometimes called an "osmoprotectant" [129,194]. Glycine-betaine accumulates in some crops under stress, like members of family *Poaceae* and *Chenopodiaceae* [195], and is absent entirely from other plants, like rice and tobacco. This directed the scientists to develop transgenic plants that have the ability to produce GB. In transgenic plants, the reproductive organs are capable of tolerating abiotic stresses if they can accumulate GB [195]. The precursor for GB is choline, and the conversion is managed by enzymes like choline monooxygenase and betaine-aldehyde dehydrogenase [151,196]. Choline supplementation to the growth media of the salt-stressed plant can act to restore the suppressed growth [197]. GB is water soluble, is not harmful at higher concentrations, and accumulates mainly in plastids and chloroplasts. Exogenous application of GB promotes salinity tolerance in plant species which do not naturally produce GB. A plant can utilize exogenously applied GB via leaves [198], as well as roots [199]. After absorption, GB is translocated in phloem [99,200]. GB is not directly involved in scavenging ROS species, but it alleviates the damaging effects of ROS by promoting enzymes responsible for the destruction or production suppression of ROS [201].

The reproductive stage of any plant during GB application is considered critical to ensure maximum yield. In various studies, it was reported that the plant reproductive organs acquire higher levels of

GB than the vegetative parts under stressed conditions. This indicates that high GB accumulation is more necessary for protecting the reproductive organs than it is for protecting the vegetative tissues from abiotic stresses, indicating that application timing is key [181,202]. The natural GB accumulating species include sugar beets, spinach, wheat, barley, and sorghum. High GB concentration is linearly linked with increased tolerance. Osmotic adjustment is the major mechanism involved in increased tolerance to abiotic stresses, especially salt stress. GB is responsible for turgor maintenance through osmotic adjustment [54,182,203]. However, this relationship is not satisfactory in some cases like *Triticum* spp. and *Agropyron* spp. [204]. Thus, this relationship varies with genotype [158]. The plant species which do not produce GB naturally can give a satisfactory yield and survival rate under salt stress conditions through the exogenous application of GB [42,205]. Exogenous GB, once applied, is transported rapidly throughout the plant. Exogenous application of GB has been reported in many plant species, including tobacco, rice, soybean, barley, and wheat. In barley, GB application improved stress tolerance by lowering water potential, which improved survivability. GB plays a role in osmotic adjustment and ionic homeostasis by maintaining high K^+ concentration compared to Na^+ ions. Exogenous application of GB also increased the K^+/Na^+ ratio [206–208]. GB also protects the photosynthetic apparatus. It enhances photosynthetic activity through increased stomatal conductance and reduced photorespiration [42,209,210].

In contrast to its positive influence, some researchers have also suggested neutral or somewhat negative responses to exogenously applied GB in some plant genotypes. For example, it appeared to have a neutral influence on growth in cotton [211], turnip, rapeseed, and tomato [212]. For the commercial application of GB, the rate, duration, timing, and frequency should be considered [158,213]. It can be used for seed treatment as well as foliar application. The application method is dependent on the plant material on which it will be applied, the timing of the application relative to plant developmental stage, and environmental conditions during the time of application [214].

Exogenously applied GB improved salt tolerance in rice by improving relative water contents in the leaves and increasing antioxidant levels, including superoxide dismutase, ascorbate peroxidase, catalase, and glutathione reductase (GR) [213]. Reduction in peroxidase activity was reported in a salt-tolerant rice genotype under salt stress. GB is also reported to reduce lipid peroxidation [215–217]. GB can prevent membrane adulterations due to osmotic stress more efficiently than proline [218]. Proline accumulation in leaves of salt-stressed plants is not reported to be correlated with exogenously applied glycine betaine [54,180]. Sugar beet is identified as the foremost source of GB [158,219]. It is appreciated as a valuable source of GB along with other beneficial compounds and is useful in inducing tolerance against salt stress in eggplant (*Solanum melongena* L.) as compared to pure GB. It has a marked influence on the morphological (growth and yield) as well as physiological and biochemical (gas exchange, photosynthetic rate, transpiration, GB accumulation) attributes [186,220].

4. Salinity and Enzymatic Attributes

Generation of reactive oxygen species (ROS) like singlet oxygen, superoxide radical, hydrogen peroxide, and hydroxyl radical as a consequence of exposure to various abiotic stresses causes injury to plants. Molecular oxygen is non-reactive and requires electron donors for the production of reactive oxygen species. These electron donors are typically metal ions [24,221]. Increased production of ROS is an indicator of plant stress [222] for proteins, lipids, pigments, DNA, and other molecules at the cellular level [223,224]. Plants combat ROS through enzymatic and non-enzymatic mechanisms. The enzymes used for scavenging ROS include superoxide dismutase (SOD), catalase (CAT), peroxidase (POD), ascorbate peroxidase (APX), glutathione reductase (GR), and glutathione-synthesizing enzymes [197,205,225]. However, increased generation of ROS scavenging enzymes does not always correspond to higher salinity tolerance. Many factors contribute to the effectiveness of antioxidant systems, including the site of antioxygenic enzyme production, enzyme action, and interaction of different antioxidant enzymes, according to Blokhina et al. [226]. Das and Roychoudhury [227] and Sairam et al. [228] reported increased SOD, APX, CAT, and GR activities in intolerant wheat plants under abiotic stresses. The site of superoxide radical synthesis is reported to be chloroplasts [229,230], mitochondria [231], and microbodies. Ascorbate peroxidase (APX) is recognized as a reducing agent for H_2O_2 to water by the utilization of ascorbate and release of monodehydroascorbate (MDHA). MDHA reductase contributes to the conversion of MDHA to AsA (ascorbate). Ascorbate peroxidase causes effective regeneration of AsA and interrupts the cascade of oxidation caused by H_2O_2 [24,232]. The antioxidant response of plants can be regulated by compounds like H_2O_2 under stress conditions. H_2O_2 contributes almost 50% of the destruction caused by oxygen radicals in photosynthetic reduction. In terms of negative impacts on plant physiology, H_2O_2 is the most harmful of all reactive oxygen radicals. APX exists in four different forms: chloroplast stromal soluble form (sAPX), chloroplast thylakoid bound form (tAPX), cytosolic form (cAPX), and glyoxysome membrane form (gmAPX) (Figure 4). H_2O_2 increased the level of other antioxidants and caused a decline in lipid peroxidation in maize under stressed conditions [210,211,233]. Saline conditions applied to potato seedlings and *Broussonetia papyrifera* showed an increase in APX activity, new POD and SOD isoenzyme activities, and alterations in isoenzyme composition [74,212]. Similarly, seed treatment of wheat by H_2O enhanced the salinity tolerance of young seedlings of wheat through prevention of oxidation damage and induction of stress proteins [234]. Exogenously applied GB has been reported to enhance antioxidant activity in terms of SOD, ascorbate peroxidase, CAT, and GR in salt-tolerant rice and wheat genotypes under salt stress conditions [197,214,235].

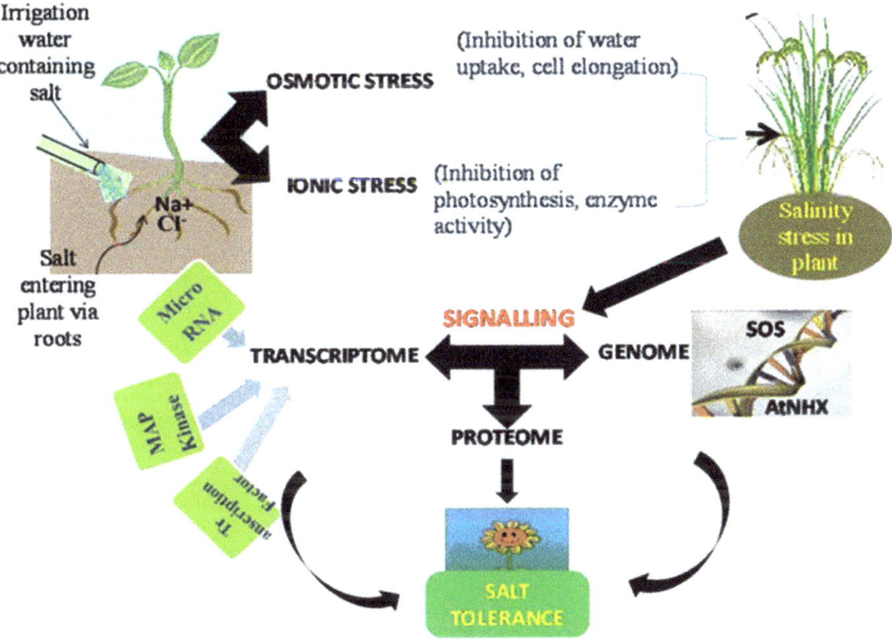

Figure 4. Schematic diagram showing routes of salt stress toxicity and various tolerance strategies in plants. Putative roles and action of genes, transcription factors, mitogen-activated protein kinases, microRNAs, and metabolites are shown [236].

Superoxide dismutase (SOD) activity is reported in almost all plant cell types. It causes disproportionation of singlet oxygen to molecular oxygen and hydrogen peroxide. The isoforms of SOD are categorized as copper- and zinc-containing superoxide dismutase (Cu/Zn-SOD), manganese-containing superoxide dismutase (Mn-SOD), and nickel and iron-containing superoxide dismutase as Ni-SOD and Fe-SOD [79]. A negative correlation was observed between proline accumulation and SOD activity in wild halophytes, so SOD activity does not necessarily induce salt tolerance in plants [216,237]. Salt treatments caused a reduction in SOD activity in potato cultivars [74]. Catalases (CATs) contribute to the plant defense system, and are synthesized in peroxisomes and glyoxysomes. CAT is responsible for the conversion of hydrogen peroxide into water and oxygen. Under saline conditions, SOD activity was reported to be more pronounced in C_3 (wheat) as compared to C_4 (maize). Both types of plants showed the same level of elevation in APX. However, an increase in GR level was more prominent in maize plants [118,238].

Exogenous application of potassium nitrate alleviated salinity effects in winter wheat by promoting some antioxidant enzymes [60,239]. Similarly, coronatine, which has properties similar to methyl jasmonate, is reported to promote the activities of antioxidant enzymes including superoxide dismutase (SOD), catalase (CAT), peroxidase (POD), and glutathione reductase (GR), and DPPH (1,1-diphenyl-2-picrylhydrazyl) scavenging behavior in cotton leaves. DPPH-radical scavenging determines non-enzymatic antioxidant activity. COR causes a decline in generation of reactive oxygen species by increasing the activity of antioxidant enzymes [240–243]. COR is functionally similar to jasmonic acid, which mitigates salt stress [244–246]. Polyamines applied to salt-stressed plants can act to ameliorate the negative impact of salinity (i.e., growth suppression) in many crop plants [247].

5. Salinity and Phytohormones

Plant hormones are signaling molecules with the ability to alter the physiological mechanisms of the plant, even if present in very minute quantities [248]. Common plant hormones are auxins, gibberellins, cytokinins, abscisic acid, and ethylene [249]. Plants growing under salt stress experience imbalances in hormonal homeostasis. Stressed conditions drastically alter physiological mechanisms of the plant, creating massive changes in endogenous hormonal contents. Higher concentrations of toxic ions are negatively correlated with the levels of plant hormones like gibberellins, auxins, and cytokinin and are positively associated with the abscisic acid level [250,251]. Exogenous plant hormone application on salt stressed plants was found to alleviate the negative effects of salinity on the morphological (leaf area, dry mass), physiological (chlorophyll content, stomatal conductance, photosynthetic rate), and yield characteristics of crops [252,253].

Abscisic acid (ABA) and ethylene are involved in signaling under stress conditions. An increase in ABA concentration in plant cells has been reported under saline conditions [254]. Carotenoids are the precursor for ABA synthesis, with roots and leaves the sites of synthesis [255]. Water deficit in the root zone causes ABA generation in roots, and ABA transport to shoots is xylem mediated. The increase in pH of xylem sap increases the transport of ABA to the guard cells, where it regulates the stomatal opening and closing through the involvement of Ca ions [256]. Alterations in ABA levels indirectly affect photosynthesis through disruption in stomatal opening and closing. The photosynthetic efficiency of the plant cell declines along with deregulation of translocation and assimilate partitioning of photosynthates [8,257].

Exogenously applied plant growth regulators have been widely reported to enhance stress tolerance in numerous plant species [258]. ABA is reported to be useful in alleviating plant salt stress under low water potential. ABA production in the plant cell is also related to ethylene synthesis under salt stress. The interaction between these two stress hormones is apparent from vegetative growth and seed germination under salt stress. During root inhibition by salt stress, ethylene regulates the ABA concentration. However, the reverse has been reported in the case of seed germination [259]. Salicylic acid has displayed a defensive role in plants experiencing stress, signaling the plant to adapt to the stressful environment [260].

Salinity and Growth Regulation

Brassinosteroids (BRs) are growth regulators which mitigate adverse growth patterns caused by salinity. It improves the germination of seeds experiencing salt stress. The improved germination rate has been reported for rice [261] and tobacco. Application of BRs as a seed treatment enhanced the growth of rice seedlings under salt stress [262]. It helps the plant to retain its green pigments and enhances nitrate reductase activity [263–265] and nitrogen-fixing capability. Brassinosteroids (28-homoBL) increased dry matter accumulation and seed yield [266]. Foliar application of brassinosteroids (24-epibrassinolide) on pepper plants grown with saline water greatly affected shoot growth parameters and leaf water contents as compared to roots. However, its effect on chlorophyll fluorescence was non-significant [267,268]. Similar patterns of brassinosteroidal effects were observed in wheat grown under salt stress. 24-epibrassinolide application on salt-stressed wheat seedlings exhibited non-significant results in terms of plant biomass, chlorophyll content, photosynthesis rate, substomatal CO_2 concentration, and water use efficiency. The incremented water use efficiency can be related to higher transpiration rate shown by salt-stressed wheat seedlings, as a result of 24-epibrassinolide application [269,270]. The efficiency of exogenously applied brassinosteroids to mitigate salinity effects varies with plant species, appropriate growth stage, dose, frequency, and method of brassinosteroidal application [271–273]. The results of BR application also vary with climatic conditions—mainly temperature, light duration, and applied fertilizers [274]. Brassinolide application to salt stressed *Vigna radiata* caused enhancement in growth, photosynthetic rate, and maximum quantum yield of PSII. Generally, brassinolide has the potential to protect the photosynthetic apparatus under salt stress. It also contributed to improving the membrane stability index and leaf water potential. However, no significant results were recorded in the case of lipid peroxidation and electrolyte leakage. Brassinolide increases antioxidant enzyme and proline contents [275–278]. Increases in proline level create a protective shield when the plant is under stress by acting as a source of carbon and nitrogen, a stabilizer of the plasma membrane, and an oxygen radical scavenger [279]. Brassinosteroids also increase pigment levels in the plant [280]. BRs also improve the nitrate and nitrite reductase activity in *Vigna radiata* under salt stress. This effect can be attributed to the ability of BRs to modulate transcription and translation at the gene level, and to increase cell nitrate uptake. Increased stress tolerance caused by brassinolide application is observable as improved growth parameters such as shoot length, root length, and plant biomass [106,281].

6. Salinity and Carbohydrate (Sugars) Metabolism in Plants

Salinity stress progressively depletes carbohydrates in plant leaves and roots [282]. Young leaves have more hexose and starch accumulation compared to older leaves. Carbon metabolism can be used to assess the salt tolerance of a plant species. Salt-tolerant genotypes accumulated more sucrose than salt-sensitive genotypes. The role of carbohydrates in osmotic adjustment has also been authenticated [231,283]. Previous studies have reported a significant correlation between sugar accumulation levels and stress tolerance in various plant species [284]. This augmentation in sugar contents may be related to the high rate of sugar hydrolysis via hydrolytic enzymes [127,285]. Sugars stabilize the plasma membrane during plant stress by interacting with phospholipids. Under salt stress, a higher accumulation of carbohydrates in leaves may contribute to osmotic adjustment [122,286]. The higher chloride contents can cause increases in carbohydrate levels in plant tissues or starch degradation in sensitive species, whereas salt-tolerant species have less starch accumulation [287]. The carbohydrate accumulation rate varies among salt-tolerant species [147,288].

7. Salinity and Root Apoplastic Barriers

Plant roots are the main organ involved in the uptake of water and nutrients from a solution, whereas xylem vessels allow nutrient transport to the aerial tissues. In addition, plant roots function as the primary site for sensing salinity levels so that the plant can respond rapidly to maintain functionality. Roots can exclude and/or counteract potentially harmful substances by modifying their anatomy [289]. In particular, the endodermis that separates the cortex from the central cylinder is characterized by the development of specific wall modifications, called "apoplastic barriers" [290], formed by a combination of Casparian strips and suberin lamellae. A significant anatomical change in the root system due to salt stress is the deposition of hydrophobic polymers such as cutin and suberin on the cell wall, polymers that are often associated with hydrophobic compounds (e.g., waxes). Rossi et al. [291] reported that different apoplastic adjustments in roots modify Na^+ fluxes to the shoots of olive trees exposed to up to 120 mM NaCl. Similarly, Krishnamurthy et al. [292] showed that the Na^+ bypass flow in rice roots was reduced by the deposition of apoplastic barriers. These findings substantiated the role of root apoplastic barriers in plants' tolerance to salt stress. Afterwards, several studies have confirmed the formation of plant root apoplastic barriers as a response to different environmental stresses such as heavy metals [268], salt, and drought stress [291]. Overall, the literature indicates that plants react to environmental constraints by developing apoplastic barriers close to the root apex to mitigate the intrusion of toxic ions. This is a specific anatomical response by roots when exposed to hostile environments.

8. Salinity and Ionic Attributes

Abiotic stress alters the patterns of nutrient availability and transport, thus causing enormous changes in plant growth. Na^+ and Cl^- ions are the chief competitors which restrict nutrients like Ca^{2+}, K^+, and NO_3^-. The increase in uptake of Na^+ and Cl^- was more prominent in mature leaves compared to young actively growing leaves [59,293]. Plants experience a deficiency of both macro and micronutrients under salt stress. However, the extent to which salinity affects micronutrient availability is determined by plant type, growing conditions, and nutrient concentration [294,295]. Hence an effective fertilization regime in salt-affected areas can be utilized to overcome the negative effects of salinity [296,297]. The increase in nutrient availability to plants enhances the plant's ability to survive under stressed conditions [1].

8.1. Salinity and Ionic Homeostasis

Salinity impairs the ionic balance of the cells. Maintaining ionic balance in plants is important for increasing plant survivability under salt stress. Plants maintain ionic balance by various processes, including cellular uptake, sequestration, and ion inclusion and exclusion [8,298]. Early maturing clones of diploid potato accumulated Na ions in their leaves while late maturing varieties excluded Na ions from their leaves (Figure 5). Late maturing varieties of potato were more salt-tolerant compared to early cultivars. The lower leaves of the potato plant accumulated more Na ions than the higher leaves in both the tested potato cultivars. Late maturing genotypes also established higher K to Na ratio, conferring salt tolerance to them [50,298]. Saline conditions alter the ionic concentration of soil solutions, with an increase in Na^+ [74] and Cl^- ions, particularly [54,299]. With the increased absorption rate of Na and chloride ions, a significant decline in other ions (e.g., K, Ca, and Mg) occurs [54,297,300].

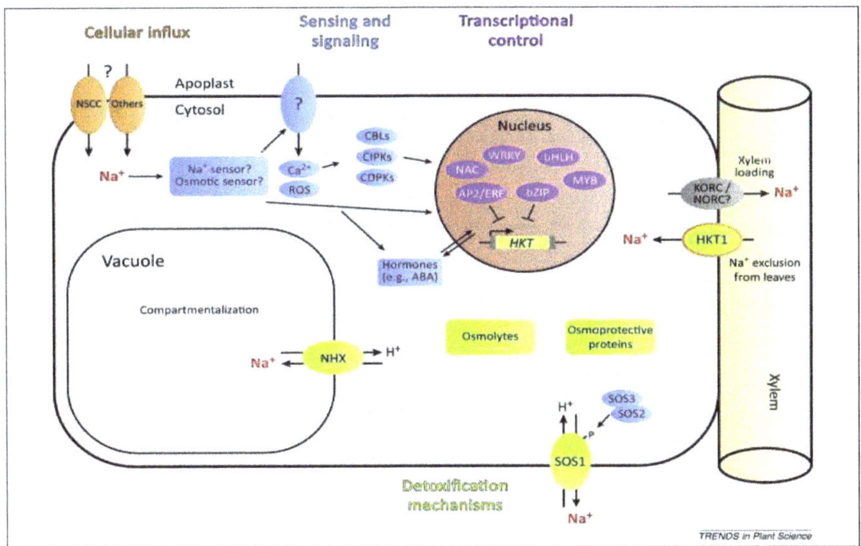

Figure 5. Overview of cellular Na$^+$ transport mechanisms and the salt stress response network in plant root cells. Na$^+$ (depicted in red) enters the cell via NSCCs and other membrane transporters (cellular Na$^+$-influx mechanisms highlighted in orange). Inside the cell, Na$^+$ is identified. This activates Ca^{2+}, ROS, and hormone signaling cascades. CBLs, CIPKs, and CDPKs are part of the Ca^{2+}-signaling pathway (sensing and signaling components highlighted in blue), which can alter the global transcriptional profile of the plant (transcription factor families in the nucleus depicted in purple; an AP2/ERF and a bZIP transcription factor that negatively regulate *HKT* gene expression are shown as an example). Ultimately, these early signaling pathways result in expression and activation of cellular detoxification mechanisms, including HKT, NHX, and the SOS Na$^+$ transport mechanisms, as well as osmotic protection strategies (cellular detoxification mechanisms highlighted in light green). Furthermore, the Na$^+$ distribution in the plant is regulated in a tissue-specific manner by unloading of Na$^+$ from the xylem. Abbreviations: NSCCs, nonselective cation channels; ROS, reactive oxygen species; CDPKs, calcium-dependent protein kinases; CBLs, calcineurin B-like proteins; CIPKs, CBL-interacting protein kinases; AP2/ERF, APETALA2/ETHYLENE RESPONSE FACTOR; bZIP, basic leucine zipper; NHX, Na$^+$/H$^+$ exchanger; SOS, salt overly sensitive [301].

Sodium is not considered an essential element for plant growth. Plants do not show any Na requirements, and they lack any particular transport system for Na. However, when the plant is exposed to high Na concentrations, Na finds its way into the plant cells following various pathways. It may gain entry into plant cell passively from soil solution osmolarity, or via voltage-dependent and independent cation channels [8]. Due to the similarity of hydrated ionic radii for Na$^+$ and K$^+$, Na$^+$ hampers K$^+$ absorption by the plant because transport proteins cannot distinguish between them. At higher a cytolic Na$^+$/K$^+$ concentration ratio, cellular processes are affected, which would otherwise be maintained by K$^+$, such as protein synthesis and enzyme activity [302]. Reduced K uptake can cause reductions in plant growth and productivity under saline conditions. Plants restrict the free cytosolic movement of Na in the cytosol by vacuole compartmentalization, in order to prevent Na from hampering the regular functioning of cytosolic enzymes. This mechanism is equally important for both glycophytes and halophytes [303]. Salt tolerance of a plant species is dependent on its ability to restrict translocation of toxic ions in shoots [304,305]. This ability is regulated by specific tissues [17], morphological features [306–308], and water use efficiency. These adaptive mechanisms alter the plant response to salinity at both cellular and entire plant levels [17,285,309].

8.2. Ionic Influx and Efflux

Sodium and chloride ions are the major ions which induce harmful effects on plant growth. The accumulation of these ions in leaves and roots determines the tolerance of a genotype to salt stress [310]. Roots are the primary organ which permits the entry of toxic ions into the plant under high saline conditions. Plant roots regulate Na and chloride contents via the extrusion mechanism. Plants respond to toxic salt levels either by exclusion to soil or by upward movement of ions through the xylem transpiration stream [311]. Phloem is a chief pathway for toxic ion transport from shoots to roots [17,278]. A passive mode of Na transport from external medium into the cell involves uniporters or ion channel type transporters like HKT, LCT1, and NSCC [45,312]. In roots, Na reaches the xylem using symplastic and apoplastic routes from the epidermis. The rate of Na transport is dependent on the barriers posed by the casparian strip and the Na extrusion from the cell. Sodium ion transport is also mediated by high-affinity potassium transporters which act as Na^+/K^+ symporters and also as Na^+-selective uniporters [313].

High salt concentration causes an increase in electrolyte leakage [106]. Sodium ion accumulation at the cellular level is a major consequence of saline environmental conditions. When Na ion levels are high, Na ions are either extruded or compartmentalized in vacuoles via Na^+/H^+ antiporters [45,314]. Ion transporters like Na^+/H^+ antiporters are involved in Na^+ ion extrusion or Na^+ ion compartmentalization in vacuoles. These antiporters may be plasma membrane-localized or tonoplast antiporters, which utilize the pH gradient, developed by P-type H^+-ATPases, respectively [7,8]. Electrochemical potential channels are the other source of ion transport across the cell membrane. These channels are ion specific and regulate ionic movement through gating, the period during which channels are open or closed [315].

Sodium extrusion and vacuolar compartmentalization are active (energy requiring) processes compared to sodium influx (passive transport). K^+ transporters are considered responsible for sodium influx into the cell. Movement of Na^+ ions from the cytosol to the vacuole is accomplished by Na^+/H^+ antiporters [316]. The plant's capacity to tolerate saline conditions depends on a high K^+/Na^+ cytosolic concentration. Plants adopt various strategies to cope with a high Na ion concentration, such as the restriction of Na^+ entry into the cell, Na^+ extrusion, and vacuole compartmentalization [279]. Sodium compartmentalized in the vacuole is an osmolytic process which provides a mechanism for water uptake. Sodium efflux is considered an adaptation of salt-sensitive plants, whereas Na compartmentalization is considered a feature of salt-tolerant species. The high salinity tolerance of halophytes can also be attributed to their ability to confine Na ions to their roots, making them Na accumulators [45,317]. Sodium exclusion from the plant body may also involve salt glands, present on the leaf surface [159].

Chloride also contributes to the undesirable effects of salt stress on plant growth. Salt-tolerant plant genotypes have the ability to inhibit chloride uptake. Chloride uptake is dependent on the shoot to root ratio and follows a passive transport system [286,318]. Higher chloride contents also induce succulence in salt-stressed plants [298], as well as a reduction in nitrate reductase activity [319]. Nitrate reductase activity inhibition is an indicator of poor nitrogen assimilation, leading to reduced protein synthesis and plant growth [320].

Adjusting the osmotic potential of the cell by K accumulation in vacuoles through different K channels and transporters appears to be a prominent plant strategy to cope with plant stress [321,322]. Potassium alters the membrane potential and turgor, maintains enzyme activities, and adjusts osmotic pressure and stomatal movement. It aids the plant in photosynthesis, protein synthesis, and oxidant metabolism [323–325].

9. Effects of Salinity on Potassium and Calcium

Potassium regulates protein synthesis, enzymatic metabolism, and photosynthesis. Low K levels were recorded in potato seedlings experiencing salt stress [31,73,326]. Potassium competes with Na under saline conditions. In saline soil, Na ions which reach the plasma membrane of a cell cause the membrane to depolarize and the K outward rectifier channels to open. This results in a loss of K from the cell [327]. Importantly, the K^+:Na^+ ratio determines the saline resistance of plants. Higher K^+ concentration is directly related to higher biomass production. Salt-tolerant plant species have the ability to retain a higher concentration of K [72,159].

Potassium and Ca occupy a critical position in the regulation of cell membrane integrity and function [328]. Calcium aids in various physiological processes, including solute movement, stomatal regulation, molecular signaling for cell defense systems, and cell repair under stress. Under saline conditions, Na^+/Ca^{2+} interactions are important because Na^+ has the ability to displace Ca^{2+} from its binding sites, thus causing a decline in Ca^{2+} availability. The cytosolic Ca concentration determines the salt sensitivity of the plant [8,329]. Calcium deficiency is evident in plants grown in saline soil conditions [330]. Calcium supplementation is beneficial in ameliorating the negative effects of salinity in beans [331,332]. The ability of plants to survive under osmotic stress is dependent on the plant's capacity to maintain high Ca^{2+}:Na^+ ratio and to exclude Na^+. Increasing Ca^{+2} (by addition of Ca^{2+} as gypsum $CaSO_4$) had an antagonistic effect on Mg^{2+} availability to the plant, as it removed Mg^{2+} from the soil complex [72,333].

Salinization management through Ca^{+2} supplementation is reported to be beneficial in enhancing the quality of celery by reducing the incidence of black heart. Calcium supplementation in the form of calcium sulfate improved the growth of tomato plants. This stabilized membrane permeability, increasing N and K concentrations in leaves [67]. Calcium sulfate has more pronounced effects on tomato growth parameters than calcium chloride, perhaps because calcium chloride can be a source of chloride ions [188]. The supplementation of K and Ca^{2+} to the salt-affected pepper plants increased vegetative growth and fruit production and decreased the incidence of blossom end rot, but caused a reduction in fresh fruit weight and marketable yield [4]. Calcium acetate positively stimulates photosynthesis and stomatal conductance. It also ameliorates the effects of low water content on osmotic potential from salt stress. However, Ca^{2+} is reported to have an inhibitory effect on proline accumulation, which affects the low osmotic potential [34,334].

Salinity and Nitrogen

Nitrogen is required in considerable quantities to satisfy the mineral needs of the plant. Moreover, nitrogen is a constituent of amino acids and nucleic acids [311]. Salt stress decreased protein contents in potatoes [335]. Nitrogen has a large effect on plant growth when there is ample water. Still, addition of nitrogen to plants, even under salt stress conditions, improves the yield in many crop plants, including tomato, millet, and wheat. Nevertheless, in a saline environment, plants have a decreased ability to uptake nitrogen [1,31]. Plants that are more susceptible to Cl^- toxicity can be managed through NO_3^- application since NO_3^- antagonizes Cl^- uptake [336]. Saline conditions reduce the regulation of nitrate uptake, metabolism, and utilization by plant species. The form of nitrogen applied affects the uptake of other nutrients like Na^+, Ca^{2+}, and K^+, and thus affects the plant's ability to tolerate saline conditions. Therefore, under saline conditions, a nitrogen fertilizer regime should be managed specifically considering the interactions between Na, NO_3, NH_4, Cl^-, etc. [337]. Potassium nitrate is a salt stress alleviating agent for melons [31].

Higher salt concentration in soil solutions interferes with the transport of nitrate in the shoots. For that reason, nitrate reductase, glutamine synthetase, and nitrite reductase enzymes all have reduced activity under stress conditions [246,338]. Nitrate reductase is responsible for the reduction of nitrate to ammonia. Nitrate reductase activity is generally less susceptible to salinity than nitrite reductase activity [339]. Increasing salinity levels decreased dry weight and protein contents of the leaves and roots of tomato seedlings. High Na and chloride ion levels suppressed the uptake of K and nitrate [340].

Nitrogen-containing compounds (NCC) accumulate in plants as a response to higher saline conditions. These compounds include amino acids, amides, quaternary ammonium compounds, and polyamines. Their reported functions are osmoprotection, osmotic adjustment, ROS scavenging, nitrogen provision, and maintenance of pH. Plant nitrogen metabolism can be regulated by phytohormone like cytokinin [29,341].

Phosphorus (P) is a prime constituent in nucleic acids, phospholipids, phosphoproteins, nucleotides, and ATP. Phosphorus uptake is reduced in plants growing under saline conditions. The availability of P is reduced due to strong ionic effects and low solubility of calcium phosphate minerals [302]. Navarro et al. [342] reported a decline in phosphorus mobility stored in vacuoles in salt-stressed melon plants. Plant growth promoting bacteria increased uptake of P and K in tomato plants grown under salt stress. P uptake may enhance the survivability of young plants in salt stress conditions [343–346].

10. Conclusions

Plants respond to salinity stress at physiological, cellular, genetic, and metabolic levels. Previous research demonstrates that among plant responses to salinity, mechanisms that control ion uptake, transport, and balance, as well as water potential, photosynthesis, cell division, osmotic adjustment, enzymatic activities, polyamine regulation, stress signaling, and regulation of root apoplastic barriers play critical roles in plant tolerance to salinity. In order to manage salt stress, significant work has been done on calcium biosensors to understand cation-sensing processes, ABA-dependent phosphorylation, changes in cell wall components, auxin and ABA associated modulations in root architecture, Na^+ exclusion mechanisms, signaling of phytohormones in roots and guard cells, and organ-specific expression of sodium transporter gene (*HKT1*). There is a need to integrate information from genomic, transcriptomic, proteomic, and metabolomics studies, as a collaborative approach for determining key pathways controlling salinity tolerance at the whole-plant level. Further studies on tissue-specific Na-sensing processes, association of Na and K biosensors with Na^+/K^+ homeostasis, ABA and other phytohormone signaling pathways, and interaction between phytohormones and ion transporters are required to illuminate the details of inter- and intracellular molecular interactions that are involved in plant stress tolerance. Further research in these areas will be helpful for mitigating salinity damage in commercially important crops.

Author Contributions: Conceived of the presented idea, M.A.S., A.S. and N.K.; developed the theory and performed the computations, M.A.S., N.K., R.M.B. and L.R.; writing—original draft preparation, M.A.S., A.S., N.K., S.A., C.G., N.M., W.N. and F.G.-S.; writing—review and editing, M.A.S., A.S., N.K. and S.A.; supervision, M.A.S. and N.K.; verified the numerical results, R.M.B., S.A., L.R., C.G., N.M., W.N. and F.G.-S. All authors have read and agreed to the published version of the manuscript.

Funding: This research received no external funding.

Conflicts of Interest: The authors confirm that there is no known conflict of interest associated with this work.

References

1. Shrivastava, P.; Kumar, R. Soil salinity: A serious environmental issue and plant growth promoting bacteria as one of the tools for its alleviation. *Saudi J. Biol. Sci.* **2015**, *22*, 123–131. [CrossRef]

2. Rogel, J.A.; Ariza, F.A.; Silla, R.O. Soil salinity and moisture gradients and plant zonation in Mediterranean salt marshes of Southeast Spain. *Wetlands* **2000**, *20*, 357–372. [CrossRef]

3. Flowers, T.J.; Garcia, A.; Koyama, M.; Yeo, A.R. Breeding for salt tolerance in crop plants—The role of molecular biology. *Acta Physiol. Plant.* **1997**, *19*, 427–433. [CrossRef]

4. Rubio, J.S.; Garcia-Sanchez, F.; Rubio, F.; Martinez, V. Yield, blossom-end rot incidence and fruit quality in pepper plants under moderate salinity are affected by K^+ and Ca^{+2} fertilization. *Sci. Hortic.* **2009**, *119*, 79–87. [CrossRef]

5. Ghassemi, F.; Jakeman, A.J.; Nix, H.A. *Salinization of Land and Water Resources*; University of New south Wales Press Ltd.: Canberra, Australia, 1995.

6. Sonowal, H.; Pal, P.B.; Shukla, K.; Ramana, K.V. Ramana, Aspalatone Prevents VEGF-Induced Lipid Peroxidation, Migration, Tube Formation, and Dysfunction of Human Aortic Endothelial Cells. *Oxid. Med. Cell. Longev.* **2017**, *11*, 2017. [CrossRef]

7. Wang, H.; Tang, X.; Shao, C.; Shao, H.; Wang, L. Molecular cloning and bioinformatics analysis of a new plasma membrane Na+/H+ antiporter gene from the halophyte Kosteletzkya virginica. *Sci. World J.* **2014**, *2014*, 141675.

8. Xiong, L.; Zhu, J.K. Molecular and genetic aspects of plant responses to osmotic stress. *Plant Cell Environ.* **2002**, *25*, 131–139. [CrossRef] [PubMed]

9. Meng, X.; Zhou, J.; Sui, N. Mechanisms of salt tolerance in halophytes: Current understanding and recent advances. *Open Llife Sci.* **2018**, *13*, 149–154. [CrossRef]

10. Ashraf, M.H.P.J.C.; Harris, P.J. Photosynthesis under stressful environments: An overview. *Photosynthetica* **2013**, *51*, 163–190. [CrossRef]

11. Siew, R.; Klein, C.R. The effect of NaCl on some metabolic and fine structural changes during the greening of etiolated leaves. *J. Cell Biol.* **1969**, *37*, 590–596. [CrossRef]

12. Slama, I.; Abdelly, C.; Bouchereau, A.; Flowers, T.; Savoure, A. Diversity, distribution and roles of osmoprotective compounds accumulated in halophytes under abiotic stress. *Ann. Bot.* **2015**, *115*, 433–447. [CrossRef] [PubMed]

13. Cushman, J.C. Molecular cloning and expression of chloroplast NADP-malate dehydrogenase during crassulacean acid metabolism induction by salt stress. *Photosynth. Res.* **1990**, *35*, 15–27. [CrossRef]

14. Sévin, D.C.; Stählin, J.N.; Pollak, G.R.; Kuehne, A.; Sauer, U. Global Metabolic Responses to Salt Stress in Fifteen Species. *PLoS ONE* **2016**, *11*, e0148888. [CrossRef] [PubMed]

15. Dschida, W.J.; Platt-Aloia, K.A.; Thomson, W.W. Epidermal peels of *Avicennia germinans* (L.) Stearn: A useful system to study the function of salt glands. *Ann. Bot. Lond.* **1992**, *70*, 501–509. [CrossRef]

16. Robert, I.L.; Frank, W.J.; Summy, K.R.; Hudson, D.Y.; Richard, S. The Biological Flora of Coastal Dunes and Wetlands. *Avicennia Germinans J. Coast. Res.* **2017**, *33*, 191–207.

17. Tester, M.; Davenport, R. Na$^+$ tolerance and Na$^+$ transport in higher plants. *Ann. Bot.* **2003**, *91*, 503–527. [CrossRef] [PubMed]

18. Shahid, M.A.; Balal, R.M.; Khan, N.; Rossi, L.; Rathinasabapathi, B.; Liu, G.; Khan, J.; Cámara-Zapata, J.M.; Martínez-Nicolas, J.J.; Garcia-Sanchez, F. Polyamines provide new insights into the biochemical basis of Cr-tolerance in Kinnow mandarin grafted on diploid and double-diploid rootstocks. *Environ. Exp. Bot.* **2018**, *1*, 248–260. [CrossRef]

19. Zhu, J.; Hasegawa, P.M.; Bressan, R.A. Molecular aspects of osmotic stress in plants. *Citri. Rev. Plant Sci.* **1997**, *16*, 253–277. [CrossRef]

20. Serrano, R.; Mulet, J.M.; Rios, G.; Marquez, J.A.; de Larrinoa, I.F.; Leube, M.P.; Mendizabal, I.; Pascual-Ahuir, A.; Proft, M.; Ros, R.; et al. A glimpse of the mechanism of ion homeostasis during salt stress. *J. Exp. Bot.* **1999**, *50*, 1023–1036. [CrossRef]

21. Flowers, T.J.; Munns, R.; Colmer, T.D. Sodium chloride toxicity and the cellular basis of salt tolerance in halophytes. *Ann. Bot.* **2015**, *115*, 419–431. [CrossRef]

22. Khan, N.; Bano, A.; Babar, M.A. The stimulatory effects of plant growth promoting rhizobacteria and plant growth regulators on wheat physiology grown in sandy soil. *Arch. Microbiol.* **2019**, *201*, 769–785. [CrossRef] [PubMed]

23. Price, A.; Hendry, G. Iron catalysed oxygen radical formation and its possible contribution to drought damage in nine native grasses and three cereals. *Plant Cell Environ.* **1991**, *14*, 477–484. [CrossRef]

24. Khan, N.; Bano, A. Effects of exogenously applied salicylic acid and putrescine alone and in combination with rhizobacteria on the phytoremediation of heavy metals and chickpea growth in sandy soil. *Int. J. Phytoremed.* **2018**, *16*, 405–414. [CrossRef] [PubMed]

25. Stevens, R.M.; Pech, J.M.; Grigson, G.J. A Short Non Saline Sprinkling Increases the Tuber Weights of Saline Sprinkler Irrigated Potatoes. *Agronomy* **2017**, *7*, 4. [CrossRef]

26. Nxele, X.; Klein, A.; Ndimba, B.K. Drought and salinity stress alters ROS accumulation, water retention, and osmolyte content in sorghum plants. *Afr. J. Bot.* **2017**, *1*, 261–266. [CrossRef]

27. Golldack, D.; Li, C.; Mohan, H.; Probst, N. Tolerance to drought and salt stress in plants: Unraveling the signaling networks. *Front. Plant Sci.* **2014**, *5*, 151. [CrossRef]

28. Naseem, H.; Ahsan, M.; Shahid, M.A.; Khan, N. Exopolysaccharides producing rhizobacteria and their role in plant growth and drought tolerance. *J. Basic Microbiol.* **2018**, *58*, 1009–1022. [CrossRef]

29. Zekri, M.; Parsons, L.R. Calcium influnces growth and leaf mineral concentration of citrus under saline conditions. *Hort. Sci.* **1990**, *25*, 784–786. [CrossRef]

30. Kumari, A.; Das, P.; Parida, A.K.; Agarwal, P.K. Proteomics, metabolomics, and ionomics perspectives of salinity tolerance in halophytes. *Front. Plant Sci.* **2015**, *6*, 537. [CrossRef]

31. Saha, J.; Brauer, E.K.; Sengupta, A.; Popescu, S.C.; Gupta, K.; Gupta, B. Polyamines as redox homeostasis regulators during salt stress in plants. *Front. Environ. Sci.* **2015**, *3*, 21. [CrossRef]

32. Mansour, M.M.F. Cell permeability under salt stress. In *Strategies for Improving Salt Tolerance in Higher Plants*; Jaiwal, P.K., Singh, R.P., Gulati, A., Eds.; Oxford and IBH: New Delhi, India, 1997; pp. 87–110.

33. Ali, S.; Xie, L. Plant Growth Promoting and Stress mitigating abilities of Soil Born Microorganisms. *Recent Pat. Food Nutr. Agric.* **2019**, *15*. [CrossRef] [PubMed]

34. Mansour, M.M.F.; Salama, K.H.A. Cellular basis of salinity tolerance in plants. *Environ. Exp. Bot.* **2004**, *52*, 113–122. [CrossRef]

35. Le Gall, H.; Philippe, F.; Domon, J.M.; Gillet, F.; Pelloux, J.; Rayon, C. Cell wall metabolism in response to abiotic stress. *Plants* **2015**, *4*, 112–166. [CrossRef] [PubMed]

36. Al-Amri, S.M. Improved growth, productivity and quality of tomato (*Solanum lycopersicum* L.) plants through application of shikimic acid. *Saudi J. Biol. Sci.* **2013**, *1*, 339–345. [CrossRef] [PubMed]

37. Mahlooji, M.; Seyed Sharifi, R.; Razmjoo, J.; Sabzalian, M.R.; Sedghi, M. Effect of salt stress on photosynthesis and physiological parameters of three contrasting barley genotypes. *Photosynthetica* **2017**, *56*, 549–556. [CrossRef]

38. Shekari, F.; Abbasi, A.; Mustafavi, S.H. Effect of silicon and selenium on enzymatic changes and productivity of dill in saline condition. *J. Saudi Soc. Agric. Sci.* **2017**, *1*, 367–374. [CrossRef]

39. Dolatabadian, A.; Sanavy, S.A.M.M.; Gholamhoseini, M.; Joghan, A.K.; Majdi, M.; Kashkooli, A. The role of calcium in improving photosynthesis and related physiological and biochemical attributes of spring wheat subjected to simulated acid rain. *Physiol. Mol. Biol. Plants.* **2013**, *19*, 189–198. [CrossRef]

40. Farooq, S.; Azam, F. The use of cell membrane stability (CMS) technique to screen for salt tolerant wheat varieties. *J. Plant Physiol.* **2006**, *163*, 629–637. [CrossRef]

41. Munns, R. Comparative physiology of salt and water stress. *Plant Cell Environ.* **2002**, *25*, 239–250. [CrossRef]

42. Dolatabadi, N.; Toorchi, M. Rapeseed (*Brassica napus* L.) genotypes response to NaCl salinity. *J. Biodiv. Environ. Sci.* **2017**, *10*, 265–270.

43. Wu, J.; Seliskar, D.M.; Gallagher, J.L. Stress tolerance in the march plant Spartina patens: Impact of NaCl on growth and root plasma membrane lipid composition. *Physiol. Plant* **1998**, *102*, 307–317. [CrossRef]

44. Balal, R.M.; Shahid, M.A.; Vincent, C.; Zotarelli, L.; Liu, G.; Mattson, N.S.; Garcia-Sanchez, F. (Kinnow mandarin plants grafted on tetraploid rootstocks are more tolerant to Cr-toxicity than those grafted on its diploids one. *Environ. Exp. Bot.* **2017**, *140*, 8–18. [CrossRef]

45. Plant, A.L.; Bray, E.A. *Regulation of Gene Expression by Abscissic Acid During Environmental Stresses*; Marcel Dekker: New York, NY, USA, 1999; pp. 303–331.

46. Delphine, S.; Alvino, A.; Zacchini, M.; Loreb, F. Consequences of salt stress on conductance to CO_2 diffusion, Rubisco characteristics and anatomy of spinach leaves. *Aust. J. Plant Physiol.* **1998**, *25*, 395–402. [CrossRef]

47. Mehmood, A.; Khan, N.; Irshad, M.; Hamayun, M. IAA Producing Endopytic Fungus Fusariun oxysporum wlw Colonize Maize Roots and Promoted Maize Growth Under Hydroponic Condition. *Eur. Exp. Biol.* **2018**, *8*, 24. [CrossRef]

48. Romero-Aranda, R.; Moya, J.L.; Tadeo, F.R.; Legaz, F.; Primo-Millo, E.; Talón, M. Physiological and anatomical disturbances induced by chloride salts in sensitive and tolerant citrus: Beneficial and detrimental effects of cations. *Plant Cell Environ.* **1998**, *21*, 1243–1253. [CrossRef]

49. Hasegawa, P.M.; Bressan, R.A.; Zhu, J.K.; Bohnert, H.J. Plant cellular and molecular responses to high salinity. *Annu. Rev. Plant Physiol.* **2000**, *51*, 463–499. [CrossRef]

50. Blumwald, E.; Aharon, G.S.; Apse, M.P. Sodium transport in plant cells. *Biochim. Biophys. Acta* **2000**, *1465*, 140–151. [CrossRef]

51. Maser, P.; Gierth, M.; Schroeder, J.I. Molecular mechanisms of potassium and sodium uptake in plants. *Plant Soil.* **2002**, *247*, 43–54. [CrossRef]

52. Khan, N.; Bano, A. Modulation of phytoremediation and plant growth by the treatment with PGPR, Ag nanoparticle and untreated municipal wastewater. *Int. J. Phytoremed.* **2016**, *18*, 1258–1269. [CrossRef]

53. Zribi, L.; Fatma, G.; Fatma, R.; Salwa, R.; Hassan, N.; Nejib, R.M. Application of chlorophyll fluorescence for the diagnosis of salt stress in tomato "*Solanum lycopersicum* (variety Rio Grande)". *Sci. Hortic.* **2009**, *120*, 367–372. [CrossRef]

54. Orlovsky, N.; Japakova, U.; Zhang, H.; Volis, S. Effect of salinity on seed germination, growth and ion content in dimorphic seeds of *Salicornia europaea* L. (Chenopodiaceae). *Plant Divers.* **2016**, *1*, 183–189. [CrossRef] [PubMed]

55. Shaterian, J.; Waterer, D.; De Jong, H.; Tanino, K.K. Differential stress responses to NaCl salt application in early and late maturing diploid potato (*Solanum* sp.) clones. *Environ. Exp. Bot* **2005**, *54*, 202–212. [CrossRef]

56. Chartzoulakis, K.; Klapaki, G. Response of two greenhouse pepper hybrids to NaCl salinity during different growth stages. *Sci. Hortic.* **2000**, *86*, 247–260. [CrossRef]

57. Kaya, C.; Tuna, A.L.; Ashrafa, M.; Altunlu, H. Improved salt tolerance of melon (*Cucumis melo* L.) by the addition of proline and potassium nitrate. *Environ. Exp. Bot* **2007**, *60*, 397–403. [CrossRef]

58. Giuffrida, F.; Graziani, G.; Fogliano, V.; Scuderi, D.; Romano, D.; Leonardi, C. Effects of nutrient and NaCl salinity on growth, yield, quality and composition of pepper grown in soilless closed system. *J. Plant Nutr.* **2014**, *37*, 1455–1474. [CrossRef]

59. Jiang, C.; Zu, C.; Lu, D.; Zheng, Q.; Shen, J.; Wang, H.; Li, D. Effect of exogenous selenium supply on photosynthesis, Na+ accumulation and antioxidative capacity of maize (*Zea mays* L.) under salinity stress. *Sci. Rep.* **2007**, *7*, 42039. [CrossRef]

60. Savvas, D.; Lenz, F. Effect of NaCl or nutrient-induced salinity on growth, yield and composition of egg plants grown in rockwool. *Sci. Hortic.* **2000**, *84*, 37–47. [CrossRef]

61. Abbas, W.; Ashraf, M.; Akram, N.A. Alleviation of salt-induced adverse effects in egg plant (*Solanum melongena* L.) by glycinebetaine and sugar beet extracts. *Sci. Hortic.* **2010**, *125*, 188–195. [CrossRef]

62. Mahjoor, F.; Ghaemi, A.A.; Golabi, M.H. Interaction effects of water salinity and hydroponic growth medium on eggplant yield, water-use efficiency, and evapotranspiration. *Int. Soil Water Conserv.* **2016**, *1*, 99–107. [CrossRef]

63. Balal, R.M.; Shahid, M.A.; Javaid, M.M.; Iqbal, Z.; Liu, G.D.; Zotarelli, L.; Khan, N. Chitosan alleviates phytotoxicity caused by boron through augmented polyamine metabolism and antioxidant activities and reduced boron concentration in *Cucumis sativus* L. Acta Physiol. Plant. **2017**, *1*, 31. [CrossRef]

64. Tabatabaie, S.J.; Nazari, J. Influence of nutrient concentrations and NaCl salinity on the growth, photosynthesis, and essential oil content of peppermint and lemon verbena. *Turk. J. Agric. For.* **2007**, *31*, 245–253.

65. Pardossi, A.; Bagnoli, G.; Malorgio, F.C.; Campiotti, A.; Tognoni, F. NaCl effects on celery (*Apium graveolens* L.) grown in NFT. *Sci. Hortic* **1999**, *81*, 229–242. [CrossRef]

66. Zheng, Y.; Jia, A.; Ning, T.; Xu, J.; Li, Z.; Jiang, G. Potassium nitrate application alleviates sodium chloride stress in winter wheat cultivars differing in salt tolerance. *J. Plant Physiol.* **2008**, *165*, 1455–1465. [CrossRef] [PubMed]

67. Saini, S.; Sharma, I.; Pati, P.K. Versatile roles of brassinosteroid in plants in the context of its homoeostasis, signaling and crosstalks. *Front. Plant Sci.* **2015**, *6*, 950. [CrossRef] [PubMed]

68. Kalhoro, N.A.; Rajpar, I.; Kalhoro, S.A.; Ali, A.; Raza, S.; Ahmed, M.; Kalhoro, F.A.; Ramzan, M.; Wahid, F. Effect of salts stress on the growth and yield of wheat (*Triticum aestivum* L.). *Am. J. Plant Sci.* **2016**, *7*, 2257. [CrossRef]

69. Xie, Z.; Duan, L.; Tian, X.; Wang, B.; Eneji, A.E.; Li, Z. Coronatine alleviates salinity stress in cotton by improving the antioxidation defence system and radical-scavenging activity. *J. Plant Physiol.* **2008**, *165*, 375–384. [CrossRef]

70. Sun, C.; Feng, D.; Mi, Z.; Li, C.; Zhang, J.; Gao, Y.; Sun, J. Impacts of Ridge-Furrow Planting on Salt Stress and Cotton Yield under Drip Irrigation. *Water* **2017**, *9*, 49. [CrossRef]

71. Caines, A.M.; Shennan, C. Interactive effects of Ca^{+2} and NaCl salinity on the growth of two tomato genotypes differing in Ca^{+2} use efficiency. *Plant Physiol. Biochem.* **1999**, *37*, 569–576. [CrossRef]

72. Evers, D.; Hemmer, K.; Hausman, J.F. Salt stress induced biometric and physiological changes in *Solanum tuberosum* L. cv. Bintje grown in vitro. *Acta Physiol. Plant* **1998**, *20*, 3–7. [CrossRef]

73. Gao, Y.; Lu, Y.; Wu, M.; Liang, E.; Li, Y.; Zhang, D.; Yin, Z.; Ren, X.; Dai, Y.; Deng, D.; et al. Ability to Remove Na$^+$ and Retain K$^+$ Correlates with Salt Tolerance in Two Maize Inbred Lines Seedlings. *Front. Plant Sci* **2016**, *7*, 1716. [CrossRef]

74. Hu, Y.; Xia, S.; Su, Y.; Wang, H.; Luo, W.; Su, S.; Xiao, L. Brassinolide Increases Potato Root Growth In Vitro in a Dose-Dependent Way and Alleviates Salinity Stress. *Biomed. Res. Int.* **2016**, 8231873. [CrossRef]

75. Alom, R.; Hasan, M.A.; Islam, M.R.; Wang, Q.F. Germination characters and early seedling growth of wheat (*Triticum aestivum* L.) genotypes under salt stress conditions. *JCSB* **2016**, *1*, 383–392. [CrossRef]

76. Tuna, A.L.; Kaya, C.; Ashraf, M.; Altunlu, H.; Yokas, I.; Yagmur, B. The effect of calcium sulphate on growth, membrane stability and nutrient uptake of tomato plants grown under salt stress. *Environ. Exp. Bot.* **2007**, *59*, 173–178. [CrossRef]

77. De Pascale, S.; Barbieri, G. Effects of soil salinity and top removal on growth and yield of broad bean as a green vegetable. *Sci. Hortic* **1997**, *71*, 147–165. [CrossRef]

78. Verma, S.; Mishra, S.N. Putrescine alleviation of growth in salt stressed *Brassica juncea* by inducing antioxidative defense system. *J. Plant Physiol.* **2005**, *162*, 669–677. [CrossRef] [PubMed]

79. Hu, Y.; Schmidhalter, U. Drought and Salinity: A comparison of their effects on mineral nutrition of plants. *J. Plant Nutr. Soil. Sci.* **2005**, *168*, 541–549. [CrossRef]

80. Rahnama, H.; Ebrahimzadeh, H. The effect of NaCl on antioxidant activities in potato seedlings. *Biol. Plant.* **2005**, *49*, 93–97. [CrossRef]

81. Korkmaz, A.; Şirikçi, R.; Kocaçınar, F.; Değer, Ö.; Demirkırıan, A.R. Alleviation of salt-induced adverse effects in pepper seedlings by seed application of glycinebetaine. *Sci. Hortic.* **2012**, *4*, 197–205. [CrossRef]

82. Roy, T.S.; Chakraborty, R.; Parvez, M.N.; Mostofa, M.; Ferdous, J.; Ahmed, S. Yield, dry matter and specific gravity of exportable potato: Response to salt. *Univ. J. Agric. Res.* **2017**, *5*, 98–103.

83. Mulholland, B.J.; Taylor, I.B.; Jackson, A.C.; Thornpson, A.J. Can ABA mediate responses of salinity have stressed tomato. *Environ. Exp. Bot.* **2003**, *50*, 17–28. [CrossRef]

84. Mayak, S.; Tirosh, T.; Glick, B.R. Plant growth promoting bacteria confer resistance in tomato plants to salt stress. *Plant Sci. Plant Sci.* **2004**, *42*, 565–572.

85. Smalle, J.; Van Der Straeten, D. Ethylene and vegetative development. *Physiol. Plant.* **1997**, *100*, 593–605. [CrossRef]

86. Akhtar, S.S.; AkhtarAktas, H.; Abak, K.; Cakmak, I. Genotypic variation in the response of pepper to salinity. *Sci. Hortic.* **2006**, *110*, 260–266.

87. Delgado, M.J.; Garrido, J.M.; Ligero, F.; Lluch, C. Nitrogen fixation and carbon metabolism by nodules and bacteroids of pea plants under sodium chloride stress. *Physiol. Plant.* **1993**, *89*, 824–829. [CrossRef]

88. Borucki, W.; Sujkowska, M. The effects of sodium chloride-salinity upon growth, nodulation, and root nodule structure of pea (*Pisum sativum* L.) plants. *Acta Physiol. Plant.* **2008**, *30*, 293–301. [CrossRef]

89. Andrzej, T.; Philip, P. Role of root microbiota in plant productivity. *J. Exp. Bot.* **2015**, *66*, 2167–2175.

90. Zahran, H.H. Rhizobium-Legume symbiosis and nitrogen fixation under severe conditions and in an arid climate. *Microbiol. Mol. Biol. R.* **1999**, *63*, 968–989. [CrossRef]

91. Batista-Santos, P.; Duro, N.; Rodrigues, A.P.; Semedo, J.N.; Alves, P.; da Costa, M.; Graça, I.; Pais, I.P.; Scotti-Campos, P.; Lidon, F.C.; et al. Is salt stress tolerance in *Casuarina glauca* Sieb. ex Spreng. associated with its nitrogen-fixing root-nodule symbiosis? An analysis at the photosynthetic level. *Plant Physiol. Biochem.* **2015**, *1*, 97–109. [CrossRef]

92. Othman, Y.; Al-karaki, G.; Al-Tawaha, A.R.; Al-Horani, A. Variation in germination and ion uptake in barley genotypes under salinity conditions. *World J. Agric. Res.* **2006**, *2*, 11–15.

93. Kandil, A.A.; Shareif, E.; Gad, M.A. Effect of Salinity on Germination and Seeding Parameters of Forage Cowpea Seed. *Res. J. Seed Sci.* **2017**, *10*, 17–26.

94. Khan, M.A.; Rizvi, Y. Effect of salinity, temperature and growth regulators on the germination and early seedling growth of *Atriplex griffithi* Var. Stocksii. *Can J. Bot.* **1994**, *72*, 475–479. [CrossRef]

95. Rajjou, L.; Duval, M.; Gallardo, K.; Catusse, J.; Bally, J.; Job, C.; Job, D. Seed germination and vigor. *Annu. Rev. Plant Biol.* **2012**, *2*, 507–533. [CrossRef]

96. Meena, K.K.; Sorty, A.M.; Bitla, U.M.; Choudhary, K.; Gupta, P.; Pareek, A.; Minhas, P.S. Abiotic Stress Responses and Microbe-Mediated Mitigation in Plants: The Omics Strategies. *Front. Plant Sci.* **2017**, *8*, 172. [CrossRef]

97. Stepien, P.; Klobus, G. Water relations and photosynthesis in *Cucumus sativus* L. leaves under salt stress. *Biol. Plant.* **2006**, *50*, 610–616. [CrossRef]

98. Dobrota, C. Energy dependant plant stress acclimation. *Rev. Environ. Sci.* **2006**, *5*, 243–251. [CrossRef]

99. Yeo, A.R.; Capron, S.J.M.; Flowers, T.J. The effect of salinity upon photosynthesis in rice (*Oryza sativa* L.) Gas exchange by individual leaves relation to their salt content. *J. Exp. Bot.* **1985**, *36*, 1240–1248. [CrossRef]

100. Muradoglu, F.; Gundogdu, M.; Ercisli, S.; Encu, T.; Balta, F.; Jaafar, H.Z.; Zia-Ul-Haq, M. Cadmium toxicity affects chlorophyll a and b content, antioxidant enzyme activities and mineral nutrient accumulation in strawberry. *Biol. Res.* **2015**, *48*, 1–7. [CrossRef]

101. El-Shintinawy, F.; El-Ansary, A. Differential Effect of Cd2+ and Ni2+ on Amino Acid Metabolism in Soybean Seedlings. *Biol. Plant.* **2000**, *43*, 79–84. [CrossRef]

102. Puniran-Hartley, N.; Hartley, J.; Shabala, L.; Shabala, S. Salinity-induced accumulation of organic osmolytes in barley and wheat leaves correlates with increased oxidative stress tolerance: In planta evidence for cross-tolerance. *Plant Physiol. Biochem.* **2014**, *1*, 32–39. [CrossRef] [PubMed]

103. Garcia-Legaz, M.F.; Ortiz, J.M.; Garcia-Lindon, A.; Cerda, A. Effect of salinity on growth, ion content and CO_2 assimilation rate in lemon varieties on different rootstocks. *Physiol. Plant.* **1993**, *89*, 427–432. [CrossRef]

104. Acosta-Motos, J.R.; Ortuño, M.F.; Bernal-Vicente, A.; Diaz-Vivancos, P.; Sanchez-Blanco, M.J.; Hernandez, J.A. Plant responses to salt stress: Adaptive mechanisms. *Agronomy* **2017**, *7*, 18. [CrossRef]

105. Hanachi, S.; Van Labeke, M.C.; Mehouachi, T. Application of chlorophyll fluorescence to screen eggplant (*Solanum melongena* L.) cultivars for salt tolerance. *Photosynthetica* **2014**, *52*, 57–62. [CrossRef]

106. Tiwari, J.K.; Munshi, A.D.; Kumar, R.; Pandey, R.N.; Arora, A.; Bhat, J.S.; Sureja, A.K. Effect of salt stress on cucumber: Na+–K+ ratio, osmolyte concentration, phenols and chlorophyll content. *Acta Physiol. Plant.* **2010**, *1*, 103–114. [CrossRef]

107. Lycoskoufis, I.H.; Savvas, D.; Mavrogianopoulos, G. Growth, gas exchange, and nutrient status in pepper (*Capsicum annuum* L.) grown in recirculating nutrient solution as affected by salinity imposed to half of the root system. *Sci. Hortic.* **2005**, *106*, 147–161. [CrossRef]

108. Zhang, T.; Gong, H.; Wen, X.; Lu, C. Salt stress induces a decrease in oxidation energy transfer from phycobilisomes to photosystem II but an increase to photosystem I in the cyanobacterium Spirulina platensis. *J. Plant Physiol.* **2010**, *12*, 20.

109. Wu, X.; He, J.; Chen, J.; Yang, S.; Zha, D. Alleviation of exogenous 6-benzyladenine on two genotypes of eggplant (*Solanum melongena* Mill.) growth under salt stress. *Protoplasma* **2014**, *251*, 169–176. [CrossRef] [PubMed]

110. Yan, K.; Shao, H.; Shao, C.; Chen, P.; Zhao, S.; Brestic, M.; Chen, X. Physiological adaptive mechanisms of plants grown in saline soil and implications for sustainable saline agriculture in coastal zone. *Acta Physiol. Plant.* **2013**, *1*, 2867–2878. [CrossRef]

111. Hayat, S.; Hasan, S.A.; Yusuf, M.; Hayat, Q.; Ahmad, A. Effect of 28- homobrassinolide on photosynthesis, fluorescence and antioxidant system in the presence or absence of salinity and temperature in *Vigna radiata*. *Environ. Exp. Bot.* **2010**, *69*, 105–112. [CrossRef]

112. Misra, A.N.; Sahu, S.M.; Misra, M.; Singh, P.; Meera, I.; Das, N.; Kar, M.; Shau, P. Sodium chloride induced changes in leaf growth, and pigment and protein contents in two rice cultivars. *Biol. Plant.* **1997**, *39*, 257–262. [CrossRef]

113. Neelesh, K.; Veena, P. Effect of Salt Stress on Growth Parameters, Moisture Content, Relative Water Content and Photosynthetic Pigments of Fenugreek Variety RMt-1. *J. Plant Sci.* **2015**, *10*, 210–221. [CrossRef]

114. Bethke, P.C.; Drew, M.C. Stomatal and non-stomatal components to inhibition of photosynthesis in leaves of *Capsicum annum* during progressive exposure to NaCl salinity. *Plant Physiol.* **1992**, *99*, 219–226. [CrossRef] [PubMed]

115. Reddy, M.P.; Vora, A.B. Changes in pigment composition, Hill reaction activity and saccharides metabolism in Bajra (*Pennisetum typhoides*) leaves under NaCl salinity. *Photosynthetica* **1986**, *20*, 50–55.

116. Gul, B.; Ansari, R.; Flowers, T.J.; Khan, M.A. Germination strategies of halophyte seeds under salinity. *Environ. Exp. Bot.* **2013**, *1*, 4–18. [CrossRef]

117. Parida, A.K.; Das, A.B. Salt tolerance and salinity affects on plant. *Ecotoxicol. Environ.* **2005**, *60*, 324–349. [CrossRef] [PubMed]

118. Cao, Y.Y.; Duan, H.; Yang, L.N.; Wang, Z.Q.; Zhou, S.C.; Yang, J.C. Effect of heat stress during meiosis on grain yield of rice cultivars differing in heat tolerance and its physiological mechanism. *Acta Agron. Sin.* **2008**, *34*, 2134–2142. [CrossRef]

119. Elhamid, E.; Sadak, M.; Tawfik, M. Alleviation of Adverse Effects of Salt Stress in Wheat Cultivars by Foliar Treatment with Antioxidant 2—Changes in Some Biochemical Aspects, Lipid Peroxidation, Antioxidant Enzymes and Amino Acid Contents. *Agric. Sci.* **2014**, *5*, 1269–1280. [CrossRef]

120. Khan, N.; Zandi, P.; Ali, S.; Mehmood, A.; Adnan Shahid, M.; Yang, J. Impact of salicylic acid and PGPR on the drought tolerance and phytoremediation potential of Helianthus annus. *Front. Microbiol.* **2018**, *9*, 2507. [CrossRef]

121. Kartashov, A.V.; Radyukina, N.L.; Ivanov, Y.V.; Pashkovskii, P.P.; Shevyakova, N.I.; Kuznetsov, V.V. Role of antioxidants systems in wild plant adaptation to salt stress. *Russ. J. Plant Physiol.* **2008**, *55*, 463–468. [CrossRef]

122. Katsumi, M. Physiological modes of brassinolide action in cucumber hypocotyl growth. In *Brassinosteriods: Chemistry, Bioactivity and Applications*; Cutler, H.G., Yokota, T., Adam, G., Eds.; American Chemical Society: Washington, DC, USA, 1991; pp. 246–254.

123. Faried, H.N.; Ayyub, C.M.; Amjad, M.; Ahmed, R. Salinity impairs ionic, physiological and biochemical attributes in potato. *Pak. J. Agric. Sci.* **2016**, *53*, 17–25.

124. Sofo, A.; Scopa, A.; Nuzzaci, M.; Vitti, A. Ascorbate Peroxidase and Catalase Activities and Their Genetic Regulation in Plants Subjected to Drought and Salinity Stresses. *Int. J. Mol. Sci.* **2015**, *16*, 13561–13578. [CrossRef] [PubMed]

125. Gupta, B.; Huang, B. Mechanism of salinity tolerance in plants: Physiological, biochemical, and molecular characterization. *Int. J. Genom.* **2014**, *2014*. [CrossRef]

126. Bhattacharyya, A.; Chattopadhyay, R.; Mitra, S.; Crowe, S.E. Oxidative Stress: An Essential Factor in the Pathogenesis of Gastrointestinal Mucosal Diseases. *Physiol. Rev.* **2014**, *94*, 329–354. [CrossRef] [PubMed]

127. Ashraf, M.; Bhatti, B. Physiological responses of some salt tolerance and salt sensitive lines of wheat (*Crthamus tinctorius* L.). *Acta Physiol. Plant.* **1999**, *117*, 23–28.

128. Adnan, A.; Erkan, Y. Effect of salinity stress on chlorophyll, carotenoid content, and proline in *Salicornia prostrata* Pall. and *Suaeda prostrata* Pall. subsp. *prostrata* (Amaranthaceae). *Braz. J. Bot.* **2016**, *39*, 101–106.

129. Kamal, A.; Qureshi, M.S.; Ashraf, M.Y.; Hussain, M. Salinity induced some changes in growth and physio-chemical aspects of two soybean [*Glycine max* (L.) merr] genotype. *Pak. J. Bot.* **2003**, *35*, 93–97.

130. Parvin, K.; Haque, M.N. Protective Role of Salicylic Acid on Salt Affected Broccoli Plants. *J. Agric. Ecol.* **2017**, *10*, 1–10. [CrossRef]

131. Stepien, P.; Klobus, G. Antioxidant defense in the leaves of C_3 and C_4 plants under salinity stress. *Physiol. Plant.* **2005**, *125*, 31–40. [CrossRef]

132. Chandrasekaran, M.; Kim, K.; Krishnamoorthy, R.; Walitang, D.; Sundaram, S.; Joe, M.M.; Selvakumar, G.; Hu, S.; Oh, S.H.; Sa, T. Mycorrhizal symbiotic efficiency on C3 and C4 plants under salinity stress—A meta-analysis. *Front. Microbiol.* **2016**, *11*, 1246. [CrossRef] [PubMed]

133. Akbar, S.; Sultan, S. Soil bacteria showing a potential of chlorpyrifos degradation and plant growth enhancement. *Braz. J. Microbiol.* **2016**, *47*, 563–570. [CrossRef]

134. Bhantana, P.; Lazarovitch, N. Evapotranspiration, Crop coefficient and growth of two young pomegranate (*Punica granatum* L.) varieties under salt stress. *Agric. Water Manag.* **2010**, *97*, 715–722. [CrossRef]

135. Gorai, M.; Ennajeh, M.; Khemira, H.; Neffati, M. Combined effect of NaCl-salinity and hypoxia on growth, photosynthesis, water relations and solute accumulation in *Phragmite saustralis* plants. *Flora* **2010**, *205*, 462–470. [CrossRef]

136. Baby, T.; Collins, C.; Tyerman, S.D.; Gilliham, M. Salinity negatively affects pollen tube growth and fruit set in grapevines and is not mitigated by silicon. *Am. J. Enol. Vitic.* **2016**, *67*, 218–228. [CrossRef]

137. Van Zelm, E.; Zhang, Y.; Testerink, C. Salt tolerance mechanisms of plants. *Ann. Rev. Plant Biol.* **2020**, *71*, 403–433. [CrossRef] [PubMed]

138. Aldesuquy, H.; Baka, Z.; Mickky, B. Kinetin and spermine mediated induction of salt tolerance in wheat plants: Leaf area, photosynthesis and chloroplast ultrastructure of flag leaf at ear emergence. *Egypt. J. Basic Appl. Sci.* **2014**, *1*, 77–87. [CrossRef]

139. Szalai, G.; Janda, K.; Darkó, É.; Janda, T.; Peeva, V.; Pál, M. Comparative analysis of polyamine metabolism in wheat and maize plants. *Plant Physiol. Biochem.* **2017**, *1*, 239–250. [CrossRef] [PubMed]

140. Syvertsen, J.; Levy, Y. *Salinity Interactions with Other Abiotic and Biotic Stresses in Citrus*; University of Florida, Indian River Research and Education Center: Ft. Pierce, FL, USA, 2005.

141. Negrão, S.; Schmöckel, S.M.; Tester, M. Evaluating physiological responses of plants to salinity stress. *Ann. Bot.* **2017**, *119*, 1–11. [CrossRef] [PubMed]

142. Moir-Barnetson, L.; Veneklaas, E.J.; Colmer, T.D. Salinity tolerances of three succulent halophytes (*Tecticornia* spp.) differentially distributed along a salinity gradient. *Funct. Plant Biol.* **2016**, *4*, 739–750. [CrossRef]

143. Khan, N.; Bano, A.; Babar, M.A. The root growth of wheat plants, the water conservation and fertility status of sandy soils influenced by plant growth promoting rhizobacteria. *Symbiosis* **2017**, *72*, 195–205. [CrossRef]

144. Bohnert, H.J.; Shen, B. Transformation and compatible solutes. *Sci. Hortic.* **1999**, *78*, 237–260. [CrossRef]

145. Shuyskaya, E.V.; Rakhamkulova, Z.F.; Lebedeva, M.P.; Kolesnikov, A.V.; Safarova, A.; Borisochkina, T.I.; Toderich, K.N. Different mechanisms of ion homeostasis are dominant in the recretohalophyte Tamarix ramosissima under different soil salinity. *Acta Physiol. Plant.* **2017**, *1*, 81. [CrossRef]

146. Hernandez, J.A.; Almansa, M.S. Short term effects of salt stress on antioxidant systems and leaf water relations of pea leaves. *Physiol. Plant.* **2002**, *115*, 251–257. [CrossRef] [PubMed]

147. Khan, N.; Ali, S.; Zandi, P.; Mehmood, A.; Ullah, S.; Ikram, M.; Ismail, M.; Babar, M.A. Role of sugars, amino acids and organic acids in improving plant abiotic stress tolerance. *Pak. J. Bot.* **1997**, *52*, 2. [CrossRef]

148. Hong, B.; Barg, R.; Ho, T.D. Developmental and organ specific expression of an ABA and stress induced protein in barley. *Plant Mol. Biol.* **1992**, *18*, 663–674. [CrossRef]

149. Hare, P.D.; Cress, W.A.; Staden, J.V. Dissecting the roles of osmolyte accumulation during stress. *Plant Cell Environ.* **1998**, *21*, 535–554. [CrossRef]

150. Galinski, E.A. Compatible solutes of halophilic eubacteria-eubacteria-molecular principles, water-solute interactions, stress protection. *Cell Mol. Life Sci.* **1993**, *49*, 487–496. [CrossRef]

151. Lutts, S.; Lefèvre, I. How can we take advantage of halophyte properties to cope with heavy metal toxicity in salt-affected areas? *Ann. Bot.* **2015**, *1*, 509–528. [CrossRef]

152. Brown, A.D. *Microbial Water Stress Physiology, Principles and Perspectives*; Wiley: Hoboken, NJ, USA, 1990.

153. Solomon, A.; Beer, S.; Waisel, Y.; Jones, G.P.; Paleg, L.G. Effects of NaCl on the Carboxylating activity of Rubisco from *Tamarix jordanis* in the presence and absence of proline-related compatible solutes. *Plant Physiol.* **1994**, *90*, 198–204. [CrossRef]

154. Papageorgiou, G.; Murata, N. The unusually strong stability effects of glycine betaine on the structure and function of the oxygen evolving photosystem II complex. *Photosynth. Res.* **1995**, *44*, 243–252. [CrossRef]

155. Bose, J.; Rodrigo-Moreno, A.; Shabala, S. ROS homeostasis in halophytes in the context of salinity stress tolerance. *J. Exp. Bot.* **2014**, *1*, 1241–1257. [CrossRef]

156. Stoop, J.M.H.; Williamson, J.D.; Pharr, D.M. Mannitol metabolism in plants: A method for coping with stress. *Trends Plant Sci.* **1996**, *1*, 139–144. [CrossRef]

157. Moran, J.F.; Becana, M.; Iturbe-Ormaetxe, I.; Frechilla, S.; Klucas, R.V.; Aparicio-Tejo, P. Drought induces oxidative stress in pea plants. *Planta* **1994**, *94*, 346–352. [CrossRef]

158. Managbanag, J.R.; Torzilli, A.P. An analysis of trehalose, glycerol, and mannitol accumulation during heat and salt stress in a salt marsh isolate of *Aureobasidium pullulans*. *Mycologia* **2002**, *1*, 384–391. [CrossRef]

159. Khan, N.; Bano, A.; Zandi, P. Effects of exogenously applied plant growth regulators in combination with PGPR on the physiology and root growth of chickpea (*Cicer arietinum*) and their role in drought tolerance. *J. Plant Interact.* **2018**, *1*, 239–247. [CrossRef]

160. Tipirdamaz, R.; Gagneul, D.; Duhazé, C.; Aïnouche, A.; Monnier, C.; Özkum, D.; Larher, F. Clustering of halophytes from an inland salt marsh in Turkey according to their ability to accumulate sodium and nitrogenous osmolytes. *Environ. Exp. Bot.* **2006**, *57*, 139–153. [CrossRef]

161. Ashraf, M.; Harris, P.J.C. Potential biochemical indicators of salinity tolerance in plants. *Plant Sci.* **2004**, *166*, 3–16. [CrossRef]

162. Sandhu, D.; Cornacchione, M.V.; Ferreira, J.F.; Suarez, D.L. Variable salinity responses of 12 alfalfa genotypes and comparative expression analyses of salt-response genes. *Sci. Rep.* **2017**, *22*, 42958. [CrossRef]

163. Liang, X.; Zhang, L.; Natarajan, S.K.; Becker, D.F. Proline Mechanisms of Stress Survival. *Antioxid. Redox Signal.* **2013**, *19*, 998–1011. [CrossRef]

164. Dar, M.I.; Naikoo, M.I.; Rehman, F.; Naushin, F.; Khan, F.A. Proline accumulation in plants: Roles in stress tolerance and plant development. In *Osmolytes and Plants Acclimation to Changing Environment: Emerging Omics Technologies*; Springer: New Delhi, India, 2016; pp. 155–166.

165. Ali, B.; Hayat, S.; Ahmad, A. 28-homobrassinolide ameliorates the saline stress in *Cicer arietinum* L. *Environ. Exp. Bot.* **2007**, *59*, 217–223. [CrossRef]

166. Mantilla, B.S.; Marchese, L.; Casas-Sánchez, A.; Dyer, N.A.; Ejeh, N.; Biran, M.; Silber, A.M. Proline Metabolism is Essential for Trypanosoma brucei brucei Survival in the Tsetse Vector. *PLoS Pathog.* **2017**, *13*, e1006158. [CrossRef]

167. Kahlaoui, B.; Hachicha, M.; Misle, E.; Fidalgo, F.; Teixeira, J. Physiological and biochemical responses to the exogenous application of proline of tomato plants irrigated with saline water. *J. Saudi Soc. Agric. Sci.* **2018**, *1*, 17–23. [CrossRef]

168. Nemoto, Y.; Sasakuma, T. Differential stress responses of early salt-stress responding genes in common wheat. *Phytochemistry* **2002**, *61*, 129–133. [CrossRef]

169. Jaarsma, R.; de Vries, R.S.; de Boer, A.H. Effect of salt stress on growth, Na+ accumulation and proline metabolism in potato (*Solanum tuberosum*) cultivars. *PLoS ONE* **2013**, *8*, e60183. [CrossRef] [PubMed]

170. Joseph, E.A.; Radhakrishnan, V.V.; Mohanan, K.V. A Study on the Accumulation of Proline— An Osmoprotectant Amino Acid under Salt Stress in Some Native Rice Cultivars of North Kerala, India. *Univ. J. Agric. Res.* **2015**, *3*, 15–22. [CrossRef]

171. Knight, H.; Anthony, J.; Knight, M.R. Calcium signaling in *Arabidopsis thaliana* responding to drought and salinity. *Plant J.* **1997**, *12*, 1067–1078. [CrossRef]

172. Misra, N.; Gupta, A.K. Interactive effects of sodium and calcium on proline metabolism in salt tolerant green gram cultivar. *Am. J. Plant Physiol.* **2006**, *1*, 1–2.

173. Ashraf, M.; Foolad, M.R. Roles of glycinebetaine and proline in improving plant abiotic stress resistance. *Environ. Exp. Bot.* **2007**, *59*, 206–216. [CrossRef]

174. Huang, Z.; Zhao, L.; Chen, D.; Liang, M.; Liu, Z.; Shao, H.; Long, X. Salt Stress Encourages Proline Accumulation by Regulating Proline Biosynthesis and Degradation in Jerusalem Artichoke Plantlets. *PLoS ONE* **2013**, *8*, e62085. [CrossRef]

175. Lutts, S.; Majerus, V.; Kinet, J.M. NaCl effects on proline metabolism in rice (*Oryza sativa*) seedlings. *Int. Rice Res.* **1999**, *25*, 39–40. [CrossRef]

176. Devnarain, N.; Crampton, B.G.; Chikwamba, R.K.; Becker, J.V.W.; O'Kennedy, M.M. Physiological responses of selected African sorghum landraces to progressive water stress and re-watering. *S. Afr. J.* **2016**, *103*, 61–69. [CrossRef]

177. Yan, H.; Gang, L.Z.; Zhao, C.Y.; Guo, W.Y. Effects of exogenous proline on the physiology of soybean plantlets regenerated from embryo in vitro and on the ultra-structure of their mitochondria under NaCl stress. *Soybean Sci.* **2000**, *19*, 314–319.

178. Hare, P.D.; Cress, W.A.; Staden, J.V. Disruptive effects of exogenous proline on chloroplast and mitochondrial ultrastructure in *Arabidopsis* leaves. *S. Afr. J.* **2002**, *68*, 393–396. [CrossRef]

179. Zeyner, A.; Romanowski, K.; Vernunft, A.; Harris, P.; Müller, A.M.; Wolf, C.; Kienzle, E. Effects of Different Oral Doses of Sodium Chloride on the Basal Acid-Base and Mineral Status of Exercising Horses Fed Low Amounts of Hay. *PLoS ONE* **2017**, *12*, e0168325. [CrossRef] [PubMed]

180. Jimenez-Bremont, J.F.; Becerra-Flora, A.; Hernandez-Lucero, E.; Rodriguez-Kessler, M.; Acosta-Gallegos, J.A.; Ramirez-Pimental, J.G. Proline accumulation in two bean cultivars under salt stress and the effect of polyamines and ornithine. *Biol. Plant.* **2006**, *50*, 763–766. [CrossRef]

181. Gill, S.S.; Tuteja, N. Polyamines and abiotic stress tolerance in plants. *Plant Signal Behav.* **2010**, *5*, 26–33. [CrossRef] [PubMed]

182. Griffith, S.M.; Banowetz, G.M. Nitrogen nutrition and flowering. In *Nitrogen Nutrition in Higher Plants*; Srivastava, H.S., Singh, R.P., Eds.; Associated Publ. Co.: New Delhi, India, 1995; pp. 385–400.

183. Kumar, R.; Khurana, A.; Sharma, A.K. Role of plant hormones and their interplay in development and ripening of fleshy fruits. *J. Exp. Bot.* **2013**, *65*, 4561–4575. [CrossRef]

184. Diao, Q.; Song, Y.; Shi, D.; Qi, H. Interaction of Polyamines, Abscisic Acid, Nitric Oxide, and Hydrogen Peroxide under Chilling Stress in Tomato (Lycopersicon esculentum Mill.) Seedlings. *Front. Plant Sci.* **2017**, *8*, 203. [CrossRef] [PubMed]

185. Shabala, S.; Cuin, T.A.; Pottosin, I. Polyamines prevent NaCl-induced K^+ efflux from pea mesophyll by blocking non-selective cation channels. *FEBS Lett.* **2007**, *581*, 1993–1999. [CrossRef]

186. Rodríguez, A.A.; Maiale, S.J.; Menéndez, A.B.; Ruiz, O.A. Polyamine oxidase activity contributes to sustain maize leaf elongation under saline stress. *J. Exp. Bot.* **2009**, *1*, 4249–4262. [CrossRef]

187. Mishra, S.N.; Sharma, I. Putrescine as a growth inducer and as a source of nitrogen for mustard seedlings under sodium chloride salinity. *Indian J. Exp. Biol.* **1994**, *32*, 916–918.

188. Lakra, N.; Tomar, P.C.; Mishra, S.N. Growth response modulation by putrescine in Indian mustard *Brassica juncea* L. under multiple stress. *Indian J. Exp. Biol.* **2016**, *54*, 262–270.

189. Maruri-López, I.; Jiménez-Bremont, J. Hetero- and homodimerization of Arabidopsis thaliana arginine decarboxylase AtADC1 and AtADC2. *Biochem. Biophys. Res. Commun.* **2017**, *11*, 508–513. [CrossRef] [PubMed]

190. Wani, S.H.; Singh, N.B.; Haribhushan, A.; Mir, J.I. Compatible solute engineering in plants for abiotic stress tolerance-role of glycine betaine. *Curr. Genom.* **2013**, *14*, 157. [CrossRef] [PubMed]

191. Baloda, A.; Madanpotra, S.; Jaiwal, P.K. Transformation of mung bean plants for abiotic stress tolerance by introducing codA gene, for an osmoprotectant glycine betaine. *J. Plant Stress Physiol.* **2017**, *3*, 5–11. [CrossRef]

192. Wei, D.; Zhang, W.; Wang, C.; Meng, Q.; Li, G.; Chen, T.H.; Yang, X. Genetic engineering of the biosynthesis of glycinebetaine leads to alleviate salt-induced potassium efflux and enhances salt tolerance in tomato plants. *Plant Sci.* **2017**, *257*, 74–83. [CrossRef] [PubMed]

193. Park, E.J.; Jeknic, Z.; Chen, T.H.H. Exogenous application of glycine betaine increases chilling tolerance in tomato plants. *Plant Cell Physiol.* **2006**, *47*, 706–714. [CrossRef] [PubMed]

194. Park, H.J.; Kim, W.Y.; Yun, D.J. A New Insight of Salt Stress Signaling in Plant. *Mol. Cells* **2016**, *39*, 447–459. [CrossRef] [PubMed]

195. Khan, N.; Bano, A. Rhizobacteria and Abiotic Stress Management. In *Plant Growth Promoting Rhizobacteria for Sustainable Stress Management*; Springer: Singapore, 2019; pp. 65–80.

196. Mohammad, N.; Naeem, K.; Zakia, A.; Abdul, G. Genetic diversity and disease response of rust in bread wheat collected from Waziristan Agency, Pakistan. *Int. J. Biodivers. Conserv.* **2011**, *3*, 10–18.

197. Shanker, A.; Venkateswarlu, B. *Abiotic Stress in Plants: Mechanisms and Adaptations. BoD–Books on Demand*; IntechOpen: Norderstedt, Germany, 2011; p. 22.

198. Nawaz, K.; Ashraf, M. Exogenous application of glycine betaine modulates activities of antioxidants in maize plants subjected to salt stress. *J. Agron. Crop Sci.* **2009**, *196*, 28–37. [CrossRef]

199. Annunziata, M.G.; Ciarmiello, L.F.; Woodrow, P.; Maximova, E.; Fuggi, A.; Carillo, P. Durum Wheat Roots Adapt to Salinity Remodeling the Cellular Content of Nitrogen Metabolites and Sucrose. *Front. Plant Sci.* **2016**, *7*, 2035. [CrossRef]

200. Wyn-Jones, R.G.; Gorham, J.; McDonnell, E. Organic and inorganic solute contents as selection criteria for salt tolerance in the Triticeae. In *Salinity Tolerance in Plants: Strategies for Crop Improvement*; Staples, R., Toennissen, G.H., Eds.; Wiley and Sons: New York, NY, USA, 1984; pp. 189–203.

201. Woodrow, P.; Ciarmiello, L.F.; Annunziata, M.G.; Pacifico, S.; Iannuzzi, F.; Mirto, A.; Carillo, P. Durum wheat seedling responses to simultaneous high light and salinity involve a fine reconfiguration of amino acids and carbohydrate metabolism. *Plant Physiol.* **2016**. [CrossRef]

202. Majeed, K.; Al-Hamzawi, A. Effect of Calcium Nitrate, Potassium Nitrate and Anfaton on Growth and Storability of Plastic Houses Cucumber (*Cucumis sativus* L. cv. Al-Hytham). *Am. J. Plant Sci.* **2010**, *5*, 278–290.

203. Kong, X.; Luo, Z.; Dong, H.; Eneji, A.E.; Li, W. Effects of non-uniform root zone salinity on water use, Na+ recirculation, and Na+ and H+ flux in cotton. *J. Exp. Bot.* **2012**, *63*, 2105–2116. [CrossRef] [PubMed]

204. Wu, H.; Shabala, L.; Liu, X.; Azzarello, E.; Zhou, M.; Pandolfi, C.; Shabala, S. Linking salinity stress tolerance with tissue-specific Na+ sequestration in wheat roots. *Front. Plant Sci.* **2015**, *6*. [CrossRef]

205. Maleka, P.; Kontturi, M.; Pehu, B.; Somersalo, S. Photosynthesis response of drought and salt stressed tomato and turnip plants to foliar applied glycinebetaine. *Physiol. Plant.* **1999**, *105*, 45–50.

206. Athar, H.U.R.; Zafar, Z.U.; Ashraf, M. Glycinebetaine Improved Photosynthesis in Canola under Salt Stress: Evaluation of Chlorophyll Fluorescence Parameters as Potential Indicators. *J. Agron. Crop Sci.* **2015**, *201*, 428–442. [CrossRef]

207. Luo, J.Y.; Zhang, S.; Peng, J.; Zhu, X.Z.; Lv, L.M.; Wang, C.Y.; Cui, J.J. Effects of Soil Salinity on the Expression of Bt Toxin (Cry1Ac) and the Control Efficiency of Helicoverpa armigera in Field- Grown Transgenic Bt Cotton. *PLoS ONE* **2017**, *12*, e0170379. [CrossRef]

208. Raza, M.A.S.; Shahid, A.M.; Saleem, M.F.; Khan, I.H.; Ahmad, S.; Ali, M.; Iqbal, R. Effects and management strategies to mitigate drought stress in oilseed rape (*Brassica napus* L.): A review. *Zemdirbyste-Agriculture* **2017**, *104*, 85–94. [CrossRef]

209. Akram, R.; Fahad, S.; Hashmi, M.Z.; Wahid, A.; Adnan, M.; Mubeen, M.; Khan, N.; Rehmani, M.I.; Awais, M.; Abbas, M.; et al. Trends of electronic waste pollution and its impact on the global environment and ecosystem *Environ. Sci. Pollut. Res. Int.* **2019**, *1*, 1–6 [CrossRef]

210. Bhuiyan, M.S.; Maynard, G.; Raman, A.; Hodgkins, D.; Mitchell, D.; Nicol, H. Salt effects on proline and glycine betaine levels and photosynthetic performance in Melilotus siculus, Tecticornia pergranulata and Thinopyrum ponticum measured in simulated saline conditions. *Funct. Plant Biol.* **2016**, *4*, 254–265. [CrossRef]

211. Demiral, T.; Türkan, I. Does exogenous glycinebetaine affect antioxidative system of rice seedlings under NaCl treatment? *J. Plant Physiol.* **2004**, *18*, 1089–1100. [CrossRef] [PubMed]

212. Talaat, N.B.; Shawky, B.T. Influence of arbuscular mycorrhizae on yield, nutrients, organic solutes, and antioxidant enzymes of two wheat cultivars under salt stress. *J. Soil Sci. Plant Nutr.* **2011**, *174*, 283–291. [CrossRef]

213. Ruiz-Lozano, J.M.; Porcel, R.; Azcón, C.; Aroca, R. Regulation by arbuscular mycorrhizae of the integrated physiological response to salinity in plants: New challenges in physiological and molecular studies. *J. Exp. Bot.* **2012**, *28*, 4033–4044. [CrossRef]

214. Ghosh, B.; Ali, M.N.; Saikat, G. Response of Rice under Salinity Stress: A Review Update. *J. Res. Rice* **2016**, *4*, 1–8. [CrossRef]

215. Mack, G.; Hoffmann, C.M.; Marlander, B. Nitrogen compounds in organs of two sugar beet genotypes (*Beta vulgaris* L.) during the season. *Field Crop. Res.* **2007**, *102*, 210–218. [CrossRef]

216. Khan, N.; Bano, A.; Rahman, M.A.; Guo, J.; Kang, Z.; Babar, M.A. Comparative physiological and metabolic analysis reveals a complex mechanism involved in drought tolerance in chickpea (*Cicer arietinum* L.) induced by PGPR and PGRs. *Sci. Rep.* **2019**, *9*, 1–19. [CrossRef]

217. Ashraf, M. Biotechnological approach of improving plant salt tolerance using antioxidants as markers. *Biotechnol. Adv.* **2009**, *27*, 84–93. [CrossRef] [PubMed]

218. Turan, S.; Tripathy, B.C. Salt and genotype impact on antioxidative enzymes and lipid peroxidation in two rice cultivars during de-etiolation. *Protoplasma* **2013**, *1*, 209–222. [CrossRef]

219. Das, M.K.; Roychoudhury, A. ROS and responses of antioxidant as ROS-scavengers during environmental stress in plants. *Front. Environ. Sci.* **2014**, *2*, 1–3. [CrossRef]

220. Elstner, E.F. Mechanisms of oxygen activation in different compartments of plant cell. In *Active Oxygen/Oxidative Stress and Plant Metabolism*; Pell, E.J., Stefen, K.L., Eds.; American society of Plant Physiol: Rockville, MD, USA, 1991; pp. 13–25.

221. Vardhini, B.V.; Anjum, N.A. Brassinosteroids make plant life easier under abiotic stresses mainly by modulating major components of antioxidant defense system. *Front. Environ. Sci.* **2015**, *12*, 67. [CrossRef]

222. Rich, P.R.; Bonner, W.D., Jr. The sites of superoxide anion generation in higher plant mitochondria. *Arch Biochem. Biophys.* **1978**, *188*, 206–213. [CrossRef]

223. Song, X.S.; Hu, W.H.; Mao, W.H.; Ogweno, J.O.; Zhou, Y.H.; Yu, J.Q. Response of ascorbate peroxidase isoenzymes and ascorbate regeneration system to abiotic stresses in *Cucumis sativus* L. *Plant Physiol. Biochem.* **2005**, *1*, 1082–1088. [CrossRef] [PubMed]

224. Mehmood, A.; Hussain, A.; Irshad, M.; Hamayun, M.; Iqbal, A.; Khan, N. In vitro production of IAA by endophytic fungus Aspergillus awamori and its growth promoting activities in Zea mays. *Symbiosis* **2019**, *77*, 225–235. [CrossRef]

225. Xia, X.J.; Gao, C.J.; Song, L.X.; Zhou, Y.H.; Shi, K.A.I.; Yu, J.Q. Role of H_2O_2 dynamics in brassinosteroid-induced stomatal closure and opening in S olanum lycopersicum. *Plant Cell Environ.* **2014**, *37*, 2036–2050. [CrossRef] [PubMed]

226. Blokhina, O.; Virolainen, E.; Fagerstedt, K.V. Antioxidants, oxidative damage and oxygen deprivation stress: A review. *Ann. Bot.* **2003**, *91*, 79–194. [CrossRef]

227. Das, S.K.; Patra, J.K.; Thatoi, H. Antioxidative response to abiotic and biotic stresses in mangrove plants: A review. *Intern. Rev. Hydrobiol.* **2016**, *101*, 3–19. [CrossRef]

228. Sairam, R.K.; Tyagi, A. Physiology and molecular biology of salinity stress tolerance in plants. *Curr. Sci.* **2004**, *86*, 407–421.

229. Zhang, M.; Fang, Y.M.; Ji, Y.H.; Jiang, Z.P.; Wang, L. Effects of salt stress on ion content, antioxidant enzymes and protein profile in different tissues of Broussonetia Papyrifera. *S. Afr. J. Bot.* **2013**, *85*, 1–9. [CrossRef]

230. Wahid, A.; Parveen, M.; Gelani, S.; Basra, S.M.A. Pretreatment of seed with H_2O_2 improves salt tolerance of wheat seedlings by alleviation of oxidative damage and expression of stress proteins. *J. Plant Physiol.* **2007**, *164*, 283–294. [CrossRef]

231. Shalata, A.; Neumann, P.M. Exogenous ascorbic acid increases resistance to salt stress and reduces lipid peroxidation. *J. Exp. Bot.* **2001**, *52*, 2207–2211. [CrossRef]

232. Khan, N.; Ali, S.; Shahid, M.A.; Kharabian-Masouleh, A. Advances in detection of stress tolerance in plants through metabolomics approaches. *Plant Omics* **2017**, *10*, 153. [CrossRef]

233. Aliniacifard, S.; Hajilou, J.; Tabatabaei, S.J.; Sifi-Kalhor, M. Effects of Ascorbic Acid and Reduced Glutathione on the Alleviation of Salinity Stress in Olive Plants. *Int. J. Fruit Sci.* **2016**, *16*, 395–409. [CrossRef]

234. Ding, X.; Jiang, Y.; He, L.; Zhou, Q.; Yu, J.; Hui, D.; Huang, D. Exogenous glutathione improves high root-zone temperature tolerance by modulating photosynthesis, antioxidant and osmolytes systems in cucumber seedlings. *Sci. Rep.* **2016**, *6*, 35424. [CrossRef] [PubMed]

235. Ahmad, B. Interactive Effects of Silicon and Potassium Nitrate in Improving Salt Tolerance of Wheat. *J. Integr. Agric.* **2013**. [CrossRef]

236. Parihar, P.; Singh, S.; Singh, R.; Singh, V.P.; Prasad, S.M. Effect of salinity stress on plants and its tolerance strategies: A review. *Environ. Sci. Pollut. Res.* **2015**, *22*, 4056–4075. [CrossRef] [PubMed]

237. Wang, N.; Qi, H.; Qiao, W.; Shi, J.; Xu, Q.; Zhou, H.; Yan, G.; Huang, Q. Cotton (*Gossypium hirsutum* L.) genotypes with contrasting K+/Na+ ion homeostasis: Implications for salinity tolerance. *Acta Physiol. Plant.* **2017**, *1*, 77. [CrossRef]

238. Walia, H.; Wilson, C.; Condamine, P.; Liu, X.; Ismail, A.M.; Close, T.J. Large scale expression profiling and physiological characterization of Jasmonic acid-mediated adaptation of barley to salinity stress. *Plant Cell Environ.* **2007**, *30*, 410–421. [CrossRef] [PubMed]

239. Qiu, Z.; Guo, J.; Zhu, A.; Zhang, L.; Zhang, M. Exogenous jasmonic acid can enhance tolerance of wheat seedlings to salt stress. *Ecotoxicol. Environ. Saf.* **2014**, *1*, 202–208. [CrossRef] [PubMed]

240. Ahmad, P.; Rasool, S.; Gul, A.; Sheikh, S.A.; Akram, N.A.; Ashraf, M.; Kazi, A.M.; Gucel, S. Jasmonates: Multifunctional Roles in Stress Tolerance. *Front. Plant Sci.* **2016**, *7*, 813. [CrossRef]

241. Khan, N.; Bano, A.; Babar, M.A. Metabolic and physiological changes induced by plant growth regulators and plant growth promoting rhizobacteria and their impact on drought tolerance in *Cicer arietinum* L. *PLoS ONE* **2019**, *14*, e0213040. [CrossRef] [PubMed]

242. Wani, S.H.; Kumar, V.; Shriram, V.; Sah, S.K. Phytohormones and their metabolic engineering for abiotic stress tolerance in crop plants. *Crop J.* **2016**, *4*, 162–176. [CrossRef]

243. Seif El-Yazal, S.A.; Seif El-Yazal, M.A.; Dwidar, E.F.; Rady, M.M. Phytohormone crosstalk research: Cytokinin and its crosstalk with other phytohormones. *Curr. Protein Pept. Sci.* **2015**, *16*, 395–405. [CrossRef] [PubMed]

244. Khan, N.; Bano, A.; Ali, S.; Babar, M.A. Crosstalk amongst phytohormones from planta and PGPR under biotic and abiotic stresses. *Plant Growth Regul.* **2020**, *90*, 189–203. [CrossRef]

245. Nafisi, M.; Fimognari, L.; Sakuragi, Y. Interplays between the cell wall and phytohormones in interaction between plants and necrotrophic pathogens. *Phytochemistry* **2015**, *1*, 63–71. [CrossRef] [PubMed]

246. Aldesuquy, H.S.; Gaber, A.M. Effect of growth regulators on leaf area, pigment content and photosynthetic activity on *Vicia faba* plants irrigated by sea water. *Biol. Plant.* **1993**, *35*, 519–527. [CrossRef]

247. Babar, S.; Siddiqi, E.H.; Hussain, I.; Hayat Bhatti, K.; Rasheed, R. Mitigating the effects of salinity by foliar application of salicylic acid in fenugreek. *Physiol. J.* **2014**, *2014*. [CrossRef]

248. Vishwakarma, K.; Upadhyay, N.; Kumar, N.; Yadav, G.; Singh, J.; Mishra, R.K.; Sharma, S. Abscisic Acid Signaling and Abiotic Stress Tolerance in Plants: A Review on Current Knowledge and Future Prospects. *Front. Plant Sci.* **2017**, *8*, 161. [CrossRef]

249. Mehmood, A.; Hussain, A.; Irshad, M.; Khan, N.; Hamayun, M.; Ismail; Afridi, S.G.; Lee, I.J. IAA and flavonoids modulates the association between maize roots and phytostimulant endophytic Aspergillus fumigatus greenish. *J. Plant Interact.* **2018**, *1*, 502–512. [CrossRef]

250. Barnawal, D.; Bharti, N.; Tripathi, A.; Pandey, S.S.; Chanotiya, C.S.; Kalra, A. ACC-deaminase-producing endophyte Brachybacterium paraconglomeratum strain SMR20 ameliorates Chlorophytum salinity stress via altering phytohormone generation. *J. Plant Growth Reg.* **2016**, *35*, 553–564. [CrossRef]

251. Beaudoin, N.; Serizet, C.; Gosti, F.; Giraudat, J. Interaction between abscisic acid and ethylene signalling cascades. *Plant Cell* **2000**, *2*, 1103–1115. [CrossRef]

252. Khan, N.; Bano, A.; Rahman, M.A.; Rathinasabapathi, B.; Babar, M.A. UPLC HRMS-based untargeted metabolic profiling reveals changes in chickpea (Cicer arietinum) metabolome following long-term drought stress. *Plant Cell Environ.* **2019**, *42*, 115–132. [CrossRef] [PubMed]

253. Anuradha, S.; Rao, S.S.R. Effect of brassinosteroids on salinity stress induced inhibition of seed germination and seedling growth of rice (*Oryza sativa* L.). *Plant Growth Regul.* **2001**, *33*, 151–153. [CrossRef]

254. Sharma, I.; Ching, E.; Saini, S.; Bhardwaj, R.; Pati, P.K. Exogenous application of brassinosteroid offers tolerance to salinity by altering stress responses in rice variety Pusa Basmati-1. *Plant Physiol. Biochem.* **2013**, *69*, 17–26. [CrossRef]

255. Anuradha, S.; Rao, S.S.R. Application of brassinosteroids to rice seeds (*Oryza sativa* L.) reduced the impact of salt stress on growth, prevented photosynthetic pigments loss and increased nitrate reductase activity. *Plant Growth Regul.* **2003**, *40*, 29–32. [CrossRef]

256. Shahid, M.A.; Balal, R.M.; Khan, N.; Simón-Grao, S.; Alfosea-Simón, M.; Cámara-Zapata, J.M.; Mattson, N.S.; Garcia-Sanchez, F. Rootstocks influence the salt tolerance of Kinnow mandarin trees by altering the antioxidant defense system, osmolyte concentration, and toxic ion accumulation. *Sci. Hortic.* **2019**, *10*, 1. [CrossRef]

257. Shahid, M.A.; Balal, R.M.; Khan, N.; Zotarelli, L.; Liu, G.D.; Sarkhosh, A.; Fernández-Zapata, J.C.; Nicolás, J.J.; Garcia-Sanchez, F. Selenium impedes cadmium and arsenic toxicity in potato by modulating carbohydrate and nitrogen metabolism. *Ecotoxicol. Environ. Saf.* **2019**, *30*, 588–599. [CrossRef] [PubMed]

258. Houimli, S.I.M.; Denden, M.; El-Hadj, S.B. Induction of salt tolerance in pepper (*Capsicum annum*) by 24-epibrassinolide. *Eurasian J. Biosci.* **2008**, *2*, 83–90.

259. Thussagunpanit, J.; Jutamanee, K.; Sonjaroon, W.; Kaveeta, L.; Chai-Arree, W.; Pankean, P.; Suksamrarn, A. Effects of brassinosteroid and brassinosteroid mimic on photosynthetic efficiency and rice yield under heat stress. *Photosynth* **2015**, *1*, 312–320. [CrossRef]

260. Qayyum, B.; Shabbaz, M.; Akram, N.A. Interacive effect of foliar application of 24—Epibrassinolide and root zone salinity on morpho – physiological attributes of wheat (*triticum aestivum* L.). *Int. J. Agric. Biol.* **2007**, *9*, 584–589.

261. Hairat, S.; Khurana, P. Improving Photosynthetic Responses during Recovery from Heat Treatments with Brassinosteroid and Calcium Chloride in Indian Bread Wheat Cultivars. *Am. J. Plant Sci.* **2015**, *6*, 1827–1849. [CrossRef]

262. Amzallag, G.N. Brassinosteroids as metahormones evidences from specific influence during critical period in sorghum development. *Plant Biol.* **2002**, *4*, 656–663. [CrossRef]

263. Gruszka, D.; Janeczko, A.; Dziurka, M.; Pociecha, E.; Oklestkova, J.; Szarejko, I. Barley Brassinosteroid Mutants Provide an Insight into Phytohormonal Homeostasis in Plant Reaction to Drought Stress. *Front. Plant Sci.* **2016**, *7*, 1824. [CrossRef]

264. Kamuro, Y.; Takatsuto, S. Practical application of brassinosteroids in agricultural fields. In *Brassinosteroids: Steroidal Plant Hormones*; Sakurai, A., Yokota, T., Clouse, S.D., Eds.; Springer: Tokyo, Japan, 1999; pp. 223–241.

265. Ali, B.; Hasan, S.A.; Hayat, S.; Hayat, Q.; Yadav, S.; Fariduddin, Q.; Ahmad, A. A role for brassinosteroids in the amelioration of aluminium stress through antioxidant system in mung bean (*Vigna radiata* L. Wilczek). *Environ. Exp. Bot.* **2008**, *62*, 153–159. [CrossRef]

266. Hasan, S.A.; Hayat, S.; Ali, B.B.; Ahmad, A. 28-Homobrassinolide protects chickpea (*Cicer arietinum*) from cadmium toxicity by stimulating antioxidants. *Environ. Pollut.* **2008**, *151*, 60–66. [CrossRef] [PubMed]

267. Rady, M.M.; Osman, A.S. Response of growth and antioxidant system of heavymetal-contaminated tomato plants to 24-epibrassinolide. *Afr. J. Agric. Res.* **2012**, *7*, 3249–3254.

268. Jain, M.; Mathur, G.; Koul, S.; Sarin, N. Ameliorative effects of proline on salt induced lipid peroxidation in Cell lines of groundnut (*Arachis hypogea* L.). *Plant Cell Rep.* **2001**, *20*, 463–468. [CrossRef]

269. Abdullahi, B.A.; Gu, X.; Gan, Q.; Yang, Y. Brassinolide amelioration of aluminum toxicity in mung bean seedling growth. *J. Plant Nutr.* **2003**, *26*, 1725–1734. [CrossRef]

270. Martínez-Cuenca, M.R.; Primo-Capella, A.; Forner-Giner, M.A. Influence of Rootstock on Citrus Tree Growth: Effects on Photosynthesis and Carbohydrate Distribution, Plant Size, Yield, Fruit Quality, and Dwarfing Genotypes. *Plant Growth* **2016**, *16*, 107.

271. Khelil, A.; Menu, T.; Ricard, B. Adaptive response to salt involving carbohydrate metabolism in leaves of a salt-sensitive tomato cultivars. *Plant Physiol. Biochem.* **2007**, *45*, 551–559. [CrossRef]

272. Bartels, D.; Sunkar, R. Drought and salt tolerance in plants. *Crit. Rev. Plant Sci.* **2005**, *24*, 23–58. [CrossRef]

273. Hoekstra, F.A.; Golovina, E.A.; Buitkink, J. Mechanisms of plant desiccation tolerance. *Trends Plant Sci.* **2001**, *6*, 431–438. [CrossRef]

274. Pierik, R.; Christa, T. The Art of Being Flexible: How to Escape from Shade, Salt, and Drought. *Top. Rev. Plast.* **2014**, *166*, 5–22. [CrossRef] [PubMed]

275. Libal-Wekseler, J.; Fernak, B.; Michael, S.T.; Zhu, J.K. Carbohydrate metabolism is affected in response to salt stress. *Horticulture* **1997**, *20*, 1043–1045.

276. Khan, N.; Bano, A. Role of plant growth promoting rhizobacteria and Ag-nano particle in the bioremediation of heavy metals and maize growth under municipal wastewater irrigation. *Int. J. Phytorem.* **2016**, *18*, 211–221. [CrossRef]

277. Schreiber, L. Transport barriers made of cutin, suberin and associated waxes. *Trends Plant Sci.* **2010**, *15*, 546–553. [CrossRef] [PubMed]

278. Lux, A.; Šottníková, A.; Opatrná, J.; Greger, M. Differences in structure of adventitious roots in Salix clones with contrasting characteristics of cadmium accumulation and sensitivity. *Physiol. Plant.* **2004**, *120*, 537–545. [CrossRef] [PubMed]

279. Shabala, S. Learning from halophytes: Physiological basis and strategies to improve abiotic stress tolerance in crops. *Ann. Bot.* **2013**, *112*, 1209–1221. [CrossRef] [PubMed]

280. White, P.J.; Broadley, M.R. Chloride in soil and its uptake and movement within the plant. A review. *Ann. Bot.* **2001**, *88*, 967–988. [CrossRef]

281. Grattan, S.R.; Grieve, C.M. Mineral nutrient acquisition and response by plants grown in saline environments. In *Hand Book of Plant and Crop Stress*; Pessarakli, M., Ed.; Marcel Dekker: New York, NY, USA, 1999; pp. 203–299.

282. Ikram, M.; Ali, N.; Jan, G.; Guljan, F.; Khan, N. Endophytic fungal diversity and their interaction with plants for agriculture sustainability under stressful condition. *Recent Pat. Food Nutr. Agric.* **2019**, *12*. [CrossRef]

283. Marzec, M.; Muszynska, A.; Gruszka, D. The Role of Strigolactones in Nutrient-Stress Responses in Plants. *Int. J. Mol. Sci.* **2013**, *14*, 9286–9304. [CrossRef] [PubMed]

284. Ali, S.; Khan, N.; Nouroz, F.; Erum, S.; Nasim, W.; Shahid, M.A. In vitro effects of GA 3 on morphogenesis of CIP potato explants and acclimatization of plantlets in field. *In Vitro Cell. Dev. Biol.* **2018**, *54*, 104–111. [CrossRef]

285. Wang, M.; Zheng, Q.; Shen, Q.; Guo, S. The critical role of potassium in plant stress response. *Int. J. Mol. Sci.* **2013**, *14*, 7370–7390. [CrossRef]

286. Ali, S.; Khan, N.; Nouroz, F.; Erum, S.; Nasim, W.A. Effects of sucrose and growth regulators on the microtuberization of cip potato (*Solanum tuberosum*) germplasm. *Pak. J. Bot.* **2018**, *1*, 763–768.

287. Maathuis, F.J.M.; Ahmad, I.; Patishtan, J. Regulation of Na$^+$ fluxes in plants. *Front. Plant Sci.* **2014**, *5*, 467. [CrossRef] [PubMed]

288. Conde, A.; Chaves, M.M.; Gerós, H. Membrane transport, sensing and signaling in plant adaptation to environmental stress. *Plant Cell Physiol.* **2011**, *1*, 1583–1602. [CrossRef]

289. Wyn Jones, R.G.; Pollard, A. Proteins, enzymes and inorganic ions. In *Encyclopedia of Plant Physiology*; New Series; Lauchli, A., Person, A., Eds.; Springer: New York, NY, USA, 1983; Volume 15B, pp. 528–562.

290. Wang, H.; Wu, Z.; Chen, Y.; Yang, C.; Shi, D. Effects of salt and alkali stresses on growth and ion balance in rice (*Oryza sativa* L.). *Plant Soil Environ.* **2011**, *8*, 286–294. [CrossRef]

291. Rossi, L.; Francini, A.; Minnocci, A.; Sebastiani, L. Salt stress modifies apoplastic barriers in olive (*Olea europaea* L.): A comparison between a salt-tolerant and a salt-sensitive cultivar. *Sci. Hortic.* **2015**, *192*, 38–46. [CrossRef]

292. Krishnamurthy, P.; Ranathunge, K.; Nayak, S.; Schreiber, L.; Mathew, M.K. Root apoplastic barriers block Na+ transport to shoots in rice (*Oryza sativa* L.). *J. Exp. Bot.* **2011**, *62*, 4215–4228. [CrossRef]

293. Kosová, K.; Vítámvás, P.; Urban, M.O.; Prášil, I.T. Plant proteome responses to salinity stress–comparison of glycophytes and halophytes. *Funct. Plant Biol.* **2013**, *23*, 775–786. [CrossRef] [PubMed]

294. Moya, J.L.; Primo-Millo, E.; Talon, M. Morphological factors determining salt tolerance in citrus seedlings: The shoot to root ratio modulates passive root uptake of chloride ions and their accumulation in leaves. *Plant Cell Environ.* **1999**, *22*, 1425–1433. [CrossRef]

295. Hussain, M.; Park, H.W.; Farooq, M.; Jabran, K.; Lee, D.J. Morphological and physiological basis of salt resistance in different rice genotypes. *Int. J. Agric. Biol.* **2013**, *15*, 113–118.

296. Cardon, D.E.; Walker, J.D.; Flowers, T.J.; Miller, A.J. Single cell measurement of the contributions of cytosolic Na$^+$ and K$^+$ to salt tolerance. *Plant Physiol.* **2003**, *131*, 676–683. [CrossRef] [PubMed]

297. Mahmud, S.; Sharmin, S.; Chowdhury, B.L.; Hussain, M.A. Research Article Effect of Salinity and Alleviating Role of Methyl Jasmonate in Some Rice Varieties. *Asian J. Plant Sci.* **2017**, *16*, 87–93. [CrossRef]

298. Manaa, A.; Ben Ahmed, H.; Valot, B.; Bouchet, J.P.; Aschi-Smiti, S.; Causse, M.; Faurobert, M. Salt and genotype impact on plant physiology and root proteome variations in tomato. *J. Exp. Bot.* **2011**, *62*, 2797–2813. [CrossRef] [PubMed]
299. Akca, Y.; Samsunlu, E. The effect of salt stress on growth, chlorophyll content, proline and nutrient accumulation and K/Na ratio in walnut. *Pak. J. Bot.* **2012**, *44*, 1513–1520.
300. Hakim, M.A.; Juraimi, A.S.; Hanafi, M.M.; Ismail, M.R.; Rafii, M.Y.; Islam, M.M.; Selamat, A. The effect of salinity on growth, ion accumulation and yield of rice varieties. *J. Anim. Plant Sci.* **2014**, *24*, 874–885.
301. Deinlein, U.; Stephan, A.B.; Horie, T.; Luo, W.; Xu, G.; Schroeder, J.I. Plant salt-tolerance mechanisms. *Trends Plant. Sci.* **2014**, *19*, 371–379. [CrossRef]
302. Adabnejad, H.; Kavousi, H.R.; Hamidi, H.; Tavassolian, I. Assessment of the vacuolar Na+/H+ antiporter (*NHX1*) transcriptional changes in *Leptochloa fusca* L. in response to salt and cadmium stresses. *Mol. Biol. Res. Commun.* **2015**, *4*, 133–142. [PubMed]
303. Guyton, A.C.; Hall, J.E. *Textbook of Medical Physiology*; Saunders: Philadelphia, PA, USA, 1986; p. 20.
304. Kumari, P.H.; Kumar, S.A.; Sivan, P.; Katam, R.; Suravajhala, P.; Rao, K.S.; Kishor, P.B.K. Overexpression of a Plasma Membrane Bound Na$^+$/H$^+$ Antiporter-Like Protein (*SbNHXLP*) Confers Salt Tolerance and Improves Fruit Yield in Tomato by Maintaining Ion Homeostasis. *Front. Plant Sci.* **2016**, *7*, 2027. [CrossRef] [PubMed]
305. Volkov, V. Salinity tolerance in plants. Quantitative approach to ion transport starting from halophytes and stepping to genetic and protein engineering for manipulating ion fluxes. *Front. Plant Sci.* **2015**, *6*, 873. [CrossRef]
306. Veraplakorn, V.; Nanakorn, M.; Kaveeta, L.; Srisom, S.; Bennett, I.J. Variation in ion accumulation as a measure of salt tolerance in seedling and callus of Stylosanthes guianensis. *Exp. Plant Physiol.* **2013**, *25*, 106–115. [CrossRef]
307. Mukhtar, I.; Shahid, M.A.; Khan, M.W.; Balal, R.M.; Iqbal, M.M.; Naz, T.; Zubair, M.; Ali, H.H. Improving salinity tolerance in chili by exogenous application of calcium and sulphur. *Soil Environ.* **2016**, *1*, 35.
308. Bao, A.; Zhao, Z.; Ding, G.; Shi, L.; Xu, F.; Cai, H. The Stable Level of *Glutamine synthetase* 2 Plays an Important Role in Rice Growth and in Carbon-Nitrogen Metabolic Balance. *Int. J. Mol. Sci.* **2015**, *16*, 12713–12736. [CrossRef]
309. Bohnert, H.J.; Sheveleva, E. Plant stress adaptations-making metabolism move. *Curr. Opin. Biol.* **1998**, *1*, 267–274. [CrossRef]
310. Bonales-Alatorre, E.; Pottosin, I.; Shabala, L.; Chen, Z.H.; Zeng, F.; Jacobsen, S.E.; Shabala, S. Differential Activity of Plasma and Vacuolar Membrane Transporters Contributes to Genotypic Differences in Salinity Tolerance in a Halophyte Species, *Chenopodium quinoa*. *Int. J. Mol. Sci.* **2013**, *14*, 9267–9285. [CrossRef] [PubMed]
311. Cherel, L. Regulation of K$^+$ channel activities in plants: From physiological to molecular aspects. *J. Exp. Bot.* **2004**, *55*, 337–351. [CrossRef] [PubMed]
312. Shabala, S.; Pottosin, I. Regulation of potassium transport in plants under hostile conditions: Implications for abiotic and biotic stress tolerance. *Physiol. Plant.* **2014**, *151*, 257–279. [CrossRef] [PubMed]
313. Humphries, E.S.A.; Dart, C. Neuronal and Cardiovascular Potassium Channels as Therapeutic Drug Targets: Promise and Pitfalls. *J. Biomol. Screen.* **2015**, *20*, 1055–1073. [CrossRef] [PubMed]
314. Machnicka, B.; Czogalla, A.; Hryniewicz-Jankowska, A.; Bogusławska, D.M.; Grochowalska, R.; Heger, E.; Sikorski, A.F. Spectrins: A structural platform for stabilization and activation of membrane channels, receptors and transporters. *Biochim. Biophys. Acta* **2014**, *1*, 620–634. [CrossRef]
315. Tavakoli, M.; Poustini, K.; Besharati, H.; Ali, S. Variable salinity responses of 25 alfalfa genotypes and comparative salt-response ion distribution. *Russ. J. Plant Physiol.* **2019**, *66*, 231–239. [CrossRef]
316. Carillo, P.; Parisi, D.; Woodrow, P.; Pontecorvo, G.; Massaro, G.; Annunziata, M.G.; Fuggi, A.; Sulpice, R. Salt-induced accumulation of glycine betaine is inhibited by high light in durum wheat. *Funct. Plant Biol.* **2011**, *22*, 139–150. [CrossRef]
317. Awada, S.; Campbell, W.F.L.; Dudle, M.; Jurinak, J.J. Interactive effects of sodium chloride, sodium sulfate, calcium sulfate and calcium chloride on snapbean growth, photosynthesis and ion uptake. *J. Plant Nutr.* **1995**, *18*, 889–900. [CrossRef]
318. Multari, S.; Stewart, D.; Russell, W.R. Potential of Fava Bean as Future Protein Supply to Partially Replace Meat Intake in the Human Diet. *Compr. Rev. Food Sci.* **2015**, *14*, 511–522. [CrossRef]

319. Hura, T.; Szewczyk-Taranek, B.; Hura, K.; Nowak, K.; Pawłowska, B. Physiological Responses of *Rosa rubiginosa* to Saline Environment. *Water Air Soil Pollut.* **2017**, *228*, 81. [CrossRef] [PubMed]

320. Gowayed, M.H.; Al-Zahrani, H.S.; Metwali, E.M. Improving the salinity tolerance in potato (*Solanum tuberosum*) by exogenous application of silicon dioxide nanoparticles. *Int. J. Agric. Biol.* **2017**, *1*, 19.

321. Pessarakli, M.; Haghighi, M.; Sheibanirad, A. Plant Responses under Environmental Stress Conditions. *Adv. Plants Agric. Res.* **2015**, *2*, 00073. [CrossRef]

322. Debouba, M.; Gouia, A.; Suzuki, M.H.; Ghorbel, M. NaCl stress effects on enzymes involved in nitrogen assimilation pathway in tomato "*Lycopersicon esculentum*" seedlings. *J. Plant Physiol.* **2006**, *163*, 1247–1258. [CrossRef]

323. Wang, H.; Zhang, M.; Guo, R.; Shi, D.; Liu, B.; Lin, X.; Yang, C. Effects of slat stress on ion balance and nitrogen metabolism of old and young leaves in rice (*Oryza sativa* L.). *BMC Plant Biol.* **2012**, *12*, 194. [CrossRef]

324. Pandey, M.; Srivastava, A.K.; D'Souza, S.F.; Penna, S. Thiourea, a ROS Scavenger, Regulates Source-to-Sink Relationship to Enhance Crop Yield and Oil Content in *Brassica juncea* (L.). *PLoS ONE* **2013**, *8*, e73921. [CrossRef]

325. Koksal, N.; Alkan-Torun, A.; Kulahlioglu, I.; Ertargin, E.; Karalar, E. Ion uptake of marigold under saline growth conditions. *Springer Plus.* **2016**, *5*, 139. [CrossRef]

326. Khan, N.; Bano, A. Role of PGPR in the Phytoremediation of Heavy Metals and Crop Growth under Municipal Wastewater Irrigation. In *Phytoremediation*; Springer: Cham, Switzerland, 2018; pp. 135–149.

327. Pravin, V.; Rosazlin, A.; Tumirah, K.; Salmah, I.; Amru, N.B. Role of Plant Growth Promoting Rhizobacteria in Agricultural Sustainability—A Review. *Molecules* **2016**, *21*, 573. [CrossRef]

328. Pal, D.; Khozin-Goldberg, I.; Cohen, Z.; Boussiba, S. The effect of light, salinity, and nitrogen availability on lipid production by *Nannochloropsis* sp. *Appl. Microbiol. Biotech.* **2011**, *90*, 1429–1441. [CrossRef] [PubMed]

329. Chen, W.; Hou, Z.; Wu, L.; Liang, Y.; Wei, C. Effects of salinity and nitrogen on cotton growth in arid environment. *Plant Soil.* **2010**, *326*, 61–73. [CrossRef]

330. Seemann, J.R.; Thomas, D.S. Salinity and nitrogen effects on photosynthesis, ribulose-1, 5-bisphosphate carboxylase and metabolite pool sizes in *Phaseolus vulgaris* L. *Plant Physiol.* **1986**, *82*, 555–560. [CrossRef] [PubMed]

331. Giblin, A.E.; Weston, N.B.; Banta, G.T.; Tucker, J.; Hopkinson, C.S. The effects of salinity on nitrogen losses from an oligohaline estuarine sediment. *Estuaries Coasts* **2010**, *33*, 1054–1068. [CrossRef]

332. Laura, R.D. Salinity and nitrogen mineralization in soil. *Soil Biol. Bioch.* **1977**, *9*, 333–336. [CrossRef]

333. Villa-Castorena, M.; Ulery, A.L.; Catalán-Valencia, E.A.; Remmenga, M.D. Salinity and nitrogen rate effects on the growth and yield of chile pepper plants. *Soil Sci. Soc. Am. J.* **2003**, *67*, 1781–1789. [CrossRef]

334. Kahn, A.E.; Durako, M.J. Thalassia testudinum seedling responses to changes in salinity and nitrogen levels. *J. Exp. Mar. Biol. Ecol.* **2006**, *335*, 1–12. [CrossRef]

335. Zhang, D.; Li, W.; Xin, C.; Tang, W.; Eneji, A.E.; Dong, H. Lint yield and nitrogen use efficiency of field-grown cotton vary with soil salinity and nitrogen application rate. *Field Crops Res.* **2012**, *138*, 63–70. [CrossRef]

336. Al-Harbi, A.R.; Wahb-Allah, M.A.; Abu-Muriefah, S.S. Salinity and nitrogen level affects germination, emergence, and seedling growth of tomato. *Int. J. Veg. Sci.* **2008**, *14*, 380–392. [CrossRef]

337. Akhtar, M.; Hussain, F.; Ashraf, M.Y.; Qureshi, T.M.; Akhter, J.; Awan, A.R. Influence of salinity on nitrogen transformations in soil. *Commun. Soil Sci. Plant Anal.* **2012**, *43*, 1674–1683. [CrossRef]

338. Esmaili, E.; Kapourchal, S.A.; Malakouti, M.J.; Homaee, M. Interactive effect of salinity and two nitrogen fertilizers on growth and composition of sorghum. *Plant Soil Environ.* **2008**, *54*, 537–546. [CrossRef]

339. Azizian, A.; Sepaskhah, A.R. Maize response to different water, salinity and nitrogen levels: Agronomic behavior. *Int. J. Plant Prod.* **2014**, *8*, 107–130.

340. Tang, Z.; Liu, Y.; Guo, X.; Zu, Y. The combined effects of salinity and nitrogen forms on Catharanthus roseus: The role of internal ammonium and free amino acids during salt stress. *J. Plant Nutr. Soil Sci.* **2011**, *174*, 135–144. [CrossRef]

341. Garg, N.; Chandel, S. The effects of salinity on nitrogen fixation and trehalose metabolism in mycorrhizal *Cajanus cajan* (L.) millsp. plants. *J. Plant Growth Reg.* **2011**, *30*, 490–503. [CrossRef]

342. Navarro, J.M.; Botella, M.A.; Cerda, A.; Martinez, V. Phosphorus uptake and translocation in salt stressed melon plants. *J. Plant Physiol.* **2001**, *158*, 375–381. [CrossRef]

343. Zakery-Asl, M.A.; Bolandnazar, S.; Oustan, S. Effect of salinity and nitrogen on growth, sodium, potassium accumulation, and osmotic adjustment of halophyte Suaeda aegyptiaca (Hasselq.) Zoh. *Arch. Agron. Soil Sci.* **2014**, *60*, 785–792. [CrossRef]

344. Zeng, W.Z.; Xu, C.; Wu, J.W.; Huang, J.S.; Zhao, Q.; Wu, M.S. Impacts of salinity and nitrogen on the photosynthetic rate and growth of sunflowers (*Helianthus annuus* L.). *Pedosphere* **2014**, *24*, 635–644. [CrossRef]

345. Cardeñosa, V.; Medrano, E.; Lorenzo, P.; Sánchez-Guerrero, M.C.; Cuevas, F.; Pradas, I.; Moreno-Rojas, J.M. Effects of salinity and nitrogen supply on the quality and health-related compounds of strawberry fruits (Fragaria× ananassa cv. Primoris). *J. Sci. Food Agric.* **2015**, *95*, 2924–2930. [CrossRef] [PubMed]

346. Min, W.; Guo, H.; Zhang, W.; Zhou, G.; Ma, L.; Ye, J.; Liang, Y.; Hou, Z. Response of soil microbial community and diversity to increasing water salinity and nitrogen fertilization rate in an arid soil. *Acta Agric. Scand. Sect. B Soil Plant. Sci.* **2016**, *66*, 117–126. [CrossRef]

Article

Isolation and Characterization of Plant Growth Promoting Endophytic Bacteria from Desert Plants and Their Application as Bioinoculants for Sustainable Agriculture

Muneera D. F. ALKahtani [1], Amr Fouda [2], Kotb A. Attia [3], Fahad Al-Otaibi [4,5], Ahmed M. Eid [2], Emad El-Din Ewais [2], Mohamed Hijri [4,6], Marc St-Arnaud [4], Saad El-Din Hassan [2], Naeem Khan [7], Yaser M. Hafez [8] and Khaled A. A. Abdelaal [8,*]

[1] Biology Department, College of Science, Princess Nourah Bint Abdulrahman University, Riyadh POX 102275-11675, Saudi Arabia; mdfkahtani@gmail.com

[2] Botany and Microbiology Department, Faculty of Science, AL-Azhar University, Nasr City, Cairo 11884, Egypt; amr_fh83@azhar.edu.eg (A.F.); aeidmicrobiology@azhar.edu.eg (A.M.E.); ewais_e@yahoo.com (E.E.-D.E.); saad.el-din.hassan@umontreal.ca (S.E.-D.H.)

[3] Center of Excellence in Biotechnology Research, King Saud University, Riyadh POX 2455-11451, Saudi Arabia; kattia1.c@ksu.edu.sa

[4] Biodiversity Centre, Institut de Recherche en Biologie Végétale, Université de Montréal and Jardin botanique de Montréal, Montréal, QC 22001, Canada; fahad.alotaibi@umontreal.ca (F.A.-O.); mohamed.hijri@umontreal.ca (M.H.); marc.st-arnaud@umontreal.ca (M.S.-A.)

[5] Department of Soil Science, King Saud University, Riyadh 11564, Saudi Arabia

[6] AgroBioSciences, Mohammed VI Polytechnic University (UM6P), 43150 Ben Guerir, Morocco

[7] Department of Agronomy, Institute of Food and Agricultural Sciences, University of Florida, Gainesville, FL 32611, USA; naeemkhan@ufl.edu

[8] Excellence Center (EPCRS), Plant Pathology and Biotechnology Laboratory, Faculty of Agriculture, Kafrelsheikh University, Kafr El Sheikh 33516, Egypt; hafezyasser@gmail.com

* Correspondence: khaled.elhaies@gmail.com

Received: 29 July 2020; Accepted: 26 August 2020; Published: 4 September 2020

Abstract: Desert plants are able to survive under harsh environmental stresses inherent to arid and semiarid regions due to their association with bacterial endophytes. However, the identity, functions, and the factors that influence the association of bacterial endophytes with desert plants are poorly known. These bacterial endophytes can be used as an untapped resource to favor plant growth and development in agro-ecosystems of arid regions. The present study is therefore focused on the isolation and identification of bacterial endophytes from two native medicinal plants (*Fagonia mollis* Delile and *Achillea fragrantissima* (Forssk) Sch. Bip.) growing spontaneously in the arid region of the South Sinai (Egypt), and characterization of their plant growth promoting (PGP) traits. Thirteen putative bacterial endophytes were isolated from the leaves of both plant species and characterized for their plant growth promoting abilities using molecular and biochemical approaches, as well as greenhouse trials. Selected endophytic bacterial strains were applied to maize plants (*Zea mays* L. var. Single cross Pioneer 30K08) to further evaluate their PGP abilities under greenhouse conditions. Isolated bacterial strains have variable plant growth promoting activities. Among these activities, isolated bacterial endophytes have the efficacy of phosphate solubilizing with clear zones ranging from 7.6 ± 0.3 to 9.6 ± 0.3 mm. Additionally, the obtained bacterial endophytes increased the productivity of indole acetic acid (IAA) in broth media from 10 to 60 µg·mL^{-1} with increasing tryptophan concentration from 1 to 5 mg·mL^{-1}. *Bacillus* and *Brevibacillus* strains were frequently isolated from the leaves of both plant species, and had significant positive effects on plant growth and shoot phosphorus (P) and nitrogen (N) contents. Results suggest that these endophytes are good candidates as plant growth promoting inoculants to help reduce chemical input in conventional agricultural practices and increase nutrient uptake and stress resilience in plant species.

Agronomy **2020**, *10*, 1325

Keywords: *Zea mays* L.; environmental stresses; endophytic bacteria; plant growth promoting ability

---

1. Introduction

The enhancement of crop productivity is required for feeding the increasing population in developing countries and often relies on the use of chemical fertilizers. However, long-term use of these fertilizers was shown to decrease bacterial diversity in soil [1,2] and can also have harmful effects on the environment, such as leaching of phosphorus and nitrogen into groundwater, and increasing soil and groundwater pollution [3]. One way to increase the sustainability of agricultural practices is the use of efficient, nutrient mobilizing microorganisms to reduce the need and dependency on chemical fertilizers [4,5]. Plant growth promoting bacteria (PGPB) that form symbiotic interactions with their host plants are crucial to improve plant productivity and health under various environmental conditions [4,6–8]. Bacterial endophytes colonize plant tissues without any apparent pathogenic symptoms and establish beneficial associations with their plant host through phytohormone synthesis, enzyme production, and nutrient mobilization and translocation, such as phosphate (PO_4^{-3}) solubilization, nitrogen fixation, and ammonia (NH_3) production [9–11]. Moreover, many endophytes display various applications such as antimicrobial mechanisms, which reduce crop losses caused by pathogens [12–16], and its metabolites integrated into different biotechnological applications [17–20].

The Sinai Peninsula is located in the Sahara-Arabian deserts and represents approximately 6% of the total land area of Egypt. The semi-arid to arid climate and winter precipitations are the main characteristics of the Sinai Peninsula desert. Plants growing in desert conditions were found to harbor a microbiome that increased the biomass during drought stress periods [21]. Medicinal plants from desert farming in Sekem (Egypt) were shown that their roots are strongly associated with bacteria [22–24]. Although the Sinai desert has diverse medicinal plants, very few studies have focused on the associated bacterial endophytes and their PGP activities. Hanna et al. [25] collect 43 different plant species from the North Sinai desert, and reported that *Fagonia mollis* was the highest plant species harboring culturable bacteria. Among these bacteria, *Gluconacetobacter diazotrophicus* was the lowest endophytic species exhibit N_2-fixing activity. In the same regards, 132 endophytic strains were isolated from 18 Egyptian medicinal plants, including nine fungal strains isolated from *Achillea fragrantissima* and exhibiting inhibitory activities against different pathogenic bacteria and yeasts [26]. Application of the bacterial endophytes (*Bacillus thuringiensis*) led to improved plant growth and increased relative water content, chlorophyll content, chlorophyll fluorescence parameter (Fv/Fm ratio), and fruit yield of sweet pepper plants [27].

Maize has become a staple food in many parts of the world, with the total production of maize surpassing that of wheat or rice. Maize crop has several uses, such as food stuff for human or as animal feed because of its high nutritional value. Maize has also been used for corn ethanol and other maize products, such as fructose, corn starch, corn oil, and corn syrup [28]. The Arab Republic of Egypt is the largest country consuming maize at the level of the African continent. However, the production of the Arab Republic of Egypt reached to about 1% of the total global production during (2005–2013), while Egypt represents the third place at the level of the African continent [29]. Therefore, the current study aimed to improve the performance maize growth under the optimal conditions and at normal habitat.

Fagonia mollis and *Achillea fragrantissima* are medicinal plants frequently found in the Sinai Peninsula, and their bacterial endophytes could partially be responsible for the production of various bioactive compounds [30], and the ability of these plants to withstand the harsh, drought condition of the Sinai Peninsula. Therefore, this study focused on the isolation and characterization of putative bacterial endophytes from *F. mollis* and *A. fragrantissima*, which are native inhabitants of the arid and extremely harsh conditions of the Sinai desert. Plant growth promoting (PGP) properties of the bacterial endophytes involving extracellular enzymes (amylase, cellulase, protease, pectinase, and xylanase) production, antimicrobial activity against selected pathogenic bacteria and fungi, indole-3-acetic acid (IAA) and NH_3 production, and P-solubilization ability were evaluated. In addition, their effect on maize growth, plant biomass production, and nutrients content in plant shoots were also investigated in order to evaluate their potentials as bioinoculants for sustainable agriculture practices.

2. Materials and Methods

2.1. Plant Sampling and Study Area

Fagonia mollis Delile (family *Zygophyllaceae*) and *Achillea fragrantissima* (Forssk.) Sch.Bip. (family *Asteraceae*) were collected from two sites, Wadi al-Zwatin (latitude 28.539290° to 28.53919° N, longitude 33.930784° to 33.92044° E) and Wadi Selebat (latitude 28.545493 to 28.543339 N, longitude 33.933707 to 33.932984 E), Saint Katherine Protectorate, South Sinai, Egypt (Figure 1). Four individual plants from each species were collected per site. The plant samples were carefully placed in sterile polyethylene bags and brought back to the laboratory in a portable cooler maintained at 4 °C using ice packs. The formal identification of the plant specimens was carried out at the herbarium of Botany and Microbiology Department of Al-Azhar University, where plant herbarium specimens were also deposited.

Figure 1. (**A**) *Fagonia mollis* Delile. and (**B**) *Achillea fragrantissima* Forssk.

2.2. Isolation of Bacterial Endophytes

On each plant, the first five leaves from a shoot tip were excised and washed under running tap water. Sterilization of leaf surfaces was done by soaking the tissues in a series of baths: sterile distilled water for 1 min, 70% ethanol for 1 min, 2.5% sodium hypochlorite for 4 min, 70% ethanol for 30 s, and a final series of rinsing thrice in sterile distilled water in three different containers. A 0.1 mL aliquot of the final rinse water was plated onto nutrient agar plates to confirm the success of surface sterilization.

The sterilized plant leaves were then cut into 5 mm segments, and twenty leaf segments per individual plant were placed in four petri dishes (9 cm; five segments/plate) containing luria broth (LB) media (tryptone 10 g·L^{-1}; yeast extract 5 g·L^{-1}; NaCl 10 g·L^{-1}; agar 15 g·L^{-1}; and 1 L dis. H_2O, adjusted to pH 7) supplemented with nystatin (25 µg·mL^{-1}) to suppress fungal growth, and incubated in the dark at 35 + 2 °C. Another twenty segments of sterilized leaves per individual plant were together crushed in 10 mL sterile saline solution using a sterile digital homogenizer (PRO25D, Pro Scentific, 120 V, Willenbrock, Oxford, CT, USA), and 1 mL of the suspension was serially diluted until 10^{-3} from which a 0.1 mL aliquot was spread onto each of the three Petri plates containing LB medium and incubated in the dark at 35 ± 2 °C [31]. The cultures were regularly observed for bacterial growth, for a period of 96 hours. Bacteria growing from the previous steps were streaked on fresh LB plates to obtain single colonies, which picked up and inoculated on LB slants and stored at 4 °C until further study.

2.3. Molecular Identification of Bacterial Endophytes

Bacterial identification was based on 16S rRNA gene sequence analysis. Genomic DNA of each isolate was extracted following the method of Miller et al. [32], with some modifications. Briefly, individual colonies from an agar plate were picked up either using a sterile toothpick or an inoculating loop and resuspended in 50 µL of sterile deionized water. The cell suspension was placed in a water bath at 97 °C and heated for 10 min, the cell lysate was centrifuged ($15,000\times g$, 10 min), and the supernatant containing the DNA was recovered. DNA concentration was determined by measuring its absorbance at UV spectrum of 260 nm using a spectrophotometer (JENWAY 6350, 230 V/50 Hz, Staffordshire, UK). A partial 16S rDNA fragment was PCR amplified using the bacterial universal primers 27f (5′-AGAGTTTGATCCTGGCTCAG-3′) and 1492r (5′-GGTTACCTTGTTACGACTT-3′) [33]. The PCR reaction contained: $1 \times$ PCR buffer, 0.5 mM $MgCl_2$, 2.5 U Taq DNA polymerase (QIAGEN), 0.25 mM dNTP, 0.5 µM of each primer, and approximately 5 ng of bacterial genomic DNA. The PCR cycling conditions were 94 °C for 3 min, followed by 30 cycles of 94 °C for 0.5 min, 55 °C for 0.5 min, and 72 °C for 1 min, followed by a final extension performed at 72 °C for 10 min. The PCR products were forward and reverse sequenced using the Applied Biosystems 3730xl DNA Analyzer technology at the Genome Quebec Innovation Center (Montreal, QC, Canada). The sequences generated in this study were deposited in GenBank under accession numbers KY555785 to KY555797. The 16S rRNA sequences were then compared against the GenBank database using the NCBI BLAST nucleotide search. A multiple sequence alignment was constructed on approximately 1200 bp of 16S rRNA gene fragments using the ClustalX 1.8 software package (http://www.clustal.org/clustal2) and a phylogenetic tree was constructed using the neighbor-joining method in the MEGA v6.1 software (www.megasoftware.net), with confidence tested by bootstrap analysis (1000 repeats).

2.4. Screening the Extracellular Enzymatic Activities of Bacterial Endophytes

The production and activity of extracellular enzymes (amylase, cellulase, protease, pectinase, and xylanase) of isolated bacterial endophytes were assessed by growing the isolates in a mineral salt (MS) media ($NaNO_3$ 5 g·L^{-1}; KH_2PO_4 1 g·L^{-1}; K_2HPO_4 2 g·L^{-1}; $MgSO_4.7H_2O$ 0.5 g·L^{-1}; KCl 0.1 g·L^{-1}; $CaCl_2$ 0,01 g·L^{-1}; $FeSO_4.7H_2O$ 0.02 g·L^{-1}; agar 15 g·L^{-1}; and 1 L dis. H_2O) complemented with various additives, depending on the enzyme being tested, as detailed below. Control treatments consisted of the same media without bacterial inoculation. After incubation for 24–48 h depending on the growth rates of the bacterial endophytes at 35 ± 2 °C, specific reagents were added (see paragraph below), and the size of the clear zone surrounding the bacterial colony was measured, indicating extracellular enzymatic activities. All assays were performed in triplicates.

Amylolytic and cellulase activity were assessed by growing the endophytic bacterial isolates on MS agar medium supplemented with 1% soluble starch and 1% cellulose or carboxy-methylcellulose (CMC) respectively. After incubation, the plates were flooded with 1% iodine. MS agar medium containing 1% gelatine was used to determine the bacterial proteolytic activity. After incubation, the degradation of gelatine was highlighted using acidic mercuric chloride as an indicator. Pectinolytic activity was determined by growing bacteria in MS medium containing 1% pectin. After the incubation period, the plates were flooded with 1% aqueous solution of hexadecyl trimethyl ammonium bromide. MS agar medium supplemented with 1% xylan from corncobs was used to measure bacterial xylanolytic activity. After the incubation period, the xylanase activity was assessed after flooding with absolute ethyl alcohol to indicate biodegradation [34].

2.5. Antimicrobial Activity of Bacterial Endophytes

To test the antimicrobial activity of the bacterial endophytes, the isolated strains were cultured in nutrient broth medium for 6 days at 35 ± 2 °C on a shaker (LABOAO, LH-2102C, Zhengzhou, China) at 180 rpm. Crude fermentation broth was blended thoroughly and centrifuged at 4000 rpm for 5 min. Liquid supernatant was extracted twice with an equal volume of ethyl acetate. The organic solvent extract was then evaporated under reduced pressure using a rotary evaporator (RE-801, BM-500 water bath (4 L), glassware C set, Yamato scientific, Tokyo, Japan). The crude extracts were dissolved in dimethyl sulfoxide (DMSO) and used for antimicrobial screening using a well diffusion method [35]. Nutrient broth media without bacterial inoculation were extracted and dissolved in DMSO and were used as controls.

Microbial strains used for antimicrobial assays were: *Staphylococcus aureus* ATCC 6538 and *Bacillus subtilis* ATCC 6633 (Gram-positive bacteria), *Escherichia coli* ATCC 8739, *Pseudomonas aeruginosa* ATCC 9027 and *Salmonella typhimurium* ATCC 14028 (Gram-negative bacteria), and *Candida albicans* ATCC 10231 (yeast). Test organisms were inoculated in Petri dishes containing Muller–Hinton agar medium (Sigma-Aldrich) for bacteria or Sabouraud agar medium (Sigma-Aldrich) for yeast [19,36,37]. Three wells of 1 cm diameter were cut in the tested organism colony using a sterile cork borer and filled with 40 µL of endophytic bacterial extract. Negative control wells were filled with 40 µL of control extract. The plates were kept at 4 °C for 4 h to allow diffusion of antimicrobial compounds, and then incubated at 35 ± 2 °C for bacteria and 28 ± 2 °C for *C. albicans* for 24 h [38,39]. The inhibition zones around the wells were measured to assess the antimicrobial activity of bacterial extracts. All antimicrobial activity assays were performed in triplicates.

2.6. Screening for In Vitro Plant Growth Promoting (PGP) Traits

2.6.1. Phosphate Solubilization

The bacterial endophytic isolates were screened for P-solubilization as follows. Pikovskaya medium (glucose 10 g·L^{-1}; Ca$_3$(PO$_4$) 2.5 g·L^{-1}; (NH$_4$)2SO$_4$ 0.5 g·L^{-1}; NaCl 0.2 g·L^{-1}; MgSO$_4$·7H$_2$O 0.1 g·L^{-1}; KCl 0.2 g·L^{-1}; FeSO$_4$·7H$_2$O 0.002 g·L^{-1}; yeast extract 0.5 g·L^{-1}; MnSO$_4$·2H$_2$O 0.002 g·L^{-1}; agar 15 g·L^{-1}; and 1 L dis. H$_2$O) was prepared and bromophenol blue was added as an indicator. The medium was inoculated with endophytic isolates and incubated for 48 h. The Pikovskaya medium without bacterial growth was used as a control. The formation of clear zones around the colony, due to the utilization of tricalcium phosphate, was measured to assess the ability of endophytes to solubilize phosphate [40].

2.6.2. Ammonia Production

The ability of the isolated endophytic bacterial strains to produce NH$_3$ was assessed after growing the bacterial strains in peptone water (peptone 10 g·L^{-1}; NaCl 5 g·L^{-1}; and 1 L dis. H$_2$O) for 72 h at 35 ± 2 °C. Peptone water without bacterial inoculation was used as a control. The addition of 1 mL of Nessler's reagent in the peptone liquid medium was used to assess the ammonia production. A color change to faint yellow indicated the minimum ammonia production while deep yellow to brownish color indicated the maximum ammonia production [41].

2.7. Quantitative Screening for Indole-3-acetic acid (IAA) Production

The ability of bacterial endophytes to produce IAA was determined in nutrient broth at 35 ± 2 °C for 24 h. One milliliter of each bacterial suspension was added to 20 mL of nutrient broth medium containing 0, 1, 2, or 5 mg·mL^{-1} tryptophan, and incubated for 14 days. Controls consisted of nutrient broth media containing 0, 1, 2, or 5 mg mL^{-1} tryptophan but without bacterial inoculation. Five milliliters of each culture were collected from the incubating broth after 14 days and centrifuged at 6000 rpm for 30 min. One milliliter of the supernatant was mixed with 1 drop of orthophosphoric acid and 2 mL of Salkowski's reagent (300 mL concentrated sulfuric acid, 500 mL distilled water, and 15 mL 0.5 M FeCl$_3$). Development of a pink color indicated IAA production. The optical density at 530 nm was measured using a spectrophotometer (Jenway 6305 UV spectrophotometer, 230 V/50 Hz, Staffordshire, UK), and the amount of IAA produced was estimated using a standard curve for authentic IAA [12].

Five isolates were chosen based on their ability to produce IAA for further analysis. Equal amount of each inoculum was added to nutrient broth medium containing 5 mg mL^{-1} tryptophan and incubated for 14 days at 35 ± 2 °C. IAA concentration was determined at 2 days intervals up to the 14th day after inoculation. Samples were centrifuged at 6000 rpm for 30 min and IAA production was determined as mentioned above. All the IAA production assays were performed in triplicates.

2.8. Effect of Bacterial Isolates on Zea mays L. Growth

2.8.1. Experimental Design

A pot experiment was conducted in a completely randomized design with five replicates of each treatment. Plants were inoculated with one of five individual bacterial isolates (*Brevibacillus* spp. Af.13, and Af.14, *Bacillus* spp. Fm.3, Fm.4, and Fm.6) or with a bacterial consortium formed of an equal amount of the five bacteria isolates. A control treatment consisted of uninoculated plants.

2.8.2. Culture Conditions

A loamy soil was collected from an agricultural field in the El-Menoufia governorate. Physical and chemical characteristics of the soil are shown in Table 1. The soil was air-dried, sieved with a 2 mm sieve, mixed with quartz sand at a soil: sand ratio of 3:1 and autoclaved twice for one hour at 121 °C. The five most potent IAA producing bacterial isolates (as listed above) were inoculated in nutrient broth and incubated at 35 ± 2 °C for 24 h on a shaker (LABOAO, LH-2102C, Zhingzhou, China) at 180 rpm. Seeds of maize (*Zea mays*, Cultivar Giza 9) were surface sterilized by soaking in 2.5% sodium hypochlorite for 3 minutes and then washed 5 times in sterile distilled water. Six groups of pregerminated seeds were separately incubated in 50 mL aliquots of the culture medium inoculated each with one of the five bacterial strains or with the bacterial consortium, and incubated at room temperature for 4 h on a shaker at 180 rpm. After incubation, the soaked seeds were sown in 1 L plastic pots filled with 900 g of sterilized soil-sand mixture. Each pot received three germinated seeds. Plants were grown in a greenhouse with a temperature of 25–30 °C and were irrigated with tap water as required without fertilization.

Table 1. Physical and chemical characteristics of the soil used in the greenhouse experiment.

Parameters	Soil Analysis
Soil Texture	Loamy Sand
Physical characters (%)	
Sand	76.8
Silt	10.9
Clay	12.2
Chemical characters (mg kg^{-1})	
P	24
K	14.075
Na	186.44
Ca	27.25
Cl	134.35

2.9. Plant Tissue Analysis

After 30 days, plants were harvested, shoot and root systems were separated, and roots were washed carefully with tap water to remove the attached soil particles. The dry weight of shoots and roots were measured after drying for 48 h at 60 °C. Phosphorus, nitrogen, and potassium contents were determined according to the methods described by AOAC international [43] and Rice et al. [44].

2.10. Statistical Analysis

Data were statistically analyzed using SPSS v17 (SPSS Inc., Chicago, IL, USA). When the normality and homogeneity of variance hypotheses were satisfied, one-way analysis of variance (ANOVA) was used to compare the bacterial isolates for extracellular enzymes production, antimicrobial activity,

IAA and ammonia production, P-solubilization ability, and the effect of these endophytes on maize growth performance. A posteriori multiple comparisons were done using Tukey's range tests at $p < 0.05$. All results are the means of three to five independent replicates, as specified above.

3. Results

Thirteen endophytic bacterial strains were isolated from the leaves of the two medicinal plants (Table 2). Nine strains were isolated from *F. mollis* plants and identified as *Bacillus* spp. (eight strains), and *Paenibacillus* sp. (one strain), while four bacterial strains were isolated from *A. fragrantissima* plants and identified as *Paenibacillus* sp. (one strain) and *Brevibacillus* sp. (three strains). The 16S rRNA gene sequences of strains Fm.2 to Fm.9 showed 96–99% sequence similarity with the sequences of *Bacillus amyloliquefaciens*, *Bacillus thuringiensis*, and *Bacillus cereus* (Figure 2 and Table 2). The 16S rRNA gene sequences of strains Fm.1 and Af.12 showed 99% of sequence similarity with *Paenibacillus barengoltzii*. Isolates Af.13 to Af.15 showed between 93 and 99% of 16S rRNA sequence similarity with *Brevibacillus agri*.

All bacterial endophytes isolated from *F. mollis* were positive for amylase, pectinase, carboxymethyl cellulase, cellulose, xylanase, and gelatinase, while those isolated from *A. fragrantissima* showed activities for only one to four enzymes (Table 3). The highest activities of cellulase and carboxymethyl cellulase were observed with *Bacillus* sp. Fm.5 (22.0 ± 1.1 and 21.3 ± 1.2 mm, respectively), while *Bacillus* sp. Fm.2 showed the highest activities of pectinase, xylanase, and gelatinase enzymes with clear zone 17.6 ± 0.6, 19.6 ± 0.3, and 22.0 ± 0.5 mm respectively. The highest gelatinase activity (22.3 ± 1.4 mm) was measured for *Bacillus* sp. Fm.3.

The antimicrobial activity of the bacterial endophytes against selected pathogenic bacterial and yeast strains are given in Table 4. The crude extract of *Brevibacillus* sp. Af.13 suppressed the growth of five tested pathogenic microorganisms, while *Bacillus* sp. Fm.8 inhibited the growth of *P. aeruginosa*, *S. typhi*, and *E. coli*. Endophytic strains *Bacillus* sp. Fm.2 and *Brevibacillus* sp. Af.13 were the only endophytes whose crude extracts showed an inhibitory effect against the pathogenic yeast *C. albicans* ATCC 10231 with clear zone 15 and 18 mm respectively. While the filtrates extracted from all strains showed some inhibition of *P. aeruginosa* ATCC 9027, the highest growth inhibition was noted from strains Fm.6, Fm.7, Fm.8, Fm.9, and Af.14 with inhibition zones ranging between 15 to 30 mm.

Table 2. The 16S rRNA sequence identification of endophytic bacterial strains from two different medicinal plants.

Plant Species	Bacterial Strain Code	Homologue Sequences (Sequence Identity %)	NCBI Accession Numbers
	Fm.1	*Paenibacillus barengoltzii* (99%)	NR_042756
	Fm.2	*Bacillus amyloliquefaciens* (98%)	NR_117946
	Fm.3	*Bacillus thuringiensis* (97%)	NR_043403
	Fm.4	*Bacillus cereus* (98%)	NR_115526
Fagonia mollis	Fm.5	*Bacillus cereus* (99%)	NR_115526
	Fm.6	*Bacillus thuringiensis* (98%)	NR_114581
	Fm.7	*Bacillus amyloliquefaciens* (97%)	NR_117946
	Fm.8	*Bacillus cereus* (97%)	NR_115526
	Fm.9	*Bacillus cereus* (96%)	NR_115526
	Af.12	*Paenibacillus barengoltzii* (99%)	NR_113900
	Af.13	*Brevibacillus agri* (95%)	NR_113767
Achillea fragrantissima	Af.14	*Brevibacillus agri* (93%)	NR_113767
	Af.15	*Brevibacillus agri* (99%)	NR_113767

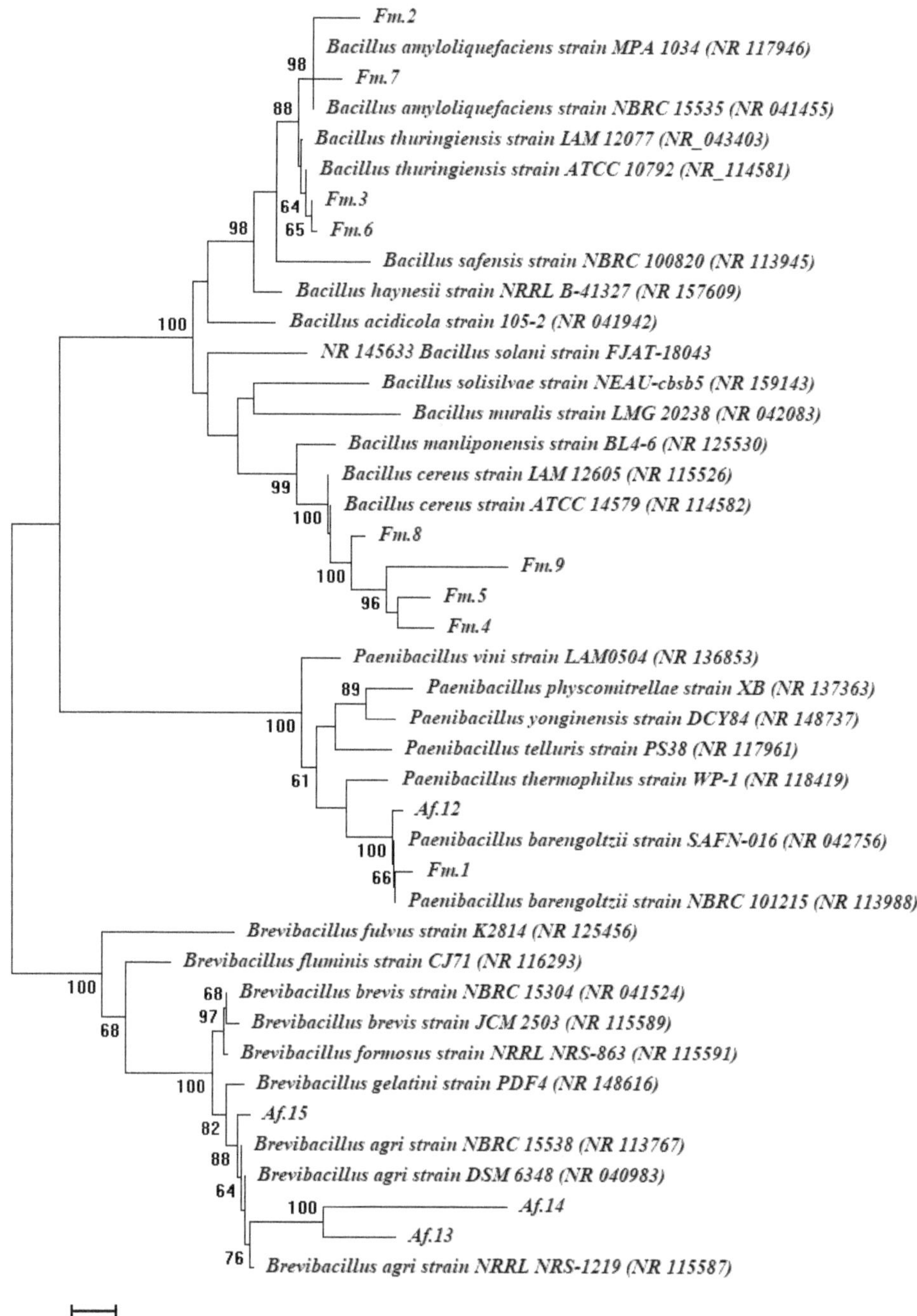

Figure 2. Phylogenetic analysis of 16S rRNA sequences of bacterial strains with reference sequences from NCBI. Fm.1–Fm.9 refers to 16S rRNA sequences of bacteria isolated from *Fagonia mollis* plants, whereas Af.13–Af.15 are the sequences from isolates from *Achillea fragrantissima*. Identity of the bacterial isolates is available in Table 2. The analysis was performed in MEGA 6 using the neighbor-joining method.

Table 3. Extracellular enzymatic activities of bacterial endophytes.

Bacterial Strains [1]	Diameter of Clear Zones (mm) [2]					
	Amylase	Pectinase	CMCase [3]	Cellulase	Xylanase	Gelatinase
C	0^d	0^d	0^e	0^d	0^f	0^f
Fm.1	17.6 ± 1.2^a	16.3 ± 0.3^b	18.6 ± 0.6^b	20.0 ± 1.7^b	17.0 ± 1.0^b	21.0 ± 0.5^a
Fm.2	17.3 ± 0.8^a	17.6 ± 0.6^a	20.6 ± 0.6^a	18.6 ± 0.8^c	19.6 ± 0.3^a	22.0 ± 0.5^a
Fm.3	17.0 ± 1.5^a	17.0 ± 0.5^a	18.0 ± 1.0^b	18.6 ± 1.2^c	16.0 ± 00.0^b	22.3 ± 1.4^a
Fm.4	17.6 ± 0.8^a	17.0 ± 0.5^a	19.0 ± 0.5^b	19.6 ± 1.7^b	14.6 ± 0.3^c	20.6 ± 0.6^b
Fm.5	17.6 ± 0.8^a	16.3 ± 0.8^b	21.3 ± 1.2^a	22.0 ± 1.1^a	18.6 ± 0.3^a	21.6 ± 0.8^a
Fm.6	17.0 ± 1.1^a	15.6 ± 0.3^c	18.3 ± 0.8^b	16.3 ± 0.3^d	16.3 ± 0.8^b	17.0 ± 1.5^d
Fm.7	17.3 ± 0.6^a	17.0 ± 0.5^a	15.6 ± 0.8^c	16.6 ± 1.2^d	14.3 ± 0.6^c	15.0 ± 1.5^e
Fm.8	17.6 ± 0.6^a	14.6 ± 0.3^c	18.0 ± 1.15^b	18.3 ± 0.3^c	14.3 ± 0.6^c	18.3 ± 1.6^c
Fm.9	15.3 ± 1.7^b	15.6 ± 0.3^c	18.6 ± 0.3^b	17.6 ± 0.8^c	9.0 ± 1.1^e	18.0 ± 1.5^c
Af.12	9.3 ± 0.3^c	0^d	19.3 ± 0.3^b	20.6 ± 0.6^b	13.6 ± 0.3^d	0^f
Af.13	0^d	0^d	0^e	0^d	0^f	19.0 ± 1.5^c
Af.14	0^d	14.0 ± 0.2^d	0^e	0^d	0^f	0^f
Af.15	0^d	17.6 ± 0.3^a	11.0 ± 1.0^d	17.33 ± 1.76^c	0^f	20.3 ± 2.3^b

[1] C: controls without bacterial inoculation. Identity of the bacterial isolates is available in Table 2. [2] Different letters between lines denote that mean values are significantly different ($p \leq 0.05$) by Tukey's test, means ± Standard Error (SE) ($n = 3$). [3] Carboxymethyl cellulase.

Table 4. Antimicrobial activities of bacterial endophytes.

Bacterial Strains [1]	Diameter of Clear Zone (mm)					
	P. aeruginosa	*S. typhi*	*E. coli*	*S. aureus*	*B. subtilis*	*C. albicans*
C	–	–	–	–	–	–
Fm.1	17	–	–	–	–	–
Fm.2	15	–	–	–	–	15
Fm.3	15	–	–	–	–	–
Fm.4	15	–	–	–	–	–
Fm.5	17	–	–	–	–	–
Fm.6	20	–	–	–	–	–
Fm.7	22	–	18	–	–	–
Fm.8	30	15	18	–	–	–
Fm.9	25	–	20	–	–	–
Af.12	15	–	–	–	–	–
Af.13	15	11	11	–	17	18
Af.14	20	–	–	–	–	–
Af.15	15	–	–	–	–	–

[1] C: controls without bacterial inoculation. Identity of the bacterial isolates is available in Table 2.

All endophytes identified as *Brevibacillus* sp. produced the highest amount of ammonia compared to *Bacillus* spp. strains (Table 5). Moreover, nine endophytes (Fm.2 to Fm.9 and Af.12) displayed significant ability to solubilize inorganic phosphate with clear zone on the Pikovskaya medium ranging from 7.6 ± 0.3 to 9.6 ± 0.3 mm. Results showed that all the isolated strains were IAA producers, with or without tryptophan (Figure 3). However, increasing tryptophan concentration from 1 to 5 mg·mL^{-1} resulted in increased bacterial ability to produce IAA from 10 to 60 µg·mL^{-1}. Strains of *Brevibacillus* spp. Af.14, Af.13, *Bacillus* sp. Fm.6, *Bacillus* sp. Fm 4, and *Bacillus* sp. Fm.3 produced the highest amount of IAA, and were selected for further analysis to measure the production of IAA at 2 day intervals in a time course over 14 days. The results indicated that the maximum IAA production with tryptophan was 5 mg mL^{-1} after 10 days. The results revealed that *Brevibacillus* sp. Af.14 produced the highest amount of IAA 59.7 µg·mL^{-1} ($p \leq 0.05$; Figure 4).

Table 5. Ammonia production and phosphate solubilization of endophytic bacterial strains.

Bacterial Strains [1]	Ammonia Production [2]	*P solubilization* Diameter of Clear Zone (mm) [3]
C	-	0 [d]
Fm.1	–	0 [d]
Fm.2	+	8.6 ± 0.3 [b]
Fm.3	++	9.0 ± 0 [a]
Fm.4	-	7.6 ± 0.3 [c]
Fm.5	++	9.6 ± 0.3 [a]
Fm.6	+	8.6 ± 0.3 [b]
Fm.7	+	8.3 ± 0.3 [b]
Fm.8	-	9.3 ± 0.3 [a]
Fm.9	++	9.3 ± 0.3 [a]
Af.12	++	9.3 ± 0.3 [a]
Af.13	++	0 [d]
Af.14	++	0 [d]
Af.15	++	0 [d]

[1] C: controls without bacterial inoculation. Identity of the bacterial isolates is available in Table 2. [2] -, +, and ++ denote no, low, and high ammonia production, respectively. [3] Different letters between columns denote that mean values are significantly different ($p \leq 0.05$) by Tukey's test, means ± SE ($n = 3$).

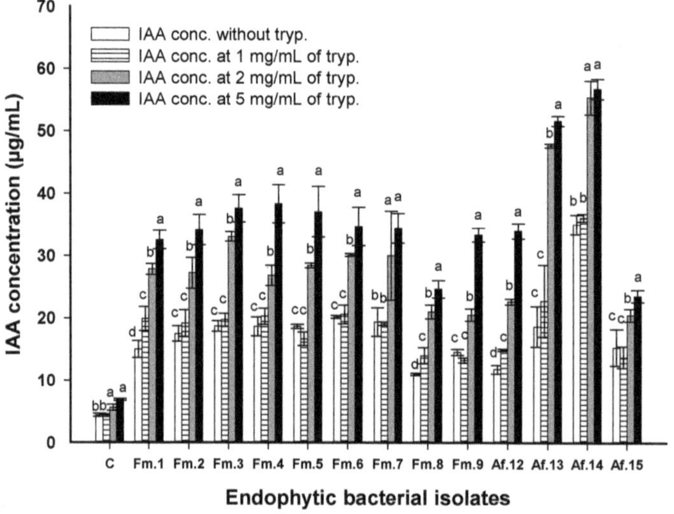

Figure 3. Quantitative production of IAA by endophytic bacterial strains with and without tryptophan. C, controls without bacterial inoculation. Identity of the bacterial isolates are available in Table 2. Data are statistically different at $p \leq 0.05$ by Tukey's test, ($n = 3$); error bars are means ± SE. Bars with the same letter for each endophytic isolate did not differ significantly, different letters on bars denote that mean values are significantly different at significant level of ($P \leq 0.05$), error bars are means ± SE.

In the greenhouse experiment, all maize plants inoculated with bacterial endophytes yielded significantly higher dry shoot weights ($F_{6,28} = 11.09$ and 10.33 respectively; $p \leq 0.001$) compared to the uninoculated control plants (Table 6). Plants inoculated with bacterial endophytes produced dry root weight higher than those recorded in control plants, but the differences were not significant ($F_{6,28} = 1.51$; $p = 0.21$).

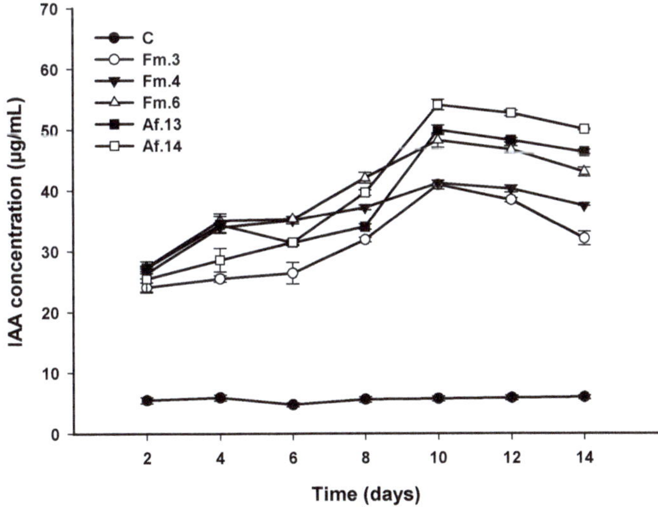

Figure 4. IAA production by the most potent bacterial strains in the presence of 5 mg mL^{-1} tryptophan and over 14 days. C, controls without any bacterial inoculation. Identity of the bacterial isolates is available in Table 2. At each time point, bars with the same letter did not differ significantly at a significant level of ($p \leq 0.05$) by Tukey's test, ($n = 3$).

Table 6. Effect of bacterial inoculations on the growth properties of maize plants.

Bacterial Strains [1]	Dry Weight (mg) [2]		Shoot Nutrients Content (mg)		
	Shoot	**Root**	**P**	**N**	**K**
C	82 ± 3.56 [c]	252.8 ± 19.5 [a]	0.42 ± 0.01 [c]	2.2 ± 0.17 [c]	8.70 ± 0.03 [b]
Fm.3	110.3 ± 5.5 [a]	322.64 ± 16.0 [a]	1.05 ± 0.07 [a]	3.9 ± 0.19 [b]	11.35 ± 0.92 [a]
Fm.4	81.8 ± 4 [c]	315.8 ± 27.9 [a]	0.72 ± 0.03 [b]	3.1 ± 0.60 [bc]	8.80 ± 0.62 [b]
Fm.6	103.2 ± 1.8 [a]	295.66 ± 30.1 [a]	0.76 ± 0.02 [b]	4.4 ± 0.18 [b]	11.11 ± 0.31 [a]
Af.13	108.2 ± 1.05 [a]	286.86 ± 19.3 [a]	0.44 ± 0.01 [c]	7.1 ± 0.18 [a]	10.57 ± 0.60 [ab]
Af.14	90.78 ± 4.55 [b]	303.8 ± 27.3 [a]	0.40 ± 0.02 [c]	7.2 ± 0.43 [a]	9.97 ± 0.13 [ab]
Mix	95.2 ± 2.8 [b]	249.6 ± 19.7 [a]	0.40 ± 0.01 [c]	3.4 ± 0.27 [bc]	10.45 ± 0.36 [ab]

[1] C: controls without bacterial inoculation. Identity of the bacterial isolates is available in Table 2. Mix, bacterial consortium consists of Fm.3, Fm.4, Fm.6, Af.13, and Af.14. [2] Different letters between columns denote that mean values are significantly different ($p \leq 0.05$) by Tukey's test, means ± SE ($n = 5$).

Inoculation of maize plants with *Bacillus* spp. Fm.3, Fm.4, and Fm.6 significantly (F6,14 = 49.07; $p \leq 0.001$) increased P shoot contents (1.05 ± 0.07, 0.72 ± 0.03, and 0.76 ± 0.02 mg respectively) as compared to the un-inoculated control plants (0.42 ± 0.01 mg), while *Brevibacillus* spp. Af.13, Af.14, and the bacterial consortium formed by a mixture of the five isolates did not affect shoot P content compared to control plants (Table 6). Analysis showed that bacterial inoculation significantly increased N shoot contents compared to the control plants (F6,14 = 35.76, $p \leq 0.001$; Table 6). Plants inoculated with *Bacillus* spp. Fm.3, and Fm.6, *Brevibacillus* spp. Af.13, and Af.14 had significantly higher ($p < 0.05$) N contents (range of 3.9 ± 0.19 to 7.2 + 0.43 mg) than those treated with other strains or uninoculated. The bacterial strains *Bacillus* spp. Fm.3 and Fm.6 significantly (F6,14 = 4.15; $p = 0.013$) increased K shoot contents (11.35 ± 0.92 and 11.11 ± 0.31 mg) in comparison with uninoculated control plants (8.70 ± 0.03 mg).

4. Discussion

In this study, 13 putative bacterial endophytic strains were isolated from two medicinal plants growing under the adverse conditions of the Sinai desert. Nine bacterial endophytes were isolated from *F. mollis* and identified as different species of *Bacillus*, and *Paenibacillus*, and four bacterial endophytes were isolated from *A. fragrantissima* and identified as *Paenibacillus* spp. and *Brevibacillus* spp. (Table 2). The plant growth promoting (PGP) activities of these bacterial strains were characterized, including extracellular enzyme production, antimicrobial action, IAA and ammonia production, and P-solubilization. In the same regard, Eida et al. [45] reported isolation of endosphere and rhizosphere bacterial groups associated with four native Saudi desert plants and proved their plant growth promotion potential including phosphate solubilization and IAA production. Based on PGP characteristics, five endophytic bacterial strains were selected to evaluate their effects on plant growth and development. Results showed that the selected endophytes have key PGP properties, and significantly increased dry weight of tissues, and P concentrations in shoots of maize plants compared with uninoculated controls. Corresponding with our results, Marag and Suman [42] isolated six bacterial endophytes including *Bacillus cereus* from two cultivars of maize, and the pot experiment indicates the efficacy of the isolates in improving biomass parameters of inoculated maize plants, in addition to compensating for approximately 25% of the NPK fertilizer input.

The bacterial endophytes exhibited different enzymatic activities involving cellulase, pectinase, xylanase, amylase, and gelatinase production [46,47]. Cellulolytic and pectinolytic activities are known to enable microorganisms to penetrate plant tissues and establish a symbiotic relationship with their host plants. The *Bacillus* spp. strain isolated in this study showed high hydrolytic activity for cellulose and pectin, as well as proteolytic activity. Similarly, different endophytic strains of *Bacillus* were shown to be strong producers of cellulase and pectinase [48]. The extracellular hydrolytic enzymes produced by endophytes contribute indirectly to plant growth promotion and protection against pathogens [49,50]. The endophytes can be described as bioproducers for amylases and xylanases based on their amylolytic and xylanolytic activities. Similarly, bacterial endophytes isolated from mangrove plants had activities associated with amylases [50]. The diverse enzymatic activities of the isolated endophytes showed their capability to catalyze different biochemical reactions and their potential for agricultural and industrial applications. Likewise, Castro et al (2014) isolated endophytic *Bacillus* from two Brazilian mangrove species, the isolates displayed extracellular amylase, esterase, lipase, protease, and endoglucanase activities and thus can be used in industrial applications [50]. Moreover, theses enzymes could enable endophytes to penetrate plant tissues and build a symbiotic relationship with their host plant, besides protecting the host from pathogens by hydrolysis of the pathogen cell wall [10].

Antimicrobial activities of the isolated endophytes were evaluated based on the suppression of microbial growth caused by the crude extracts. The estimation of antimicrobial activity of crude extracts is the initial step required for the discovery of new antimicrobial compounds. Selection of bacterial isolates as inoculants based on their PGP traits, and on their inhibitory effect against different pathogens, has received attention and has been suggested as an approach to enhance plant growth and protect plants against diseases [51]. In the current study, the isolated endophytes showed a significant antagonistic effect against different pathogenic microorganisms. Endophytic bacteria can indirectly assist plant growth through the production of substances, which inhibit plant pathogens [52,53]. Endophytes isolated from other medicinal plants have also produced novel bioactive compounds [49,54]. Hassan [10] isolated six bacterial endophytes including *Bacillus cereus* and *Bacillus subtilis* from the native desert medicinal plant *Teucrium polium* L., the isolates manifested variable broad-spectrum activity against *Pseudomonas aeruginosa*, *Salmonella typhimurium*, *Escherichia coli*, *Staphylococcus aureus*, *Bacillus subtilis*, and *Candida albicans*. Accordingly, suggesting their application as biocontrol agents [10].

Sun et al. [55] showed that 10 endophytic bacterial strains of *Bacillus* and *Streptomyces* isolated from *Polygonum cuspidatum* exhibited antagonistic effects against different plant pathogens. *Bacillus licheniformis* and *Bacillus pumilus* endophytic isolates from *Platycodon grandiflorum* roots also exhibited a significant antifungal action against *Phytophthora capsici*, *Fusarium oxysporum*, *Rhizoctonia solanic*, and *Pythium ultimum*. The endophytic bacterial strain *Paenibacillus* sp. IIRAC-30 was isolated from cassava and suppressed the growth of *Rhizoctonia solani* [56]. *Bacillus amyloliquefaciens* was isolated from the Chinese medicinal plant *Scutellaria baicalensis* Georgi. A crude extract of this strain exhibited antagonistic effects against some plant pathogens, food-borne pathogenic and spoilage microorganisms [53].

The PGP properties of bacteria have been investigated to select bacteria with high potential to be used as biofertilizers. These tests are critical in light of the fact that they identify bacteria with higher benefits for plants before testing them in field traits [57]. Ammonia and IAA production, as well P-solubilization, are among various mechanisms exhibited by bacteria that enhance plant growth [58]. Here, most endophytic bacterial isolates were able to produce different amounts of ammonia. It is often found that ammonia-producing bacteria can supply ammonia as a nitrogen source for plant growth [59]. Bacterial endophytes can enhance plant growth through the production of ammonia through the hydrolysis of urea into ammonia and carbon dioxide [60]. With regard to P-solubilization, most of the isolated endophytes showed variable capacity to solubilize phosphate. Rodrigues et al. [61] found that about 47% of bacterial endophytes isolated from sugarcane have low P-solubilizing indices. At soil with low phosphate supply, inoculation of P-solubilizing endophytic bacteria leading to increase of pant growth performance.

Indole-3-acetic acid (IAA) is a phytohormone that can be produced by plants and various microorganisms. This hormone not only enhances plant growth but also contributes in the interaction between plants and microorganisms [62]. In this study, all endophytic bacterial strains had the ability to produce IAA in the absence and presence of tryptophan, the precursor for IAA production. Although most microorganisms utilize tryptophan in IAA synthesis [63,64], the advantage of bacterial endophytes is that they can produce IAA without tryptophan supplementation. Rodrigues et al. [61] showed that 57% of bacterial endophytes secreted high IAA concentration of 21.05–139.21 μg mL^{-1} in 72 h in the presence of 5 mM tryptophan. Endophytic bacterial strains were shown to produce higher IAA concentrations than rhizospheric strains, suggesting a closer link, and potential symbiosis, between endophytes and their hosts [12]. Thus, in the current study, a higher capacity to produce IAA was used to select five bacterial strains to determine their effect on maize growth performance. Bokhari et al. [65] isolated *Bacillus circulans* PK3-138 from plants grown in Pakistan desert, reported the potency of this isolate for IAA production. Similarly, four bacterial endophytes (*Sphingomonas* sp., *Bacillus* sp., *Pantoea* sp., and *Enterobacterc* sp.) isolated from the roots of elephant grass showed valuable PGP traits including IAA production at a range of 10.50–759.19 mg/L, and ammonia production capacity. So, these inoculants could be used for increasing crop yield in a sustainable mode [58].

We found that inoculated plants produced more biomass than uninoculated plants. Plant–microbe interactions are well known to influence nutrient transfer between microorganisms and plants [66]. Therefore, it is possible that plant biomass production varied with different microbial taxa assemblages in the roots due to their various abilities to supply nutrients to their host. The results showed that the shoot P concentration was significantly increased in plants inoculated with Fm.3, Fm.4, and Fm.6 compared to the uninoculated plants. P-solubilizing bacteria help plants to access insoluble forms of phosphate, such as apatite, through excretion of protons and organic acids, mainly gluconic acid, rendering phosphate available to plants for uptake [11,67]. These bacteria can also produce enzymes that mineralize organic phosphorus, which also render it available for plants [67]. The capacity of microorganisms to absorb immobile nutrients such as P from soils and transfer it to their host plants is one of the main effects of microbial symbiosis; however, microbial capacity for nutrient transfer varies with different microorganisms [68]. Basically, plant roots can be colonized simultaneously by multiple microorganisms, which can positively or negatively benefit the host plant [69,70].

Importantly, not only did the isolated endophytic *Bacillus* and *Brevibacillus* species display the highest level of IAA and ammonia production, but they also had various plant growth-promoting traits. Bacteria that were isolated and characterized in the present study are potential candidates for plant bioinoculation in agricultural practices, in particular those that inhibited pathogens and harbored the highest levels of IAA production and of nutrient uptake.

5. Conclusions

Very few isolates were obtained herein to claim that, bacterial endophytes inhabiting the two studied medicinal plants, *F mollis* and *A. fragrantissima*, mainly belong to *Bacillus* and *Brevibacillus* spp. Bacterial endophytes characterization including extracellular enzymatic activity, antimicrobial actions, P-solubilization activity, ammonia, and IAA production were performed in terms of their plant growth-promoting abilities in-vitro and in plants. Five bacterial strains identified as *Brevibacillus agri* Af.13, and Af.14, *Bacillus* sp. Fm3, *Bacillus* sp. Fm.4, and *Bacillus* sp. Fm.6 were selected and inoculated into maize plant to increase their growth performance under normal conditions. These endophytic bacterial isolates significantly promote plant growth and increase P and N shoot contents of maize plant. However, in order to demonstrate the beneficial role of these bacterial endophytes in plant growth promotion of their host plants, particularly under real field conditions, further investigation of their mechanisms of colonization and competition against other soil microbial communities will be required.

Author Contributions: Conceptualization, A.F., A.M.E., E.E.-D.E., S.E.-D.H., K.A.A.A., Y.M.H., K.A.A. and M.D.F.A.; methodology, A.F., A.M.E., E.E.-D.E., S.E.-D.H., K.A.A.A., Y.M.H., K.A.A. and M.D.F.A.; software, A.F., A.M.E., F.A.-O., M.H., M.S.-A., E.E.-D.E., S.E.-D.H., K.A.A.A., Y.M.H., K.A.A. and M.D.F.A.; formal analysis, A.F., A.M.E., E.E.-D.E., S.E.-D.H., F.A.-O., M.H., M.S.-A., K.A.A.A., Y.M.H., K.A.A., A.M.E. and N.K.; investigation, A.F., A.M.E., E.E.-D.E., S.E.-D.H., F.A.-O., M.H., M.S.-A., K.A.A.A., and N.K.; resources, A.F., A.M.E., E.E.-D.E., S.E.-D.H., K.A.A.A., Y.M.H.; data curation, A.F., A.M.E., E.E.-D.E., S.E.-D.H., F.A.-O., M.H., M.S.-A., K.A.A.A., Y.M.H. and K.A.A.; writing—original draft preparation, A.F., A.M.E., E.E.-D.E., F.A.-O., M.H., M.S.-A., S.E.-D.H. and K.A.A.A.; writing—review and editing, A.F., A.M.E., E.E.-D.E., S.E.-D.H., K.A.A.A., Y.M.H., K.A.A., A.M.E., M.A. and N.K.; visualization, A.F., A.M.E., E.E.-D.E., S.E.-D.H., K.A.A.A., Y.M.H. and K.A.A.; funding acquisition, M.D.F.A., K.A.A., Y.M. and K.A.A.A. All authors have equal contribution. All authors have read and agreed to the published version of the manuscript.

Funding: This research was funded by the Deanship of Scientific Research at Princess Nourah bint Abdulrahman University through the Fast-track Research Funding Program.

Acknowledgments: The authors extend their appreciation to the Deanship of Scientific Research at Princess Nourah bint Abdulrahman University for funding this research through the Fast-track Research Funding Program.

Conflicts of Interest: The authors declare no conflict of interest.

References

1. Sun, R.; Zhang, X.; Guo, X.; Wang, D.; Chu, H. Bacterial diversity in soils subjected to long-term chemical fertilization can be more stably maintained with the addition of livestock manure than wheat straw. *Soil Biol. Biochem.* **2015**, *88*, 9–18. [CrossRef]

2. Zhou, J.; Jiang, X.; Zhou, B.; Zhao, B.; Ma, M.; Guan, D.; Li, J.; Chen, S.; Cao, F.; Shen, D.; et al. Thirty four years of nitrogen fertilization decreases fungal diversity and alters fungal community composition in black soil in northeast China. *Soil Biol. Biochem.* **2016**, *95*, 135–143. [CrossRef]

3. Rafi, M.M.; Krishnaveni, M.S.; Charyulu, P.B.B.N. Phosphate-Solubilizing Microorganisms and Their Emerging Role in Sustainable Agriculture. In *Recent Developments in Applied Microbiology and Biochemistry*; Buddolla, V., Ed.; Academic Press: Cambridge, MA, USA; Elsevier: Amsterdam, The Netherlands, 2019; Volume 17, pp. 223–233. [CrossRef]

4. De La, T.; Neyser, V.R.; Clara, I.R.; Martha, R.; Carlos, A.; Federico, A.G.; Héctor, P.; Reiner, R. Effect of plant growth-promoting bacteria on the growth and fructan production of *Agave americana* L. *Brazilian J. Microbiol.* **2016**, *47*, 587–596. [CrossRef] [PubMed]

5. Gamez, R.; Cardinale, M.; Montes, M.; Ramirez, S.; Schnell, S.; Rodriguez, F. Screening, plant growth promotion and root colonization pattern of two rhizobacteria (*Pseudomonas fluorescens* Ps006 and *Bacillus amyloliquefaciens* Bs006) on banana cv. Williams (*Musa acuminata* Colla). *Microbiol. Res.* **2019**, *220*, 12–20. [CrossRef] [PubMed]

6. Abdelaal, K.A.A.; Tawfik, S.F. Response of Sugar Beet Plant (*Beta vulgaris* L.) to Mineral Nitrogen Fertilization and Bio-Fertilizers. *Int. J. Curr. Microbiol. Appl. Sci.* **2015**, *4*, 677–688.

7. Abdelaal, K.A.A. Pivotal Role of Bio and Mineral Fertilizer Combinations on Morphological, Anatomical and Yield Characters of Sugar Beet Plant (*Beta vulgaris* L.). *Middle East J. Agric. Res.* **2015**, *4*, 717–734.

8. Abdelaal, K.A.A.; Badawy, S.A.; Abdel Aziz, R.M.; Neana, S.M.M. Effect of mineral nitrogen levels and PGPR on morphophysiological characters of three sweet sorghum varieties (*Sorghum bicolor* L. Moench). *J. Plant Prod.* **2015**, *6*, 189–203. [CrossRef]

9. Hassan, S.E.; Salem, S.S.; Fouda, A.; Awad, M.A.; El-Gamal, M.S.; Abdo, A.M. New approach for antimicrobial activity and bio-control of various pathogens by biosynthesized copper nanoparticles using endophytic actinomycetes. *J. Radiation Res. Appl. Sci.* **2018**, *11*, 262–270. [CrossRef]

10. Hassan, S.E. Plant growth-promoting activities for bacterial and fungal endophytes isolated from medicinal plant of *Teucrium polium* L. *J. Adv. Res.* **2017**, *8*, 687–695. [CrossRef]

11. Naseem, H.; Ahsan, M.; Shahid, M.A.; Khan, N. Exopolysaccharides producing rhizobacteria and their role in plant growth and drought tolerance. *J. Basic Microbiol.* **2018**, *58*, 1009–1022. [CrossRef]

12. Taktek, S.; St-Arnaud, M.; Yves Piché, J.; Fortin, A.; Antoun, H. Igneous phosphate rock solubilization by biofilm-forming mycorrhizobacteria and hyphobacteria associated with *Rhizoglomus irregulare* DAOM 197198. *Mycorrhiza* **2017**, *27*, 13–22. [CrossRef] [PubMed]

13. Abbamondi, G.R.; Tommonaro, G.; Weyens, N.; Thijs, S.; Sillen, W.; Gkorezis, P.; Iodice, C.; Rangel, M.; Nicolaus, B.; Vangronsveld, J. Plant growth-promoting effects of rhizospheric and endophytic bacteria associated with different tomato cultivars and new tomato hybrids. *Chem. Biol. Technol. Agric.* **2016**, *3*, 1. [CrossRef]

14. Hafez, Y.M.; Attia, K.A.; Kamel, S.; Alamery, S.; El-Gendy, S.; Al-Dosse, A.; Mehiar, F.; Ghazy, A.; Abdelaal, K.A.A. *Bacillus subtilis* as a bio-agent combined with nano molecules can control powdery mildew disease through histochemical and physiobiochemical changes in cucumber plants. *Physiol. Mol. Plant Path.* **2020**, *111*, 101489. [CrossRef]

15. Hafez, Y.; Emeran, A.; Esmail, S.; Mazrou, Y.; Abdrabbo, D.; Abdelaal, K.A.A. Alternative treatments improve physiological characters, yield and tolerance of wheat plants under leaf rust disease stress. *Fresenius Environ. Bull.* **2020**, *29*, 4738–4748.

16. St-Arnaud, M.; Vujanovic, V. Effect of the arbuscular mycorrhizal symbiosis on plant diseases and pests. In *Mycorrhizae in Crop Production*; Hamel, C., Plenchette, C., Eds.; Haworth Press: Binghampton, NY, USA, 2007; pp. 67–122.

17. Hafez, Y.M.; Abdelaal, K.A.A.; Badr, M.M.; Esmaeil, R.A. Control of Puccinia triticina the causal agent of wheat leaf rust disease using safety resistance inducers correlated with endogenously antioxidant enzymes up-regulation. *Egyptian J. Biol. Pest Control* **2017**, *27*, 1–10.

18. Fouda, A.; Abdel-Maksoud, G.; Abdel-Rahman, M.A.; Eid, A.M.; Barghoth, M.G.; El-Sadany, M.A. Monitoring the effect of biosynthesized nanoparticles against biodeterioration of cellulose-based materials by Aspergillus niger. *Cellulose* **2019**, *26*, 6583–6597. [CrossRef]

19. Fouda, A.; Abdel-Maksoud, G.; Abdel-Rahman, M.A.; Salem, S.S.; Hassan, S.E.; El-Sadany, M.A. Eco-friendly approach utilizing green synthesized nanoparticles for paper conservation against microbes involved in biodeterioration of archaeological manuscript. *Int. Biodet. Biodegr.* **2019**, *142*, 160–169. [CrossRef]

20. Fouda, A.; Hassan, S.E.; Salem, S.S.; Shaheen, T.I. In-Vitro cytotoxicity, antibacterial, and UV protection properties of the biosynthesized Zinc oxide nanoparticles for medical textile applications. *Microb. Pathog.* **2018**, *125*, 252–261. [CrossRef]

21. Cherif, H.; Marasco, R.; Rolli, E.; Ferjani, R.; Fusi, M.; Soussi, A.; Mapelli, F.; Blilou, I.; Borin, S.; Boudabous, A.; et al. Oasis desert farming selects environment-specific date palm root endophytic communities and cultivable bacteria that promote resistance to drought. *Environ. Microb. Rep.* **2015**, *7*, 668–678. [CrossRef]

22. Khan, N.; Bano, A.; Rahman, M.A.; Rathinasabapathi, B.; Babar, M.A. UPLC-HRMS-based untargeted metabolic profiling reveals changes in chickpea (*Cicer arietinum*) metabolome following long-term drought stress. *Plant Cell Environ.* **2019**, *42*, 115–132. [CrossRef]

23. Köberl, M.; Müller, H.; Ramadan, E.M.; Berg, G. Desert Farming Benefits from Microbial Potential in Arid Soils and Promotes Diversity and Plant Health. *PLoS ONE* **2011**, *6*, e24452. [CrossRef] [PubMed]

24. Khan, N.; Bano, A.; Babar, M.A. The stimulatory effects of plant growth promoting rhizobacteria and plant growth regulators on wheat physiology grown in sandy soil. *Arch. Microbiol.* **2019**, *201*, 769–785. [CrossRef] [PubMed]

25. Hanna, A.L.; Youssef, H.H.; Amer, W.M.; Monib, M.; Fayez, M.; Hegazi, N.A. Diversity of bacteria nesting the plant cover of north Sinai deserts, Egypt *J. Adv. Res.* **2013**, *4*, 13–26. [CrossRef] [PubMed]

26. Selim, K.A.; El-Beih, A.A.; AbdEl-Rahman, T.M.; El-Diwany, A.I. Biodiversity and antimicrobial activity of endophytes associated with Egyptian medicinal plants. *Mycosphere* **2011**, *2*, 669–678. [CrossRef]

27. ALKahtani, M.D.F.; Attia, K.A.; Hafez, Y.M.; Khan, N.; Eid, A.M.; Ali, M.A.M.; Abdelaal, K.A.A. Chlorophyll Fluorescence Parameters and Antioxidant Defense System Can Display Salt Tolerance of Salt Acclimated Sweet Pepper Plants Treated with Chitosan and Plant Growth Promoting Rhizobacteria. *Agronomy* **2020**, *10*, 1180. [CrossRef]

28. Kiarie, S.; Nyasani, J.O.; Gohole, L.S.; Maniania, N.K.; Subramanian, S. Impact of Fungal Endophyte Colonization of Maize (*Zea mays* L.) on Induced Resistance to Thrips and Aphid-Transmitted Viruses. *Plants* **2020**, *9*, 416. [CrossRef]

29. Abd el Fatah, H.Y.; Mohamed, E.A.; Hassan, M.B.; Mohamed, K.A. An Economic Analysis for Maize Market in Egypt. *Middle East J. Agric. Res.* **2015**, *4*, 873–878.

30. Berg, G.; Grube, M.; Schloter, M.; Smalla, K. Unraveling the plant microbiome: Looking back and future perspectives. *Front. Microbiol.* **2014**, *5*, 148. [CrossRef]

31. Arora, S.; Patel, P.N.; Vanza, M.J.; Rao, G.G. Isolation and characterization of endophytic bacteria colonizing halophyte and other salt tolerance plant species from coastal Gujarta. *Afr. J. Microb. Res.* **2014**, *8*, 1779–1788.

32. Miller, D.N.; Bryant, J.E.; Madsen, E.L.; Ghiorse, W.C. Evaluation and optimization of DNA extraction and purification procedures for soil and sediment samples. *Appl. Environ. Microbiol.* **1999**, *65*. [CrossRef]

33. Lane, D.J. 16S/23S rRNA sequencing. In *Nucleic Acids Techniques in Bacterial Systematics*; Stackebrandt, E., Goodfellow, M., Eds.; John Wiley & Sons: Chichester, UK, 1991.

34. Lv, Y.; Zhang, F.; Chen, J.; Cui, J.; Xing, Y.; Li, X.; Guo, S. Diversity and antimicrobial activity of endophytic fungi associated with the alpine plant *Saussurea involucrata*. *Biol. Pharm. Bull.* **2010**, *33*, 1300–1306. [CrossRef]

35. Yadav, R.; Chauhan, N.; Chouhan, A.S.; Soni, V.K.; Omray, L. Antimicrobial screening of various extracts of *Aphanmixis polystachya* stems bark. *Int. J. Adv. Pharm. Sci.* **2010**, *1*, 147–150.

36. Mohamed, A.A.; Fouda, A.; Abdel-Rahman, M.A.; Hassan, S.E.; El-Gamal, M.S.; Salem, S.S.; Shaheen, T.I. Fungal strain impacts the shape, bioactivity and multifunctional properties of green synthesized zinc oxide nanoparticles. *Biocat. Agric. Biotech.* **2019**, *19*, 101103. [CrossRef]

37. Sharaf, O.M.; Al-Gamal, M.S.; Ibrahim, G.A.; Dabiza, N.M.; Salem, S.S.; El-ssayad, M.F.; Youssef, A.M. Evaluation and characterization of some protective culture metabolites in free and nano-chitosan-loaded forms against common contaminants of Egyptian cheese. *Carbohyd. Polym.* **2019**, *223*, 115094. [CrossRef] [PubMed]

38. Alsharif, S.M.; Salem, S.S.; Abdel-Rahman, M.A.; Fouda, A.; Eid, A.M.; Hassan, S.E.; Awad, M.A.; Mohamed, A.A. Multifunctional properties of spherical silver nanoparticles fabricated by different microbial taxa. *Heliyon* **2020**, *6*, 3943. [CrossRef]

39. Fouda, A.; Hassan, S.E.-D.; Abdo, A.M.; El-Gamal, M.S. Antimicrobial, Antioxidant and Larvicidal Activities of Spherical Silver Nanoparticles Synthesized by Endophytic *Streptomyces* spp. *Biol. Trace Element Res.* **2020**, *195*, 707–724. [CrossRef]

40. Jasim, B.; John, C.; Jyothis, M.; Radhakrishnan, E.K. Plant growth promoting potential of endophytic bacteria isolated from *Piper nigrum*. *Plant Growth Regul* **2013**, *71*, 1–11. [CrossRef]

41. Singh, P.; Kumar, V.; Agrawal, S. Evaluation of phytase producing bacteria for their plant growth promoting activities. *Intern. J. Microbiol.* **2014**, 426483. [CrossRef]

42. Marag, P.S.; Suman, A. Growth stage and tissue specific colonization of endophytic bacteria having plant growth promoting traits in hybrid and composite maize (*Zea mays* L.). *Microbiol. Rese* **2018**, *214*, 101–113. [CrossRef]

43. AOAC. International Official Methods of Analysis of AOAC International. In *Official Methods of Analysis of AOAC International*; George, W., Ed.; Association of Official Analytical Chemists: Rockville, MD, USA, 2016; p. 3172.

44. Rice, E.W.; Baird, R.B.; Eaton, A.D. *Standard Methods for the Examination of Water and Wastewater*, 23rd ed.; American Public Health Association, American Water Works Association: Washington, DC, USA, 2017.

45. Eida, A.A.; Ziegler, M.; Lafi, F.F.; Michell, C.T.; Voolstra, C.R.; Hirt, H.; Saad, M.M. Desert plant bacteria reveal host influence and beneficial plant growth properties. *PLoS ONE* **2018**, *13*, e0208223. [CrossRef] [PubMed]

46. Eid, A.M.; Salim, S.S.; Hassan, S.S.; Ismail, M.A.; Fouda, A. Role of Endophytes in Plant Health and Abiotic Stress Management. In *Microbiome in Plant Health and Disease: Challenges and Opportunities*; Kumar, V., Prasad, R., Kumar, M., Choudhary, D.K., Eds.; Springer: Singapore, 2019; pp. 119–144.

47. Fouda, A.; Hassan, S.E.; Eid, A.M.; Ewais, E.E. The Interaction Between Plants and Bacterial Endophytes Under Salinity Stress. In *Endophytes and Secondary Metabolites*; Jha, S., Ed.; Springer Nature: Cham, Switzerland, 2019; pp. 591–607.

48. Choi, Y.W.; Hodgkiss, I.; Hyde, K. Enzyme production by endophytes of *Brucea javanica*. *J. Agric. Tech.* **2005**, *1*, 55–66.

49. Glick, B.R. Plant growth-promoting bacteria: Mechanisms and applications. *Scientifica* **2012**, 1–15. [CrossRef] [PubMed]

50. Castro, R.A.; Quecine, M.C.; Lacava, P.T.; Batista, B.D.; Luvizotto, D.M.; Marcon, J.; Ferreira, A.; Melo, I.S.; Azevedo, J.L. Isolation and enzyme bioprospection of endophytic bacteria associated with plants of Brazilian mangrove ecosystem. *SpringerPlus* **2014**, *3*, 382. [CrossRef] [PubMed]

51. Vurukonda, S.S.P.; Giovanardi, D.; Stefani, E. Plant growth promoting and biocontrol activity of *Streptomyces* spp. as endophytes. *Intern. J. Mol. Sci.* **2018**, *19*, 952. [CrossRef] [PubMed]

52. Chow, Y.Y.; Rahman, S.; Ting, A.S.Y. Understanding colonization and proliferation potential of endophytes and pathogen in planta via plating, polymerase chain reaction, and ergosterol assay. *J. Adv. Res.* **2017**, *8*, 13–21. [CrossRef]

53. Khan, N.; Zandi, P.; Ali, S.; Mehmood, A.; Adnan Shahid, M.; Yang, J. Impact of salicylic acid and PGPR on the drought tolerance and phytoremediation potential of Helianthus annus. *Front. Microbiol.* **2018**, *9*, 2507. [CrossRef]

54. Asraful, I.; Md, S.; Math, R.K.; Kim, J.M.; Yun, M.G.; Cho, J.J.; Kim, E.J.; Lee, Y.H.; Yun, H.D. Effect of plant age on endophytic bacterial diversity of balloon flower (*Platycodon grandiflorum*) root and their antimicrobial activities. *Curr. Microbiol.* **2010**, *61*, 346–356. [CrossRef]

55. Sun, L.; Lu, Z.; Bie, X.; Lu, F.; Yang, S. Isolation and characterization of a co-producer of fengycins and surfactins, endophytic *Bacillus amyloliquefaciens* ES-2, from *Scutellaria baicalensis* Georgi. *World J. Microbiol. Biotechnol.* **2006**, *22*, 1259–1266. [CrossRef]

56. Canova, S.P.; Petta, T.; Reyes, L.F.; Zucchi, T.; Moraes, L.A.; Melo, I. Characterization of lipopeptides from *Paenibacillus* sp. (IIRAC30) suppressing *Rhizoctonia solani*. *World J. Microbiol. Biotechnol.* **2010**, *26*, 2241–2247. [CrossRef]

57. Biasolo, G.; Kucmanski, D.A.; Salamoni, S.P.; Gardin, J.P.; Minotto, E.; Baratto, C.M. Isolation, characterization and selection of bacteria that promote plant growth in grapevines (*Vitis* sp.). *J. Agric. Sci.* **2017**, *9*, 184–194. [CrossRef]

58. Li, X.; Geng, X.; Xie, R.; Fu, L.; Jiang, J.; Gao, L.; Sun, J. The endophytic bacteria isolated from elephant grass (*Pennisetum purpureum* Schumach) promote plant growth and enhance salt tolerance of hybrid Pennisetum. *Biotech. Biofuels* **2016**, *9*, 190. [CrossRef] [PubMed]

59. Banik, A.; Dash, G.K.; Swain, P.; Kumar, U.; Mukhopadhyay, S.K.; Dangar, T.K. Application of rice (*Oryza sativa* L.) root endophytic diazotrophic *Azotobacter* sp. strain Avi2 (MCC 3432) can increase rice yield under green house and field condition. *Microbiol. Res.* **2019**, *219*, 56–65. [CrossRef] [PubMed]

60. Khan, N.; Bano, A.; Curá, J.A. Role of Beneficial Microorganisms and Salicylic Acid in Improving Rainfed Agriculture and Future Food Safety. *Microorganisms* **2020**, *8*, 1018. [CrossRef] [PubMed]

61. Rodrigues, A.A.; Forzani, M.V.; Soares, R.; Sibov, S.T.; Vieira, J. Isolation and selection of plant growth-promoting bacteria associated with sugarcane. *Agric. Res. Tropics* **2016**, *46*, 149–158. [CrossRef]

62. Lin, L.; Zhengyi, L.; Chunjin, H.; Zhang, X.; Chang, S.; Yang, L.; Yangrui, L.; Qianli, A. Plant growth-promoting nitrogen-fixing *Enterobacteria* are in association with sugarcane plants growing in Guangxi, China. *Microbes Environ.* **2012**, *27*, 391–398. [CrossRef]

63. Fouda, A.; Hassan, S.E.; Eid, A.M.; Ewais, E.E. Biotechnological applications of fungal endophytes associated with medicinal plant *Asclepias sinaica* (Bioss). *Ann. Agric. Sci.* **2015**, *60*, 95–104. [CrossRef]

64. Hassan, S.E.; Fouda, A.; Radwan, A.A.; Salem, S.S.; Barghoth, M.G.; Awad, M.A.; Abdo, A.M.; El-Gamal, M.S. Endophytic actinomycetes Streptomyces spp mediated biosynthesis of copper oxide nanoparticles as a promising tool for biotechnological applications. *JBIC* **2019**, *24*, 377–393. [CrossRef]

65. Bokhari, A.; Essack, M.; Lafi, F.F.; Andres-Barrao, C.; Jalal, R.; Alamoudi, S.; Saad, M.M. Bioprospecting desert plant Bacillus endophytic strains for their potential to enhance plant stress tolerance. *Sci. Rep.* **2019**, *9*. [CrossRef]

66. Orozco, M.; Ma del, C.; Ma del, C.R.; Bernard, R.G.; Gustavo, S. Microbiome engineering to improve biocontrol and plant growth-promoting mechanisms. *Microbiol. Res.* **2018**, *208*, 25–31. [CrossRef]

67. Khan, N.; Bano, A.; Babar, M.A. The root growth of wheat plants, the water conservation and fertility status of sandy soils influenced by plant growth promoting rhizobacteria. *Symbiosis* **2017**, *72*, 195–205. [CrossRef]
68. Gyaneshwar, P.; Naresh, K.G.; Parekh, L.J.; Poole, P.S. Role of soil microorganisms in improving P nutrition of plants. *Plant Soil* **2002**, *245*, 83–93. [CrossRef]
69. Hijri, M. Analysis of a large dataset of mycorrhiza inoculation field trials on potato shows highly significant increases in yield. *Mycorrhiza* **2016**, *26*, 209–214. [CrossRef] [PubMed]
70. Read, A.; Louise, F.; Taylor, H. The Ecology of Genetically Diverse Infections. *Science* **2001**, *292*, 1099. [CrossRef] [PubMed]

Article

PGPR Modulation of Secondary Metabolites in Tomato Infested with *Spodoptera litura*

Bani Kousar [1], Asghari Bano [2,*] and Naeem Khan [3,*] 

[1] Department of Plant Sciences Faculty of Biological Sciences, Quaid-i-Azam University, Islamabad 45320, Pakistan; bani.kousar@gmail.com
[2] Department of Biosciences, University of Wah, Wah Cantt 47040, Pakistan
[3] Department of Agronomy, Institute of Food and Agricultural Sciences, University of Florida, Gainesville, FL 32608, USA
* Correspondence: banoasghari@gmail.com (A.B.); naeemkhan@ufl.edu (N.K.)

Received: 3 May 2020; Accepted: 28 May 2020; Published: 30 May 2020

Abstract: The preceding climate change demonstrates overwintering of pathogens that lead to increased incidence of insects and pest attack. Integration of ecological and physiological/molecular approaches are imperative to encounter pathogen attack in order to enhance crop yield. The present study aimed to evaluate the effects of two plant growth promoting rhizobacteria (*Bacillus endophyticus* and *Pseudomonas aeruginosa*) on the plant physiology and production of the secondary metabolites in tomato plants infested with *Spodoptera litura* (Fabricius) (Lepidoptera: Noctuidae). The surface sterilized seeds of tomato were inoculated with plant growth promoting rhizobacteria (PGPR) for 3–4 h prior to sowing. Tomato leaves at 6 to 7 branching stage were infested with *S. litura* at the larval stage of 2nd instar. Identification of secondary metabolites and phytohormones were made from tomato leaves using thin layer chromatography (TLC) and high performance liquid chromatography (HPLC) and fourier-transform infrared spectroscopy (FTIR). Infestation with *S. litura* significantly decreased plant growth and yield. The PGPR inoculations alleviated the adverse effects of insect infestation on plant growth and fruit yield. An increased level of protein, proline and sugar contents and enhanced activity of superoxide dismutase (SOD) was noticed in infected tomato plants associated with PGPR. Moreover, p-kaempferol, rutin, caffeic acid, p-coumaric acid and flavonoid glycoside were also detected in PGPR inoculated infested plants. The FTIR spectra of the infected leaf samples pre-treated with PGPR revealed the presence of aldehyde. Additionally, significant amounts of indole-3-acetic acid (IAA), salicylic acid (SA) and abscisic acid (ABA) were detected in the leaf samples. From the present results, we conclude that PGPR can promote growth and yield of tomatoes under attack and help the host plant to combat infestation via modulation in IAA, SA, ABA and other secondary metabolites.

Keywords: *Spodoptera litura* (Fabricius) (Lepidoptera: Noctuidae); *Solanum lycopersicum* L.; secondary metabolites; plant insect interactions

1. Introduction

Lycopersicon esculentum Mill. (Tomato) is one of the widely used vegetables cultivated all over the world. It is the important source of vitamin C and vitamin A [1], lycopene (carotenoids), pro-vitamin A, β-carotene and flavonoids [2]. In the recent years, its yield is significantly reduced by the infestation of leaf caterpillars.

Leaf caterpillar *S. litura* (Fabricius) (Lepidoptera: Noctuidae), also known as tropical armyworm, is among the main pests of cultivated crops that can cause significant damage to tomato crop. To this date, *S. litura* has infected about 290 plant species, belonging to 99 families [3,4]. It grows throughout the year, and mounts nearly 7 to 8 generations per year. The larvae of *S. litura* feed initially on plant leaves and latterly feed on almost every part of the plant. The larvae can cause 12 to 23% damage to tomatoes in the monsoon and 9.4 to 27.4% in winter [5]. This insect had shown strong resistance to all conventional and some new chemically synthesized insecticides [6,7]. To combat this notorious insect attack, one can develop new insect resistant cultivars. The main drawbacks of the new cultivars' development are time and expenses. Alternatively, the use of plant growth promoting rhizobacteria having biocontrol properties is a sustainable and eco-friendly approach.

Rhizosphere bacteria form a close association with the roots of plants, they nourish on the soil nutrients and root-exudates of plants; in return they protect the host against the biotic and abiotic stresses and help in host growth [8,9]. Plant growth promoting rhizobacteria (PGPR) boost plant growth directly through the production of phytohormones and indirectly as biocontrol agents [10]. PGPR employs different mechanisms to promote plant growth and control phyto-pathogens. One of the widely recognized mechanisms is the production of inhibitory allelo-chemicals, the production of antibiotics, siderophore, lytic enzymes and the induction of systemic resistance (ISR) in host plants against a broad spectrum of pathogens [11]. Induced systemic resistance (ISR) protects the plant against a broad range of diseases [12,13], triggered by a wide variety of beneficial microbes [14].

PGPR consortium of *S. marcescens*, *B. amyloliquefaciens*, *P. putida*, *P. fluorescens* and *B. cereus* significantly increased the number of fruit/plant [15]. The three bacterial species viz. *B. amyloliquefaciens*, *B. subtilis* and *B. brevis* have significantly improved the activity of defense related enzymes in tomato plants infected with bacterial canker [16]. Several bacterial species *(Pseudomonas, Azotobacter, Azospirillum, Pseudomonas + Azotobacter, Pseudomonas + Azospirillum, Azotobacter + Azospirillum and Pseudomonas + Azotobacter + Azospirillum)* played a key role in nutrient uptake by tomato plants. Also, the rhizospheric bacteria significantly improved shoot and root dry weights, enhanced and modulated production of secondary metabolites [17] and induced resistance to various diseases [18]. *Pseudomonas aeruginosa* is an aerobic, gram-negative rod-shaped bacterium of *Pseudomonadaceae* [19] that was reported to have antifungal activity against *Fusarium moniliforme* [20]. Both *Pseudomonas aeruginosa* and *Bacillus endophyticus* were catalase and oxidase positive, solubilize phosphorus and produce bacteriocin. These bacterial strains showed significant ($p < 0.05$) increase in dry matter production, plant height and root length of maize [21]. They were found positive for the production of antibiotics [22] and had a protruding impact on plant metabolism and plant defense against environmental stresses [23,24].

The present investigation was based on the hypothesis that rhizobacteria isolated from stressed habitats can induce tolerance to plants against environmental stresses in a much better way than those from normal conditions [25]. The rhizobacteria *Bacillus endophyticus* strain Y5 (Accession no. JQ792035) and *Pseudomonas aeruginosa* JYR (Accession no JQ792038) were isolated from the semiarid areas of Yousaf wala Sahiwal (15% soil moisture) and arid areas of Jhang (9% soil moisture), where maize is grown as a main crop. Soil sampling was done at the tasseling stage of maize. The role of those two PGPRs used as bioinoculants was studied on growth and yield of tomato (*Solanum lycopersicum* L.) infested with *S. litura*.

2. Materials and Methods

2.1. Plant Material

The experiment was conducted in the green house of Quaid-i-Azam University, Islamabad. Seeds of *Solanum lycopersicum* L. cv. Rio Grande was obtained from the National Agricultural Research Centre (NARC) Islamabad. Prior to sowing the seeds were surface sterilized with 70% ethanol for 2–3 min, followed by shaking in 10% clorox for 2–3 min. The seeds were finally washed with autoclaved distilled water to remove the traces of treated chemicals [13].

2.2. Preparation of Inocula and Method of Inoculation

Fresh cultures (24 h old) of *Bacillus endophyticus* and *Pseudomonas aeruginosa* were used to inoculate Luria-Bertani (LB) broth, incubated on a rotary shaker for 48 h at 28 °C. The cultures were centrifuged at 3000 rpm for 10 min. Supernatant was discarded, and the pellet containing the bacterial cells was suspended in the autoclaved distilled water to adjust the optical density ($\lambda = 1$) at 660 nm. The inoculum prepared was found to have 10^6 cells/mL. Sterilized seeds were soaked in the bacterial inoculum for 3 to 4 h. The seeds soaked in autoclaved distilled water for the same period were treated as a control [5].

2.3. Growing Conditions and the Treatments

Seeds were sown in pots containing autoclaved sand and soil mixed in 1:3 ratio [26]. Pots were kept in the greenhouse of Quaid-i-Azam University using randomized complete block design with four replicates per treatment. The growing conditions were: photoperiod 16 h, temp 22–28 °C and humidity 60–80%.

The treatments included: Tomato seeds uninoculated uninfested control (C); Tomato seeds inoculated with *Bacillus endophyticus* (T1); Tomato seeds inoculated with *Pseudomonas aeruginosa* (T2); plants infested with *S. litura* (T3); Tomato seeds inoculated with *Bacillus endophyticus* and latterly infested the leaves at 6 to 7 branching stage with *S. litura* (T4); Tomato seeds inoculated with *Pseudomonas aeruginosa* and infested the leaves at 6 to 7 branching stage with *S. litura* (T5).

The tropical armyworm was obtained from the Insectary department, National Agricultural Research Centre (NARC), Islamabad. The leaves of tomato seedlings at 6 to 7 branching stage were infested with larvae of *S. litura* at larval stage of 2nd instar. The larvae were starved for 2 h prior to infestation.

2.4. Height and Weight of Plants and Weight of Tomato Fruit

At the time of harvesting, four plants were marked from each treatment to measure the average height (cm) of the plant and their fresh and dry weights were recorded. After 180 days of sowing, the red ripened fruits were harvested and their fresh weight was measured [27].

2.5. Physiological and Biochemical Attributes of Plants

The physiological and biochemical parameters of leaves were measured after insect infestation.

2.5.1. Leaf Protein Content

Protein content of fresh leaves of tomato plant was estimated following the method of Lowry et al. [28], using Bovine Serum Albumin (BSA) as a standard. Fresh leaves (0.1 g) were grinded in 1 mL of phosphate buffer (pH 6.8) and centrifuged for 10 min at 3000 rpm. The supernatant (0.1 mL) was poured into the test tube and a total volume of 1 mL was made with distilled water. A mixture of 50 mL of Na_2CO_3, NaOH and Na-K tartrate and 1mL of $CuSO_4.5H_2O$ was added. After shaking for 10 min, 0.1 mL of Folin phenol reagent was added. The absorbance of each sample was recorded at 650 nm after 30 min incubation. The concentration of protein was determined using the following formula:

$$Protein\left(\frac{mg}{g}\right) = \frac{K - value \times dilution\ factor \times absorbance}{weight\ of\ sample}$$

K value = 19.6

Dilution factor = 2

Weight of leaf sample = 100 mg

2.5.2. Chlorophyll and Carotenoids Content

Estimation of chlorophyll contents was made according to the method of Arnon [29]. The tomato leaves (0.05 g) were grinded in 10 mL dimethyl sulfoxide (DMSO). The tubes were incubated at 65 °C for 4 h and then the optical density of the sample was recorded at 665 nm and 645 nm. The carotenoids content was determined following the method of Lichtenthaler and Welburn [30].

$$Chloropyll\ a \left(\frac{mg}{g}\right) = 1.07(OD_{663}) - 0.09(OD_{645})$$

$$Chloropyll\ b \left(\frac{mg}{g}\right) = 1.77(OD_{645}) - 0.28(OD_{663})$$

$$Carotenoids \left(\frac{mg}{g}\right) = Absorbance\ (OD_{663}) \times 4$$

2.5.3. Proline Content of Leaves (µg/g)

Free proline content in tomato plant leaves was estimated following the method of Bates et al. [31]. Fresh plant leaf (0.5 g) was grounded in 3% sulfosalicylic acids and kept overnight at 4 °C. The extract was centrifuged at 3,000 rpm for 10 min. The supernatant was mixed with acidic ninhydrin and boiled for 1 h. The solution was then cooled and toluene was added. The absorbance of the toluene layer was recorded at 520 nm against toluene blank. The content of free proline was estimated on fresh weight basis following the formula:

$$Proline \left(\frac{\mu g}{g}\right) = \frac{K - value \times dilution\ factor \times absorbance}{leaf\ weight}$$

Value of K= 17.52

Dilution factor= 2

Weight of leaf sample= 100 mg

2.5.4. Sugar Estimation

The colorimetric determination of total sugar (simple sugar, oligosaccharides and reducing sugar) was done following the method of Dubois et al. [32]. Fresh tomato leaves (500 mg) were grinded with 10 mL of distilled water in autoclaved mortar and pestle, centrifuged at 3000 rpm for 5 min. To the supernatant (100 µL), 1 mL of 80% (w/v) phenol and 5 mL concentrated sulfuric acid was added. The mixture was heated in a water bath till boiling and then incubated for 4 h at room temperature. The absorbance of each sample was finally measured at 420 nm.

$$Sugar \left(\frac{mg}{g}\right) = \frac{K - value \times dilution\ factor \times absorbance}{leaf\ weight}$$

Value of K = 20

Dilution factor = 10

Weight of leaf sample = 500mg

2.5.5. Superoxide Dismutase (SOD) Assay

The SOD activity was estimated following the method of Beauchamp and Fridovich [33].
The activity of Superoxide dismutase was expressed as units/100 g F.W.
Superoxide dismutase was calculated by the following formula:

R4 = R_3-R_2

SOD activity = R_4/A

R_1 = O.D of Reference, R_2 = O.D of Blank, R_3 = O.D of Sample

$A = R_1$ (50/100)

2.5.6. Determination of Indole acetic acid (IAA), Gibberellic acid (GA) and Abscisic acid (ABA) Contents

The extraction and purification for above mentioned phytohormones were made following the method of Kettner and Doerffling [34]. Plant leaves (1g) were grinded in 80% methanol at 4 °C with butylated hydroxytoluene (BHT) used as antioxidant. The extract was centrifuged and the supernatant was reduced by using a rotary thin film evaporator (RFE). The aqueous phase was partitioned 4 times at pH 2.5–3 with $\frac{1}{2}$ volume of ethyl acetate. The ethyl acetate was evaporated by a rotary thin film evaporator. The residue was re-dissolved in 1 mL of methanol (100%) and examined on HPLC (LC-8A Shimadzu, C-R4A Chromatopac; SCL-6B system controller) using UV detector and C-18 column (39 × 300 mm). The wavelength used for the detection of IAA was 280 nm and for GA was 254 nm. For ABA, the samples were injected onto a C_{18} column and eluted at 254 nm with a linear gradient of methanol (30–70%), containing 0.01% acetic acid, at a flow rate of 0.8 mL min^{-1} [35].

2.5.7. Determination of Salicylic Acid (SA) Content of Leaves

Enyedi et al. [36] and Seskar et al. [37] method was employed for salicylic acid detection. After crushing the fresh leaves (1 g) of tomato in 10 mL of 80% methanol at 4 °C. The sample was kept for 3 days with subsequent change in methanol after 24 h. The methanol was then evaporated using RFE and the residue was dissolved again in methanol, filtered and subjected to high-performance liquid chromatography (HPLC) (Agilent Technologies USA) equipped with S-1121 dual piston solvent delivery system and S-3210 UV/VIS diode array detector. Detection of SA was done at 280 nm by co-chromatography with 2-hydroxybenzoic acid as standard. The peak areas were recorded and calculated with SRI peak simple chromatography data acquisition and integration software (SRI instruments, Torrance, CA, USA).

2.5.8. Measurement of Shoot and Root Fresh and Dry Weights and Root Area

Shoots of 4 plants per treatment were cut at the base and weighed immediately by using the electronic balance, to measure the fresh weight of shoot. The chopped shoot was then dried at 70 °C for 72 h and dry weight was measured. The roots of the same plants were washed thoroughly with running tap water to remove soil debris. The water was absorbed on filter paper and weighed to measure the fresh weight of the root. The same root samples were used for determination of root dry weight after drying in the oven till constant weight was obtained [13]. The root area was calculated by using root law Software, Washington State University [38].

2.6. Thin Layer Chromatography of Methanolic Extract of Tomato Leaves

Leaves of tomato plant were harvested 24 h after infestation (80 DAS); shade dried at room temperature and grinded to fine powder. Powdered leaves (20 g) was extracted in 400 mL methanol for 72 h. The methanolic extract was dried using rotary evaporator (RFE), the residue (3 mg) was dissolved in 500 µL methanol and collected in eppendorf tube and stored at −4 °C.

Extract was spotted on a TLC plate (20 × 20 cm) coated with silica gel HF (250-350 nm). The mobile phase used was chlorotorm: methanol (95.5 v/v). The bands, representing various compounds were visualized under UV (254 nm and 380 nm) [39]. The Rf value of each band was calculated and identification of the compound from each band at specific Rf was made from the literature documented.

2.7. FTIR Spectroscopy

All spectra were obtained with the help of an OMNI-sampler attenuated total reflectance (ATR) accessory on a Nicolet FTIR spectrophotometer followed by the method of Lu et al. [40] and Liu et al. [41] with some modifications. Small amount of TLC eluent corresponding to the Rf value of major bands were placed directly on the germanium piece of the infrared spectrometer with constant pressure applied and data of infrared absorbance, collected over the wave number ranged from 4000 cm^{-1} to 675 cm^{-1} and computerized for analyses by using the Omnic software [42].

2.8. Statistical Analysis of Data

The data was subjected to analysis of variance using Statistix 8.1 software. The differences among various treatment means were compared using the least significant differences test (LSD) at $p \leq 0.05$ probability level (Table S1).

3. Results

3.1. Plant Growth Attributes

The plant spread, which is a measurement of plant width, was significantly (31%) higher in PGPR treated plants under unstressed condition over control (Figure 1). Insect infestation decreased the plant spread by 41%, the decrease was ameliorated by the PGPRs and the value was even greater than the control. The plant height was significantly increased in PGPR inoculated plants (Figure 1). The insect infestation significantly reduced ($p \leq 0.05$) the height of the plant by 40%, and root area by 50% of the control (Figure 1). The PGPR inoculated plants alleviated the inhibitory effects of insect infestation on plant height and root area such that the root area was significantly higher than the control. Both the shoot and root fresh weights were significantly (44% and 34%) increased in PGPR inoculated plants (Figures 2 and 3). Infestation with the insect decreased the fresh weights of both the root and shoot, the shoot fresh weight was more adversely affected. The PGPR inoculation had ameliorated the insect-induced decrease in the root and shoot fresh weight.

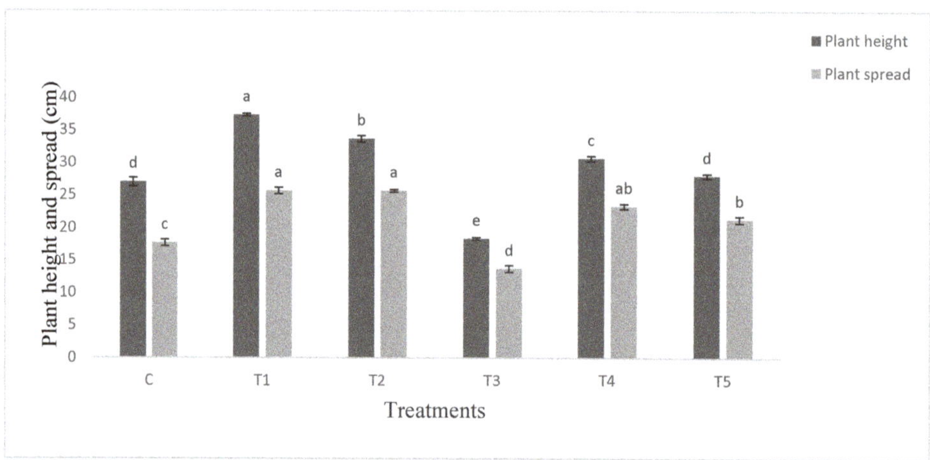

Figure 1. Mean plant height and plant spread (cm) of tomato under control and infested conditions. Data are means of four replicates along with standard error bars. Different letters on the bar represent significant differences ($p < 0.05$) among treatments.

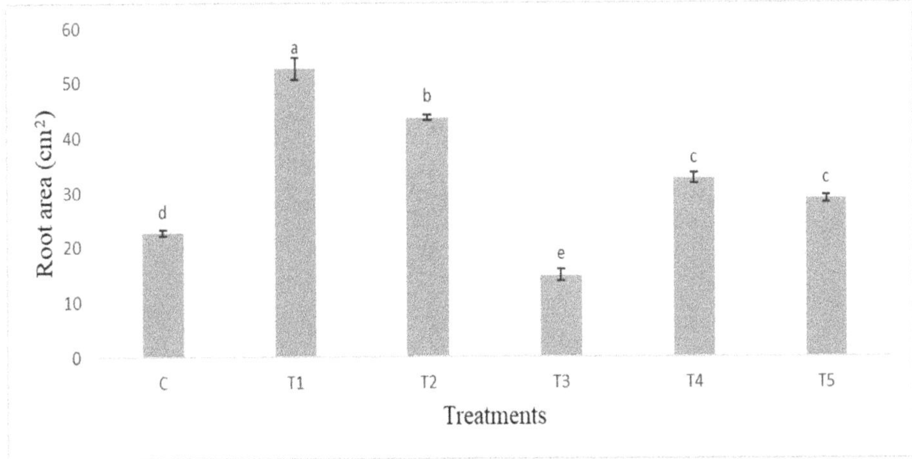

Figure 2. Root area (cm^2) of tomato plant infested with *S. litura* and under control condition. Data are means of four replicates along with standard error bars. Different letters are indicating significant differences ($p < 0.05$) among treatments.

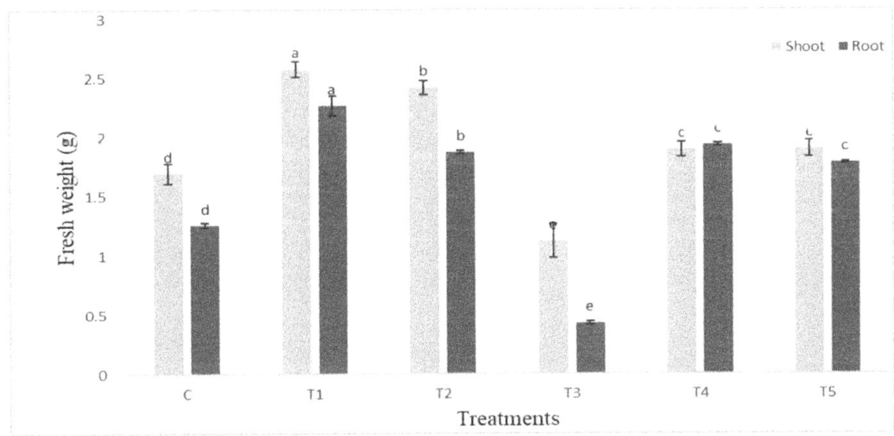

Figure 3. Fresh weight of shoot and root (g) of tomato plant infested with *S. litura* and under control condition. Data are means of four replicates along with standard error bars. Different letters are indicating significant differences ($p < 0.05$) among treatments.

C-uninoculated uninfested control, T1-Seeds inoculated with *Bacillus endophyticus*, T2-Seeds inoculated with *Pseudomonas aeruginosa*, T3-Plants infested with *S. litura*, T4-Seeds inoculated with *Bacillus endophyticus* and plants infested with *S. litura*, T5-Seeds inoculated with *Pseudomonas aeruginosa* and plants infested with *S. litura*.

C-uninoculated uninfested control, T1-Seeds inoculated with *Bacillus endophyticus*, T2-Seeds inoculated with *Pseudomonas aeruginosa*, T3-Plants infested with *S. litura*, T4-Seeds inoculated with *Bacillus endophyticus* and plants infested with *S. litura*, T5-Seeds inoculated with *Pseudomonas aeruginosa* and plants infested with *S. litura*.

C-uninoculated uninfested control, T1-Seeds inoculated with *Bacillus endophyticus*, T2-Seeds inoculated with *Pseudomonas aeruginosa*, T3-Plants infested with *S. litura*, T4-Seeds inoculated with

Bacillus endophyticus and plants infested with *S. litura*, T5-Seeds inoculated with *Pseudomonas aeruginosa* and plants infested with *S. litura*.

The dry weight of root and shoot was also higher ($p \leq 0.05$) in PGPR inoculated plants (Figure 4). The root was more responsive and the % increase in root weight was greater. The leaves were almost eaten by the insect and the shoot weight was significantly decreased to 81% whereas root weight was decreased by 38% over the control.

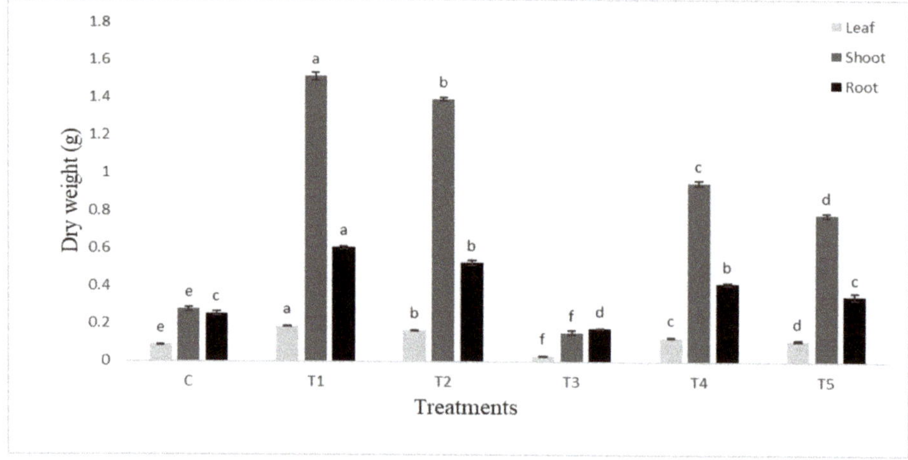

Figure 4. Dry weight of leaf, shoot and root (g) of tomato plant infested with *S. litura* and under control condition. Data are means of four replicates along with standard error bars. Different letters are indicating significant differences ($p < 0.05$) among treatments.

C-uninoculated uninfested control, T1-Seeds inoculated with *Bacillus endophyticus*, T2-Seeds inoculated with *Pseudomonas aeruginosa*, T3-Plants infested with *S. litura*, T4-Seeds inoculated with *Bacillus endophyticus* and plants infested with *S. litura*, T5-Seeds inoculated with *Pseudomonas aeruginosa* and plants infested with *S. litura*.

3.2. Physiological Parameters

The proline production was lower ($p \leq 0.05$) in the untreated control plants (Figure 5). Under unstressed conditions the PGPR treatments stimulated proline content of leaves by 18% over control. Similar percent of increase was recorded in plants infested with *S. litura*. Both the PGPR inoculated plants infested with *S. litura* exhibited marked increase in proline content of leaves over infested plants. The maximum (59%) increase was recorded in the *Bacillus endophyticus* inoculated plants infested with *S. litura*. Chlorophyll a, b and carotenoids followed the similar pattern of response to PGPR and *S. litura* infestation (Figure 6). The response of PGPR was higher ($p \leq 0.05$) particularly for carotenoids content. Both the protein and the sugar contents were higher ($p \leq 0.05$) in PGPR inoculated plants (Figure 7) under unstressed conditions. *Pseudomonas aeruginosa* showed maximum (1.4 fold) increase in sugar content over infested plants. The infestation with *S. litura* had increased sugar and protein contents significantly higher than the control. The inoculated plants receiving insect infestation exhibited up to 2.25 fold increase in sugar content as compared to that of infested plants.

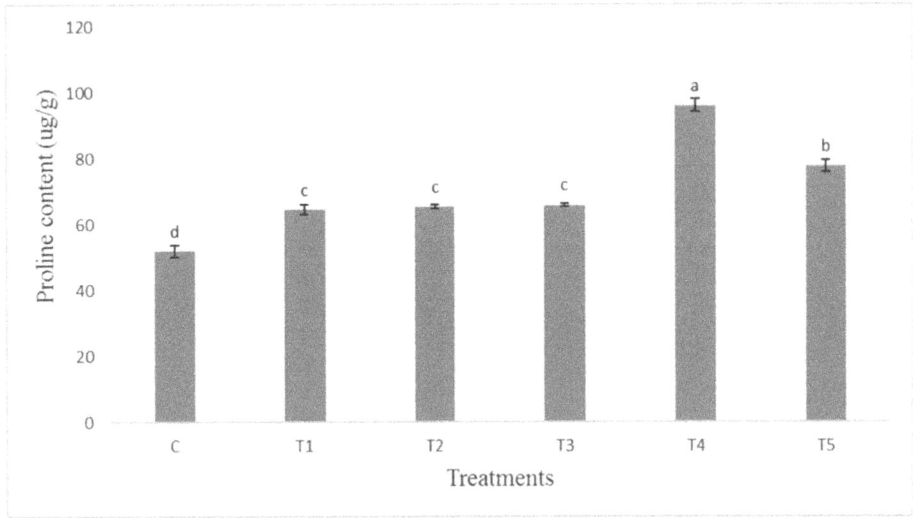

Figure 5. Proline content (µg/g) of tomato leaves infested with *S. litura* and under control condition. Data are means of four replicates along with standard error bars. Different letters are indicating significant differences ($p < 0.05$) among treatments.

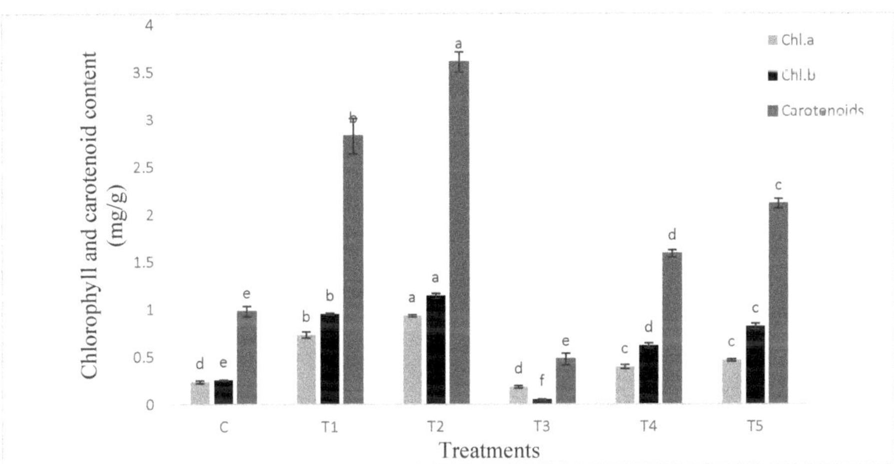

Figure 6. Chlorophylls and carotenoids content (mg/g) of tomato leaves infested with *S. litura* and under control condition. Data are means of four replicates along with standard error bars. Different letters are indicating significant differences ($p < 0.05$) among treatments.

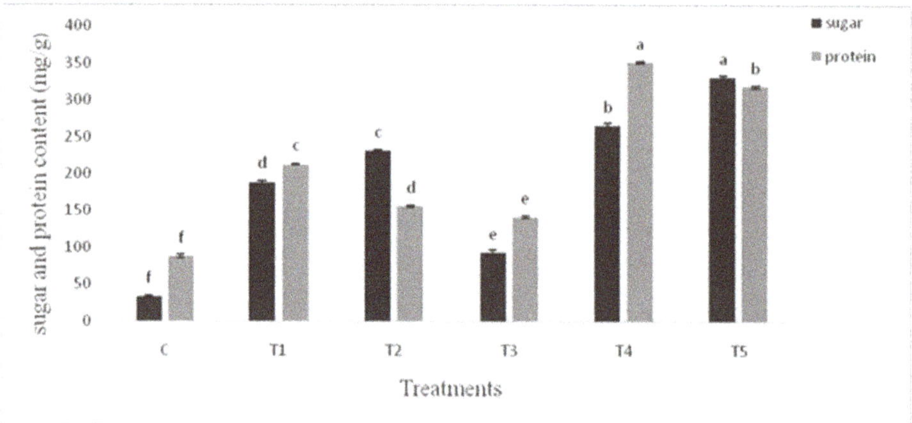

Figure 7. Sugar and protein content (mg/g) of tomato leaves infested with *S. litura* and under control condition. Data are means of four replicates along with standard error bars. Different letters are indicating significant differences ($p < 0.05$) among treatments.

The weight of tomato fruit was about 35% greater in plants inoculated with *Bacillus endophyticus* while *Pseudomonas aeruginosa* inoculated plants exhibited 44% increase over control. There was 26% decrease in the weight of tomato fruit in infested plants (Figure 8). The PGPR inoculated plants ameliorated the inhibitory effect of the insect and showed up to 78% increase in the fruit weight over infested plants.

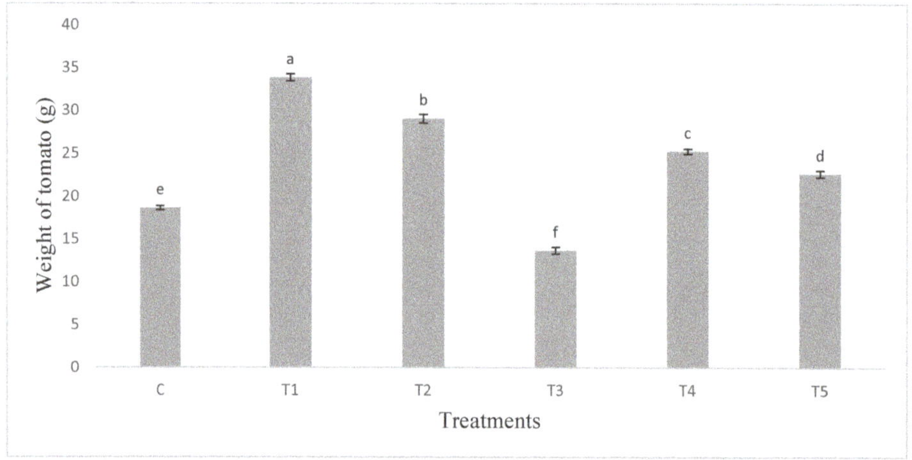

Figure 8. Weight of tomato fruits/plant (g) infested with *S. litura* and under control condition. Data are means of four replicates along with standard error bars. Different letters are indicating significant differences ($p < 0.05$) among treatments.

The infestation with insects enhanced the SOD activity. The SOD activity was three fold higher in leaves of plants inoculated with *Bacillus endophyticus* (T1). Plants inoculated with *Pseudomonas aeruginosa* (T2) on infestation further augmented SOD (Figure 9).

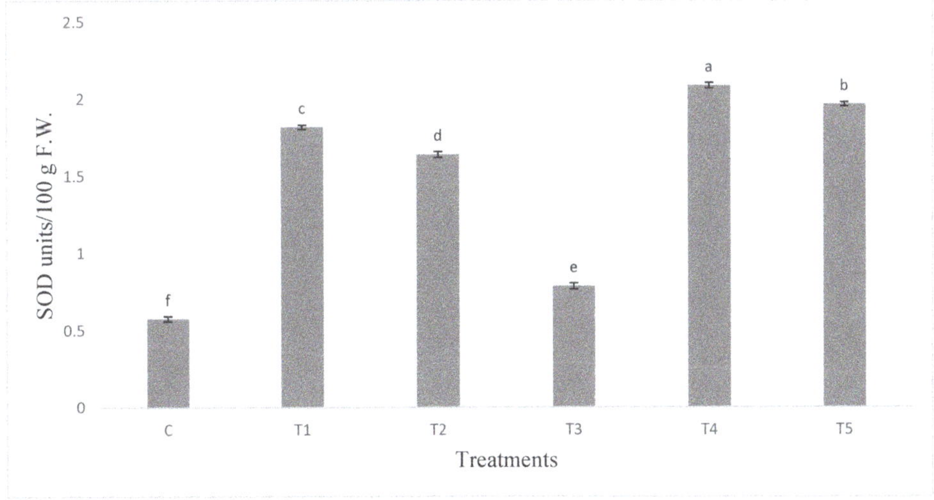

Figure 9. Superoxide dismutase (SOD) activity in tomato leaves infested with *S. litura* and under control condition. Data are means of four replicates along with standard error bars. Different letters are indicating significant differences ($p < 0.05$) among treatments.

3.3. Phytohormones Contents of Leaves

The data in Figure 10 revealed that uninoculated uninfested control leaves of tomato had traceable amounts of Salicylic acid. Insect infestation produced very little amounts of SA. Both the PGPR produced significantly higher amounts of SA in plants, *Pseudomonas* sp. being more efficient. The SA was 1.8 folds greater than infested plant leaves. In *Pseudomonas* inoculated plants, this was further augmented and significantly higher (3.6 fold) SA was recorded in infested plant leaves pretreated with *Pseudomonas aeruginosa*. IAA was not detected in the control and insect infested plants but both the PGPR produced significant amount of IAA in the leaves of inoculated plant which was further augmented and up to 449 µg IAA/g leaves was detected in the leaves of plants infested with *S. litura* and pretreated with *Pseudomonas aeruginosa* (Figure 10). Insect infestation increased the GA content of leaves significantly over control. Several fold increases in GA production were recorded in both the PGPR inoculated plants: *Pseudomonas aeruginosa* being most efficient. Both the PGPR inoculated plants overcame the insect infestation induced decrease in GA content (Figure 10). The ABA content was significantly lower in the infested plant leaves as compared to control. *Bacillus endophyticus* inoculation showed significantly higher ABA production under controlled conditions and the value was several times greater than control in the inoculated plant infested with *S. litura*.

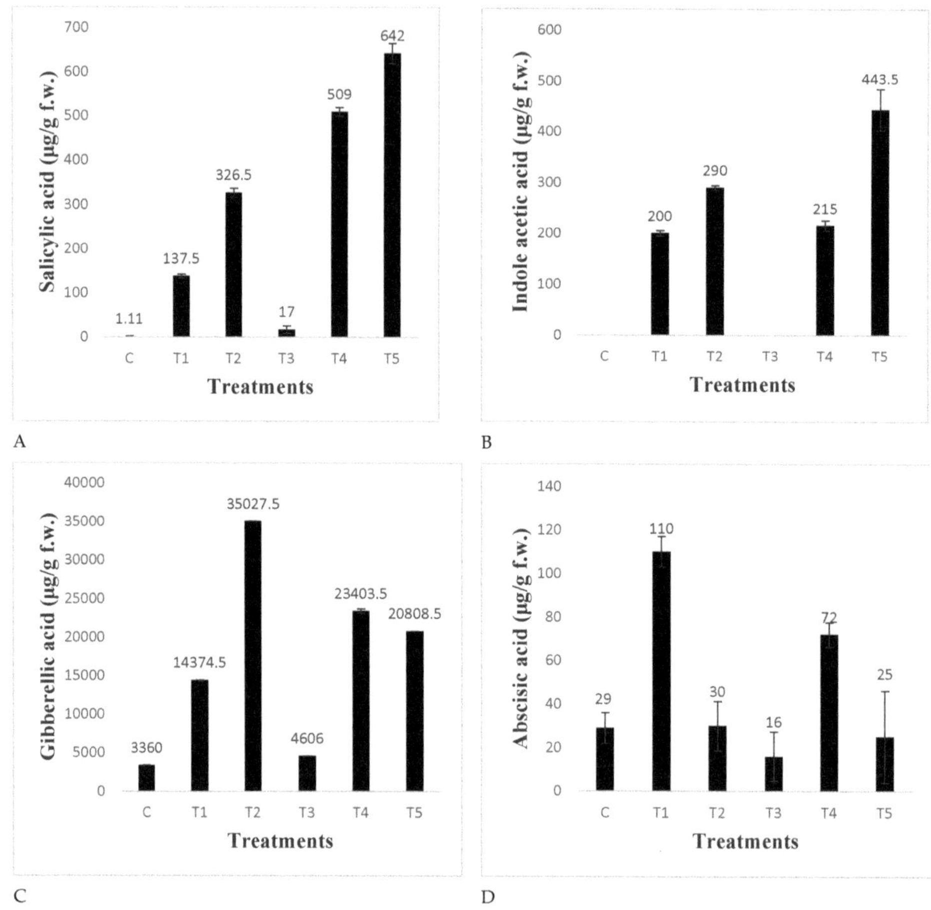

Figure 10. Phytohormone content in the leaves of tomato plants infested with *S. litura* and under control condition. (**A**): Salicylic acid; (**B**): Indole acetic acid; (**C**): Gibberellic acid; (**D**): Abscisic acid.

3.4. Detection of Secondary Metabolites from Extract of Tomato Leaves

Thin layer chromatography of tomato leaf extract showed 29 bands of different colors under UV light (Table 1). Calculated Rf values of leaf extract were compared with Rf values of standard compounds ferulic acid (0.72), salicylic acid (0.60), o-coumeric acid (0.74), trans-cinnamic acid (0.74), caffeic acid (0.85), p-coumaric acid (0.77).

The un-inoculated non infested control plant leaves extract contained caffeic acid (Rf 0.85) and quercetin (Rf 0.88). This was in contrast to *Bacillus endophyticus* inoculated plant leaves which exhibited some unidentified compounds at Rf 0.50 in addition to myricitin (Rf 0.73) o-coumaric (Rf 0.74) whereas, *Pseudomonas aeruginosa* inoculated plants showed the presence of flavonoids, ferulic acid, o-coumaric, kaempferol-7-neoheps-eridiside-glycosides in addition to some unidentified compounds of low polarity. Infestation with *S. litura* resulted in the production of caffeic acid and o-coumaric acid in addition to low and high polarity unidentified compounds. PGPR inoculated plants on infestation produced salicylic acid, rutin and kaempferol in addition to p-Coumaric acid and some unidentified compounds.

Table 1. Putative secondary metabolites identified on the basis of the Rf values in the extract of tomato leaves of different treatments.

Treatments	Rf Values	Color	Compounds
Control	0.85	Red	Caffeic acid
	0.85	Red	Quercitin
T1	0.50	Red	Unidentified
	0.73	Red	Myricitin
	0.79	Red	o-coumaric acid
T2	0.21	Red	Flavonoid-glycoside
	0.39	Red	Unidentified
	0.50	Red	Unidentified
	0.71	Red	Ferulic acid
	0.79	Red	o-coumaric acid
T3	0.55	Yellow	Kampferol-7-neoheps-eridiside
	0.14	Red	Unidentified
	0.41	Red	Unidentified
	0.84	Red	Caffeic acid
	0.76	Yellow	p-Cumaric acid
T4	0.16	Red	Unidentified
	0.23	Red	Unidentified
	0.43	Red	Rutin
	0.60	Red	Salicylic acid
	0.82	Red	Kaempferol
	0.76	Yellow	p-Cumaric acid
T5	0.16	Red	Unidentified
	0.23	Red	Unidentified
	0.43	Red	Rutin
	0.60	Red	Salicylic acid
	0.82	Red	Kaempferol
	0.77	Yellow	p-Cumaric acid

3.5. Fourier Transform Infrared Spectrometry (FTIR) of Tomato Leaves

The data presented in Table 2 revealed that control plant (uninfested and uninoculated) leaves extract had shown the presence of amines and amides with N-H stretch and bend, aliphatic amines stretching with C-N, alkenes with C-H bend and alkyl halides with C-Cl stretch. Plants inoculated with *Pseudomonas aeruginosa* exhibited an additional bonding indicating the presence of alkynes (at frequency of 638.51 with –C≡C–H: C–H bend) which were absent in uninoculated un-infested plants. While extract of plant leaves infested with *S. litura* exhibited alkanes with C–H stretch. In addition to amines and amides with N-H stretch, aliphatic amines stretching with C-N, alkenes with C-H bend and alkyl halides with C-Cl stretching. This was in contrast to plant leaves extract previously inoculated with *Bacillus endophyticus* (T4) or *Pseudomonas aeruginosa* and infested with *S. litura* (T5) which exhibited higher frequency of N-H stretch and =C-H bend and additional bonding indicating the presence of aldehyde and amine with H–C=O: C–H stretch, N–H bend (at frequencies of 2827.91 and 1630.78) as compared to plant extract infested with *S. litura* (T5)

Table 2. Fourier-transform infrared spectroscopy (FTIR) of thin-layer chromatography (TLC) eluent of tomato leaves under different treatments.

Treat.	Frequency	Bond	Functional Group	Characteristics of Peak
C	3408.9	N–H- stretch	1°, 2° amines/amides	Medium
	1632.5	N–H- bend	1° amines	Medium
	1068.3	C–N- stretch	aliphatic amines	Medium
	967.6	=C–H- bend	Alkenes	Strong
	799.5	C–Cl stretch	alkyl halides	Medium
T1	3412.90	N–H stretch	1°, 2° amines, amides	Medium
	1633.25	N–H- bend	1° amines	Medium
	1066.14	C–N- stretch	aliphatic amines	Strong
	967.48	=C–H- bend	Alkenes	Medium
	799.21	C–Cl stretch	alkyl halides	Broad, strong
T2	3410.75	N–H- stretch	1°, 2° amines, amides	Medium
	1633.24	C–N- stretch	aliphatic- amines	Strong
	1068.66	C–N- stretch	aliphatic amines	Strong
	967.13	=C–H- bend	Alkenes	Medium
	799.17	C–Cl- stretch	alkyl halides	Broad, strong
	638.51	–C≡C–H:C–H- bend	Alkynes	Broad, strong
T3	3376.84	N–H stretch	1°, 2° amines, amides	Medium
	2922.79	C–H- stretch	Alkanes	Medium
	1071.40	C–N- stretch	aliphatic amines	Strong
	966.62	=C–H- bend	Alkenes	Medium
	799.60	C–Cl- stretch	alkyl halides	Broad, strong
T4	3412.33	N–H- stretch	1°, 2° amines, amides	Medium
	2827.91	H–C=O: C–H- stretch	Aldehydes	Medium
	1630.78	N–H- bend	1° amines	Medium
	1066.63	C–N- stretch	aliphatic amines	Strong
	967.01	=C–H- bend	Alkenes	Medium
	799.50	C–Cl- stretch	alkyl halides	Broad, strong
T5	3405.01	N–H- stretch	1°, 2° amines, amides	Medium
	1632.25	N–H- bend	1° amines	Medium
	1066.10	C–N- stretch	aliphatic amines	Strong
	967.23	=C–H- bend	Alkenes	Medium
	799.56	C–Cl- stretch	alkyl halides	Broad, strong

Values are mean of 4 replications per treatment. Small amount of TLC eluent corresponding to the Rf-value of key bands were placed directly on the germanium piece of the infrared spectro-meter with persistent pressure and the infrared absorbance was collected over the wave number ranged from 4000 cm^{-1} –675 cm^{-1} and computerized for analyses by using the Omnic software.

4. Discussion

This paper evaluates the effect of PGPR as a growth promoter as well as a biocontrol agent. Although the study deals with tomatoes only in one season and also limited by the lack of behavioral study of the insect.

The growth parameters of tomato were considerably amplified after PGPR inoculation, under uninfested condition; the *Pseudomonas aeruginosa* being more effective. Of note, the effectivity of PGPR were higher under infested conditions and produce higher proline as osmoregulant, more defense hormone e.g., SA and higher level of growth promoting hormone, e.g., IAA contents and for inducing antioxidant enzyme, SOD. The PGPR inoculation not only overcame the infestation induced decrease in root and shoot weight but also increased the root and shoot weight. The PGPR effect was more pronounced on shoot dry weight. Similar results were reported by Avis et al. [43] and Babalola [44]. Shannag and Abadneh [45] reported that fresh and dry weight of shoot and root were decreased by *Aphis fabae Scopoli* in Faba Bean as compared to its respective control. Yadav et al. [46]

reported a marked increase in shoot and root dry weight in chickpea treated with PGPR. The increased weight of tomato fruit was correlated with the number of flowers, branches, plant biomass concomitant with the osmotic balance and alleviating oxidative stress. Both the PGPR were effective and significantly enhanced (≥35%) the fruit fresh weight. This increase in fruit fresh weight by the PGPR may be attributed to the fact that PGPR significantly improves the root growth and plant vigor which lead to enhanced fruit production. Fabro et al. [47] reported that tomato plants treated with PGPR showed the highest number of branches when compared to infested control. PGPR has positive effects on tomato fruit quality attributes, particularly on size and texture [48]. Widnyana [49] reported that inoculation of tomato plants with *Pseudomonas* and *Bacillus* sp. speeded up the plant growth and yield and protection against plant pathogens. Similar results that PGPR inoculation enhances the plant growth, yield and fruit weight were also reported by Almaghrabi et al. [50] and Murphy et al. [51].

Results revealed that PGPR alleviated the osmotic imbalance by increasing, proline content in the insect infested plants. The different PGPR behaved differently, both for proline production and antioxidant enzymes. It is demonstrated that *Pseudomonas aeruginosa* combats osmotic stress in infested plants through increase in sugar content as osmoregulant whereas, *Bacillus endophyticus* enhances proline content to combat osmotic imbalance. The osmotic stress is one of the secondary stresses caused by insect infestation. Proline is revealed to be an osmoregulant that accumulates in plants under a wide range of stress conditions [52,53]. It is well known that free proline accumulation in vascular plants demonstrated stresses including pathogen attack [54,55]. The accumulation of cellular osmolytes such as proline, sugar alcohols, glucosinolates etc. and soluble sugars and the expression of antioxidant systems help plants in sustaining cellular function, crucial for physiological stability of plants under stress. Ullah et al. [56] indicated that application of PGPR to plants displayed substantial increase in proline content as compared to untreated plants. Phenolics are produced by many plant species for protection against biotic or abiotic stress growth conditions and their accumulation correlates with antioxidant capacity of plants in a number of species [57,58].

Compatible solutes are used for osmotic adjustment under adverse environmental conditions [59,60]. The soluble carbohydrates in plants attacked by a fungal pathogen, as well as proportions of individual sugars, may be variously modified, both by plant regulatory mechanisms and by pathogen interference. There are several causes for quantitative and qualitative changes of sugars at the infection site. The level of sugars is reduced by their consumption for both energy and structural purposes, their uptake by the pathogen, while in autotrophic tissues it happens due to the inhibition of photosynthesis [61]. The results further demonstrate the PGPR induced changes in chlorophyll and carotenoids in normal and insect infested plants and the *P. aeruginosa* being most effective. Similar results were reported by Wang et al. [62] that PGPR isolates increased chlorophyll content significantly in tomatoes. Inoculation with *Pseudomonas* B-25 resulted in greater synthesis of chlorophyll than the diseased control. Botha et al. [63] showed that *Diuraphis noxia* feeding caused decrease in chlorophyll content in Tugela and decreased levels of chlorophyll a upon infestation [64,65].

It was demonstrated that the *P. aeruginosa* adjust osmotic stress following infestation by stimulation in antioxidant SOD activity. The PGPR effectively enhanced the SOD activity to scavenge the ROS and prevent oxidative stress in plant cells. The observed enhancement in PGPR induced SOD activity in infested plants is a mechanism to combat insect induced oxidative stress. Recently it has been reported by Sharma and Mathur [66] that PGPR alone and/or in association with fungi significantly enhanced the antioxidant enzyme activities in *Brassica juncea* infested with *Spodoptera litura* that lead to enhanced immune system against herbivory. Similarly, Zhao et al [67] reported Aphid resistance in plants infested with *B. tabaci* nymphs, associated with enhanced antioxidant activities. They concluded that this resistance probably acted via interactions with SA-mediated defense responses.

PGPR promoted growth by nutrient acquisition and by producing bioactive compounds [68,69]. They also improve the nutrient uptake in plants by modulating plant hormones level, thereby increasing root proliferation [70]. However, response of the 2 PGPR differed substantially; *Bacillus endophyticus* exhibited lower IAA but higher GA than that of *Pseudomonas aeruginosa*. PGPR can control plant disease directly, through the production of antagonistic compounds, and indirectly, through the elicitation of a plant defense response [71]. Fernandez-Aunion et al. [72] also reported that PGPR enhances plant growth by synthesis of bioactive compounds and activating plant defense system.

While the PGPR-elicited ISR has been studied extensively in the model plant *Arabidopsis*, it is not well characterized in crop plants. The induction of ISR was investigated by *Bacillus cereus* strain BS107 against *Xanthomonas axonopodis* pv. *vesicatoria* in pepper leaves. Choudhary and Jobri [73] demonstrated the induction of ISR elicited by *Bacillus* spp. against several fungal bacterial and viral pathogens including root knot nematodes. Yang et al. [74] reported genetic evidence of the priming effect of a *rhizobacterium* on the expression of defense genes involved in ISR in pepper. A stronger negative effect of the PGPR on the performance of leaf folder larvae was noted in rice and found that combined treatment of PGPR is more effective than individually. Several plant secondary compounds such as glucosinolates and cyanogenic glycosides yield toxic products after hydrolysis by enzymes stored and liberated during attack by chewing insects [75,76].

It is demonstrated from the present findings that SA and ABA are both involved in inducing tolerance to plants, but the mechanism of inducing tolerance against the insect varied among the PGPR used. For example, *Bacillus endophyticus* inoculation ameliorated the adverse effects of insect infestation by significantly increasing SA and ABA many folds higher than infested plants whereas, *Pseudomonas aeruginosa* ameliorated the infestation by increasing SA higher than the former strain. Salicylic acid is the integral part of signal transduction pathways initiating resistance to disease and infection [77–79]. Plant defense in response to microbial attack is controlled by signaling molecules including SA, JA and ethylene [80]. SA is an important director of pathogen stimulated systemic acquired resistance (SAR), whereas JA and ET are compulsory for rhizobacteria-mediated induced systemic resistance (ISR) [81]. Branch et al. [82] found that SA is a vital constituent of motioning the induced resistance to root-knot nematodes [83].

Plant phenolics comprises a wide array of secondary metabolites including flavonoids, Cinnamic acid, Kaempferol, Coumaric acid as well as salicylic acid synthesized to provide resistance to plants. Their number, type and concentration increase under insect attack [84] and appear to be stimulated following PGPR application. *Pseudomonas aeruginosa* produced both flavonoid glycoside and kaempferol in addition to coumaric acid whereas, *Bacillus endophyticus* had only myricitin and lack kaempferol and coumaric acid but on infestation both produced similar bioactive metabolites in plant. The chromatographic separation of leaf extract also revealed the presence of bands corresponding to Rf value of SA as well as phenolic compounds e.g., Kaempferol and coumaric acid thereby demonstrating the induction of ISR by PGPR in the inoculated plants infested with *S. litura*. Generally, the role of phenolic compounds in defence is related to their antibiotic, antinutritional or unpalatable properties. Besides their involvement in plant- animal or plant-microbe interactions; plant phenolics also play a key role as antioxidants and stress signaling [85–87]. Hammerschmidt [88] reported that phenolic metabolites are related to the resistance phenomenon of plants against their enemies.

The FTIR spectrum was used to identify the functional group of the active components based on the peak value in the region of infrared radiation. Production of alkynes and aldehyde in plants inoculated with *Bacillus endophyticus* and *Pseudomonas aeruginosa* on infestation with *S. litura* demonstrate the PGPR induced defense strategy against insects. Similar results reported by Panda and Khush [89] that chemical derived substances e.g., alkanes, aldehydes, ketones, waxes are involved in host-plant resistance to insects' function as a protective layer to save the plant. Whereas, Shavit [90] reported that inoculation of tomato plants with *P. fluorescens* WCS417r enhanced the performance of the phloem feeding insect *Bemicia tabaci*. Previously, the FTIR has been applied to classify the actual structure of certain plant secondary metabolites [91]. FTIR is one of the extensively used approaches to categorize the chemical ingredients and clarify the compounds structures [92]. The FTIR of the leaves extract revealed the presence of additional peaks of aldehyde in the FTIR of leaves of infested plants pretreated with PGPR. Chehab et al. [93] reported that aldehydes play a positive role in plant defense. Plants defend themselves from pathogens attack by producing secondary metabolites and proteins [94,95].

5. Conclusions

The *Bacillus endophyticus* and *Pseudomonas aeruginosa* can be used to combat oxidative and osmotic stresses induced by *S. litura* infestation. Both the PGPRs combat insects induced adverse effects on plant growth and productivity through the production of phenolics, SA and ABA. *Bacillus endophyticus* was more effective in the improved defense strategy induction through the modulation of phytohormones and secondary metabolites. These PGPR are more effective under uninfested conditions and can be implicated as bioinoculant to endure the plants to cope better with insect infestation. Since there is alteration in the functional group and presence of aldehyde predominantly detected in plants treated with PGPR and infested with the insect armyworm. Further investigations using nuclear magnetic resonance (NMR) and liquid chromatography-mass spectrometry (LC-MS) are needed to unveil the secondary metabolites produced in PGPR inoculated plants versus uninoculated insect infested plants. Finally, an integrated approach of molecular mechanism of PGPR induced defense in plants against pests and parasites needs thorough investigation.

Supplementary Materials: The following are available online at http://www.mdpi.com/2073-4395/10/6/778/s1, Table S1: Results of the statistical analysis (ANOVA) for the performed parameters.

Author Contributions: Conceptualization, A.B.; Methodology, B.K. and N.K.; Software, B.K. and N.K.; Validation, B.K., A.B. and N.K.; Formal Analysis, N.K.; Investigation, B.K. and A.B.; Resources, A.B.; Data Curation, N.K.; Writing—Original Draft Preparation, B.K.; Writing—Review & Editing, A.B. and N.K; Visualization, B.K.; Supervision, A.B.; Project Administration, A.B. All authors have read and agreed to the published version of the manuscript.

Funding: This research received no external funding.

Conflicts of Interest: The authors declare no conflict of interest.

References

1. Tyssandier, V.; Feillet-Coudray, C.; Caris-Veyrat, C.; Guilland, J.C.; Coudray, C.; Bureau, S.; Reich, M.; Amiot-Carlin, M.J.; Bouteloup-Demange, C.; Boirie, Y.; et al. Effect of tomato product consumption on the plasma status of antioxidant microconstituents and on the plasma total antioxidant capacity in healthy subjects. *J. Am. Coll. Nutr.* **2004**, *23*, 148–156. [CrossRef]

2. Friedman, M. Anticarcinogenic, cardioprotective, and other health benefits of tomato compounds lycopene, α-tomatine, and tomatidine in pure form and in fresh and processed tomatoes. *J Agr Food Chem.* **2013**, *61*, 9534–9550. [CrossRef]

3. Wu, C.J.; Fan, S.Y.; Jiang, Y.H.; Yao, H.H.; Zhang, A.B. Inducing gathering effect of taro on *Spodoptera lituraFabricius*. *Chin. J. Eco.* **2004**, *23*, 172–174. [CrossRef]

4. Ahmad, M.; Ghaffar, A.; Rafiq, M. Host plants of leaf worm, Spodoptera litura (Fabricius) (Lepidoptera: Noctuidae) in Pakistan. *Asian J. Agric. Biol.* **2013**, *1*, 23–28.

5. Khan, N.; Bano, A.; Zandi, P. Effects of exogenously applied plant growth regulators in combination with PGPR on the physiology and root growth of chickpea (*Cicer arietinum*) and their role in drought tolerance. *J. Plant Interact.* **2018**, *13*, 239–247. [CrossRef]

6. Sayyed, A.H.; Ahmad, M.; Saleem, M.A. Cross-resistance and genetics of resistance to indoxacarb in *Spodoptera litura* (Lepidoptera: Noctuidae). *J. Econ. Entomol.* **2008**, *101*, 472–479. [CrossRef]

7. Saleem, M.A.; Ahmad, M.; Aslam, M.; Sayyed, A.H. Resistance to selected organochlorine, organophosphate, carbamates and pyrethroid insecticides in *Spodoptera litura* (Lepidoptera: Noctuidae) from Pakistan. *J. Econ. Entomol.* **2008**, *101*, 1667–1675. [CrossRef]

8. Bais, H.P.; Weir, T.L.; Perry, L.G.; Gilroy, S.; Vivanco, J.M. The role of root exudates in rhizosphere interactions with plants and other organisms. *Annu. Rev. Plant Biol.* **2006**, *57*, 233–266. [CrossRef] [PubMed]

9. Khan, N.; Bano, A. Growth and Yield of Field Crops Grown Under Drought Stress Condition Is Influenced by the Application of PGPR. In *Field Crops: Sustainable Management by PGPR*; Springer Nature: Berlin/Heidelberg, Germany, 2019; pp. 337–349.

10. Nihorimbere, V.; Ongena, M.; Smargiassi, M.; Thonart, P. Beneficial effect of the rhizosphere microbial community for plant growth and health. *Biotechnol. Agron. Soc. Environ.* **2011**, *15*, 327.

11. Compant, S.; Duffy, B.; Nowak, J.; Clément, C.; Barka, E.A. Use of plant growth-promoting bacteria for biocontrol of plant diseases: Principles, mechanisms of action, and future prospects. *Appl. Environ. Microbiol.* **2013**, *71*, 4951–4959. [CrossRef]

12. War, A.R.; Paulraj, M.G.; War, M.Y.; Ignacimuthu, S. Differential defensive response of groundnut to *Helicoverpa armigera* (Hubner) (Lepidoptera: Noctuidae). *J. Plant Interact.* **2011**, *7*, 45–55. [CrossRef]

13. Khan, N.; Bano, A.; Rahman, M.A.; Rathinasabapathi, B.; Babar, M.A. UPLC-HRMS-based untargeted metabolic profiling reveals changes in chickpea (*Cicer arietinum*) metabolome following long-term drought stress. *Plant Cell Environ.* **2019**, *42*, 115–132. [CrossRef] [PubMed]

14. Naseem, H.; Ahsan, M.; Shahid, M.A.; Khan, N. Exopolysaccharides producing rhizobacteria and their role in plant growth and drought tolerance. *J. Basic Microbiol.* **2018**, *58*, 1009–1022. [CrossRef] [PubMed]

15. Vaikuntapu, P.R.; Dutta, S.; Samudrala, R.B.; Rao, V.R.; Kalam, S.; Podile, A.R. Preferential promotion of Lycopersicon esculentum (Tomato) growth by plant growth promoting bacteria associated with tomato. *Indian J. Microbiol.* **2014**, *54*, 403–412. [CrossRef]

16. Passari, A.K.; Upadhyaya, K.; Singh, G.; Abdel-Azeem, A.M.; Thankappan, S.; Uthandi, S.; Hashem, A.; Abd_Allah, E.F.; Malik, J.A.; Alqarawi, A.S.; et al. Enhancement of disease resistance, growth potential, and photosynthesis in tomato (*Solanum lycopersicum*) by inoculation with an endophytic actinobacterium, Streptomyces thermocarboxydus strain BPSAC147. *PLoS ONE* **2019**, *14*, e0219014. [CrossRef]

17. Moradali, M.F.; Ghods, S.; Rehm, B.H. *Pseudomonas aeruginosa* lifestyle: A paradigm for adaptation, survival, and persistence. *Front. Microbiol.* **2017**, *7*, 39. [CrossRef]

18. Khan, N.; Zandi, P.; Ali, S.; Mehmood, A.; Adnan Shahid, M.; Yang, J. Impact of salicylic acid and PGPR on the drought tolerance and phytoremediation potential of Helianthus annus. *Front. Microbiol.* **2018**, *9*, 2507. [CrossRef]

19. Quais, M.K.; Munawar, A.; Ansari, N.A.; Zhou, W.W.; Zhu, Z.R. Interactions between brown planthopper (Nilaparvata lugens) and salinity stressed rice (Oryza sativa) plant are cultivar-specific. *Sci. Rep.* **2020**, *10*, 1–14. [CrossRef]

20. Farooq, U.; Bano, A. Screening of indigenous bacteria from rhizosphere of maize (Zea mays L.) for their plant growth promotion ability and antagonism against fungal and bacterial pathogens. *J. Anim. Plant Sci.* **2013**, *23*, 1642–1652.

21. Naz, R.; Bano, A.; Yasmin, H.; Samiullah, H.; Farooq, U. Antimicrobial potential of the selected plant species against some infectious microbes used. *J. Med. Plants Res.* **2011**, *5*, 5247–5253.

22. Egamberdieva, D.; Wirth, S.J.; Alqarawi, A.A.; Abd_Allah, E.F.; Hashem, A. Phytohormones and beneficial microbes: Essential components for plants to balance stress and fitness. *Front. Microbiol.* **2017**, *8*, 2104. [CrossRef] [PubMed]

23. Khan, N.; Bano, A.; Ali, S.; Babar, M.A. Crosstalk amongst phytohormones from planta and PGPR under biotic and abiotic stresses. *Plant. Growth Regul.* **2020**, *90*, 189–203. [CrossRef]

24. Kudoyarova, G.; Arkhipova, T.N.; Korshunova, T.; Bakaeva, M.; Loginov, O.; Dodd, I.C. Phytohormone mediation of interactions between plants and non-symbiotic growth promoting bacteria under edaphic stresses. *Front. Plant Sci.* **2019**, *10*, 1368. [CrossRef] [PubMed]

25. Kumar, A.; Patel, J.S.; Meena, V.S.; Ramteke, P.W. Plant growth-promoting rhizobacteria: Strategies to improve abiotic stresses under sustainable agriculture. *J. Plant Nutr.* **2019**, *42*, 1402–1415. [CrossRef]

26. Mahmoudi, T.R.; Yu, J.M.; Liu, S.; Pierson III, L.S.; Pierson, E.A. Drought-Stress Tolerance in Wheat Seedlings Conferred by Phenazine-Producing Rhizobacteria. *Front. Microbiol.* **2019**, *10*, 1590. [CrossRef]

27. Baliyan, S.P.; Rao, M.S. Evaluation of tomato varieties for pest and disease adaptation and productivity in Botswana. *Int. J. Agric. Food Res.* **2013**, *2*, 2. [CrossRef]

28. Lowry, O.H.; Rosebrough, N.J.; Farr, A.L.; Randall, R.J. Protein measurement with the Folin phenol reagent. *J. Biol. Chem.* **1951**, *193*, 265–275.

29. Arnon, D.I. Copper enzymes in isolated chloroplasts. Polyphenoloxidase in Beta vulgaris. *Plant Physiol.* **1949**, *24*, 1–15. [CrossRef]

30. Lichtenthaler, H.K.; Wellburn, A.R. Determinations of total carotenoids and chlorophylls a and b of leaf extracts in different solvents. *Biochem. Soc. Trans.* **1983**, *11*, 591–592. [CrossRef]

31. Bates, L.S.; Waldren, R.P.; Teare, I.D. Rapid determination of free proline for water-stress studies. *Plant Soil* **1973**, *39*, 205–207. [CrossRef]
32. Dubois, S.M.; Giles, K.A.; Hamilton, J.K.; Rebers, P.A.; Smith, F. Calorimetric method for determination of sugar and related substance. *Anal. Chem.* **1956**, *28*, 350. [CrossRef]
33. Beauchamp, C.; Fridovich, I. Superoxide dismutase: Improved assays and an assay applicable to acrylamide gels. *Anal. Biochem.* **1971**, *44*, 276–287. [CrossRef]
34. Kettner, J.; Dörffling, K. Biosynthesis and metabolism of abscisic acid in tomato leaves infected with Botrytis cinerea. *Planta* **1995**, *196*, 627–634. [CrossRef]
35. Hansen, H.; Dörffling, K. Root-derived trans-zeatin riboside and abscisic acid in drought-stressed and rewatered sunflower plants: Interaction in the control of leaf diffusive resistance? *Funct. Plant Biol.* **2003**, *30*, 365–375. [CrossRef]
36. Enyedi, A.J.; Raskin, I. Induction of UDP-glucose: Salicylic acid glucosyltransferase activity in tobacco mosaic virus-inoculated tobacco (*Nicotiana tabacum*) leaves. *Plant Physiol.* **1993**, *101*, 1375–1380. [CrossRef]
37. Sesker, M.; Shulaev, V.; Raskin, I. Endogenous methyl salicylate in pathogen-inoculated tobacco plants. *Plant Physiol.* **1998**, *116*, 387–392. [CrossRef]
38. Khan, N.; Bano, A. Effects of exogenously applied salicylic acid and putrescine alone and in combination with rhizobacteria on the phytoremediation of heavy metals and chickpea growth in sandy soil. *Int. J. Phytoremediat.* **2018**, *20*, 405–414. [CrossRef]
39. Subramanian, S.; Ramakrishnan, N. Chromatographic finger print analysis of Naringi crenulata by HPTLC technique. *Asian Pac. J. Trop. Biomed.* **2011**, *1*, S195–S198. [CrossRef]
40. Lu, X.; Wang, J.; Al-Qadiri, H.M.; Ross, C.F.; Powers, J.R.; Tang, J.; Rasco, B.A. Determination of total phenolic content and antioxidant capacity of onion (*Allium cepa*) and shallot (*Allium oschaninii*) using infrared spectroscopy. *Food Chem.* **2011**, *129*, 637–644. [CrossRef]
41. Liu, Y.; Hu, T.; Wu, Z.; Zeng, G.; Huang, D.; Shen, Y.; He, X.; Lai, M.; He, Y. Study on biodegradation process of lignin by FTIR and DSC. *Environ. Sci. Pollut. Res.* **2014**, *21*, 14004–14013. [CrossRef]
42. Lammers, K.; Arbuckle-Keil, G.; Dighton, J. FT-IR study of the changes in carbohydrate chemistry of three New Jersey pine barrens leaf litters during simulated control burning. *Soil Bio Biochem.* **2009**, *41*, 340–347. [CrossRef]
43. Avis, T.J.; Gravel, V.; Antoun, H.; Tweddell, R.J. Multifaceted beneficial effects of rhizosphere microorganisms on plant health and productivity. *Soil Biol. Biochem.* **2008**, *40*, 1733–1740. [CrossRef]
44. Babalola, O.O. Beneficial bacteria of agricultural importance. *Biotechnol. Lett.* **2010**, *32*, 1559–1570. [CrossRef] [PubMed]
45. Shannag, H.K.; Abadneh, J.A. Influence of Black Bean Aphids, *Aphis fabae Scopoli*. On growth rates of Faba Bean. *World J. Agric. Sci.* **2007**, *3*, 344–349.
46. Yadav, J.; Jay, P.V.; Kavindra, N.T. Plant growth promoting activities of fungi and their effect on chickpea plant growth. *Asian J. Biol. Sci.* **2011**, *4*, 291–299. [CrossRef]
47. Ali, N.; Hadi, F. Phytoremediation of cadmium improved with the high production of endogenous phenolics and free proline contents in Parthenium hysterophorus plant treated exogenously with plant growth regulator and chelating agent. *Environ Sci Pollut.* **2015**, *22*, 13305–13318. [CrossRef]
48. Mena-Violante, H.G.; Olalde-Portugal, V. Alteration of tomato fruit quality by root inoculation with plant growth-promoting rhizobacteria (PGPR): Bacillus subtilis BEB-13bs. *Sci. Hortic.* **2007**, *113*, 103–106. [CrossRef]
49. Widnyana, I.K. PGPR (Plant Growth Promoting Rizobacteria) Benefits in Spurring Germination, Growth and Increase the Yield of Tomato Plants. In *Recent Advances in Tomato Breeding and Production*; IntechOpen: London, UK, 2018.
50. Almaghrabi, O.A.; Massoud, S.I.; Abdelmoneim, I.S. Influence of inoculation with plant growth promoting rhizobacteria (PGPR) on tomato plant growth and nematode reproduction under greenhouse conditions. *Saudi J. Biol. Sci.* **2013**, *20*, 57–61. [CrossRef]
51. Murphy, J.F.; Reddy, M.S.; Ryu, C.M.; Kloepper, J.W.; Li, R. Rhizobacteria-mediated growth promotion of tomato leads to protection against Cucumber mosaic virus. *Phytopathol* **2003**, *93*, 1301–1307. [CrossRef]
52. Lu, H.F.; Cheng, C.G.; Tang, X.; Hu, Z.H. Spectrum of *Hypericum* and *Triadenum* with reference to their identification. *J. Integr. Plant Biol.* **2004**, *46*, 401–406. [CrossRef]

53. Fabro, G.; Kovács, I.; Pavet, V.; Szabados, L.; Alvarez, M.E. Proline accumulation and AtP5CS2 gene activation are induced by plant-pathogen incompatible interactions in Arabidopsis. *Mol. Plant-Microbe Interact.* **2004**, *17*, 343–350. [CrossRef] [PubMed]

54. Janardan, Y.; Verma, J.P.; Tiwari, K.N. Effect of plant growth promoting rhizobacteria on seed germination and plant growth chickpea (*Cicer arietinum L.*) under in vitro conditions. *Biol. Forum* **2010**, *2*, 15–18.

55. Isah, T. Stress and defense responses in plant secondary metabolites production. *Biol. Res.* **2009**, *52*, 39. [CrossRef] [PubMed]

56. Ullah, S.; Ashraf, M.; Asghar, H.N.; Iqbal, Z.; Ali, R. Review Plant growth promoting rhizobacteria-mediated amelioration of drought in crop plants. *Soil Environ.* **2019**, *38*, 1–20. [CrossRef]

57. Abideen, Z.; Qasim, M.; Rasheed, A.; Adnan, M.Y.; Gul, B.; Khan, M.A. Antioxidant activity and polyphenolic content of *Phragmites karka* under saline conditions. *Pak. J. Bot.* **2015**, *47*, 813–818.

58. Welbaum, G.E.; Sturz, A.V.; Dong, Z.; Nowak, J. Managing soil microorganisms to improve productivity of agro-ecosystems. *Crit. Rev. Plant Sci.* **2004**, *23*, 175–193. [CrossRef]

59. Ullah, S.; Asema, M.; Asghari, B. Effect of PGPR on growth and performance of *Zea mays*. *Res. J. Agric. Environ. Manag.* **2013**, *2*, 434–447.

60. Willcox, J.K.; Catignani, G.L.; Lazarus, S. Tomatoes and cardiovascular health. *Crit. Rev. Food Sci. Nutr.* **2003**, *43*, 1–18. [CrossRef]

61. Chen, Y.; Cao, C.; Guo, Z.; Zhang, Q.; Li, S.; Zhang, X.; Gong, J.; Shen, Y. Herbivore exposure alters ion fluxes and improves salt tolerance in a desert shrub. *Plant Cell Environ.* **2020**, *43*, 400–419. [CrossRef]

62. Berger, S.; Sinha, A.K.; Roitsch, T. Plant physiology meets phytopathology: Plant primary metabolism and plant–pathogen interactions. *J. Exp. Bot.* **2007**, *58*, 4019–4026. [CrossRef]

63. Wang, C.J.; Yang, W.; Wang, C.; Gu, C.; Niu, D.D.; Liu, H.X.; Wang, Y.P.; Guo, J.H. Induction of drought tolerance in cucumber plants by a consortium of three plant growth-promoting rhizobacterium strains. *PLoS ONE* **2012**, *7*, e52565. [CrossRef] [PubMed]

64. Botha, A.M.; Lacock, L.; van Niekerk, C.; Matsioloko, M.T.; du Preez, F.B.; Loots, S.; Venter, E.; Kunert, K.J.; Cullis, C.A. Is photosynthetic transcriptional regulation in Triticum aestivum L. cv. 'TugelaDN' a contributing factor for tolerance to Diuraphis noxia (Homoptera: Aphididae)? *Plant Cell Rep.* **2006**, *25*, 41–54. [CrossRef] [PubMed]

65. Burd, J.D.; Elliott, N.C. Changes in chlorophyll a fluorescence induction kinetics in cereals infested with Russian wheat aphid (Homopetra: Aphididea). *J. Econ. Entomol.* **1996**, *89*, 1332–1337. [CrossRef]

66. Ni, X.; Quisenberry, S.S.; Heng-Moss, T.; Markwell, J.; Higley, L.; Baxendale, F.; Sarath, G.; Klucas, R. Dynamic change in photosynthetic pigments and chlorophyll degradation elicited by cereal aphid feeding. *Entomol. Exp. Appl.* **2002**, *105*, 43–53. [CrossRef]

67. Sharma, G.; Mathur, V. Modulation of insect-induced oxidative stress responses by microbial fertilizers in Brassica juncea. *FEMS Microbiol. Ecol.* **2020**, *96*, 40. [CrossRef]

68. Zhao, Z.H.; Hui, C.; Hardev, S.; Ouyang, F.; Dong, Z.; Ge, F. Responses of cereal aphids and their parasitic wasps to landscape complexity. *J. Econ. Entomol.* **2014**, *107*, 630–637. [CrossRef]

69. Mehboob, I.; Zahir, Z.A.; Arshad, M.; Tanveer, A.; Khalid, M. Comparative effectiveness of different Rhizobium sp. for improving growth and yield of maize (*Zea mays L.*). *Soil Environ.* **2012**, *31*, 37–46.

70. Zhang, S.; Moyne, A.L.; Reddy, M.S.; Kloepper, J.W. The role of salicylic acid in induced systemic resistance elicited by plant growth-promoting rhizobacteria against blue mold of tobacco. *Biol. Control.* **2002**, *25*, 288–296. [CrossRef]

71. Bobby, M.N.; Wesely, E.G.; Johnson, M. FT-IR studies on the leaves of *Albizia lebbeck* benth. *Int. J. Pharm. Pharm. Sci.* **2012**, *4*, 293–296.

72. Mansour, M.F. Nitrogen containing compounds and adaptation of plants to salinity stress. *Biol. Plant.* **2000**, *43*, 491–500. [CrossRef]

73. Fernandez-Aunión, C.; Hamouda, T.B.; Iglesias-Guerra, F.; Argandoña, M.; Reina-Bueno, M.; Nieto, J.J.; Aouani, M.E.; Valrgas, C. Biosynthesis of compatible solutes in rhizobial strains isolated from Phaseolus vulgaris nodules in Tunisian fields. *BMC Microbiol.* **2010**, *10*, 192. [CrossRef] [PubMed]

74. Choudhary, D.K.; Johri, B.N. Interactions of Bacillus spp. and plants–with special reference to induced systemic resistance (ISR). *Microbiol. Res.* **2009**, *164*, 493–513. [CrossRef] [PubMed]

75. Yang, J.; Yen, H.E. Early salt stress effects on the changes in chemical composition in leaves of ice plant and Arabidopsis. A Fourier transform infrared spectroscopy study. *Plant Physiol.* **2000**, *130*, 1032–1042. [CrossRef] [PubMed]

76. Liu, H.; Sun, S.; Lv, G.; Chan, K.C. Study on Angelica and its different extracts by Fourier transform infrared spectroscopy and two-dimensional correlation IR spectroscopy. *Spectrochim. Acta Part A* **2006**, *64*, 321–326. [CrossRef]

77. Pineda, A.; Zheng, S.J.; Van Loon, J.J.; Pieterse, C.M.; Dicke, M. Helping plants to deal with insects: The role of beneficial soil-borne microbes. *Trends Plant Sci.* **2010**, *15*, 507–514. [CrossRef]

78. Schoonhoven, L.M.; Van Loon, B.; van Loon, J.J.; Dicke, M. *Insect-Plant Biology*; Oxford University Press on Demand: Oxford, UK, 2005.

79. Natikar, P.K.; Balika, R.A. Tobacco caterpillar Spodoptera litura (Fabricus) toxicity, ovicidal action, oviposition deterrent activity, ovipositional preference and its management. *Biochem. Cell. Arch* **2015**, *15*, 383–389.

80. Brimecombe, M.J.; De Leij, F.A.; Lynch, J.M. Rhizodeposition and microbial populations. In *the Rhizosphere Biochemistry and Organic Substances at the Soil-Plant Interface*; Pinton, R., Veranini, Z., Nannipieri, P., Eds.; Taylor and Francis: New York, NY, USA, 2007; pp. 73–110.

81. Tsukanova, K.A.; Ch botar, V..; Meyer, J.J.M.; Bibikova, T.N. Effect of plant growth-promoting Rhizobacteria on plant hormone homeostasis. *South Afri. J. Bot.* **2017**, *113*, 91–102. [CrossRef]

82. Mendes, R..; Garbeva, P..; Raaijmakers, J.M. The Rhizosphere Microbiome: Significance of Plant Beneficial, Plant Pathogenic, and Human Pathogenic Microorganisms. *FEMS Microbiol. Rev.* **2013**, *37*, 634–663. [CrossRef]

83. Branch, C.; Hwang, C.F.; Navarre, D.A.; Williamson, V.M. Salicylic acid is part of the Mi-1-mediated defense response to root-knot nematode in tomato. *Mol. Plant-Microbe Interact.* **2004**, *17*, 351–356. [CrossRef]

84. Ton, J.; Van-Pelt, J.A.; Van Loon, L.C.; Pieterse, C.M. Differential effectiveness of salicylate-dependent and jasmonate/ethylene-dependent induced resistance in *Arabidopsis*. *Mol. Plant-Microbe Interact.* **2002**, *15*, 27–34. [CrossRef]

85. Dixit, G.; Srivastava, A.; Rai, K.M.; Dubey, R.S.; Srivastava, R.; Verma, P.C. Distinct defensive activity of phenolicsandphenylpropanoid pathway genesin different cotton varieties toward chewing pests. *Plant. Signal. Behav.* **2020**, *15*, 1747689. [CrossRef] [PubMed]

86. Lattanzio, V.; Lattanzio, V.M.; Cardinali, A. Role of phenolics in the resistance mechanisms of plants against fungal pathogens and insects. *Phytochem. Adv. Res.* **2006**, *661*, 23–67.

87. Jacobo-Velázquez, D.A.; González-Agüero, M.; Cisneros-Zevallos, L. Cross-talk between signaling pathways: The link between plant secondary metabolite production and wounding stress response. *Sci. Rep.* **2015**, *5*, 8608. [CrossRef] [PubMed]

88. Weber, D.; Egan, P.A.; Muola, A.; Ericson, L.E.; Stenberg, J.A. plant resistance does not compromise parasitoid-based biocontrol of a strawberry pest. *Sci. Rep.* **2020**, *10*, 1–10. [CrossRef]

89. Hammerschmidt, R. Phenols and plant–pathogen interactions: The saga continues. *Physiol. Mol. Plant Pathol.* **2005**, *66*, 77–78. [CrossRef]

90. Panda, N.; Khush, G.A. *Host Plant Resistance to Insects*; CAB International: Wallingford, UK, 1995.

91. Shavit, R.; Lalzar, M.O.; Saul, B.; Morin, S. Inoculation of tomato plants with rhizobacteria enhances the performance of the phloem-feeding insect *Bemisia tabaci*. *Front. Plant Sci.* **2013**, *4*, 1941–1947. [CrossRef]

92. Subramanion, L.J.; Zakaria, Z.; Sreenivasan, S. Phytochemicals screening, DPPH free radical scavenging and xanthine oxidase inhibitiory activities of *Cassia fistula* seeds extract. *J. Med. Plants Res.* **2011**, *5*, 1941–1947.

93. Khan, N.; Bano, A. Modulation of phytoremediation and plant growth by the treatment with PGPR, Ag nanoparticle and untreated municipal wastewater. *Int. J. phytoremediat.* **2016**, *18*, 1258–1269. [CrossRef]

94. Mintenig, S.M.; Int-Veen, I.; Löder, M.G.; Primpke, S.; Gerdts, G. Identification of microplastic in effluents of waste water treatment plants using focal plane array-based micro-Fourier-transform infrared imaging. *Water Res.* **2017**, *108*, 365–372. [CrossRef]

95. Chehab, E.W.; Eich, E.; Braam, J. Thigmomorphogenesis: A complex plant response to mechano-stimulation. *J. Exp. Bot.* **2009**, *60*, 43–56. [CrossRef]

 *agronomy*

Article

Large Scale Screening of Rhizospheric Allelopathic Bacteria and Their Potential for the Biocontrol of Wheat-Associated Weeds

Tasawar Abbas [1], Zahir Ahmad Zahir [1], Muhammad Naveed [1,*], Sana Abbas [2], Mona S. Alwahibi [3], Mohamed Soliman Elshikh [3] and Adnan Mustafa [4]

1   Soil Microbiology and Biochemistry Laboratory, Institute of Soil and Environmental Sciences, University of Agriculture, Faisalabad 38040, Pakistan; tasawarabbasuaf@gmail.com (T.A.); zazahir@yahoo.com (Z.A.Z.)
2   Department of Chemistry, Government College Women University Faisalabad, Faisalabad 38040, Pakistan; sanabas.gc@gmail.com
3   Department of Botany and Microbiology, College of Science, King Saud University, Riyadh 11451, Saudi Arabia; malwahibi@ksu.edu.sa (M.S.A.); Melshikh@ksu.edu.sa (M.S.E.)
4   National Engineering Laboratory for Improving Quality of Arable Land, Institute of Agricultural Resources and Regional Planning, Chinese Academy of Agricultural Sciences, Beijing 100081, China; adnanmustafa780@gmail.com
*   Correspondence: muhammad.naveed@uaf.edu.pk; Tel.: +92-3336560577

Received: 21 August 2020; Accepted: 22 September 2020; Published: 25 September 2020

Abstract: Conventional weed control practices have generated serious issues related to the environment and human health. Therefore, there is a demand for the development of alternative techniques for sustainable agriculture. The present study performed a large-scale screening of allelopathic bacteria from the rhizosphere of weeds and wheat to obtain biological weed control inoculants in the cultivation of wheat. Initially, around 400 strains of rhizobacteria were isolated from the rhizosphere of weeds as well as wheat that grows in areas of chronic weed invasions. A series of the screen was performed on these strains, including the release of phytotoxic metabolites, growth inhibition of sensitive *Escherichia coli*, growth inhibition of indicator plant of lettuce, agar bioassays on five weeds, and agar bioassay on wheat. Firstly, 22.6% (89 strains) of the total strains were cyanogenic, and among the cyanogenic strains, 21.3% (19 strains) were inhibitory to the growth of sensitive *E. coli*. Then, these 19 strains were tested using lettuce seedling bioassay to show that eight strains suppressed, nine strains promoted, and two strains remained ineffective on the growth. These 19 strains were further applied to weeds and wheat on agar bioassays. The results indicated that dry matter of broad-leaved dock, wild oat, little seed canary grass, and common lambs' quarter were reduced by eight strains (23.1–68.1%), seven strains (38.5–80.2%), eight strains (16.5–69.4%), and three strains (27.5–50.0%), respectively. Five strains suppressed the growth of wheat, nine strains increased its dry matter (12.8–47.9%), and five remained ineffective. Altogether, the strains that selectively inhibit weeds, while retaining normal growth of wheat, can offer good opportunities for the development of biological weed control in the cultivation of wheat.

Keywords: allelopathic bacteria; antimetabolites; biological control; phytotoxic metabolites; rhizobacteria; weed invasion

---

1. Introduction

Dramatic increases in food production have been observed in the latter half of the twentieth century owing to the use of agro-chemicals, mechanization, irrigation, high yielding varieties, and post-harvest technology. The production of wheat in Pakistan has increased to ~25 m ton from 4.55 m ton in 1965 [1,2]. The pest attacks continue to incur losses to crop production owing to the diversity of pests and their resistance to prevailing control practices. The use of pesticides has increased from 15 to 20-fold over the last fifty years [3]. Chemical herbicides have gained importance in crop production in the face of a shortage of labor and limited application of mechanical control [4]. The mechanical control is known to contribute to soil erosion and its degradation [5]. Herbicides have led to the emergence of resistant biotypes of weeds, making the herbicide compounds useless to control these weeds [6]. Hence, the discovery of new compounds with novel modes of action is needed to replace these herbicides with more effective compounds to control such weeds. The discovery of such compounds, having herbicidal properties, has reduced over time. Further, the control of one type of weeds with herbicides has provided space to the proliferation of other weed species, which were less problematic for crop production in the past [7]. They have caused losses of biodiversity in the environment. It has deprived the ecosystems of some of their vital functions. Herbicides have aggravated the loss of biodiversity by killing the susceptible species, restricting the growth of others and the degradation of natural resources [8]. Poisoning, growth retardation, sterility, and deaths of wildlife owing to herbicide exposure have been reported by [9]. The residues of herbicides, apart from polluting the natural resources and destroying life forms, may also accumulate in the edible portions of plants, which facilitate their entry to the food chain and bodies of humans. It causes poisoning and chronic diseases in human beings, leading to deaths [10]. Human health disorders caused by herbicides include disorders of the nervous system, malformation of the embryo, loss of fertility, loss of immunity, kidney disorders, and liver disorders [11].

Farmers pay only the costs of manufacturing and marketing of herbicides, which provides economic access to farmers to adopt chemical weed control. The additional costs incurred on the treatment of human illnesses, degradation of natural resources and environment, and loss of biodiversity also need to be paid by farmers, society, or governments. Hence, the scenario of economic, environmental, and biological costs of chemical weed control pushes the researchers towards finding out safer weed control techniques. The importance of biological control has dramatically increased in the present situation. It presents a safer, inexpensive, and easier solution to the above-discussed issues of other control practices. It relies on increasing the strength, population, and activities of the organisms, resulting in growth reduction of weeds [12].

The past efforts in this area were focused on pathogens causing diseases in weeds [13] and insects feeding on weeds [14]. The success of insect biocontrol agents is limited by the existence of multiple hosts of insects in nature, which may cause the emergence of new pests of crops [15]. The pathogens of weeds used for biocontrol wait for suitable environmental conditions to cause infections and diseases in weed plants [13]. It may usually lead to delayed disease development, even after the weeds have caused economic losses of crops. Plant allelochemicals have also been investigated for biological weed control [16]. Their efficacy for weed control is reduced owing to the soil reactions, biodegradation, and mobility. It reduces their bioavailability and phytotoxicity on weeds [17]. These limitations of conventional biological weed control have discouraged researchers of this field, and the popularity of chemical weed control has increased dramatically.

The low success rate in conventional biological weed control has driven scientists to explore the characteristics of the rhizosphere inhabiting bacteria of weeds and crops for the development of novel weed biocontrol techniques. However, researchers have made efforts to explore the type of rhizobacteria, which produce substances inhibitory to the growth of weeds and are the least explored candidates for biological weed control. They release their secondary metabolites (phytotoxic in nature) in the rhizosphere, which is followed by their absorption in weeds. It results in a growth reduction of these weeds. The nature of this interaction between plants and microorganisms may be termed as plant-microbe allelopathy, and the bacteria responsible for these interactions may be called as allelopathic bacteria (AB) [18]. The discovery of host specificity in such microbial interactions with plants by [19] has opened ways for their potential application in crops for weed control. It reflects the properties of non-inhibition or even promotion of growth of crops among these rhizobacteria [20]. Therefore, the present study was conducted to explore such bacteria from the rhizosphere of weeds and wheat growing in fields facing weed invasions chronically, characterize them for the biological weed control, and evaluate their effects on the growth of wheat and weeds species of wheat.

2. Materials and Methods

2.1. Isolation of Rhizobacteria

We collected a large pool of samples of wheat and five weeds along with earth ball across the District of Faisalabad, Punjab, Pakistan. The sampling field was selected based on chronic weed invasions over the last 5 years. The weed species sampled included field bindweed, little seed canary grass, common lambs' quarter, wild oat, and broad-leaved dock. The scientific names of these weeds are *Convolvulus arvensis*, *Phalaris minor*, *Chenopodium album*, *Avena fatua*, and *Rumex dentatus*, respectively. These samples were transferred to the laboratory in an icebox and stored at 4 °C. The rhizosphere soil of these samples was used for the isolation of rhizobacteria using the dilution plating technique. A hundred microliters of each of the serial dilutions (10^{-1}–10^{-8}) were spread on the sterilized King's B agar media in Petri plates aseptically. This media was prepared by adding 1.5-g K_2HPO_4, 10 mL glycerol, 20 gm proteose peptone, 1.5 gm $MgSO_4.7H_2O$, and 20-g agar and making up the volume of one liter with distilled water following King et al. [21]. The growth of rhizobacterial colonies was obtained after 48 h of incubation of these plates at 28 ± 1 °C. The fast-growing colonies were picked and transferred to other Petri plates containing sterilized King's B agar media. These colonies were, hence, purified after some streaking. In this way, 393 strains were purified and preserved at −20 °C in 40% glycerol.

2.2. Cyanide Production Assay on Strains of Rhizobacteria

The method given by Bakker and Schipper [22] was followed for the qualitative determination of the production of hydrogen cyanide (HCN) by the isolated strains of rhizobacteria. The pieces of filter paper to the sizes of Petri plates were made, autoclaved for sterilization, and soaked in a 1% solution of picric acid for 12 h. These soaked filter papers were dried aseptically. Glycine amended media was prepared by adding 0.35 gm K_2HPO_4, 2.5 mL glycerol, 5 gm proteose peptone, 0.35 gm $MgSO_4 \cdot 7H_2O$, 5 gm glycine, and 20-g agar and making up the volume to one liter with distilled water. It gave out quarter strength media with glycine amendment. It was autoclaved and poured in Petri plates. The fresh culture of the strains was used to make a layer on the surface of the media and placing the picric acid-soaked paper on the inner side of the Petri plate lid. The paper was fastened with the help of a 10% solution of Na_2CO_3. The plates were closed and tightened with parafilm to avoid the leakage of gas. The plates were incubated at 28 °C and periodically observed for a change in the color of filter paper. The turning of color to brown indicated the production of HCN, while the intensity of brown color indicated the level of its production (Figure 1).

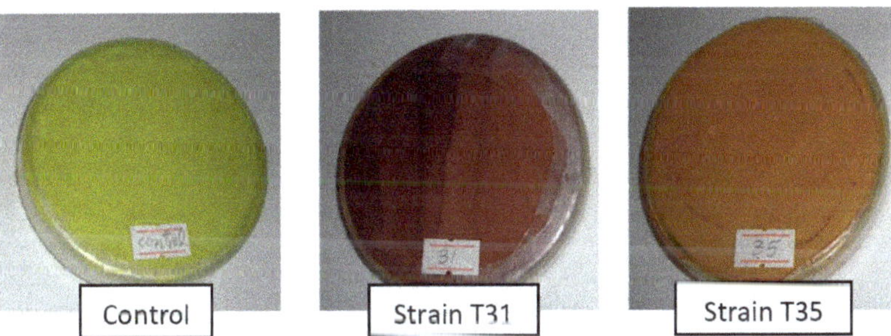

Figure 1 Pictorial view of cyanide production by rhizobacteria.

2.3. Antimetabolite Assay on E. coli

The bacterial production of toxic metabolites in extracellular spaces can be tested in a simple test based on the growth retardation of sensitive bacteria, *E. coli* [23]. All the strains (393) were tested for HCN production, while this assay was performed on only those strains that produced HCN to any level in step 1. These were 89 strains. Strain K12 of *E. coli* was cultured on LB agar media and placed in an incubator at 28 °C. After 2 days, the gentle rubbing of the surface and mixing with sterilized 0.01 M MgSO$_4$ solution formed the culture suspension of *E. coli*. The population of cells of bacteria in the suspension was maintained at 10^8 cells mL^{-1} through the measurement of optical density at 600 nm and the addition of 0.01 M MgSO$_4$ to get the value around 0.55–0.6. A layer of the harvested cell suspension was made on the Petri plates containing sterilized media (King's B). The culture of strains of cyanogenic rhizobacteria was spot inoculated at 3 points of equal distance on the plates pre-inoculated with *E. coli*. The plates were placed in an incubator at <40 °C. The production and release of toxic substances by the strains were evident from the zone of clearing around the spot of inoculation of strain. It indicated that the extracellular release of toxic compounds by the strains killed the growth of *E. coli* around its growth. The diameters of the zone of the clearing were recorded.

2.4. Antimetabolite Assay on Lettuce (Lectuca sativa L.) Seedlings

Nineteen strains restricted the growth of *E. coli* in the previous test. These strains were tested on the seedlings of lettuce as lettuce is considered sensitive to any type of phytotoxic substances and, hence, can be used as an indicator plant [24]. The fresh culture of the selected strains was prepared in Petri plates on KB media. This culture was suspended with the help of a sterilized buffer solution of MgSO$_4$ (0.01 M) by shaking gently. The suspension was collected in test tubes, and the cell population was maintained using optical density measurement at 600 nm with a value of 0.33. It established the population at 10^6 cells mL^{-1}.

The seeds of lettuce were disinfected on their surface in a parallel activity. The surface disinfection process comprised of seed dipping in ethanol for a moment, followed by the treatment with sodium hypochlorite (5%) for three minutes and complete rinsing of the seed with autoclaved water [25]. These seeds were allowed to germinate in the growth chamber.

Water agar was used as a medium for the growth of lettuce seedlings, where agar was added into the water at the rate of 1%. It was sterilized and poured in large-sized Petri plates, having a diameter of 15 cm. Seeds with good germination were picked up and transferred to the surface of these plates aseptically. Twenty germinating seeds of lettuce were placed on each plate.

Thirty microliters of the bacterial cell suspension were dispensed to each seed for inoculation. Three Petri plates were prepared for each strain in the same way. The control plates were treated with 30 μL buffer (0.01 M MgSO$_4$) per seed. The plates were placed at ambient temperature in the dark for 4 days. Then, the seedlings were removed from the plates and blotted. The measurements of masses and lengths of roots and shoots were done. The data were analyzed statistically to determine the significant differences [26].

2.5. Antimetabolite Assay on Weeds Using Presumed Allelopathic Bacteria

The strains of rhizobacteria obtained after the above-mentioned steps of the screening process were now called as presumed allelopathic bacteria. These strains were, now, used for testing on weeds. We selected four weeds of wheat for this assay i.e., wild oat, broad-leaved dock, common lambs' quarter, and little seed canary grass. These weeds cause maximum economic losses in the wheat crop in Pakistan. Nineteen strains were used to conduct this study in an experimental set up similar to the one used for bioassay on lettuce seedlings in Section 2.4. The culture of each strain was prepared in King's B broth. The culture was centrifuged to get the supernatant and form the bacterial pellets. These pellets were mixed in a sterilized buffer (0.01 M MgSO$_4$) to adjust the optical density value of 0.55 at 600 nm. It gave out the bacterial cell population at 10^8 cells mL^{-1}.

Water agar was prepared by adding 10 g of agar in 1 L distilled water and sterilizing in an autoclave at 121 °C and 15 PSI pressure for 20 min. The water agar was poured on large-sized Petri plates. It served as a medium for the growth of seedlings (Figure 2).

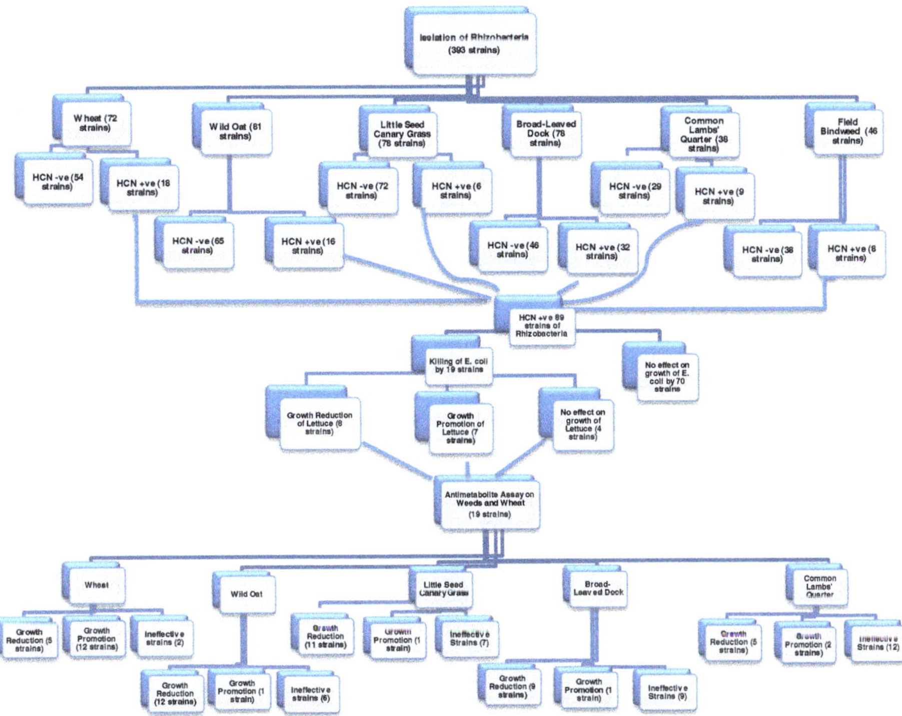

Figure 2. Flow chart of isolation and large-scale screening of allelopathic bacteria for the biocontrol of wheat-associated weeds.

The seeds of the selected weeds were surface disinfected by washing with ethanol (70%) momentarily, followed by washing with sodium hypochlorite (5%) and rinsing of seeds in plenty of sterilized water [25]. These seeds were placed in the growth chamber for germination.

Twenty germinated seeds were placed inside each prepared Petri plates aseptically. The culture suspension of each strain was applied at the rate of 30 µL per seed. For the control treatment, the sterilized buffer (0.01 M $MgSO_4$) was applied at the same rate. The plates were placed at ambient temperature in the dark. Each treatment in the experiment was replicated four times. After 7 days, the seedlings were uprooted from the water agar plates and blotted. These seedlings were measured for the lengths and weights of roots and shoots. The data were analyzed statistically to determine the significant differences following Steel et al. [26].

2.6. Antimetabolite Assay on Wheat Using Presumed Allelopathic Bacteria

The same nineteen strains were also tested for their effects on the growth of seedlings of wheat in a similar agar bioassay (Figure 3). The culture suspension of the strains was prepared following the same method as above. The large-sized Petri plates containing water agar were prepared as in previous bioassays. The surface of seeds of wheat was disinfected following Abd-Alla et al. [25]. Then, the seeds were placed for germination. The germinated seeds were placed on the already prepared Petri plates aseptically. The culture suspension of each strain was dispensed at the rate of 30 µL per seed. For the control treatment, the sterilized buffer (0.01 M MgSO$_4$) was dispensed to each seed at the rate of 30 μL. Each treatment was replicated four times. The seedlings were uprooted after five days and blotted. The data of lengths and weights of roots and shoots were taken and analyzed statistically to determine the significant differences following Steel et al. [26]. These analyses were carried out using *Statistix 8.1* software. All the data were first subjected to analysis of variance (ANOVA) test in this software, followed by multiple comparisons of means using the linear model. The least significant difference (LSD) test was then applied to determine the significant difference among treatments at $p < 0.05$.

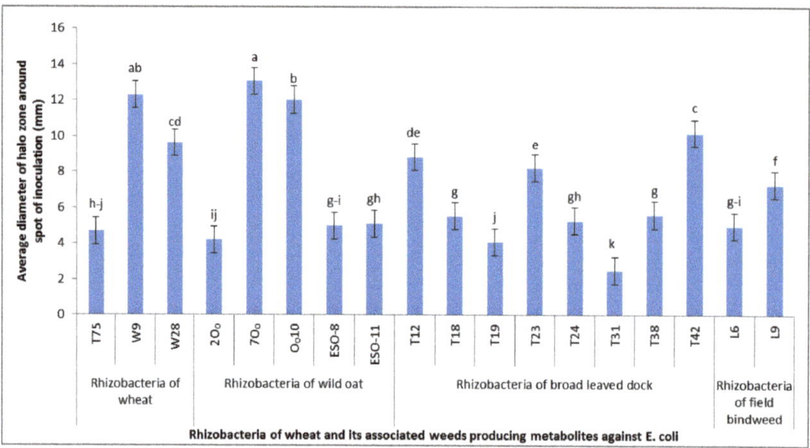

Figure 3. Cyanogenic rhizobacteria of weeds and wheat-producing metabolites against *E. coli* in the antimetabolite assay.

2.7. Cluster Analysis for the Screening of Biological Weed Control Agents

Cluster analysis was carried out for the grouping of strains applied in antimetabolite assays on weeds and wheat. The strains were categorized as non-selective biological weed control agents (the strains that reduced the growth of all the tested weeds and wheat), selective (the strains that reduced the growth of some of the tested plants and also wheat), selective (the strains that reduced the growth of some of the weeds but not wheat), and selective (the strains that reduced the growth of one more weed but promoted the growth of wheat). Five most efficient strains of allelopathic bacteria obtained from this study were identified through 16s rDNA sequencing as *Pseudomonas* strain T42 as *Pseudomonas putida*, strains L9 and 7O$_0$ as *P. fluorescens*, strain O$_0$10 as *P. aeruginosa*, and strain W9 as *P. alcaligenes*.

3. Results

The present study explored the rhizosphere of wheat and five weeds of wheat in search of allelopathic bacteria for the development of biological weed control agents. The selected weeds cause huge economic losses to the production of wheat in Pakistan annually [27]. These weed species were wild oat, common lambs' quarter, little seed canary grass, broad-leaved dock, and field bindweed. The job was carried out by the isolation of a large number of strains of rhizobacteria (393) from the weeds and wheat growing in areas of high weed invasion. Multiple bioassays were conducted on these strains to evaluate if they produced some phytotoxic substances, whether the release of such substances resulted in growth suppression of weeds, and if they were selective to inhibit the growth of weeds but not crop. The screening process of rhizobacteria to find out allelopathic bacteria from the rhizosphere of weeds and wheat is shown in the form of a flow chart (Figure 2).

3.1. Isolation of Rhizobacteria

We isolated 78 strains from the rhizosphere of wild oat, 81 from the broad-leaved dock, 78 from common lambs' quarter, 46 from field bindweed, 38 from little seed canary grass, and 72 from wheat. The total number of strains was 393. Multiple screening tests were conducted on these strains to characterize weed suppressive allelopathic bacteria.

3.2. Production of HCN by Rhizobacteria

The proportion of strains producing cyanide to various levels is shown in Table 1. We got 89 strains, which could produce cyanide to any level. Among these, 33 strains produced a low amount of cyanide, 25 medium, 20 high, and 11 very high, depending upon the intensity of change of color of picrate-treated filter paper inside the Petri plates and the time taken to change the color. The proportion of cyanogenic strains in the rhizosphere of the broad-leaved dock was calculated to be 41.0%, that of wild oat was 19.8%, of little seed canary grass was 7.7%, of common lambs' quarter was 23.7%, of field bindweed was 17.4%, and that of wheat was 25.0%. However, the majority of strains (77.6%) did not produce HCN in this study. These counted to 304 in number out of 393. The pictorial view of this assay is given in (Figure 1).

Table 1. The proportion of cyanogenic rhizobacteria in the rhizosphere of wheat and its associated weeds. The cyanide production by the strains was indicated after 48, 36, 24, and 12 h of incubation for low, medium, high, and very high cyanide production activity, respectively.

Category	Rhizosphere of						Total Strains
	Wheat	Broad-Leaved Dock	Wild Oat	Little Seed Canary Grass	Field Bindweed	Common Lambs' Quarter	
Non-cyanogenic strains	54	46	65	72	38	29	304
Low cyanide activity strains	8	3	8	5	3	6	33
Medium cyanide activity strains	6	12	3	0	2	2	25
High cyanide activity strains	2	14	0	1	2	1	20
Very high cyanide activity strains	2	3	5	0	1	0	11
Total strains	72	78	81	78	46	38	393

3.3. Antimetabolite Assay on E. coli

Clearing zones were produced around the inoculation spot of some strains, while the growth of most of the strains was mixed with the growth of *E. coli*, i.e., mutualistic strains. The clearing zones indicated the killing of *E. coli*, which occurred with nineteen strains. The diameter of these clearing or halo zones indicated the level of inhibition of growth of *E. coli* (Figure 3). Strain $7O_0$ produced the maximum diameter of the halo zone, which was followed by strains W9, $O_0 10$, T42, W28, T12, T23, and L9. The average diameter of zones produced by these strains was measured to be 1.3 ± 0.08, 1.23 ± 0.13, 1.21 ± 0.08, 1.01 ± 0.10, 0.96 ± 0.09, 0.88 ± 0.06, 0.82 ± 0.06, and 0.72 ± 0.07 cm, respectively. The remaining strains showed positive interaction with the growth of *E. coli*.

3.4. Antimetabolite Assay on Lettuce Seedlings

Results indicated that the strains imparted mixed effects on the growth of lettuce seedlings (Table 2). Five of the application strains significantly reduced the dry matter, root length, and shoot length of lettuce seedlings from 18.8 to 38.9%, 19.7 to 36.3%, and 17.3 to 24.3%, respectively. These strains were T18, T12, W9, W28, and $O_0 10$. The strains L6 and T31 caused a significant reduction in root length only. However, the strain T38 caused a significant reduction in the length of root and shoot. There were seven strains, which increased the dry matter, root length, and shoot length of lettuce seedlings from 15.7 to 41.5%, 16.7 to 61.4%, and 26.2 to 43.4%, respectively. These strains were T23, T42, T19, $2O_0$, T24, L9, and $7O_0$. The strains B11 and ESO-8 increased the shoot length only. The other strains remained ineffective on the growth of seedlings of lettuce.

Table 2. The effect of cyanogenic *E. coli* inhibiting rhizobacteria on lettuce seedlings in agar bioassay. Values sharing the same letter(s) in a column do not differ significantly from each other at $p < 0.05$. Values in a column indicate mean ± standard error.

Treatments	Root Length (cm)	Shoot Length (cm)	Dry Matter (mg)
Control	5.08 ± 0.14 [d,e]	4.01 ± 0.13 [f,g]	51.49 ± 0.006 [f,g]
T12	4.07 ± 0.16 [g,h]	3.19 ± 0.10 [i]	38.94 ± 0.006 [i,j]
T18	3.79 ± 0.16 [h,i]	3.32 ± 0.09 [i]	41.81 ± 0.006 [h,i,j]
T19	7.1 ± 0.13 [b]	5.39 ± 0.03 [b,c]	67.13 ± 0.006 [a,b]
T23	6.08 ± 0.12 [c]	5.22 ± 0.13 [c]	61.7 ± 0.010 [b,c,d]
T24	5.92 ± 0.27 [c]	5.07 ± 0.15 [c]	59.57 ± 0.007 [c,d,e]
T31	4.36 ± 0.20 [f,g]	3.68 ± 0.12 [g,h]	44.2 ± 0.007 [h,i]
T38	4.06 ± 0.16 [g,h]	3.37 ± 0.08 [h,i]	42.07 ± 0.007 [hij]
T42	6.35 ± 0.22 [c]	4.63 ± 0.15 [d]	58.42 ± 0.007 [d,e]
T75	4.8 ± 0.26 [e,f]	4.02 ± 0.15 [f]	50.7 ± 0.003 [f,g]
$2O_0$	6.19 ± 0.20 [c]	5.16 ± 0.14 [c]	65.26 ± 0.007 [b,c]
$7O_0$	8.19 ± 0.17 [a]	5.76 ± 0.11 [a]	72.19 ± 0.006 [a]
$O_0 10$	3.4 ± 0.14 [i,j]	3.18 ± 0.09 [i]	39.78 ± 0.006 [hij]
ESO-8	5.37 ± 0.20 [d]	4.43 ± 0.17 [d]	55.0 ± 0.003 [e,f]
ESO-11	5.03 ± 0.25 [d,e]	4.06 ± 0.11 [e,f]	52.0 ± 0.007 [f,g]
L6	4.48 ± 0.13 [f,g]	3.79 ± 0.11 [f,g]	45.79 ± 0.003 [g,h]
L9	8.01 ± 0.19 [a]	5.73 ± 0.05 [a,b]	72.88 ± 0.007 [a]
B11	5.26 ± 0.11 [d,e]	4.38 ± 0.18 [d,e]	54.53 ± 0.006 [e,f]
W9	3.34 ± 0.12 [i,j]	3.11 ± 0.08 [i]	35.99 ± 0.003 [j]
W28	3.231 ± 0.19 [j]	3.04 ± 0.14 [i]	38.08 ± 0.003 [i,j]
LSD	0.517	0.345	6.42

3.5. Antimetabolite Assay on Broad-Leaved Dock

The effects of the applied strains on the growth of the seedling of the broad-leaved dock were mixed, i.e., inhibiting, promoting, and neutral (Table 3). The dry matter, root length, and germination rate of the broad-leaved dock were significantly reduced by eight of the applied strains from 23.1 to 68.1%, 23.9 to 61.8%, and 26.7 to 64.4% than control, respectively. These strains were T42, $O_0 10$, L9, T38, $7O_0$, ESO-11, W9, and W28. The strain T19 caused a reduction in root length and germination rate only. The strain T31 caused a significant increase in root length and germination rate of the dock. The other strains remained ineffective on the growth of the seedlings of the dock.

3.6. Antimetabolite Assay on Wild Oat

Seven strains significantly reduced the dry matter, root length, and germination rate of wild oat from 38.5 to 80.2%, 19.4 to 60.2%, and 25.4 to 70.9%, respectively (Table 3). These strains were $2O_0$, ESO-8, $O_0 10$, T42, W28, W9, and $7O_0$. The strains T18, T12, ESO-11, and T75 significantly inhibited the germination rate from 14.5 to 25.4% but no other parameters. The strain T24 only reduced the root length of wild oat. The root length and germination rate of the weed were significantly increased by strain T19 up to 13.3 and 14.5%, respectively. The other strains remained ineffective on the growth of the seedlings of wild oat.

3.7. Antimetabolite Assay on Little Seed Canary Grass

Eight of the nineteen applied strains caused a significant reduction in dry matter, root length, and germination rate of little seed canary grass from 16.5 to 69.4%, 24.2 to 63.6%, and 20 to 52.7%, respectively (Table 4). These eight strains were T75, 7O$_0$, T42, ESO-11, O$_0$10, W9, L9, and W28. The strains T18 and T12 reduced only the root length (10.5–20%) and germination rate (18.2–25.4%). The strain 2O$_0$ significantly reduced the dry matter (21.2%) and root length (10.8%) of the weed. However, the strain T19 significantly increased the dry matter (23.5%) and root length (10.4%) of the weed. Other strains remained ineffective on the growth of the seedlings of this weed. The pictorial view of the assay is available in (Figure 4).

Table 3. The effect of presumed allelopathic bacteria on the germination and seedling growth of broad-leaved dock and wild oat in agar bioassay. Values sharing the same letter(s) in a column do not differ significantly from each other at $p < 0.05$. Values in a column indicate mean ± standard error.

Treatment	Broad-Leaved Dock			Wild Oat		
	Germination Rate (%)	Root Length (cm)	Dry Matter (g)	Germination Rate (%)	Root Length (cm)	Dry Matter (g)
Control	75.0 ± 0.58 [b,c]	3.52 ± 0.13 [b]	0.307 ± 0.014 [a,b,c]	73.3 ± 0.33 [b,c]	6.0 ± 0.16 [b,c,d]	0.32 ± 0.03 [b,c,d]
T12	73.4 ± 0.88 [b,c]	3.48 ± 0.12 [b]	0.29 ± 0.035 [a,b,c,d]	62.7 ± 0.88 [d,e,f]	5.6 ± 0.17 [c,d,e]	0.29 ± 0.02 [b,c,d]
T18	80.0 ± 1.00 [a,b]	3.5 ± 0.10 [b]	0.303 ± 0.026 [a,b,c]	58.7 ± 0.67 [f,g]	5.52 ± 0.24 [d,e]	0.28 ± 0.02 [c,d]
T19	63.4 ± 0.67 [d,e]	2.98 ± 0.18 [c,d]	0.247 ± 0.013 [c,d,e]	84.0 ± 0.58 [a]	6.79 ± 0.13 [a]	0.41 ± 0.03 [a]
T23	68.3 ± 0.33 [c,d]	3.33 ± 0.11 [b,c]	0.277 ± 0.018 [b,c,d]	66.7 ± 0.33 [c,d,e]	6.16 ± 0.16 [b]	0.30 ± 0.02 [b,c,d]
T24	71.6 ± 0.33 [b,c,d]	3.46 ± 0.11 [b]	0.297 ± 0.023 [a,b,cd]	68.0 ± 0.58 [c,d]	5.28 ± 0.14 [e,f]	0.29 ± 0.02 [b,c,d]
T31	86.6 ± 0.33 [a]	4.1 ± 0.23 [a]	0.353 ± 0.014 [a]	73.3 ± 0.33 [b,c]	5.75 ± 0.15 [b,c,d,e]	0.32 ± 0.02 [b,c,d]
T38	55.0 ± 0.58 [e,f]	2.68 ± 0.28 [d,e]	0.233 ± 0.017 [d,e,f]	77.3 ± 0.33 [a,b]	5.96 ± 0.08 [b,c,d]	0.33 ± 0.02 [b,c]
T42	33.3 ± 0.67 [i,j]	1.73 ± 0.29 [h,i]	0.13 ± 0.020 [h,i]	32.0 ± 0.58 [k]	2.74 ± 0.18 [j]	0.11 ± 0.01 [h,i]
T75	70.0 ± 1.00 [c,d]	3.33 ± 0.06 [b,c]	0.28 ± 0.036 [b,c,d]	60.0 ± 1.00 [e,f,g]	5.88 ± 0.33 [b,c,d]	0.28 ± 0.02 [b,c,d]
2O$_0$	73.4 ± 0.33 [b,c]	3.61 ± 0.14 [b]	0.303 ± 0.022 [a,b,c]	54.7 ± 0.88 [g,h]	4.55 ± 0.18 [g]	0.20 ± 0.01 [e,f]
7O$_0$	41.7 ± 0.33 [h,i]	1.96 ± 0.06 [g,h]	0.17 ± 0.023 [f,g,h]	49.3 ± 0.67 [h,i]	3.77 ± 0.22 [h,i]	0.16 ± 0.003 [f,g,h]
O$_0$10	51.6 ± 0.33 [f,g]	2.17 ± 0.10 [f,g]	0.203 ± 0.027 [e,f,g]	45.3 ± 0.67 [i,j]	3.6 ± 0.19 [i]	0.15 ± 0.03 [f,g,h]
ESO-8	73.4 ± 0.67 [b,c]	3.41 ± 0.21 [b]	0.29 ± 0.026 [a,c,d]	60.0 ± 0.58 [e,f,g]	4.83 ± 0.18 [f,g]	0.20 ± 0.01 [f,g]
ESO-11	48.4 ± 1.20 [f,g,h]	2.53 ± 0.06 [e,f]	0.203 ± 0.018 [e,f,g]	54.7 ± 0.88 [g,h]	5.56 ± 0.29 [d,e]	0.26 ± 0.02 [d,e]
L6	56.6 ± 0.67 [e,f]	2.77 ± 0.08 [d,e]	0.24 ± 0.020 [c,d,e]	82.7 ± 0.33 [a]	6.13 ± 0.17 [b,c]	0.35 0.03 [a,b]
L9	41.6 ± 0.67 [h,i]	1.88 ± 0.10 [g,h]	0.157 ± 0.022 [g,h,i]	21.3 ± 0.33 [l]	2.39 ± 0.23 [j]	0.63 ± 0.01 [i]
B11	76.6 ± 0.67 [b,c]	3.65 ± 0.11 [b]	0.317 ± 0.033 [a,b]	70.7 ± 0.33 [b,c]	6.0 ± 0.19 [b–d]	0.32 ± 0.02 [b,c,d]
W9	26.6 ± 0.67 [j]	1.34 ± 0.12 [i]	0.097 ± 0.018 [i]	41.3 ± 0.33 [j]	3.42 ± 0.13 [i]	0.13 ± 0.03 [g,h]
W28	43.4 ± 0.88 [g,h]	1.41 ± 0.08 [i]	0.14 ± 0.029 [g,h,i]	44.0 ± 1.16 [i,j]	4.3 ± 0.13 [g,h]	0.17 ± 0.02 [f,g,h]
LSD	9.8205	0.429	0.068	7.32	0.545	0.0642

Table 4. The effect of presumed allelopathic bacteria on the germination and seedling growth of little seed canary grass and common lambs' quarter. Values sharing the same letter(s) in a column do not differ significantly from each other at $p < 0.05$. Values in a column indicate mean ± standard error.

Treatments	Little Seed Canary Grass			Common Lambs' Quarter		
	Germination Rate (%)	Root Length (cm)	Dry Matter (g)	Germination Rate (%)	Root Length (cm)	Dry Matter (g)
Control	73.3 ± 1.20 [a,b]	4.59 ± 0.22 [b,c]	0.283 ± 0.026 [b,c]	63.3 ± 1.00 [c,d,e]	2.87 ± 0.19 [b,c,d,e]	0.27 ± 0.01 [a,b,c]
T12	54.7 ± 1.45 [d,e,f,g]	4.11 ± 0.06 [d,e]	0.287 ± 0.024 [b,c]	61.0 ± 0.67 [c,d,e]	2.57 ± 0.12 [d,e,f,g]	0.25 ± 0.02 [b,c,d]
T18	60.0 ± 1.53 [c,d,e,f]	3.68 ± 0.08 [e,f]	0.267 ± 0.026 [b,c,d]	61.5 ± 0.88 [c,d,e]	2.29 ± 0.11 [f,g,h]	0.24 ± 0.01 [b,c,d]
T19	81.3 ± 0.67 [a]	5.07 ± 0.11 [a]	0.35 ± 0.021 [a]	63.2 ± 1.00 [c,d,e]	2.50 ± 0.17 [e,f,g,h]	0.27 ± 0.01 [a,b,c]
T23	66.7 ± 0.33 [b,c]	4.64 ± 0.15 [a,b]	0.26 ± 0.015 [b,c,d]	58.7 ± 1.45 [c,d,e]	2.73 ± 0.11 [c,d,e,f]	0.25 ± 0.02 [b,c,d]
T24	68.0 ± 0 [b,c]	4.28 ± 0.16 [b,c,d]	0.237 ± 0.013 [c,d,e]	66.7 ± 1.00 [b,c]	3.14 ± 0.22 [a,b,c]	0.29 ± 0.01 [a,b]
T31	74.7 ± 0.33 [a,b]	4.32 ± 0.08 [b,c,d]	0.287 ± 0.014 [b,c]	57.7 ± 0.33 [d,e,f]	2.9 ± 0.22 [b,c,d,e]	0.28 ± 0.06 [a,b]
T38	74.7 ± 0.67 [a,b]	4.58 ± 0.11 [b,c]	0.277 ± 0.024 [b,c]	57.2 ± 0.33 [d,e,f]	2.61 ± 0.28 [d,e,f]	0.24 ± 0.03 [b,c,d]
T42	54.7 ± 0.88 [d,e,f,g]	2.89 ± 0.10 [g,h]	0.153 ± 0.017 [g,h,i]	65.7 ± 0.67 [b,c,d]	3.04 ± 0.19 [a,b,c,d]	0.24 ± 0.04 [b,c,d]
T75	53.3 ± 1.67 [e,f,g,h]	2.34 ± 0.09 [i]	0.175 ± 0.005 [f,g,h]	47.7 ± 0.67 [g]	2.04 ± 0.17 [h,i]	0.19 ± 0.02 [d,e,f]
2O$_0$	64.0 ± 1.00 [b,c,d,e]	4.1 ± 0.24 [d,e]	0.223 ± 0.007 [d,e,f]	75.6 ± 0.33 [a]	3.51 ± 0.09 [a]	0.32 ± 0.01 [a]
7O$_0$	42.7 ± 0.67 [h,i]	2.37 ± 0.18 [i]	0.100 ± 0.006 [j]	55.7 ± 0.33 [e,f,g]	2.90 ± 0.17 [b,c,d,e]	0.21 ± 0.01 [b,c,d,e]
O$_0$10	34.7 ± 0.67 [i]	1.8 ± 0.19 [j]	0.087 ± 0.019 [j]	63.3 ± 0.58 [c,d,e]	2.93 ± 0.11 [b,c,d,e]	0.24 ± 0.02 [b,c,d]
ESO-8	65.3 ± 0.88 [b,c,d]	4.16 ± 0.21 [c,d]	0.237 ± 0.012 [c,d,e]	64.3 ± 0.67 [b,c,d]	2.93 ± 0.27 [b,c,d,e]	0.28 ± 0.02 [a,b,c]
ECO 11	49.0 ± 1.53 [g,h]	3.48 ± 0.08 [f]	0.237 ± 0.022 [c,d,e]	49.0 ± 0.67 [g]	1.69 ± 0.14 [i,j]	0.17 ± 0.02 [e,f]
L6	80.0 ± 0.58 [a]	4.57 ± 0.11 [b,c]	0.293 ± 0.013 [b]	72.3 ± 0.88 [a,b]	3.32 ± 0.23 [a,b]	0.30 ± 0.01 [a,h]
L9	49.3 ± 0.67 [f,g,h]	2.54 ± 0.20 [h,i]	0.123 ± 0.007 [i,j]	63.3 ± 1.53 [c,d,e]	2.69 ± 0.12 [c,d,e,f]	0.27 ± 0.04 [a,b,c]
B11	74.7 ± 0.67 [a,b]	4.52 ± 0.11 [b,c,d]	0.257 ± 0.012 [b,c,d]	49.0 ± 0.88 [g]	2.09 ± 0.11 [g,h,i]	0.21 ± 0.01 [c,d,e]
W9	58.7 ± 0.33 [c,d,e,f,g]	3.26 ± 0.27 [f,g]	0.197 ± 0.032 [e,f,g]	59.0 ± 0.67 [c,d,e]	2.98 ± 0.29 [b,c,d,e]	0.26 ± 0.03 [a,b,c,d]
W28	49.3 ± 1.77 [f,g,h]	1.67 ± 0.09 [j]	0.137 ± 0.003 [h,i,j]	50.0 ± 1.16 [f,g]	1.29 ± 0.13 [i]	0.13 ± 0.02 [f]
LSD	11.432	0.445	0.0507	8.127	0.524	0.073

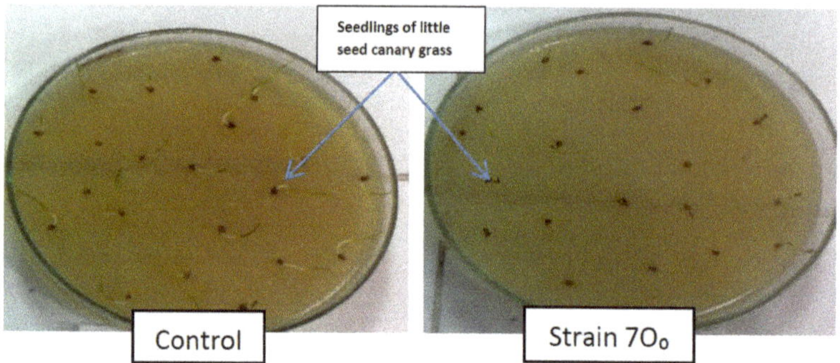

Figure 4. The pictorial view of seedlings of little seed canary grass growing on water agar in agar bioassay.

3.8. Antimetabolite Assay on Common Lambs' Quarter

The present study reported a decrease in dry matter, root length, and germination rate of common lambs' quarter by three of the applied strains from 27.5 to 50.0%, 29.0 to 55.0%, and 21.0 to 24.6%, respectively (Table 4). These strains were W28, ESO-11, and T75. The strain B11 caused a reduction in root length (27.3%) and germination rate (22.8%) only. The strain T18 caused a reduction in root length only, which was 20.3% lesser than the control. However, a significant increase in root length (13.0%) and germination rate (19.3%) was observed with the inoculation of strain 2O$_0$. The strain L6 increased the germination rate of the weed by 14%. The other strains remained ineffective on the growth of the seedlings of this weed.

3.9. Antimetabolite Assay on Wheat

There were three strains in the whole lot, which significantly reduced the dry matter, shoot length, root length, and germination rate of wheat from 23.4 to 34%, 21.0 to 38.5%, 27.2 to 52.8%, and 8.3 to 10.4%, respectively (Figure 5, Table 5). These three strains were ESO-11, W28, and T18. Two strains (T75 and T12) reduced the dry matter (23.4 and 26.6%), root length (24.8 and 50.1%), and shoot length (18.9 and 35.5%) of the crop. However, there were six strains, which significantly increased the dry matter, shoot length, root length, and germination rate of wheat from 24.5 to 47.9%, 14.6 to 29.7%, 19.4 to 37.7%, and 12.5 to 18.8%, respectively. These strains were T23, 7O$_0$, 2O$_0$, L9, T24, and T19. The strains L6, O$_0$10, and B11 caused an increment in dry matter of the crop up to 13.8, 12.8, and 27.7% than control, respectively. The strains T38 and T31 caused a significant increase in shoot length of the crop up to 18.9 and 18.7% than control, respectively. The strain T42, however, increased the germination rate of the crop up to 8.3%. The other strains remained ineffective on the growth of the seedlings of wheat.

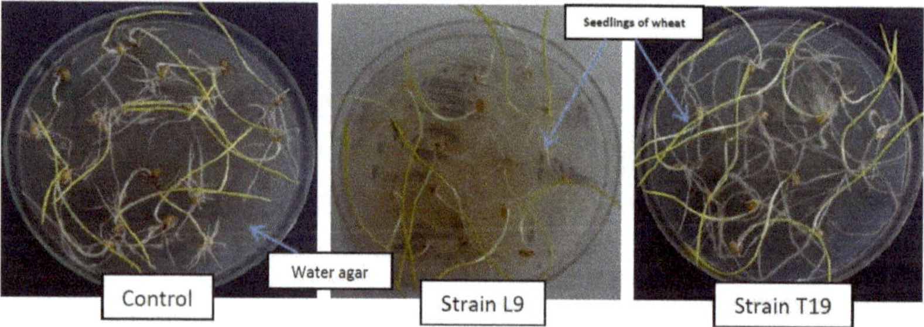

Figure 5. The pictorial view of seedlings of wheat growing on water agar in agar bioassay.

Table 5. The effect of presumed allelopathic bacteria on the germination and seedling growth of wheat. Values sharing the same letter(s) in a column do not differ significantly from each other at $p < 0.05$. Values in a column indicate mean ± standard error.

Treatments	Germination Rate (%)	Root Length (cm)	Shoot Length (cm)	Dry Matter (g)
Control	80.0 ± 0.58 [d,e]	6.60 ± 0.50 [c,d,e]	8.58 ± 0.22 [c,d]	0.313 ± 0.018 [f,g]
T12	75.0 ± 0.58 [e,f]	4.97 ± 0.08 [f]	6.96 ± 0.27 [e]	0.233 ± 0.003 [h]
T18	73.35 ± 0.33 [f]	4.81 ± 0.10 [f]	6.78 ± 0.16 [e]	0.240 ± 0.015 [h]
T19	93.4 ± 0.33 [a]	8.22 ± 0.15 [b]	9.83 ± 0.37 [b]	0.407 ± 0.012 [c]
T23	90.0 ± 0.58 [a,b]	8.13 ± 0.14 [b]	10.13 ± 0.34 [b]	0.390 ± 0.006 [c,d]
T24	90.0 ± 0 [a,b]	7.92 ± 0.13 [b]	10.05 ± 0.21 [b]	0.400 ± 0.015 [c]
T31	80.0 ± 0.58 [d,e]	6.47 ± 0.23 [d,e]	10.19 ± 0.20 [b]	0.300 ± 0.020 [g]
T38	80.0 ± 0.58 [d,e]	6.67 ± 0.27 [c,d,e]	10.21 ± 0.17 [b]	0.310 ± 0.015 [f,g]
T42	86.5 ± 0.33 [b,c]	7.15 ± 0.25 [c]	9.097 ± 0.38 [c]	0.343 ± 0.012 [e,f]
T75	75.0 ± 0 [e,f]	3.3 ± 0.17 [g]	5.53 ± 0.10 [f]	0.230 ± 0.012 [h]
$2O_0$	93.4 ± 0.33 [a]	7.89 ± 0.26 [b]	10.03 ± 0.18 [b]	0.413 ± 0.003 [b,c]
$7O_0$	93.4 ± 0.33 [a]	9.09 ± 0.21 [a]	11.13 ± 0.25 [a]	0.450 ± 0.010 [a,b]
O_010	85 ± 0 [b–d]	7.17 ± 0.15 [c]	8.69 ± 0.19 [c,d]	0.353 ± 0.003 [d,e]
ESO-8	81.5 ± 0.67 [c,d]	6.53 ± 0.17 [c–e]	8.52 ± 0.16 [c,d]	0.347 ± 0.014 [e,f]
ESO-11	73.4 ± 0.67 [f]	3.38 ± 0.33 [g]	5.56 ± 0.15 [f]	0.220 ± 0.006 [h]
L6	83.4 ± 0.33 [c,d]	6.41 ± 0.29 [e]	8.6 ± 0.14 [c,d]	0.357 ± 0.019 [d,e]
L9	95 ± 0 [a]	9.09 ± 0.12 [a]	11.13 ± 0.20 [a]	0.463 ± 0.020 [a]
B11	83.4 ± 0.67 [c,d]	6.32 ± 0.23 [e]	8.42 ± 0.17 [d]	0.400 ± 0.006 [c]
W9	81.7 ± 0.67 [c,d]	7.14 ± 0.31 [c,d]	8.62 ± 0.38 [c,d]	0.337 ± 0.014 [e,f,g]
W28	71.6 ± 0.67 [f]	3.12 ± 0.26 [g]	5.28 ± 0.17 [f]	0.207 ± 0.014 [h]
LSD	6.565	0.682	0.673	0.0376

3.10. Cluster Analysis

The cluster analysis was performed to categorize the tested strains of this study based on the objectives of this study (Table 6). All the strains were categorized into four groups: the first group of two strains (W28 and ESO-11) comprised of non-selective strains, which reduced the growth of seedlings of all the tested plants; the second group of three strains (T75, T18, and T12) comprised of selective strains, which reduced the growth of little seed canary grass, wild oat, common lambs' quarter, and wheat but not of the broad-leaved dock; the third group of three strains (W9, ESO-8, and T38) comprised of selective strains, which reduced the growth of seedlings of wild oat, broad-leaved dock, and little seed canary grass but not of wheat and common lambs' quarter; and the fourth group of nine strains (T24, $2O_0$, O_010, L9, B11, T19, T42, $7O_0$, and L6) comprised of selective strains, which reduced the growth of seedlings of little seed canary grass, broad-leaved dock, and wild oat but increased the growth of seedlings of wheat. The remaining two strains of this study (T23 and T31) did not suppress the growth of any weed or wheat.

Table 6. Cluster analysis for the selection of bioherbicidal agents based on the response of rhizobacteria in wheat and its associated weeds in agar bioassays. Candidate strains for biological weed control in wheat are indicated in bold.

Category of Strains	Strain	Effects on Weeds and Wheat		
		Inhibition	Promotion	No Effect
Non-selective	ESO-11	All the tested weeds and wheat	–	–
	W28			
Selective and inhibitory to wheat	T12	Wheat, wild oat, and little seed canary grass	–	Broad-leaved dock and common lambs' quarter
	T18	Wheat, wild oat, little seed canary grass, and common lambs' quarter	–	Broad-leaved dock
	T75			
Selective and non-inhibitory to wheat	**T38**	Broad-leaved dock	–	Wheat, wild oat, little seed canary grass, and common lambs' quarter
	ESO-8	Wild oat	–	Wheat, little seed canary grass, broad-leaved dock, and common lambs' quarter
	W9	Wild oat, little seed canary grass, and broad-leaved dock	–	Wheat and common lambs' quarter
Selective and promotory to wheat	T19	Broad-leaved dock	Wheat, wild oat, and little seed canary grass	Common lambs' quarter
	T24	Wild oat	Wheat	Little seed canary grass, broad-leaved dock, and common lambs' quarter
	T42	Wild oat, little seed canary grass, and broad-leaved dock	Wheat	Common lambs' quarter
	$7O_0$			
	O_010			
	L9			
	$2O_0$	Wild oat and little seed canary grass	Wheat and common lambs' quarter	Broad-leaved dock
	L6	Broad-leaved dock	Wheat, wild oat, and common lambs' quarter	Little seed canary grass
	B11	Common lambs' quarter	Wheat	Wild oat, little seed canary grass, and broad-leaved dock
–	T23	–	Wheat	Wild oat, little seed canary grass, broad-leaved dock, and common lambs' quarter
	T31	–	Broad-leaved dock	Wheat, wild oat, little seed canary grass, and common lambs' quarter

4. Discussion

The study of diverse forms of soil-inhabiting microorganisms and their activities may be helpful in resolving many agricultural, environmental, and ecological issues created by unsustainable farming practices. The invasion of weeds in crops reduces their yields, and farmers adopt unsustainable and unhealthy practices to reduce the losses of their crops. The harmful impacts of tillage and chemicals have been established. Therefore, the present study explored an alternative, inexpensive, sustainable, and environmentally and ecologically safe technique for weed control in crops. It was aimed at finding out the natural mechanisms of rhizobacteria, which function to limit the growth of weeds, alleviate the biotic stress of weeds on crops, and produce a vigorous crop stand. Strengthening such natural processes through augmentation, inoculation, or other processes is required for the development of biological weed control in crops. This may help us to resolve the above-mentioned issues created by conventional control practices [11].

The rhizosphere inhabiting bacteria, which release phytotoxic metabolites in the rhizosphere and result in germination/growth reduction of weeds, are called allelopathic bacteria [18]. The present study is the pioneering work executed in Pakistan, which is aimed at searching such rhizobacteria with their novel characteristics to develop biological weed control. The probability of the existence of such bacteria has been speculated in the rhizosphere of weeds and crops, which are growing together over many years or where the weed invasions occur more frequently [28]. Therefore, we collected the samples of weeds and wheat from areas/fields across the District of Faisalabad, Pakistan, where the weed invasions were more frequent. The findings of this work support the above-mentioned finding of Schippers et al. [28]. They also reported that growth inhibitory rhizobacteria grew, strengthened, and increased their activities in the agricultural crops, where a single crop is grown year after year. It resulted in the reduction of yields of crops. They reported the increase in cyanogenic bacteria and cyanide production in the rhizosphere of potatoes when this crop was continuously grown over a field for 3 years. Their findings increased the importance of crop rotation.

We isolated 393 strains of rhizobacteria from the rhizosphere of five weeds and wheat in this study. These strains were passed through a comprehensive screening process based on the production of phytotoxic metabolites in vitro, suppression of indicator bacteria and plants, in vivo suppression of weeds, and their effects on wheat crops. The protocols followed for these purposes obtained support from the findings of Bakker and Schipper [22], Kremer and Souissi [29], and Kremer [24]. The first test conducted on these strains was the qualitative production of HCN. It was considered a major substance responsible for the growth inhibition of some plants by Kremer and Souissi [29]. This study obtained 22.6% of strains (89) to have produced cyanide at various levels. The distribution of cyanogenic strains in different weeds and wheat was also variable. This was synonymous with the findings of Kremer and Souissi [29]. The proportion of cyanogenic strains in their study (32%) was, however, higher than in our study. This difference might be due to differences in agro-ecological conditions and prevalent agricultural practices. Zeller et al. [19] found that the sensitivity of different weeds and crops to cyanide was variable, and the cyanogenic bacteria might cause suppression of some weeds without imparting harmful effects on the accompanying crop in certain cases. They applied various levels of cyanide to five weeds and wheat and reported that this characteristic of rhizobacteria might be used for the selective suppression of three seeds (*C. jacea*, *G. mollugo*, and *H. murinum*), invading the wheat crop without disturbing the growth of wheat.

The cyanogenic strains of our study were further tested for the production of toxic metabolites using the indicator of sensitive bacteria (*E. coli* strain K12). The relevance of this assay for the screening of rhizobacteria weed control agents was reported by Kremer et al. [30]. We got 21.3% of the cyanogenic strains to suppress the growth of sensitive bacteria. As all the cyanogenic strains did not suppress the growth of sensitive bacteria in this study, one may speculate that the strains inhibiting the growth of bacteria may also have possessed the characteristics of production of some other toxic compounds along with cyanide. This assay indicated that the strains inhibiting the growth of sensitive bacteria might be producing multiple growth inhibitory compounds, collectively termed as antimetabolites, and could be more suitable for testing on weeds and wheat in the next screening studies.

Nineteen strains, obtained from the above screening procedures, were tested on sensitive plant species, i.e., lettuce. The effects of these strains on the growth of the seedlings of lettuce were variable. Some strains inhibited, some promoted, and others remained ineffective. Hence, all the strains inhibiting the growth of *E. coli* did not inhibit the growth of lettuce in our study. This finding agreed with Kremer et al. [30]. Kremer and Kennedy [31] also reported the growth reduction of lettuce by such rhizobacteria. As the strains tested on lettuce were all cyanogenic in nature, Zermane et al. [32] also reported mixed effects of cyanogenic rhizobacteria on lettuce. The non-inhibition of lettuce by some strains may be due to the non-host interactions, where these strains needed to grow with their host in order to express their characteristics [33]. There also exist differences in the metabolic functions of *E. coli* and lettuce, the former being a prokaryote, and the latter being a eukaryotic plant species. There may also be the difference of compounds, causing antibiosis against bacteria and plants. The results obtained in our study reflected the release of diverse types of metabolites and their functions by these strains, which affected the growth of bacteria and plants. For similar reasons, we tested all the above-mentioned strains on weeds and wheat in the further screening process. This decision in our study has grounds in Souissi and Kremer [34]. They reported the reduction in the growth of weeds by those strains of rhizobacteria, which did not reduce the growth of lettuce. In other words, the growth reduction of lettuce and weeds by rhizobacteria could not be correlated in their study.

Stability or consistency in the characteristics of strains of our study may be evident from the above studies. It increased our reliance on these strains for further studies regarding their effects on weeds and wheat. We found all type of effects of the strains on weeds and wheat, i.e., there were strains inhibitory to all the weeds and wheat, suppressive to one or more weeds and wheat, suppressive to one or more weeds but not to wheat, and suppressive to one or more weeds but promoted the growth of wheat. This array of responses by the strains of allelopathic bacteria has multiple applications if further studies on their characterization and response under natural conditions are carried out. These may be developed for application to control weeds and strengthen crop in poor agricultural systems (selective strains) and control weeds in non-agricultural systems (non-selective strains). The reasons for selectivity may be a difference of tolerance to toxic metabolites in weeds and crop, release of toxic metabolites by these strains only in the rhizosphere of their host plants, the difference in availability of substrates required for the production of toxic metabolites in the rhizosphere of weeds and wheat, a difference of survival, colonization, and establishment in the rhizosphere of weeds and wheat, and difference of mechanisms in the rhizosphere of host and non-host plants [19,20,35]. The findings of our study became more evident when the strains were further characterized by the production of indole-3-acetic acid, exopolysaccharides, siderophores, catalases, chitinases, oxidases, and P solubilization. The most prominent strains were identified as pseudomonads. The effects of the five most efficient strains on weeds and wheat were tested under axenic conditions in Abbas et al. [36]. The strains inhibiting one or more weeds and promoting wheat may be more successful for weed control under natural conditions. These may strengthen the weak crop plants, increase their competitive ability, and, hence, increase the scale of weed control by allelopathic bacteria. The non-selective strains inhibitory to wheat may be tested for their effects on other crops to explore opportunities for their application in other cropping systems. The efforts on augmentation of effects of allelopathic bacteria under natural conditions may be helpful to realize the dream of biological weed control. The strains of allelopathic bacteria obtained from this study can be further tested for their effects on weeds and wheat under field conditions. Further efforts may be required to improve their efficiency of weed control under natural conditions. Application methods of allelopathic bacteria may also be needed to be optimized. This will produce a bioherbicide for the control of weeds in an environmentally friendly and sustainable manner.

5. Conclusions

The rhizosphere of five weeds and wheat, growing in areas of high weed invasion, was explored for the allelopathic bacteria. A large collection of strains of rhizobacteria was passed through a comprehensive screening process for this purpose. We got 22.6% strains cyanogenic in nature, 21.3% of which (19 strains) inhibited the growth of sensitive bacteria. These strains were applied to lettuce, which showed mixed effects. These strains were later tested on four weeds and wheat. We got strains inhibitory to all these weeds (eight for the broad-leaved dock, seven for wild oat, eight for little seed canary grass, and three for common lambs' quarter). They reduced the dry matter of these weeds from

23.1 to 68.1%, 38.5 to 80.2%, 16.5 to 69.4%, and 27.5 to 50.0%, respectively. Only five of these strains were inhibitory to wheat; the others either remained neutral (five strains) or improved the growth of wheat (nine strains). These strains offer opportunities for the development of biological weed control.

Author Contributions: Conceptualization: M.N., Z.A.Z.; Data curation: T.A., A.M., S.A.; Formal analysis and software: A.M., M.S.A., M.S.E.; Investigation: M.N., Z.A.Z., Methodology: T.A., A.M., and M.S.E.; Supervision: M.N., Z.A.Z., Writing—original draft: T.A.; Writing—review and editing: M.N., Z.A.Z., M.S.A. All authors have read and agreed to the published version of the manuscript.

Funding: Researchers Supporting Project number (RSP-2020/173), King Saud University, Riyadh, Saudi Arabia.

Acknowledgments: This research was financially supported by the Higher Education Commission of Pakistan under the Indigenous Ph.D. Fellowship Program for 5000 scholars Phase II Batch II (Grant # 213-65052-2AV2-105. The authors would like to extend their sincere appreciation to the Researchers Supporting Project number (RSP-2020/173), King Saud University, Riyadh, Saudi Arabia.

Conflicts of Interest: Authors declare that they have no conflicting/competing interests.

References

1. Anonymous. *Pakistan Economic Survey*; Economic Adviser, Government of Pakistan, Ministry of Finance: Rawalpindi, Pakistan, 1966. Available online: http://www.irispunjab.gov.pk/StatisticalReport/Pakistan%20Economic%20Surveys/Economic%20Survey%201996-97.pdf (accessed on 24 September 2020).

2. Anonymous. *Economic Survey of Pakistan 2015–16*; Economic Advisor's Wing, Finance Division, Government of Pakistan: Islamabad, Pakistan, 2016. Available online: http://www.finance.gov.pk/survey_1516.html (accessed on 24 September 2020).

3. Oerke, C.E. Centenary review on crop losses to pests. *J. Agric. Sci.* **2006**, *144*, 31–43. [CrossRef]

4. Shad, R.A. Weeds and weed control. In *Crop Production*; Nazir, S., Bashir, F., Bantel, R., Eds.; National Book Foundation: Islamabad, Pakistan, 2015; pp. 175–204.

5. Ghorbani, R.; Leifert, C.; Seel, W. Biological control of weeds with antagonistic plant pathogens. *Adv. Agron.* **2005**, *86*, 191–225.

6. Gaines, T.A.; Duke, S.O.; Morran, S.; Rigon, C.A.G.; Tranel, P.J.; Kupper, A.; Dayan, F.E. Mechanisms of evolved herbicide resistance. *J. Biol. Chem.* **2020**, *295*, 10307–10330. [CrossRef] [PubMed]

7. Quimby, P.C.; King, L.R.; Grey, W.E. Biological control as a means of enhancing the sustainability of crop/land management systems. *Agric. Ecosyst. Environ.* **2002**, *88*, 147–152. [CrossRef]

8. Geiger, F.; Bengtsson, J.; Berendse, F.; Weisser, W.W.; Emmerson, M.; Morales, M.B.; Ceryngier, P.; Liira, J.; Tschantke, T.; Winqvist, C.; et al. Persistent negative effects of pesticides on biodiversity and biological control potential on European farmland. *Basic Appl. Ecol.* **2010**, *11*, 97–105. [CrossRef]

9. Pimentel, D. Environmental and economic costs of the application of pesticides primarily in the United States. *Environ. Dev. Sustain.* **2005**, *7*, 229–252. [CrossRef]

10. Alavanja, M.C.R.; Hoppin, J.A.; Kamel, F. Health effects of chronic pesticide exposure: Cancer and neurotoxicity. *Annu. Rev. Public Health* **2004**, *25*, 155–197. [CrossRef]

11. Abbas, T.; Zahir, Z.A.; Naveed, M.; Kremer, R.J. Limitations of existing weed control practices necessitate the development of alternative techniques based on biological approaches. *Adv. Agron.* **2017**, *147*, 239–280.

12. Quimby, P.C.; Birdsall, J.L. Fungal agents for biological control of weeds: Classical and augmentative approaches. In *Novel Approaches to Integrated Pest Management*; Reuveni, R., Ed.; CRC Press: Boca Raton, FL, USA, 1995; pp. 293–308.

13. Boyette, C.D.; Hoagland, R.E. Bioherbicidal potential of a strain of *Xamthomonas* spp. for control of common cocklebur (*Xanthium strumarium*). *Biocontrol. Sci. Technol.* **2013**, *23*, 183–196. [CrossRef]

14. Coombs, E.M.; Clark, J.K.; Piper, G.L.; Cofrancesco, A.F., Jr. *Biological Control of Invasive Plants in the United States*; Oregon State University Press: Corvallis, OR, USA, 2004.

15. Denslow, J.S.; D'Antonio, C.M. After biocontrol: Assessing indirect effects of insect releases. *Biol. Control* **2005**, *35*, 307–318. [CrossRef]

16. Farooq, M.; Bajwa, A.A.; Cheema, S.A.; Cheema, Z.A. Application of allelopathy in crop production. *Int. J. Agric. Biol.* **2013**, *15*, 1367–1378.

17. Kobayashi, K. Factors affecting phytotoxic activity of allelochemicals in soil. *Weed Biol. Manag.* **2004**, *4*, 1–7. [CrossRef]

18. Kremer, R.J. The role of allelopathic bacteria in weed management. In *Allelochemicals: Biological Control of Plant Pathogens and Diseases*; Inderjit, M.K.G., Ed.; Springer: Dordrecht, The Netherlands, 2006; pp. 143–156.

19. Zeller, S.L.; Brandl, H.; Schmid, B. Host-plant selectivity of rhizobacteria in a crop/weed model system. *PLoS ONE* **2007**, *2*, 846–858. [CrossRef]

20. Kennedy, A.C.; Johnson, B.N.; Stubbs, T.L. Host range of a deleterious rhizobacterium for biological control of downy brome. *Weed Sci.* **2001**, *49*, 792–797. [CrossRef]

21. King, E.; Ward, M.; Raney, D. Two simple media for the demonstration of pycyanin and Xuorescein. *J. Lab. Clin. Med.* **1954**, *44*, 301–307. [PubMed]

22. Bakker, A.W.; Schipper, B. Microbial cyanide production in the rhizosphere in relation to potato yield reduction and *Pseudomonas* spp. mediated plant growth stimulation. *Soil Biol. Biochem.* **1987**, *19*, 451–457. [CrossRef]

23. Gasson, M.J. Indicator technique for antimetabolic toxin production by phytopathogenic species of *Pseudomonas*. *Appl. Environ. Microbiol.* **1980**, *39*, 25–29. [CrossRef] [PubMed]

24. Kremer, R.J. Interactions between the plants and microorganisms. *Allelopath. J.* **2013**, *31*, 51–70.

25. Abd-Alla, M.H.; Morsy, F.M.; El-Enany, A.E.; Ohyama, T. Isolation and characterization of a heavy metal resistant isolate of *Rhizobium leguminosarum* bv. *viciae* potentially applicable for biosorption of Cd^{+2} and Co^{+2}. *Int. Biodeterior. Biodegrad.* **2012**, *67*, 48–55. [CrossRef]

26. Steel, R.G.D.; Torrie, J.H.; Dicky, D.A. *Principles and Procedures of Statistics—A Biometrical Approach*, 3rd ed.; McGraw Hill Book International Co.: Singapore, 1997.

27. Matloob, A.; Safdar, M.E.; Abbas, T.; Aslam, F.; Khaliq, A.; Tanveer, A.; Rehman, A.; Chadhar, A.R. Challenges and prospects for weed management in Pakistan: A review. *Crop Prot.* **2020**, *134*, 104724. [CrossRef]

28. Schippers, B.; Bakker, A.W.; Bakker, P.A.H.M. Interactions of deleterious and beneficial rhizosphere microorganisms and the effect of cropping practices. *Annu. Rev. Phytopathol.* **1987**, *25*, 339–358. [CrossRef]

29. Kremer, R.J.; Souissi, T. Cyanide production of rhizobacteria and potential for suppression of weed seedling growth. *Curr. Microbiol.* **2001**, *43*, 182–186. [CrossRef] [PubMed]

30. Kremer, R.J.; Begonia, M.F.T.; Stanley, L.; Lanham, E.T. Characterization of rhizobacteria associated with weed seedlings. *Appl. Environ. Microbiol.* **1990**, *56*, 1649–1655. [CrossRef] [PubMed]

31. Kremer, R.J.; Kennedy, A.C. Rhizobacteria as biocontrol agents of weeds. *Weed Technol.* **1996**, *10*, 601–609. [CrossRef]

32. Zermane, N.; Souissi, T.; Kroschel, J.; Sikora, R. Biocontrol of broomrape (*Orobanche crenata* Forsk. and *Orobanche foetida* Poir.) by *Pseudomonas fluorescens* isolate Bf7-9 from the faba bean rhizosphere. *Biocontrol Sci. Technol.* **2007**, *17*, 483–497. [CrossRef]

33. Weiland, G.W.; Neumann, R.; Backhaus, H. Variation of microbial communities in soil, rhizosphere and rhizoplane in response to crop species, soil type and crop development. *Appl. Environ. Microbiol.* **2001**, *67*, 5849–5854. [CrossRef]

34. Souissi, T.; Kremer, R.J. A rapid microplate callus bioassay for assessment of rhizobacteria for biocontrol of leafy spurge (*Euphorbia esula* L.). *Biocontrol Sci. Technol.* **1998**, *8*, 83–92. [CrossRef]

35. Owen, A.; Zdor, R. Effect of cyanogenic rhizobacteria on the growth of velvetleaf (*Abutilon theophrasti*) and corn (*Zea mays* L.) in autoclaved soil and the influence of supplemental glycine. *Soil Biol. Biochem.* **2001**, *33*, 801–809. [CrossRef]

36. Abbas, T.; Zahir, Z.A.; Naveed, M. Bioherbicidal activity of allelopathic bacteria against weeds associated with wheat and their effects on growth of wheat under axenic conditions. *BioControl* **2017**, *62*, 719–730. [CrossRef]

 *agronomy*

Review

Relevance of Plant Growth Promoting Microorganisms and Their Derived Compounds, in the Face of Climate Change

Judith Naamala and Donald L. Smith *

Department of Plant Science, McGill University, Lakeshore Road, Ste. Anne de Bellevue, 21111, Montreal, QC H9X3V9, Canada; naamala.judith@mail.mcgill.ca
* Correspondence: Donald.Smith@McGill.Ca

Received: 29 May 2020; Accepted: 6 August 2020; Published: 12 August 2020

Abstract: Climate change has already affected food security in many parts of the world, and this situation will worsen if nothing is done to combat it. Unfortunately, agriculture is a meaningful driver of climate change, through greenhouse gas emissions from nitrogen-based fertilizer, methane from animals and animal manure, as well as deforestation to obtain more land for agriculture. Therefore, the global agricultural sector should minimize greenhouse gas emissions in order to slow climate change. The objective of this review is to point out the various ways plant growth promoting microorganisms (PGPM) can be used to enhance crop production amidst climate change challenges, and effects of climate change on more conventional challenges, such as: weeds, pests, pathogens, salinity, drought, etc. Current knowledge regarding microbial inoculant technology is discussed. Pros and cons of single inoculants, microbial consortia and microbial compounds are discussed. A range of microbes and microbe derived compounds that have been reported to enhance plant growth amidst a range of biotic and abiotic stresses, and microbe-based products that are already on the market as agroinputs, are a focus. This review will provide the reader with a clearer understanding of current trends in microbial inoculants and how they can be used to enhance crop production amidst climate change challenges.

Keywords: plant growth promoting microorganisms; climate change; abiotic stress; biotic stress

---

1. Introduction

The world is at a point where we can no longer prevent all of the effects of climate change (because some of it is already here), but can only slow its further progress. The purpose of this paper is therefore to give the reader an understanding of why plant growth promoting organisms, or their products, are relevant, amidst climate change challenges, by showing how they can be used to mitigate the effects of climate change on crop production. The paper also highlights the various ways in which this approach can be used, and the role that inoculant formulation plays in maintaining the efficacy, durability and handling of microbial inoculants. The major drivers of climate change are human driven [1–3]. Burning of fossil fuels for energy, agriculture and industrialisation all contribute to emission of greenhouse gases (GHGs), such as: methane, carbon dioxide and nitrous oxide (N_2O). Agriculture is a major contributor to greenhouse gas emissions [4,5], especially through the use of N based fertilizers, methane emissions from animals and animal manure, deforestation to acquire more land for crop production, etc. According to the intergovernmental panel on climate change (IPCC) report on GHG emissions, energy consumption contributes about 35%, agriculture, forestry and related land use 24%, industry 21% and transport 14% [6]. The greenhouse gases then trap heat radiating from the earth's surface, causing global warming. Unfortunately, climate change also adversely affects agriculture [6,7], especially because, along with increases in global temperature, comes the increased prevalence of biotic and abiotic stresses that are detrimental to agriculture production, such as: pests, pathogens, nutrient deficiencies, salinity and weather extremes [1,8–10], some of which may encourage the further use of chemicals to correct, while there is little that can be done about others such as high temperatures and floods. Unmanaged, such factors affect plant growth and render arable land unproductive. This puts us in a challenging situation, especially because world population is growing so that there is a need to increase food production [5], both through increasing yield per unit area and reclaiming more land for crop production [11]. Therefore, while we strive hard to hold greenhouse gas emissions to 'bearable' levels, there is also a need for sustainable approaches that will ensure increased food production in the face of climate change. The use of agrochemicals has boosted crop productivity and contributed to food security, especially in developed countries. However, shortcomings related to their improper and continuous use, such as: increased greenhouse gas emissions (which is a major contributor to global warming), surface and ground water contamination, residual contamination of crop harvest, which poses health concerns to both humans and animals, as well as high costs related to their use. These circumstances have created a need for a more ecofriendly and sustainable approach for enhancing crop productivity in the face of climate change [11–13].

Several approaches have been suggested; the use of plant growth promoting microorganisms and compounds that they produce is perhaps the most promising [14]. The holobiont refers to plants and their associated microbes, which probably coexisted since the colonization of land by the first terrestrial plants [15–17]. This association is referred to as the holobiont [18], and it is dynamic, with the plant asserting a great influence on the nature of phytomicrobiome, especially in its rhizosphere [19], which is mainly attributed to the composition of their root exudates. The rhizosphere, endosphere and phyllosphere may be comprised of pathogenic, neutral and beneficial microbes, in relation to the plant [18,20]. Microbes that exert beneficial effects on the plant are termed plant growth promoting microorganisms (PGPM). These microbes may inhabit the rhizosphere, rhizoplane, phyllosphere, endosphere, etc. [19] For decades, PGPM such as rhizobia, mycorrhizae and plant growth promoting bacteria (PGPR, first defined by Kloepper and Schroth, in 1978) have been reported to enhance plant growth under stressed and non-stressed conditions. The use of microbial inoculants is an old practice [21] that has recently gained more prominence during the last three decades. Much research has been done on rhizobia, and currently a lot is being done on plant growth promoting rhizobacteria and PGPR derived compounds. The ability of microbes to suppress plant pathogens, as well as mitigate the effect of abiotic stress on plants, has been investigated by many researchers, and the findings are promising.

Although they occur naturally in the rhizosphere, and plant tissue, PGPM populations are often insufficient to induce desired effects, hence, it is recommendable to isolate them from their natural environments and multiply their populations before reintroduction into the soil or onto the plant as microbial inoculants [14]. Products in the form of microbe-produced compounds are currently gaining popularity among researchers, although they are less well known among farmers, in comparison to microbial cell inoculants, packaged as either single microbial strains or consortia, which have been commercialised for quite some time [21,22]. Microbe based inoculants are generally from the bacteria (such as Bacillus and Rhizobia) and fungi (especially Trichordema) subgroups [19,22,23], although some groups of archea have also been reported to enhance plant growth. Microbially produced compounds, such as lipochitooligosaccharides (LCO), as plant growth enhancers, on the other hand, are only gaining attention recently, which may explain their lesser availability on the agro-input market. Figure 1 below summarizes some of the mechanisms PGPM employ to mitigate the effects of biotic and abiotic stress on plants, which are later discussed in detail.

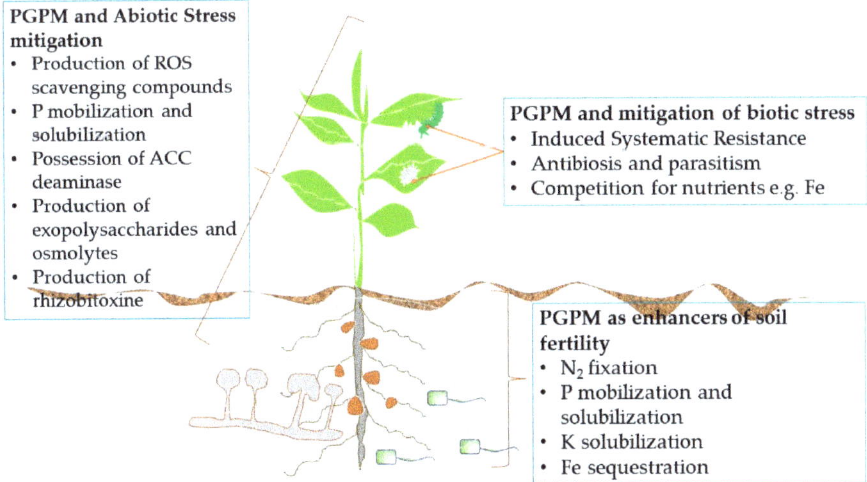

Figure 1. Mechanisms employed by plant growth promoting microorganisms (PGPM) to mitigate effects of biotic and abiotic stress on plants.

2. PGPM as Enhancers of Soil Fertility

For proper growth and development, plants need enough supply of essential macro (Nitrogen, Phosphorus, Potasium, Magnesium, Calcium, etc.) and micro (iron, manganese, boron, zinc, molybdenum, copper) nutrients. N, P and K are the most limiting as far as plant growth is concerned. Unfortunately, with climate change comes abiotic stresses like high temperature, drought and salinity, which influence the biogeochemical transformation of nutrients like P, K, and N, making them either available or less available for plant uptake [24–26]. While the lack of bioavailable macro and microelements is natural in the soil, it could be worsened by climate change. Nitrogen, phosphorus and potassium as the most plant growth limiting elements and their biogeochemical cycle, are affected by temperature and rainfall amongst other abiotic factors, which happen to be affected by climate change. Processes like decomposition, mineralisation, immobilisation, etc. are largely influenced by temperature and rainfall. Processes like soil erosion should also be noted, which is majorly due to run off and wind affect soil fertility as the nutrient rich topsoil is washed away.

Alkalinity affects the availability of Fe, Cu, Zn and Mn, while very low pH is associated with Al toxicity. Processes such as mineralization and nitrogen fixation are affected by moisture, temperature and pH, because they are driven by soil microorganisms like rhizobia, nitrifying bacteria, etc., and enzymes [24,26], which are also affected by abiotic stress. A study by DaMatta et al. [27] showed a decrease in leaf N content of *Coffea canephora* due to water stress. For PGPM technology to be relevant, amidst climate change, it is paramount that stress tolerant strains are identified and used. At the same time, the availability of these nutrients is essential, because they play a key role in minimizing the effects of other abiotic stresses like drought, salinity and high temperature on crops. The roles N, P, K, Ca, Mg and Fe play in the mitigation of abiotic stress have been reported by many researchers [27–33]. For instance, N and P have been reported to minimize the effects of drought stress [24–26,34,35]. K plays a major role in drought stress as well, since it is involved in the opening and closing of the stomata. Agricultural soils have been degraded due to continuous and intense cropping. Agricultural practices like continuous cropping, especially monocropping of non-leguminous crops, without application of fertiliser, is one way of depleting soils of nutrients [36]. This is a common practice of many smallholder farmers, especially in sub-Saharan Africa, due to the inaccessibility and cost of fertiliser [37]. Climate change is only further degrading the situation, because factors such as high temperatures, drought, flooding, salinity, extreme pH, etc. may cause changes in the physiochemical properties of essential soil nutrients such as N, Fe, P and K, thereby limiting their mobility and/or affect their availability for plant uptake, while enhancing the accumulation of toxic elements such as aluminum (Al^{3+}). The role of stress tolerant beneficial microbes in maintaining/increasing crop production amidst climate change challenges cannot be ignored. In order to reclaim land that has been abandoned due to inadequate nutrients for crop growth, considering the financial and environmental costs related to synthetic fertilisers, stress tolerant plant growth promoting organisms can be a cheaper and sustainable approach. With the need to reclaim more land for crop production, emphasis on enhancing soil fertility is inevitable, because nutrients can enhance plant tolerance to abiotic stress. Therefore, there is a need to address the issue using more sustainable approaches. With limited alternatives, and research output so far, microbial inoculants are a promising approach to enhance soil fertility, particularly in conjunction with the various challenges associated with climate change. Microbial inoculants may be defined as formulations comprised of microorganisms, such as bacteria and fungi, as the active ingredients, which once applied on plants, can enhance their growth [19,22,38]. They may also enhance plant quality through the increased concentration of essential nutrients such as proteins [14], and valuable metabolites such as flavonoids, phenolics, alkaloids and carotenoids [23]. Microbial inoculants may also enhance soil biodiversity and properties such as soil structure [22]. As biofertilizers, microbial inoculants enhance the availability and uptake of essential plant nutrients, such as: nitrogen, phosphorus, iron, zinc, and potassium [11–13], which, if lacking or available in inadequate quantities, could limit plant growth.

2.1. Nitrogen Fixation

Some free-living and symbiotic bacteria fix atmospheric dinitrogen into plant usable forms, initially ammonium, through biological nitrogen fixation. Symbionts such as *Rhizobia*, *Bradyrhizobium*, *Sinorhizobium*, *Frankia*, *Actinobacteria* and *Bukholderia* form specialized structures called nodules on their host plants, where they obtain nourishment and shelter, and in turn, fix nitrogen [38,39]. The process is referred to as symbiotic nitrogen fixation and it occurs in both legumes and non-leguminous plants, although that of legumes is the most studied. Communication in the form of molecular signals from both the microbe and host plant, as well as a complex of enzymes (e.g., nitrogenase) and genes (nif and/or symbiotic genes), are involved in the process of nitrogen fixation. On the other hand, free-living nitrogen fixing bacteria such as *Azotobacter* do not need to occupy plant tissue to fix nitrogen. Because of its high energy requirement, plants tend to prefer applied N fertilizer to biological nitrogen fixation (BNF), hence, for effectiveness, synthetic N should not be used along with biological nitrogen fixing organisms, because the plant may suppress the nitrogen fixing symbiosis. Where a starter dose of synthetic N is necessary, it should be applied cautiously, because high N supply can have an inhibitory effect on nodulation (nodule dry weight and number of nodules) and nitrogenase activity [24,40,41]. Arbuscular mycorrhizal fungi, through their hyphae, can enhance the acquisition of soil N by the plant [42], although there are wide variabilities as to the degree of this, whose causes are not yet known [22]. The efficiency and effectiveness of nitrogen fixing bacteria varies among and within plant species, and, in the agricultural context, are largely limited to members of the fabaceae family. Other crops can benefit from the symbiosis by including legumes in crop rotation regimes. There is also a need for more research on how to extend such modifications to non-leguminous plants. Approaches such as genetic engineering to enable non-legume nitrogen fixation and enhance effective communication with N fixing microorganisms can be further researched. Although genetic engineering is questionable, especially its ecological impact, some of the questions are likely from a lack of adequate information on the technology. Extensive research to address most of the questions can be very helpful.

2.2. Phosphate Mobilisation and Solubilisation

Although phosphorus is an abundant element in most soils, it frequently occurs in forms unavailable for plant use. The application of external sources of P fertiliser, such as single super phosphate, diammonium phosphate, etc., can help meet plants' P requirements, but this too may be immobilised shortly after application, making it largely unavailable for plant uptake [43]. mobilization (chemical solubilization and mineralization), which results in plant available forms of the respective nutrients and solubilization, which is a more general term and does not necessarily result in readily plant available forms. For instance, the solubilization of organic P does not necessarily mean that the P is already plant available, as it may still be bound in unavailable organic forms (e.g., phytates). PGPM may enhance soil phosphorus availability for plant uptake through solubilisation and/or mobilisation of inorganic phosphorus. A PGPM may possess both or either mechanisms. The terms, phosphorus solubilisation and phosphorus mobilisation are often used synonymously by many researchers, although they are not necessarily the same thing. P solubilisation is the broader term, which may entail P mobilisation. Goldstein and Krishnaraj [44] described phosphate solubilising microorganisms as those that convert sparing soluble organic or mineral P, into soluble orthophosphate, in a way that significantly increases P availability to a specific plant or plant population within the microorganism's native soil ecosystem. The same author defined phosphate mobilising microorganisms as those that convert sparingly soluble organic or mineral P, into soluble orthophosphate P, in a way that significantly contributes to pool of available orthophosphate (Pi) in the native soil ecosystem. Phosphorus solubilising bacteria, such as: *Pseudomonas*, *Bacillus*, *Burkholderia* and *Rhizobium*, and some fungal species solubilise inorganic phosphates from sparingly soluble forms such as: tricalcium phosphate, dicalcium phosphate and aluminum phosphate, to forms such as hydrogen phosphate ($HPO_4{}^{-2}$), or dihydrogen phosphate ($H_2PO_4{}^{-1}$), which plants can utilise [37,42–46] through the production of low molecular weight organic acid anions, such as gluconate, lactate, glycolate and oxalate. Phosphorus mobilisers, on the other hand, produce enzymes (such as phosphatase, phytase and phosphonoacetate hydrolase) that chelate cations, bind phosphates and dephosphorylate organic phosphates [22,24]. Dephosphorylation is catalyzed by hydrolase enzymes such as phosphonoacetate, which some PGPM can produce. For instance, ectomycorrhiza and ericoid mycorrhizal fungi produce extracellular acid phosphatases and phytases, which catalyse the mineralisation of P from organic complexes in the soil [42,47]. Other fungal species, such as *Aspergillus niger*, also produce organic acids which aid the process of P solubilisation [48,49]. Through the possession of hyphae, some mycorrhizae such as arbuscular mycorrhizae can deliver up to 80% of the phosphorus taken up by the host plant [27,40].

2.3. Sequestering of Iron

Some PGPM, like *Pseudomonas fluorescens* and *Rhizobia meliloti*, sequester iron through the production of siderophores, which can be grouped into four, namely: hydroxamates, catecholates, carboxylates and pyoverdines [50]. Currently, about 500 siderophores have been reported by researchers. Although plants cannot absorb Fe^{3+}, siderophores have a high affinity for Fe^{3+}, which results in an iron-siderophore complex that is then absorbed by plants [51], into their tissues, hence, aiding plants in meeting their iron requirements [14,17,43,52]. In 2013, study findings of Radzki et al. [53] showed an increase in iron content at 12 weeks for iron deficient tomato plants, following inoculation of siderophore producing bacteria, evidence that microbial siderophores can be a source of iron for plants. A study by Sharma and Johri [54] also showed an increase in maize plant growth following inoculation with siderophore producing PGPR. The uptake of Fe-microbial siderophore complexes by strategy II plants, via ligand exchange, between ferrated microbes and a phyto siderophore, was also reported by Yehuda et al. [55] It should also be noted that some plant species can also produce siderophores which bind Fe^{3+}, to form a complex that can be taken up by the plant with the aid of ligands. Production of siderophores is also a benefit in the context of biocontrol in a sense that potential plant pathogens, especially fungal pathogens, are outcompeted for iron sources, which may lead to their death, or ineffectiveness.

2.4. Potassium Solubilisation

Microbes such as: *Arthrobacter* sp. *Bacillus edaphicus*, *Bacillus circulans* and *Bacillus mucilaginosus* convert sparingly soluble and mineral potassium to soluble forms available for plant use [56]. Through the release of H^+ and organic anions, such as citrate, malate and oxalate, arbuscular mycorrhiza can also increase the solubility of mineral K [57], thereby increasing the availability of potassium anions for plant uptake, although the increase in K^+ availability is sometimes related to the increase in phosphorus availability [22,58].

Some PGPR can also directly influence plant growth through the production of phytohormones such as auxins and gibberellins, which enhance plant growth when plant phytohormones are at suboptimal concentrations [37]. They may also produce enzymes which regulate hormone concentration in plant tissue. For instance, some plant growth promoting microorganisms can produce an enzyme, ACC deaminase, which breaks down ACC, a precursor of ethylene, into an alpha keto butyrate and ammonium, hence lowering ethylene concentration in plant tissues [56,59–62]. With more research and proper manipulation, PGPM, with the ability to enhance plant growth, may not necessarily fully replace chemical fertilizer, but lower their use, directly and indirectly, through increasing the plants' nutrient uptake efficiency from applied chemical fertilizers [63]. Manipulations such as developing an effective consortium of microbes that are able to make available key elements in the soil would greatly reduce the need to use chemical fertilizers. For instance, rhizobial species require iron for good growth, and in their nitrogenase complex, hence co-inoculation of rhizobia and siderophore producing PGPM could enhance nodulation and nitrogen fixation [64]. Use of biofertilizers can lower the need to burn fossil fuels for fertilizer production, and the associated contribution to greenhouse gas emissions.

3. PGPM and Control of Plant Pests and Diseases

With global warming comes new species of pests, weeds and pathogens currently prevalent in warmer environments. The use of chemicals to suppress such plant growth inhibitors is effective but with negative outcomes related to improper use, cost, and increasing evolution of tolerance to the chemical. The antagonist properties of biocontrols against such plant growth suppressors have been reported by many researchers, and the results are promising. A diversity of PGPM with biocontrol properties has been identified by researchers, conferring benefits to a variety of crop species [65–73]. Berendsen [74] showed that plants, when exposed to pathogen attack, can recruit specific plant growth promoting microbes with biocontrol activity, against the pathogen in question. It is believed that manipulating plant recruited PGPM for inoculant production could be more effective in controlling targeted pathogens, than PGPM isolated from places with no pathogen attack. Biocontrols have the potential to minimise the use of industrially manufactured chemicals in agricultural production. This would mean a decline in burning of fossil fuels, and hence a reduction in greenhouse gas emissions. This is because some pesticides are synthesized in laboratories using hydrocarbons like petroleum, which is a fossil fuel. Reduction in their use can mean a reduction in burning fossil fuels, hence less CO_2 emission to the atmosphere. It would also reduce effects on non-targeted members in the ecosystems, which are sometimes affected by chemical use.

Biocontrols may act directly to inhibit growth of biotic agents through hyper parasitism and production of bioactive substances, such as: antibiotics, hydrogen cyanide and phenazines [73], or indirectly through competition for nutrients and active sites on plants, as well as inducing the plant's systemic resistance against the harmful biotic factor [22,38]. Siderophore producing PGPM tend to outcompete other microorganisms for iron sources, which causes inefficiencies in terms of pathogen activities, especially for fungal pathogens, which eventually leads to their death [64]. Induced systemic resistance is triggered by microbe associated molecular patterns (MAMPS), such as lipopolysaccharides, that plants recognise and respond to by turning on their defence systems [75]. Since MAMPS differ among PGPM, it is believed that microbial consortia made up of more than one microbe may induce stronger systemic resistance than single strains [72], although further research needs to be done for a clearer understanding of this potential. PGPM can not only mitigate crop, but also suppress crop pests, such as spidermites [76], moths [77], aphids [78], nematodes [67,79], leaffolder pest [80] and cutworms [81], which greatly contribute to losses incurred in crop production, right from planting to harvesting and storage, if not managed well. PGPR control pests through mechanisms such as production of volatile compounds, such as compounds such as β-ocimene and β-caryophyllene [76], that attract natural enemies of the pest in question. For example, a study by Pangesti et al. [77] showed an increase in the concentration of parasitoid *Microplitis mediator*, a natural predator of *Mamestra brassicae* following the inoculation of *Arabidopsis thaliana* roots with the rhizobacterium *Pseudomonas fluorescens* WCS417r. Other mechanisms through which PGPM mitigate the effects of pests include, increased activity of antioxidant enzymes and increased content of proteins and phenolics in plants, etc. In other cases, the biocontrol agent may not have an effect on the biotic antagonist, but will enhance plant yield in the presence of the antagonist [78]. This particular strategy seems very useful, especially in cases where biotic stress factors such as weeds and pests become resistant or unresponsive to other control strategies. It may also enhance/preserve species diversity, hence maintaining ecosystem functionality.

Some PGPM are efficient against pathogens as single strains, while others perform better as a consortium. Details of single strains vs. consortia are discussed later in this review. Table 1 lists PGPM with potential biocontrol activities against the pathogens of various crop species. With the increasing campaign against the use of chemicals, as a means of combating climate change, such strains are a promising substitute for chemicals that are currently prevalent in agricultural production. Currently, the global biocontrol market is approximately 2 billion USD [79], and is expected to grow further. More research on existing microbial species or microbe-produced compounds with biocontrol properties is still desirable, as is the identification of new ones.

Table 1. Biocontrol species against biotic stressors of different crop species.

PGPM	Biotic Stress	Host Plant	Reference
Bacillus amyloliquefaciens LY-1	*Peronophythora litchii*	Litchi (*Litchi chinensis* Sonn.)	[71]
Burkholderia cepacia	*Fusarium oxysporum*	*Solanum tuberosum*	[65]
Pseudomonas fluorescens	*Fusarium graminearum*	*Triticum aestivum*(wheat) cv. Tabuki	[66]
Pseudomonas fluorescens CHAO	*Gaeumannomyces graminis var. tritici*	*Triticum* sp.	[64]
Pseudomonas fluorescens CHAO	*Thielaviopsis basicola*	*Nicotiana tabacum*	[64]
Bacillus spp.	*Heterodera glycines*	*Glycine max.*	[82]
Serratia proteamaculans	*Meloidogyne incognita*	*Solanum lycopersicum* L.	[72]
Bacillus aryabhattai A08	*Meloidogyne incognita*	*Solanum lycopersicum* L.	[83]
Serratia plymuthica HRO-C48	*Botrytis cinerea*	_	[84]
Serratia plymuthica strain C-1, *Chromobacterium* sp. strain C-61 and *Lysobacter enzymogenes* strain C-3 consortium	*Phytophthora capsici*	*Cupsicum* spp.	[85]
Paenibacillus sp. 300 + *Streptomyces* sp. 385,	*Oxysporum f.* sp. *Cucumerinum*	*Cucumis sativus*	[86]
Pseudomonas fluorescens WCS 358	*Fusarium oxysporum f* sp. *Raphani*	*Raphanus sativus*	[87]
Pseudomonas fluorescens	*Macrophomina phseolina*	*Coleus forskohlii* Briq.	[68]
Pseudomonas aeruginosa 7NSK2	*Pythium splendens*	*Lycopersicon esculentum*	[88]
Pseudomonas fluorescens	*Pythium* spp.	*Triticum* sp.	[64]
Pseudomonas fluorescens	*Pythium ultimum*	*Gossypium* sp.	[64]
Bradyrhizobium japonicum NCIM 2746	*Rhizopus* sp. and, *Fusarium* sp.	*Glycine max* L.	[89]
Paenibacillus lentimorbus B30488	*Scelerotium rolfsii*	*Solunum lycopersicum* L.	[69]

Table 1. *Cont.*

PGPM	Biotic Stress	Host Plant	Reference
Pseudomonas putida UW4	*Agrobacterium tumefaciens*	*Solanum lycopersicum*	[90]
Burkholderia phytofirmans PsJN	*Agrobacterium tumefaciens*	*Solanum lycopersicum*	[90]
Bacillus cereus PX35, Bacillus subtilis SM21 and Serrati asp. XY2	*Meloidogyne incognito*	*S. lycopersicum*	[91]
Pseudomonas fluorescens strain S35	*Phytophthora infestans*	*Solanum tuberosum*	[73]
Pseudomonas frederiksbergensis strain 49 and Pseudomonas fluorescens strain 19 consortium	*Phytophthora infestans*	*Solanum tuberosum*	[73]
Pseudomonas putida strain R32	*Phytophthora infestans*	*Solanum tuberosum*	[73]
Pseudomonas chlororaphis spp. *strain R47*	*Phytophthora infestans*	*Solanum tuberosum*	[73]
Pseudomonas spp. *strain S49*	*Phytophthora infestans*	*Solanum tuberosum*	[73]
Bacillus and Pseudomonas spp. *consortium*	*Fusarium oxysporum U3 and Alternaria sp. U10*	*Nicotiana attenuata*	[92]
Chaetomium sp. C72 and Oidodendron sp. Oi3 consortium	*Fusarium oxysporum U3 and Alternaria sp. U10*	*Nicotiana attenuata*	[92]
Pseudomonas chlororaphis R47	*Phytophthora infestans*	*Solanum tuberosum*	[69,93]
Pseudomonas fluorescens strain LBUM 636	*Phytophthora infestans*	*Solanum tuberosum*	[94]
Agrobacterium radiobacter var radiobacter	*Crown gall*	*Solunum lycopersicon*	[95]
Tricoderma koningiopsis Th003 WP	*Fusarium oxysporum*	*Physalis peruviana*	[67]
Trichoderma harzianum Tr6 + Pseudomonas sp. Ps14	*Fusarium oxysporum f. sp. radicis cucumerinum*	*Cucumis sativus*	[96]
Pseudomonas sp. Ps14	*Botrytis cinerea*	*Arabidopsis thaliana*	[96]
Trichoderma harzianum Tr6	*Botrytis cinerea*	*Arabidopsis thaliana*	[96]
Pseudomonas putida	*Spodoptera litura*	*Solanum lycopersicum L.*	[96]
Pseudomnas flourescences Pf1, Bacillus subtilis Bs and Trichoderma viridae Tv consortium	*Lasiodiplodia theobromae*	*Polianthes tuberosa L.*	[97]
Pseudomonas sp. 23S	*Clavibacter michiganensis*	*Solanum lycopersicum L.*	[98]
Peanibacillus lentimorbus B-30488	*cucumber mosaic virus*	*Nicotiana tabacum cv White burley*	[99]
Serratia liquefaciens MG1	*Alternaria alternate*	*Solanum lycopersicum*	[100]
Xanthomonas sp. WCS2014-23, Stenotrophomonas sp. WCS2014-113 and Microbacterium sp. WCS2014-259	*Hyaloperonospora arabidopsidis*	*Arabidopsis thaliana*	[74]
Lactobacillus plantarum SLG17 and Bacillus amyloliquefaciens FLN13	*Fusarium* spp.	*Triticum durum*	[101]
Fusarium oxysporum strain Fo162	*Aphis gossypii Glover*	*Cucurbita pepo*	[102]
Rhizobium etli strain G12	*Aphis gossypii Glover*	*Cucurbita pepo*	[102]
Bacillus subtilis strain BEB-DN	*Bemisia tabaci*	*Solanum lycopersicum*	[103]
Bacillus amyloliquefaciens (SN13)	*Rhizoctonia solani*	*Rice (Oryza sativa)*	[104]
Pseudomonas fluorescens Migula strains Pf1 and AH1	*Desmia funeralis*	*Oryza sativa*	[80]
Pseudomonas putida and Rothia sp.	*Spodoptera litura*	*Solanum lycopersicum*	[81]

4. PGPM and Abiotic Stress

With climate change, the occurrence of extreme abiotic stresses, such as floods, salinity, high temperature and drought are expected to increase [3,8–10,105]. In fact, much of this is already being experienced in some parts of the world. Winters are becoming warmer in some regions; rainfall is becoming scarcer and more erratic, causing droughts and desertification [1,2,104] in other regions. With less rainfall, salinity is more likely to occur, either through irrigation or natural causes [106–111].

All these factors affect crop production, and their management inputs are sufficiently costly that many farmers may not be able to afford them. Factors such as high temperatures can generally not be managed under field conditions. Therefore, there is the need for a strategy that is ecofriendly and manageable by the majority of crop producers. PGPM have been reported to mitigate effects of abiotic stress on plants, hence, allowing the plant to grow and yield relatively well under stress conditions [112–115]. Various researchers have reported the ability of a wide range of PGPM to enhance plant growth, in the presence of abiotic stressors, such as salinity [116,117], drought [114,118,119], heavy metals and acidity. In fact, the ability of some PGPMs to enhance plant growth is only triggered in the presence of stress [118]. They employ mechanisms such as the production of ROS scavenging compounds, possession of ACC deaminase (an enzyme that lowers ethylene concentration in plants exposed to stress), and the production of exopolysaccharides and osmolytes. For example, Akhtar et al. [120] observed an increase in the antioxidant activity of catalase (CAT) in the roots of drought stressed maize plants treated with *Bacillus licheniformis* (FMCH001). Treated plants also exhibited a higher dry weight and higher water use efficiency. Yang et al. [121] also reported the increased activity of catalase and dehydroascobate reductase enzymes in salinity stressed Quinoa plants treated with an endophytic bacterium known as *Burkholderia Phytofirmans* PsJN, compared to the untreated plants. The former also exhibited a higher shoot biomass, grain weight and grain yield compared to the latter. Some rhizobia spp. produce the compound rhizobitoxine, which inhibits the activity of ACC synthetase, hence lowering ethylene activity that would otherwise inhibit nitrogen fixation. A PGPM may possess one or more of these mechanisms, all of which act to help a plant thrive under stress conditions. Like plants, PGPM can also be affected by abiotic stress, such as salinity, high temperature and drought, which can lower their efficacy in promoting plant growth, or even death of the microbe, in cases of prolonged exposure to extremes of such conditions [122]. Therefore, it is essential that the strains chosen for use are tolerant to the abiotic stress, whose effect in plants they tend to mitigate. Strains isolated from areas affected by abiotic stress may have an edge over those isolated under normal conditions, although this may not always be the case. The use of microbial consortia may be helpful, especially in areas where more than one factor inhibits crop growth (which is almost always the case under field conditions). However, more research needs to be conducted, for the better deployment of PGPM technology. The exploitation of such microbes has a definite potential to maintain crop production amidst increasing abiotic stresses that are rendering some currently arable land unfit for crop production. Table 2, below, shows some PGPM strains that have been discovered and characterized by researchers, with the potential to mitigate the effects of abiotic stress on a range of plant species.

Table 2. Examples of PGPM that enable plants to withstand abiotic stress.

PGPM	Abiotic Stress	Host Plant	Reference
Pseudomonas putida MTCC5279	Drought	chickpea (*Cicer arietinum*)	[114]
Pseudomonas fluorescens REN1	Flooding	Rice (*Oryza sativa*)	[123]
Variovorax paradoxus 5C-2,	Salinity	Peas	[115]
Bacillus amyloliquefaciens SQR9	salinity	Maize	[112]
Dietzia natronolimnaea	Salinity	Wheat (*Triticum aestivum*)	[116]
Serratia nematodiphila	Low temperature	pepper (*Capsicum annum*)	[124]
Burkholderia phytofirmans PsJN	Low temperature	grapevine (*Vitis vinifera*)	[125]
Pseudomonas vancouverensis	Low temperature	Tomato (*Solanum lycopersicum*)	[113]
Pseudomonas sp. S1	drought	*Capsicum annum*	[118]

Table 2. *Cont.*

PGPM	Abiotic Stress	Host Plant	Reference
Pseudomonas sp. S1	drought	*Vitis vinifera*	[118]
Achromobacter xylosoxidans	Flooding stress	*Ocimumsanctum*	[126]
Pseudomonas sp. 54RB + *Rhizobium* sp. Thal-8	Salinity	*Zea mays cv. Agaiti 2002*	[127]
Pseudomonas putida KT2440, *Sphingomonas* sp. OF178, *Azospirillum brasilense* Sp7 and *Acinetobacter* sp. EMM02) consortium	drought	Zea mays	[119]
Achromobacter xylosoxidans	salinity	*Catharanthus roseus*	[128]
Burkholdera cepacia SE4	salinity	*Cucumis sativus* L.	[124]
Pseudomonas putida (W2)	salinity	*Triticum aestivum* L.	[56]
Pseudomonas fluorescens (W17)	salinity	*Triticum aestivum* L.	[56]
Kocuria flava AB402	Arsenic toxicity	*Oryza sativum*	[129]
Bacillus vietnamensis AB403	Arsenic toxicity	*Oryza sativum*	[129]
Trichoderma spp. strain, M-35	Arsenic toxicity	*Cicer arietinum*	[130]
Burkholderia cepacia and *Penicillium chrysogenum* consortium	waste motor oil toxicity	*Sorghum bicolor*	[131]
Bacillus safensis	High temperature	*Triticum aestivum* L.	[132]
Pseudomonas aeruginosa	Zn-induced oxidative stress	*Triticum aestivum* L.	[133]
Bacillus licheniformis (FMCH001)	oxidative stress Drought	*Zea mays* L. cv. Ronaldinho	[120]
Burkholderia phytofirmans PsJN	Salinity	*Chenopodium quinoa* Willd	[121]

5. Commercialisation of Microbial Inoculants

Making PGPM technology available for farmers is key to ensuring their adaptation as agricultural inputs. Commercialisation of promising strains is one way of making promising strains accessible by farmers. Although various strains that possess desirable properties under laboratory and greenhouse conditions may be isolated, developing a commercial product, effective under field conditions, is not an easy task, especially because numerous factors determine the efficiency of introduced species. Characteristics such as: possession of multiple mechanisms of enhancing plant growth, ability to compete favorably and establish populations in the rhizosphere, persistence in the rhizosphere over seasons, and ability to be cultured in artificial environments [15,61,89] are desired for potential PGPM strains. However, many plant and soil factors, such as plant species, soil temperature, composition and prevalence of native microbes, soil pH, etc., may work together against a strain which is otherwise excellent under controlled environment conditions. Even before introduction into the field, factors such as formulation play a major part concerning a product's efficacy. For instance, solid inoculant formulations are desired for their longer shelf life, however, the process of drying microbes often results in lower microbial cell counts, hence lowering their competitiveness, since number contributes greatly to their ability to compete with native microbes [134]. Exposing a potential PGPM to some level of stress before formulation may increase its survival rates during formulation and after field application [134]. Before introducing a potential PGPM inoculant into the market, a series of events, such as greenhouse and field trials, characterization, toxicology profiling, etc. occur, most of which are intended to increase strain survival and efficacy in the field.

Formulation of Microbial Inoculants for Commercial Purposes and Their Mode of Application

Microbial inoculants are usually a combination of microbial cells and/or their parts/compounds and a nonliving carrier that may be in form of a liquid or solid material [15,24,38]. Microbial cells may be either active or dormant; in the latter case, they have to be activated before or after inoculation [15]. They may also be pure cultures (single strains) or a combination of microbial strains (microbial consortia) [15,38]. Formulation is a major contributor to the variation in performance of inoculants observed in farmers' fields and at research stations. Formulation can shield the microbe from adverse environmental conditions, increase their shelf life and also supply their nutritional requirements, hence enhancing their chances of survival in the field [23,134]. Normally, a group of microbes are isolated from their natural habitat (soil or plant tissue), tested for their ability to promote plant growth under a range of conditions, and the superior strains are selected for commercialisation purposes. The strains are multiplied and formulated under controlled environment conditions, after which the efficiency of the inoculant is evaluated under field conditions [23]. The method of formulation ought to consider the target crop, target market and mode of application, the latter because the type of formulation often dictates the mode of application of the inoculant. For instance, solid formulations are mainly applied through seed dressing, or broadcasting onto the field, while liquid formulations have a wide range of application methods [15,24,38]. Liquid carriers are mostly water and/or organic solvents (other than microbial media), such as glycerol and carboxymethyl cellulose that are added to increase properties such as stickiness and dispersal abilities [23]. There are several types of solid carriers, such as clay, vermiculite, peat and charcoal [15]. Care should be taken, when selecting microbial carriers, to ensure they have no negative impact on the environment or the microbe itself [15,38]. Although they are easy to handle and work with, liquid carriers may require specialised storage conditions (cool conditions that necessitate a cooling mechanism) for a long shelf life [23], which makes their marketing and use in developing countries difficult, due to limited and unstable power supply on most farms. Solid formulations, on the other hand, are bulky and may require larger storage facilities, when compared to liquids. However, materials such as peat have an outstanding reputation as inoculant carriers, and are successfully used in both North and South America [23]. The formulation method opted for should ensure the affordability of the final product by the target market, since a very expensive product is likely to meaningfully increase production costs, which is undesirable. For instance, sterile carriers are preferred over nonsterile carriers [23], however the former are costlier than the latter, which may make them unaffordable to many farmers across the globe. The formulation method should also ensure the compatibility of the inoculant with agronomic practices, such as weed control methods, irrigation, etc.

Once a formulated product exhibits positive responses, in field and greenhouse trials, it is put on the market for accessibility by farmers. While the isolation and characterisation of microbial strains from their natural habitats is largely done by academic research institutions, the production of microbial inoculants for commercial purposes is dominated by registered companies, which obtain patents and rights over specific inoculants. Table 3 below shows such microbial based products on the market as plant growth stimulants.

Table 3. Examples of microbial inoculants currently available on the market, and their producing companies.

Inoculant	Country	Producer	Use	Reference
Bacillus megaterium	*Sri Lanka*	*BioPowerLanka*	*Phosphorus solubilisation*	[135]
Pseudomonas striata, B. Polymyxa and B.megaterium consortium	*India*	*AgriLife*	*Phosphorus solubilisation*	[135]
Acidithiobacillus ferrooxidans	*India*	*AgriLife*	*Iron mobilization*	[135]
Trichoderma and Bradyrhizobium spp. *(Excalibre-SA) consortium*	*USA*	*ABM®*	*N fixation Growth stimulation*	[18]
BIODOZ® (B. japonicum)	*Denmark*	*Novozymes*	*Nitrogen fixation*	[134]
Cell-Tech® (B.japonicum)	*Belgium*	*Monsanto (Bayer)*	*Nitrogen fixation*	[134]
Nitragin® S. meliloti	*Belgium*	*Monsanto BioAg^{TM} (Bayer)*	*Nitrogen fixation*	[134]
Cedomon® Pseudomonas chlororaphis	*Sweden*	*BioAgriAB*	*Biopesticide*	[134]
Sheathguard^{TM} Pseudomonas fluorescens	*India*	*AgriLife*	*Biopesticide*	[134]
Galltrol® -A Agrobacterium radiobacter	*USA*	*AgBioChem*	*Biopesticide*	[134]
HISTICK® Bradyrhizobium japonicum	*Germany*	*BASF SE*	*Nitrogen fixation*	[135]
Bacillus + Pseudomonas + Lactobacillus + Saccharomyces spp.	*Canada*	*EVL Inc*	*Biostimulant*	
Xen Tari (Bacillus thuringiensis)	*USA*	*Valent USA*	*Biopesticide*	[136]
VOTIVO FS seed treatment (Bacillus firmus)	*USA*	*Bayer*	*Biopesticide*	[136]
VectoLex FG (Bacillus sphaericus)	*USA*	*Valent Biosciences*	*Biopesticide*	[136]
Venerate XC (Burkholderia rinojensis)	*USA*	*Marrone Bio Innovations*	*Biopesticide*	[136]
Zequanox (Pseudomonas fluorescens)	*USA*	*Marrone Bio Innovations*	*Biopesticide*	[136]
BotaniGard ES/WP, Mycotrol (Beauveria bassiana)	*USA*	*Lam International*	*Biopesticide*	[136]
Naturalis L (Beauveria bassiana)	*USA*	*Troy BioSciences*	*Biopesticide*	[136]
BioCeres WP (Beauveria bassiana)	*USA*	*BioSafe*	*Biopesticide*	[136]
Met-52 EC and Met-52 G (Metarhizium brunneum (anisopliae s.L.)	*USA*	*Novozymes*	*Biopesticide*	[136]
MeloCon WG (Purpureocillium lilacinum)	*USA*	*Bayer*	*Biopesticide*	[136]
Cyd-X, Cyd-X HP (Cydia pomonella (CpGV)	*USA*	*Certis USA*	*Biopesticide*	[136]
FruitGuard (Plodia interpunctella GV	*USA*	*Agrivir*	*Biopesticide*	[136]
Serenade (Bacillus subtilis QST 713)		*Agraquest*	*Biocontrol*	[79]
Bacillus firmus I-1582 WP5 (B. firmus I-1582)		*Bayer Crop Science*	*Biocontrol*	[79]
Cedomon (Pseudomonas chlororaphis MA342)		*Bioagri*		[79]
Proradix (Pseudomonas sp. DSMZ 13134)		*Sourcon–Padena Germany, Itary*	*Biocontrol*	[79]
Novodor (B. thuringiensis ssp. tenebrionis NB 176)	*USA*	*Valent Bioscience*	*Biocontrol*	[79]

6. Limitations to Global Use of Microbial Inoculants

Although microbial inoculants are viewed as the most viable hope, with regard to sustainable agriculture in the face of climate change, their use and adoption globally are still wanting, due to a range of reasons, that vary between developed and developing countries. Adaptation to use of microbial inoculants is developing at a relatively faster pace [24] in the developed world than in developing areas, such as Africa, where their use is restricted by limited availability of resources and knowledge, among other factors. In the developed world, microbial use is slowed largely by inconsistencies in enhancing plant growth, in which case crop producers opt for chemicals, which generally provide stable results. There are many cases where the excellent performance of an inoculant observed during pre-commercialisation trials does not translate to efficiency on farmers' fields. Even when it does, sometimes the results are not consistent, which frustrates the farmers. Some of these inconsistences may be attributed to biotic and abiotic soil factors and plant factors which directly or indirectly affect the introduced microorganism(s) [23]. For instance, some inoculants are cultivar and species specific, in that applying them outside the target species will yield no results. Soil factors such as salinity and temperature are dynamic and affect the survival and effectiveness of the applied microbial strains. This implies that soil conditions should always be favorable for the introduced microbe, otherwise inconsistencies are bound to prevail. Therefore, there is a need to sensitise farmers regarding the proper use of microbial products to minimise such inconsistencies. Unless sensitisation is properly conducted, we cannot rule out inappropriate practices such as farmers applying rhizobial inoculants together with high doses of nitrogen fertilizer, expecting better results than the inoculant or fertilizer used alone. In fact, nitrogen fertilizer will inhibit biological nitrogen fixation. Similarly, applying a biocontrol to a soil or plant that lacks the pathogen it can antagonise/suppress may not yield results. It is also important to understand the status of the soil/plant as the application of microbial inoculants may inhibit plant growth where the soil/plant already contains optimal concentrations of the compound that the microbe produces to enhance plant growth. For instance, application of IAA producing PGPM on plants with an already optimal concentration of IAA may yield negative effects on the plant, due to excess IAA [43]. Understanding soil conditions will also guide the farmer regarding how often to apply the inoculant. Some require seasonal, annual or even twice in a season application, while after some time, application may not be necessary, especially where the microbe establishes reasonable populations in the soil. Successful microbial inoculants employ mechanisms that give them a competitive advantage over the native strains. For instance, rhizobia and mycorrhizal fungi have a signaling system with their host plants, which gives them an advantage over their competitors. Introduced microbes may also outcompete native microbes through the production of antimicrobials, which may kill or deter other microbes, as well as the production of siderophores that give them a competitive advantage over other microbes for iron resources in the soil, hence proliferating better, especially in iron limited soils [14]. Nevertheless, it is important to increase the competitive advantage of introduced microbes, by ensuring high microbial concentrations in the inoculant and use of adequate formulations [18]. With approaches such as metagenomics, the microbial population of the target environment can be studied, and potential PGPM studied for their ability to out compete the latter in field, greenhouse and laboratory conditions. However, this may not be an easy task, given that microbial populations in crop production fields may differ meaningfully due to a wide range of factors. Location and plant specific nature of some phytomicrobiome elements for inoculant production should also be prioritised, since such microbes, to a great extent, are more adapted to the environment and/or plant conditions, which may increase their chances of survival and persistence in the soil. The idea of using microbial consortia may also work to our benefit, as will be discussed below. This does not, however, disqualify single strain inoculants; their advantages are also discussed below.

In less developed countries, especially in sub-Saharan Africa, reasons for low adoption also vary between large- and small-scale farmers. For large-scale farmers, such as those in Zimbabwe, South Africa and Kenya, the ineffectiveness of many microbe-based products in the field contributes meaningfully to the low adoption of microbial inoculants [137,138]. For small scale farmers, costs and inadequate knowledge of such products are the major drivers. These two factors, especially costs, also limit the use of other agricultural inputs, such as high-quality seed. Exceptions can be made for a smaller group of small scale farmers, whose farms' researchers run experiments/field trials, because then, they can obtain access to the inputs from researchers largely free of charge, otherwise, they mostly depend on crop rotations (which are sometimes not properly done) and animal manures, while others just grow their preferred crops year in year out. The lack of knowledge can be attributed to the large gap between research and extension. Researchers achieve good findings, but due to poor funding and poor dissemination techniques, this knowledge never reaches the farmer [138]. Publications do not help much, because many small-scale farmers are illiterate, and even those who can read have limited access to technologies such as smart phones, computers and the internet. It should be noted that many small-scale farmers are also low-income earners, who struggle to meet their basic needs. In countries where governments are not directly involved in the distribution of agricultural inputs, dealers may not be willing to extend products to people who they well know cannot afford them, which leads to unavailability of and/or inaccessibility of the products by the farmers. In such cases, intervention strategies should definitely be at least a bit different and more vigorous. First and foremost, the knowledge of existence of PGPM technology needs to be spread to these largely small scale farmers. Projects like N2 Africa have done a good job in trying to spread the BNF technology, although more effort is still needed. Extension officers should be updated on new findings and products, and be properly facilitated to extend this knowledge to the farmers. Governments may consider subsidizing products and getting directly involved in their distribution to the farmers. Promiscuous soybean varieties are already a good strategy of eliminating the need for inoculation. It would be better to develop strategies that enable the use of farm-based PGPM inoculants, as many farmers have limited access to agro-input markets, in part due to poor transport networks. Locally made cooling facilities such as charcoal based refrigerators and unglazed clay pots may also be helpful. However, the former would be a contradictory measure, given that it would encourage deforestation. The whole sensitisation process should involve all stakeholders, such as governments, extension officers, agricultural schools, and private companies that contract small scale farmers to grow crops for them for use as raw materials. The latter, especially, provide the farmer with chemicals such as pesticides and fertilizers; therefore, their involvement cannot be ignored.

7. Microbial Consortia

In order to address issues associated with the use of single strains as inoculants, microbial consortia have gained popularity. This may be relevant, especially now that the prevalence of both biotic and abiotic stresses due to climate change are likely to increase. Microbial consortia technology involves the use of more than one microbial species in a single inoculant product. The microbes may have the same or different modes of action [18,70,89], and may be from different phyla, genera or even other groupings, for example, a combination of bacterial and fungal strains. Microbial consortia may have an advantage over single strains when the species synergistically interact and confer benefits to each other [70,71,89,117]. For instance, one strain may breakdown a substrate, unavailable to other species, converting it into forms that the other members of the consortium can utilise as a source of nutrients [14], or produce exopolysaccharides which offer protection against stress to all members of the consortium [134], produce compounds which are signals that activate plant growth promotion capability of other members of the consortium, through the production of plant growth stimulating compounds, that they would otherwise not produce, for instance, in pure culture. In cases where microbes with the same mode of action are used, members may have varying tolerance to different biotic and abiotic stresses, which enhances survival of at least a member that will confer intended benefits to the plant. In the case of different modes of action, these complement each other and confer a more effective benefit to the plant. It could also be that some members of the consortium are simply helpers of the strains meant to benefit the plant. Such helper strains, for instance mycorrhiza helper bacteria, should facilitate the target strain in plant colonisation, conferring benefits to the plant.

Researchers have reported inefficient strains that became efficient in a consortium. For example, Santhanam et al. [89] observed that the inclusion of two bacterial strains with insignificant effects on mortality of sudden wilt pathogens in tobacco, in a consortium with three other bacteria improved resistance of plants to the same pathogen, in comparison to the consortium of 3 used alone. Mycorrhizal fungi, in association with a helper bacterium, may have better established mycelia and plant root colonisation, if the bacterium produces substances that directly enhance the germination of fungal spores, or indirectly enhance the establishment of mycorrhiza through the production of antimicrobials that reduce competition from other microbes or minor pathogens [14].

Because of the interaction advantage, some microbes perform better in microbial consortia than when applied individually [70,89]. However, the reverse is true for some PGPM species, as reported by other researchers [70,92,100]. Therefore, the role that single strain inoculants play cannot be written off easily, especially because microbial consortia also have their shortcomings. Coming up with effective compatible combinations in which all members actively benefit the plant can be challenging, practically given that some members of the consortium may produce compounds lethal to other members [133]. Even if the produced compounds do not go to the extreme of killing other members, they may cause a shutdown of their plant growth promoting system, or interfere with their growth, as de Vrieze et al. [70] observed in a consortium of five *Pseudomonas* strains. In such cases, it is probable that only a subset of the consortium members will actively benefit the plant, the rest being "dormant" or dead. Difficulties concerning the formulation of microbial consortia may also be associated with the variations in optimal growth conditions. For more than one species, or even genus, creating conditions that will favour all members while retaining their ability to promote plant growth may not be easy. Finally, manufacturing consortia can be challenging, as very small changes at the outset can result in very different levels of consortium members in the final product, resulting in product inconsistencies.

8. Microbial Compounds as "Inoculants"

The use of microbial compounds as "inoculants" is slowly gaining popularity after successful trials [139–145]. To be a true inoculant, the material must contain living cells that colonize the plant. In this case, the technology may be the product of microbial growth and may be more valuable as a result of climate change were biotic and abiotic factors may lower or completely halt the effectiveness of microbial cell based inoculants. This practice involves the separation of cell free supernatant from microbial cells, and the subsequent separation and purification of the compound from the cell-free supernatant, mainly through high pressure liquid chromatography (HPLC). The pure compound is then tested for its ability to promote plant growth under greenhouse and field conditions, prior to commercialisation. Before commercialisation, other tests, such as the effect of the compound on non-target organisms and humans, as well as checks regarding legal regulations, are usually carried out. The effect of the compound on non-target organisms such as plants, humans and animals ought to be substantially understood too, as with studying the residual effects of the compound (how much of it remains in the edible parts of the plant, and in the soil, following application). Therefore, before any compound can be commercialised, its ability to be purified, and produced on a large scale, should be verified [143]. The compound should be identified and characterised based on its physiological and biological properties.

The efficacy and type of microbial bioactive compounds produced are influenced by microbial species and conditions to which the PGPM is exposed. Slight alterations in growth conditions may result in different compounds, produced at different levels, and with varying degrees of efficacy. For instance, varying the pH, a *Pseudomonas* species culture caused it to produce different phenazine compounds with varying efficacy against *Fusarium oxysporum* f. sp. radicis-lycopersici [143]. Sometimes, the PGPM has to be exposed to stressful conditions before it will produce bioactive compounds, as such compounds may only be produced to enhance the survival of the microorganism under stressful conditions. Therefore, it is important to have an adequate understanding of the conditions under which a certain PGPM will produce plant growth stimulating compounds.

So far, not many (compared to microbial strains) bioactive compounds have been identified for use in crop production. The Smith laboratory at McGill University has thuricin17 and lipochitooligosaccharide (LCO). Thuricin 17 is a bacteriocin secreted by *Bacillus thuringiensis*, a non-symbiotic endophytic bacterium. The compound is known to have anti-microbial properties, which gives *Bacillus thuringiensis* a competitive advantage over other bacteria of the same grouping [140]. After a series of experiments, thuricin 17 was discovered to have growth promoting properties for tomato, soybean, canola, arabidopsis, and rapeseed and switch grass [117,140–142,144,145]. More trials are on-going, and the technology has yet to be commercialised. Lipo-chitooligosaccharide, on the other hand, is produced by rhizobia, as a signal to its host plants [139]. Formerly extensively studied for its role in the nodulation process, the compound is currently patented and being marketed by Novozymes as a plant growth stimulant, where its effects are greatest under abiotic stress conditions. Other compounds such as phenazine-1-carboxylic acid (PCA) have also been commercialised [143,146–153]. Table 4 shows the various compounds with potential use as agro-inputs. Some of them are already commercialised.

Table 4. Microbial compounds of agricultural importance.

Compound	Producing Microbe	Function	Comment	Reference
LCO	*Bradyrhizobium japonicum*	Biostimulant	*Stimulates plant growth under stressed and non stressed conditions.*	[117,146]
Thuricin17	*Bacillus thuringiensis*	Biostimulant	*Enhances growth of different crops eg Soybean in stressed and non stressed conditions*	[141,142]
Anisomycin	*Streptomyces* sp.	herbicide	*Effective against Digitaria spp.*	[149]
Phenazine-1-carboxyamide (PCN)	*Pseudomonas* spp.	biocontrol	*It is effective against; Fusarium oxysporum f. sp. Radicis-lycopersici, Xanthomonas oryzae pv. Oryzae, Rhizoctonia solani, Botrytis cinerea*	[143,148,149,151]
Phenazine-1-carboxylic acid (PCA)	*Pseudomonas* spp.	biocontrol	*It is effective against Fusarium oxysp.orum f. sp.Radicis-lycopersici, Colletotrichum orbiculare, Gaeumannomyces graminis var. tritici, Phytophthora capsici*	[143,146,149,152, 153]
Pyocyanin (PYO)	*Pseudomonas* spp.	biocontrol	*Effective against: Sclerotium rolfsii, Macrophomina phaseolina*	[154–156]
Pyrrolnitrin	*Burkholderia pyrrocinia 2327*	biocontrol	*It has antifungal properties against; Ralstonia solani, Phytophthora capsici, and Fusarium oxysporum*	[157,158]
Phencomycin	*Burkholderia glumae 411gr-6*	biocontrol	*Effective against; Alternaria brassicicola, Aspergillus oryzae, Cladosporium cucumerinum, Colletotrichum gloeosporioides*	[159]
Ornibactin	*Burkholderia contaminans MS14*	biocontrol	*Siderophore with biocontrol activity against Erwinia amylovora, Ralstonia solanacearum, Pseudomonas syringae B301, Clavibacter michiganensis subsp. michiganensis*	[160]
Iturin A2	*Bacillus subtilis B47*	biocontrol	*Effective against fungi; Bipolaris maydis*	[161]
Mycosubtilin	*Bacillus subtilis*	biocontrol	*Has anti fungal properties, effective against; Bremia lactucae*	[162]
Herboxidiene	*Streptomyces* sp. *A7847*	herbecide	*Effective on a number of weed sp.*	[163]
Phosphinothricin	*Streptomyces hygroscopicus*	herbecide		

Table 4. *Cont.*

Compound	Producing Microbe	Function	Comment	Reference
Cyanobacterin	*Scytonema hofmanni*	*herbecide*	*Effective on cynobacteria, algae and higher plants*	[164]
Avermectin	*Streptomyces avermitilis*	*Insectide nematocide*	*Effective against Spider mites, Citrus red mite, horn worms, army worms, etc.*	[165]

9. Microbial Cells or Microbial Compounds?

Given the current understanding, a question would be, what should a crop producer adopt, given a choice between the microbial cells and microbial compound based products. The answer to such a question cannot be as definite as that specific factors may call for either of the two, or even the use of both simultaneously. Before one reaches the level of farmer preferences, soil and environmental factors as well as economic implications, intended use and handling may be major considerations. For instance, in the reclamation of areas heavily affected by abiotic stress, use of microbial cells may not be a good idea, if they are not able to survive some harsh conditions. Even if they did, the efficacy of their plant growth promotion capacity may be greatly affected. Compounds, on the other hand, are less affected by such abiotic stresses, and hence have a greater chance of being successful under such conditions. The use of compounds or both compound(s) and microbial cells may be desirable, especially when an abiotic stress such as drought interrupts signaling between plant and PGPM. In such a case, external application of the signal may rectify the disruption. Prudent et al. [142] observed a 17% increase in soybean biomass under drought conditions following co-inoculation with *Bradyrhizobium japonicum* and thuricin17, compared to inoculation with the rhizobial cells alone. The use of microbial compounds may also be a better choice in cases where the microbe is a facultative pathogen, such as *Agrobacterium* spp. [92]. In such cases, the pathogen effect of the microbe on plants is minimised. Application of microbial compounds may also benefit a wider range of crop species compared to microbial cells, given that many microbes can be at least somewhat species specific. A case would be that of lipochitooligosaccharides (LCOs), which can be utilised to enhance growth of legumes and non-leguminous crops [148], under stressed and non-stressed conditions [115,137], but to a greater extent, under stress conditions. For instance, LCOs enhanced fruit and flower production in tomato (*Lycopersicon esculentum*) plants [148], and stimulated the growth of soybean and corn plants [139]. The compound was also reported to enhance the germination of soybean seeds subjected to high NaCl concentrations [117] and canola [145]. Such benefits from LCO would not be provided to these crops had *Bradyrhizobium japonicum* been applied. Compounds are also less bulky and less costly, in most cases requiring small doses to be efficient. This relieves crop producers of storage and transportation concerns.

However, there are scenarios where the use of microbial cells is inevitable. For instance, the role that rhizobia play in nitrogen fixation, or mycorrhizae in P mobilisation and acquisition by plant roots could not be fulfilled by microbial compounds. Nitrogen fixing bacteria cannot be substituted by compounds in areas were N is limiting. Microbial cells have the potential to establish microbial populations in the rhizosphere, which may eliminate the need for further inoculation, a characteristic most farmers would desire, since it not only has positive financial implications, but also saves labour. Based on this, it is safe to assume that marketing companies would opt for compounds, since they guarantee continuous sales. However, the long and laborious process of isolating and purifying microbial compounds may also contribute to their scarcity and willingness of some researchers and companies to take that route.

10. Way Forward and Recommendations

With climate change conditions increasing, and the desperate need to come up with sustainable approaches of enhancing crop productivity to meet the food demand of the growing population, microbes are a prominent source of hope. However, a great deal still needs to be done to bridge the gap between their use in developed and developing countries. More research should be done to address issues of inconsistencies observed on crop producers' fields, following the use of microbial inoculants. It is obvious that single strains and consortia, or microbial cells and microbial compounds are issues that need to be evaluated on a case-by-case basis. Therefore, a better suggestion would be that more research be done to provide consumers with options that can address their unique needs, while being economically viable.

11. Conclusions

Lowering the effects of climate change on crop production, through reducing greenhouse gas emissions, is one of the major focuses of researchers in recent times. With proper manipulation, plant growth promoting microorganisms and compounds, they produce have potential to enhance growth and yield of plants exposed to biotic and biotic stress(es). This can complement other strategies, such as conservation farming and breeding for stress tolerant crop cultivars, to create an integrated approach of enhancing crop production in the face of climate change. Given that the prevalence of stress is predicted to increase with climate change, more research is needed to come up with better and more effective alternatives of utilising PGPM technology; not only to enhance plant growth, but also to reduce greenhouse gas emissions from the agricultural sector, which is a meaningful contributor.

Author Contributions: J.N. gathered all reading material and wrote the review; D.L.S. did all the editing and guidance on scientific knowledge. All authors have read and agreed to the published version of the manuscript.

Funding: This research was funded by the Natural Science and Engineering Research Council of Canada grant number RGPIN 2020-07047.

Conflicts of Interest: There is no conflict of interest.

References

1. Lott, F.C.; Christidis, N.; Stott, P.A. Can the 2011 East African drought be attributed to human-induced climate change? *Geophys. Res. Lett.* **2013**, *40*, 1177–1181. [CrossRef]
2. Rossi, F.; Olguın, E.J.; Diels, L.; de Philippis, R. Microbial fixation of CO_2 in waterbodies and in drylands to combat climate change, soil loss and desertification. *New Biotechnol.* **2015**, *32*, 109–120. [CrossRef]
3. Bradley, B.A.; Curtis, C.A.; Chambers, J.C. Bromus Response to Climate and Projected Changes with Climate Change. In *Exotic Brome-Grasses in Arid and Semiarid Ecosystems of the Western US*; Germino, M., Chambers, J., Brown, C., Eds.; Springer Series on Environmental Management; Springer: Cham, Switzerland, 2016; pp. 257–274. [CrossRef]
4. Richards, M.B.; Wollenberg, E.; van Vuuren, D. National contributions to climate change mitigation from agriculture: Allocating a global target. *Clim. Policy* **2018**, *18*, 1271–1285. [CrossRef]
5. Loboguerrero, A.M.; Campbell, B.M.; Cooper, P.J.M.; Hansen, J.W.; Rosenstock, T.; Wollenberg, E. Food and Earth Systems: Priorities for Climate Change Adaptation and Mitigation for Agriculture and Food Systems. *Sustainability* **2019**, *11*, 1372. [CrossRef]
6. IPCC. *2014: Climate Change 2014: Synthesis Report. Contribution of Working Groups I, II and III to the Fifth Assessment Report of the Intergovernmental Panel on Climate Change*; Pachauri, R.K., Meyer, L.A., Eds.; IPCC: Geneva, Switzerland, 2014; p. 151.
7. Porter, J.R.; Xie, L.; Challinor, A.J.; Cochrane, K.; Howden, S.M.; Iqbal, M.M.; Lobell, D.B.; Travasso, M.I. Food security and food production systems. In *Climate Change 2014: Impacts, Adaptation, and Vulnerability Part A: Global and Sectoral Aspects. Contribution of Working GroupII to the Fifth Assessment Report of the Intergovernmental Panel on Climate Change*; Field, C.B., Barros, V.R., Dokken, D.J., Mach, K.J., Mastrandrea, M.D., Bilir, T.E., Chatterjee, M., Ebi, K.L., Estrada, Y.O., Genova, R.C., et al., Eds.; Cambridge University Press: Cambridge, UK; New York, NY, USA, 2014; pp. 485–533.
8. Dawson, T.P.; Perryman, A.H.; Osborne, T.M. Modelling impacts of climate change on global food security. *Clim. Chang.* **2016**, *134*, 429–440. [CrossRef]

9.   Bouwer, L.M.; Bubeck, P.; Aerts, J.C. Changes in future flood risk due to climate and development in a Dutch polder area. *Glob. Environ. Chang.* **2010**, *20*, 463–471. [CrossRef]

10. Mirza, M.M.Q. Climate change, flooding in South Asia and implications. *Reg. Environ. Chang.* **2011**, *11*, 95–107. [CrossRef]

11. Nam, W.; Hayes, M.J.; Svoboda, M.D.; Tadesse, T.; Wilhite, D.A. Drought hazard assessment in the context of climate change for South Korea. *Agric. Water Manag.* **2015**, *160*, 106–117. [CrossRef]

12. Barea, J.M. Future challenges and perspectives for applying microbial biotechnology in sustainable agriculture based on a better understanding of plant-microbiome interactions. *J. Soil Sci. Plant Nutr.* **2015**, *15*, 261–282. [CrossRef]

13. Gupta, G.; Parihar, S.S.; Ahirwar, N.K.; Snehi, S.K.; Singh, V. Plant Growth Promoting Rhizobacteria (PGPR): Current and Future Prospects for Development of Sustainable Agriculture. *Microb. Biochem. Technol.* **2015**, *7*, 96–102. [CrossRef]

14. Bender, S.F.; Wagg, C.; van der Heijden, M.G.A. An Underground Revolution: Biodiversity and Soil Ecological Engineering for Agricultural Sustainability. *Trends Ecol. Evol.* **2016**, *31*, 440–452. [CrossRef] [PubMed]

15. Babalola, O.O.; Glick, B.R. The use of microbial inoculants in African agriculture: Current practice and future prospects. *J. Food Agric. Environ.* **2012**, *10*, 540–549.

16. Smith, D.L.; Subramanian, S.; Lamont, J.R.; Bywater-Ekegärd, M. Signaling in the phytomicrobiome: Breadth and potential. *Front. Plant Sci.* **2015**, *6*, 709. [CrossRef] [PubMed]

17. Smith, D.L.; Gravel, V.; Yergeau, E. Signaling in the Phytomicrobiome. *Front. Plant Sci.* **2017**, *8*, 611. [CrossRef] [PubMed]

18. Backer, R.; Rokem, J.S.; Ilangumaran, G.; Lamont, J.; Praslickova, D.; Ricci, E.; Subramanian, S.; Smith, D.L. Plant Growth-Promoting Rhizobacteria: Context, Mechanisms of Action, and Roadmap to Commercialization of Bio stimulants for Sustainable Agriculture. *Front. Plant Sci.* **2018**, *9*, 1473. [CrossRef]

19. Hartmann, A.; Rothballer, M.; Hense, B.A.; Schröder, P. Bacterial quorum sensing compounds are important modulators of microbe-plant interactions. *Front. Plant Sci.* **2014**, *5*, 131. [CrossRef]

20. Sánchez-Cañizares, C.; Jorrín, B.; Poole, P.S.; Tkacz, A. Understanding the holobiont: The interdependence of plants and their microbiome. *Curr. Opin. Microbiol.* **2017**, *38*, 188–196. [CrossRef]

21. Compant, S.; Samad, A.; Faist, H.; Sessitsch, A. A review on the plant microbiome: Ecology, functions and emerging trends in microbial applications. *J. Adv. Res.* **2019**, *19*, 29–37. [CrossRef]

22. Berg, M.; Koskella, B. Nutrient and dose dependent protection against a plant pathogen. *Curr. Biol.* **2018**, *28*, 2487–2492. [CrossRef]

23. Bashan, Y.; de-Bashan, L.E.; Prabhu, S.R.; Hernandez, J. Advances in plant growth-promoting bacterial inoculant technology: Formulations and practical perspectives (1998–2013). *Plant Soil* **2014**, *378*, 1–33. [CrossRef]

24. Alori, E.T.; Dare, M.O.; Babalola, O.O. Microbial Inoculants for Soil Quality and Plant Health. In *Sustainable Agriculture Reviews*; Lichtfouse, E., Ed.; Springer: Cham, Switzerland, 2017; Volume 22, pp. 281–307. [CrossRef]

25. Malusa, E.; Sas-Paszt, L.; Ciesielska, J. Technologies for Beneficial Microorganisms Inocula Used as Biofertilizers. *Sci. World J.* **2012**, *2012*, 491206. [CrossRef]

26. Silva, E.C.; Nogueira, R.J.M.C.; Silva, M.A.; Albuquerque, M.B. Drought Stress and Plant Nutrition. *Plant Stress* **2011**, *5*, 32–41.

27. DaMatta, F.; Loos, R.A.; Silva, E.A.; Loureiro, M.E.; Ducatti, C. Effects of soil water déficit and nitrogen nutrition on water relations and photosynthesis of pot-grown *Coffea canephora* Pierra. *Trees* **2002**, *16*, 555–558. [CrossRef]

28. Tietema, A.; De Boer, W.; Riemer, L.; Verstraten, J.M. Nitrate production in nitrogen saturated acid forest soils: Vertical distributions and characteristics. *Soil Biol. Biochem.* **1992**, *24*, 235–240. [CrossRef]

29. Tripathi, D.K.; Singh, S.; Gaur, S.; Singh, S.; Yadav, V.; Liu, S.; Singh, V.P.; Sharma, S.; Srivastava, P.; Prasad, S.M.; et al. Acquisition and Homeostasis of Iron in Higher Plants and Their Probable Role in Abiotic Stress Tolerance. *Front. Environ. Sci.* **2018**, *5*, 86. [CrossRef]

30. Wang, W.; Zheng, Q.; Shen, Q.; Guo, S. The Critical Role of Potassium in Plant Stress Response. *Int. J. Mol. Sci.* **2013**, *14*, 7370–7390. [CrossRef]

31. Waraich, E.A.; Ahmad, R.; Halim, A.; Aziz, T. Alleviation of temperature stress by nutrient management in crop plants: A review. *J. Soil Sci. Plant Nutr.* **2012**, *12*, 221–244. [CrossRef]

32. Waraich, E.A.; Ahmad, R.; Ashraf, M.Y.; Saifullah; Ahmad, M. Improving agricultural water use efficiency by nutrient management in crop plants. Acta Agriculturae Scandinavica. *Sect. B Plant Soil Sci.* **2011**, *61*, 291–304. [CrossRef]

33. Karmakar, R.; Das, I.; Dutta, D.; Rakshit, A. Potential Effects of Climate Change on Soil Properties: A Review. *Sci. Int.* **2016**, *4*, 51–73. [CrossRef]

34. Wu, F.U.; Bao, W.; Li, F.L.; Wu, N. Effects of water stress and nitrogen supply on leaf gas exchange and fluorescence parameters of *Sophora davidii* seedlings. *Photosynthetica* **2008**, *46*, 40–48. [CrossRef]

35. Faye, I.; Diouf, O.; Guisse', A.; Se'ne, M.; Diallo, N. Characterizing Root Responses to Low Phosphorus in Pearl Millet [*Pennisetum glaucum* (L.) R. Br.]. *Agron. J.* **2006**, *98*, 1187–1194. [CrossRef]

36. Alori, E.T.; Fawole, O.B. Impact of chemical inputs on arbuscular mycorrhiza spores in soil: Response of AM Spores to fertilizer and herbicides. *Alban J. Agric. Sci.* **2017**, *16*, 10–13.

37. Lal, R. Restoring Soil Quality to Mitigate Soil Degradation. *Sustainability* **2015**, *7*, 5875–5895. [CrossRef]

38. Alori, E.T.; Babalola, O.O. Microbial Inoculants for Improving Crop Quality and Human Health in Africa. *Front. Microbiol.* **2018**, *9*, 2213. [CrossRef] [PubMed]

39. Naamala, J.; Jaiswal, S.K.; Dakora, F.D. Microsymbiont diversity and phylogeny of native Bradyrhizobia associated with soybean (*Glycine max* L. Merr.) nodulation in South African soils. *Syst. Appl. Microbiol.* **2016**, *39*, 336–344. [CrossRef]

40. Graham, J.H.; Leonard, R.T.; Menge, J.A. Membrane-Mediated Decrease in Root Exudation Responsible for Phorphorus Inhibition of Vesicular-Arbuscular Mycorrhiza Formation. *Plant. Physiol.* **1981**, *68*, 548–552. [CrossRef]

41. Sprent, J.I.; Stephens, J.H.; Rupela, O.P. Environmental effects on nitrogen fixation. In *World Crops: Cool Season Food Legumes*; Summerfield, R.J., Ed.; Current Plant Science and Biotechnology in Agriculture; Springer: Dordrech, The Netherlands, 1988; Volume 5, pp. 801–810. [CrossRef]

42. Marschner, H.; Dell, B. Nutrient uptake in mycorrhizal symbiosis. *Plant Soil* **1994**, *159*, 89–102. [CrossRef]

43. Glick, B.R. Plant growth promoting bacteria: Mechanisms and applications. *Scientifica* **2012**, *2012*, 1–15. [CrossRef]

44. Goldstein, A.H.; Krishnaraj, P.U. Phosphate solubilizing microorganisms vs. phosphate mobilizing microorganisms: What separates a phenotype from a trait? In *First International Meeting on Microbial Phosphate Solubilization. Developments in Plant and Soil Sciences*; Velázquez, E., Rodríguez-Barrueco, C., Eds.; Springer: Dordrecht, The Netherlands, 2007; Volume 102, pp. 203–213. [CrossRef]

45. Meding, S.M.; Zasoski, R.J. Hyphal-mediated transfer of nitrate, arsenic, cesium, rubidium, and strontium between arbuscular mycorrhizal forbs and grasses from a California oak woodland. *Soil Biol. Biochem.* **2008**, *40*, 126–134. [CrossRef]

46. Hayat, R.; Ali, S.; Amara, U.; Khalid, R.; Ahmed, I. Soil beneficial bacteria and their role in plant growth promotion: A review. *Ann. Microbiol.* **2010**, *60*, 579–598. [CrossRef]

47. Straker, C.J.; Mitchell, D.T. The activity and characterization of acid phosphatases in endomycorrhizal fungi of the Ericaceae. *New Phytol.* **1986**, *104*, 243–256. [CrossRef]

48. Khan, M.S.; Zaidi, A.; Ahemad, M.; Oves, M.; Wani, P.A. Plant growth promotion by phosphate solubilizing fungi—Current perspective. *Arch. Agron. Soil Sci.* **2010**, *56*, 73–98. [CrossRef]

49. Elias, F.; Woyessa, D.; Muleta, D. Phosphate Solubilisation Potential of Rhizosphere Fungi Isolated from Plants in Jimma Zone, Southwest Ethiopia. *Int. J. Microbiol.* **2016**, *2016*, 5472601. [CrossRef] [PubMed]

50. Daly, D.H.; Velivelli, S.L.S.; Prestwich, B.D. The Role of Soil Microbes in Crop Biofortification. In *Agriculturally Important Microbes for Sustainable Agriculture*; Meena, V., Mishra, P., Bisht, J., Pattanayak, A., Eds.; Springer: Singapore, 2017. [CrossRef]

51. Khan, A.; Singh, J.; Upadhayay, V.K.; Singh, A.V.; Shah, S. Microbial Biofortification: A Green Technology Through Plant Growth Promoting Microorganisms. In *Sustainable Green Technologies for Environmental Management*; Shah, S., Venkatramanan, V., Prasad, R., Eds.; Springer: Singapore, 2019. [CrossRef]

52. Bhatti, T.M.; Yawar, W. Bacterial solubilization of phosphorus from phosphate rock containing sulfur-mud. *Hydrometallurgy* **2010**, *103*, 54–59. [CrossRef]

53. Radzki, W.; Gutierrez Manero, F.J.; Algar, E.; Lucas Garcıa, J.A.; Garcıa-Villaraco, A.; Solano, B.R. Bacterial siderophores efficiently provide iron to iron-starved tomato plants in hydroponics culture. *Antonie Van Leeuwenhoek* **2013**, *104*, 321–330. [CrossRef]

54. Sharma, A.; Johri, B.N. Growth promoting influence of siderophore-producing *Pseudomonas* strains GRP3A and PRS9 in maize (*Zea mays* L.) under iron limiting conditions. *Microbiol. Res.* **2003**, *158*, 243–248. [CrossRef]

55. Yehuda, Z.; Shenker, M.; Romheld, V.; Marschner, H.; Hadar, Y.; Chen, Y. The Role of Ligand Exchange in the uptake of Iron from Microbial Siderophores by Cramineous Plants. *Plant Physiol.* **1996**, *112*, 1273–1280. [CrossRef]

56. Nadeem, S.M.; Zahir, Z.A.; Naveed, M.; Ashghar, H.N.; Arshad, M. Rhizobacteria capable of producing ACC deaminase may mitigate salt stress in wheat. *Soil Biol. Biochem.* **2010**, *74*, 533–542. [CrossRef]

57. Meena, V.S.; Maurya, B.R.; Prakash, J. Does a rhizospheric microorganism enhance K^+ availability in agricultural soils? *Microbiol. Res.* **2014**, *169*, 337–347. [CrossRef]

58. Cardoso, I.M.; Kuyper, T.W. Mycorrhizas and tropical soil fertility. *Agric. Ecosyst. Environ.* **2006**, *116*, 72–84. [CrossRef]

59. Jalili, F.; Khavazi, K.; Pazira, E.; Nejati, A.; Rahmani, H.A.; Sadaghiani, H.R.; Miransari, M. Isolation and characterization of ACC deaminase-producing fluorescent pseudomonads, to alleviate salinity stress on canola (*Brassica napus* L.) growth. *J. Plant Physiol.* **2008**, *166*, 667–674. [CrossRef] [PubMed]

60. Ali, S.; Charles, T.C.; Glick, B.R. Amelioration of high salinity stress damage by plant growth promoting bacterial endophytes that contain ACC deaminase. *Plant Physiol. Biochem.* **2014**, *80*, 160–167. [CrossRef] [PubMed]

61. Pérez-Montano, F.; Alías-Villegas, C.; Bellogín, R.A.; del Cerro, P.; Espuny, M.R.; Jiménez-Guerrero, I.; López-Baena, F.J.; Ollero, F.J.; Cubo, T. Plant growth promotion in cereal and leguminous agricultural important plants: From microorganism capacities to crop production. *Microbiol. Res.* **2014**, *169*, 325–336. [CrossRef] [PubMed]

62. Jha, C.K.; Saraf, M. Plant growth promoting Rhizobacteria (PGPR): A review. *E3 J. Agric. Res. Dev.* **2015**, *5*, 0108–0119.

63. Dodd, I.C.; Ruiz-Lozano, J.M. Microbial Enhancement of crop resource use efficiency. *Curr. Opin. Biotechnol.* **2012**, *23*, 236–242. [CrossRef]

64. Hassen, A.I.; Bopape, F.L.; Sanger, L.K. Microbial Inoculants as Agents of Growth Promotion and Abiotic Stress Tolerance in Plants. In *Microbial Inoculants in Sustainable Agricultural Productivity*; Singh, D., Singh, H., Prabha, R., Eds.; Springer: New Delhi, India, 2016; pp. 23–36. [CrossRef]

65. Recep, K.; Fikrettin, S.; Erkol, D.; Cafer, E. Biological control of the potato dry rot caused by *Fusarium* species using PGPR strains. *Biol. Control* **2009**, *50*, 194–198. [CrossRef]

66. Moussa, T.A.A.; Almaghrabi, O.A.; Abdel-Moneim, T.S. Biological control of the wheat root rot caused by *Fusarium graminearum* using some PGPR strains in Saudi Arabia. *Ann. Appl. Biol.* **2013**, *163*, 72–81. [CrossRef]

67. Díaz, A.; Smith, A.; Mesa, P.; Zapata, J.; Caviedes, D.; Cotes, A.M. *Control of Fusarium Wilt in Cape Gooseberry by Trichoderma koningiopsis and PGPR*; Pertot, I., Elad, Y., Barka, E.A., Clément, C., Eds.; Working Group Biological Control of Fungal and Bacterial Plant Pathogens; IOBC Bulletin: Dijon, France, 2013; Volume 86, pp. 89–94.

68. Vanitha, S.; Ramjegathesh, R. Bio Control Potential of *Pseudomonas fluorescens* against Coleus Root Rot Disease. *J. Plant Pathol. Microb.* **2014**, *5*, 216. [CrossRef]

69. Dixit, R.; Agrawal, L.; Gupta, S.; Kumar, M.; Yadav, S.; Chauhan, P.S.; Nautiyal, C.S. Southern blight disease of tomato control by 1-aminocyclopropane1-carboxylate (ACC) deaminase producing *Paenibacillus lentimorbus* B30488. *Plant Signal. Behav.* **2016**, *11*, e1113363. [CrossRef]

70. Li, X.L.; George, E.; Marschner, H. Extension of the phosphorus depletion zone in VA-mycorrhizal white clover in a calcareous soil. *Plant Soil* **1991**, *136*, 41–18. [CrossRef]

71. Wu, Y.; Lin, H.; Lin, Y.; Shi, J.; Xue, S.; Hung, Y.; Chen, Y.; Wang, H. Effects of biocontrol bacteria *Bacillus amyloliquefaciens* LY-1 culture broth on quality attributes and storability of harvested litchi fruit. *Postharvest Biol. Technol.* **2017**, *132*, 81–87. [CrossRef]

72. Zhao, D.; Zhao, H.; Zhao, D.; Zhua, X.; Wang, Y.; Duan, Y.; Xuan, Y.; Chen, L. Isolation and identification of bacteria from rhizosphere soil and their effect on plant growth promotion and root-knot nematode disease. *Biol. Control* **2018**, *119*, 12–19. [CrossRef]

73. de Vrieze, M.; Germanier, F.; Vuille, N.; Weisskopf, L. Combining Different Potato-Associated *Pseudomonas* Strains for Improved Biocontrol of *Phytophthora infestans*. *Front. Microbiol.* **2018**, *9*, 2573. [CrossRef] [PubMed]

74. Berendsen, R.L.; Vismans, G.; Yu, K.; Song, Y.; de Jonge, R.; Burgman, W.P.; Burmølle, M.; Herschend, J.; Bakker, P.A.; Pieterse, C.M. Disease-induced assemblage of a plant-beneficial bacterial consortium. *ISME J.* **2018**, *12*, 1496–1507. [CrossRef] [PubMed]

75. Gadhave, K.R.; Hourston, J.E.; Gange, A.C. Developing Soil Microbial Inoculants for Pest Management: Can One Have Too Much of a Good Thing? *J. Chem. Ecol.* **2016**, *42*, 348–356. [CrossRef]

76. Schausberger, P.; Peneder, S.; Juerschik, S.; Hoffmann, D. Mycorrhiza changes plant volatiles to attract spidermite enemies. *Funct. Ecol.* **2012**, *26*, 441–449. [CrossRef]

77. Pangesti, N.; Weldegergis, B.T.; Langendorf, B.; van Loon, J.J.; Dicke, M.; Pineda, A. Rhizobacterial colonization of roots modulates plant volatile emission and enhances the attraction of a parasitoid wasp. to host-infested plants. *Oecologia* **2015**, *178*, 1169–1180. [CrossRef]

78. Herman, M.A.B.; Nault, B.A.; Smart, C.D. Effects of plant growth–promoting rhizobacteria on bell pepper production and green peach aphid infestations in New York. *Crop Prot.* **2008**, *27*, 996–1002. [CrossRef]

79. Velivelli, S.L.S.; Sessitsch, A.; Prestwich, B.D. The Role of Microbial Inoculants in Integrated Crop Management Systems. *Potato Res.* **2014**, *57*, 291–309. [CrossRef]

80. Karthiba, L.; Saveetha, K.; Suresh, S.; Raguchander, T.; Saravanakumar, D.; Samiyappan, R. PGPR and entomopathogenic fungus bioformulation for the synchronous management of leaffolder pest and sheath blight disease of rice. *Pest. Manag. Sci.* **2010**, *66*, 555–564. [CrossRef]

81. Bano, A.; Muqarab, R. Plant defence induced by PGPR against *Spodoptera litura* in tomato (*Solanum lycopersicum* L.). *Plant Biol.* **2017**, *19*, 406–412. [CrossRef] [PubMed]

82. Xiang, N.; Lawrence, K.S.; Kloepper, J.W.; Donald, P.A.; McInroy, J.A. Biological control of *Heterodera glycines* by spore-forming plant growth-promoting rhizobacteria (PGPR) on soybean. *PLoS ONE* **2017**, *12*, e0181201. [CrossRef] [PubMed]

83. Viljoen, J.F.; Labuschagne, N.; Fourie, H.; Sikora, R.A. Biological control of the root-knot nematode *Meloidogyne incognita* on tomatoes and carrots by plant growth-promoting rhizobacteria. *Trop Plant Pathol.* **2019**, *44*, 284–291. [CrossRef]

84. Frankowski, J.; Lorito, M.; Scala, F.; Schmid, R.; Berg, G.; Bahl, H. Purification and properties of two chitinolytic enzymes of *Serratia plymuthica* HRO-C48. *Arch. Microbiol.* **2001**, *176*, 421–426. [CrossRef]

85. Kim, Y.C.; Jung, H.; Kim, K.Y.; Park, S.K. An effective biocontrol bioformulation against *Phytophthora* blight of pepper using growth mixtures of combined chitinolytic bacteria under different field conditions. *Eur. J. Plant Pathol.* **2008**, *120*, 373–382. [CrossRef]

86. Singh, P.P.; Shin, Y.C.; Park, C.S.; Chung, Y.R. Biological control of Fusarium wilt of cucumber by chitinolytic bacteria. *Phytopathology* **1999**, *89*, 92–99. [CrossRef]

87. Leeman, M.; Ouder, F.M.D.; Pelt, J.A.V.; Dirk, F.P.M.; Steij, H.; Bakker, P.A.; Schippers, B. Iron availability affects induction of systemic resistance to Fusarium wilt of radishes by *Pseudomonas fluorescens*. *Phytopathology* **1996**, *86*, 149–155. [CrossRef]

88. Buysens, S.; Heungens, K.; Poppe, J.; Höftee, M. Involvement of pyochelin and pyoverdin in suppression of Pythium-induced damping-off of tomato by *Pseudomonas aeruginosa* 7NSK2. *Appl. Environ. Microbiol.* **1996**, *62*, 865–871. [CrossRef]

89. Khandelwal, S.R.; Manwar, A.V.; Chaudhari, B.L.; Chincholkar, S.B. Siderophorogenic bradyrhizobia boost yield of soybean. *Appl. Biochem. Biotechnol.* **2002**, *102*, 155–168. [CrossRef]

90. Toklikishvili, N.; Dandurishvili, M.; Tediashvili, N.; Lurie, G.S.; Szegedi, E.; Glick, B.R.; Chermin, L.N. Inhibitory effect of ACC deaminase-producing bacteria on crown gall formation in tomato plants infected by *Agrobacterium tumefaciens* or A. vitis. *Plant Pathol.* **2010**, *59*, 1023–1030. [CrossRef]

91. Niu, D.D.; Zheng, Y.; Zheng, L.; Jiang, C.H.; Zhou, D.M.; Guo, J.H. Application of PSX biocontrol preparation confers root-knot nematode management and increased fruit quality in tomato under field conditions. *Biocontrol. Sci. Technol.* **2016**, *26*, 174–180. [CrossRef]

92. Santhanam, S.; Luu, V.T.; Weinhold, A.; Goldberg, J.; Oh, Y.; Baldwin, I.T. Native root-associated bacteria rescue a plant from a sudden-wilt disease that emerged during continuous cropping. *Proc. Natl. Acad. Sci. USA* **2015**, *112*, E5013–E5020. [CrossRef] [PubMed]

93. Hunziker, L.; Bönisch, D.; Groenhagen, U.; Bailly, A.; Schulz, S.; Weisskopf, L. *Pseudomonas* strains naturally associated with potato plants produce volatiles with high potential for inhibition of *Phytophthora infestans*. *Appl. Environ. Microbiol.* **2015**, *81*, 821–830. [CrossRef] [PubMed]

94. Guyer, A.; de Vrieze, M.; Bönisch, D.; Gloor, R.; Musa, T.; Bodenhausen, N.; Bailly, A.; Weisskopf, L. The Anti-Phytophthora Effect of Selected Potato-Associated Pseudomonas Strains: From the Laboratory to the Field. *Front. Microbiol.* **2015**, *6*, 1309. [CrossRef] [PubMed]

95. New, P.B.; Kerr, A. Biological Control of Crown Gall: Field Measurements and Glasshouse Experiments. *J. Appl. Bact.* **1972**, *35*, 279–287. [CrossRef]

96. Allizadeh, H.; Behboudi, K.; Masoud, A.; Javan-Nikkhah, M.; Zamioudis, C.; Corné, M J P.; Bakker, A.H.M. Induced systemic resistance in cucumber and *Arabidopsis thaliana* by the combination of *Trichoderma harzianum* Tr6 and *Pseudomonas* sp. Ps14. *Biol. Control* **2013**, *65*, 14–23. [CrossRef]

97. Durgadevi, D.; Srivignesh, S.; Sankaralingam, A. Effect of Consortia Bioformulation of Rhizobacteria on Induction of Systemic Resistance in Tuberose against Peduncle Blight Disease. *Int. J. Bio Resour. Stress Manag.* **2018**, *9*, 510–517. [CrossRef]

98. Takishita, Y.; Charron, J.B.; Smith, D.L. Biocontrol Rhizobacterium *Pseudomonas* sp. 23S Induces Systemic Resistance in Tomato (*Solanum lycopersicum* L.) Against Bacterial Canker *Clavibacter michiganensis* subsp. michiganensis. *Front. Microbiol.* **2018**, *9*, 2119. [CrossRef]

99. Kumar, S.; Chauhan, P.S.; Agrawal, L.; Raj, R.; Srivastava, A.; Gupta, S.; Mishra, S.K.; Yadav, S.; Singh, P.C.; Raj, S.K.; et al. *Paenibacillus lentimorbus* Inoculation Enhances Tobacco Growth and Extenuates the Virulence of Cucumber mosaic virus. *PLoS ONE* **2016**, *11*, e0149980. [CrossRef]

100. Schuhegger, R.; Ihring, A.; Gantner, S.; Bahnweg, G.; Knappe, C.; Vogg, G.; Hutzler, P.; Schmid, M.; Breusegem, F.V.; Eberl, L.; et al. Induction of systemic resistance in tomato by N-acyl-L-homoserine lactone-producing rhizosphere bacteria. *Plant Cell Environ.* **2006**, *29*, 909–918. [CrossRef]

101. Baffoni, L.; Gaggia, F.; Dalanaj, N.; Prodi, A.; Nipoti, P.; Pisi, A.; Biavati, B.; Gioia, D.D. Microbial inoculants for the biocontrol of *Fusarium* spp. in durum wheat. *BMC Microbiol.* **2015**, *15*, 242. [CrossRef] [PubMed]

102. Martinuz, A.; Schouten, A.; Menjivar, R.; Sikora, R. Effectiveness of systemic resistance toward *Aphis gossypii* (Hom., Aphididae) as induced by combined applications of the endophytes *Fusarium oxysporum* Fo162 and *Rhizobium etli* G12. *Biol. Control* **2012**, *62*, 206–212. [CrossRef]

103. Valenzuela-Soto, J.H.; Estrada-Hernandez, M.G.; Ibarra-Laclette, E.; Delano-Frier, J.P. Inoculation of tomato plants (*Solanum lycopersicum*) with growth-promoting *Bacillus subtilis* retards whitefly *Bemisia tabaci* development. *Planta* **2010**, *231*, 397–410. [CrossRef] [PubMed]

104. Srivastava, S.; Bist, V.; Srivastava, S.; Singh, P.C.; Trivedi, P.K.; Asif, M.H.; Chauhan, P.S.; Nautiyal, C.S. Unraveling aspects of *Bacillus amyloliquefaciens* mediated enhanced production of rice under biotic stress of *Rhizoctonia solani*. *Front. Plant Sci.* **2016**, *7*, 587. [CrossRef] [PubMed]

105. Collins, M.; Knutti, R.; Arblaster, J.; Dufresne, L.; Fichefet, T.; Friedlingstein, P.; Gao, X.; Gutowski, W.J.; Johns, T.; Krinner, G.; et al. Long-term Climate Change: Projections, Commitments and Irreversibility. In *Climate Change 2013: The Physical Science Basis. Contribution of Working Group I to the Fifth Assessment Report of the Intergovernmental Panel on Climate Change*; Stocker, T.F., Qin, D., Plattner, G.-K., Tignor, M., Allen, S.K., Boschung, J., Nauels, A., Xia, Y., Bex, V., Midgley, P.M., Eds.; Cambridge University Press: New York, NY, USA, 2013; pp. 1029–1136.

106. Xu, D.Y.; Kang, X.W.; Zhuang, D.F.; Pan, J.J. Multi-scale quantitative assessment of the relative roles of climate change and human activities in desertification—A case study of the Ordos Plateau, China. *J. Arid Environ.* **2010**, *74*, 498–507. [CrossRef]

107. Tank, N.; Saraf, M. Salinity-resistant plant growth promoting rhizobacteria ameliorates sodium chloride stress on tomato plants. *J. Plant Interact.* **2010**, *5*, 51–58. [CrossRef]

108. Rousk, J.; Elyaagubi, F.K.; Jones, D.L.; Godbold, D.L. Bacterial salt tolerance is unrelated to soil salinity across an arid agroecosystem salinity gradient. *Soil Biol. Biochem.* **2011**, *43*, 1881–1887. [CrossRef]

109. Egamberdieva, D.; Lugtenberg, B. Use of Plant Growth-Promoting Rhizobacteria to Alleviate Salinity Stress in Plants. In *Use of Microbes for the Alleviation of Soil Stresses*; Miransari, M., Ed.; Springer Science + Business Media: New York, NY, USA, 2014; pp. 73–96. [CrossRef]

110. Shrivastava, P.; Kumar, R. Soil salinity: A serious environmental issue and plant growth promoting bacteria as one of the tools for its alleviation. *Saudi J. Biol. Sci.* **2015**, *22*, 123–131. [CrossRef]

111. Yan, N.; Marschner, P.; Cao, W.; Zuo, C.; Qin, W. Influence of salinity and water content on soil microorganisms. *Int. Soil Water Conserv. Res.* **2015**, *3*, 316–323. [CrossRef]

112. Chen, L.; Liu, Y.; Wu, G.; Njeri, K.V.; Shen, Q.; Zhang, N.; Zhang, R. Induced maize salt tolerance by rhizosphere inoculation of *Bacillus amyloliquefaciens* SQR9. *Physiol. Plant* **2016**, *158*, 34–44. [CrossRef]

113. Subramanian, P.; Mageswari, A.; Kim, K.; Lee, Y.; Sa, T. Psychrotolerant endophytic *Pseudomonas* sp. strains OB155 and OS261 induced chilling resistance in tomato plants (*Solanum lycopersicum* mill.) by activation of their antioxidant capacity. *Mol. Plant Microbe Interact.* **2015**, *28*, 1073–1081. [CrossRef] [PubMed]

114. Tiwari, S.; Lata, C.; Chauhan, S.P.; Chandra Shekhar Nautiyal, C.P. *Pseudomonas putida* attunes morphophysiological, biochemical and molecular responses in *Cicer arietinum* L. during drought stress and recovery. *Plant Physiol. Biochem.* **2016**, *99*, 108–117. [CrossRef] [PubMed]

115. Wang, Q.; Dodd, I.C.; Belimov, A.A.; Jiang, F. Rhizosphere bacteria containing 1-aminocyclopropane-1-carboxylate deaminase increase growth and photosynthesis of pea plants under salt stress by limiting Na^+ accumulation. *Funct. Plant Biol.* **2016**, *43*, 161–172. [CrossRef] [PubMed]

116. Bhartirt, N.; Pandey, S.S.; Barnawal, D.; Patel, V.K.; Kalra, A. Plant growth promoting rhizobacteria *Dietzia natronolimnaea* modulates the expression of stress responsive genes providing protection of wheat from salinity stress. *Sci. Rep.* **2016**, *6*, 34768. [CrossRef]

117. Subramanian, S.; Ricci, E.; Souleimanov, A.; Smith, D.L. A proteomic approach to lip-chitooligosaccharide and thuricin 17 effects on soybean germination under salt stress. *PLoS ONE* **2016**, *11*, e0160660. [CrossRef] [PubMed]

118. Rolli, E.; Marasco, R.; Vigani, G.; Ettoumi, B.; Mapelli, F.; Deangelis, M.L.; Gandolfi, C.; Casati, F.; Previtali, F.; Gerbino, R.; et al. Improved plant resistance to drought is promoted by the root-associated microbiome as a water stress-dependent trait. *Environ. Microbiol.* **2015**, *17*, 316–331. [CrossRef]

119. Molina-Romero, D.; Baez, A.; Quintero-Hernández, V.; Castañeda-Lucio, M.; Fuentes-Ramírez, L.E.; Bustillos-Cristales, M.D.R.; Rodríguez-Andrade, O.; Morales-García, Y.E.; Munive, A.; Muñoz-Rojas, J. Compatible bacterial mixture, tolerant to desiccation, improves maize plant growth. *PLoS ONE* **2017**, *12*, e0187913. [CrossRef]

120. Akhtar, S.S.; Amby, D.B.; Hegelund, J.N.; Fimognari, L.; Großkinsky, D.K.; Westergaard, J.C.; Muller, R.; Melba, L.; Liu, F.; Roitsch, T. *Bacillus licheniformis* FMCH001 increases water use efficiency via growth stimulation in both normal and drought conditions. *Front. Plant Sci.* **2020**, *11*, 297. [CrossRef]

121. Yang, A.; Akhtar, S.S.; Fu, Q.; Naveed, M.; Iqbal, S.; Roitsch, T.; Jacobsen, S.E. *Burkholderia Phytofirmans* PsJN Stimulate Growth and Yield of Quinoa under Salinity Stress. *Plants* **2020**, *9*, 672. [CrossRef]

122. Zahran, H.H. Rhizobium-Legume Symbiosis and Nitrogen Fixation under Severe Conditions and in an Arid Climate. *Microbiol. Mol. Biol. Rev.* **1999**, *63*, 968–989. [CrossRef]

123. Etesami, H.; Mirseyedhosseini, H.; Alikhani, H.A. Bacterial biosynthesis of 1-aminocyclopropane-1-caboxylate (ACC) deaminase, a useful trait to elongation and endophytic colonization of the roots of rice under constant flooded conditions. *Physiol. Mol. Biol. Plants* **2014**, *20*, 425–434. [CrossRef] [PubMed]

124. Kang, S.M.; Khan, A.L.; Waqs, M.; You, Y.H.; Hamayun, M.; Joo, G.; Shahzad, R.; Choi, K.; Lee, I.J. Gibberellin-producing *Serratia nematodiphila* PEJ1011 ameliorates low temperature stress in *Capsicum annuum* L. *Eur. J. Soil Biol.* **2015**, *68*, 85–93. [CrossRef]

125. Fernandez, O.; Theocharis, A.; Bordiec, S.; Feil, R.; Jacquens, L.; Clément, C.; Fontaine, F.; Barka, E.A. *Burkholderia phytofirmans* PsJN Acclimates Grapevine to Cold by Modulating Carbohydrate Metabolism. *Mol. Plant Microbe Interact.* **2012**, *25*, 496–504. [CrossRef]

126. Barnawal, D.; Bharti, N.; Maji, D.; Chanotiya, C.S.; Kalra, A. 1-Aminocyclopropane-1-carboxylic acid (ACC) deaminase containing rhizobacteria protect *Ocimum sanctum* plants during water logging stress via reduced ethylene generation. *Plant Physiol. Biochem.* **2012**, *58*, 227–235. [CrossRef] [PubMed]

127. Bano, A.; Fatima, M. Salt tolerance in *Zea mays* (L.) following inoculation with Rhizobium and Pseudomonas. *Biol. Fertil. Soils* **2009**, *45*, 405–413. [CrossRef]

128. Karthikeyan, B.; Joe, M.M.; Islam, M.D.R.; Sa, T. ACC deaminase containing diazotrophic endophytic bacteria ameliorate salt stress in *Catharanthus roseus* through reduced ethylene levels and induction of antioxidative defense systems. *Symbiosis* **2012**, *56*, 77–86. [CrossRef]

129. Mallick, I.; Bhattacharyya, C.; Mukherji, S.; Sarkar, S.C.; Mukhopadhyay, U.K.; Ghosh, A. Effective rhizoinoculation and biofilm formation by arsenic immobilizing halophilic plant growth promoting bacteria (PGPB) isolated from mangrove rhizosphere: A step towards arsenic rhizoremediation. *Sci. Total Environ.* **2018**, *610*, 1239–1250. [CrossRef] [PubMed]

130. Tripathi, P.; Singh, P.C.; Mishra, A.; Srivastava, S.; Chauhan, R.; Awasthi, S.; Mishra, S.; Dwivedi, S.; Tripathi, P.; Kalra, A.; et al. Arsenic tolerant *Trichoderma* sp. reduces arsenic induced stress in chickpea (*Cicer arietinum*). *Environ. Pollut.* **2017**, *223*, 137–145. [CrossRef]

131. Sánchez-Yáñez, J.M.; Alonso-Bravo, J.N.; Dasgupta-Schuber, N.; Márquez-Benavides, L. Bioremediation of soil contaminated by waste motor oil in 55,000 and 65,000 and phytoremediation by *Sorghum bicolor* inoculated with *Burkholderia cepacia* and *Penicillium chrysogenum*. *J. Selva Andina Biosph.* **2015**, *3*, 86–94.

132. Sarkar, J.; Chakraborty, B.; Chakraborty, U. Plant Growth Promoting Rhizobacteria Protect Wheat Plants Against Temperature Stress Through Antioxidant Signalling and Reducing Chloroplast and Membrane Injury. *J. Plant Growth Regul.* **2018**, *37*, 1396–1412. [CrossRef]

133. Islam, F.; Yasmeen, T.; Ali, O.; Ali, S.; Arif, S.M.; Sabir Hussain, S.; Rizv, H. Influence of *Pseudomonas aeruginosa* as PGPR on oxidative stress tolerance in wheat under Zn stress. *Ecotoxicol. Environ. Saf.* **2014**, *104*, 285–293. [CrossRef] [PubMed]

134. Berninger, T.; Lopez, O.G.; Bejarano, A.; Preininger, C.; Sessitsch, A. Maintenance and assessment of cell viability in formulation of non-sporulating bacterial inoculants. *Microb. Biotechnol.* **2018**, *11*, 277–301. [CrossRef]

135. Mehnaz, S. An Overview of Globally Available Bioformulations. In *Bioformulations: For Sustainable Agriculture*; Arora, N., Mehnaz, S., Balestrini, R., Eds.; Springer: New Delhi, India, 2016; pp. 267–281. [CrossRef]

136. Arthur, S.; Dara, S.K. Microbial biopesticides for invertebrate pests and their markets in the United States. *J. Invertebr. Pathol.* **2018**, *165*, 13–21. [CrossRef]

137. Babalola, O.O.; Glick, B.R. Indigenous African agriculture and plant associated microbes: Current practice and future transgenic prospects. *Sci. Res. Essays* **2012**, *7*, 2431–2439.

138. Aremu, B.R.; Alori, E.T.; Kutu, R.F.; Babalola, O.O. Potentials of Microbial Inoculants in Soil Productivity: An Outlook on African Legumes. In *Microorganisms for Green Revolution. Microorganisms for Sustainability*; Panpatte, D., Jhala, Y., Vyas, R., Shelat, H., Eds.; Springer: Singapore, 2017; Volume 6. [CrossRef]

139. Souleimanov, A.; Prithiviraj, B.; Smith, D. The major Nod factor of *Bradyrhizobium japonicum* promotes early growth of soybean and corn. *J. Exp. Bot.* **2002**, *53*, 1929–1934. [CrossRef] [PubMed]

140. Gray, E.J.; Lee, K.D.; Souleimanov, A.M.; Di Falco, M.R.; Zhou, X.; Ly, A.; Charles, T.C.; Driscoll, B.T.; Smith, D.L. A novel bacteriocin, thuricin 17, produced by plant growth promoting rhizobacteria strain *Bacillus thuringiensis* NEB17: Isolation and classification. *J. Appl. Microbiol.* **2006**, *100*, 545–554. [CrossRef]

141. Subramanian, S.; Souleimanov, A.; Smith, D.L. Proteomic Studies on the Effects of Lipochitooligosaccharide and Thuricin 17 under Unstressed and Salt Stressed Conditions in *Arabidopsis thaliana*. *Front. Plant Sci.* **2016**, *7*, 1314. [CrossRef]

142. Arunachalam, S.; Schwinghamer, T.; Dutilleul, P.; Smith, D.L. Multi-Year Effects of Biochar, Lipo-Chitooligosaccharide, Thuricin 17, and Experimental Bio-Fertilizer for Switchgrass. *Agron. J.* **2018**, *110*, 77–84. [CrossRef]

143. Navarro, M.O.P.; Piva, A.C.M.; Simionato, A.S.; Spago, F.R.; Modolon, F.; Emiliano, J.; Azul, A.M.; Chryssafidis, A.L.; Andrade, G. Bioactive Compounds Produced by Biocontrol Agents Driving Plant Health. In *Microbiome in Plant Health and Disease*; Kumar, V., Prasad, R., Kumar, M., Choudhary, D., Eds.; Springer: Singapore, 2019; pp. 337–374. [CrossRef]

144. Prudent, M.; Salon, C.; Souleimanov, S.A.; Emery, R.J.N.; Smith, D.L. Soybean is less impacted by water stress using *Bradyrhizobium japonicum* and thuricin-17 from *Bacillus thuringiensis*. *Agron. Sustain. Dev.* **2015**, *35*, 749–757. [CrossRef]

145. Schwinghamer, T.; Souleimanov, A.; Dutilleul, P.; Smith, D.L. Supplementation with solutions of lipo-chitooligosaccharide Nod Bj V (C18:1, MeFuc) and thuricin 17 regulates leaf arrangement, biomass, and root development of canola (*Brassica napus* [L.]). *Plant Growth Regul.* **2015**, *78*, 31–41. [CrossRef]

146. Yuan, L.; Li, Y.; Wang, Y.; Zhang, X.; Xu, Y. Optimization of critical medium components using response surface methodology for phenazine 1 carboxylic acid production by *Pseudomonas* sp. M-18Q. *J. Biosci. Bioeng.* **2008**, *3*, 232–237. [CrossRef] [PubMed]

147. Xu, S.; Pan, X.; Luo, J.; Wu, J.; Zhou, Z.; Liang, X.; He, Y.; Zhou, M. Effects of phenazine-1-carboxylic acid on the biology of the plant-pathogenic bacterium *Xanthomonas oryzae* pv. Oryzae. *Pestic. Biochem. Physiol.* **2015**, *117*, 39–46. [CrossRef] [PubMed]

148. Chen, C.; Mciver, J.; Yang, Y.; Bai, Y.; Schultz, B.; Mciver, A. Foliar application of lipochitooligosaccharides (Nod factors) to tomato (*Lycopersicon esculentum*) enhances flowering and fruit production. *Can. J. Plant Sci.* **2007**, *87*, 365–372. [CrossRef]

149. Duke, S.O.; Lydon, J. Herbicides from Natural Compounds. *Weed Technol.* **1987**, *1*, 122–128. [CrossRef]

150. Zhang, Z.K.; Huber, D.J.; Qu, H.X.; Yun, Z.; Wang, H.; Huang, Z.H.; Huang, H.; Jiang, Y.M. Enzymatic browning and antioxidant activities in harvested litchi fruit as influenced by apple polyphenols. *Food Chem.* **2015**, *171*, 191–199. [CrossRef]
151. Shanmugaiah, V.; Mathivanan, N.; Varghese, B. Purification, crystal structure and antimicrobial activity of phenazine-1-carboxamide produced by a growth-promoting biocontrol bacterium, *Pseudomonas aeruginosa* MML2212. *J. Appl. Microbiol.* **2010**, *108*, 703–711. [CrossRef]
152. Huang, H.; Sun, L.; Bi, K.; Zhong, G.; Hu, M. The effect of phenazine-1-carboxylic acid on the morphological, physiological, and molecular characteristics of *Phellinus noxius*. *Molecules* **2016**, *21*, 613. [CrossRef]
153. Puopolo, G.; Masi, M.; Raio, A.; Andolfi, A.; Zoina, A.; Cimmino, A.; Evidente, A. Insights on the susceptibility of plant pathogenic fungi to phenazine-1-carboxylic acid and its chemical derivatives. *Nat. Prod. Res.* **2013**, *27*, 956–966. [CrossRef]
154. Gheorghe, I.; Popa, M.; Marutescu, L.; Saviuc, C.; Lazar, V.; Chifiriuc, M.C. Lessons from interregn communication for development of novel, ecofriendly pesticides. In *New Pesticides and Soil Sensors*; Grumezescu, A.M., Ed.; Academic Press: London, UK, 2017; pp. 1–46. [CrossRef]
155. Rane, M.R.; Sarode, P.D.; Chaudhari, B.L.; Chincholkar, S.B. Exploring antagonistic metabolites of established biocontrol agent of marine origin. *Appl. Biochem. Biotechnol.* **2008**, *151*, 665–675. [CrossRef]
156. Kare, E.; Arora, N.K. Dual activity of pyocyanin from *Pseudomonas aeruginosa*—Antibiotic against phytopathogen and signal molecule for biofilm development by rhizobia. *Can. J. Microbiol.* **2011**, *57*, 708–713. [CrossRef]
157. Jung, B.K.; Hong, S.J.; Park, G.S.; Kim, M.C.; Shin, J.H. Isolation of *Burkholderia cepacia* JBK9 with plant growth–promoting activity while producing pyrrolnitrin antagonistic to plant fungal diseases. *Appl. Biol. Chem.* **2018**, *61*, 173–180. [CrossRef]
158. Okada, A.; Banno, S.; Ichiishi, A.; Kimura, M.; Yamaguchi, I.; Fujimura, M. Pyrrolnitrin interferes with osmotic signal transduction in *Neurospora crassa*. *J. Pestic. Sci.* **2005**, *30*, 378–383. [CrossRef]
159. Han, J.W.; Kim, J.D.; Lee, J.M.; Ham, J.H.; Lee, D.; Kim, B.S. Structural elucidation and antimicrobial activity of new phencomycin derivatives isolated from *Burkholderia glumae* strain 411gr6. *J. Antibiot.* **2014**, *67*, 721. [CrossRef]
160. Deng, P.; Foxfire, A.; Xu, J.; Baird, S.M.; Jia, J.; Delgado, K.H.; Shin, R.; Smith, L.; Lu, S.E. Siderophore product ornibactin is required for the bactericidal activity of Burkholderia contaminans MS14. *Appl. Environ. Microbiol.* **2017**, *83*, e00051-17. [CrossRef]
161. Ye, Y.; Li, Q.; Fu, G.; Yuan, G.; Miao, J.; Lin, W. Identification of antifungal substance (Iturin A2) produced by *Bacillus subtilis* B47 and its effect on southern corn leaf blight. *J. Integr. Agric.* **2012**, *11*, 90–99. [CrossRef]
162. Deravel, J.; Lemière, S.; Coutte, F.; Krier, F.; Hese, N.V.; Béchet, M.; Sourdeau, N.; Höfte, M.; Leprêtre, A.; Jacques, P. Mycosubtilin and surfactin are efficient, low ecotoxicity molecules for the biocontrol of lettuce downy mildew. *Appl. Microbiol. Biotechnol.* **2014**, *98*, 6255–6264. [CrossRef]
163. Isaac, B.G.; Ayer, S.W.; Elliott, R.C.; Stonard, R.J. Herboxidiene: A potent phytotoxic polyketide from Streptomyces sp. A7847. *J. Org. Chem.* **1992**, *57*, 7220–7226. [CrossRef]
164. Saxena, S.; Pandey, A.K. Microbial metabolites as eco-friendly agrochemicals for the next millennium. *Appl. Microbiol. Biotechnol.* **2001**, *55*, 395–403. [CrossRef]
165. Tanaka, Y.; Omura, S. Agroactive compounds of microbial origin. *Annu. Rev. Microbiol.* **1993**, *47*, 57–87. [CrossRef]

Article

An Assessment of Seaweed Extracts: Innovation for Sustainable Agriculture

El Chami Daniel and Galli Fabio *

Timac Agro Italia S.p.A., S.P.13—Località Ca' Nova, I-26010 Ripalta Arpina (CR), Italy; daniel.elchami@roullier.com
* Correspondence: fabio.galli@roullier.com; Tel.: +39-0373-669-111; Fax: +39-0373-669-291

Received: 19 August 2020; Accepted: 19 September 2020; Published: 21 September 2020

Abstract: Plant growth regulators (PGRs) are described in the literature as having a significant role in securing crop management of modern agriculture in conditions of abiotic and biotic stressors. A joint field experiment was carried out to assess the role of seaweed-based extracts in pear trees and to test the "less for more" theory, which consists of getting more and better agricultural produce using fewer innovative inputs. The trials took place on two production seasons (from March till September 2018–2019) and the selected case study was on a pear orchard (*Pyrus communis* L. cv. Abate Fètel) in Emilia Romagna (Italy) by Fondazione Navarra and Timac Agro Italia S.p.A. Results demonstrate that, depending on the yearly climate conditions, it was possible to substantially reduce the primary nutrients by 35–46% and total fertilisation units applied by 13% and significantly improve quantitative and qualitative production indicators (average weight of fruits (5%) and total yield (19–55%)). Results also confirm a positive correlation between plant growth regulators and agronomic efficiency of pears which increased between five and nine times compared to the conventional nutrition programme. These outcomes constitute scientific evidence for decision making in farm management.

Keywords: pear trees; PGR; sustainable development; crop nutrition; fertiliser; Timac Agro Italia

1. Introduction

For decades, plant nutrition has been under scrutiny for concerns about negative externalities generated by the use of fertilisers in agriculture, which emerged in the late 1960s [1]. Since then, a clear correlation has been found between plant nutrition, the eutrophication of surface water, the accumulation of nitrate in water bodies and energy consumption. Even more recently, global studies have warned about unprecedented nitrate contamination of water [2], which is creating direct irreversible damage to natural ecosystems and human health [3]. Further, the most universal form of deteriorated water quality in the world in recent decades is freshwater eutrophication from phosphorus loss [4,5].

Looking at the glass half-full, the importance of fertilisers in agriculture has been extensively documented in the literature for over 150 years of research and experiments. The relevance of plant nutrition is fundamental for (i) normal growth and reproduction of crops [6], (ii) average crop yield increase [7] and (iii) improving soil fertility [8]. However, the fertilising rates have reached the optimum in the developed world, and the new direction is to reduce them. This has been one of the European Green Deal recommendations, for example, as expressed by the "farm to fork" strategy (The Farm to Fork (F2F) Strategy is at the heart of the European Green Deal set out in 2019 to make Europe the first climate-neutral continent by 2050. The strategy comprehensively addresses the challenges of sustainable food systems and recognises the inextricable links between healthy people, healthy societies and a healthy planet.) with a target of diminishing nutrient losses by at least 50% and reducing fertiliser use by at least 20% by 2030 [9].

The focus of scientific innovation is currently on crop biostimulants to activate natural plant processes, which, according to the documented literature, improve nutrient uptake and efficiency, crop quality and yield and build plant tolerance to abiotic and biotic stressors [10–12]. A statutory definition of biostimulants was provided in 2018 by the primary agricultural and food policy tool of the United States federal government (Farm Bill: https://www.congress.gov/115/bills/hr2/BILLS-115hr2enr.pdf). This definition is consistent with the one currently proposed by the European Bio-stimulant Industry Council (EBIC) (http://www.bio-stimulants.eu/) and in line with the definition under review by the European Union in the context of revising the existing EU regulation (EC) No. 2019/1009 relating to fertilisers (https://eur-lex.europa.eu/legal-content/EN/TXT/HTML/?uri=OJ:L:2019:170:FULL&from=EN).

The definition sums up the scientific aspects raised in the literature and describes a plant biostimulant as "a substance or micro-organism that, when applied to seeds, plants, or the rhizosphere, stimulates natural processes to enhance or benefit nutrient uptake, nutrient efficiency, tolerance to abiotic stress, or crop quality and yield".

Nevertheless, du Jardin [13] identified in a review study seven categories of biostimulants: (i) humic and fulvic acids, (ii) protein hydrolysates and other N-containing compounds, (iii) seaweed extracts and botanicals, (iv) chitosan and other biopolymers, (v) inorganic compounds, (vi) beneficial fungi and (vii) beneficial bacteria. This emerging field of research is very promising and represents one of the fundamental management aspects of agro-systems to reach sustainable agriculture that is more resilient to climate change and able to feed the increasing population [14].

Therefore, the literature still needs to explore different research aspects related to the biostimulant categories and their use in agriculture to answer evolving enquiries that arise with the technological advances in this field. Recently, algae have proved to contain natural active compounds with biostimulation and/or bioregulation effects [15], e.g., phytohormones, hormone-like substances, vitamins, antibiotics, amino acids, and primary, secondary and micro-nutrients.

Even though several forms of applying algal constituents have been reported in the literature, algal extracts from seaweed have proved to be the most efficient in terms of growth enhancement and stress tolerance [16–18]. Indeed, Mutale-joan et al. [16] have tested 15 different Crude Bio-Extracts (CBEs) obtained from acid hydrolysis of microalgae on tomato plant growth, chlorophyll content, nutrient uptake and metabolite profile. The authors have recorded positive effects on plant development, particularly, significant root and shoot length improvement and increased nutrients uptake. Further, Shukla et al. [17] have reviewed the ability of Ascophyllum Nodosum Extracts (ANE) to improve plant growth and agricultural productivity and have confirmed the plant growth promotion, the improvement of root/microbe interactions and nutrient use efficiency in plants and enhancement of plant tolerance to abiotic and/or biotic stresses. Finally, Michalak et al. [18] have successfully tested different seaweed extracts to enhance carotenoid and chlorophyll content in plant shoots and develop root thickness and above-ground biomass.

In this context, this paper proposes to explore the category of seaweed extracts produced by Timac Agro Italia, the Italian holding of the French multinational Groupe Roullier, a world leader in the field of plant nutrition, with the largest private research centre in Europe dedicated to plant physiology and nutrition and investing in these technologies. The selection of the trial crop also has significance, because pears are one of the major fruits of temperate climates, grown in almost all four corners of the world, reaching a total harvested area of 1.5 million hectares in 2018 and over 23.5 million tons of production according to FAOSTAT [19] (Figure 1). The tree belongs to two species: the common pear cultivated mainly in Europe, the Near East, America and Australia and known as the European pear (*Pyrus communis L.*) given its European descendants and the Nashi or Oriental pear (*Pyrus pyrifolia*) widely grown in Asia.

Area Harvested (% of total hectares)

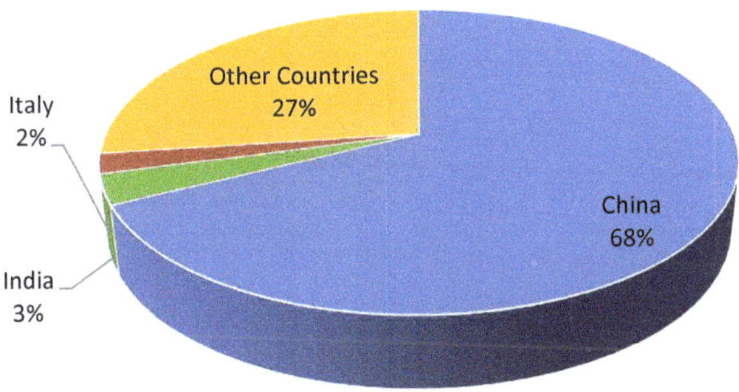

Total Production (% of total tons)

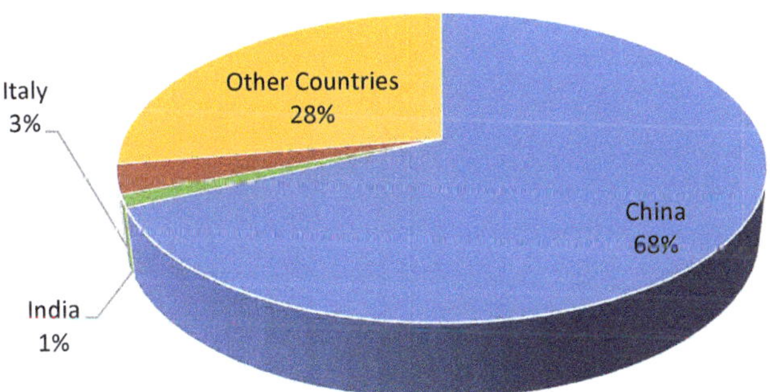

Figure 1. Harvested area of pear (**top**) and total production (**bottom**) in the top 3 countries and the world.

2. Materials and Methods

The experiment was carried out in Emilia Romagna (Italy) at an Abate Fetel orchard for the relevance of this cultivar in Italy, which happens to be the main producer of pears in Europe [20], the third producer in the world in terms of area harvest and the second after China in terms of total production.

There are over 3000 identified pear cultivars worldwide [21], and in Italy, Abate Fetel (also known as Abbé Fetel) and 3 others (Conference, Beurrè Bosc, Doyenne du Comice) are the major cultivars commercially grown, providing more than 70% of the total annual production [22]. Further, the selection of Emilia Romagna has local importance, given that this region, in terms of fruit trees, is the first ranked in harvest area, production and the average size of farms (Table 1).

Table 1. Top 5 regions in fruit production and farm size in Italy.

Region	Area Harvested		Number of Farms	
	(ha)	**(%)**	**(N)**	**(%)**
Emilia Romagna (EMR)	67,454.3	15.9	18,355	7.8
Campania (CAM)	58,836.7	13.9	32,133	13.6
Sicily (SIC)	54,295.5	12.8	36,055	15.3
Piedmont (PIE)	43,673.3	10.3	20,168	8.5
Lazio (LAZ)	36,318.8	8.6	15,323	6.5
Sum of top 5 regions	260,578.5	61.4	122,034	51.7
Other regions	163,725	38.6	114,206	48.3
Total Italian fruit farms	424,303.5	100	236,240	100
Average Italian fruit farm size	1.8 ha			
Average fruit farm size in EMR	3.7 ha			
Farm size	Area Harvested		Number of Farms	
	(ha)	**(%)**	**(N)**	**(%)**
Small (<10 ha)	222,270.4	52.4	201,324	85.2
Medium (10–50 ha)	150,171.9	35.4	30,674	13.0
Large (>50 ha)	51,861.5	12.2	4242	1.8
Total Italian fruit farms	424,303.8	100	236,240	100

Source: Istat [23]

2.1. Case Study

The experiment took place at the experimental field of the Navarra Foundation, a reference of agricultural knowledge for the Navarra agricultural technical institute and farmers of the north-east of Italy, given its contribution to the development of the region's agri-food sector through research, experiments, innovation and knowledge transfer.

The experimental field has a total area of ≈2.5 ha, similar to the average size of fruit tree farms in the area, and is located in Ferrara (Table 2), characterised by a warm and temperate climate classified as Cfa by the Köppen-Geiger system. The historic precipitation and temperature measured at a weather station in Ferrara (Table 2) between 1961 and 1990 revealed a yearly average temperature of 13.1 °C and rainfall of around 689.5 mm [24], with considerable rain at high temperatures in the driest months (Figure 2).

The trials took place over two consecutive seasons in 2018–2019 on a V-shaped orchard system planted in 2005 using 3.8 m spacing between rows and variable in-row spacing of 0.5 m, with a tree density of 5263 trees per hectare. The orchard was evenly irrigated with a drip system without any variation between the rows and was covered with black anti-hail netting.

The soil structure is silty clay loam according to the classification system of the United States Department of Agriculture (USDA) and silt clay according to the International Soil Sciences Society (ISSS). The general composition of the soil is about 60% silt, 30% clay and 10% sand; it presents low compaction risk, high fertility indicators (organic matter 2.21%, C/N = 8.87) and a high content of available nutrients, given that the field is experimental with continuous trials carried out yearly. Soil tests were carried out before and during the experiment to guide the definitions of annual fertilisation programmes.

Table 2. Geospatial coordinates of experimental field and weather station of reference.

Location	Latitude	Longitude	Altitude
Experimental field	44.857 N	11.653 E	5 m
Weather station	44.861 N	11.656 E	4 m

Source: [24,25].

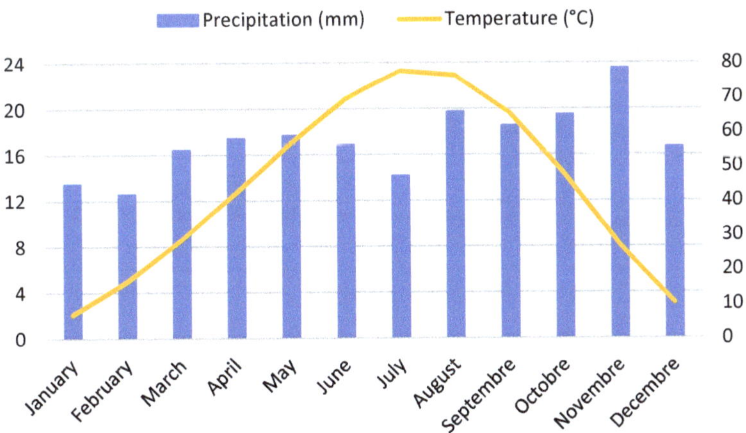

Figure 2. Historic average of precipitation and temperature in Ferrara (1961–1990) (after [24]).

The fertilisation programme was divided into 3 treatments: control, grown without any fertilisation; conventional treatment (CF), representing empirical nutritional treatment (primary, secondary and micro-nutrients and organic matter of animal or vegetable origin) conceived from the available products in the region (to simulate a conventional nutritional programme); and Timac Agro treatment (TIMAC), corresponding to a programme based on integrating conventional nutrients (primary, secondary and micro-nutrients and organic matter of animal or vegetable origin) with cutting-edge technologies created to reduce the environmental burden of fertilisers, increase the profit of farms and improve their well-being. The total fertilisation units per hectare for each treatment are reported in Table 3, which shows a great difference in the quantity of fertiliser applied in the 2 years. This difference is mainly due to the general climate conditions during the year, which considerably determines the quality and quantity of agricultural yield.

Table 3. Fertilisation units (FU) applied per treatment and hectare in 2018 and 2019 seasons.

Type	Element	2018 FU ha^{-1}			2019 FU ha^{-1}		
		CF	TIMAC	*TIMAC/CF*	CF	TIMAC	*TIMAC/CF*
	Nitrogen (N)	205.1	141.3	**68.9%**	174.9	120.7	**69.0%**
Primary Nutrients	Phosphorus (P$_2$O$_5$)	184.4	79.0	**42.8%**	103.1	77.6	**75.3%**
	Potassium (K$_2$O)	292.7	145.1	**49.6%**	246.0	140.8	**57.2%**
	Total Primary Nutrients	**682.2**	**365.4**	**53.6%**	**524**	**339.1**	**64.7%**
	Calcium (CaO)	42.5	46.4	**109.2%**	4.8	43.2	**900.0%**
Secondary Nutrients	Magnesium (MgO)	21.6	24.1	**111.6%**	3.0	35.8	**1193.3%**
	Sulphur (SO$_3$)	109	200.2	**183.7%**	49.5	197.6	**399.2%**
	Total Secondary Nutrients	**173.1**	**270.7**	**156.4%**	**57.3**	**276.6**	**482.7%**
	Boron (B)	0.45	0.88	**197.5%**	0	1.45	–
	Copper (Cu)	0.43	0.06	**13.9%**	0	0.05	–
	Iron (Fe)	5.83	3.90	**66.9%**	2.25	1.50	**66.7%**
Micro-Nutrients	Manganese (Mn)	0.24	0	–	0.03	0.07	**233.3%**
	Molybdenum (Mo)	0.01	0.04	**655.7%**	0	0.30	–
	Zinc (Zn)	0.31	0.10	**32.3%**	0.02	0.11	**550.0%**
	Total Micro-Nutrients	**7.26**	**4.98**	**68.6%**	**2.30**	**3.48**	**151.3%**
OM	**Total Organic Matter**	**43.8**	**44.4**	**101.3%**	**48.7**	**41.2**	**106.1%**

A supplement of complexed seaweed-based extracts was added to the TIMAC treatment (58 L ha^{-1}) in different growth stages and concentrations (Table 4) of three technologies: Fertiactyl®, NMX®, Seactiv®. These technologies are registered in the European Patent Office (EPO) under the numbers EP0609168, EP1147706 and EP0855375, respectively. The species involved in the extraction are *Lithothamnion corallioides*, *Lithothamnion glaciale*, *Lithothamnion tophiforme* and *Phymatolithon calcareum* and the general composition of the technologies are as follows:

Table 4. Quantity of seaweed-based technologies applied in Timac Agro treatment (TIMAC) treatment and corresponding growth stage.

Technology	Quantity (L ha^{-1})	Growth Stage
Fertiactyl®	8	
NMX®	3	Vegetative growth
Seactiv®	3	
NMX®	6	
Seactiv®	6	Fruit set
Seactiv®	3	Post-harvest

Fertiactyl®: based on organic substrates acids whose fulvic and humic acids have been mobilized through the formation of soluble inorganic salts, phenolic and polyphenolic acids and zeatin;

NMX®: based on precursor, inhibitor and simulating enzymes (e.g., precursor compound of cyclic AMP; inhibitor compound of the enzymes of the Phosphodiesterases, stimulating compound of the enzymes of the Adenyl-Cyclase, etc.), mixed with mineral fertilisers and plant growth regulators (PGRs) (e.g., auxins, cytokinins, gibberellins, n-ethanolamines, polyamines, sugars, etc.);

Seactiv®: based on adenine derivatives to help, as foliar treatment agent, the migration and distribution within plants of nitrogenous nutrients and/or micro-nutrients.

The concentration in the different stages did not vary between 2018 and 2019.

2.2. Statistical Analysis

The experimental design was a randomised block design to minimise the effects of systematic errors. This design consisted of dividing the experimental block into 3 fertilisation treatments randomly selected within the block, with 2 replicates of 5 trees for each treatment. In total, 60 trees were used for data collection (all performed manually) and statistical analysis to determine whether mean scores differed significantly across the treatments. The measurements performed were divided into 3 pillars as follows:

Total harvest (t ha^{-1});

Average fruit weight (g);

$$\text{Flower density} = \text{FBT} = \text{Number of Floral Buds per Tree} \tag{1}$$

$$\text{Fruit density} = \text{NFT} = \text{Number of Fruits per Tree} \tag{2}$$

$$\text{Fruit set} = \text{FS} = (\text{NFT}/\text{FBT}) \times 100 \tag{3}$$

$$\begin{aligned} &\text{Agronomic efficiency AE} \left(\text{kg kg}^{-1}\right) \\ &= \text{Yield}_{\text{fertilised}} \left(\text{kg ha}^{-1}\right) - \text{Yield}_{\text{not fertilised}} \left(\text{kg ha}^{-1}\right)/\text{N Applied} \left(\text{kg ha}^{-1}\right) \end{aligned} \tag{4}$$

The collected field data were statistically examined, separately for each year, using analysis of variance (one-way ANOVA) with statistical probability ($p - value \le 0.05$) and Tukey's honestly significant difference (HSD) test, which is a single-step multiple comparison procedure to find significantly different means. An excel spreadsheet has been used for this purpose.

The assumptions of both tests are essentially the same and they are three: normality (experimental errors of the data are normally distributed), homogeneity (equal variances between treatments) and independence (each sample is randomly selected and independent).

3. Results

Plants within a population often vary in the numbers of open flowers and fruits. The correlation between these two indicators is calculated by the fruit set, a ratio defined as the transition from flower to young fruit. These quantitative indicators in the development process of any plant are correlated to the rate of pollination [26] and determine the final yield quantities (or total harvest).

Field data for two consecutive years demonstrated an increase in all quantitative indicators under the TIMAC treatment compared to the conventional treatment and the control, which generated the highest harvest for TIMAC treatment. The literature has mentioned different abiotic and biotic stressors which could lead to flower drop in the early season [27]. In this experiment, the difference between the two years of trials is substantial and can be explained by the yearly climate variability in Italy. Indeed, 2019 was a dry winter with many clear-sky mornings generating diamond dust and leading to frost flowers, this generated a lower flower density trend in 2019 compared to 2018. Yet, TIMAC treatment has participated in avoiding further damage with higher flower density compared to the other treatment.

Even though the numerical difference is considerable, statistical significance is present only between the control and TIMAC treatments. Complete statistical results are listed in a final table (Table 5).

Table 5. Statistical results of selected indicators.

Indicator	Treatment	First Year, 2018			Second Year, 2019		
		Mean	Std Dev.	Variance	Mean	Std Dev.	Variance
Flower Density	Control	223.5 [a]	25.4	681.1	61.0 [a]	33.9	1211.0
	CF	219.4 [a]	28.7	865.8	53.1 [a]	28.3	844.6
	TIMAC	223.6 [a]	27.7	807.6	74.0 [a]	36.3	1384.2
Fruit Density	Control	38.2 [a]	10.5	115.4	11.1 [a]	6.1	39.8
	CF	42.4 [a,b]	18.1	343.2	13.9 [a]	9.4	93.7
	TIMAC	50.4 [b]	17.1	308.4	25.1 [b]	11.4	136.2
Fruit Set	Control	17.4 [a]	5.3	29.4	25.2 [a]	22.9	551.6
	CF	20.1 [a]	10.2	109.5	30.1 [a]	17.7	328.1
	TIMAC	22.3 [a]	6.2	40.9	37.9 [a]	18.7	368.0
Fruit Weight	Control	238.9 [a]	29.0	882.9	225.2 [a]	25.5	684.6
	CF	237.1 [a]	31.8	1064.4	228.6 [a]	24.7	643.9
	TIMAC	242.7 [a]	24.5	633.8	230.6 [a]	30.1	952.4
Total Harvest	Control	47.5 [a]	13.1	180.7	13.5 [a]	8.1	69.7
	CF	51.6 [a,b]	19.8	413.5	16.1 [a]	9.9	104.0
	TIMAC	63.3 [b]	19.1	384.6	29.8 [b]	12.7	169.3

[a,b] Letters indicate statistically significant means of different indicators.

However, high plant fertility would negatively affect the quality of the harvest, which, in conventional agriculture, would require flower removal intervention to boost the quality [28–30]. The role of plant regulators has been limited mainly to this growth stage of plants to control fruit set [31] and simultaneously boost the quantity and quality of yield [32].

While the average fruit weight of different pear cultivars can vary according to the genetic characteristics [33], within the same variety, fruit fresh weight is considered to be one of the most important quality indicators [34] determining the market value of the harvest. The effect of climate variability between 2018 and 2019, which initially generated a reduction in fruit density, led to a drop in the total harvest; however, the yield gap between the two treatments (CF vs. TIMAC) confirms the role of plant growth regulators (PGRs) in reducing crop response to abiotic stress. Further, fruit quality did not undergo any variation. Figure 3 shows the qualitative and quantitative improvement of yield in the TIMAC treatment compared to the control and CF.

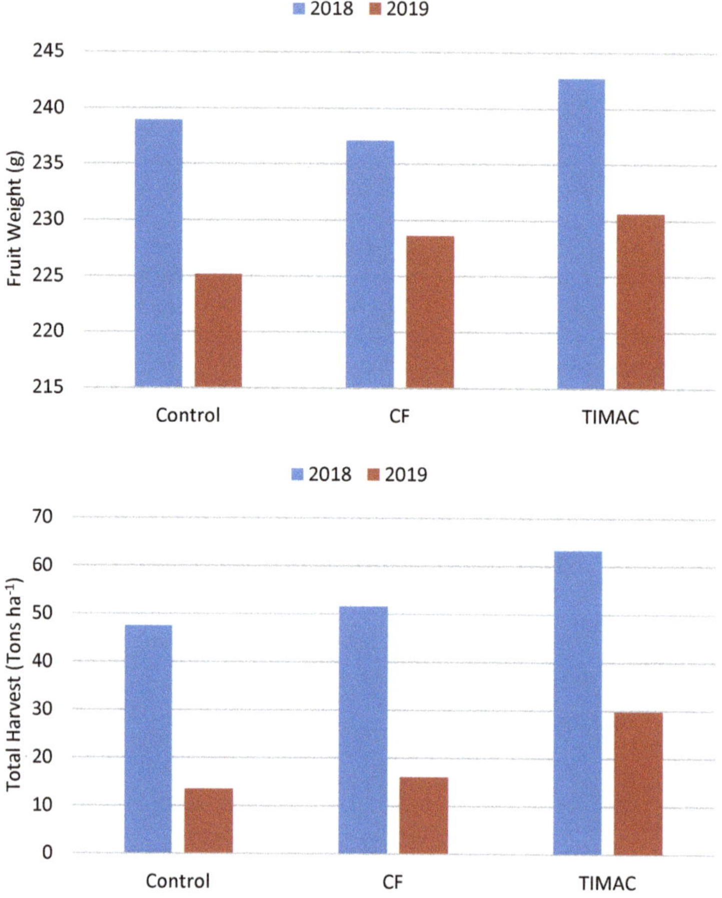

Figure 3. Improvement of fruit weight (**top**) and total harvest (**bottom**) under TIMAC treatment.

The outcomes of this experiment confirm the results of An et al. [31], showing a positive correlation between phytohormones application (auxin and ethylene) and fruit quantity and quality, and Bons and Kaur [32], who reviewed the effects of different PGRs on fruit set. The review included the impacts of 6-Benzyladenine (BA), 6-Benzylaminopurine (BAP), Gibberellic acid (GA3) and Naphthalene acetic acid (NAA).

Nutrient agronomic efficiency measures the technical performance of a crop and is calculated based on the yield difference between fertilised and unfertilised treatments, divided by the fertiliser units applied. Specifically, AE estimates productivity improvement gained by the use of nutrient input. Initially, it was used to evaluate nitrogen performance [35,36] and to improve the environmental and economic performance of agriculture, then it was extended to include the performance of phosphorus nutrition [37] to have a broader meaning and use agronomic efficiency correlated inputs for agro-system performance as an indicator of a transition to sustainable agriculture [38].

In this study, we calculated nitrogen, phosphorus and total nutrient efficiency for conventional fertilisation and TIMAC treatments (Table 6). The results show that the efficiency of the TIMAC treatment varied from 5.18 to 9.37 times higher than the conventional treatment (CF).

Table 6. Nutrient agronomic efficiency of different treatments during experimental years.

Year	Treatment	$AE_{(N)}$	$AE_{(P_2O_5)}$	$AE_{(Total)}$
	CF	20.1	22.3	4.8
2018	TIMAC	112.1	200.4	24.7
	TIMAC/CF	**5.58**	**8.98**	**5.18**
	CF	14.4	24.4	4.3
2019	TIMAC	134.9	209.8	26.3
	TIMAC/CF	**9.37**	**8.59**	**6.10**

4. Discussion

The results over two consecutive years of experiments show the role of PGRs in increasing crop tolerance to abiotic stress and improving physiological activities such as nutrient uptake and assimilation, reducing the total fertilisation units (FU) to around 13% ($\approx$35.6 FU), which is an encouraging outcome towards the reduction of fertilisers according to the European Farm-to-Fork Strategy. Extrapolating this result to the total area of pears cultivated worldwide ($\approx$1.4 million ha), this means saving a minimum of 49.8 million fertilisation units annually, without considering overfertilisation still practised in many countries around the world.

The results also reveal a substantial reduction in P_2O_5 use (over 45%), which is a significant result impacting the AE of phosphorus, and global efforts to reduce and/or substitute the use of phosphate rock, a mineral fundamental for food security, is expected to end in a short time [39,40]. The agronomic efficiency was higher in the second experimental, which presented higher abiotic stress due to climate conditions, affecting total yield. This confirms studies describing the important role of biostimulants as abiotic stress alleviators [41,42].

The field experiment confirmed the reviewed literature in Bons and Kaur [32], which assessed the positive correlation between plant growth regulators and the quality and quantity of harvests, as the TIMAC treatment improved both the quality and quantity of pears. Therefore, these results disproved the results of Dicenta et al. [43], which did not show a correlation between fruit set and total harvest.

There is not a consensus in the scientific community on one process to explain this because the mechanisms underlining seaweed extract-induced stimulation are still not completely revealed, though, several factors can be attributed to their activity. Literature has demonstrated that components within seaweed extracts may modulate innate pathways for the biosynthesis of phytohormones in plants [44]. It is also proved that marine bioactive substances (IPA extracts) enhance nutrient flux in [45] and increase endogenous antioxidant activity in plants [46]. Furthermore, the presence of phytohormones in seaweed extracts, as regulators of various cellular processes and responses, is presumably another factor that may have improved metabolic activity of crops [47] leading to better nutrition and yields.

Some questions that the research has raised and some future recommendations are mainly related to the importance of a balanced nutrition programme for sustainable management of crops. This can be defined by Liebig's law of the minimum, which is a fundamental principle in plant nutrition. This research partially demonstrates the importance of this law to the overall agronomic efficiency of crops (AE was not assessed in this study). Furthermore, it is recommended to follow the framework suggested by El Chami et al. [14], who proposed a methodology to reached sustainable agro-systems based on a life cycle study [48]; future studies will be intensified and address these questions, and will implement that methodology towards fulfilling the European Farm-to-Fork Strategy and the United Nations Sustainable Development Goals.

5. Conclusions

In conclusion, the experiment has confirmed the role of PGRs in reducing the total fertilisation units by 13% while improving the harvest between 19 and 55% and the quality of the fruit by about 5%. This improvement has positively affected the agronomic efficiency from five to nine times compared to the conventional nutrition programme. PGRs are, indeed, the precursors of a new agricultural revolution which will transform the sector to sustainably face the present and future challenges of humanity. Thus, trials will continue to explore all aspects raised in the discussion to add applicable high-impact scientific evidence about the agronomic performance of PGRs.

Author Contributions: Conceptualisation, methodology, analysis and writing, E.C.D. Experiment and data curation G.F. All authors have read and agreed to the published version of the manuscript.

Funding: This research received no external funding.

Acknowledgments: The authors would like to thank the Navarra Foundation, with special recognition of Alessandro Zago and Michele Mariani for their support in the experimental field.

Conflicts of Interest: The authors declare no conflict of interest.

References

1. Viets, F.G.; Lunin, J. The environmental impact of fertilizers. *Crit. Rev. Env. Control* **1975**, *5*, 423–453. [CrossRef]

2. Zhou, Z. *A Global Assessment of Nitrate Contamination in Groundwater*; Internship Report; Wageningen University: Wageningen, The Netherlands, 2015.

3. Ward, M.H.; Jones, R.R.; Brender, J.D.; de Kok, T.M.; Weyer, P.J.; Nolan, B.T.; Villaneuva, C.M.; van Breda, S.G. Drinking Water Nitrate and Human Health: An Updated Review. *Int. J. Environ. Res. Public Health* **2018**, *15*, 1557. [CrossRef] [PubMed]

4. MacDonald, G.K.; Bennett, E.M.; Potter, P.A.; Ramankutty, N. Agronomic phosphorus imbalances across the world's croplands. *Proc. Natl. Acad. Sci. USA* **2011**, *108*, 3086–3091. [CrossRef] [PubMed]

5. Sharpley, A.N.; McDowell, R.W.; Kleinman, P.J.A. Phosphorus loss from land to water: Integrating agricultural and environmental management. *Plant Soil* **2001**, *237*, 287–307. [CrossRef]

6. Barker, A.V.; Pilbeam, D.J. Introduction. In *Handbook of Plant Nutrition*, 2nd ed.; Barker, A.V., Pilbeam, D.J., Eds.; CRC Press: Boca Raton, FL, USA, 2007; pp. 3–18.

7. Colla, G.; Cardarelli, M.; Bonini, P.; Rouphael, Y. Foliar applications of protein hydrolysate, plant and seaweed extracts increase yield but differentially modulate fruit quality of greenhouse tomato. *Hortscience* **2017**, *52*, 1214–1220. [CrossRef]

8. Dong, W.; Zhang, X.; Wang, H.; Dai, X.; Sun, X.; Qiu, W.; Yang, F. Effect of Different Fertilizer Application on the Soil Fertility of Paddy Soils in Red Soil Region of Southern China. *PLoS ONE* **2012**, *7*, e44504. [CrossRef] [PubMed]

9. EU. *Farm to Fork Strategy: For a Fair, Healthy and Environmentally-Friendly Food System*; EU: Brussels, Belgium, 2020; p. 22. Available online: https://ec.europa.eu/food/sites/food/files/safety/docs/f2f_action-plan_2020_strategy-info_en.pdf (accessed on 12 June 2020).

10. Carillo, P.; Ciarmiello, L.F.; Woodrow, P.; Corrado, G.; Chiaiese, P.; Rouphael, Y. Enhancing sustainability by improving plant salt tolerance through macro- and micro-algal biostimulants. *Biology* **2020**, *9*, 253. [CrossRef]

11. Drobek, M.; Frąc, M.; Cybulska, J. Plant Bio-stimulants: Importance of the Quality and Yield of Horticultural Crops and the Improvement of Plant Tolerance to Abiotic Stress—A Review. *Agronomy (Basel)* **2019**, *9*, 335.

12. Battacharyya, D.; Zamani, M.; Babgohari, M.Z.; Rathor, P.; Prithiviraj, B. Seaweed extracts as biostimulants in horticulture. *Sci. Hortic.* **2015**, *196*, 39–48. [CrossRef]

13. Du Jardin, P. Plant bio-stimulants: Definition, concept, main categories and regulation. *Sci. Hortic.* **2015**, *196*, 3–14. [CrossRef]

14. El Chami, D.; Daccache, A.; El Moujabber, M. How can sustainable agriculture increase climate resilience? A systematic review. *Sustainability (Basel)* **2020**, *12*, 3119. [CrossRef]

15. Michalak, I.; Chojnacka, K.; Saeid, A. Plant Growth Biostimulants, Dietary Feed Supplements and Cosmetics Formulated with Supercritical CO_2 Algal Extracts. *Molecules (Basel)* **2017**, *22*, 66. [CrossRef] [PubMed]

16. Mutale-joan, C.; Redouane, B.; Elmernissi, N.; Kasmi, Y.; Lyamlouli, K.; Sbabou, L.; Zeroual, Y.; El Arroussi, H. Screening of microalgae liquid extracts for their bio-stimulant properties on plant growth, nutrient uptake and metabolite profile of *Solanum lycopersicum* L. *Sci. Rep.* **2020**, *10*, 2820. [CrossRef] [PubMed]

17. Shukla, P.S.; Mantin, E.G.; Adil, M.; Bajpai, S.; Critchley, A.T.; Prithiviraj, B. Ascophyllum nodosum-Based Biostimulants: Sustainable Applications in Agriculture for the Stimulation of Plant Growth, Stress Tolerance, and Disease Management. *Front. Plant Sci.* **2019**, *29*, 655. [CrossRef]

18. Michalak, I.; Górka, B.; Wieczorek, P.P.; Rój, E.; Lipok, J.; Łęska, B.; Messyasz, B.; Wilk, R.; Schroeder, G.; Dobrzyńska-Inger, A.; et al. Supercritical fluid extraction of algae enhances levels of biologically active compounds promoting plant growth. *Eur. J. Phycol.* **2016**, *51*, 243–252. [CrossRef]

19. FAOSTAT. *Crop Production Data*; The United Nations Food and Agricultural Organisation (FAO): Rome, Italy, 2020; Available online: http://www.fao.org/faostat/en/#data (accessed on 12 June 2020).

20. Eurostat. *Database—Agricultural Production*; European Statistics (Eurostat): Brussels, Belgium, 2020; Available online: https://ec.europa.eu/eurostat/data/database (accessed on 12 June 2020).

21. Elzebroek, A.T.G.; Wind, K. Edible fruits and nuts. In *Guide to Cultivated Plants*; Elzebroek, A.T.G., Wind, K., Eds.; CAB International: Wallingford, UK; Boston, MA, USA, 2008; pp. 25–131.

22. Eccher, T.; Pontiroli, R. Old pear varieties in Northern Italy. *Acta Hortic.* **2005**, *671*, 243–246. [CrossRef]

23. Istat. *The 2010 Agricultural Census*; The National Institute of Statistics (Istat): Rome, Italy, 2020; Available online: https://www.istat.it/it/censimenti-permanenti/censimenti-precedenti/agricoltura/agricoltura-2010 (accessed on 12 June 2020).

24. Harris, I.C.; Jones, P.D. *CRU TS4.01: Climatic Research Unit (CRU) Time-Series (TS) Version 4.01 of High-Resolution Gridded Data of Month-by-Month Variation in Climate (Jan. 1901–Dec. 2016)*; University of East Anglia, Climatic Research Unit, Centre for Environmental Data Analysis: Norwich, UK, 2017.

25. Google Earth. Navarra Foundation and Weather station in Ferrara (44.857 N, 11.653 E and 44.861 N, 11.656 E). Google Earth Pro. v. 7.3, 3 D Map, Building Data Layer. 2020. Available online: http://www.google.com/earth/index.html (accessed on 12 June 2020).

26. Essenberg, C.J. Explaining Variation in the Effect of Floral Density on Pollinator Visitation. *Am. Nat.* **2012**, *180*, 153–166. [CrossRef]

27. Alburquerque, N.; Burgos, L.; Egea, J. Influence of flower bud density, flower bud drop and fruit set on apricot productivity. *Sci. Hortic.* **2004**, *102*, 397–406. [CrossRef]

28. Balkic, R.; Gunes, E.; Altinkaya, L.; Gubbuk, H. Effect of male bud flower removal on yield and quality of 'Dwarf Cavendish' banana. *Acta Hortic.* **2016**, *1139*, 587–590. [CrossRef]

29. Dong, H.; Zhang, D.; Tang, W.; Li, W.; Li, Z. Effects of planting system, plant density and flower removal on yield and quality of hybrid seed in cotton. *Field Crop. Res.* **2005**, *93*, 74–84. [CrossRef]

30. Daugaard, H. The effect of flower removal on the yield and vegetative growth of A+ frigo plants of strawberry (Fragaria×ananassa Duch). *Sci. Hortic.-Amsterdam* **1999**, *82*, 153–157. [CrossRef]

31. An, J.; Althiab Almasaud, R.; Bouzayen, M.; Zouine, M.; Chervin, C. Auxin and ethylene regulation of fruit set. *Plant Sci.* **2020**, *292*, 110381. [CrossRef]

32. Bons, H.K.; Kaur, M. Role of plant growth regulators in improving fruit set, quality and yield of fruit crops: A review. *J. Hortic. Sci. Biotechnol.* **2019**, *95*, 137–146. [CrossRef]

33. Lāce, B.; Lācis, G.; Blukmanis, M. Average fruit weight variability of Pear cultivars under growing conditions of Latvia. *Acta Hortic.* **2015**, *1094*, 189–195. [CrossRef]

34. Zhang, C.; Tanabe, K.; Wang, S.; Tamura, F.; Yoshida, A.; Matsumoto, K. The Impact of Cell Division and Cell Enlargement on the Evolution of Fruit Size in *Pyrus pyrifolia*. *Ann. Bot.-London* **2006**, *98*, 537–543. [CrossRef] [PubMed]

35. Craswell, E.T.; Godwin, D.C. The efficiency of nitrogen fertilisers applied to cereals in different climates. *Adv. Plant. Nutr.* **1984**, *1*, 1–55.

36. Novoa, R.; Loomis, R.S. Nitrogen and plant production. *Plant Soil* **1981**, *58*, 177–204. [CrossRef]

37. Casanova, E.F. Agronomic evaluation of fertilisers with special reference to natural and modified phosphate rock. *Fertil. Res.* **1995**, *41*, 211–218. [CrossRef]

38. De Koeijer, T.J. *Efficiency Improvement for a Sustainable Agriculture: The Integration of Agronomic and Farm Economics Approaches*; Thesis Wageningen University: Wageningen, The Netherlands, 2002; p. 143.

39. Shepherd, J.G.; Kleemann, R.; Bahri-Esfahani, J.; Hudek, L.; Suriyagoda, L.; Vandamme, E.; van Dijk, K.C. The future of phosphorus in our hands. *Nutr. Cycl. Agroecosyst.* **2016**, *104*, 281–287. [CrossRef]

40. Herring, J.R.; Fantel, R.J. Phosphate rock demand into the next century: Impact on world food supply. *Nat. Resour. Res.* **1993**, *2*, 226–246. [CrossRef]

41. Bulgari, R.; Franzoni, G.; Ferrante, A. Biostimulants Application in Horticultural Crops under Abiotic Stress Conditions. *Agronomy (Basel)* **2019**, *9*, 306. [CrossRef]

42. Van Oosten, M.J.; Pepe, O.; De Pascale, S.; Silletti, S.; Maggio, A. The role of biostimulants and bioeffectors as alleviators of abiotic stress in crop plants. *Chem. Biol. Technol. Agric.* **2017**, *4*, 5. [CrossRef]

43. Dicenta, F.; Ortega, E.; Egea, J. Influence of flower density on fruit set rate and production in almond. *Acta Hortic.* **2006**, *726*, 307–310. [CrossRef]

44. Wally, O.S.D.; Critchley, A.T.; Hiltz, D.; Craigie, J.S.; Han, X.; Zaharia, L.I.; Abrams, S.R.; Prithiviraj, B. Regulation of phytohormone biosynthesis and accumulation in *Arabidopsis* following treatment with commercial extract from the marine macroalga *Ascophyllum nodosum*. *J. Plant Growth Regul.* **2013**, *32*, 324–339. [CrossRef]

45. Mancuso, S.; Azzarello, E.; Mugnai, S.; Briand, X. Marine bioactive substances (IPA extract) improve foliar ion uptake and water stress tolerance in potted Vitis vinifera plants. *Adv. Hort. Sci.* **2006**, *20*, 156–161.

46. Fan, D.; Hodges, D.M.; Zhang, J.; Kirby, C.W.; Ji, X.; Locke, S.J.; Critchley, A.T.; Prithiviraj, B. Commercial extract of the brown seaweed Ascophyllum nodosum enhances phenolic antioxidant content of spinach (Spinacia oleracea L.) which protects Caenorhabditis elegans against oxidative and thermal stress. *Food Chem.* **2011**, *124*, 195–202. [CrossRef]

47. Fahad, S.; Hussain, S.; Matloob, A.; Khan, F.A.; Khaliq, A.; Saud, S.; Hassan, S.; Shan, D.; Khan, F.; Ullah, N.; et al. Phytohormones and plant responses to salinity stress: A review. *Plant Growth Regul.* **2015**, *75*, 391–404. [CrossRef]

48. El Chami, D.; Daccache, A. Assessing sustainability of winter wheat production under climate change scenarios in a humid climate—An integrated modelling framework. *Agric. Syst.* **2015**, *140*, 19–25. [CrossRef]

Review

Flavonoids in Agriculture: Chemistry and Roles in, Biotic and Abiotic Stress Responses, and Microbial Associations

Ateeq Shah and Donald L. Smith *

Department of Plant Science, Macdonald Campus, McGill University,
Sainte-Anne-de-Bellevue, QC H9X 3V9, Canada; Ateeq.Shah@mail.mcgill.ca
* Correspondence: donald.smith@mcgill.ca

Received: 26 June 2020; Accepted: 11 August 2020; Published: 17 August 2020

Abstract: The current world of climate change, global warming and a constantly changing environment have made life very stressful for living entities, which has driven the evolution of biochemical processes to cope with stressed environmental and ecological conditions. As climate change conditions continue to develop, we anticipate more frequent occurrences of abiotic stresses such as drought, high temperature and salinity. Living plants, which are sessile beings, are more exposed to environmental extremes. However, plants are equipped with biosynthetic machinery operating to supply thousands of bio-compounds required for maintaining internal homeostasis. In addition to chemical coordination within a plant, these compounds have the potential to assist plants in tolerating, resisting and escaping biotic and abiotic stresses generated by the external environment. Among certain biosynthates, flavonoids are an important example of these stress mitigators. Flavonoids are secondary metabolites and biostimulants; they play a key role in plant growth by inducing resistance against certain biotic and abiotic stresses. In addition, the function of flavonoids as signal compounds to communicate with rhizosphere microbes is indispensable. In this review, the significance of flavonoids as biostimulants, stress mitigators, mediators of allelopathy and signaling compounds is discussed. The chemical nature and biosynthetic pathway of flavonoid production are also highlighted.

Keywords: flavonoids; biotic and abiotic stress; symbiosis; signaling; rhizobium; AMF; salinity; allelopathy

---

1. Introduction

Climate change is the most serious threat to current human culture. Escalating global food demand and ever-increasing global warming put humanity in jeopardy. According to ongoing global temperature analysis carried out by NASA's Goddard Institute for Space Studies (GISS) scientists, the average global temperature has increased by about 1 °C since 1880 [1], and it is estimated that every 2 °C rise in global temperature will cause on hundred million human deaths and bring millions of species to the brink of extinction [2]. After fossil fuel burning for energy generation, agriculture is the second-largest contributor to climate change through the emission of greenhouse gases (GHGs) including carbon dioxide (CO_2), methane (CH_4) and nitrous oxide (N_2O) [3]. Commercial fertilizers are very convenient, easy to handle and a rapid source of soil nutrient recharge, however, the toxic and residual effects of synthetic chemicals have altered thinking around this. It is estimated that reductions in mineral fertilizer use could lead to a 20% reduction in GHG emissions [4]. As the world warms, there is an immediate need to adjust what have become inadequate and inappropriate policies. There is an urgency to develop ecofriendly land practices and more sustainable agriculture. The implementation of biobased products, for example, ushering in the use of organic farming, biofertilizers and biocontrol techniques, will be a progressive step towards sustainable global food security. In this review, we focused on a specific type of biostimulant, flavonoids, and their role in sustainable agriculture. Flavonoids are examples of a versatile set of low molecular weight secondary metabolites with a polyphenolic structure, involved in plant physiological functions, often demonstrating protective effects against biotic and abiotic stresses including UV-B radiation [5], salt stress [6] and drought [7], at least in part by detoxifying the Reactive Oxygen Species (ROS) produced under stress conditions in plants [8,9]. Flavonoids also play a crucial role in plant–microbe associations, predominantly plant–rhizobia and arbuscular mycorrhizal symbioses [10]. Certain flavonoids act as signaling compounds triggering nodule induction by inducing transcription of *nod* genes in rhizobia, the first step in legume–rhizobia symbiotic relationships [11]. In addition, some flavonoids act to combat certain pests and pathogens [12]. Some classes of flavonoids act as color pigments, producing specific hues in leaves and flower petals, helping plants attract pollinators [13]. Moreover, flavonoids have indirect effects on nutrient supply and availability by enhancing mycorrhizal symbioses and colonization of the rhizosphere by beneficial microorganisms [14].

2. Biosynthesis and Classification of Flavonoids

The biosynthesis of distinct flavonoid-based compounds is the result of condensation of one molecule of 4-coumaroyl-CoA (6-carbon) and three molecules of malonyl-CoA, carried out by the enzyme chalcone synthase (CHS). The two major precursors originate from two different pathways of cellular metabolism: the acetate pathway and shikimate pathway providing ring A and ring B, respectively, with chain linkages forming ring C. Ring A is generated from malonyl-CoA synthesized by carboxylation of acetyl-CoA via the acetate pathway, however, ring B along with the linking chain (ring C) is synthesized from coumaroyl-CoA via the shikimate pathway (Figure 1). Coumaroyl-CoA is generated directly from the amino acid phenylalanine by three enzymatic reactions of the phenylpropanoid pathway [15].

The condensation of these aromatic rings by these pathways results in the synthesis of chalcone which will then form flavanone after isomerase-catalyzed cyclization. The later compounds undergo further modifications such as hydroxylation, glycosylation or methylation resulting in the enormous range of flavonoid colors we see today.

Flavonoids are the largest family of natural products; more than nine thousands of these phenolic substances have been found in various plants [16]. Flavonoids have a basic structure containing three phenolic rings, namely A (6 carbon) and B (6-carbon) linked with the central C (3-carbon) ring; C_6-C_3-C_6 which can produce several derivatives and sub-class compounds with distinct substitutions in the basic structure [17]. The major subgroups of flavonoids are; flavonols, flavones, flavanones, flavanonols, flavanols, anthocyanins, isoflavonoids and chalcones [18]. However, based on the attachment of the B ring to the C ring, flavonoids have been classified into three major subgroups: Flavonoids (2-phenylbenzopyrans): The B ring is attached on 2-position of ring C), Isoflavonoids (3-benzopyrans): The B ring is attached on 3-position of ring C) and Neoflavonoids (4-benzopyrans: unlike isoflavonoids; the B ring is attached at 4-position of C ring) [19].

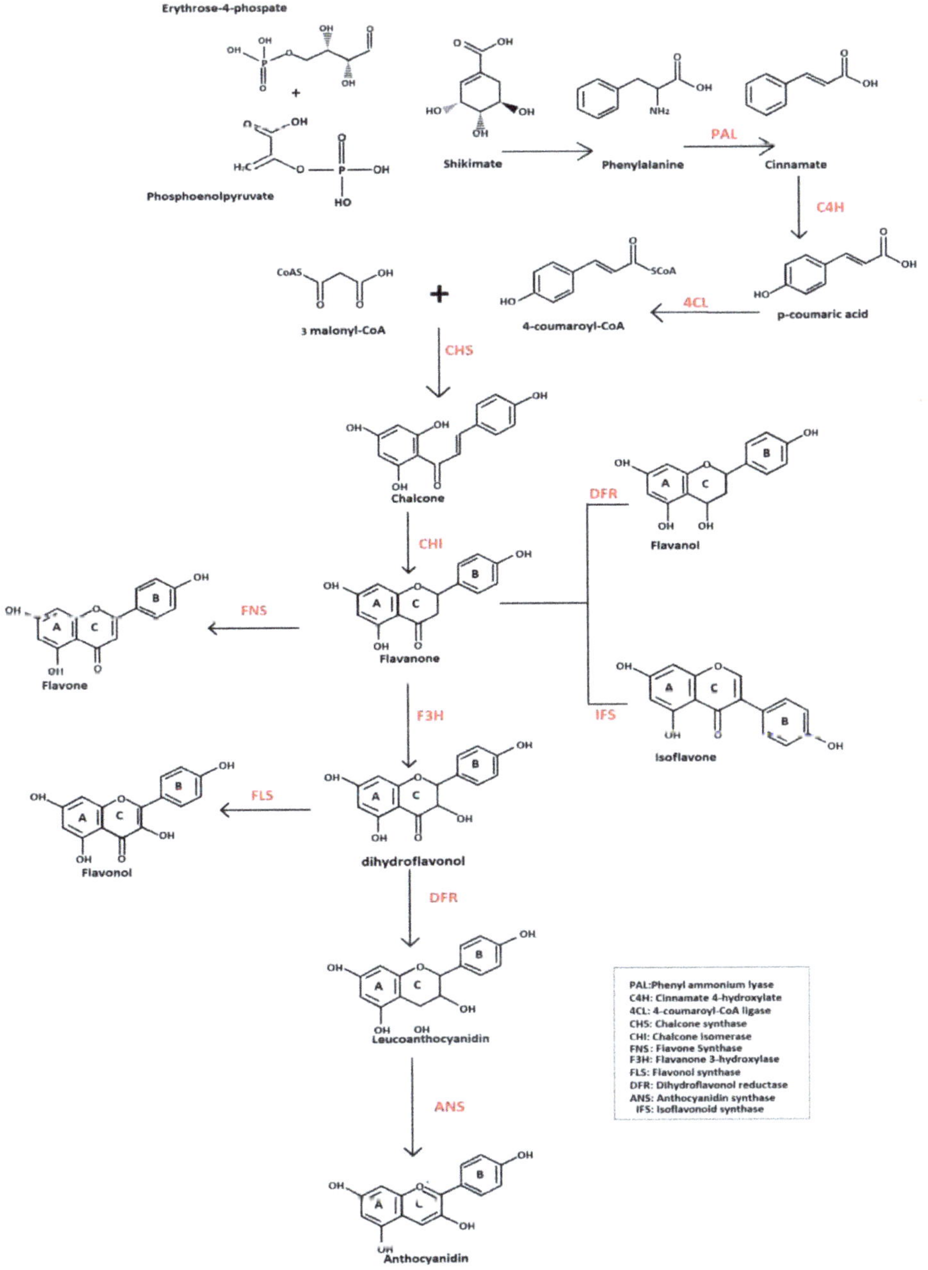

Figure 1. Biosynthesis of flavonoids.

2.1. Flavonols

Flavonols are the most abundant flavonoids in plants. The most studied subclasses of flavonols are the quercetins, kaempferols, myricetins and fisetins; distinctions in the structures of each subclass are shown in Figure 2. The substitution patterns in quercetins and kaempferols are 3,5,7,3',4'-OH and 3,5,7,4'-OH, respectively. These are very often found in plants as glycosides. The major dietary sources of flavonols are fruits and vegetables, predominantly onions, but also including the apple, strawberry, lettuce and other leafy vegetables. In addition, black and green tea and red wine are also rich sources of flavonols. In general, soft fruits, leaves of medicinal plants and green leafy vegetables have greater levels of flavonoids than other vegetable and fruit plants [20]. However, cooking may lower the concentration of flavonols in vegetables such as tomato and onion [21].

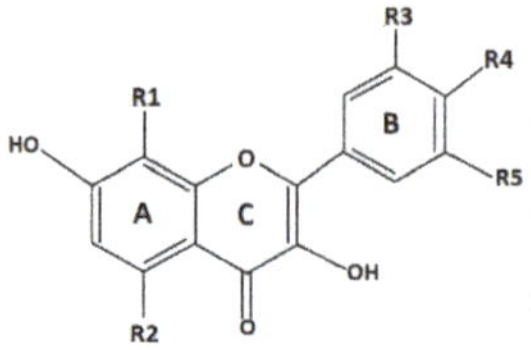

Subclass	R1	R2	R3	R4	R5
Kaempferol	H	OH	H	OH	H
Quercetin	H	OH	OH	OH	H
Myricetin	H	OH	OH	OH	OH

Flavonols

Figure 2. Flavonols: chemical structures, types and substitution positions in the basic skeleton.

2.2. Flavones

Flavones are one of the major subgroups of flavonoids. Fruits and vegetables including parsley, carrot, pepper, celery, olive oil, and peppermint are the main dietary sources of flavones [22]. Chrysin, apigenin, rutin (glycoside) and luteolin are the most studied subclasses of flavones. Substitution patterns in the basic structure of flavones are 5,7-OH (chrysin), 5,7,4'-OH (apigenin), and 5,7,3',4'-OH (luteolin) [23] (Figure 3).

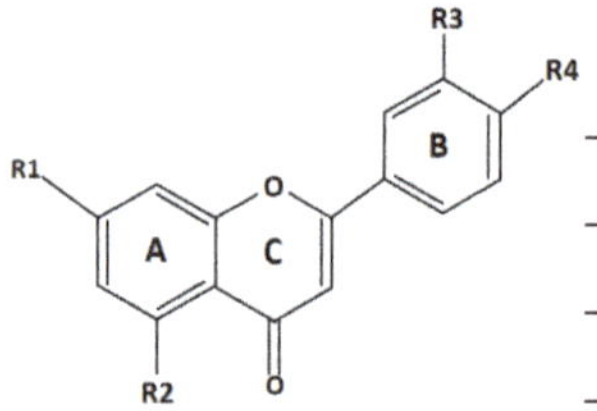

Subclass	R1	R2	R3	R4
Chrysin	OH	OH	H	H
Luteolin	OH	OH	OH	OH
Apigenin	OH	OH	H	OH

Flavones

Figure 3. Flavones: chemical structure, types and substitution positions in the basic skeleton.

2.3. Flavanones

Flavanones are different from flavones through their possession of a single bond between C2 and C3 of the C ring. Flavanones are most abundant in solid tissues of citrus fruits such as orange lemon and grape. The most studied types of flavanones are hesperidin and naringenin (Figure 4) The hydroxylation and substitution patterns in flavanone are 5,7,4'-OH (naringenin) and 5,3'-OH, 4'OMe, 7-rutinose (hesperidin) [23].

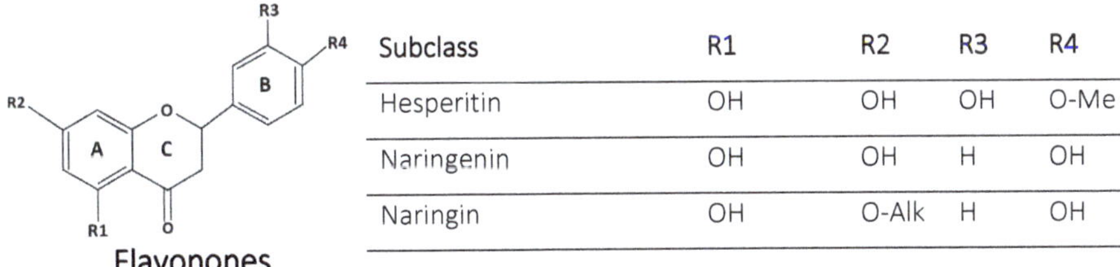

Subclass	R1	R2	R3	R4
Hesperitin	OH	OH	OH	O-Me
Naringenin	OH	OH	H	OH
Naringin	OH	O-Alk	H	OH

Figure 4. Flavonones: chemical structures, types and substitution positions in the basic skeleton.

2.4. Isoflavonoids

Isoflavonoids, with a B ring attached at 3-position of the C ring, are structurally different from other flavonoid classes. Isoflavonoids are found to be very helpful in microbial signaling and nodule induction in legume–rhizobia symbioses. Common examples of isoflavonoids are aglycone and glycosides of genistein and daidzein. The main natural sources of isoflavonoids are legumes such as soybean, as they are reported to exude these as signaling compounds to communicate with microbial symbionts [24]. The subclasses of isoflavonoids are shown in Figure 5.

Compound	R1	R2
Daidzein	H	H
Genistein	OH	H
Glycitein	H	OCH$_3$

Figure 5. Isoflavonoids: chemical structure, types and substitution positions in the basic skeleton.

2.5. Anthocyanidins

Anthocyanidins are responsible for the coloration of many fruits and vegetables. The red and blue colors in apple, grape and berries are due to anthocyanin or anthocyanin glycoside pigments [25]. The color depends on the structure of the compound which, usually changes due to hydroxylation and methylation at specific positions of the A and B rings [18]. Unlike other flavonoids, except flavanols, it carries no ketone group at the 4-position of C ring. Some of the anthocyanidin subclasses are shown in Figure 6.

Compound	R1	R2
Cyanidin	OH	H
Delphinidin	OH	OH
Pelargonidin	H	H

Figure 6. Anthocyanidins: chemical structures, types and substitution positions in the basic skeleton.

2.6. Flavanols

Flavanols have a missing ketone group at the 4-position of the heterocyclic ring, C, like anthocyanidins. The major sources of flavanols are grape (seed, pulp, stem and skin), berries, tea, wine, apple, pear and peach [26,27]. The most common examples of flavanols are (+)-catechin and (−)-epicatechin (Figure 7).

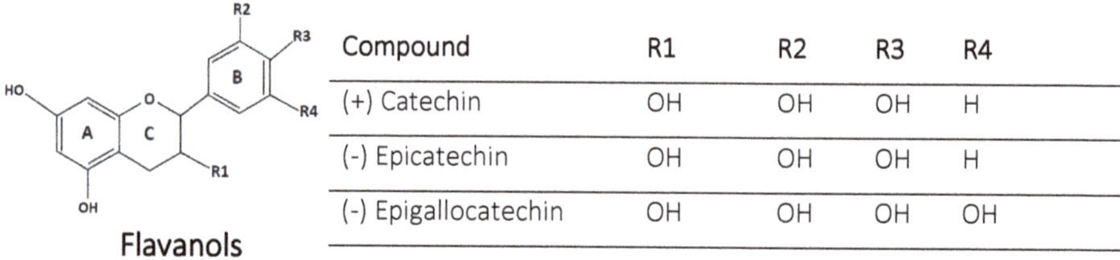

Compound	R1	R2	R3	R4
(+) Catechin	OH	OH	OH	H
(-) Epicatechin	OH	OH	OH	H
(-) Epigallocatechin	OH	OH	OH	OH

Figure 7. Flavanols: chemical structure, types and substitution positions in the basic skeleton.

2.7. Chalcones

Chalcone is the only class with an open ring; it serves as a precursor for various flavonoid classes. The missing C ring in the structure makes it quite different from other flavonoids. The major dietary sources of chalcones are apple, hops or beer [28], berries, tomato and certain wheats [18]. The most studied chalcone is chalco naringenin (Figure 8).

Compound	R1	R2
Naringenin chalcone	OH	H
Eriodictyol chalcone	OH	OH

Figure 8. Chalcones: chemical structure, types and substitution positions in the basic skeleton.

3. Role of Flavonoids in Plant Growth and Crop Yield

Plant growth and production are directly associated with soil productivity. In agronomic terms, it refers to the ability of soil to produce a certain yield of agricultural crops. However, the ability of soil is determined by numerous factors including soil physicochemical properties and management-related factors. Generally, it is the measure of inputs versus outputs, which in agronomic situations are related to water and nutrient supply (inputs) versus crop yield (output) [29]. In addition to these factors, the phytomicrobiome is becoming recognized as a key pillar and a major contributor to crop production. The soil is home to vast numbers of microbial species including free-living, symbionts, host-specific and non-host specific which could be either beneficial or harmful to plant growth and productivity. It is estimated that 1 g of soil contains 1 billion bacterial cells with 10,000 distinct genomes [30,31]. These microbial entities in phytomicrobial associations aid plants in their growth in both stressed and unstressed conditions, by a range of mechanisms. In the phytomicrobiome, associated entities trade for their benefits by aiding one another; microbes get reduced carbon-rich food materials exuded from plant roots and in turn, microbes aid plants in nutrient acquisition, disease resistance and stress mitigation by direct and indirect means. The contribution of the phytomicrobiome in relation to crop yield and production has been demonstrated by numerous studies [32–34]. However, this productive association is the result of crosstalk between trading partners, which is carried out by a complex signal cascade in the rhizosphere. In phytomicrobial associations, signaling is a key activity, and any interruption in signal transmission may disturb and/or reduce the interaction and efficacy of the symbionts.

Plants are equipped with biosynthetic machinery operating to supply thousands of bio-compounds that are required to perform vital functions. There is an array of compounds exuded from plant roots including sugars, amino acids, organic acids, phenolics, enzymes and growth regulators. These secretions serve as a source of carbon for many microorganisms associated with the plant root system. Besides providing a carbon-rich environment, plants generate chemical signals: a mode of communication with microbes in the rhizosphere to initiate colonization and other specific activities for mutual benefit [35,36]. The changing environment and anthropogenic global warming, and demand for sustainable agriculture for future security, have increased the significance of the phytomicrobiome in agricultural practices. Research on this area is becoming more prevalent in the agricultural scientific literature, and scientists are working to better understand plant–microbial associations and signaling to increase their efficacy in order to obtain substantial yield enhancements. In this review, we considered publications highlighting plant-microbial association in relation to signal compounds: flavonoids affecting plant growth and their production. Flavonoids have direct and indirect positive impacts on plant growth and development. The production of specific flavonoids and their accumulation in certain plant tissues following external or internal stimulation is largely unexplained. However, flavonoids are very effective in certain plant–microbe interactions in the rhizosphere and can have effects in combating biotic and abiotic stresses [37]. The roles of flavonoids, flavonoid subclasses and substitution patterns are extensively demonstrated in Table 1.

Table 1. Flavonoid subclasses, dietary sources, major functions in plants and structural substitutions in the basic skeleton.

	Compound	Flavonoid Class	Structure Substitution	Food Sources	Role in Plants	References
1	Quercetin	Flavonols	3,5,7,3′,4′-OH	Onions, apples, berries	Antioxidant, Allelopathic	[38,39]
2	Kaempferol	Flavonols	3,5,7,4′-OH	Tea, broccoli, cabbage, beans, tomato, strawberries and grapes	Antioxidant, antibacterial, insect repellent, abiotic stress mitigation	[40–43]
3	Chrysin	Flavones	5,7-OH	Honey, propolis	Antioxidant, UV-A/B Resistance, AMF symbiosis	[40,44,45]
4	Apigenin	Flavones	5,7,4′-OH	Parsley, Pepper	Antioxidant, AMF spore germination (symbiosis), phytoalexin	[46,47]
5	Naringenin	Flavanone	5,7,4′-OH	Grape, apple, orange	AntioxidantAMF Hyphal growth (Symbiosis)	[25,48]
6	Hesperidin	Flavanone	5,3′-OH, 4′OMe, 7-rutinose	Citrus, orange juice	Antioxidant	[25,49]
7	Genistein	Isoflavonoid	5,7,4′-OH	Currants, raisin, legumes	Nodule induction, signaling	[24,46]
8	Daidzein	Isoflavonoid	7,4′-OH	Currants, raisin, legumes	Nodule induction, signaling, chelation	[46,50]
9	Apigeninidin	Anthocyanin	7,4′-OH	Flowers, fruits	Color pigmentation, pollinator attractant, UV-B absorber	[19,25,51]
10	Fisetin	Flavonols	3,7,4′,5′-OH	Apple, strawberry, onion, cucumber	Antioxidant	[25,52]
11	Myricetin	Flavonols	3,5,7,3′,4′,5′-OH	Berries, tea, wine	Antioxidant	[53]
12	Luteolin	Flavones	5,7,3′,4′-OH	Broccoli, chilli, onion leaves bilimbi fruit and leaves, carrot, local celery	*Nod* gene inducer	[54]
13	Rutin	Flavones	5,7,3′,4′-OH, 3-rutinose	Parsley, Pepper, carrot	Mycorrhizae symbiosis, abiotic stress mitigation	[22,55]
14	(+)-catechin	Flavanol	3,5,7,3′,5′-OH	Grapes, pears, apples	Antioxidant, ROS scavengers	[26,56,57]
15	(−)-epicatechin	Flavanol	3,5,7,3′,5′-OH	Strawberry, apple	Antioxidant	[56,57]

3.1. Flavonoids in the Rhizosphere

The rhizosphere is the most complex and intensive place for the interaction of plants with the external environment. It is the area of maximum biological community activity, nutrient acquisition (mobility, solubility and diffusion) and plant–microbial interaction, which may depend on the secretion of exudates containing large and small molecular weight organic and inorganic compounds including ions, phenolics, enzymes, secondary metabolites and carbohydrates [58].

Flavonoids (secondary metabolites), very often in both aglycone and glycoside forms, are likely to be exuded from root systems and have indirect effects on plant growth by mediating belowground interactions, including attracting compatible rhizosphere-dwelling rhizobia, stimulating mycorrhizal growth and hyphal branching, enhancing solubility of nutrients including phosphorus and iron and repulsion of pests and root pathogens [50,59]. Studies suggested that flavonoid secretions from roots are carried out through active transport (ATP dependent), catalyzed by ABC transporters [60]. However, flavonoid secretion can also be passive, through degrading of root cap and epidermal cells [61]. In the rhizosphere, flavonoid persistence and mobility may be influenced by solubility, structure, availability of microbes and binding sites, as these compounds can be adsorbed to cation binding sites of soil or cell walls. Flavonoid glycosides are sparingly soluble in water and expected to be less adsorbed to binding sites, improving mobility and availability [10]. In addition, flavonoid secretion may be influenced by environmental stresses, including nutrient supply (nitrogen and phosphorus) in the soil [62].

3.2. Flavonoids and Legume-Rhizobium Interaction

The evolution of intimate relationships that enable plants and microbes to coexist has been the subject of many studies; these have attempted to explain and simplify the interactions that occur between an individual or specific plants and their symbionts. However, in reality, these interactions are far more complex and involve a range of microbes associated with a single plant, exchanging chemical signals [63]. However, these interactions aid plants in many ways. Soil fertility and/or nutrient acquisition is one of the major services provided by soil microbes. Nitrogen deficiency, due to rapid nitrogen loss from soil by leaching, denitrification and immobilization, is a leading problem in crop production. Atmospheric nitrogen is fixed and becomes part of the soil nitrogen recharge, by biological and artificial means, of fixation from the atmosphere. However, the natural biological fixation of atmospheric nitrogen (N_2) contributes about 60% of the total atmospheric nitrogen fixation, which has gained the attention of researchers and growers. Below-ground interactions leading to the establishment of legume nitrogen-fixing symbioses are carried out by signal exchange, as a mode of communication between host and symbiont, for mutual benefit.

The plants release chemo-attractants, in the form of flavonoids, to initiate the symbiotic process [64]. Flavonoids are very often exuded in greatest concentrations from root tips, which is near the site of rhizobium attachment and infection [65]. These secondary metabolites serve as signaling compounds to attract rhizobia toward plant roots and to activate *nod* genes in the rhizobia, which then start the nodulation process in legumes [66]. The *nod* genes are responsible for making lipo-chitooligosaccharides (LCOs), referred to as nod factors, which are released in response to chemical stimulus (generally isoflavonoids) from plant roots. LCOs or other nod factors initiate root hair curling, the formation of infection threads and bacterial entrance into the host plant root cells [67]. In legume crops, almost half of the nitrogen required is fixed by nitrogen-fixing microbes, predominantly, *Rhizobium* and *Bradyrhizobium*; the rest is supplied by fertilizer supplements [68].

Isoflavones are considered to be very active and effective in plant–microbe interactions; they are very operative in signaling and enhancing nodulation by inducing *nod* gene systems. *Nod* gene inducing flavonoids (quercetin and luteolin) released from the seeds and roots of *Medicago sativa* L were investigated. Many, but not all, of the flavonoids, were found to induce *nod* genes in *Rhizobium melioti* [69], indicating that rhizobia is responsive to selective flavonoid signals. Nodulation related gene induction in *R. melioti* by flavonoids released from alfalfa was investigated. A chalcone (4,4′-dihydroxy-2′-methoxychalcone) released, was reported to be the primary *nod* gene inducer in the group. Moderate inducing activities for 4′,7-dihydroxyflavone and 4′-7-dihydroxyflavanone were also reported [70]. In addition, genistein (isoflavonoid), was tested under salt stress with inoculation of *Bradyrhizobium* to evaluate effects on nodulation, N_2-fixation and physiological changes. The results revealed that genistein increased photosynthesis levels, nodulation and nitrogen fixation under saline and non-saline conditions [71]. Isoflavonoids are hypothesized to induce *nod* gene expression and to control the concentration of auxin in soybean roots. Results from similar work provided genetic evidence of isoflavonoid involvement in soybean nodulation and assumed this to be essential for nodule induction, by inducing the *nod* genes in *Bradyrhizobium japonicum* [72].

These biological signals are crucial factors in plant–microbe associations and are very often disrupted by known and unknown causes. Studies on subtropical legume (soybean) nodulation indicated that root zone temperatures (RZTs) below 25 °C decrease nodule induction and N_2-fixation [73,74]. The appropriate range of RZT required for optimal nodulation is 25 to 30 °C. Below optimal RZT (25 °C), with each degree decrease in temperature there is a 2-day delay in the onset of N_2-fixation; below 17 °C each degree RZT decrease delays the onset of N_2-fixation by about one week [75]. However, suboptimal RZTs hinder root hair infection to a greater degree than nodule initiation and development [76]. In addition, N_2-fixation activity by the nitrogenase enzyme complex is delayed, as is nitrogen assimilation [77]. However, the interorganismal signaling disruption by suboptimal RZTs (17.5 and 15 °C) could be minimized by genistein application. Preincubation of *Bradyrhizobium japonicum* with genistein increased the number of nodules, N_2-fixation and plant total dry weight at suboptimal RZT. It is, at least in part, because rhizobial *nod* gene induction is temperature-dependent, making bradyrhizobia less sensitive to signal molecules. However, nodulation events began earlier at suboptimal RZT following genistein application and this stimulated the production of nod factors (LCO), nodule formation and nitrogen fixation [78].

The environmental growth conditions, in terms of nodulation and N_2-fixation, may affect the efficacy of applied signaling compounds and/or inoculated microbial strain, for example, under field conditions, the plant root system is surrounded by an array of phytomicrobiome members, which may compete for plant-supplied reduced carbon; in addition, other factors such as temperature, CO_2 limitation, water and nutrient supply may alter nodulation and the onset of nitrogen fixation [79,80]. However, similar results were observed by preincubation of *B. Japonicum* with genistein application under field conditions. Genistein application improved N_2-fixation (40%) and total nitrogen yield through increased nodule numbers and accelerated onset of N_2-fixation; however, these effects were greater in N-stressed plants. In addition, genistein preincubation of *B. japonicum* has meaningful impacts on yield components [81]. Genistein preincubated inoculum (*B. japonicum*) increased soybean grain yield and protein content by 16 and 70% respectively, as compared to those receiving only inoculum [82]. Similar studies indicated an increase in yield and protein content of soybean by 25.5 and 21.6%, respectively [78]. However, crop responses to genistein application vary with the genetic makeup of the crop cultivars. It was suggested that high yield potential cultivars respond more to genistein application [83]. In addition to crop responses, the cultivar differences also determine the concentration and accumulation of a range of secondary metabolites. The influences of soybean cultivars and selection for yield on the concentration of health beneficial compounds, including isoflavonoids (examples being genistein and daidzein) were determined. The findings revealed a positive correlation between yield and isoflavonoid concentration and suggested that breeders selecting for higher soybean yield may select for higher isoflavonoid concentration. However, isoflavonoid concentrations were negatively correlated with protein content, unlike oil content [84].

3.3. Flavonoids and Mycorrhizal Associations

Phosphorus (P) is the second most important plant nutrient, after nitrogen. However, P is one of the most deficient and least accessible primary nutrients and is often limiting to plant growth. Phosphorus solubilization and affinity for the soil matrix and organic complexes are crucial reasons for its unavailability for plant utilization. The rhizospheric microbial community plays a vital role in phosphorus availability and solubility by producing enzymes that can mobilize adsorbed phosphorus and enhance the process of mineralization. Moreover, beneficial soil microbial species have positive impacts on root growth and development and may increase root surface area, which ultimately enhances the phosphorus depletion zone and effective phosphorus mobility [85]. In plant–mycorrhizal symbioses, mycorrhizal fungi regulate plants to reduce root growth while increasing root extension through hyphal outgrowth (100 times longer than root hairs) and the depletion zone for phosphorus, increasing its availability by extending and proliferating in the soil far beyond the reach of root surfaces [86]. Arbuscular mycorrhizal (AM) associations are probably the most taxonomically extensive associations, formed by 70–90% of plant species. Arbuscular mycorrhizal associations are formed by a monophyletic group of fungi from the phylum Glomeromycota which is estimated to utilize about 20% of plant photosynthates, which is approximately equal to 5 billion t of carbon per year [87]. The mycorrhizal symbiosis is very effective at enhancing plant growth in drought conditions and nutrient-deficient soils, specifically those that are phosphorus-deficient. It was observed that AM inoculation increased potato yield by 9.5% in inoculated fields, as compared to uninoculated fields, with yields of 42.2 and 38.3 t ha^{-1}, respectively. Arbuscular mycorrhizal associations have substantial potential to increase crop productivity, but their efficacy and association in large-scale crop production systems are not yet fully investigated [34].

Root exudation, chemo-stimulation and presence of other microbial communities in the rhizosphere may affect mycorrhizal symbioses and their colonization. Root exuded flavonoids have been shown to enhance the mycorrhizal symbioses by stimulating fungal spore germination, hyphal growth and colonization of roots. Flavonoids are considered to be universal signaling compounds [88]. However, some scientists have a different perception regarding AM symbiotic signaling. The normal development of mycorrhizal relationships in the absence of flavonoid-based signaling compounds in carrot root extracts and deficient activity of chalcone synthase (necessary for flavonoid biosynthesis) brought them to a conclusion that "Flavonoids are not necessary plant signal compounds in Arbuscular Mycorrhizal symbiosis". However, if present, their influence greatly stimulates AM development [89] and plant growth depending on concentration, spore density and plant growth stage [90].

The effects of the AM stimulating flavonoid, formononetin, on potato yield mediated by native AMF were examined. The results revealed an increase in plant dry matter, tuber development and phosphorus use efficiency. Perhaps the effects were more prominent at low phosphorus (P) levels. In addition, formononetin increased soil sporulation more than 3 fold [91]. However, the variation in response of different cultivars to formononetin application indicated that signaling responses by participatory symbionts may depend on genetic characteristics [92]. Similar results suggested that formononetin seed application to soybean may reduce the need for phosphorus (P) fertilizer by 50% [93]. A similar study was conducted on the hyphal growth and root colonization of arbuscular mycorrhizal fungi (AMF) on tomato as affected by flavones and flavonols. A correlation was observed between the number of entry points and root colonization percentage for the specific tested AMF (*Gigaspora rosea*, *Gigaspora margarita*, *Glomus mosseae*, and *Glomus intraradices*). Flavones (chrysin and luteolin) and flavonols (morin) enhanced the colonization and number of entry points, whereas kaempferol and rutin have no effect on presymbiotic growth of AMF and subsequent root colonization [94]. The flavonoids (apigenin, hesperetin and naringenin) enhanced spore germination, hyphal growth and root colonization of *Gigaspora margarita* [95]. In addition, quercetin glycosides, exuded from alfalfa, were found to be effective in enhancing AMF symbioses by increasing hyphal growth and branching, and spore germination of *G. macrocarpum* and *G. etunicatum* [50].

Moreover, flavonoids have been found to solubilize phosphorus by enhancing mycorrhizal colonization of root systems and may help in nutrient availability and mineralization of nitrogen and other nutrients [96]. They may also act as metal chelating agents, making certain micronutrients more available for plants [97]. Flavonoids released from roots of white lupin cause significant increases in phosphorus acquisition [98]. An Isoflavonoid present in root exudates of alfalfa was shown to dissolve phosphates of iron, making both iron and phosphorus more available to plants. In addition, flavonoids, including genistein, quercetin and kaempferol, can make iron available by chelating and reducing iron oxides in the rhizosphere from Fe^{+3} to Fe^{+2} [99].

4. Flavonoids and Plant Abiotic Stresses

Plants, as they are sessile, are exposed to an array of unfavorable environmental conditions. Ecological variations and intense growth conditions affect plant hemostasis, physiology and growth, leading to diminished and stunted plants. A wide range of unfavorable biotic and abiotic stresses threaten sustainable agriculture and are often responsible for diminished crop production [100]. The mechanisms plants use to cope with abiotic stresses are coordinated among plant organs and tissues through chemical signals [101]. Most plant responses to stress conditions are not fully understood. However, the biosynthetic flavonoids and their ability to induce resistance against biotic and abiotic stresses have acquired some attention. Flavonoids are found to be supportive in abiotic stresses, including UV radiation, salt and drought stress.

4.1. Flavonoids as UV Scavengers

UV radiation is invisible, short wavelength and highly energetic radiation. UV radiation is divided into three segments based on the wavelength of the light: UV-A, B and C with a wavelength of 315–400 nm, 280–315 nm and less than 280 nm, respectively [19]. These wavelengths have enough energy to cause damage and abnormalities in plants by breaking chemical bonds through photochemical reactions [62]. The energy of the photon depends on the wavelength of the radiation; the shorter the wavelength, the more energetic the radiation will be. UV-C (less than 280 nm) is most energetic and can ionize certain molecules. UV-B can cause severe metabolic disruption in plants by negatively affecting photosynthesis, starch concentration and transpiration, and promoting cellular damage; it may also increase disease susceptibility by making defense mechanisms weak, affecting the process of cell division and inhibiting overall plant growth [65,102,103]. However, the absorbance of UV-B radiation by flavonoids permits little radiation in this wavelength range to pass through leaf epidermal cells [62]. Plant resistance to UV radiation is due to flavonoids (anthocyanins) filtering UV-B by absorbing such radiations and detoxifying the ROS produced by photochemical reactions [104].

The significance of UV-B scavenging flavonoids on two apple varieties (Granny Smith and Braeburn) exposed to sunlight and UV-B radiation was estimated. The results revealed different contents and compositions of UV-B absorbing compounds in the two varieties. The Granny Smith (anthocyanin-free) fruit showed significant decreases in chlorophyll and carotenoid contents. Conversely, Braeburn exposed to sunlight had higher contents of chlorophyll and carotenoids. However, quercetin glycosides were the principal compounds absorbing UV-A and UV-B radiations. [105]. Similar work has investigated tomato under controlled environment conditions; the findings demonstrated an increase of flower/fruit synchronization under high radiation, with minimal effect on vegetative plant parts. In addition, an increase of UV-B receptors and chlorophyll content was also observed, along with phenylpropanoid compounds responsible for UV absorption by-products of antioxidant pathways. [106]. UV-A/B, as an abiotic stress, can be used to enhance fruit quality by activating oxidation pathways in plants [107].

The effect of UV light on flavonoid content in barley leaves was tested, and results revealed significant increases in flavonoid (saponarin, lutonarin) content when leaves were exposed to UV-B radiation, as compared to control conditions (absence of UV-B) [108]. In addition, the increased flavonoid accumulation in response to UV-B radiation may reduce the damage in exposed leaves by absorbing specific radiation wavelengths. [109]. Therefore, plants grown in open-environment conditions, exposed to full sunlight, have greater flavonoid contents than plants grown in greenhouses.

4.2. Flavonoids in Managing Salt and Drought Stress

Salinity, one of the most concerning abiotic stressors, is a major constraint to global crop productivity. The worldwide extent of soil affected by salts is about 955 M ha, while 77 M ha are affected with secondary salinization, and of these 58% are irrigated lands. Almost 20% of irrigated land is affected by excess soluble salts [110]. Soil salinization and increased accumulation of soluble salts in the root zone, predominantly NaCl, is caused by natural and/or human activities which have resulted in degraded and abandoned formally fertile and productive agriculture lands [111]. Excess soluble salts in soil solution may limit plant growth, primarily through two mechanisms: osmotic stress and ion toxicity. First, low solute/osmotic potential due to increased ion concentration (NaCl) in soil water reduces the total soil-water potential (Ψ) which in turn reduces the ability of plant roots to uptake water, eventually leading to diminished plant growth. Second, ion toxicity in plant tissues, more frequently due to sodium accumulation, causes cellular damage by membrane disruption and disturbs plant physiological processes, including photosynthesis, respiration, transpiration and osmoregulation, resulting in necrosis or chlorosis, leading to reduced plant growth [112–114]. Depending on sensitivity and tolerance to salinity, plants are classified as either glycophytes or halophytes [115]. Most of the agricultural crops are glycophytes (low tolerance) and tend to exclude Na^+ and Cl^- from roots. Unlike glycophytes, halophytes are often native to saline growth conditions and tolerate salt concentrations that kill 99% of other vegetation. The salinity tolerance of halophytes relies on ionic homeostasis by controlled uptake and compartmentalization of ions (Na^+, K^+, and Cl^-) and accumulation of metabolically compatible solutes (organic) in the cytoplasm to balance the solute potential of ions accumulated in the vacuole [116,117]. Salt responses in plants follow a biphasic mechanism. The first, osmotic phase (rapid), begins immediately after root zone salinization increases to a threshold level, resulting in reduced shoot growth and leaf area, and causes stomatal closure. The second, ionic phase, begins with increased accumulation of ions (Na^+) to toxic levels in the cytoplasm, leading to chlorosis followed by leaf death. Osmotic stress not only exerts immediate effects on plant growth but is also more chronic than the ionic phase [118]. Increases in an array of compatible organic solutes is proposed to balance solute potential including sucrose, proline, glycine-betaine and sorbitol [119,120]. In addition to compartmentalization, some plants can prevent salt accumulation (whole plant or cellular level) and avoid toxic effects of ions on crop physiology including photosynthesis [113]. In response to salt stress, plants undergo several morphological, physiological and metabolic changes to cope with the stress conditions. These adaptations involve several biochemical pathways, sustained osmotic potential, ion compartmentalization and exclusion of toxin ions. Subsequent to ionic toxicity, specific toxic substances, ROS, including superoxide, singlet oxygen and hydrogen peroxide, cause oxidative damage in cells, which is considered a secondary effect of salinity [121,122]. However, plants are equipped with specific defense mechanisms to cope with such stress conditions, by initiating antioxidant pathways including enzymes and antioxidant agents; flavonoids, carotenoids and specific vitamins [121,123,124]. Reactive oxygen species or free radicals are molecular species that contain at least one unpaired electron in their atomic shells, making them highly reactive. Reactive species (RS) are quite unstable, most of them exist not more than 10^{-6} s in biological systems, and to be more stable they react with biomolecules by either donating or receiving an electron [125].

Oxygen is poisonous; aerobes are equipped with defense mechanisms mediated by antioxidants, which is how they survive such toxicity. The antioxidant defenses in biological systems are the result of several strategies [126]:

1. Suppressing RS formation either by uncoupling proteins triggered by superoxide, indicating that it may reduce mitochondrial ROS formation [127] or inhibition of enzymes involved in RS formation, for example, inhibition of cyclooxygenase, lipoxygenase and NADH oxidase by flavonoids [128]
2. Substitution of biomolecules vulnerable to oxidative damage with resistant ones
3. Antioxidants acting as "sacrificial agents" by reacting with reactive species to prevent them from reacting with important biomolecules [129].

Flavonoids have been found to play an important role as antioxidants by detoxifying and scavenging of ROS produced [130] as by-products of oxidative metabolism [121] during abiotic stresses including salt and drought. However, the accumulation of such metabolites advances when plants face any environmental uncertainty. It was observed that anthocyanin accumulation in response to salt stress increased by 40%, which could be a phytochemical strategy to combat salt stress and subsequent toxic reactions [131]. In addition, the protective nature of anthocyanin was also compared in two groups of rice genotypes: salt sensitive and salt resistant. The total anthocyanin content in salt-tolerant genotypes was higher than in salt-sensitive varieties, with antioxidant activities of 125–199% and 106–113%, respectively. It was therefore concluded that anthocyanins in rice contribute to cellular protection by detoxifying accumulated salts [132]. The effects of applying genistein (an isoflavonoid) to rhizobial culture on signal production and subsequent growth and yield of soybean have been investigated. Results demonstrated a significant increase in plant growth with increased leaf area and number of nodules. Genistein application enhanced crop yield by 21% under salt stress [133]. Salinity may inhibit signal exchange between host and symbiont, which is very important for initiating a functional symbiotic nitrogen-fixation relationship. The interaction of soybean and *B. japonicum* was evaluated under salt stress. The findings are consistent with similar studies. Genistein application enhanced the stimulation of growth and signaling between the symbiotic partners and, hence, increased nodulation and growth of the plant. The results may help in cultivating soybean in a more efficient and productive way under unfavorable environmental conditions [134]. Likewise, an increased concentration of flavonoids has been found in tomato plants when exposed to salt or drought stress, however, plant growth and chlorophyll content were significantly reduced, indicating no and/or insignificant effect of flavonoids on plant growth and physiology [121]. In contrast, findings from a similar study on flavonoid biosynthesis and accumulation in wheat leaves under drought stress suggested that drought resistance in wheat is closely related to increased flavonoid accumulation [135]. Similar results were observed in two native shrubs from Argentina. Flavonoid accumulation was observed throughout the year; however, a significant increase was noted during times of intensive drought [136]. Drought mitigation by flavonoids and flavonoid derivatives has been confirmed in *Arabidopsis thaliana*. The role of individual flavonoids was unclear, however, increased production of flavonoids in plants and associated drought resistance was confirmed [95].

5. Flavonoids against Plant Biotic Stress

Plants, as they are sessile, have no possible way to physically remove themselves from invading pests and pathogens. In nature, plants are exposed to an array of pathogenic fungi, bacteria and herbivore pests. However, plants have evolved strategies to combat such unwanted guests. Pathogens do not generally succeed in infecting plants that are not host species (non-host resistance) and/or resistant varieties (incompatible interaction), but intense damage after infection may be caused in susceptible plants (compatible). The damage caused by pathogens in most cases is inversely correlated with the hypersensitive response including reinforcement of cell wall, induction of lytic enzymes and production of phytoalexins [137]. One of the defense strategies adopted by plants in response to invading pests and pathogens is formation, accumulation and secretion of phytoalexins. Phytoalexins are chemicals released by plants in response to pests and pathogens, to ward off the disease and disease-causing agents. Flavonoids are the most-described secondary metabolites in plant defense systems [132]. Their role in plant physiology, morphology and communication and defense mechanisms is considerable. Certain flavonoids are found to be strong phytoalexins against pathogenic bacteria, fungi and nematodes, and may act as insect repellents. External morphological modifications in plants may also act as protective mechanisms against invading pests, predominantly feeding animals including insects, however the chemical tool of insect repellent is more effective. The synthesis of these antibiotic secondary metabolites in plants is due to infection caused by pathogens, bacteria, fungi and nematodes, and may also be induced by feeding insects [138,139].

5.1. Phytoalexin Flavonoids as Nematicides

Parasitic nematodes cause tremendous crop yield loss by forming cysts or galls on roots. Plants, in response to nematode invasion, produce several chemicals to increase resistance to or minimize the effect of, nematode presence. The synthesis and accumulation of flavonoids within plant root systems is often stimulated by nematode invasion. The induced synthesis of such phytoalexins assists plants in coping with nematode infections.

Coumestrol (an isoflavonoid) can act as a phytoalexin; it is synthesized in lima bean as nematicide when infected by *Pratylenchus penetrans*. Glyceollin is a protecting isoflavonoid synthesized in soybean roots when infected by the root-knot nematode; *Meloidogyne incognita*. The results demonstrated a decrease in nematode mobility and O_2 uptake [140]. The synthesis of glyceollin, resistance inducing flavonoids in soybean, in response to *Meloidogyne penetrans* infection minimized crop damage [141]. *Heterodera glycines* (soybean cyst nematode) is the most destructive parasitic nematode of soybean. Accumulation of the phytoalexin glyceollin in root systems of soybean after cyst nematode invasion was determined by HPLC (High Performance Liquid Chromatography): A form of liquid chromatography, used to separate, identify and quantify compounds in a solution. No glyceollin was found in control plants, however, the content of the phytoalexin glyceollin increased at the 2nd, 4th and 6th days after inoculation, by 12, 19 and 23 μg g^{-1} root, respectively [142]. The major phytoalexin in oat, when infected with major nematodes of cereals, was identified as O-methyl-apigenin-C-deoxyhexoside-O-hexoside (a flavone-C-glycoside). The phytoalexin flavone, induced by nematode invasion, was extracted from oat roots and shoots, and was significantly reduced invasion by major cereal nematodes: *H. avenae* and *P. neglectus* [143]. The interaction of flavonoids with parasitic nematodes as defense mechanisms is unclear, however, flavonoids were found to be protecting agents as they inhibit nematode motility and chemotaxis [140].

5.2. Flavonoids against Pathogenic Fungi

Fungi, as the most dominant disease-causing agent in plants, adversely affect agricultural crop production. The wide range of diseases caused by fungi decreases crop production dramatically. However, plants have adapted themselves to produce resistance mechanisms against biotic stresses. The production of phytoalexins in response to pathogenic invasion, predominantly fungi, is an effective tool used by plants for combatting biotic stress. The effect of cucumber powdery mildew and subsequent biochemical changes in response to invading pathogen was investigated. Results revealed that silicon can contribute to powdery mildew resistance in cucumber by increasing the accumulation of a fungi-toxic phytoalexin, which was identified as the flavonol aglycone rhamnetin (a flavonoid) [144]. Brown rot lesion is a disease caused by the fungus *Phytophthora citrophthora* in citrus fruits. The correlation of infection caused by pathogen and level of phytoalexin flavones accumulation in host plant was examined. The increased accumulation of heptamethoxyflavone, nobiletin, sinensetin, and tangeretin was confirmed along with the antifungal effects of phytoalexin flavonoids. The most effective flavonoids against *P. citrophthora* were naringenin and hesperetin [145]. Similar results were found in tangelo fruit defense mechanisms against *P. citrophthora*. The accumulation of isoflavonoids was induced by 6-benzylaminopurine application, which enhanced fruit resistance to the pathogenic fungus by 60%. The most inhibiting of the accumulated phytoalexins were nobiletin, sinensetin, heptamethoxyflavone, followed by tangeretin [146]. Further research was carried out to evaluate phytoalexin accumulation in soybean cotyledons using four species of *Aspergillus*. All the pathogenic species induced accumulation of phytoalexins in soybean cotyledons, however the phytoalexins glyceollin at 955 μg g^{-1} (fw) and coumestrol at 27.2 μg g^{-1} (fw), following inoculation with *A. sojae* and *A. niger*, accumulated to the greatest degrees [147].

5.3. Antibacterial Effects of Flavonoids

The study of natural defense mechanisms of plants related to synthesizing antimicrobial phytoalexins in response to biotic stress demonstrated that phytoalexin level is increased as part of the resistance to phytopathogenic agents [148]. Information regarding phytoalexin accumulation in response to fungal invasion is considerable, however very little is known about phytoalexin synthesis in response to phytopathogenic bacterial invasion [149]. The production of antibacterial phytoalexins in bean leaves was studied by inoculating bean plants with *Pseudomonas spp.* Coumestrol, a phytoalexin isoflavonoid, was accumulated in infected bean leaves and inhibited the growth of two pathogenic bacterial species: *P. mars-prunorum and P. phaseolicola*. The coumestrol was obtained from hypersensitive and susceptible lesions at 1 and 5 days after inoculation. Their accumulation explains the inhibition of bacterial colonization in hypersensitive and susceptible lesions of bean leaves [150]. The accumulation of isoflavonoid in soybean leaves in response to *Pseudomonas glycinea* invasion was investigated. Coumestrol and daidzein were identified as the major phytoalexins accumulated in response to pathogenic (*P. glycinea*) and non-pathogenic (*P. lachrymans*) inoculations of soybean leaves. The data demonstrated inhibiting properties of coumestrol against pathogenic bacterial colonization and suggested that the resistance in soybean leaves against *P. glycinea* was due to induced accumulation of isoflavonoids [151]. In contrast, coumestrol was found ineffective against pathogenic bacterial strains when tested with another five isoflavonoids on twenty isolates of pathogenic and saprophytic bacteria, including species of *Pseudomonas, Xanthomonas* and *Achromobacter.* However, phaseollinisoflavan and kievitone showed antibacterial activity by strongly inhibiting the population of *Xanthomonas* and *Achromobacter* species [152]. Recent research toward finding natural solutions against biotic stress introduced new products against certain disease-causing agents. A new compound isolated from the roots of *Erythrina poeppigiana*, identified as an isoflavonoid was isolated against *Staphylococcus aureus,* and compared with five other root isolates. Results revealed strong inhibiting activities against inoculated pathogens. The minimum inhibitory concentration was 12.5 g mL^{-1} against thirteen (13) strains of S. *aureus*. It was also assumed that new compounds could act as potent antibiotics against infections caused by S. *aureus* [153].

5.4. Flavonoids as Insect/Herbivore Repellents

Plants, as sessile beings, act as an available food source for herbivores, including insect pests. Plants have evolved defense strategies to avoid and/or deal with such biotic stresses, by secreting and accumulating repellent molecules or signals, which are either plant constitutes, in some cases produced inductively in response to pest invasion [154]. The defense mechanism may initiate from undamaged tissues by secreting phytoalexins in response to chemical signals from wounded tissues, which may repel or intoxicate insects [155]. Most of the insect repellent or antifeedant molecules are alkaloids, flavonoids and other secondary metabolites [156]. Morphological modifications like thorns and waxes can make feeding difficult for insect pests. Plants usually use two different strategies: direct and indirect methods, as defense mechanisms against insect herbivory. Direct methods involve the accumulation of insect repellents or toxic substances to minimize the level of damage. On the other hand, plants release signaling compounds as chemoattractants to signal predators, which may feed on the pest and minimize the plant damage [157]. Rotenone (isoflavonoid) is a very effective botanical insecticide used as a basis for insect repellents. Rotenone is a major component of insecticidal resins which may be extracted from roots of some legumes, including those in the genera *Lonchocarpus, Derris* and *Tephrosia*. Pyrethrum is another bioactive material against insect pests, also containing flavonoids as a major constituent. Pyrethrum can be extracted from flowers of *Chrysanthemum* species [158]. Likewise, the antifeeding effects of four isoflavonoids (genistein, formononetin, daidzein and biochanin A), isolated from two red clover cultivars, were investigated against clover root borer. The isoflavonoids decreased insect weight and activity. Genistein and formononetin had high anti-feeding activity against *Hylastinus obscurus* (clover root borer). The results could be utilized in controlling curculionid [159]. In addition, the insecticidal effects of phenolics of pea plants were tested against *Acyrthosiphon pisum*. The high concentration of phenolics and flavonoids in infested plants, as compared to controls, suggested the induced accumulation of antifeedants in response to insect pest presence. In addition, flavonoids (luteolin and genistein) added as supplements to artificial diets prolonged the stylet probing, onset of salivation and passive ingestion. Salivation and passive ingestion completely stopped at higher concentrations of flavonoids. such measures could be employed to induce resistance against certain invading pests [160]. Additionally, the role of phenolics and flavonoids as insect repellents is illustrated by several lines of research [161–163]

The antifeedant and toxic effects of four isolated flavonoids (isoglabratephrin (þ)-glabratephrin, tephroapollin-F and lanceolatin-A) from aerial parts of *Tephrosia apollinea L.* were determined. The flavonoids exhibited toxic effects against insect pests: *Sitophilus oryzae, Rhyzopertha dominica* and *Tribolium*, with mortality percentages of 78.6, 64.6 and 60.7%, respectively, at 3.5 mg mL^{-1}. [164]

6. Allelopathic/Phytotoxic Behavior of Flavonoids

Weeds are a very significant challenge to crop plants as they are constantly competing for light, nutrients and water, interfering with crop functioning and causing tremendous yield loss directly or indirectly. Reductions in crop yield are generally much greater due to weeds than other pests. It is estimated that about 34% of yield loss among the major crops is caused by weeds [165]. Some of the major crops affected by weeds are wheat, soybean, rice, maize, cotton and potato with yield reductions of 23, 37, 37, 40, 36 and 30% respectively [166]. Weed management strategies have always been a significant part of agricultural systems but have changed significantly based on the accessibility of tools and techniques, environmental and sustainability concerns, starting from ancient techniques such as pulling by hand and soil tilling with simple tools, to current use of herbicides and mechanized conventional tillage, which are the most recent and, so far, most effective techniques available [167]. However, despite being very effective, commercial herbicides are finding themselves eschewed by growers because of their toxic and residual effects which contravene the principles of sustainability, eventually contributing to climate change, which is a consequence of such unsustainable human activities. In addition, the continuous use of chemical herbicides induces herbicide resistance in weed populations, which is a crucial long-term consideration in weed management. In contrast, biopesticides are gaining significant popularity among crop scientists because of their environmentally friendly behavior, as they contain biochemicals with no, or minimal, residual effects. The concept of using plant-derived biochemicals as "weedicides" originated from the allelopathic effects mediated by certain plants by employing allelochemicals released into the environment.

The term allelopathy was first defined by Molisch [168], indicating that the effects that result (directly or indirectly) from exuded biochemicals transferred from one plant to another. This definition, suggested by Molisch, implies only plant activity. However, the term "allelopathy" was later refined to include microorganisms (bacteria, algae, fungi and viruses) in his definition, as a significant part of allelopathic processes [169]. Allelopathy is an interference process in which either plants or their dead parts exude phytotoxic chemicals which interfere with the physiology and growth of other plants [170]. The allelopathic behavior of certain entities (plants and microorganisms) has been demonstrated in the literature, however, given the extensive uncertainty, this area needs more exploration in order to understand allelochemical behavior, including the formation of allelopathic compounds and their chemical nature, viability, efficiency and mode of action in plant-plant and plant–microbe interactions, to improve their practical implementation in the field.

Several plants are known to have allelopathic natures through allelochemical exudation including wheat, rice, rye, barley, sorghum and sugarcane. These plants can be manipulated to suppress weeds through an allelopathic approach within crop rotations, intercropping, cover crops and mulch [171]. The phytotoxic effect of sunflower cultivars was evaluated against weed species in wheat either by growing with weeds or applied as residues over a wheat crop and weeds. Sunflower cultivars have shown strong allelopathic effects on weeds, however, variation among the cultivars was observed, indicating that allelochemical exudation or phytotoxic effects and weed suppression vary with cultivar/genotype.

In addition, the sunflower cultivars suppressed total weed density and biomass by 10–87% and 34–81%, respectively [172].

There is an array of biochemicals, produced as secondary metabolites in plants, or released during their decomposition by microbes, which act as active allelochemicals in plant ecosystems. These phytotoxic substances, based on their chemical nature, are classified into 14 categories including cinnamic acid and its derivatives: coumarin, flavonoids, tannins, terpenoids and steroids. Recent publications regarding flavonoids have evidenced their phytotoxicity and growth inhibitory effects, which could be a sustainable approach toward integrated weed management [173].

In early plants, bryophytes and ferns, some of the allelochemicals found are synthesized in the early stages of flavonoid biosynthesis, however, additional flavonoid classes accumulate in angiosperms and gymnosperms, reflecting the employment of genes beyond just those involved in flavonoid biosynthesis [137]. Flavonoids have been reported in the literature for over 50 years as allelochemicals in the rhizosphere [174]. They are either exuded from roots or released from decomposed plant tissues as leachates, persist for days in the soil and their activity (inhibitory or stimulatory) depends on their concentration and solubility. In addition, phytotoxic compounds can also accumulate in leaves and pollen, which eventually inhibit seed germination of other plants after falling onto the soil [12].

It has been observed that flavonoids are produced by many legumes; quercetin and kaempferol (aglycon and glycosylated) released from seeds and roots, possess phyto-inhibitory effects. If present in lower concentrations such compounds may stimulate seed germination while in higher concentrations they may inhibit seedling growth [175]. Flavonoids isolated from roots of *Stellera chamaejasme* L., a toxic and ecologically threatening weed, showed strong phytotoxic activity against *Arabidopsis thaliana*. The isolated flavonoids reduced seedling growth and root development. In addition, endogenous auxin distribution in *Arabidopsis thaliana* was also influenced, indicating a critical factor in phytotoxicity [176]. Spotted knapweed is one of the more noxious and economically devastating weeds of North America, destroying crops and other weeds by phytotoxicity. An allelochemical identified was flavan-3-ol (¬-)-catechin (flavonoid), was shown to be responsible for the invasive behavior and phytotoxicity of *Centaurea maculosa* (spotted knapweed) [103].

Recently, the herbicidal effects of 10 crude extracts obtained from Tunisian plants were assessed. Among the five phenolic compounds, three of the flavonoids had significant herbicidal effects on *Trifolium incarnatum*. Flavonoids inhibited seed germination and seedling growth and caused severe necrosis and chlorosis. Based on their efficiency, flavonoids were formulated into a natural herbicide and interestingly, the extracts showed the same herbicidal effects as an industrial biopesticide containing pelargonic acid [177]. A similar study was conducted, indicating the phytotoxicity of *Plantago major* extracts on germination and seedling growth of purslane (*Portulaca oleracea*). It was observed that the level of phytotoxicity or inhibition was directly proportional to extract concentration. Phytotoxicity of a higher extract dose (40 mg mL^{-1}) was greater than the lowest one evaluated (2.5 mg mL^{-1}) and these concentrations inhibited germination by 30.24 and 4.60%, respectively. In addition, the highest concentrations significantly inhibited radical and plumule growth. The biologically active organic compounds in plant extracts were analyzed and found to be phenolics, tannins, alkaloids, flavonoids and saponins [178]. However, the compounds were not tested alone in this study, leaving no evidence of individual phytotoxic intensity of biological compounds. The need for sustainable and eco-friendly approaches in agricultural systems fosters great interest in bioproducts and biological control agents. However, further studies are required to obtain a better understanding of the many phytotoxic flavonoids.

7. Conclusions

The indispensable role of flavonoids in stress mitigation and signaling behavior in plants is highlighted. More specifically, we reviewed the protective nature of flavonoids in plants against certain biotic and abiotic stress conditions. The polyphenolic structure and diverse chemical nature of flavonoids facilitate multiple mechanisms of action, favoring plant survival under a range of harsh conditions. Despite current knowledge of this matter, the use of flavonoids in agriculture is very limited. Soil microbiota are ecofriendly contributors in sustainable agriculture, and the iconic role of flavonoids in improving phyto-microbial associations is the "icing on the cake". Still, however, we know very little about the chemo-communications between plants and the vast number of microbial strains in the rhizosphere (phytomicrobiome members); much is left to be explored and elucidated. In addition, flavonoids play an indispensable role against plant biotic and abiotic stresses. Flavonoids could be employed as an ecofriendly and sustainable approach towards stress mitigation. The phytotoxic and pesticidal effects of flavonoids provide insight regarding how effective these biochemicals could be in the field if practically implemented. The use of bioflavonoids as natural herbicides is an area of growing interest in integrated weed management. Further research and investigations are required to understand the full range of activity of flavonoids produced naturally and/or applied artificially.

Author Contributions: Authors have contributed equally to this work. All authors have read and agreed to the published version of the manuscript.

Funding: The authors would like to acknowledge the support for this review paper was provided through the Biomass Canada Cluster (BMC), which is funded through Agriculture and Agri-Food Canada's AgriScience program and industry partners.

Acknowledgements: The authors would like to acknowledge the support for this review paper was provided through the Biomass Canada Cluster (BMC), which is funded through Agriculture and Agri-Food Canada's AgriScience program and industry partners.

Conflicts of Interest: The authors declare no conflict of interest.

References

1. Observatory Earth. World of Change: Global Temperatures. Available online: https://earthobservatory.nasa.gov/world-of-change/global-temperatures (accessed on 24 March 2020).
2. Richardson, Y.; Blin, J.; Julbe, A. A short overview on purification and conditioning of syngas produced by biomass gasification: Catalytic strategies, process intensification and new concepts. *Prog. Energy Combust. Sci.* **2012**, *38*, 765–781. [CrossRef]
3. EPA. Global Emissions by Gas. Available online: https://www.epa.gov/ghgemissions/global-greenhouse-gas-emissions-data (accessed on 24 March 2020).
4. Scialabba, N.E.-H.; Müller-Lindenlauf, M. Organic agriculture and climate change. *Renew. Agric. Food Syst.* **2010**, *25*, 158–169. [CrossRef]
5. Chalker-Scott, L. Environmental significance of anthocyanins in plant stress responses. *Photochem. Photobiol.* **1999**, *70*, 1–9. [CrossRef]
6. Chen, S.; Wu, F.; Li, Y.; Qian, Y.; Pan, X.; Li, F.; Wang, Y.; Wu, Z.; Fu, C.; Lin, H.; et al. NtMYB4 and NtCHS1 are critical factors in the regulation of flavonoid biosynthesis and are involved in salinity responsiveness. *Front. Plant Sci.* **2019**, *10*. [CrossRef] [PubMed]
7. Shojaie, B.; Mostajeran, A.; Ghanadian, M. Flavonoid dynamic responses to different drought conditions: Amount, type, and localization of flavonols in roots and shoots of *Arabidopsis thaliana* L. *Turk. J. Biol.* **2016**, *40*, 612–622. [CrossRef]
8. Brunetti, C.; Di Ferdinando, M.; Fini, A.; Pollastri, S.; Tattini, M. Flavonoids as antioxidants and developmental regulators: Relative significance in plants and humans. *Int. J. Mol. Sci.* **2013**, *14*, 3540–3555. [CrossRef] [PubMed]
9. Cushnie, T.T.; Lamb, A.J. Antimicrobial activity of flavonoids. *Int. J. Antimicrob. Agents* **2005**, *26*, 343–356. [CrossRef]
10. Singla, P.; Garg, N. Plant flavonoids: Key players in signaling, establishment, and regulation of rhizobial and mycorrhizal endosymbioses. In *Mycorrhiza-Function, Diversity, State of the Art*; Springer: Berlin/Heidelberg, Germany, 2017; pp. 133–176.
11. Liu, C.W.; Murray, J.D. The role of flavonoids in nodulation host-range specificity: An update. *Plants (Basel)* **2016**, *5*, 33. [CrossRef] [PubMed]
12. Mierziak, J.; Kostyn, K.; Kulma, A. Flavonoids as important molecules of plant interactions with the environment. *Molecules* **2014**, *19*, 16240–16265. [CrossRef]

13. Dudek, B.; Warskulat, A.-C.; Schneider, B. The Occurrence of Flavonoids and Related Compounds in Flower Sections of Papaver nudicaule. *Plants* **2016**, *5*, 28. [CrossRef]
14. Cesco, S.; Mimmo, T.; Tonon, G.; Tomasi, N.; Pinton, R.; Terzano, R.; Neumann, G.; Weisskopf, L.; Renella, G.; Landi, L.; et al. Plant-borne flavonoids released into the rhizosphere: Impact on soil bio-activities related to plant nutrition. A review. *Biol. Fertil. Soils* **2012**, *48*, 123–149. [CrossRef]
15. Nabavi, S.M.; Samec, D.; Tomczyk, M.; Milella, L.; Russo, D.; Habtemariam, S.; Suntar, I.; Rastrelli, L.; Daglia, M.; Xiao, J.; et al. Flavonoid biosynthetic pathways in plants: Versatile targets for metabolic engineering. *Biotechnol. Adv.* **2020**, *38*, 107316. [CrossRef] [PubMed]
16. Wang, Y.; Chen, S.; Yu, O. Metabolic engineering of flavonoids in plants and microorganisms. *Appl. Microbiol. Biotechnol.* **2011**, *91*, 949–956. [CrossRef] [PubMed]
17. Aherne, S.A.; O'Brien, N.M. Dietary flavonols: Chemistry, food content, and metabolism. *Nutrition* **2002**, *18*, 75–81. [CrossRef]
18. Panche, A.N.; Diwan, A.D.; Chandra, S.R. Flavonoids: An overview. *J. Nutr. Sci.* **2016**, *5*, e47. [CrossRef]
19. Samanta, A.; Das, G.; Das, S.K. Roles of Flavonoids in Plants. *Int. J. Pharm. Sci. Technol.* **2011**, *6*, 12–35.
20. Sultana, B.; Anwar, F. Flavonols (kaempferol, quercetin, myricetin) contents of selected fruits, vegetables and medicinal plants. *Food Chem.* **2008**, *108*, 879–884. [CrossRef]
21. Crozier, A.; Lean, M.E.; McDonald, M.S.; Black, C. Quantitative analysis of the flavonoid content of commercial tomatoes, onions, lettuce, and celery. *J. Agric. Food Chem.* **1997**, *45*, 590–595. [CrossRef]
22. López-Lázaro, M. Distribution and biological activities of the flavonoid luteolin. *Mini Rev. Med. Chem.* **2009**, *9*, 31–59. [CrossRef]
23. Heim, K.E.; Tagliaferro, A.R.; Bobilya, D.J. Flavonoid antioxidants: Chemistry, metabolism and structure-activity relationships. *J. Nutr. Biochem.* **2002**, *13*, 572–584. [CrossRef]
24. Del Valle, I.; Webster, T.M.; Cheng, H.-Y.; Thies, J.E.; Kessler, A.; Miller, M.K.; Ball, Z.T.; MacKenzie, K.R.; Masiello, C.A.; Silberg, J.J.; et al. Soil organic matter attenuates the efficacy of flavonoid-based plant-microbe communication. *Sci. Adv.* **2020**, *6*, eaax8254. [CrossRef] [PubMed]
25. Erlund, I. Review of the flavonoids quercetin, hesperetin, and naringenin. Dietary sources, bioactivities, bioavailability, and epidemiology. *Nutr. Res.* **2004**, *24*, 851–874. [CrossRef]
26. Andersen, O.M.; Markham, K.R. *Flavonoids: Chemistry, Biochemistry and Applications*; CRC Press: Boca Raton, FL, USA, 2005.
27. Arts, I.C.; van de Putte, B.; Hollman, P.C. Catechin contents of foods commonly consumed in The Netherlands. 1. Fruits, vegetables, staple foods, and processed foods. *J. Agric. Food Chem.* **2000**, *48*, 1746–1751. [CrossRef] [PubMed]
28. Tsao, R. Chemistry and biochemistry of dietary polyphenols. *Nutrients* **2010**, *2*, 1231–1246. [CrossRef]
29. Hillel, D.; Hatfield, J.L. *Encyclopedia of Soils in the Environment*; Elsevier: Amsterdam, The Netherlands, 2005; Volume 3.
30. Dykhuizen, D. Species numbers in bacteria. *Proc. Calif. Acad. Sci.* **2005**, *56*, 62.
31. Lal, R.; Francaviglia, R. *Sustainable Agriculture Reviews 29: Sustainable Soil Management: Preventive and Ameliorative Strategies*; Springer: Berlin/Heidelberg, Germany, 2019; Volume 29.
32. Mia, M.B.; Shamsuddin, Z. Rhizobium as a crop enhancer and biofertilizer for increased cereal production. *Afr. J. Biotechnol.* **2010**, *9*, 6001–6009.
33. Afzal, A.; Bano, A. Rhizobium and phosphate solubilizing bacteria improve the yield and phosphorus uptake in wheat (Triticum aestivum). *Int. J. Agric. Biol.* **2008**, *10*, 85–88.
34. Hijri, M. Analysis of a large dataset of mycorrhiza inoculation field trials on potato shows highly significant increases in yield. *Mycorrhiza* **2016**, *26*, 209–214. [CrossRef]
35. Winkel-Shirley, B. Biosynthesis of flavonoids and effects of stress. *Curr. Opin. Plant Biol.* **2002**, *5*, 218–223. [CrossRef]
36. Dixon, R.A.; Pasinetti, G.M. Flavonoids and isoflavonoids: From plant biology to agriculture and neuroscience. *Plant Physiol.* **2010**, *154*, 453–457. [CrossRef]
37. Cetinkaya, H.; Kulak, M.; Karaman, M.; Karaman, H.S.; Kocer, F. Flavonoid accumulation behavior in response to the abiotic stress: Can a uniform mechanism be illustrated for all plants. In *Flavonoids—From Biosynthesis to Human Health*; Intechopen: London UK, 2017 [CrossRef]

38. Trichopoulou, A.; Vasilopoulou, E.; Hollman, P.; Chamalides, C.; Foufa, E.; Kaloudis, T.; Kromhout, D.; Miskaki, P.; Petrochilou, I.; Poulima, E.; et al. Nutritional composition and flavonoid content of edible wild greens and green pies: A potential rich source of antioxidant nutrients in the Mediterranean diet. *Food Chem.* **2000**, *70*, 319–323. [CrossRef]

39. Parvez, M.; Tomita-Yokotani, K.; Fujii, Y.; Konishi, T.; Iwashina, T. Effects of quercetin and its derivatives on the growth of Arabidopsis thaliana and Neurospora crassa. *Biochem. Syst. Ecol.* **2004**, *32*, 631–635. [CrossRef]

40. Škerget, M.; Kotnik, P.; Hadolin, M.; Hraš, A.R.; Simonič, M.; Knez, Ž. Phenols, proanthocyanidins, flavones and flavonols in some plant materials and their antioxidant activities. *Food Chem.* **2005**, *89*, 191–198. [CrossRef]

41. Díaz, M.; Rossini, C. Bioactive natural products from Sapindaceae deterrent and toxic metabolites against insects. In *Insecticides–Pest Engineering*; Perveen, F., Ed.; InTech: Rijeka, Croatia, 2012; pp. 287–308.

42. Tatsimo, S.J.N.; Tamokou, J.d.D.; Havyarimana, L.; Csupor, D.; Forgo, P.; Hohmann, J.; Kuiate, J.-R.; Tane, P. Antimicrobial and antioxidant activity of kaempferol rhamnoside derivatives from Bryophyllum pinnatum. *BMC Res. Notes* **2012**, *5*, 158. [CrossRef] [PubMed]

43. Mendki, P.; Salunke, B.; Kotkar, H.; Maheshwari, V.; Mahulikar, P.; Kothari, R. Antimicrobial and Insecticidal Activities of Flavonoid Glycosides from Calotropis procera L. for Post-harvest Preservation of Pulses. *Biopestic. Int.* **2005**, *1*, 193–200.

44. Manach, C.; Scalbert, A.; Morand, C.; Rémésy, C.; Jiménez, L. Polyphenols: Food sources and bioavailability. *Am. J. Clin. Nutr.* **2004**, *79*, 727–747. [CrossRef]

45. Wu, N.-L.; Fang, J.-Y.; Chen, M.; Wu, C.-J.; Huang, C.-C.; Hung, C.-F. Chrysin protects epidermal keratinocytes from UVA- and UVB-induced damage. *J. Agric. Food Chem.* **2011**, *59*, 8391–8400. [CrossRef]

46. Liggins, J.; Bluck, L.; Runswick, S.; Atkinson, C.; Coward, W.; Bingham, S. Daidzein and genistein contents of vegetables. *Br. J. Nutr.* **2000**, *84*, 717–725. [CrossRef]

47. Pei, Y.; Siemann, E.; Tian, B.; Ding, J. Root flavonoids are related to enhanced AMF colonization of an invasive tree. *AoB Plants* **2020**, *12*. [CrossRef]

48. Cavia-Saiz, M.; Busto, M.D.; Pilar-Izquierdo, M.C.; Ortega, N.; Perez-Mateos, M.; Muñiz, P. Antioxidant properties, radical scavenging activity and biomolecule protection capacity of flavonoid naringenin and its glycoside naringin: A comparative study. *J. Sci. Food Agric.* **2010**, *90*, 1238–1244. [CrossRef]

49. Wilmsen, P.K.; Spada, D.S.; Salvador, M. Antioxidant activity of the flavonoid hesperidin in chemical and biological systems. *J. Agric. Food Chem.* **2005**, *53*, 4757–4761. [CrossRef] [PubMed]

50. Tsai, S.M.; Phillips, D.A. Flavonoids released naturally from alfalfa promote development of symbiotic glomus spores in vitro. *Appl. Environ. Microbiol.* **1991**, *57*, 1485–1488. [CrossRef]

51. Harborne, J.B. The flavonoids: Advances in research since 1986 (Harborne, J.B.). *J. Chem. Educ.* **1995**, *72*, A73. [CrossRef]

52. Naeimi, A.F.; Alizadeh, M. Antioxidant properties of the flavonoid fisetin: An updated review of in vivo and in vitro studies. *Trends Food Sci. Technol.* **2017**, *70*, 34–44. [CrossRef]

53. Ong, K.C.; Khoo, H.-E. Biological effects of myricetin. *Gen. Pharmacol. Vasc. Syst.* **1997**, *29*, 121–126. [CrossRef]

54. Peters, N.; Frost, J.; Long, S. A plant flavone, luteolin, induces expression of Rhizobium meliloti nodulation genes. *Science* **1986**, *233*, 977–980. [CrossRef]

55. Garcia, K.; Delaux, P.M.; Cope, K.R.; Ane, J.M. Molecular signals required for the establishment and maintenance of ectomycorrhizal symbioses. *New Phytol.* **2015**, *208*, 79–87. [CrossRef]

56. Rice-evans, C.A.; Miller, N.J.; Bolwell, P.G.; Bramley, P.M.; Pridham, J.B. The relative antioxidant activities of plant-derived polyphenolic flavonoids. *Free Radic. Res.* **1995**, *22*, 375–383. [CrossRef]

57. Tsanova-Savova, S.; Ribarova, F.; Gerova, M. (+)-Catechin and (−)-epicatechin in Bulgarian fruits. *J. Food Compos. Anal.* **2005**, *18*, 691–698. [CrossRef]

58. Pathan, S.I.; Ceccherini, M.T.; Sunseri, F.; Lupini, A. Rhizosphere as hotspot for plant-soil-microbe interaction. In *Carbon and Nitrogen Cycling in Soil*; Springer: Berlin/Heidelberg, Germany, 2020; pp. 17–43.

59. Mandal, S.M.; Chakraborty, D.; Dey, S. Phenolic acids act as signaling molecules in plant-microbe symbioses. *Plant Signal. Behav.* **2010**, *5*, 359–368. [CrossRef]

60. Sugiyama, A.; Shitan, N.; Yazaki, K. Involvement of a soybean ATP-binding cassette-type transporter in the secretion of genistein, a signal flavonoid in legume-Rhizobium symbiosis. *Plant Physiol.* **2007**, *144*, 2000–2008. [CrossRef]

61. Shaw, L.J.; Morris, P.; Hooker, J.E. Perception and modification of plant flavonoid signals by rhizosphere microorganisms. *Environ. Microbiol.* **2006**, *8*, 1867–1880. [CrossRef] [PubMed]

62. Kovács, E.; Keresztes, Á. Effect of gamma and UV-B/C radiation on plant cells. *Micron* **2002**, *33*, 199–210. [CrossRef]

63. Chaparro, J.M.; Sheflin, A.M.; Manter, D.K.; Vivanco, J.M. Manipulating the soil microbiome to increase soil health and plant fertility. *Biol. Fertil. Soils* **2012**, *48*, 489–499. [CrossRef]

64. Munoz Aguilar, J.M.; Ashby, A.M.; Richards, A.J.M.; Loake, G.J.; Watson, M.D.; Shaw, C.H. Chemotaxis of Rhizobium leguminosarum biovar phaseoli towards flavonoid inducers of the symbiotic nodulation genes. *Microbiology* **1988**, *134*, 2741–2746. [CrossRef]

65. Hassan, S.; Mathesius, U. The role of flavonoids in root-rhizosphere signalling: Opportunities and challenges for improving plant-microbe interactions. *J. Exp. Bot.* **2012**, *63*, 3429–3444. [CrossRef]

66. Bolaños-Vásquez, M.C.; Werner, D. Effects of Rhizobium tropici, R. etli, and R. leguminosarum bv. phaseoli on nod gene-inducing flavonoids in root exudates of Phaseolus vulgaris. *Mol. Plant Microbe Interact.* **1997**, *10*, 339–346. [CrossRef]

67. Stambulska, U.Y.; Bayliak, M.M. Legume-rhizobium symbiosis: Secondary metabolites, free radical processes, and effects of heavy metals. In *Bioactive Molecules in Food*; Springer: Berlin/Heidelberg, Germany, 2019; pp. 1–32. [CrossRef]

68. Davidson, E.A. The contribution of manure and fertilizer nitrogen to atmospheric nitrous oxide since 1860. *Nat. Geosci.* **2009**, *2*, 659–662. [CrossRef]

69. Hartwig, U.A.; Joseph, C.M.; Phillips, D.A. Flavonoids released naturally from alfalfa seeds enhance growth rate of Rhizobium meliloti. *Plant Physiol.* **1991**, *95*, 797–803. [CrossRef]

70. Maxwell, C.A.; Hartwig, U.A.; Joseph, C.M.; Phillips, D.A. A chalcone and two related flavonoids released from alfalfa roots induce *nod* genes of *Rhizobium meliloti*. *Plant Physiol.* **1989**, *91*, 842–847. [CrossRef]

71. Dolatabadian, A.; Sanavy, S.A.M.M.; Ghanati, F.; Gresshoff, P.M. Morphological and physiological response of soybean treated with the microsymbiont Bradyrhizobium japonicum pre-incubated with genistein. *S. Afr. J. Bot.* **2012**, *79*, 9–18. [CrossRef]

72. Subramanian, S.; Stacey, G.; Yu, O. Endogenous isoflavones are essential for the establishment of symbiosis between soybean and Bradyrhizobium japonicum. *Plant J.* **2006**, *48*, 261–273. [CrossRef] [PubMed]

73. Jones, F.R.; Tisdale, W. Effect of soil temperature upon the development of nodules on the roots of certain legumes. *J. Agric. Res.* **1921**, *22*, 17–37.

74. Lynch, D.; Smith, D. Soybean (Glycine max) modulation and N2-fixation as affected by exposure to a low root-zone temperature. *Physiol. Plant.* **1993**, *88*, 212–220. [CrossRef]

75. Pan, B.; Zhang, F.; Smith, D.L. Genistein addition to the rooting medium of soybean at the onset of nitrogen fixation increases nodulation. *J. Plant Nutr.* **1998**, *21*, 1631–1639. [CrossRef]

76. Gibson, A. Factors in the physical and biological environment affecting nodulation and nitrogen fixation by legumes. *Plant Soil* **1971**, *35*, 139–152. [CrossRef]

77. Layzell, D.; Rochman, P.; Canvin, D. Low root temperatures and nitrogenase activity in soybean. *Can. J. Bot.* **1984**, *62*, 965–971. [CrossRef]

78. Zhang, F.; Smith, D.L. Preincubation of bradyrhizobium japonicum with genistein accelerates nodule development of soybean at suboptimal root zone temperatures. *Plant Physiol.* **1995**, *108*, 961–968. [CrossRef]

79. Kasper, S.; Christoffersen, B.; Soti, P.; Racelis, A. Abiotic and biotic limitations to nodulation by leguminous cover crops in South Texas. *Agriculture* **2019**, *9*, 209. [CrossRef]

80. Homida, M.; Issa, A.A.; Ohyam, T. Impact of harsh environmental conditions on nodule formation and dinitrogen fixation of legumes. In *Advances in Biology and Ecology of Nitrogen Fixation*; Intechopen: London, UK, 2014. [CrossRef]

81. Zhang, F.; Smith, D.L. Application of genistein to inocula and soil to overcome low spring soil temperature inhibition of soybean nodulation and nitrogen fixation. *Plant Soil* **1997**, *192*, 141–151. [CrossRef]

82. Belkheir, A.M.; Zhou, X.; Smith, D.L. Variability in yield and yield component responses to genistein pre-incubated Bradyrhizobium japonicum by soybean [Glycine max (L.) Merr] cultivars. *Plant soil* **2001**, *229*, 41–46. [CrossRef]

83. Belkheir, A.; Zhou, X.; Smith, D. Response of soybean [Glycine max (L.) Merr.] cultivars to genistein-preincubated bradyrhizobium japonicum: Nodulation and dry matter accumulation under Canadian short-season conditions. *J. Agron. Crop Sci.* **2000**, *185*, 167–175. [CrossRef]

84. Morrison, M.; Cober, E.; Saleem, M.; McLaughlin, N.; Fregeau-Reid, J.; Ma, B.; Yan, W.; Woodrow, L. Changes in isoflavone concentration with 58 years of genetic improvement of short-season soybean cultivars in Canada. *Crop Sci.* **2008**, *48*, 2201–2208. [CrossRef]

85. Hinsinger, P. Bioavailability of soil inorganic P in the rhizosphere as affected by root-induced chemical changes: A review. *Plant soil* **2001**, *237*, 173–195. [CrossRef]

86. Mohammadi, K.; Khalesro, S.; Sohrabi, Y.; Heidari, G. A review: Beneficial effects of the mycorrhizal fungi for plant growth. *J. Appl. Environ. Biol. Sci.* **2011**, *1*, 310–319.

87. Parniske, M. Arbuscular mycorrhiza: The mother of plant root endosymbioses. *Nat. Rev. Microbiol.* **2008**, *6*, 763–775. [CrossRef]

88. Vierheilig, H.; Bago, B.; Albrecht, C.; Poulin, M.-J.; Piché, Y. Flavonoids and arbuscular-mycorrhizal fungi. In *Flavonoids in the Living System*; Manthey, J.A., Buslig, B.S., Eds.; Springer: Boston, MA, USA, 1998; pp. 9–33. [CrossRef]

89. Becard, G.; Taylor, L.P.; Douds, D.D.; Pfeffer, P.E.; Doner, L.W. Flavonoids are not necessary plant signal compounds in arbuscular mycorrhizal symbioses. *MPMI Mol. Plant Microbe Interact.* **1995**, *8*, 252–258. [CrossRef]

90. Siqueira, J.; Safir, G.; Nair, M. Stimulation of vesicular-arbuscular mycorrhiza formation and growth of white clover by flavonoid compounds. *New Phytol.* **1991**, *118*, 87–93. [CrossRef]

91. Davies, F.T.; Calderón, C.M.; Huaman, Z. Influence of arbuscular mycorrhizae indigenous to peru and a flavonoid on growth, yield, and leaf elemental concentration of 'yungay'potatoes. *HortScience* **2005**, *40*, 381–385. [CrossRef]

92. Davies, F.T., Jr.; Calderón, C.M.; Huaman, Z.; Gómez, R. Influence of a flavonoid (formononetin) on mycorrhizal activity and potato crop productivity in the highlands of Peru. *Sci. Hortic.* **2005**, *106*, 318–329. [CrossRef]

93. De Almeida Ribeiro, P.R.; dos SANTOS, J.V.; de Carvalho, T.S.; da Silva, J.S.; de Resende, P.M.; de Souza Moreira, F.M. Formononetin associated with phosphorus influences soybean symbiosis with mycorrhizal fungi and Bradyrhizobium. *Biosci. J.* **2016**, *32*.

94. Scervino, J.M.; Ponce, M.A.; Erra-Bassells, R.; Bompadre, J.; Vierheilig, H.; Ocampo, J.A.; Godeas, A. The effect of flavones and flavonols on colonization of tomato plants by arbuscular mycorrhizal fungi of the genera Gigaspora and Glomus. *Can. J. microbiol.* **2007**, *53*, 702–709. [CrossRef] [PubMed]

95. Nakabayashi, R.; Mori, T.; Saito, K. Alternation of flavonoid accumulation under drought stress in Arabidopsis thaliana. *Plant Signal. Behav.* **2014**, *9*, e29518. [CrossRef]

96. Dakora, F.D.; Phillips, D.A. Root exudates as mediators of mineral acquisition in low-nutrient environments. *Plant Soil* **2002**, *245*, 35–47. [CrossRef]

97. Gupta, R.; Chakrabarty, S. Gibberellic acid in plant: Still a mystery unresolved. *Plant Signal. Behav.* **2013**, *8*, e25504. [CrossRef]

98. Tomasi, N.; Weisskopf, L.; Renella, G.; Landi, L.; Pinton, R.; Varanini, Z.; Nannipieri, P.; Torrent, J.; Martinoia, E.; Cesco, S. Flavonoids of white lupin roots participate in phosphorus mobilization from soil. *Soil Biol. Biochem.* **2008**, *40*, 1971–1974. [CrossRef]

99. Cesco, S.; Neumann, G.; Tomasi, N.; Pinton, R.; Weisskopf, L. Release of plant-borne flavonoids into the rhizosphere and their role in plant nutrition. *Plant Soil* **2010**, *329*, 1–25. [CrossRef]

100. Cramer, G.R.; Urano, K.; Delrot, S.; Pezzotti, M.; Shinozaki, K. Effects of abiotic stress on plants: A systems biology perspective. *BMC Plant Biol.* **2011**, *11*, 163. [CrossRef]

101. Haak, D.C.; Fukao, T.; Grene, R.; Hua, Z.; Ivanov, R.; Perrella, G.; Li, S. Multilevel regulation of abiotic stress responses in plants. *Front. Plant Sci.* **2017**, *8*, 1564. [CrossRef]

102. Nihorimbere, V.; Ongena, M.; Smargiassi, M.; Thonart, P. Beneficial effect of the rhizosphere microbial community for plant growth and health. *Biotechnol. Agron. Soc. Environ.* **2011**, *15*, 327–337.

103. Walker, T.S.; Bais, H.P.; Grotewold, E.; Vivanco, J.M. Root exudation and rhizosphere biology. *Plant Physiol.* **2003**, *132*, 44–51. [CrossRef] [PubMed]

104. Chen, H.; Gao, H.; Fang, X.; Ye, L.; Zhou, Y.; Yang, H. Effects of allyl isothiocyanate treatment on postharvest quality and the activities of antioxidant enzymes of mulberry fruit. *Postharvest Biol. Technol.* **2015**, *108*, 61–67. [CrossRef]

105. Solovchenko, A.; Schmitz-Eiberger, M. Significance of skin flavonoids for UV-B-protection in apple fruits. *J. Exp. Bot.* **2003**, *54*, 1977–1984. [CrossRef] [PubMed]

106. Mariz-Ponte, N.; Mendes, R.; Sario, S.; de Oliveira, J.F.; Melo, P.; Santos, C. Tomato plants use non-enzymatic antioxidant pathways to cope with moderate UV-A/B irradiation: A contribution to the use of UV-A/B in horticulture. *J. Plant Physiol.* **2018**, *221*, 32–42. [CrossRef]

107. Merzlyak, M.N.; Solovchenko, A.E.; Gitelson, A.A. Reflectance spectral features and non-destructive estimation of chlorophyll, carotenoid and anthocyanin content in apple fruit. *Postharvest Biol. Technol.* **2003**, *27*, 197–211. [CrossRef]

108. Schmitz-Hoerner, R.; Weissenböck, G. Contribution of phenolic compounds to the UV-B screening capacity of developing barley primary leaves in relation to DNA damage and repair under elevated UV-B levels. *Phytochemistry* **2003**, *64*, 243–255. [CrossRef]

109. Tossi, V.; Lombardo, C.; Cassia, R.; Lamattina, L. RETRACTED: Nitric oxide and flavonoids are systemically induced by UV-B in maize leaves. *Plant Sci.* **2012**, *193–194*, 103–109. [CrossRef]

110. Pathan, M.S.; Lee, J.-D.; Shannon, J.G.; Nguyen, H.T. Recent advances in breeding for drought and salt stress tolerance in soybean. In *Advances in Molecular Breeding Toward Drought and Salt Tolerant Crops*; Jenks, M.A., Hasegawa, P.M., Jain, S.M., Eds.; Springer: Dordrecht, The Netherlands, 2007; pp. 739–773. [CrossRef]

111. Ilangumaran, G.; Smith, D.L. Plant growth promoting rhizobacteria in amelioration of salinity stress: A systems biology perspective. *Front. Plant Sci.* **2017**, *8*, 1768. [CrossRef]

112. Greenway, H.; Munns, R. Mechanisms of salt tolerance in nonhalophytes. *Ann. Rev. Plant Physiol.* **1980**, *31*, 149–190. [CrossRef]

113. Chaves, M.M.; Flexas, J.; Pinheiro, C. Photosynthesis under drought and salt stress: Regulation mechanisms from whole plant to cell. *Ann. Bot.* **2009**, *103*, 551–560. [CrossRef]

114. Perri, S.; Entekhabi, D.; Molini, A. Plant osmoregulation as an emergent water-saving adaptation. *Water Resour. Res.* **2018**, *54*, 2781–2798. [CrossRef]

115. Bernstein, N. Plants and salt: Plant response and adaptations to salinity. In *Model Ecosystems in Extreme Environments*; Elsevier: Amsterdam, The Netherlands, 2019; pp. 101–112.

116. Jones, G.W.; Gorham, J. Intra-and Inter-Cellular Compartmentation of Ions. In *Salinity: Environment-Plants-Molecules*; Springer: Dordrecht, The Netherlands, 2002; pp. 159–180.

117. Flowers, T.J.; Colmer, T.D. Salinity tolerance in halophytes. *New Phytol.* **2008**, *179*, 945–963. [CrossRef] [PubMed]

118. Munns, R.; Tester, M. Mechanisms of salinity tolerance. *Ann. Rev. Plant Biol.* **2008**, *59*, 651–681. [CrossRef] [PubMed]

119. Koyro, H.-W. Effect of salinity on growth, photosynthesis, water relations and solute composition of the potential cash crop halophyte *Plantago coronopus* (L.). *Environ. Exp. Bot.* **2006**, *56*, 136–146. [CrossRef]

120. Rhodes, D.; Nadolska-Orczyk, A.; Rich, P. Salinity, osmolytes and compatible solutes. In *Salinity: Environment-Plants-Molecules*; Springer: Dordrecht, The Netherlands, 2002; pp. 181–204.

121. Al Hassan, M.; Martínez Fuertes, M.; Ramos Sánchez, F.J.; Vicente, O.; Boscaiu, M. Effects of salt and water stress on plant growth and on accumulation of osmolytes and antioxidant compounds in cherry tomato. *Not. Bot. Horti Agrobot. Cluj Napoca* **2015**, *43*, 1–11. [CrossRef]

122. AbdElgawad, H.; Zinta, G.; Hegab, M.M.; Pandey, R.; Asard, H.; Abuelsoud, W. High salinity induces different oxidative stress and antioxidant responses in maize seedlings organs. *Front. Plant Sci.* **2016**, *7*, 276. [CrossRef]

123. Wang, H.-M.; Xiao, X.-R.; Yang, M.-Y.; Gao, Z.-L.; Zang, J.; Fu, X.-M.; Chen, Y.-H. Effects of salt stress on antioxidant defense system in the root of Kandelia candel. *Bot. Stud.* **2014**, *55*, 57. [CrossRef]

124. Abogadallah, G.M. Antioxidative defense under salt stress. *Plant Signal. Behav.* **2010**, *5*, 369–374. [CrossRef]

125. Engwa, G.A. Free radicals and the role of plant phytochemicals as antioxidants against oxidative stress-related diseases. In *Phytochemicals: Source of Antioxidants and Role in Disease Prevention*. BoD–Books on Demand; Intechopen: London, UK, 2018; pp. 49–74.
126. Halliwell, B.; Gutteridge, J.M.C. *Free Radicals in Biology and Medicine*, 5th ed.; Oxford University Press: New York, NY, USA, 2015.
127. Smith, A.M.; Ratcliffe, R.G.; Sweetlove, L.J. Activation and function of mitochondrial uncoupling protein in plants. *J. Biol. Chem.* **2004**, *279*, 51944–51952. [CrossRef]
128. Pietta, P.-G. Flavonoids as antioxidants. *J. Nat. Prod.* **2000**, *63*, 1035–1042. [CrossRef]
129. Lü, J.M.; Lin, P.H.; Yao, Q.; Chen, C. Chemical and molecular mechanisms of antioxidants: Experimental approaches and model systems. *J. Cell. Mol. Med.* **2010**, *14*, 840–860. [CrossRef] [PubMed]
130. Di Ferdinando, M.; Brunetti, C.; Fini, A.; Tattini, M. Flavonoids as antioxidants in plants under abiotic stresses. In *Abiotic Stress Responses in Plants*; Springer: New York, NY, USA, 2012; pp. 159–179. [CrossRef]
131. Kaliamoortiiy, S.; Rao, A. Effect of salinity on anthocyanin accumulation in the root of maize. *Science* **1994**, *248*, 1637–1638.
132. Sugiyama, A.; Yazaki, K. Flavonoids in plant rhizospheres: Secretion, fate and their effects on biological communication. *Plant Biotechnol.* **2014**, *31*, 431–443. [CrossRef]
133. Miransari, M.; Smith, D. Overcoming the stressful effects of salinity and acidity on soybean nodulation and yields using signal molecule genistein under field conditions. *J. Plant Nutr.* **2007**, *30*, 1967–1992. [CrossRef]
134. Miransari, M.; Smith, D. Alleviating salt stress on soybean (Glycine max (L.) Merr.)–Bradyrhizobium japonicum symbiosis, using signal molecule genistein. *Eur. J. Soil Biol.* **2009**, *45*, 146–152. [CrossRef]
135. Ma, D.; Sun, D.; Wang, C.; Li, Y.; Guo, T. Expression of flavonoid biosynthesis genes and accumulation of flavonoid in wheat leaves in response to drought stress. *Plant Physiol. Biochem.* **2014**, *80*, 60–66. [CrossRef]
136. Varela, M.C.; Arslan, I.; Reginato, M.A.; Cenzano, A.M.; Luna, M.V. Phenolic compounds as indicators of drought resistance in shrubs from Patagonian shrublands (Argentina). *Plant Physiol. Biochem.* **2016**, *104*, 81–91. [CrossRef]
137. Koes, R.E.; Quattrocchio, F.; Mol, J.N. The flavonoid biosynthetic pathway in plants: Function and evolution. *BioEssays* **1994**, *16*, 123–132. [CrossRef]
138. Maddox, C.E.; Laur, L.M.; Tian, L. Antibacterial activity of phenolic compounds against the phytopathogen Xylella fastidiosa. *Curr. Microbiol.* **2010**, *60*, 53. [CrossRef]
139. Rattan, R.S. Mechanism of action of insecticidal secondary metabolites of plant origin. *Crop Prot.* **2010**, *29*, 913–920. [CrossRef]
140. Kaplan, D.; Keen, N.; Thomason, I. Studies on the mode of action of glyceollin in soybean incompatibility to the root knot nematode, Meloidogyne incognita. *Physiol. Plant Pathol.* **1980**, *16*, 319–325. [CrossRef]
141. Chin, S.; Behm, C.A.; Mathesius, U. Functions of flavonoids in plant(-)nematode interactions. *Plants (Basel)* **2018**, *7*, 85. [CrossRef] [PubMed]
142. Huang, J.-S.; Barker, K.R. Glyceollin I in soybean-cyst nematode interactions: Spatial and temporal distribution in roots of resistant and susceptible soybeans. *Plant Physiol.* **1991**, *96*, 1302–1307. [CrossRef] [PubMed]
143. Soriano, I.; Asenstorfer, R.; Schmidt, O.; Riley, I. Inducible flavone in oats (Avena sativa) is a novel defense against plant-parasitic nematodes. *Phytopathology* **2004**, *94*, 1207–1214. [CrossRef] [PubMed]
144. Fawe, A.; Abou-Zaid, M.; Menzies, J.; Bélanger, R. Silicon-mediated accumulation of flavonoid phytoalexins in cucumber. *Phytopathology* **1998**, *88*, 396–401. [CrossRef]
145. del Río, J.A.; Gómez, P.; Baidez, A.G.; Arcas, M.C.; Botía, J.M.; Ortuño, A. Changes in the levels of polymethoxyflavones and flavanones as part of the defense mechanism of citrus sinensis (Cv. Valencia Late) fruits against phytophthora citrophthora. *J. Agric. Food Chem.* **2004**, *52*, 1913–1917. [CrossRef]
146. Ortuno, A.; Arcas, M.; Botia, J.; Fuster, M.; Del Río, J. Increasing resistance against Phytophthora citrophthora in tangelo Nova fruits by modulating polymethoxyflavones levels. *J. Agric. Food Chem.* **2002**, *50*, 2836–2839. [CrossRef]
147. Boué, S.M.; Carter, C.H.; Ehrlich, K.C.; Cleveland, T.E. Induction of the soybean phytoalexins coumestrol and glyceollin by Aspergillus. *J. Agric. Food Chem.* **2000**, *48*, 2167–2172. [CrossRef]
148. Cushnie, T.P.; Lamb, A.J. Recent advances in understanding the antibacterial properties of flavonoids. *Int. J. Antimicrob. Agents* **2011**, *38*, 99–107. [CrossRef]

149. Gnanamanickam, S.S.; Patil, S.S. Accumulation of antibacterial isoflavonoids in hypersensitively responding bean leaf tissues inoculated with Pseudomonas phaseolicola. *Physiol. Plant Pathol.* **1977**, *10*, 159–168. [CrossRef]

150. Lyon, F.M.; Wood, R. Production of phaseollin, coumestrol and related compounds in bean leaves inoculated with Pseudomonas spp. *Physiol. Plant Pathol.* **1975**, *6*, 117–124. [CrossRef]

151. Keen, N.; Kennedy, B. Hydroxyphaseollin and related isoflavanoids in the hypersensitive resistance reaction of soybeans to Pseudomonas glycinea. *Physiol. Plant Pathol.* **1974**, *4*, 173–185. [CrossRef]

152. Wyman, J.G. Antibacterial activity of selected isoflavonoids. *Phytopathology* **1978**, *68*, 583. [CrossRef]

153. Tanaka, H.; Sato, M.; Oh-Uchi, T.; Yamaguchi, R.; Etoh, H.; Shimizu, H.; Sako, M.; Takeuchi, H. Antibacterial properties of a new isoflavonoid from Erythrina poeppigiana against methicillin-resistant Staphylococcus aureus. *Phytomedicine* **2004**, *11*, 331–337. [CrossRef] [PubMed]

154. War, A.R.; Paulraj, M.G.; Ahmad, T.; Buhroo, A.A.; Hussain, B.; Ignacimuthu, S.; Sharma, H.C. Mechanisms of plant defense against insect herbivores. *Plant Signal. Behav.* **2012**, *7*, 1306–1320. [CrossRef] [PubMed]

155. Gao, Q.-M.; Zhu, S.; Kachroo, P.; Kachroo, A. Signal regulators of systemic acquired resistance. *Front. Plant Sci.* **2015**, *6*. [CrossRef]

156. Tridiptasari, A.; Brawijaya, U.; Leksono, A.S.; Siswanto, D. Antifeedant Effect of *Moringa oleifera* (L.) Leaf and Seed Extract on Growth and Feeding Activity of Spodoptera litura (Fab.) (Lepidoptera: Noctuidae). *J. Exp. Life Sci.* **2019**, *9*, 25–31. [CrossRef]

157. Aljbory, Z.; Chen, M.S. Indirect plant defense against insect herbivores: A review. *Insect Sci.* **2018**, *25*, 2–23. [CrossRef]

158. Rosell, G.; Quero, C.; Coll, J.; Guerrero, A. Biorational insecticides in pest management. *J. Pestic. Sci.* **2008**, *33*, 103–121. [CrossRef]

159. Quiroz, A.; Mendez, L.; Mutis, A.; Hormazabal, E.; Ortega, F.; Birkett, M.A.; Parra, L. Antifeedant activity of red clover root isoflavonoids on Hylastinus obscurus. *J. Soil Sci. Plant Nutr.* **2017**, *17*, 231–239. [CrossRef]

160. Goławska, S.; Łukasik, I. Antifeedant activity of luteolin and genistein against the pea aphid, Acyrthosiphon pisum. *J. Pest. Sci. (2004)* **2012**, *85*, 443–450. [CrossRef] [PubMed]

161. Johnson, E.T.; Dowd, P.F. Differentially enhanced insect resistance, at a cost, in Arabidopsis thaliana constitutively expressing a transcription factor of defensive metabolites. *J. Agric. Food Chem.* **2004**, *52*, 5135–5138. [CrossRef] [PubMed]

162. Simmonds, M.S. Importance of flavonoids in insect–plant interactions: Feeding and oviposition. *Phytochemistry* **2001**, *56*, 245–252. [CrossRef]

163. Lattanzio, V.; Lattanzio, V.M.; Cardinali, A. Role of phenolics in the resistance mechanisms of plants against fungal pathogens and insects. *Phytochem. Adv. Res.* **2006**, *661*, 23–67.

164. Nenaah, G.E. Toxic and antifeedant activities of prenylated flavonoids isolated from Tephrosia apollinea L. against three major coleopteran pests of stored grains with reference to their structure–activity relationship. *Nat. Prod. Res.* **2014**, *28*, 2245–2252. [CrossRef]

165. Weston, L.A.; Duke, S.O. Weed and crop allelopathy. *Crit. Rev. Plant Sci.* **2003**, *22*, 367–389. [CrossRef]

166. Oerke, E.C. Crop losses to pests. *J. Agric. Sci.* **2005**, *144*, 31–43. [CrossRef]

167. Young, S.L.; Pierce, F.J. *Automation: The Future of Weed Control*; Springer Science and Business Media: Berlin/Heidelberg, Germany, 2013.

168. Molisch, H. *Der Einfluss Einer Pflanze Auf die Andere, Allelopathie*; Fischer: Jena, Germany, 1937.

169. Rice, E. *Allelopathy*, 2nd ed.; Acad. Press Inc.: Orlando, FL, USA, 1984; p. 422.

170. Rob, M.M.; Hossen, K.; Iwasaki, A.; Suenaga, K.; Kato-Noguchi, H. Phytotoxic activity and identification of phytotoxic substances from Schumannianthus dichotomus. *Plants* **2020**, *9*, 102. [CrossRef]

171. Jabran, K.; Mahajan, G.; Sardana, V.; Chauhan, B.S. Allelopathy for weed control in agricultural systems. *Crop Prot.* **2015**, *72*, 57–65. [CrossRef]

172. Alsaadawi, I.S.; Sarbout, A.K.; Al-Shamma, L.M. Differential allelopathic potential of sunflower (*Helianthus annuus* L.) genotypes on weeds and wheat (*Triticum aestivum* L.) crop. *Arch. Agron. Soil Sci.* **2012**, *58*, 1139–1148. [CrossRef]

173. Palma-Tenango, M.; Soto-Hernández, M.; Aguirre-Hernández, E. Flavonoids in agriculture. In *Flavonoids—From Biosynthesis to Human Health*; Intech: Rijeka, Croatia, 2017. [CrossRef]

174. Weston, L.A.; Mathesius, U. Flavonoids: Their structure, biosynthesis and role in the rhizosphere, including allelopathy. *J. Chem. Ecol.* **2013**, *39*, 283–297. [CrossRef] [PubMed]

175. Zhang, H.; Gao, J.-M.; Liu, W.-T.; Tang, J.-C.; Zhang, X.-C.; Jin, Z.-G.; Xu, Y.-P.; Shao, M.-A. Allelopathic substances from walnut (Juglans regia L.) leaves. *Allelopath. J.* **2008**, *21*, 425–431.
176. Yan, Z.; Guo, H.; Yang, J.; Liu, Q.; Jin, H.; Xu, R.; Cui, H.; Qin, B. Phytotoxic flavonoids from roots of Stellera chamaejasme L. (Thymelaeaceae). *Phytochemistry* **2014**, *106*, 61–68. [CrossRef] [PubMed]
177. Kaab, S.B.; Rebey, I.B.; Hanafi, M.; Hammi, K.M.; Smaoui, A.; Fauconnier, M.L.; De Clerck, C.; Jijakli, M.H.; Ksouri, R. Screening of Tunisian plant extracts for herbicidal activity and formulation of a bioherbicide based on Cynara cardunculus. *S. Afr. J. Bot.* **2020**, *128*, 67–76. [CrossRef]
178. Al-obaidi, A.F. Phytotoxicity of plantago major extracts on germination and seedling growth of purslane (portulaca oleracea). In *Seed Dormancy and Germination*; IntechOpen: London, UK, 2020.

 *agronomy*

Article

Agricultural Utilization of Unused Resources: Liquid Food Waste Material as a New Source of Plant Growth-Promoting Microbes

Waleed Asghar, Shiho Kondo, Riho Iguchi, Ahmad Mahmood  and Ryota Kataoka *

Department of Environmental Sciences, Faculty of Life and Environmental Sciences, University of Yamanashi, Kofu, Yamanashi 400-8510, Japan; waleedasghar978@gmail.com (W.A.); g19lr004@yamanashi.ac.jp (S.K.); lustiness_17@yahoo.co.jp (R.I.); ahmadmahmood91@gmail.com (A.M.)
* Correspondence: rkataoka@yamanashi.ac.jp

Received: 15 June 2020; Accepted: 30 June 2020; Published: 2 July 2020

Abstract: Organic amendment is important for promoting soil quality through increasing soil fertility and soil microbes. This study evaluated the effectiveness of using liquid food waste material (LFM) as a microbial resource, by analyzing the microbial community composition in LFM, and by isolating plant growth-promoting bacteria (PGPB) from the material. High-throughput sequencing of LFM, collected every month from May to September 2018, resulted in the detection of >1000 bacterial operational taxonomic units (OTUs) in the LFM. The results showed that *Firmicutes* was abundant and most frequently detected, followed by *Proteobacteria* and *Actinobacteria*. Of the culturable strains isolated from LFM, almost all belonged to the genus *Bacillus*. Four strains of PGPB were selected from the isolated strains, with traits such as indole acetic acid production and 1-aminocyclopropane-1-carboxylic acid deaminase activity. Lettuce growth was improved via LFM amendment with PGPB, and *Brassica rapa* showed significant differences in root biomass when LFM amendment was compared with the use chemical fertilizer. Field experiments using LFM showed slight differences in growth for *Brassica rapa,* lettuce and eggplant, when compared with the use of chemical fertilizer. LFM is a useful microbial resource for the isolation of PGPB, and its use as fertilizer could result in reduced chemical fertilizer usage in sustainable agriculture.

Keywords: bacterial community composition; liquid food waste materials (LFM); plant growth-promoting bacteria (PGPB); plant growth-promoting (PGP) traits

1. Introduction

Plant growth-promoting microorganisms (PGPM) are broadly accepted to enhance crop production [1]. Plant growth-promoting microorganisms enhance plant growth and development through a variety of functions, encompassing the increase of macro-nutrient availability to the host plant by assembly of growth-promoting chemicals [2], nitrogen fixation [3], solubilization of inorganic phosphate and mineralization of organic phosphate [4], production of different types of phytohormones-like organic compounds [5,6] and biological control of phytopathogens by synthesizing antibiotics and/or competing with harmful microorganisms [7,8]. Therefore, the continuous use of PGPM could lead to it replacing pesticides and chemical fertilizers [9].

On another front, the overuse of chemical fertilizers and continuous agricultural activities results in the deterioration of soil quality [10,11] The associated loss of soil health, fertility and nutrient status leads to continuous input requirements. Crop nutrition needs can be met through the provision of inorganic as well as organic fertilizers and biofertilizers. Overreliance on inorganic fertilizers stretches the economics of the farming community, and also leads to consumption of available non-renewable nutrient resources, compromises the potential plant-beneficial microbiome, and can have a severe environmental impact [12]. In contrast, the concerns around the use of organic fertilizers include that they are bulky, slow release, have inconsistent composition and can spread weed seeds, among other things. Therefore, sustainable solutions must be sought for crop production, while focusing on the utilization of all available resources. Organic waste production, which can be animal- or plant-based, including food leftovers, vegetable and fruit peels and market refuse, is a worldwide issue, and its disposal and treatment is increasingly important in developing countries [13]. The large amount of this waste that is produced is a major economic, social and environmental challenge [14], which is associated with extensive handling costs. There are great potential benefits to recycling and reusing this material in agriculture. Recent efforts have led to up to a 25% reduction in food waste in some parts of the world, however, in Japan, although food waste legislation has helped to reduce the volume of food waste produced, more needs to be done in addressing this issue, as reviewed by [15]. General waste, other than that generated by food processing industries and households, contains about 60% organic matter [16]. Hence, the separation of organic matter from general waste streams should be targeted, and treated as a resource rather than a problem [11,17].

Organic waste contains fatty acids, proteins and carbohydrates [18,19] among other constituents, which can be utilized as a source of crop nutrition. The application of organic waste materials in agriculture has been reported to reduce runoff, improve soil structure and increase soil biological activity [20]. In addition, some research has showed that local effective microorganisms (LEM) are a beneficial inoculant for the nitrogen mineralization of organic materials [21,22]. Therefore, the better management of organic waste materials could lead to preservation of soil quality and sustainable crop production [9]. Previous studies have explored the potential of the utilization of food and organic wastes in domestic, agricultural and industrial applications [11,23–25]. Among the variety of waste processing and manipulation procedures prior to their application in agriculture, most have had associated physical, chemical or biological problems. In the effective utilization of food waste, quick manipulation, easy operation and little or no reduction in the nutritional composition of the waste products are all considered publicly acceptable, and could increase the waste's potential for wide application and dissemination. Under this scenario, nutrient retention can be ensured, and minimal damage to the plant-beneficial microbes present in the food waste would be achieved. A food waste recycling facility started operating in 2014 in Kai-City, Yamanashi Prefecture, Japan, which collects food waste from school restaurants in the vicinity, processes the waste using lactic acid fermentation, and supplies the final product in liquid form to farmers (Kai City Biomass [26]). The food waste recycling facility has a structure divided into four phases. In the first phase, food wastes and water, along with an inoculum of microorganisms, such as lactic acid bacteria, are added, and the mixture is agitated and gradually moved to phases 2, 3 and 4. During that time, the pH drops to 3 and the temperature rises above 50 °C to promote fermentation. This liquid food waste material (LFM) has been used as a crop nutrient source by many farmers in the area. Although LFM has been mainly employed for use in agriculture and/or for energy production, the microbiological potential of the plant growth-promoting microbes in LFM has not been studied, to the best of our knowledge. In this study, we explored the microbial community composition of the final form of the recycled waste materials, and studied the ecology of those microorganisms, while also investigating the plant growth-promoting traits of the culturable bacterial isolates.

2. Materials and Methods

2.1. Liquid Food Waste Materials

The LFM was obtained in March 2017 and March 2018 for pot and field experiments, respectively, and in May, June, July, August and September 2018 for bacterial composition analysis (and in October for isolation of microbes), from the Biomass Center at Kai City, Yamanashi, Japan. A portion (50 mL) of the material each month was stored at −80 °C for DNA extraction and high-throughput sequencing.

The Kai City facility produces LFM from residues of local school-provided lunches using the lactic acid fermentation process at the rate of approximately 90 L/day^{-1}. The total carbon and nitrogen of the final form of fertilizer were 31.7% and 1.41%, respectively. The C/N ratio of LFM was 22.4. In addition, the pH of LFM was 3.42 because of the lactic acid fermentation process. Electrical conductivity (EC), nitrate–nitrogen ($NO_3{}^-$-N), ammonium-N ($NH_4{}^+$-N) and available phosphate (Trough-P) were at the values of 6.65 mS/cm^{-1}, 0.95 mg/L^{-1}, 14.8 mg/L^{-1} and 0.69 mg/L^{-1}, respectively.

2.2. Assay of Liquid Food Waste Material (LFM) Utilization

An incubation experiment was carried out to assess the mineralization of $NO_3{}^-$-N from the LFM according to a modified Soil Environmental Analytical Method, 1997. A total of 100 mL of the LFM material was weighed and mixed with 300 g of soil obtained from University of Yamanashi (UofY) Research Farm (hereinafter referred to as UofY farm soil); soil type is gray lowland soil (pH 6.79 ± 0.33; EC (mS/cm^{-1}) 0.11 ± 0.08; $NO_3{}^-$-N 23.1 ± 3.13 mg/kg^{-1}; available phosphate 421 ± 86.8 mg/kg^{-1}). The pots were covered by aluminium foil and incubated for 14 weeks at 25 °C. Since rapeseed cake is used as an organic fertilizer, it was used as a control for nitrogen release after application to soil. The $NO_3{}^-$-N content was measured via the alkali reduction diazo dye method (Soil Environmental Analytical Method, 1997).

2.3. Isolation of Bacteria from LFM

Bacterial isolation from LFM was performed through the dilution plating technique. A total of 1 mL of LFM and 4.0 mL of sterile distilled water was placed in a test tube and mixed thoroughly using a vortex mixer. Subsequently, 50-µL dilutions were taken from the first tube and spread onto Reasoner's 2A agar (R2A) media (Eiken Chemical Co. Ltd., Tochigi, Japan) using a disposable spreader; plates of each dilution were incubated at 25 °C for 3 days. Colonies appearing after 3 days were re-streaked until a single pure colony type per plate was achieved.

2.4. DNA Extraction and PCR Amplification for Culturable Bacteria

DNA was extracted from the isolated strains using the ZR Fungal/Bacterial DNA MiniPrep Kit™ (Zymo Research Corp., Irvine, CA, USA). 16S rRNA gene sequencing was carried out for identification of the strains. Extracted DNA from isolated strains was mixed with prior to PCR amplification. The universal primers 341F (5′-CCTACGGGAGGCAGCAG-3′) and 1378R (5′-TGTGCAAGGAGCAGGGAC-3′) were used to amplify the 16S rRNA gene on a TaKaRa PCR Thermal Cycler Dice® Series Gradient (Takara, Shiga, Japan). The PCR amplification conditions were as follows: 95 °C for 5 min, followed by 30 cycles of 94 °C for 30 s, 58 °C for 30 s, and 72 °C for 1 min, followed by a final extension at 72 °C for 7 min [27]. The amplification mixture for PCR (total volume: 25 µL) contained 1 µL of DNA template, 1 µL of each primer, 9.5 µL of sterilized distilled water and 12.5 µL of GoTaq Green Master mix (Promega, Madison, WI, USA). The amplification products (5 µL) were subjected to electrophoresis on a 1% (*w/v*) agarose gel in tris-acetate-ethylenediaminetetraacetic acid buffer at 100 V for 25 min, and visualized by GelRed™ staining (1:20,000 dilution; Biotium, Fremont, CA, USA). The DNA sequences obtained were compared with those previously reported in the DNA Data Bank of Japan (http://blast.ddbj.nig.ac.jp/), and the nearest neighbor was noted. The sequences of numbers 2, 4, 6 and 11 were submitted to DNA Data Bank of Japan (DDBJ)

2.5. High-Throughput DNA Sequencing

DNA was isolated from the stored LFM samples using the FastDNA™ Spin Kit for Soil (MP Biomedicals Japan, Tokyo, Japan). The DNA concentration was measured using a nano-spectrophotometer and DNA was diluted to 1 ng/µL^{-1} using sterile water. The V4 region of the 16S rRNA gene was amplified using the primers 515F (5′-GTGCCAGCMGCCGCGGTAA-3′) and 806R (5′-GGACTACHVGGGTWTCTAAT-3′) with additional barcode sequences. All PCR reactions were carried out with Phusion® High-Fidelity PCR Master Mix (New England Biolabs

Japan Inc., Tokyo, Japan). The quality and quantity of PCR products was assessed by mixing equal volumes of a loading buffer (containing SYBR green) with PCR products and electrophoresing the samples on 2% (*w/v*) agarose gel. Samples with a bright main strip between 400 and 450 bp were chosen for further experiments. The PCR products were mixed in equal density ratios. Thereafter, the mixed PCR products were purified with a Qiagen Gel Extraction Kit (Qiagen, Hilden, Germany). The libraries—250 bp paired-end reads generated with NEBNext® UltraTM DNA Library Prep Kit for Illumina (New England Biolabs Japan Inc.,Tokyo, Japan) and quantified via Qubit and quantitative PCR—were sequenced on an Illumina HiSeq 2500 platform. Quality control was performed at each step of the procedure. Paired-end reads were assigned to samples based on their unique barcode and truncated by cutting off the barcode and primer sequence. Paired-end reads were merged using FLASH (V1.2.7, http://ccb.jhu.edu/software/FLASH/) [28]. Quality filtering of the raw tags was performed under specific filtering conditions to obtain high-quality clean tags [29] according to the QIIME (V1.7.0, http://qiime.org/scripts/split_libraries_fastq.html) quality control process [30]. The tags were compared with the reference database (Gold database, http://drive5.com/uchime/uchime_download.html) using the UCHIME algorithm (UCHIME Algorithm, http://www.drive5.com/usearch/manual/uchime_algo.html) [31] to detect chimera sequences (http://www.drive5.com/usearch/manual/chimera_formation.html). Next, the chimera sequences were removed [32], and the effective tags were finally obtained. Sequence analysis was performed via Uparse software (Uparse v7.0.1001 http://drive5.com/uparse/) using all the effective tags [33]. Sequences with ≥97% similarity were assigned to the same operational taxonomic units (OTUs). A representative sequence for each OTU was screened for further annotation. For each representative sequence, Mothur software was used against the small subunit rRNA database of SILVA (http://www.arb-silva.de/) [34] for species annotation at each taxonomic rank (Threshold: 0.8–1) [35]. The phylogenetic relationship of the representative sequences of all OTUs was obtained by using MUSCLE software (Version 3.8.31, http://www.drive5.com/muscle/) for rapid comparison of multiple sequences [36]. The abundance of OTUs was normalized using a standard sequence number corresponding to the sample with the least sequences. Subsequent analyses were all performed based on this output normalized data.

The reads were submitted to the DDBJ Sequence Read Archive (https://www.ddbj.nig.ac.jp/dra/index-e.html) under Bioproject, and are available under accession number DRA010367.

2.6. Plant Growth-Promoting Traits of Isolates

Indole acetic acid production: The isolated strains were tested for indole-3-acetic acid (IAA) production. Cultures of each isolate were grown at 25 °C for 4 days in IAA production media (2 g beef extract, 3 g $CaCO_3$, 30 g glucose, pH 7 in 1 L of distilled water) with or without 1 mM (final concentration) tryptophan. The cultures were centrifuged at 10,000 g for 10 min. IAA production was measured in 300 μL of supernatant using 1.2 mL of Salkowski's reagent [37,38]; absorbance was measured at 535 nm in a spectrophotometer, and the concentration was estimated from a standard curve. Control/blank samples were prepared without bacterial inoculation.

1-aminocyclopropane-1-carboxylic acid (ACC) deaminase and nitrogen fixation: DNA was extracted using ZR Bacterial/Fungal DNA MiniPrep Kit™ (Zymo Research Corp., Irvine, CA, USA). The PCR amplification conditions for ACC deaminase and *nif*H genes were as follows: 1 cycle of 95 °C for 5 min, then 30 cycles of 94 °C for 30 s, 55 °C for 30 s, and 72 °C for 1 min, followed by a final extension at 72 °C for 7 min using a T100™ Thermal Cycler (Bio-rad, Hercules, CA, USA). The PCR mixture (total volume: 25 μL) contained 1 μL of DNA template, 1 μL of 10 mM primers (PolF (5′ TGCGAYCCSAARGCBGACTC 3′) and PolR (5′ ATS GCC ATCATY TCR CCG GA 3′) [39] for *nif*H genes; ACCF (5′ GCCAARCGBGAVGACTGCAA 3′) and ACCR (5′ TGCATSGAYTTGCCYTC 3′) [40] for ACC deaminase), 12.5 μL of GoTaq® Green Master Mix and 9.5 μL sterilized distilled water. The PCR amplification products were checked via 1.0% of agarose gel electrophoresis, staining and visualization.

Siderophore production: A slightly modified Chrome Azurol S (CAS) method was used for determination of siderophore production by bacterial isolates [41,42]. A total of 100 mL of medium was prepared as follows: 7.3 mg of hexadecyl trimetly ammonium bromide, 6.04 mg of CAS, 3.04 g of piperazine-1,4-bis(2-ethanesulfonic acid) and 1 mL of 1 mM $FeCl_3 \cdot 6H_2O$. A quantity of 10 mL of the siderophore production medium was applied over the surface of agar plates containing cultivated microorganisms. The blue CAS agar changed to light yellow or orange if siderophores were produced by the bacteria; the siderophore production was evaluated by the following index: + color change, − no color change and ++ color change detected over the entire medium.

Phosphate solubilization: The medium developed by Pikovskaya [43] was used for qualitative estimation of calcium phosphate solubilization by the isolates. Selected strains were inoculated into the media and incubated for 7 days at 25 °C. Zones of clearance around the bacterial colonies were indicative of phosphate solubilization; the results were compiled on the basis of the following index: − No clear zone, ± detectable clear zone but very weak activity, + detectable clear zone.

2.7. Pot Experiments

Two pot experiments were conducted using UofY farm soil. The first compared selected bacterial isolates with an uninoculated control to evaluate the role of specific isolates, and the second compared LFM with an untreated control and a fertilizer control to determine the role of LFM in plant growth promotion. For the first experiment, 11 bacterial isolates, that were selected based on PGP traits, were compared with an uninoculated control in a *Lactuca sativa* var. crispa (lettuce) growth trial. Pots (size: 100 cm^2; Fujiwara Seisakusho, Ltd., Tokyo, Japan) were filled with 300 g of soil (dry weight) and soil moisture was maintained at 60% of water holding capacity daily. A suspension of each bacterial strain (grown for 48 h (stationary phase) at 25 °C with shaking in PDB medium) was applied to the pots while the same volume of uninoculated PDB was applied as the control. Subsequently, a lettuce seedling germinated on a petri dish was transferred to each pot. The plants were grown for five weeks, harvested, and the dry weight of aboveground and belowground parts was recorded after being put into the dry oven set at 60 °C. From the results of the pot experiment using lettuce, four isolates (numbers 2, 4, 6 and 11), which showed the maximum growth-enhancement of lettuce, were selected (data not shown), and then these strains were tested under similar growth conditions for *Brassica campestris* (brassica) and the same parameters were recorded. In the second experiment, the response to LFM was compared with that of chemical fertilizer and control treatments. To achieve this goal, 100 mL of LFM and chemical fertilizer (HYPONeX Japan Corp., Ltd., Osaka, Japan. Liquid Fertilizer, N:P:K = 6:10:5) was mixed with soil to achieve a final concentration of 200 mg/kg^{-1} soil N, while there was no amendment in the control pots. Similar growing conditions and parameters were recorded as for the first experiment.

2.8. Field Experiment Using LFM

A field experiment was conducted comparing chemical fertilizer with LFM at the University of Yamanashi Research Farm (35°60′39.5″ N, 138°57′82.9″ E). The field experimental plots (4 m × 2 m) were treated with LFM and chemical fertilizer. In this field, the chemical fertilizer plots have been continuously treated with chemical fertilizer, and cow compost was applied every two years in all subplots until the year before the study. Soil chemical properties were as follows: pH (H_2O) 7.0, EC 0.12 (mS/cm^{-1}), ex Ca 2940 mg/kg^{-1}, ex-Mg 874 mg/kg^{-1}, ex-K 381 mg/kg^{-1}, CEC 14 ($cmolc/kg^{-1}$), Trough-P 344 mg/kg^{-1}, NH_4-N 5.8 mg/kg^{-1}, NO_3-N 59.4 mg/kg^{-1}. Two replicates were prepared for each of the test vegetables: *Brassica rapa* var., *Lactuca sativa* var. crispa (lettuce) and *Solanum melongena* (eggplant). LFM was input at the rate of 200 kg/ha^{-1}, 200 kg/ha^{-1} and 75 kg/ha^{-1} to the final concentration of soil N for brassica, lettuce and eggplant, respectively, whereas chemical fertilizer input was 200 kg/ha^{-1} soil N for all crops. *Brassica rapa* and lettuce were planted at 20 to 30 plants/plot, and eggplant was cultivated at 9 plants/plot. *Brassica rapa* was grown for 29 days and plant height was measured upon harvest. Lettuce was grown for 56 days; plant height and dry weight of the edible part were measured at harvest. The eggplants were harvested when the fruits grew to a suitable size (around 120 g/fruit); the quantity and weight of the fruits were measured. Fruit harvest began on 12 August 2018 and continued until 29 September 2018.

2.9. Statistical Analysis

Analysis of variance (ANOVA) was carried out to determine the statistical effects of treatments in Statistix 8.0 (Analytical Software, Tallahassee, FL, USA) followed by pairwise comparison of treatment means using Tukey's honestly significant difference (HSD) test and multiple comparisons through Dunnett's test. For the veracity of sequencing data analysis, raw data was merged and filtered to obtain clean data. Effective data was used for operational taxonomic unit (OTUs) clustering. The clustering analysis was applied, and a clustering tree was constructed to study the similarities among different samples. The unweighted pair–group method with arithmetic means (UPGMA) Clustering was performed as a type of hierarchical clustering method to interpret the distance matrix using average linkage, and was conducted using QIIME software (Version 1.7.0).

3. Results

3.1. Isolation of Plant Growth Promoting Bacteria (PGPB)

The bacterial isolation from the LFM was performed via standard methods, through the serial dilution plating technique. Various different strains appeared in the media. The number of culturable bacteria in the LFM was 3.5×10^4 colony forming units/mL^{-1} of LFM. After isolation, 31 strains were randomly selected, and 11 out of those 31 strains were examined for plant growth-promoting (PGP) traits. The sequence of strain numbers 2, 4, 6 and 11 (approximately 940 nt; GenBank accession No. LC553393, LC553394, LC553395, LC553396) was compared with other bacterial nucleotide sequences in GenBank. All strains exhibited a high sequence similarity with *Bacillus* spp.

3.2. Identification of Culturable Bacteria

All 31 isolates that were identified belonged to genus *Bacillus* (Figure S2); these were type A—closely related to *Bacillus velezensis* strain FZB42 (frequency: 3.3%), type B—closely related to *Bacillus amyloliquefaciens* strain MPA 1034 (frequency: 56.7%), type C—closely related to *Bacillus vallismortis* strain NRRL B-14890 (frequency: 26.7%), type D—closely related to *Bacillus subtilis* subsp. *inaquosorum* strain BGSC 3A28 (frequency: 3.3%), type E—closely related to *Bacillus wiedmannii* strain FSL W8-0169 (frequency: 3.3%), type F—closely related to *Bacillus velezensis* strain NTGB-29 (frequency: 3.3%), and type G—closely related to *Bacillus vallismortis* strain DSM 11031 (frequency: 3.3%).

3.3. PGP Traits

Because of the importance of microbial IAA production in influencing the root architecture and initial plant growth, IAA production was examined for 14 of the 31 isolates. Strain numbers 2, 4, 6 and 11 were positive for IAA production with tryptophan (Table 1). Amplification of the *nifH* gene confirmed N fixation potential in strain number 6, whereas amplification of the ACC deaminase gene was positive for all four selected strains (Table 1). Only strain number 11 showed zones of clearance on the Pikovskaya agar plates, indicating the phosphate solubilization ability of this strain (Table 1). Strain number 11 was also positive for siderophore production, with complete color change from blue to yellow, when compared with other non-siderophore-producing strains and the control that had a negative reaction.

Table 1. Plant growth-promoting traits, where + indicates the possession of the following trait, and – indicates the lack of the trait.

Strain No.	Indole-3-Acetic Acid (IAA)	Phosphate Solubilization	Nitrogen Fixation	1-Aminocyclopropane-1-Carboxylic Acid Deaminase	Siderophore
1	–	–	–	–	–
2	+	–	–	+	–
3	–	–	–	–	+
4	+	–	–	+	–
5	-	–	–	+	–
6	+	–	+	+	–
7	–	+	–	–	–
8	–	+	–	–	–
9	–	–	-	+	+
10	–	+	–	–	–
11	+	+	–	+	+

3.4. High-Throughput DNA Sequencing of LFM

In total, 192,355 reads were obtained; the average number of observed species per sample was 1013 ± 170 (max: 1306, min: 892), and the coefficient of variation was 0.17. The average bacterial composition was shown via the integration of the clustering results and the relative abundance of each sample by phylum (Figures 1a,b and S3). *Proteobacteria* were most frequently detected, followed by *Firmicutes* and *Actinobacteria*. The species composition at the phylum level was different in August when compared with that from the other months (Figure 1a). At the genus level, when the top 10 genera were compared between different months, the genera composition of the LFM in May was different from that from the other months (Figures 1a,b and S3). This is because there were few *Lactobacillus* spp. at this time, and the fermentation was in the early stages.

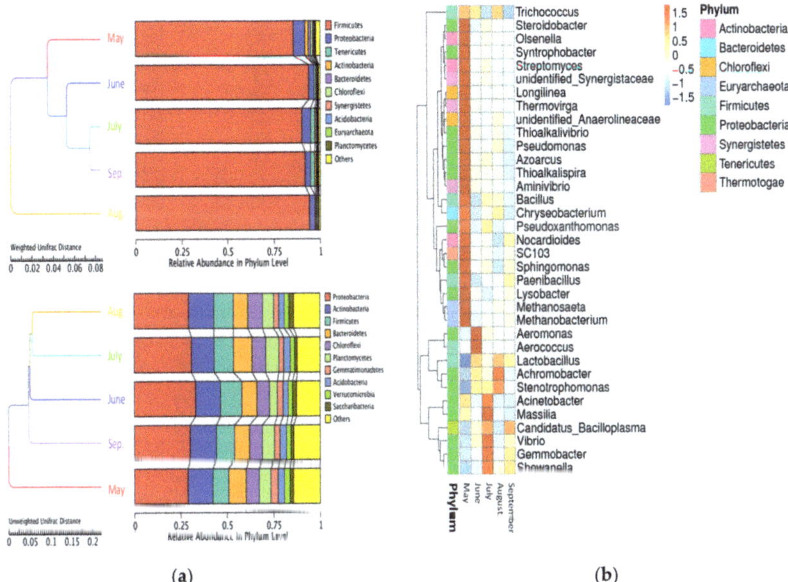

Figure 1. (a) Unweighted pair-group method with arithmetic means cluster tree based on unweighted and weighted unifrac distance. This was displayed with the integration of clustering results and the relative abundance of each sample by phylum. (b) The top 10 taxa at the genus level were selected to form the distribution histogram of relative abundance.

3.5. Incubation Study of LFM Utilization

The release of NO_3^--N and phosphate was used to determine the potential for nutrient provision from LFM. The release of NO_3^--N started from 14 days in the LFM, and approximately 200 mg/kg^{-1} of NO_3^--N had accumulated by the end of 14 weeks of incubation (Figure S1a). Phosphate availability followed the same trend as that of NO_3^--N release (Figure S1b). Available phosphate was rapidly released from rapeseed cake, whereas no phosphate was detected from LFM until week 2. From week 3, the availability of phosphates increased slightly in the LFM.

3.6. Pot Experiments

A pot experiment was conducted to examine the effect of selected strains on the growth of brassica and lettuce. In the initial experiment on lettuce, 11 strains were tested; 4 of these strains showed high activity when compared with the control. Therefore, strains 2, 4, 6 and 11 were tested on brassica, and showed significant differences in their growth-promoting effect (Figure 2). A pot experiment was also conducted to assess the effect of LFM on the growth of both brassica and lettuce. There was no significant difference in the growth characteristics of lettuce between the LFM and chemical fertilizer treatments (Figure 3a); however, brassica exhibited significant differences in its root biomass with LFM amendment (Figure 3b, Tukey's HSD, $p < 0.05$).

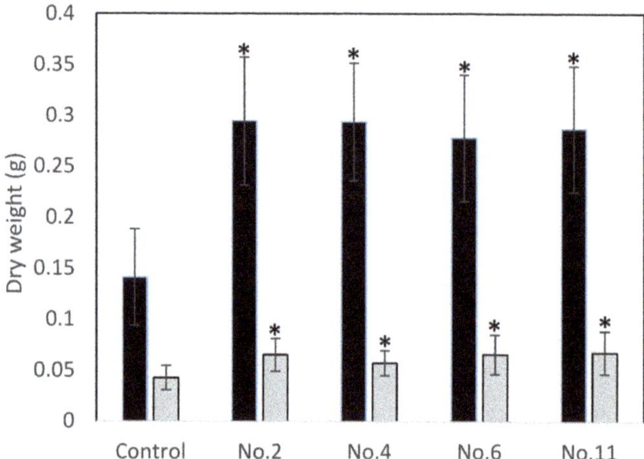

Figure 2. Growth response of *Brassica campestris* with plant growth-promoting bacteria selected in this study. Control ($n = 7$), strain number 2 ($n = 10$), strain number 4 ($n = 10$), strain number 6 ($n = 7$), strain number 11 ($n = 10$), Dunnett test ($p < 0.05$). The vertical bar indicates the standard error. * indicates significance differences between treatments when compared with the control ($p < 0.05$).

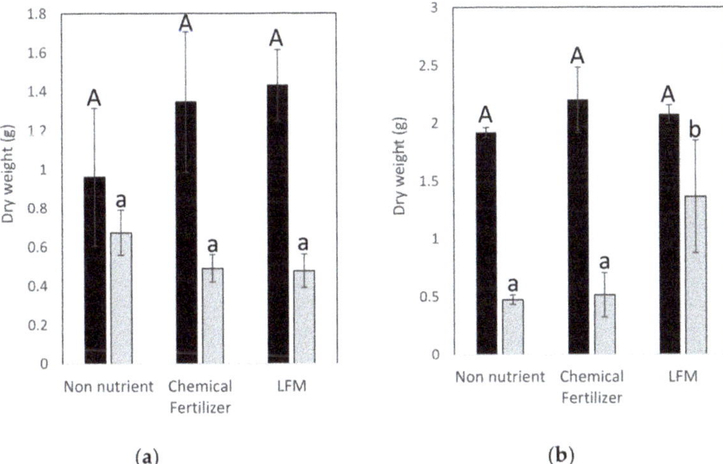

(a)　　　　　　　　　　　　　(b)

Figure 3. Pot experiment testing lettuce (**a**) and *Brassica rapa* (**b**) growth using liquid food waste materials (LFM). Values presented are means and standard error (*n* = 3). Closed bar and Gray bar mean edible part and root, respectively. Treatments of the same crop with different letters are significantly different by Tukey's HSD ($p < 0.05$).

3.7. Field Experiments

Field experiments were conducted to assess the effect of LFM on the growth of *Brassica rapa*, lettuce and eggplant. The growth of *Brassica rapa* and lettuce in the field was similar to that in the pot experiment. The heights achieved by *Brassica rapa* were 33.0 ± 0.67 cm and 32.5 ± 0.78 cm, with LFM and chemical fertilizer, respectively. The lettuce grown with LFM amendment was slightly larger than that grown with chemical fertilizer, but not significantly so (Figure 4a). Eggplant also grew slightly better with LFM than with chemical fertilizer, but the differences were not significant (Figure 4b,c for eggplant).

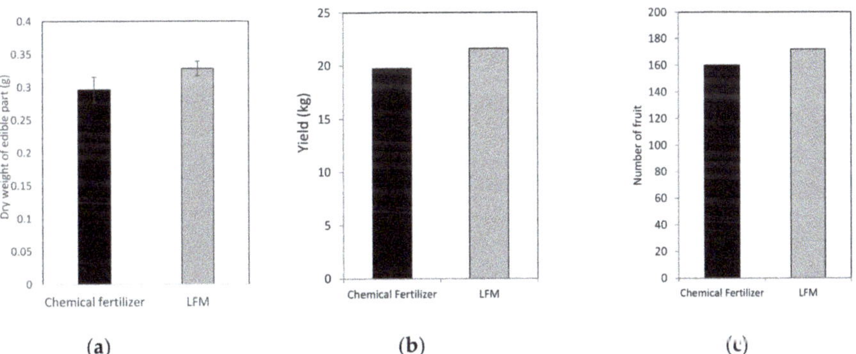

(a)　　　　　　　　　　　　(b)　　　　　　　　　　　　(c)

Figure 4 Field experiment testing the effect of liquid food waste materials (LFM) on the growth of Lettuce (**a**) and eggplant (**b**,**c**).

4. Discussion

This study showed that useful microorganisms, such as PGPB, were present in LFM produced from the recycling of unused resources. This is an important finding that leads to the promotion of recycling, and also indicates that unused resources are useful as microbial resources. LFM can be used as a fertilizer, and has other positive effects on vegetable growth. However, there are limitations to the plant growth-enhancement functions of LFM. If we apply isolated strains to LFM in order to promote a plant growth function more effective than that of the original LFM, some supplementary nutrients are needed to produce an effect. This point still remains an issue.

High-throughput sequencing analysis of the bacterial community composition in the LFM revealed substantial differences between the sampling months May and August. A similar bacterial community composition was observed for the samples from June, July and September. The May sample was still in the early stages of fermentation, and there was no school-provided lunch in August; therefore, LFM was stored in the tank, which might help to explain the differences in the community composition of these samples.

Most of the bacteria isolated from LFM were *Bacillus* spp., which is a spore-forming bacteria. This was because the LFM pH was reduced to 3.0 through the process of lactic acid fermentation, and the temperature exceeded 50 °C. However, this result was obtained because we used R2A media for isolation, and we detected the family of *Lactobacillaceae* in the LFM through high-throughput sequencing (Figure S3). The most frequently isolated *Bacillus* spp. strains were *Bacillus amyloliquefaciens* (56.7%) and *Bacillus vallismortis* (26.7%). *Bacillus* species are known to produce dormant spores [44], and enact an anti-pathogen activity through the assembly of non-ribosomal cyclic lipopeptides [45]. In addition, *Bacillus* species are considered PGPB because of their potential for antibiotic production, biofilm formation on the plant root surface, and production of plant hormones [46,47]. Furthermore, seed treatment with *Bacillus* species has been shown to significantly enhance shoot fresh and dry weight, as well as plant height, in various crops [48,49]. From the results of this study, >1000 bacterial OTUs were identified in LFM; therefore, it might be possible to isolate other useful strains, other than *Bacillus* spp., under a range of isolation conditions, including increased pH.

Organic materials slowly release nutrients, but they are still a promising alternative to chemical fertilizers, as their application can reduce nutrient leaching, volatilization and problems of toxicity [50]. In the present study, LFM was used as an alternative to chemical fertilizers in order to investigate the release of $NO_3{}^-$-N and available phosphate (Figure S1). Low amounts of available phosphate were released from LFM during the incubation study because of the low total phosphate concentration in LFM. LFM released $NO_3{}^-$-N from the third week of incubation, and released approximately 200 kg/ha^{-1} $NO_3{}^-$-N during the 13 weeks of incubation. Moreover, the biomass richness of soil fertilized with LFM was higher than that treated with chemical fertilizer [51], and LFM did not change the soil pH after treatment through our study.

The growth-promoting effect of strain number 11, with an IAA-producing ability and an ACC deaminase activity, was confirmed in a pot experiment; growth was significantly promoted with inoculation by strain number 11, when compared with the control (Figure 2). The growth of lettuce in a pot was the same with both LFM and chemical fertilizer. For *Brassica rapa*, the growth of the edible (aboveground) part was the same with both LFM and chemical fertilizer, whereas the root biomass was significantly increased with LFM, when compared with chemical fertilizer (Figure 3b). These results indicate that the PGPB in LFM contributed to the increase in root biomass of *Brassica rapa*. A field experiment was conducted to assess the effect of LFM on the growth of *Brassica rapa*, lettuce and eggplant. Although the N input by LFM was less than half that of the chemical fertilizer, the growth of the eggplant with LFM was the same as that with chemical fertilizer. The yield with LFM was higher than, but not significantly different from, that achieved with chemical fertilizer. These results indicate the abundance and activity of PGPB in LFM, and their efficacy in supporting eggplant growth under the conditions tested. LFM could be a viable alternative to commercially available chemical fertilizers, without an adverse effect on soil and vegetable growth. A previous study showed the positive effect of PGPB inoculation on vegetable growth and yield [52].

All the selected strains that showed growth-enhancement in the pot experiment had an IAA-producing ability (Table 1). IAA is a type of plant hormone that promotes root elongation and enhances root growth. Many PGPBs with the IAA-production ability have been isolated in previous studies [1,53]. Furthermore, all strains that were positive for IAA production also showed ACC deaminase activity (strain numbers 2, 4, 6 and 11). This suggests that IAA production and ACC deaminase activity contribute greatly to enhancing the plant growth in our isolated strains, while *Caulobacter* sp. had a negative impact on plant growth, even though it produced higher levels of IAA [54]. In addition, previous studies have shown that PGP microbes and PGPB can promote plant growth indirectly or directly, through the production of ACC deaminase and through reducing the ethylene level in the developing plants through the roots [52,55], by generating plant growth hormones like IAA [56]. It is likely that ACC deaminase and IAA production promote root growth in a similar fashion [57,58]. Of the selected strains, only strain number 11 showed phosphorus solubilizing potential and siderophore production (Table 1). Phosphate solubilization is effective in soils with low available phosphoric acid, and siderophore production chelates the iron in soils with high pH to help plant uptake [52,59]. However, the detailed mechanism of plant growth-enhancement is complex, and further investigation is needed [54,60].

The selection of PGPB strains from LFM was important to confirm the positive effect of the inoculants on plant growth, and to optimize their application for maximum impact on vegetable crops. The main aim of this study was to reduce the commercial use of chemical fertilizers, by utilizing LFM as an alternative fertilizer with the maximum impact on crop growth and soil, and minimal environmental impact. Further investigation into LFM use as an organic fertilizer should evaluate any adverse impact of its application to the soil environment.

Supplementary Materials: The following are available online at http://www.mdpi.com/2073-4395/10/7/954/s1, Figure S1: Nitrate nitrogen (**a**) and available phosphate (**b**) release from organic materials, ● rapeseed oil cake, ▲ Liquid food waste material (LFM). Figure S2: 31 strains isolated from liquid food waste material belong to genus *Bacillus*. Type A closely related to *Bacillus velezensis* strain FZB42, type B closely related to *Bacillus amyloliquefaciens* strain MPA 1034, type C closely related to *Bacillus vallismortis* strain NRRL B-14890, type D closely related to *Bacillus subtilis* subsp. inaquosorum strain BGSC 3A28, type E closely related to *Bacillus wiedmannii* strain FSL W8-0169, type F closely related to *Bacillus velezensis* strain NTGB-29, and type G closely related to *Bacillus vallismortis* strain DSM 11031. Figure S3: Top 30 bacterial compositions, in different taxonomic levels.

Author Contributions: Conceptualization, R.K.; methodology, S.K., R.I., R.K., W.A., A.M.; writing—original draft preparation, R.K., W.A., A.M.; writing—review and editing, R.K.; visualization, R.K., W.A.; supervision, R.K.; project administration, R.K.; funding acquisition, R.K. All authors have read and agreed to the published version of the manuscript.

Funding: This research was funded by Strategic research project in university of yamanashi, grant number 14 and the Japan Society for the Promotion of Science (JSPS), Ministry of Education, Culture, Sports, Science and Technology, Japan [KAKENHI Grant Number 19K12379].

Acknowledgments: The authors are grateful to Mr. Yuta KOBAYASHI for supporting to manage the field conditions. We also thank Kai city for their many supports.

Conflicts of Interest: The authors declare no conflict of interest.

References

1. Wang, J.; Li, R.; Zhang, H.; Wei, G.; Li, Z. Beneficial bacteria activate nutrients and promote wheat growth under conditions of reduced fertilizer application. *BMC Microbiol.* **2020**, *20*, 1–12.

2. Khan, N.; Bano, A.; Rahman, M.A.; Guo, J.; Kang, Z.; Babar, M.A. Comparative physiological and metabolic analysis reveals a complex mechanism involved in drought tolerance in chickpea (*Cicer arietinum* L.) induced by PGPR and PGRs. *Sci. Rep.* **2019**, *9*, 1–19.

3. Lin, L.; Li, Z.; Hu, C.; Zhang, X.; Chang, S.; Yang, L.; Li, Y.; An, Q. Plant growth-promoting nitrogen-fixing enterobacteria are in association with sugarcane plants growing in Guangxi, China. *Microbes Environ.* **2012**, *27*, 391–398. [CrossRef] [PubMed]

4. Zaheer, A.; Malik, A.; Sher, A.; Qaisrani, M.M.; Mehmood, A.; Khan, S.U.; Ashraf, M.; Mirza, Z.; Karim, S.; Rasool, M. Isolation, characterization, and effect of phosphate-zinc-solubilizing bacterial strains on chickpea (*Cicer arietinum* L.) growth. *Saudi J. Biol. Sci.* **2019**, *26*, 1061–1067. [CrossRef]

5. Sivasankari, B.; Anandharaj, M.; Daniel, T. Effect of PGR producing bacterial strains isolated from vermisources on germination and growth of *Vigna unguiculata* (L.) Walp. *J. Biochem. Technol.* **2014**, *5*, 808–813.

6. Grobelak, A.; Kokot, P.; Hutchison, D.; Grosser, A.; Kacprzak, M. Plant growth-promoting rhizobacteria as an alternative to mineral fertilizers in assisted bioremediation-sustainable land and waste management. *J. Environ. Manag.* **2018**, *227*, 1–9. [CrossRef]

7. Sun, G.; Yao, T.; Feng, C.; Chen, L.; Li, J.; Wang, L. Identification and biocontrol potential of antagonistic bacteria strains against Sclerotinia sclerotiorum and their growth-promoting effects on Brassica napus. *Biol. Control.* **2017**, *104*, 35–43. [CrossRef]

8. Etesami, H.; Emami, S.; Alikhani, H.A. Potassium solubilizing bacteria (KSB): Mechanisms, promotion of plant growth, and future prospects A review. *J. Soil Sci. Plant. Nutr.* **2017**, *17*, 897–911. [CrossRef]

9. Padmini, O.S.; Wuryani, S.; Aryani, R. Application of Organic Fertilizer and Plant Growth-Promoting Rhizobacteria (PGPR) to Increase Rice Yield and Quality. In *ICoSI: Proceedings of the 2nd International Conference on Sustainable Innovation 2014*; Taufik, T., Prabasari, I., Rineksane, I.A., Yaya, R., Widowati, R., Rosyidi, S.A.P., Riyadi, S., Harsanto, P., Eds.; Springer: Berlin/Heidelberg, Germany, 2017; pp. 3–11.

10. Komatsuzaki, M.; Ohta, H. Soil management practices for sustainable agro-ecosystems. *Sustain. Sci.* **2007**, *2*, 103–120. [CrossRef]

11. Mahmood, A.; Iguchi, R.; Kataoka, R. Multifunctional food waste fertilizer having the capability of Fusarium-growth inhibition and phosphate solubility: A new horizon of food waste recycle using microorganisms. *Waste Manag.* **2019**, *94*, 77–84. [CrossRef]

12. Mahmood, A.; Kataoka, R. Metabolite profiling reveals a complex response of plants to application of plant growth-promoting endophytic bacteria. *Microbiol. Res.* **2020**, *234*, 126421. [CrossRef] [PubMed]

13. Sinha, R.K.; Herat, S.; Bharambe, G.; Patil, S.; Bapat, P.; Chauhan, K.; Valani, D. Human waste-as potential resource: Converting trash into treasure by embracing the 5 r's philosophy for safe and sustainable waste management. *Environ. Res. J.* **2009**, *13*, 111–171.

14. Gustavsson, J.; Cederberg, C.; Sonesson, U.; Van Otterdijk, R.; Meybeck, A. *Global Food Losses and Food Waste*; FAO: Rome, Italy, 2011.

15. Liu, C.; Hotta, Y.; Santo, A.; Hengesbaugh, M.; Watabe, A.; Totoki, Y.; Allen, D.; Bengtsson, M. Food waste in Japan: Trends, current practices and key challenges. *J. Clean. Prod.* **2016**, *133*, 557–564. [CrossRef]

16. Lin, C.S.K.; Pfaltzgraff, L.A.; Herrero-Davila, L.; Mubofu, E.B.; Abderrahim, S.; Clark, J.H.; Koutinas, A.A.; Kopsahelis, N.; Stamatelatou, K.; Dickson, F. Food waste as a valuable resource for the production of chemicals, materials and fuels. Current situation and global perspective. *Energy Environ. Sci.* **2013**, *6*, 426–464. [CrossRef]

17. Baroutian, S.; Munir, M.T.; Sun, J.Y.; Eshtiaghi, N.; Young, B.R. Rheological characterisation of biologically treated and non-treated putrescible food waste. *Waste Manag.* **2018**, *71*, 494–501. [CrossRef]

18. Yan, S.; Li, J.; Chen, X.; Wu, J.; Wang, P.; Ye, J.; Yao, J. Enzymatical hydrolysis of food waste and ethanol production from the hydrolysate. *Renew. Energy* **2011**, *36*, 1259–1265. [CrossRef]

19. Leung, C.C.J.; Cheung, A.S.Y.; Zhang, A.Y.-Z.; Lam, K.F.; Lin, C.S.K. Utilisation of waste bread for fermentative succinic acid production. *Biochem. Eng. J.* **2012**, *65*, 10–15. [CrossRef]

20. Marinari, S.; Masciandaro, G.; Ceccanti, B.; Grego, S. Influence of organic and mineral fertilisers on soil biological and physical properties. *Bioresour. Technol.* **2000**, *72*, 9–17. [CrossRef]

21. Ney, L.; Franklin, D.; Mahmud, K.; Cabrera, M.; Hancock, D.; Habteselassie, M.; Newcomer, Q.; Dahal, S. Impact of inoculation with local effective microorganisms on soil nitrogen cycling and legume productivity using composted broiler litter. *Appl. Soil Ecol.* **2020**, *154*, 103567. [CrossRef]

22. Ney, L.; Franklin, D.; Mahmud, K.; Cabrera, M.; Hancock, D.; Habteselassie, M.; Newcomer, Q. Examining trophic-level nematode community structure and nitrogen mineralization to assess local effective microorganisms' role in nitrogen availability of swine effluent to forage crops. *Appl. Soil Ecol.* **2018**, *130*, 209–218. [CrossRef]

23. Truong, L.; Morash, D.; Liu, Y.; King, A. Food waste in animal feed with a focus on use for broilers. *Int. J. Recycl. Org. Waste Agric.* **2019**, *8*, 417–429. [CrossRef]

24. Torres-León, C.; Ramírez-Guzman, N.; Londoño-Hernandez, L.; Martinez-Medina, G.A.; Díaz-Herrera, R.; Navarro-Macias, V.; Alvarez-Pérez, O.B.; Picazo, B.; Villarreal-Vázquez, M.; Ascacio-Valdes, J.; et al. Food waste and byproducts: An opportunity to minimize malnutrition and hunger in developing countries. *Front. Sustain. Food Syst.* **2018**, *2*. [CrossRef]

25. Girotto, F.; Alibardi, L.; Cossu, R. Food waste generation and industrial uses: A review. *Waste Manag.* **2015**, *45*, 32–41. [CrossRef]

26. Center, K.-C.B. Food waste recycle. Available online: https://www.city.kai.yamanashi.jp/kurashi_tetsuduki/gomi_kankyo_pet/kankyo/3704.html (accessed on 27 May 2020).

27. Mahmood, A.; Takagi, K.; Ito, K.; Kataoka, R. Changes in endophytic bacterial communities during different growth stages of cucumber (*Cucumis sativus* L.). *World J. Microbiol. Biotechnol.* **2019**, *35*, 104. [CrossRef]

28. Magoč, T.; Salzberg, S.L. FLASH: Fast length adjustment of short reads to improve genome assemblies. *Bioinformatics* **2011**, *27*, 2957–2963. [CrossRef]

29. Bokulich, N.A.; Subramanian, S.; Faith, J.J.; Gevers, D.; Gordon, J.I.; Knight, R.; Mills, D.A.; Caporaso, J.G. Quality-filtering vastly improves diversity estimates from Illumina amplicon sequencing. *Nat. Methods* **2013**, *10*, 57–59. [CrossRef]

30. Caporaso, J.G.; Kuczynski, J.; Stombaugh, J.; Bittinger, K.; Bushman, F.D.; Costello, E.K.; Fierer, N.; Peña, A.G.; Goodrich, J.K.; Gordon, J.I.; et al. QIIME allows analysis of high-throughput community sequencing data. *Nat. Methods* **2010**, *7*, 335. [CrossRef]

31. Edgar, R.C.; Haas, B.J.; Clemente, J.C.; Quince, C.; Knight, R. UCHIME improves sensitivity and speed of chimera detection. *Bioinformatics* **2011**, *27*, 2194–2200. [CrossRef]

32. Haas, B.J.; Gevers, D.; Earl, A.M.; Feldgarden, M.; Ward, D.V.; Giannoukos, G.; Ciulla, D.; Tabbaa, D.; Highlander, S.K.; Sodergren, E.; et al. Chimeric 16S rRNA sequence formation and detection in Sanger and 454-pyrosequenced PCR amplicons. *Genome Res.* **2011**, *21*, 494–504. [CrossRef]

33. Edgar, R.C. UPARSE: Highly accurate OTU sequences from microbial amplicon reads. *Nat. Methods* **2013**, *10*, 996. [CrossRef]

34. Wang, Q.; Garrity, G.M.; Tiedje, J.M.; Cole, J.R. Naïve Bayesian Classifier for Rapid Assignment of rRNA Sequences into the New Bacterial Taxonomy. *Appl. Environ. Microbiol.* **2007**, *73*, 5261–5267. [CrossRef] [PubMed]

35. Quast, C.; Pruesse, E.; Yilmaz, P.; Gerken, J.; Schweer, T.; Yarza, P.; Peplies, J.; Glöckner, F.O. The SILVA ribosomal RNA gene database project: Improved data processing and web-based tools. *Nucleic Acids Res.* **2013**, *11*, D590–D596. [CrossRef] [PubMed]

36. Edgar, R.C. Muscle: Multiple sequence alignment with high accuracy and high throughput. *Nucleic Acids Res.* **2004**, *32*, 1792–1797. [CrossRef]

37. Acuña, J.; Jorquera, M.; Martínez, O.; Menezes-Blackburn, D.; Fernández, M.; Marschner, P.; Greiner, R.; Mora, M. Indole acetic acid and phytase activity produced by rhizosphere bacilli as affected by pH and metals. *J. Soil Sci. Plant. Nutr.* **2011**, *11*, 1–12.

38. Patten, C.L.; Glick, B.R. Role of Pseudomonas putida indoleacetic acid in development of the host plant root system. *Appl. Environ. Microbiol.* **2002**, *68*, 3795–3801. [CrossRef]

39. Poly, F.; Ranjard, L.; Nazaret, S.; Gourbière, F.; Monrozier, L.J. Comparison of nifH gene pools in soils and soil microenvironments with contrasting properties. *Appl. Environ. Microbiol.* **2001**, *67*, 2255–2262. [CrossRef]

40. Jha, B.; Gontia, I.; Hartmann, A. The roots of the halophyte Salicornia brachiata are a source of new halotolerant diazotrophic bacteria with plant growth-promoting potential. *Plant. Soil* **2012**, *356*, 265–277. [CrossRef]

41. Schwyn, B.; Neilands, J. Universal chemical assay for the detection and determination of siderophores. *Anal. Biochem.* **1987**, *160*, 47–56. [CrossRef]

42. Pérez-Miranda, S.; Cabirol, N.; George-Téllez, R.; Zamudio-Rivera, L.; Fernández, F. O-CAS, a fast and universal method for siderophore detection. *J. Microbiol. Methods* **2007**, *70*, 127–131. [CrossRef]

43. Pikovskaya, R. Mobilization of phosphorus in soil in connection with vital activity of some microbial species. *Mikrobiologiya* **1948**, *17*, 362–370.

44. Piggot, P.J.; Hilbert, D.W. Sporulation of Bacillus subtilis. *Curr. Opin. Microbiol.* **2004**, *7*, 579–586. [CrossRef] [PubMed]

45. Zhang, B.; Qin, Y.; Han, Y.; Dong, C.; Li, P.; Shang, Q. Comparative proteomic analysis reveals intracellular targets for bacillomycin L to induce Rhizoctonia solani Kühn hyphal cell death. *Biochim. Biophys. Acta* **2016**, *1864*, 1152–1159. [CrossRef] [PubMed]

46. Santos, C.A.; Nobre, B.; da Silva, T.L.; Pinheiro, H.; Reis, A. Dual-mode cultivation of Chlorella protothecoides applying inter-reactors gas transfer improves microalgae biodiesel production. *J. Biotechnol.* **2014**, *184*, 74–83. [CrossRef] [PubMed]

47. Rahman, A.; Uddin, W.; Wenner, N.G. Induced systemic resistance responses in perennial ryegrass against M agnaporthe oryzae elicited by semi-purified surfactin lipopeptides and live cells of B acillus amyloliquefaciens. *Mol. Plant. Pathol.* **2015**, *16*, 546–558. [CrossRef]

48. Yuan, J.; Ruan, Y.; Wang, B.; Zhang, J.; Waseem, R.; Huang, Q.; Shen, Q. Plant growth-promoting rhizobacteria strain Bacillus amyloliquefaciens NJN-6-enriched bio-organic fertilizer suppressed Fusarium wilt and promoted the growth of banana plants. *J. Agric. Food Chem.* **2013**, *61*, 3774–3780. [CrossRef] [PubMed]

49. Gowtham, H.; Murali, M.; Singh, S.B.; Lakshmeesha, T.; Murthy, K.N.; Amruthesh, K.; Niranjana, S. Plant growth promoting rhizobacteria-Bacillus amyloliquefaciens improves plant growth and induces resistance in chilli against anthracnose disease. *Biol. Control.* **2018**, *126*, 209–217. [CrossRef]

50. Machado, D.L.M.; Lucena, C.C.d.; Santos, D.d.; Siqueira, D.L.d.; Matarazzo, P.H.M.; Struiving, T.B. Slow-release and organic fertilizers on early growth of Rangpur lime. *Rev. Ceres* **2011**, *58*, 359–365. [CrossRef]

51. Cheong, J.C.; Lee, J.T.; Lim, J.W.; Song, S.; Tan, J.K.; Chiam, Z.Y.; Yap, K.Y.; Lim, E.Y.; Zhang, J.; Tan, H.T. Closing the food waste loop: Food waste anaerobic digestate as fertilizer for the cultivation of the leafy vegetable, xiao bai cai (*Brassica rapa*). *Sci. Total Environ.* **2020**, *715*, 136789. [CrossRef]

52. Sheirdil, R.A.; Hayat, R.; Zhang, X.-X.; Abbasi, N.A.; Ali, S.; Ahmed, M.; Khattak, J.Z.K.; Ahmad, S. Exploring potential soil bacteria for sustainable wheat (*Triticum aestivum* L.) production. *Sustainability* **2019**, *11*, 3361. [CrossRef]

53. Zhang, Z.; Zhou, W.; Li, H. The role of GA, IAA and BAP in the regulation of in vitro shoot growth and microtuberization in potato. *Acta Physiol. Plant.* **2005**, *27*, 363–369. [CrossRef]

54. Berrios, L.; Ely, B. Plant growth enhancement is not a conserved feature in the Caulobacter genus. *Plant. Soil* **2020**, *449*, 81–95. [CrossRef]

55. Dey, R.; Pal, K.; Bhatt, D.; Chauhan, S. Growth promotion and yield enhancement of peanut (*Arachis hypogaea* L.) by application of plant growth-promoting rhizobacteria. *Microbiol. Res.* **2004**, *159*, 371–394. [CrossRef] [PubMed]

56. Mishra, M.; Kumar, U.; Mishra, P.K.; Prakash, V. Efficiency of plant growth promoting rhizobacteria for the enhancement of *Cicer arietinum* L. growth and germination under salinity. *Adv. Biol Res.* **2010**, *4*, 92–96.

57. Glick, B.R.; Cheng, Z.; Czarny, J.; Duan, J. Promotion of Plant Growth by ACC Deaminase-Producing Soil Bacteria. In *New Perspectives and Approaches in Plant Growth-Promoting Rhizobacteria, Research*; Bakker, P.A.H.M., Raaijmakers, J.M., Bloemberg, G., Höfte, M., Lemanceau, P., Cook, B.M., Eds.; Springer: Berlin/Heidelberg, Germany, 2007; pp. 329–339.

58. Pandey, S.; Gupta, S. ACC deaminase producing bacteria with multifarious plant growth promoting traits alleviates salinity stress in French bean (*Phaseolus vulgaris*) plants. *Front. Microbiol.* **2019**, *10*, 1506.

59. Vansuyt, G.; Robin, A.; Briat, J.-F.; Curie, C.; Lemanceau, P. Iron acquisition from Fe-pyoverdine by Arabidopsis thaliana. *Mol. Plant. Microbe Interact.* **2007**, *20*, 441–447. [CrossRef]

60. Khan, N.; Bano, A.; Babar, M.A. Metabolic and physiological changes induced by plant growth regulators and plant growth promoting rhizobacteria and their impact on drought tolerance in *Cicer arietinum* L. *PLoS ONE* **2019**, *14*, e0213040. [CrossRef] [PubMed]

 *agronomy*

Article

Salicylic Acid Improves Boron Toxicity Tolerance by Modulating the Physio-Biochemical Characteristics of Maize (*Zea mays* L.) at an Early Growth Stage

Muhammad Nawaz [1,*], Sabtain Ishaq [2], Hasnain Ishaq [1], Naeem Khan [3], Naeem Iqbal [1], Shafaqat Ali [4,5,*], Muhammad Rizwan [4], Abdulaziz Abdullah Alsahli [6] and Mohammed Nasser Alyemeni [6]

[1] Department of Botany, Government College University Faisalabad, Lahore 54000, Pakistan; hasnain.official1@gmail.com (H.I.); drnaeem@gcuf.edu.pk (N.I.)

[2] Department of Botany, University of Agriculture Faisalabad, Faisalabad 38000, Pakistan; sabtainishaq28@gmail.com

[3] Department of Agronomy, Institute of Food and Agricultural Sciences, University of Florida, Gainesville, FL 32611, USA; naeemkhan@ufl.edu

[4] Department of Environmental Sciences and Engineering, Government College University, Lahore 54000, Pakistan; mrazi1532@yahoo.com

[5] Department of Biological Sciences and Technology, China Medical University, Taichung 40402, Taiwan

[6] Botany and Microbiology Department, College of Science, King Saud University, P.O. Box. 2460, Riyadh 11451, Saudi Arabia; aalshenaaalshenaifi@ksu.edu.sa (A.A.A.); mnyemeni@ksu.edu.sa (M.N.A.)

[*] Correspondence: muhammadnawaz@gcuf.edu.pk (M.N.); shafaqataligill@yahoo.com or shafaqataligill@gcuf.edu.pk (S.A.)

Received: 14 November 2020; Accepted: 16 December 2020; Published: 21 December 2020

Abstract: The boron (B) concentration surpasses the plant need in arid and semi-arid regions of the world, resulting in phyto-toxicity. Salicylic acid (SA) is an endogenous signaling molecule responsible for stress tolerance in plants and is a potential candidate for ameliorating B toxicity. In this study, the effects of seed priming with SA (0, 50, 100 and 150 µM for 12 h) on the growth, pigmentation and mineral concentrations of maize (*Zea mays* L.) grown under B toxicity were investigated. One-week old seedlings were subjected to soil spiked with B (0, 15 and 30 mg kg^{-1} soil) as boric acid. Elevating concentrations of B reduced the root and shoot length, but these losses were significantly restored in plants raised from seeds primed with 100 µM of SA. The B application decreased the root and shoot fresh/dry biomasses significantly at 30 mg kg^{-1} soil. The chlorophyll and carotenoid contents decreased with increasing levels of B, while the contents of anthocyanin, H_2O_2, ascorbic acid (ASA) and glycinebetaine (GB) were enhanced. The root K and Ca contents were significantly increased, while a reduction in the shoot K contents was recorded. The nitrate concentration was significantly higher in the shoot as compared to the root under applied B toxic regimes. However, all of these B toxicity effects were diminished with 100 µM SA applications. The current study outcomes suggested that the exogenously applied SA modulates the response of plants grown under B toxic conditions, and hence could be used as a plant growth regulator to stimulate plant growth and enhance mineral nutrient uptake under B-stressed conditions.

Keywords: biomass reduction; cereal crops; growth regulators; metal stress

1. Introduction

Abrupt changes in climate along with the potential abiotic and biotic stresses are serious challenges for plant growth and production worldwide [1]. Environmental stresses negatively influence the germination, growth and yield of the crop plants. The continuous yield losses caused by abiotic stresses are one of the important reasons for socioeconomic imbalance [2]. Drought reduces the yield of staple food crops throughout the world up to 70% [3], and the effects of drought and salt stress on plant growth mechanisms and patterns have been discussed [3,4]. In the last few decades, soil and water resources are being contaminated with toxic elements due to industrial revolution and urbanization together with the use of artificial fertilizers [5,6]. Increasing levels of these toxic elements are imposing harmful effects on plants, plant-dependent animals and ultimately human health [7].

Boron is an important micronutrient in many plants for their normal functioning [8]. It is also considered to be an essential element for vascular plants according to the defined criteria for essentiality. The indirect association of B with photosynthesis has been reported in crop plants—e.g., soybean [9]. However, the rate of emergence and productivity is also decreased in many plants, including tomato, maize, wheat, alfalfa and carrot under B toxicity [10]. The B toxicity significantly reduces the yield of crop plants in relatively dry areas of the world [11]. Some of the factors contributing to the elevating levels of B are the use of fertilizers, mining and irrigation [12,13]. The B-induced toxicity occurs more commonly in saline soil in semi-arid geographical zones [14]. The interplay between salt stress and B nutrition in plants has been described, with contrasting results showing antagonistic and synergistic relations even within the same plant species [15]. It has been observed that salinity increases B toxicity [16], but the interaction of salinity and B is not fully understood [17], making it an important area of research in plant physiology and ecotoxicology.

Oxidative stress may result from a deficiency or excess of B, which triggers the over-production of reactive oxygen species (ROS). The ROS and their derivatives are highly toxic agents and damage cellular membranes due to lipid peroxidation, causing protein denaturation and mutations in DNA [18]. Different nutrients such as silicon (Si) [19], zinc (Zn) [20,21], potassium (K) [22] and calcium (Ca) [23] can ameliorate B toxicity in different crop plants. The SA signal molecule [24] plays an important role in reducing the hazardous effects posed by biotic and abiotic stresses. Thus, SA has been used by many researchers to reduce the hazardous effects of different stresses such as osmotic stress [25], heat, saline and B toxicity in wheat [26].

Among the most important staple foods, maize holds an important position after wheat and rice [27]. Maize is well known for its high potential of extracting heavy metals from soil [28]. Despite this phytoextraction ability, maize is affected by various environmental stresses along with the high metal concentrations. The abiotic stress effects on maize growth and yield have been studied [29,30]. In the current study, the main objective was to assess the effects of high B toxicity under the remodeling effects of SA in terms of physio-biochemical improvements in the maize cultivar Gohar-19.

2. Results

For assessing the effects of SA on mitigating the effects of B toxicity, plants were supplied with 0, 50, 100 and 150 µM of SA. The B toxicity levels were 0, 15 and 30 mg kg^{-1} soil. Roots transport B via passive diffusion or facilitate transport [30] in the plant body through transpiration streams and it is accumulated in older shoots without being translocated [31], therefore the study parameters include both the root and shoot data of maize cv. Gohar-19.

2.1. Root and Shoot Length

The B toxicity significantly reduced the root and shoot length of maize seedlings. High B concentrations in soil inhibit the root and shoot growth due to the decreased photosynthetic activity and net plant productivity. Elevating the B concentration in soil decreased the root and shoot length up to 21.77% and 25.25%, respectively, which are significant reductions (Table 1, Figure 1). The priming of seeds with SA reduced the B toxic effects and retained the root and shoot lengths. Plant seeds that were primed with various concentrations (0, 50, 100 and 150 µM) of SA improved the root and shoot lengths. Significant increases in the root and shoot lengths were observed at 100 µM SA (Figure 2, Table 1). A 23.8% increase in root length was observed with the application of 100 µM of SA in 30 mg kg^{-1} of

B-treated plants, while a 26.7% decrease was observed in the shoot length of 30 mg kg^{-1} B-treated plants as compared with the control. The SA application at 100 μM was found to be the best treatment and caused increases in the shoot length in 30 mg kg^{-1} B-treated plants up to 31.8%.

Table 1. Effects of SA (0, 50, 100 and 150 μM) on the plant root and shoot length of maize cultivar Gohar-19 under different B toxicity levels (0, 15 and 30 mg kg^{-1}).

		Root Length (cm)		
		0 mg kg^{-1} B	15 mg kg^{-1} B	30 mg kg^{-1} B
	0 μM	27.1 ± 0.89 [c]	24.3 ± 1.05 [b]	21.2 ± 0.88 [c]
SA	50 μM	27.8 ± 1.01 [b]	24.2 ± 0.97 [b]	21.34 ± 1.03 [c]
	100 μM	29 ± 0.98 [a]	28.2 ± 0.87 [a]	26.4 ± 0.77 [a]
	150 μM	26.4 ± 1.12 [d]	23 ± 1.24 [c]	21.8 ± 1.02 [b]
		Shoot Length (cm)		
	0 μM	30.3 ± 0.69 [b]	28.5 ± 0.85 [b]	22.65 ± 1.25 [c]
SA	50 μM	30.2 ± 1.13 [b]	28.7 ± 0.77 [b]	22.10 ± 1.02 [d]
	100 μM	32.0 ± 0.99 [a]	30.5 ± 0.98 [a]	29.00 ± 0.84 [a]
	150 μM	28.0 ± 1.21 [c]	27.4 ± 0.66 [c]	27.00 ± 0.96 [b]

LSD 5% = 0.44. Values in the same column with different letters in superscript differ significantly.

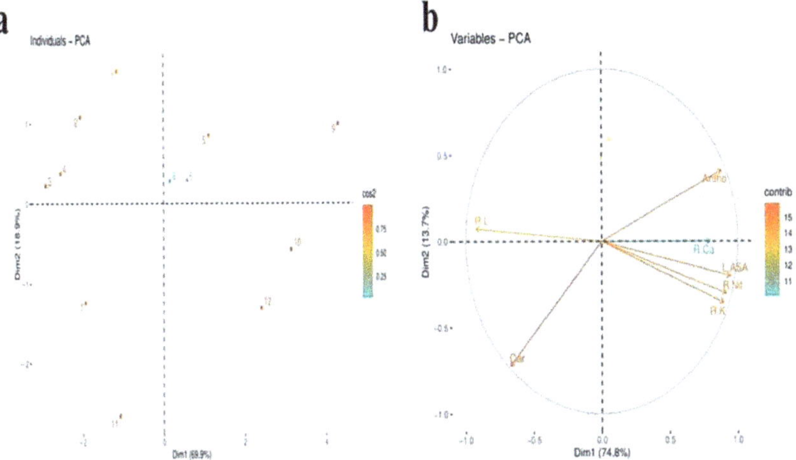

Figure 1. Score (**a**) and loading plot (**b**) of principal component analysis (PCA) on different attributes of maize cultivar Gohar-19 plants supplemented with and without SA while grown under B stress. Score plot represents the separation of treatments as T1: 0 mg B without SA; T2: 0 mg B with 50 μM SA; T3: 0 mg B with 100 μM SA; T4: 0 mg B with 150 μM SA; T5: 15 mg/kg B without SA; T6: 15 mg/kg B with 50 μM SA; T7: 15 mg/kg B with 100 μM SA; T8: 15 mg/kg B with 150 μM SA; T9: 30 mg/kg B without SA; T10: 30 mg/kg B with 50 μM SA, T11. 30 mg/kg B with 100 μM SA; T12: 30 mg/kg B with 150 μM SA. Attributes evaluated include R L = root length; Car = carotenoids; R Nit = root nitrate; R K = root potassium; R Ca = root calcium, Antho = anthocyanin; L ASA – leaf ascorbic acid.

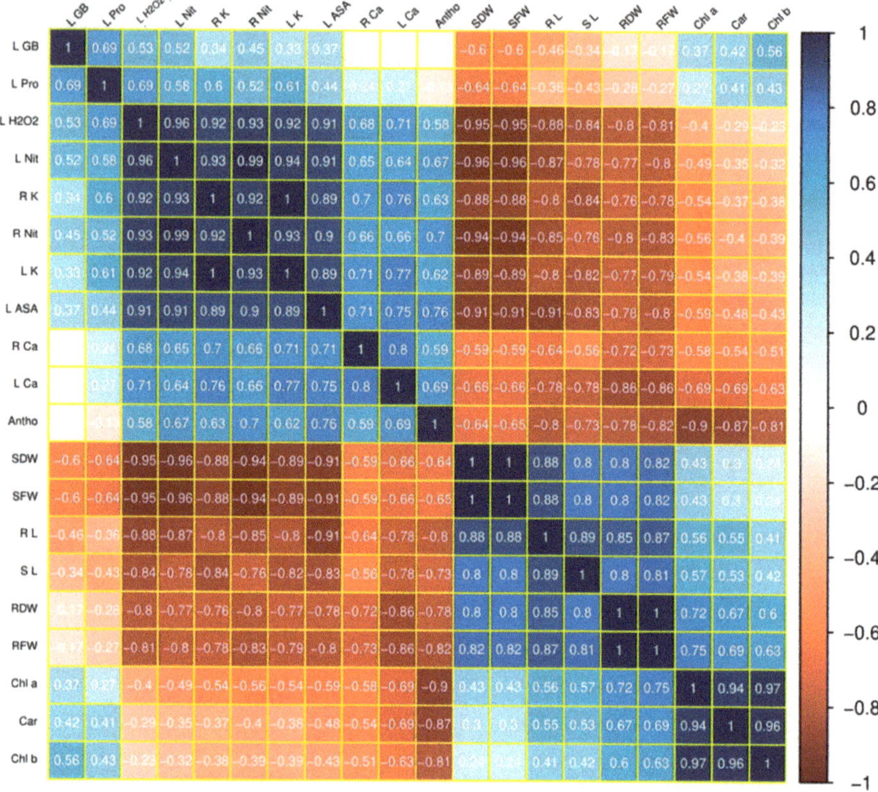

Figure 2. Correlations (r values) among the different studied parameters of maize cultivar Gohar-19 grown under different B stress levels and fertigated with and without SA. R L = root length; S L = shoot length; RFW=root fresh weight; RDW = root dry weight; SFW = shoot fresh weight; SDW = shoot dry weight; Chl a = chlorophyll a; Chl b = chlorophyll b; Car = carotenoids; Antho = anthocyanin; L ASA = leaf ascorbic acid; L H₂O₂ = leaf hydrogen peroxide; L Pro = leaf proline; L GB = leaf glycine betaine; R K = root potassium; L K = leaf potassium; R Ca = root calcium; L Ca = leaf calcium; R Nit = root nitrate; L Nit = leaf nitrate.

2.2. Plant Biomass

Plant fresh and dry biomass were also reduced in response to increasing the B treatment levels. The results obtained exhibited a positive correlation with the root and shoot lengths. Reducing and increasing patterns in plant fresh and dry biomass were observed in the root and shoot lengths. We observed 30% and 32.89% decreases in the root and shoot fresh biomass, respectively. The maximum increase in plant biomass was noted in plants raised from seeds primed with 100 µM of SA, as shown in Table 2. This gain in the plant growth biomarkers was due to the enhanced photosynthetic activity and improved antioxidant status of the plant body (Figure 2).

Table 2. Effect of SA (0, 50, 100 and 150 μM) on the plant fresh and dry weight of maize cultivar Gohar-19 under varying B toxicity levels (0, 15 and 30 mg kg^{-1}).

Treatment	Root Fresh Weight (g)	Root Dry Weight (g)	Shoot Fresh Weight (g)	Shoot Dry Weight (g)
00 μM SA + 00 mg/g^{-1} B	1.00 ± 0.89 [abc]	0.75 ± 0.21 [bc]	3.80 ± 0.33 [abc]	2.53 ± 1.01 [abcd]
00 μM SA + 15 mg kg^{-1} B	0.89 ± 0.95 [bc]	0.65 ± 0.34 [d]	3.10 ± 0.41 [abc]	2.07 ± 0.85 [abcd]
00 μM SA + 30 mg kg^{-1} B	0.70 ± 0.55 [c]	0.5 ± 0.21 [d]	2.55 ± 0.55 [c]	1.70 ± 0.33 [d]
50 μM SA + 00 mg kg^{-1} B	1.25 ± 0.45 [ab]	1.02 ± 0.35 [ab]	3.85 ± 1.01 [abc]	2.57 ± 0.65 [abcd]
50 μM SA + 15 mg kg^{-1} B	1.15 ± 0.75 [abc]	0.95 ± 0.34 [ab]	3.00 ± 0.95 [abc]	2.00 ± 0.35 [abcd]
50 μM SA + 30 mg kg^{-1} B	0.85 ± 0.65 [bc]	0.62 ± 0.32 [d]	2.70 ± 0.55 [bc]	1.80 ± 0.45 [cd]
100 μM SA + 00 mg kg^{-1} B	1.50 ± 0.76 [ab]	1.26 ± 0.22 [a]	4.30 ± 0.25 [a]	2.87 ± 0.27 [a]
100 μM SA + 15 mg kg^{-1} B	1.35 ± 0.55 [ab]	1.09 ± 0.36 [abc]	3.60 ± 0.97 [abc]	2.40 ± 0.85 [abc]
100 μM SA + 30 mg kg^{-1} B	1.15 ± 0.75 [abc]	0.9 ± 0.23 [abc]	3.00 ± 0.85 [abc]	2.00 ± 0.33 [abcd]
150 μM SA + 00 mg kg^{-1}B	1.25 ± 0.82 [ab]	0.99 ± 0.45 [abc]	3.90 ± 0.21 [abc]	2.60 ± 0.43 [abc]
150 μM SA + 15 mg kg^{-1} B	1.00 ± 071 [abc]	0.75 ± 0.35 [bc]	3.25 + 0.85 [abc]	2.17 ± 0.55 [abcd]
150 μM SA + 30 mg kg^{-1} B	0.95 ± 0.66 [bc]	0.71 ± 0.32 [c]	2.85 ± 0.79 [abc]	1.90 ± 0.65 [bcd]
LSD 5%	0.51	0.49	1.46	0.98

Values in the same column with different letters in superscript differ significantly.

2.3. Photosynthetic Pigments

Elevated B levels significantly reduced the photosynthetic pigment contents of maize seedlings. It was observed that the chl *a* contents were reduced with increasing B treatment levels. The 30 mg kg^{-1} B imposed deteriorative effects and reduced the chl *a* contents effectively (Figure 3a). An improvement in the chl *a* concentration was recorded through priming seeds with SA. Seeds primed with 100 μM of SA expressed the maximum chl *a* content, which suggests reduced toxicity effects.

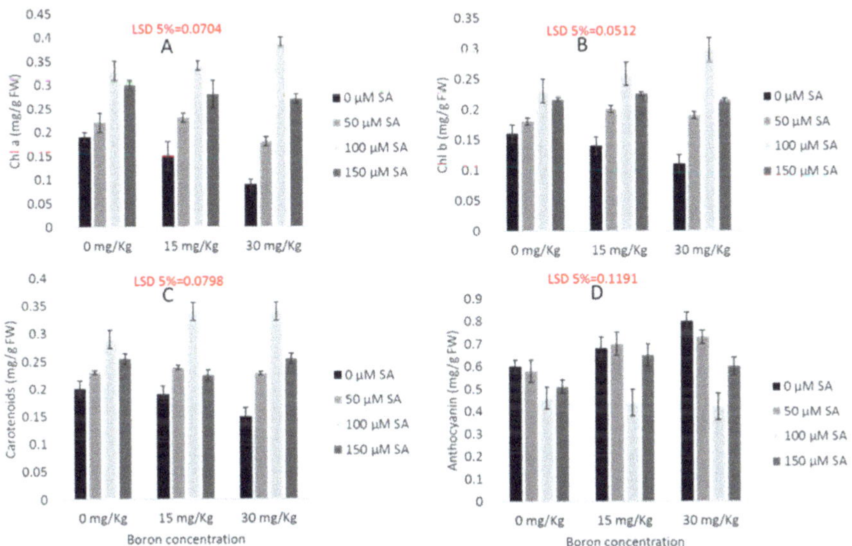

Figure 3. Effect of SA on chlorophyll a (**A**), chlorophyll b (**B**), carotenoids (**C**) and anthocyanin (**D**) contents of maize cultivar Gohar-19 under varying B toxicity levels.

The chl *b* contents were also reduced under B toxicity as compared to the control. An increase in the chl *b* contents was observed with respect to the control in the plants emerging from primed seeds. An increase of 30.4% in the chl *b* contents was observed at 100 μM SA treatment, while a non-significant increase was observed at 150 μM SA treatment as compared to the control (Figure 3b). The carotenoid contents were reduced effectively under 30 mg kg^{-1} B. The B application at 30 mg Kg^{-1} caused reductions of 52.6%, 31.3% and 45% in the chl*a*, chl*b* and carotenoids, respectively. The SA priming improved the carotenoid contents by reducing the drastic effects of B toxicity. A non-significant change in the carotenoid contents was observed in 50 and 150 μM SA primed seeds, while 100 μM SA significantly enhanced the carotenoid contents as compared to the control (Figures 2 and 3c).

2.4. Anthocyanin

The anthocyanin contents increased with increasing the levels of B toxicity. Significant increases in the anthocyanin contents were observed in plants treated with 15 and 30 mg kg^{-1} B. There was a 33.33% increase in anthocyanin contents when 30 mg kg^{-1} soil B was applied, as compared to the control. The SA priming reduced the anthocyanin contents overall, but only 100 μM of SA caused a 47.5% reduction in the anthocyanin contents (Figure 2 andFigure 3c).

2.5. Ascorbic Acid

The toxic effects of B increased the ASA contents of maize seedlings. The B treatment of 30 mg kg^{-1} significantly increased the ASA content up to 44% as compared with the control (Figure 4). Priming with SA reduced the B toxic effects. Only 100 and 150 μM of SA effectively mitigated the toxic effects on plants grown in pots containing 15 mg kg^{-1} B. However, under a high boron toxicity, only 100 μM of SA significantly reduced the ASA content up to 36% as compared to the control, as shown in Table 3, Figure 2.

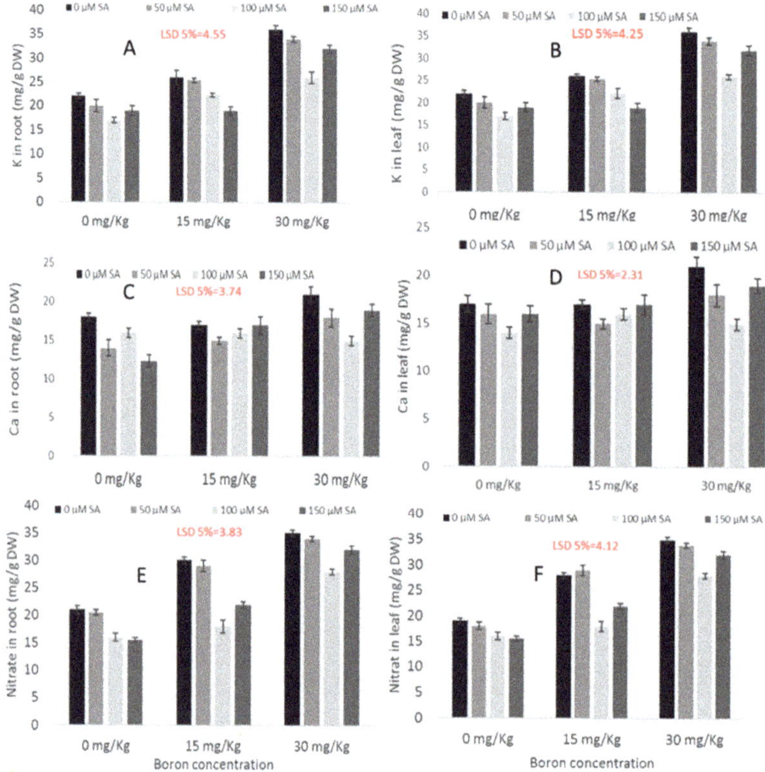

Figure 4. Effect of SA on the K (**A**,**B**), Ca (**C**,**D**) and nitrate (**E**,**F**) contents in the roots and leaves of maize cultivar Gohar-19 under varying B toxicity levels.

Table 3. Effect of SA (0, 50, 100 and 150 µM) on the leaf ascorbic acid, H_2O_2, proline and glycine betaine contents of maize cultivar Gohar-19 under varying B toxicity levels (0, 15 and 30 mg kg^{-1}).

Treatment	Leaf ASA (µmoles/g FW)	Leaf H_2O_2 (mg/g FW)	Leaf Proline (µMole/g FW)	Leaf GB (µg/g FW)
00 µM SA + 00 mg kg^{-1} B	210 ± 1.53	0.80 ± 0.15	32.00 ± 0.5	1.60 ± 0.05
00 µM SA + 15 mg kg^{-1} B	320 ± 1.15	1.80 ± 0.06	36.50 ± 0.5	1.80 ± 0.03
00 µM SA + 30 mg kg^{-1} B	375 ± 0.58	2.50 ± 0.10	46.00 ± 0.5	1.90 ± 0.03
50 µM SA + 00 mg kg^{-1} B	209 ± 1.00	0.70 ± 0.05	33.00 ± 0.5	1.70 ± 0.05
50 µM SA + 15 mg kg^{-1} B	300 ± 0.58	1.34 ± 0.03	37.00 ± 0.58	2.50 ± 0.06
50 µM SA + 30 mg kg^{-1} B	360 ± 0.58	2.40 ± 0.03	47.00 ± 0.50	2.50 ± 0.08
100 µM SA + 00 mg kg^{-1} B	200 ± 1.00	0.64 ± 0.02	33.00 ± 0.29	1.80 ± 0.05
100 µM SA + 15 mg kg^{-1} B	260 ± 0.58	1.00 ± 0.03	46.00 ± 0.76	2.00 ± 0.05
100 µM SA + 30 mg kg^{-1} B	240 ± 0.58	1.90 ± 0.03	58.00 ± 0.29	2.80 ± 0.05
150 µM SA + 00 mg kg^{-1} B	211 ± 1.15	0.78 ± 0.02	37.00 ± 0.29	1.90 ± 0.05
150 µM SA + 15 mg kg^{-1} B	276 ± 0.76	1.45 ± 0.05	37.00 ± 0.58	2.50 ± 0.09
150 µM SA + 30 mg kg^{-1} B	335 ± 0.29	2.20 ± 0.20	48.00 ± 0.29	2.40 ± 0.08
LSD 5%	0.51	0.49	1.46	0.98

2.6. H_2O_2 Concentration

Increasing the B toxicity enhanced the H_2O_2 content effectively. The H_2O_2 content was highly affected by 30 mg kg^{-1} B in soil. The effective treatment in term of reducing the H_2O_2 content was 100 µM SA, which decreased the H_2O_2 content up to 84% (Table 3, Figure 2).

2.7. Proline Content

The toxic effects of B significantly increased the proline contents. The B treatment of 30 mg kg^{-1} increased the proline contents up to 43.75% as compared to the control. Priming with SA remained productive in reducing the B toxic effects. Only 100 and 150 µM SA effectively mitigated the toxic effects in plants grown in pots containing 15 mg kg^{-1} B. However, under a high boron toxicity, only 150 µM SA significantly reduced the toxic effects by increasing the proline content in comparison with the control, as shown in Table 3, Figure 2.

2.8. Glycine Betaine

An outstanding improvement was noted in the leaf GB contents of plants grown under B stress. Exogenous applications of SA further increased the GB content in the leaves of plants experiencing B toxicity stress. All the treatments of SA affected the GB level, however 100 µM SA increased the GB contents up to 100% under 30 mg kg^{-1} of B treatment as compared to the control, as indicated in Table 3, Figure 2.

2.9. Potassium Content

An increase in the K contents was observed in response to the B toxicity. Applications of SA reduced the K contents and a maximum reduction of up to 27.8% was noted in the plants primed with 100 µM of SA. The K uptake and accumulation exhibited quite similar patterns in the plant root and shoot (Figures 2 and 4a,b).

2.10. Calcium Content

Boron toxicity significantly influenced the Ca accumulation in the root of the maize cultivar Gohar-19. The Ca content was reduced at a lower SA treatment level as compared to the control. With increasing the SA concentration up to 100 and 150 µM, higher increments in the Ca contents relative to the control were recorded. A total of 100 µM SA treatment was found to be effective in reducing toxic effects by lowering the Ca content up to 28.7% in the 30 mg kg^{-1} B group. The Ca accumulation in the plant leaves was reduced due to the B toxicity even at high concentrations of SA applications. No significant effects of SA on Ca contents were observed (Figures 2 and 4a,b).

2.11. Nitrate Concentration

An increase in the nitrate contents was observed with an increasing B level. Various SA concentrations (0, 50, 100 and 150 µM) reduced the nitrate content. The application of 100 µM of SA reduced the nitrate content up to 20% in the most destructive B treatment of 30 mg kg^{-1} (Figures 2 and 4a,b).

The relationship between B toxicity and the morpho-physiological attributes of maize under SA application is illustrated in Figure 2.

A Pearson correlation analysis was conducted to quantify the interactive effects of B toxicity and SA application on plant growth and biomass, chlorophyll contents, lipid peroxidation and the antioxidant and nutrient uptake of maize (Figure 2). B toxicity was negatively correlated with plant growth and biomass, photosynthetic pigments, oxidative stress and antioxidative response. Chlorophyll contents were positively correlated with plant biomass accumulation. Positive correlations were also identified among growth attributes and K, Ca and nitrate contents.

2.12. Principal Component Analysis

The combinatorial effect of B toxicity and SA application was evaluated on important attributes of maize plants by the synthesis of the score and loading plots of PCA, as presented in Figure 4. All the three applied B treatments with and without SA were successfully dispersed by the first two principal components (Figure 1a). The maximum variance among all the components was based on extracted components—i.e., PC1 (Dim1) and PC2 (Dim2), where component Dim1 contributed 69.9% while the contribution of Dim2 was 18.9% (Figure 1b).

3. Discussion

The only non-metallic element of group 13 of the periodic Table is B, which exhibits a trivalent oxidation state. Naturally, B is present in the form of borate, boric acid and borosilicate mineral. In the Earth crust, the B level varies between 1–500 mg kg^{-1} and 2–100 mg kg-1 as per the geographical region and soil composition status [32].

B has a considerable importance due to its supportive role in plant development and growth. It helps in the processes of cell division, the formation of cell wall and the elongation of cells [33]. However, B causes toxic effects at very high or very low levels. B toxicity mostly co-exists together with some other abiotic stresses—e.g., salt and drought stress [34]. A high B toxicity reduces the plant growth and other attributes.

El-Shazoly [25] conducted a study to describe the SA effects on B toxicity stress in wheat. The results of such a study were in agreement with those of the present study. The SA application also enhanced the root and shoot length, supporting the findings of the previous works [35]. It has been reported that a high level of B causes abnormal cell division in the root meristematic zone [35], hypodermis formation and suberin deposition [36], thus limiting plant growth and development. The excess of B also causes cytotoxic effects during mitosis, which in turn reduces the root and shoot biomass [37,38]. In the present study, 100 µM of SA significantly enhanced the plant biomass by mitigating the B toxicity. Sarafi et al. [39] reported that the B toxicity reduces the plant dry weight up to 48%, the number of leaves and the root dry weight in the pepper plant (*Capsicum annuum*). In this study, the applications of melatonin (MEL) and resveratrol (RES) were studied, where a treatment of 100 µM of RES and 1 µM of MEL effectively reversed the reductions in fresh and dry weights under B toxic effects, respectively. Eser and Aydemir [22] reported that kinetin application prevented the B-induced reductions in the plant fresh and dry weight of wheat plants under B stress. Moreover, the high B content (50 mg kg^{-1}) in soil reduced the shoot fresh and dry weight of tomato plants [40]. It has been particularized that B toxicity causes the down-regulation of the photosystem biochemical components and the inhibition of the electron transport rate [41], thus lowering the activity of carbon fixation enzymes [41,42]. High levels of B can also cause the root growth inhibition, accompanied by a decrease in plant dry weight [43]. The reduction in root growth may be due to the intense lignification of cell wall [44]. However, it has been reported that lignification is not mainly responsible for root growth inhibition, but is rather a defensive attribute for reducing B uptake [36].

The high B level (30 mg kg^{-1}) reduced the photosynthetic pigments biosynthesis. However, SA application reversed these negative effects, and the most effective treatment was 100 µM of SA. The findings of present study are in line with those of EI-Shazoly [25]. Plant growth and development are considerably dependent on photosynthetic pigments. It has been reported that the inhibition of plant growth by B stress is associated with reduced photosynthetic pigments. Indeed, the present study indicated that the biosynthesis of chlorophyll and carotenoid was negatively affected by B toxicity stress. Our results depicted a negative relationship between the biosynthesis of photosynthetic pigments and the increasing applied B stress regimes. This decline in photosynthetic pigments might be owing to H_2O_2 accumulation, which damages the photosynthetic reaction centers. Papadakis et al. [45] reported that one of the possible reasons for the reduction in photosynthetic activity in plants grown under excess of B was the structural damage of thylakoids. In general, SA, being a versatile molecule, interacts with other hormones to promote the induction of enzymes and antioxidants to alleviate the toxic effects of stress [46].

Regarding the mineral contents, Kaya and Ashraf [46] described that B toxicity significantly reduces the N, K and Ca contents in tomato. However, nitric oxide application induced the level of minerals and minimized the B toxicity effects. EI-Shazoly [25] described that a low level of boron (3 mg kg^{-1} soil) does not affect the K content in wheat plants, however a high level could decrease it. Moreover, the Ca level was reduced due to the B toxicity (3 mg kg^{-1} soil), but increased upon SA and thiamin application.

High levels of B increased the anthocyanin contents in sweet basil (*Ocimum basilicum* L.) plants, indicating possible stress responses or poor nutrient mobilization from the plant root [47] We also found that B stress elevated the anthocyanin levels in the root and shoot of maize cultivar Gohar-19. The application of SA at higher levels reduced the stress level. Additionally, the reduction in anthocyanin content in plants treated with SA predicts a reduction in stress severity.

Ascorbic acid is an important antioxidant and scavenger of ROS [48–50]. The ASA content was significantly affected by B stress in the present study. Abiotic stresses result in a higher accumulation of ASA than that of other stress. Increased ROS scavenging enzymatic and nonenzymatic activity by excessive B concentrations has already been reported in barley, chickpea, tomato and grapes [51–54].

In general, plants up-regulate the synthesis of different osmolytes in cytosol and other organelles to cope with the deleterious effects of environmental stresses. Proline and GB are considered to be key osmolytes for the osmotic adjustment. Proline, being a secondary metabolite, plays a key role in stress tolerance as an antioxidant and osmoprotectant [55]. Stress-related genes are activated by GB to detoxify ROS and protect photosynthetic machinery under stressed conditions [56]. In the current study, SA applications triggered the accumulation of proline and GB to cope with the B toxicity effects through scavenging ROS and the activation of the antioxidant defense systems.

The PCA results depicted that the application of salicylic acid had a significant ameliorative effect for B toxicity on the studied parameters of maize plants. The same effects of SA have been reported in salt-stressed sunflower plants [57]. Overall, the applied B stress exerted hazardous effects on the growth and ecophysiological attributes of maize. These results were in accordance with the findings of previous reports which have reported decreases in the growth traits of various plant species grown under environmental stress conditions [58–66]. Based on the findings of the current study, we conclude that SA applications improved the growth of B-stressed maize plants at the seedling stage through increasing the biosynthesis of photosynthetic pigments, osmolytes and antioxidants. The high level of B deteriorates photosystem II centers, as the low levels of chlorophyll and carotenoids are linked with biomass reduction caused by B toxicity. High levels of osmoprotectants such as proline may act as signaling molecules for scavenging ROS, thus stabilizing the membrane structures as well as cascading the stress-tolerant gene expression. Further studies at the molecular level may elaborate the comprehensive understanding lying behind these modulations of SA against B toxicity in maize. The induction of B toxicity tolerance in maize plants after SA application is also associated with antioxidant defense system improvement.

4. Materials and Methods

4.1. Plant Material and Experimental Design

Seeds of maize (cultivar Gohar-19) were obtained from the Maize & Millets Research Institute, Yusafwala, Sahiwal, Pakistan, and the experiments were conducted at the Department of Botany, Government College University Faisalabad, Pakistan. The seeds were surface-sterilized using a 1% sodium hypochloride solution for 5 min. Surface-sterilized seeds were thoroughly washed with distilled water and air-dried for 12 h. These seeds were soaked in 0, 50, 100 and 150 μM of SA solution for 12 h. Plastic pots were filled with 1 kg of washed and air-dried sand at the botanical garden, Government College University Faisalabad, and a 100% field capacity was maintained in pots by adding boron-free water. The experiment was carried out in a completely randomized design (CRD) with three replicates. Ten seeds were sown in each pot. After 5 days of germination, 5 seedlings were selected based on their similarity in size and vigor. B stress was applied using Nable's solution containing boric acid (H_3BO_3) by maintaining pH at 5.7. The final B concentrations of 0, 15 and 30 mg Kg^{-1} soil were maintained in each pot for one week. After one week of B stress application, the plants were harvested and stored in a freezer for further analysis.

4.2. Morphological Parameters and Plant Biomass

The root and shoot lengths were measured for individual plants using a meter scale. The root and shoot fresh and dry weight, after drying at 70 °C for 72 h, was calculated using the same weight balance.

4.3. Physiological and Biochemical Analysis

Different physio-biochemical analyses were carried out as described below.

4.3.1. Photosynthetic Pigments

Contents of chlorophyll a, b and carotenoids were determined using a 0.5 g fresh leaf sample. The Arnon [67] method with minor modifications was used for the determination of photosynthetic pigments. The collected sample was ground in 15 mL of 85% acetone and centrifuged at 10,000× *g* for 15 min. The absorbance of the supernatant was measured at 480, 645 and 663 nm using a spectrophotometer (Hitachi U-2001, Tokyo, Japan).

4.3.2. Anthocyanin Content

The anthocyanin content was measured as reported previously [68]. Fresh root and shoot samples (0.1 g each) were ground separately in 2 mL of 1% acidified methanol (1 mL HCL and 99 mL methanol), then the extract was heated up to 50 °C for one in a water bath. The anthocyanin content was then quantified using a spectrophotometer (Hitachi U-2001, Tokyo, Japan) at 535 nm.

4.3.3. Ascorbic Acid Content

The protocol of Mukherjee and Choudhuri [68] was followed to determine the ascorbic acid content. Root and shoot fresh samples (0.1 g) were taken and ground in 5 mL of trichloroacetic acid (TCA) using a pestle and mortar. The extract was filtered, and 4 mL of the homogenate sample was allowed to react with 2 mL of 2% dinitrophenyl hydrazine in an acidified medium. One drop of 10% thiourea (prepared in 70% ethanol) was added and the mixture was then allowed to boil at 100 °C for 15 min. The absorbance at 530 nm was recorded through UV-spectrophotometer (Hitachi U-2001, Tokyo, Japan) for calculating the ascorbic acid content.

4.3.4. H_2O_2 Content Determination

Shoot samples were extracted in a cold acetone for H_2O_2 content determination. One milliliter of extract was mixed with 1 mL of 0.1% titanium dioxide in 20% sulfuric acid and centrifuged at 8000 rpm for 15 min. The supernatant was then used to measure the absorbance at 415 nm. H_2O_2 content was calculated using a standard curve plotted in a range of 0.5–5 mM H_2O_2 and was expressed as mg g^{-1} FW.

4.3.5. Potassium Content

Dry samples of plant root and shoot (1 mg) were dissolved in 9 mL of distilled water. Flame photometer was used for the determination of potassium content, as reported previously [69].

4.3.6. Calcium Content

Dry samples of plant root and shoot (1 mg) were added to 9 mL of distilled water. Flame photometer method was used for the determination of the calcium content [69].

4.3.7. Nitrate Content

Dry root and shoot samples were dissolved in 1 mL of TCA (1.24 g TCA+ 500 mL H_2SO_4) and 1 mL of distilled water. For determining the nitrate oxide, the absorbance was recorded at 530 nm using a UV vis spectrophotometer (Hitachi U-2001, Tokyo, Japan).

5. Statistical Analysis

Collected data were subjected to an analysis of variance (ANOVA) using the statistical software Co-Stat version 6.2, Cohorts Software, 2003 (Monterey, CA, USA). The treatment means were equated by the least significant difference method (Fisher's LSD) at p value of ≤0.05 level. Before applying ANOVA, the data were standardized using means of inverse or logarithmic transformations wherever necessary. The correlations and PCA of the mean values of all variables were found using XL-STAT 2010.

Author Contributions: Conceptualization, M.N., N.K., N.I., S.A., M.R., A.A.A. and M.N.A.; Data curation, M.N. and S.I.; Formal analysis, M.N. and S.I.; Funding acquisition, A.A.A. and M.N.A.; Investigation, S.I. and N.K.; Methodology, M.N., S.I. and N.I.; Project administration, M.N., N.I. and S.A.; Resources, M.N., H.I., S.A. and A.A.A.; Software, H.I., N.I., S.A., M.R. and M.N.A.; Supervision, M.N. and S.A.; Validation, H.I.; Visualization, H.I.; Writing–original draft, M.N., N.K., M.R. and A.A.A.; Writing–review & editing, N.K., S.A., M.R. and M.N.A. All authors have read and agreed to the published version of the manuscript.

Funding: This study was financially supported by the Higher Education Commission (HEC) of Pakistan. The authors would like to extend their sincere appreciation to the Researchers Supporting Project Number (RSP-2020/236), King Saud University, Riyadh, Saudi Arabia.

Acknowledgments: This study was financially supported by the Higher Education Commission (HEC) of Pakistan. The authors would like to extend their sincere appreciation to the Researchers Supporting Project Number (RSP-2020/236), King Saud University, Riyadh, Saudi Arabia.

Conflicts of Interest: The authors declare no conflict of interest.

References

1. Jaradat, A.; Simulated, A. Climate change deferentially impacts phenotypic plasticity and stoichiometric homeostasis in major food crops. *Emir. J. Food Agr.* **2018**, *30*, 429–442.
2. Khan, N.; Ali, S.; Shahid, M.A.; Kharabian-Masouleh, A. Advances in detection of stress tolerance in plants through metabolomics approaches. *Plant Omics* **2017**, *10*, 153. [CrossRef]
3. Mantri, N.; Patade, V.; Penna, S.; Ford, R.; Pang, E. Abiotic stress responses in plants: Present and future. In *Abiotic Stress Responses in Plants*; Springer: New York, NY, USA, 2012; pp. 1–19.

4.   Bakhsh, A. Engineering crop plants against abiotic stress: Current achievements and prospects. *Emir. J. Food Agr.* **2015**, *27*, 24–39. [CrossRef]

5.   Cheng, S. Heavy metal pollution in China: Origin, pattern and control. *Environ. Sci. Pollut. Res.* **2003**, *10*, 192–198. [CrossRef]

6.   Kelepertzis, E. Accumulation of heavy metals in agricultural soils of Mediterranean: Insights from Argolida basin, Peloponnese, Greece. *Geoderma* **2014**, *221*, 82–90. [CrossRef]

7.   Zhao, F.J.; Ma, Y.; Zhu, Y.G.; Tang, Z.; McGrath, S.P. Soil contamination in China: Current status and mitigation strategies. *Environ. Sci. Technol.* **2014**, *49*, 750–759. [CrossRef]

8.   Hussain, S.; Zhang, J.H.; Zhong, C.; Zhu, L.F.; Cao, X.C.; YU, S.M.; Jin, Q.Y. Effects of salt stress on rice growth, development characteristics and the regulating ways. *J. Integr. Agric.* **2017**, *16*, 2357–2374. [CrossRef]

9.   Liu, P.; Yang, Y.S.; Xu, G.D.; Fang, Y.H.; Yang, Y.A.; Kalin, R.M. The effect of molybdenum and boron in soil on the growth and photosynthesis of three soybean varieties. *Plant Soil Environ.* **2005**, *51*, 197–205. [CrossRef]

10.   Khan, N.; Bano, A.; Babar, M.A. The stimulatory effects of plant growth promoting rhizobacteria and plant growth regulators on wheat physiology grown in sandy soil. *Arch. Microbiol.* **2019**, *201*, 769–785. [CrossRef]

11.   Ayvaz, M.; Avcı, M.K.; Yamaner, C.; Koyuncu, M.; Güven, A.; Fagerstedt, K. Does excess boron affect the malondialdehyde levels of potato cultivars? *Eurasia J. Biosci.* **2013**, *7*, 47–53. [CrossRef]

12.   Stiles, A.R.; Liu, C.; Kayama, Y.; Wong, J.; Doner, H.; Funston, R.; Terry, N. Evaluation of the boron tolerant grass, Puccinellia distans, as an initial vegetative cover for the phytorestoration of a boron-contaminated mining site in southern California. *Environ. Sci. Technol.* **2011**, *45*, 8922–8927. [CrossRef] [PubMed]

13.   Princi, M.P.; Lupini, A.; Araniti, F.; Longo, C.; Mauceri, A.; Sunseri, F.; Abenavoli, M.R. Boron toxicity and tolerance in plants: Recent advances and future perspectives. In *Plant Metal Interaction*; Elsevier: Amsterdam, The Netherlands, 2016; pp. 115–147.

14.   Kayıhan, C.; Öz, M.T.; Eyidogan, F.; Yucel, M.; Oktem, H.A. Physiological, biochemical and transcriptomic responses to boron toxicity in leaf and root tissues of contrasting wheat cultivars. *Plant Mol. Biol. Rep.* **2017**, *35*, 97–109. [CrossRef]

15.   Yermiyahu, U.; Ben-Gal, A.; Keren, R.; Reid, R.J. Combined effect of salinity and excess boron on plant growth and yield. *Plant Soil* **2008**, *304*, 73–87. [CrossRef]

16.   Diaz, F.J.; Grattan, S.R. Performance of tall wheatgrass (*Thinopyrum ponticum* cv. Jose) irrigated with saline-high boron drainage water: Implications on ruminant mineral nutrition. *Agric. Ecosyst. Environ.* **2009**, *131*, 128–136. [CrossRef]

17.   Zhang, B.; Chu, G.; Wei, C.; Ye, J.; Li, Z.; Liang, Y. The growth and antioxidant defense responses of wheat seedlings to omethoate stress. *Pestic. Biochem. Phys.* **2011**, *100*, 273–279. [CrossRef]

18.   Khan, N.; Zandi, P.; Ali, S.; Mehmood, A.; Adnan Shahid, M.; Yang, J. Impact of salicylic acid and PGPR on the drought tolerance and phytoremediation potential of Helianthus annus. *Front. Microbiol.* **2018**, *9*, 2507. [CrossRef]

19.   Tavallali, V. Interactive effects of zinc and boron on growth, photosynthesis and water relations in pistachio. *J. Plant Nutr.* **2017**, *40*, 1588–1603. [CrossRef]

20.   Nasim, M.; Rengel, Z.; Aziz, T.; Regmi, B.D.; Saqib, M. Boron toxicity alleviation by zinc application in two barley cultivars differing in tolerance to boron toxicity. *Pak. J. Agric. Sci.* **2015**, *52*, 151–158.

21.   Samet, H.; Cikili, Y.; Dursun, S. The role of potassium in alleviating boron toxicity and combined effects on nutrient contents in pepper (*Capsicum annuum* L.). *Bulg. J. Agric. Sci.* **2015**, *21*, 64–70.

22.   Siddiqui, M.H.; Al-Whaibi, M.H.; Sakran, A.M.; Ali, H.M.; Basalah, M.O.; Faisal, M.; Al-Amri, A.A. Calcium-induced amelioration of boron toxicity in radish. *J. Plant Growth Regul.* **2013**, *32*, 61–71. [CrossRef]

23.   Moustafa-Farag, M.; Mohamed, H.I.; Mahmoud, A.; Elkelish, A.; Misra, A.N.; Guy, K.M.; Kamran, M.; Ai, S.; Zhang, M. Salicylic Acid Stimulates Antioxidant Defense and Osmolyte Metabolism to Alleviate Oxidative Stress in Watermelons under Excess Boron. *Plants* **2020**, *9*, 724. [CrossRef] [PubMed]

24.   Maghsoudia, K.; Arvinb, M.J. Salicylic acid and osmotic stress effects on seed germination and seedling growth of wheat (*Triticum aestivum* L.) cultivars. *Plant Ecophysiol.* **2010**, *2*, 7–11.

25.   El-Shazoly, R.M.; Metwally, A.A.; Hamada, A.H. Salicylic acid or thiamin increases tolerance to boron toxicity stress in wheat. *J. Plant Nutr.* **2019**, *42*, 702–722. [CrossRef]

26.   Hussain, H.A.; Men, S.; Hussain, S.; Zhang, Q.; Ashraf, U.; Anjum, S.A.; Ali, I.; Wang, L. Maize Tolerance against Drought and Chilling Stresses Varied with Root Morphology and Antioxidative Defense System. *Plants* **2020**, *9*, 720. [CrossRef] [PubMed]

27. Aliyu, H.G.; Adamu, H.M. The potential of maize as phytoremediation tool of heavy metals. *Eur. Sci. J.* **2014**, *6*, 32–33.

28. Wang, B.; Liu, C.; Zhang, D.; He, C.; Zhang, J.; Li, Z. Effects of maize organ-specific drought stress response on yields from transcriptome analysis. *BMC Plant Biol.* **2019**, *19*, 335. [CrossRef]

29. Hussain, H.A.; Men, S.; Hussain, S.; Chen, Y.; Ali, S.; Zhang, S.; Zhang, K.; Li, Y.; Xu, Q.; Liao, C.; et al. Interactive effects of drought and heat stresses on morpho-physiological attributes, yield, nutrient uptake and oxidative status in maize hybrids. *Sci. Rep.* **2019**, *9*, 3890. [CrossRef]

30. Khan, N.; Bano, A. Effects of exogenously applied salicylic acid and putrescine alone and in combination with rhizobacteria on the phytoremediation of heavy metals and chickpea growth in sandy soil. *Int. J. Phytoremediation* **2018**, *16*, 405–414. [CrossRef]

31. Brown, P.H.; Bellaloui, N.; Wimmer, M.A.; Bassil, E.S.; Ruiz, J.; Hu, H.; Pfeffer, H.; Dannel, F.; Römheld, V. Boron in plant biology. *Plant Biol.* **2002**, *4*, 205–223. [CrossRef]

32. Parks, J.L.; Edwards, M. Boron in the environment. *Crit. Rev. Environ. Sci. Technol.* **2005**, *35*, 81–114. [CrossRef]

33. Blevins, D.G.; Lukaszewski, K.M. Boron in plant structure and function. *Annu. Rev. Plant Biol.* **1998**, *49*, 481–500. [CrossRef] [PubMed]

34. Zafar-ul-Hye, M.; Munir, K.; Ahmad, M.; Imran, M. Influence of boron fertilization on growth and yield of wheat crop under salt stress environment. *Soil Environ.* **2016**, *35*, 181–186.

35. Liu, D.; Jiang, W.; Zhang, L.; Li, L. Effects of boron ions on root growth and cell division of broad bean (*Vicia Faba* L.). *Isr. J. Plant Sci.* **2000**, *48*, 47–58. [CrossRef]

36. Ghanati, F.; Morita, A.; Yokota, H. Deposition of suberin in roots of soybean induced by excess boron. *Plant Sci.* **2005**, *168*, 397–405. [CrossRef]

37. Konuk, M.; Liman, R.; Cigerci, I.H. Determination of genotoxic effect of boron on *Allium cepa* root meristematic cells. *Pak. J. Bot.* **2007**, *39*, 73–79.

38. Sarafi, E.; Tsouvaltzis, P.; Chatzissavvidis, C.; Siomos, A.; Therios, I. Melatonin and resveratrol reverse the toxic effect of high boron (B) and modulate biochemical parameters in pepper plants (*Capsicum annuum* L.). *Plant Physiol. Biochem.* **2017**, *112*, 173–182. [CrossRef]

39. Eser, A.; Aydemir, T. The effect of kinetin on wheat seedlings exposed to boron. *Plant Physiol. Biochem.* **2016**, *108*, 158–164. [CrossRef]

40. Khan, N.; Bano, A.; Ali, S.; Babar, M.A. Crosstalk amongst phytohormones from planta and PGPR under biotic and abiotic stresses. *Plant Growth Regul.* **2020**, *90*, 189–203. [CrossRef]

41. Chen, M.; Mishra, S.; Heckathorn, S.A.; Frantz, J.M.; Krause, C. Proteomic analysis of *Arabidopsis thaliana* leaves in response to acute boron deficiency and toxicity reveals effects on photosynthesis, carbohydrate metabolism and protein synthesis. *J. Plant Physiol.* **2013**, *171*, 235–242. [CrossRef]

42. Turan, M.; Taban, N.; Taban, S. Effect of calcium on the alleviation of boron toxicity and localization of boron and calcium in cell wall of wheat. *Not. Bot. Horti Agrobot. Cluj-Napoca* **2009**, *37*, 99–103.

43. Ghanati, F.; Morita, A.; Yokota, H. Induction of suberin and increase of lignin content by excess boron in tobacco cells. *J. Plant Nutr. Soil Sci.* **2002**, *48*, 357–364. [CrossRef]

44. Sharma, A.; Sidhu, G.P.S.; Araniti, F.; Bali, A.S.; Shahzad, B.; Tripathi, D.K.; Brestic, M.; Skalicky, M.; Landi, M. The Role of Salicylic Acid in Plants Exposed to Heavy Metals. *Molecules* **2020**, *25*, 540. [CrossRef] [PubMed]

45. Papadakis, I.; Dimassi, K.; Bosabalidis, A.; Therios, I.; Patakas, A.; Giannakoula, A. Boron toxicity in 'Clementine' mandarin plants grafted on two root stocks. *Plant Sci.* **2004**, *166*, 539–547. [CrossRef]

46. Kaya, C.; Ashraf, M. Exogenous application of nitric oxide promotes growth and oxidative defense system in highly boron stressed tomato plants bearing fruit. *Sci. Hortic.* **2015**, *185*, 43–47. [CrossRef]

47. Pardossi, A.; Romani, M.; Carmassi, G.; Guldl, L.; Landi, M.; Incrocci, L.; Maggini, R.; Puccinelli, M.; Vacca, W.; Ziliani, M. Boron accumulation and tolerance in sweet basil (*Ocimum basilicum* L.) with green or purple leaves. *Plant Soil* **2015**, *395*, 375–389. [CrossRef]

48. Foyer, C.H.; Noctor, G. Ascorbate and glutathione: The heart of the redox hub. *Plant Physiol.* **2011**, *155*, 2–18. [CrossRef]

49. Qian, H.F.; Peng, X.F.; Han, X.; Ren, J.; Zhan, K.Y.; Zhu, M. The stress factor, exogenous ascorbic acid, affects plant growth and the antioxidant system in *Arabidopsis thaliana*. *Russ. J. Plant Physiol.* **2014**, *61*, 467–475. [CrossRef]

50. Karabal, E.; Yucel, M.; Huseyin, A.O. Antioxidant responses of tolerant and sensitive barley cultivars to boron toxicity. *Plant Sci.* **2003**, *164*, 925–933. [CrossRef]

51. Gunes, A.; Soylemezoglu, G.; Inal, A.; Bagci, E.G.; Coban, S.; Sahin, O. Antioxidant and stomatal responses of grapevine (*Vitis vinifera* L.) to boron toxicity. *Sci. Hortic.* **2006**, *110*, 279–284. [CrossRef]

52. Cervilla, L.M.; Blasco, B.; Ríos, R.; Romero, L.; Ruiz, J. Oxidative stress and antioxidants in tomato (*Solanum lycopericum*) plants subjected to boron toxicity. *Ann. Bot.* **2007**, *100*, 747–756. [CrossRef]

53. Ardic, M.; Sekmen, A.H.; Tokur, S.; Ozdemir, F.; Turkan, I. Antioxidant responses of chickpea plants subjected to boron toxicity. *Plant Biol.* **2009**, *11*, 328–338. [CrossRef] [PubMed]

54. Khan, N.; Bano, A.; Curá, J.A. Role of Beneficial Microorganisms and Salicylic Acid in Improving Rainfed Agriculture and Future Food Safety. *Microorganisms* **2020**, *8*, 1018. [CrossRef] [PubMed]

55. Shaki, F.; Maboud, H.E.; Niknam, V. Effects of salicylic acid on hormonal cross talk, fatty acids profile, and ions homeostasis from salt-stressed safflower. *J. Plant Interact.* **2019**, *14*, 340–346. [CrossRef]

56. Ahmad, P.; Abdel Latef, A.A.; Hashem, A.; Abd-Allah, E.F.; Gucel, S.; Tran, L.S.P. Nitric oxide mitigates salt stress by regulating levels of osmolytes and antioxidant enzymes in chickpea. *Front. Plant Sci.* **2016**, *7*, 347. [CrossRef]

57. Hossain, M.A.; Hoque, M.A.; Burritt, D.J.; Fujita, M. Proline Protects Plants against Abiotic Oxidative Stress: Biochemical and Molecular Mechanisms. In *Oxidative Damage to Plants*; Academic Press: Cambridge, MA, USA, 2014; pp. 477–522.

58. El-Esawi, M.A.; Elkelish, A.; Soliman, M.; Elansary, H.O.; Zaid, A.; Wani, S.H. *Serratia marcescens* BM1 Enhances Cadmium Stress Tolerance and Phytoremediation Potential of Soybean Through Modulation of Osmolytes, Leaf Gas Exchange, Antioxidant Machinery, and Stress-Responsive Genes Expression. *Antioxidants* **2020**, *9*, 43. [CrossRef]

59. Abdelaal, K.A.; EL-Maghraby, L.M.; Elansary, H.; Hafez, Y.M.; Ibrahim, E.I.; El-Banna, M.; El-Esawi, M.; Elkelish, A. Treatment of Sweet Pepper with Stress Tolerance-Inducing Compounds Alleviates Salinity Stress Oxidative Damage by Mediating the Physio-Biochemical Activities and Antioxidant Systems. *Agronomy* **2020**, *10*, 26. [CrossRef]

60. Alhaithloul, H.A.; Soliman, M.H.; Ameta, K.L.; El-Esawi, M.A.; Elkelish, A. Changes in Ecophysiology, Osmolytes, and Secondary Metabolites of the Medicinal Plants of *Mentha piperita* and *Catharanthus roseus* Subjected to Drought and Heat Stress. *Biomolecules* **2020**, *10*, 43. [CrossRef]

61. El-Esawi, M.A.; Alaraidh, I.A.; Alsahli, A.A.; Alzahrani, S.M.; Ali, H.M.; Alayafi, A.A.; Ahmad, M. *Serratia liquefaciens* KM4 Improves Salt Stress Tolerance in Maize by Regulating Redox Potential, Ion Homeostasis, Leaf Gas Exchange and Stress-Related Gene Expression. *Int. J. Mol. Sci.* **2018**, *19*, 3310. [CrossRef]

62. Zafar-ul-Hye, M.; Naeem, M.; Danish, S.; Khan, M.J.; Fahad, S.; Datta, R.; Brtnicky, M.; Kintl, A.; Hussain, G.S.; El-Esawi, M.A. Effect of Cadmium-Tolerant Rhizobacteria on Growth Attributes and Chlorophyll Contents of Bitter Gourd under Cadmium Toxicity. *Plants* **2020**, *9*, 1386. [CrossRef]

63. Naveed, M.; Bukhari, S.S.; Mustafa, A.; Ditta, A.; Alamri, S.; El-Esawi, M.A.; Rafique, M.; Ashraf, S.; Siddiqui, M.H. Mitigation of Nickel Toxicity and Growth Promotion in Sesame through the Application of a Bacterial Endophyte and Zeolite in Nickel Contaminated Soil. *Int. J. Environ. Res. Public Health* **2020**, *17*, 8859. [CrossRef]

64. Imran, M.; Hussain, S.; El-Esawi, M.A.; Rana, M.S.; Saleem, M.H.; Riaz, M.; Ashraf, U.; Potcho, M.P.; Duan, M.; Rajput, I.A.; et al. Molybdenum Supply Alleviates the Cadmium Toxicity in Fragrant Rice by Modulating Oxidative Stress and Antioxidant Gene Expression. *Biomolecules* **2020**, *10*, 1582. [CrossRef] [PubMed]

65. Ali, Q.; Shahid, S.; Ali, S.; El-Esawi, M.A.; Hussain, A.I.; Perveen, R.; Iqbal, N.; Rizwan, M.; Nasser Alyemeni, M.; El-Serehy, H.A.; et al. Fertigation of Ajwain (*Trachyspermum ammi* L.) with Fe-Glutamate Confers Better Plant Performance and Drought Tolerance in Comparison with FeSO₄. *Sustainability* **2020**, *12*, 7119. [CrossRef]

66. Soliman, M.; Alhaithloul, H.A.; Hakeem, K.R.; Alharbi, B.M.; El-Esawi, M.; Elkelish, A. Exogenous Nitric Oxide Mitigates Nickel-Induced Oxidative Damage in Eggplant by Upregulating Antioxidants, Osmolyte Metabolism, and Glyoxalase Systems. *Plants* **2019**, *8*, 562. [CrossRef] [PubMed]

67. Arnon, D.I. Copper enzymes in isolated chloroplasts. Polyphenoloxidase in *Beta vulgaris*. *Plant Physiol.* **1949**, *24*, 1–15. [CrossRef] [PubMed]

68. Mukherjee, S.P.; Choudhuri, M.A. Implications of water stress-induced changes in the levels of endogenous ascorbic acid and hydrogen peroxide in Vigna seedlings. *Physiol. Plant* **1983**, *58*, 166–170. [CrossRef]

69. Stark, D.; Wray, V. Anthocyanins. In *Methods in Plant Biology*; Volume 1: Plant Phenolics; Harborne, J.B., Ed.; Academic Press: London, UK; Harcourt Brace Jovanovich: London, UK, 1989; pp. 325–356.

Publisher's Note: MDPI stays neutral with regard to jurisdictional claims in published maps and institutional affiliations.

Article

Identification and Quantification of Plant Growth Regulators and Antioxidant Compounds in Aqueous Extracts of *Padina durvillaei* and *Ulva lactuca*

Israel Benítez García [1], Ana Karen Dueñas Ledezma [1], Emmanuel Martínez Montaño [1,2], Jesús Aarón Salazar Leyva [1], Esther Carrera [3] and Idalia Osuna Ruiz [1,*]

[1] Maestría en Ciencias Aplicadas, Unidad Académica de Ingeniería en Biotecnología, Universidad Politécnica de Sinaloa (UPSIN), Carretera Municipal Libre Mazatlán Higueras Km 3., Mazatlán, SIN 82199, Mexico; ibenitez@upsin.edu.mx (I.B.G.); 2017031067@upsin.edu.mx (A.K.D.L.); emartinez@upsin.edu.mx (E.M.M.); jsalazar@upsin.edu.mx (J.A.S.L.)

[2] Cátedras CONACYT, Consejo Nacional de Ciencia y Tecnología, Ciudad de México 03940, Mexico

[3] Instituto de Biología Molecular y Celular de Plantas (UPV-CSIC), Instituto de Biología Molecular y Celular de Plantas (IBMCP), Universidad Politécnica de Valencia-Consejo Superior de Investigaciones Científicas (CSIC), 46022 Valencia, Spain; ecarrera@ibmcp.upv.es

* Correspondence: iosuna@upsin.edu.mx

Received: 15 May 2020; Accepted: 15 June 2020; Published: 18 June 2020

Abstract: Aqueous seaweed extracts have diverse compounds such as Plant-Growth Regulators (PGRs) which have been utilized in agricultural practices for increasing crop productivity. Algal biomass of *Padina durvillaei* and *Ulva lactuca* have been suggested for use as biofertilizers because of plant growth-enhancing properties. This work aimed to identify the main PGRs and antioxidant properties in *P. durvillaei* and *U. lactuca* extracts, such as abscisic acid, auxins, cytokinins, gibberellins, jasmonates, and salicylates, to assess their potential use as biofertilizers that improve plant growth and crop yield. Phytochemical analyses of two seaweed extracts showed a significantly higher content of sulfates, flavonoids, and phenolic compounds in *P. durvillaei* extract, which could be linked to its higher antioxidant activity (DPPH, ABTS, and FRAP) compared to *U. lactuca* extract. The identification and quantification of PGRs showed two gibberellins (GA1 and GA4), abscisic acid (ABA), indoleacetic acid (IAA), three cytokinins (tZ, IP, and DHZ), jasmonic acid (JA), and salicylic acid (SA) in two seaweed extracts. However, GA4, tZ, and DHZ contents were significantly higher in *P. durvillaei* compared to *U. lactuca* extracts. These findings evidence that *P. durvillaei* and *U. lactuca* extracts are suitable candidates for use as biofertilizers.

Keywords: seaweed extract; phytohormone profiling; fertilizers; antioxidant; plant growth regulators; brown seaweed; green algae

---

1. Introduction

Seaweed extracts have been used in agricultural activities due to their content of macroelements (alginate, agar, carrageenan, etc.), which activate the synthesis of endogenous hormones in plants [1,2] and contribute microelements (N, Ca, Mg, Mn, B, Br, I, Zn, Cu, and Co), amino acids, and vitamins that enrich soil in plant crops [3]. Besides, seaweed extracts contain biochemical compounds such as chlorophylls, carotenoids, and phenolics that confer antioxidant protection [4,5]. The antioxidant properties of algae extracts have been widely evaluated and attributed to sulfated polysaccharides, pigments, and phenolic compounds [4–8], which provide desirable characteristics for their potential use in crops, since, in addition to conferring antioxidant protection, compounds such as polysaccharides have been linked to growth promoting activities [8].

Some environmentally-friendly extraction methods generally include boiling or soaking with distilled water, which have been used as biostimulants for plant growth [9]. Phytohormone-like Plant-Growth Regulators (PGRs) have been identified in algal extracts, such as abscisic acid, auxins, cytokinins, gibberellins, jasmonates, or salicylates, all of which regulate plant cell metabolism and boost production and growth [10–12]. For this reason, marine algae have been used in agriculture as organic fertilizers to achieve sustainable crop production [13,14] and counter the excessive use of fertilizers and synthetic hormones (e.g., 2.4-dichlorophenoxyacetic acid and naphthaleneacetic acid) that may potentially affect both the environment and humans [12,15].

Several authors have reported that algal extracts induce physiological processes in treated plants, such as germination, emergence, root growth, nutrient mobilization, maturation, tolerance to stress, and disease resistance; these responses are similar to those observed in plant crops treated with synthetic hormones [3,10,16–18]. Some seaweed extracts are marketed as liquid biofertilizers or biostimulants [3,16,19], mostly enriched with biomass of *Ascophyllum nodosum*, *Sargassum* spp., and *Macrocystis pyrifera* [3,20,21]. However, the different species of algae show variations regarding PGR biosynthesis [22]; thus, algal extracts exert variable physiological effects on different crops.

In Latin America, algal extracts have recently been used as biostimulants. Some reports demonstrate the benefits of the application of seaweed extracts harvested in coastal areas on various crops. The macroalgae *M. pyrifera*, *Gelidium robustum*, *Chondracanthus canaliculatus*, *Sargassum* spp., *Ulva lactuca*, and *Padina gymnospora*, have been used as biostimulants, fertilizers, and root promoters, as well as to stimulate growth and increase antifungal protection in tomato plants (*Solanum lycopersicum*) [20,23]. However, the exploitation of marine algae, mainly those involved in massive arrivals, is still incipient; moreover, it is not well known if improvements in yield and production of crops fertilized with algal extracts are due to the presence of PGRs in algal organic matter and/or if a possible contribution of other metabolites contribute to the biostimulant effects. Therefore, it is necessary to study the chemical and bioactive composition of a new algal extract when it is prepared to consider its potential use as plant grow stimulant.

In addition, PGRs in seaweeds have been insufficiently studied. Therefore, information is needed to support the use of marine sources, such as algae, to achieve sustainable agriculture practices in the future. This will reduce the environmental impact associated with the excessive use of chemical fertilizers, and also the potential risks to consumers resulting from the indiscriminate application of synthetic PGRs, along with the fact that algae provide other bioactive compounds that enhance protection against stress oxidative, improving plant health.

The objective of the present investigation was to characterize the chemical and bioactive composition (antioxidant activity, PGR identification and content) of aqueous extracts of the macroalgae *Ulva lactuca* and *Padina durvillaei*, and evaluate the use of such aqueous extracts as a potential biofertilizer.

2. Material and Methods

2.1. Seaweed Collection and Reagents

Specimens of the seaweeds *Padina durvillaei* (Bory Saint-Vicent, 1957) and *Ulva lactuca* (Linnaeus 1753) were collected in Mazatlan Bay, Sinaloa, Mexico (23°1′2 9.1″ LN, 106°25′29.7′ LW), in March 2017. Fresh samples were rinsed with distilled water, lyophilized, ground with a commercial grinder, and stored at −20 °C until used. All chemicals used in this research were analytical grade and supplied from Sigma (Sigma-Aldrich Co., St. Louis, MO, USA), unless otherwise specified.

2.2. Seaweed Extracts

Seaweed extracts were obtained using distilled water according to Tierney et al. [24], modified as follows: dried algal material was mixed with water at 21 °C (1:10, *w:v*) with stirring for 3 h; then, the extract was filtered through a fiber glass filter (1.2 μm pore size) and the algal residue extracted again (twice). Filtrates were pooled and centrifuged at 12,000× g and 4 °C for 20 min; then, the supernatant was collected. Finally, the aqueous extract was lyophilized and stored at −20 °C until analyzed.

The extraction yield was calculated according to Equation (1):

$$\text{Extraction yield (\%)} = (\text{grams of dry aqueous extract/grams of dry seaweed}) \times 100 \tag{1}$$

2.3. Chemical Composition

Carbohydrate content was measured using the phenol–sulfuric acid method [25] using D-glucose as standard. Soluble protein content was determined with Bradford's method using Bovine Serum Albumin (BSA) as standard [26].

Sulfate content was measured with the barium chloride-gelatin assay using potassium sulfate as standard [27]. The uronic acid content was determined with the sulfuric acid-carbazole colorimetric method using D-glucuronic acid as standard [28].

2.3.1. Total Phenolic Content (TPC)

Total soluble phenolic content was determined using the Folin–Ciocalteu method [29]. Dry samples were reconstituted with acetone (1 mg/mL); then, a 100 mL of each sample was mixed with 150 mL of Folin solution (previously diluted 1:1 with deionized water) followed by the addition of 1 mL of 2% sodium carbonate in 0.4% sodium hydroxide. The mixture was incubated in the dark at room temperature for 20 min. The resulting blue complex was read in a spectrophotometer at 750 nm. Phenolic content was expressed as mg of gallic acid equivalent (GAE) per g of sample (dry weight). A gallic acid standard curve was constructed at the concentration range of 0–0.25 mg/mL.

2.3.2. Total Flavonoids Content (TFC)

Total flavonoid content was assessed according to Luximon-Ramma et al. [30]. Samples of solutions (1 mL) were diluted in equal volumes of a 2% aluminum chloride solution (2 g of $AlCl_3 \cdot 6H_2O$ in 100 mL of methanol). The mixture was incubated at room temperature for 10 min. Absorbance was read at 367 nm. The results were expressed in mg of quercetin equivalents (QE) per gram of sample (dry weight). A quercetin standard curve was constructed at the concentration range of 0–0.5 mg/mL.

2.4. Antioxidant Evaluation

2.4.1. DPPH Free-Radical Scavenging Activity

The free-radical scavenging potential of the seaweed extracts was analyzed according to the method proposed by Mensor et al. [31], modified as follows: a 100 mL aliquot of each extract (at concentrations of 0.0015 to 1.5 mg/mL) was mixed with 900 mL of an ethanol solution of 0.3 mM 2,2-diphenyl-1-picrylhydrazyl (DPPH); the mixture was incubated for 30 min in the dark at room temperature. Then, absorbance was measured at 518 nm. Trolox was used as standard, whereas the DPPH solution served as control to calculate the degree of radical scavenging by samples as well as the reference standard.

The percentage of DPPH scavenging was calculated with Equation (2):

$$\% \text{ DPPH scavenging} = [(1 - \text{Absorbance of sample})/\text{Absorbance of Control}] \times 100 \tag{2}$$

2.4.2. ABTS Free-Radical Scavenging Activity

The scavenging activity of 2,2′-azinobis [3-ethylbenzthiazoline]-6-sulphonic acid (ABTS) was determined according to the method by Przygodzka et al. [32], modified as follows: the ABTS radical was previously activated for 12–16 h at room temperature in the dark; the resulting ABTS radical solution was diluted with ethanol and its absorbance read at 734 nm, yielding a value of 0.80. A 100 μL aliquot of each sample (at a concentration of 0.0015 to 1.5 mg/mL) was mixed with 2.9 mL of ABTS solution and the absorbance was read 10 min after mixing. Trolox was used as reference standard, whereas the ABTS radical solution served as control to calculate the degree of radical scavenging by samples as well as the reference standard.

The percentage of ABTS scavenging was calculated using Equation (3):

$$\% \ ABTS \ scavenging = [(A - B)/A] \times 100 \tag{3}$$

where A is absorbance of the *ABTS* control solution and B is absorbance of the test solution.

2.4.3. Ferric Reducing Antioxidant Power (FRAP) Assay

The FRAP assay was performed according to the methods of Benzie and Strain [33], with the minor modification reported by Szôllôsi and Varga [34]. The FRAP reagent was made from three different solutions: Solution A: 300 mM acetate buffer, pH 3.6; Solution B: 10 mM TPTZ dissolved in 40 mM HCl; and Solution C: 20 mM $FeCl_3 \cdot 6H_2O$. The work solution was prepared by mixing A, B, and C in a 10:1:1 ratio (by volume). For the assay, 100 µL of sample were mixed with 1400 µL of FRAP, and then incubated at room temperature for 30 min in the dark. Finally, absorbance was read at 593 nm. Trolox was used as reference standard.

2.5. Identification and Quantification of Plant Growth Regulator Profiles

PGRs for acid hormones (gibberellins, GAs; indolacetic acid, IAA; jasmonic acid, JA; abscisic acid, ABA; and salicylic acid, SA), and cytokinins or basic hormones (dihydrozeatine, DHZ; isopentyladenine, iP; and t-zeatine, tZ) were identified and quantified by ultra-high performance liquid chromatography–mass spectrometry (UHPLC-MS) using a Thermo Scientific™ Q Exactive™ Hybrid Quadrupole-Orbitrap mass spectrometer at Institute for Plant Molecular and Cell Biology (IBMCP), Spain. Extraction and separation of plant hormone profiles were performed as described by Seo et al. [35]. The lyophilized extract was suspended in 80% methanol (MeOH) containing 1% acetic acid, mixed by stirring for 1 h at 4 °C and centrifuged at 14 000× g at 4 °C for 4 min. The supernatant extract was stored at −20 °C overnight and then centrifuged at 14 000× g at 4 °C for 4 min. Then, the supernatant was dried in a vacuum evaporator. The dry residue was dissolved in 1% (*v/v*) acetic acid and passed consecutively through an Oasis HLB reverse-phase column (30 mg; Waters) and an Oasis MCX cation exchanger. Acid hormones were eluted with MeOH and basic hormones with 60% MeOH containing 5% aqueous ammonia [35]. The final residues were dried and dissolved in 5% (*v/v*) acetonitrile, 1% (*v/v*) MeOH, and 1% (*v/v*) acetic acid. Then, hormones were separated by UHPLC with a reverse Accucore C18 column (2.6 mm inner diameter, 100 mm length; Thermo Fisher Scientific) with a 2% to 55% (*v/v*) acetonitrile gradient containing 0.05% (*v/v*) acetic acid at 400 mL for 21 min. The plant hormones were analyzed with a Q-Exactive mass spectrometer (Orbitrap detector; Thermo Fisher Scientific) by targeted selected ion monitoring (capillary temperature, 300 °C; S-lens RF level, 70; resolution, 70,000) and electrospray ionization (spray voltage, 3 kV; heater temperature, 150 °C; sheath gas-flow rate, 1.90 mL/min; auxiliary gas-flow rate, 0.42 mL/min). The concentrations of plant hormones in extracts were determined using embedded calibration curves and the Xcalibur 2.2 SP1 build 48 and TraceFinder programs. The internal standards for quantification of each plant hormone were D6-ABA, D2-GA1, D2-GA4, D5-tZ, D3-DHZ, D6-iP, D2-IAA, D6-SA, and D2-JA (Olchemim Ltd., Olomouc, Czech).

2.6. Statistical Analyses

The results of all assays are reported as mean ± standard deviation. Comparisons between groups were performed using the Student's *t*-test. Statistical significance was set at $p < 0.05$.

All statistical analyses were performed using the statistical program Sigma Plot version 11.0 (2018 Systat Software, Inc.; Erkrath, Germany).

3. Results

3.1. Chemical Composition and Antioxidant Capacity Evaluation of Seaweed Extracts

The highest extraction yield was achieved in the *Ulva lactuca* extract (Table 1), which was nearly twice the yield for *Padina durvillaei* (7.55 ± 4.05% and 3.34 ± 1.20%, respectively). With regard to the chemical composition of the soluble components in both extracts, similar contents of carbohydrates and uronic acids were found; however, the percentage of sulfates and the content of polyphenols and flavonoids varied, with higher values in the *P. durvillaei* extract (6.63 ± 0.7%, 34.26 ± 1.39 GAE/g, and 16.16 ± 2.87 QE/g, respectively). Soluble protein was found in the *P. durvillaei* extract (1.28 ± 0.5%), but not in the *U. lactuca* extract.

Table 1. Chemical composition of seaweed extracts.

	Padina Durvillaei	*Ulva Lactuca*
Extraction Yield (%)	3.34 ± 1.20	7.55 ± 4.05
Chemical Composition (dwt.)		
Soluble Protein (%)	1.28 ± 0.56	N. D.
Carbohydrates (%)	16.36 ± 0.08	16.19 ± 0.07
Uronic Acids (%)	8.79 ± 0.60	8.01 ± 0.21
Sulfates (%)	6.63 ± 0.76 a	4.05 ± 1.13 b
Total Polyphenols Content (mg GAE/g)	34.26 ± 1.39 a	27.29 ± 1.57 b
Total Flavonoids Content (mg QE/g)	16.16 ± 2.87 a	10.22 ± 0.96 b

All chemicals components are expressed on dry weight basis (dwt.). Values represented are mean of triplicates; values followed by different letters are significantly different at $p < 0.05$. N.D., not detected; GAE/g, mg of Gallic acid equivalent per gram of seaweed extract; QE/g, mg of Quercetin equivalent per gram of seaweed extract.

Regards to the antioxidant activity in extracts (Figure 1), greater activity was recorded in the *P. durvillaei* extract for the DPPH and ABTS tests, with 0.37 and 0.07 TEAC per gram of extract, respectively (Figure 1A,B). The antioxidant capacity, as measured by DPPH, was three-fold for the *P. durvillaei* extract versus the *U. lactuca* extract (0.13 TEAC/g). The ferric reducing antioxidant power (FRAP) was only recorded in *P. durvillaei* extracts, with TEAC values of 0.02/g (Figure 1C).

3.2. PGR Contents in Seaweed Extracts

Two groups of PGRs (acidic and basic) were identified by UHPLC-MS (Figure 2); their identification and quantification in aqueous algal extracts correspond to abcisic acid (ABA), an auxin (IAA), three cytokinins (dihydrozeatine, DHZ; isopentyladenine, IP; and t-zeatine, tZ), two gibberellins (GA1 and GA4), jasmonic acid (JA), and salicylic acid (SA). The quantification of acidic PGRs showed that SA and IAA are the main chemical groups in both algal extracts. SA concentration was 88.8 ± 52.9 ng/g in the *P. durvillaei* extract and 67.6 ± 4.2 ng/g in the *U. lactuca* extract (Figure 2A). IAA concentration was slightly higher in *U. lactuca* (49.3 ± 5.2 ng/g) relative to *P. durvillaei* (39.0 ± 12.8 ng/g), but these differences were not statistically significant (Figure 2A). On the other hand, the *U. lactuca* extract contained the highest ABA concentration (17.8 ± 5.2 ng/g) compared to the *P. durvillaei* extract (1.5 ± 0.5 ng/g), JA concentrations were similar in both extracts (Figure 2A). With regard to gibberellins, the highest GA4 concentration (1.51 ± 0.64 ng/g) was noted in the *U. lactuca* extract, compared to the *P. durvillaei* extract (0.37 ± 0.08 ng/g), while GA1 showed similar levels in both extracts (Figure 2B).

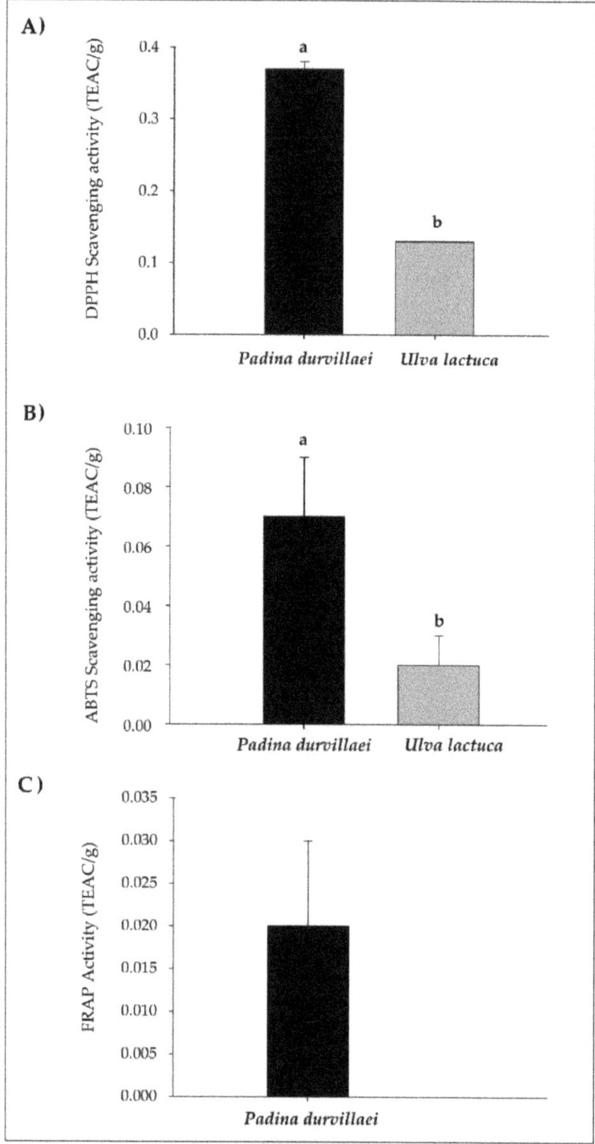

Figure 1. In vitro antioxidant activity of seaweed extracts of *Padina durvillaei* and *Ulva lactuca* assessed by different methods: (**A**) DPPH scavenging activity; (**B**) ABTS scavenging activity; and (**C**) FRAP activity. Bars represent mean of triplicates ± SD; different letters are significantly different at $p < 0.05$. TEAC/g, mmol of Trolox equivalent of antioxidant activity per gram of seaweed extract.

The quantification of basic PGRs (Figure 2C) showed that tZ is the most abundant hormone in the *U. lactuca* extract (10.95± 4.31 ng/g). On the other hand, IP concentrations were similar in both extracts (4.41 1.85 ng/g in *P. durvillaei* and 4.57 ± 0.38 ng/g in *U. lactuca*). DHZ concentrations were low in both extracts, with 1.09 ± 0.44 ng/g in *P. durvillaei* and 0.77 ± 0.09 ng/g in *U. lactuca*.

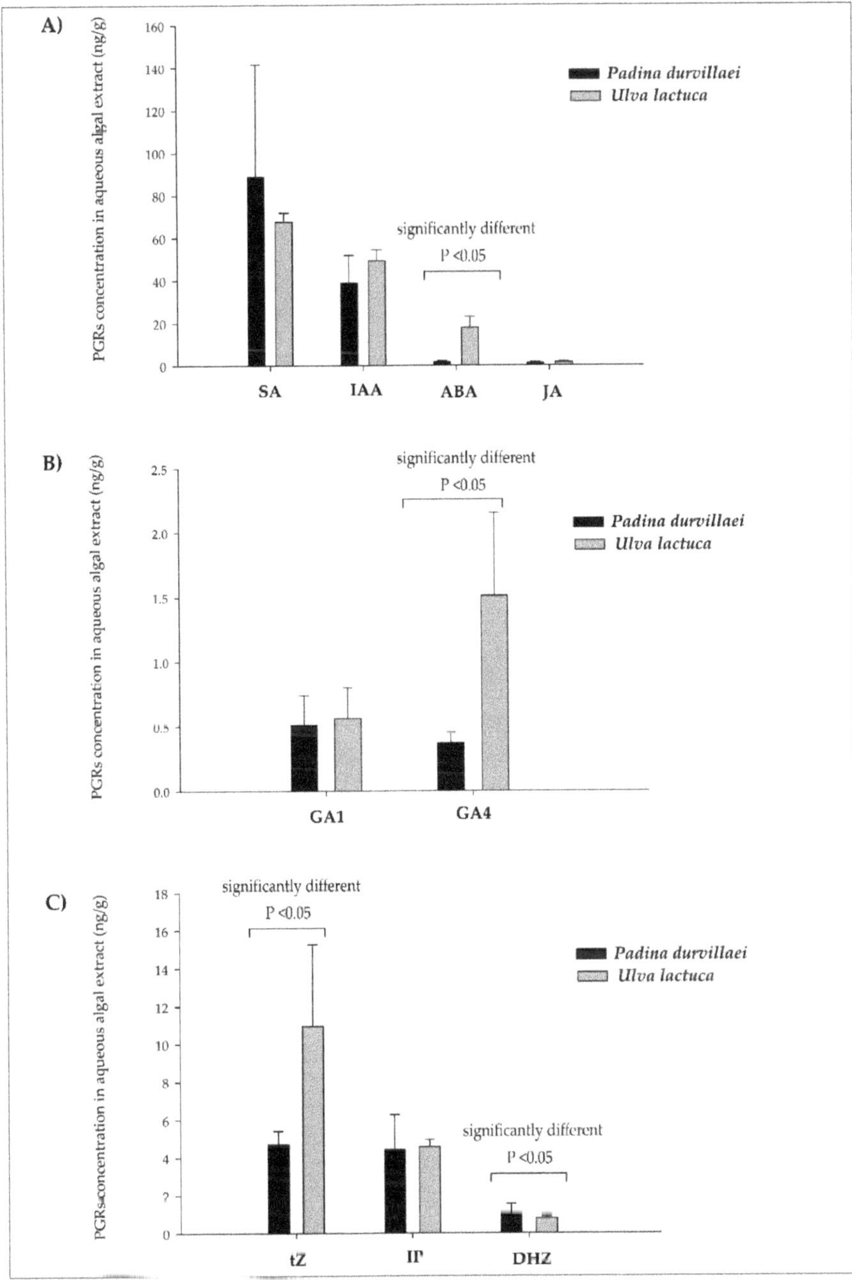

Figure 2. PGRs quantification by UHPLC-MS in aqueous algal extracts of *Padina durvillaei* and *Ulva lactuca*: (**A**) acid PGRs (salicylic acid, SA; indolacetic acid, IAA; abscisic acid, ABA; and jasmonic acid, JA); (**B**) gibberellins (GA1 and GA4); and (**C**) cytokinins or basic hormones (t-zeatine, tZ; isopentyladenine, iP; and dihydrozeatine, DHZ). Bars represent mean of triplicates ± SD. Significant differences at $p < 0.05$ of some type of PGR between algal extracts have been indicated.

4. Discussion

In this study, we used two species of algae to obtain aqueous extracts. Both species are frequently observed as floating seaweed mats that reach the northeast Pacific coasts; they belong to different taxonomic groups: *P. durvillaei* to the class Phaeophyceae (brown algae) and *U. lactuca* to the class Chlorophyceae (green algae). This leads to potential chemo-taxonomic differences, coupled with spatiotemporal variations associated with growth and environmental adaptation. Ultimately, these differences result in the variability in chemical composition, as observed in the aqueous extracts analyzed in the present study. Such variability has also been observed in other studies reporting taxonomic compositional differences in marine algae thriving in temperate waters (e.g., *Caulerpa sertularioides*, *Rhizoclonium riparium*, *Gracilaria vermoculophylla*, and *Spyridia filamentosa*), including the presence of biochemical compounds associated with bioactive characteristics, such as fucosterol, β-sitosterol, omega-3 fatty acids, and various photosynthetic pigments [36]. An example is the case of temperate macroalgae in Denmark, where aqueous extracts were analyzed from 16 species of macroalgae in different taxonomic families and whose total polyphenolics content showed no statistically significant differences [6].

From the compositional differences observed in macroalgae, their extracts exhibit different levels of biological activity, particularly antioxidant activity; the present study recorded a higher activity of the *P. durvillaei* extract. This extract has a higher content of polyphenols and flavonoids, compounds with a known ability to scavenge synthetic radicals in in vitro tests (DPPH and ABTS), as well as the presence of sulfated polysaccharides. Note that no ferric reducing antioxidant power (FRAP) was detected in *Ulva* extracts, while it was very weak in the *Padina* extract; this may indicate the influence of compositional characteristics of sulfated polysaccharides in extracts. Besides, substances that interfere in the antioxidant tests used were found, such as proteins and uronic acids. High contents of sulfates, protein, and uronic acids were also observed, which have been associated with the low antioxidant activity observed in polysaccharide fractions isolated from aqueous extracts of green algae such as *Ulva fasciata* [37].

It was also observed that the presence of fucose in sulfated polysaccharides from brown algae conferred a higher ferric reducing activity relative to polysaccharides from green algae that do not contain this sugar (e.g., *Sargassum wightii* vs. *Ulva lactuca*) [7]; a similar effect is likely to influence the behavior observed in our study. In general, secondary metabolites such as polyphenols are more abundant in brown algae, with some, e.g., 2,4,6-trihydroxybenzoate (a benzoic acid derivative), being unique to this group of macroalgae [38]. It should be stressed that polyphenols are compounds with electrons that can be donated, thus conferring a higher antioxidant response when classified methods are used, including electron transfer such as ABTS and FRAP [39].

On the other hand, a higher antioxidant activity was observed in both extracts with the DPPH method. This indicates that the antioxidant chemicals in extracts function mainly through a reaction mechanism involving the transfer of a hydrogen atom from these compounds [6]; this sort of antioxidant capacity was about three times higher for the *P. durvillaei* extract relative to the *U. lactuca* extract. Some studies have shown that aqueous extracts from brown algae have a higher polyphenolics content relative to red and green algae, which confer on them greater antioxidant activity. For example, the antioxidant activity (assessed with the DPPH test) of aqueous extracts from the brown algae *Fucus serratus* and *F. vesiculus* was 42 times higher than in extracts from the green algae *Enteromorpha intestinalis* and 89 times higher than in *Ulva lactuca* [6]. The same behavior was observed for the FRAP test, since the reducing power from brown algae extracts was 3–5 times lower relative to green algae; total polyphenols content and type of polyphenols (ferulic, vanillic, coumaric, and gallic acids) contribute highly to the antioxidant and reducing activities [6]. Likewise, Wang et al. [8] reported higher total polyphenols contents in aqueous extracts of brown algae relative to red and green algae, which led to a high in vitro antioxidant activity measured with DPPH for extracts obtained from *Fucus vesiculus*, *F. serratus*, *Ascophyllum nodosum*, and *Laminaria hyperborea*, with antioxidant values representing up to 200 times the antioxidant activity observed for *Ulva lactuca*.

Originally, PGRs in plants have been found in trace amounts (fmol to pmol per gram, wet weight), with gibberellins, abscisic acid, cytokinins, indoleacetic acid, ethylene, jasmonates, and salicylates as those most studied PGRs [40]. Unlike plants, algae accumulate higher concentrations of PGRs (pmol to μmol per gram fresh weight), mainly auxins and cytokinins; some authors hypothesize that bioactive gibberellins inducing germination and growth may also accumulate, as well as jasmonates and salicylates. However, few species of macroalgae contain gibberellins, jasmonate, and salicylates [3].

The presence and number of PGRs in algae and terrestrial plants differ according to the species or variety [12,22]. For example, Mori et al. [41] reported the identification of auxins, cytokinins, and salicylic acid (SA), but not AG3 and jasmonates, in extracts of two red algae (*Pyropia yezoensis* and *Bangia fuscopurpurea*). Another study involving fourteen seaweeds in the Turkey coast reported the presence of five PGRs (t-zeatine (t-Z), IAA, GA3, ABA, and 6-benzyl amino purine (BAP) in two algae, namely *Petalonia fascia* (brown algae) and *Caulerpa racemosa* var. cylindracea (green algae), and the absence of GA3 and ABA in the eleven remaining seaweeds, including *Sargassum vulgare* and *Ulva rigida*, which are used to produce biofertilizers [22]. By contrast, this work identified nine PGRs (SA, IAA, ABA, JA, t-Z, IP, DHZ, GA4, and GA1) in extracts from two algae (*P. durvillaei* and *U. lactuca*) distributed along the Mexican Pacific coasts. The presence of SA and JA is worth noting, as these have not been identified in any other algae; these PGRs strengthen the defense capabilities of plants by inducing acquired systemic resistance (SAR).Gibberellins GA4 and GA1 are also present, which show biological activity in plants.

Mexico produces and markets fertilizers based on extracts of seaweeds such as *Macrocystis pyrifera*, *Sargassum* spp, *Ascophyllum nodosum*, *Laminaria* spp, *Egregia menziesii*, and *Gelidium robustum*, using methods that involve hydrothermal treatment under acidic, neutral, and alkaline conditions [20]. These extracts boost germination, rooting, and plant growth, likely resulting from the content of polysaccharides, macro- and microelements [20,42–44]. However, although studies on algal extracts show evident effects on plant growth, few reports demonstrate the presence of PGRs in such extracts, and these can only be inferred from the physiological effects shown [42]. Unlike marketed products based on seaweed extracts containing polysaccharides obtained using methods under alkaline, acidic, and neutral conditions, this study used aqueous extraction, which ensured the extraction of nine PGRs in high concentrations for use in plant crops.

Studies related to the identification and quantification of PGRs in algae, such as *Ascophyllum nodosum* and *Sargassum muticum* (Phaeophyceae) used in the development of biofertilizers [45], report results differing from those obtained in this work. For example, *A. nodosum*, whose extracts are used in products such as Phylgreen®, contains concentrations of IAA, ABA, and IP of 7.53, 17.63, and 16.11 pmol/g dry weight (DW), respectively [46], while extracts of *Sargassum heterophyllum* accumulate IP and t-Z at 48.2 and 2.4 pmol/g DW, respectively [47]. In contrast, this study reports concentrations in extracts of *P. durvillaei* (IAA 39.0 ng/g, ABA 1.5 ng/g, IP 4.41 ng/g DW, and t-Z 4.7 ng/g DW) and *U. lactuca* (IAA 49.3 ng/g, ABA 17.8 ng/g, IP 4.57 ng/g, and t-Z 10.95 ng/g DW) that are higher than those reported by Jannin et al. [46] and Stirk et al. [47]. This finding suggests that the algae studied here are potential alternatives suitable for use as biofertilizers based on their high PGR content.

On the other hand, the identification of GA1 and GA4 in this study contrasts with reports by Dumale et al. (2018) [48], who identified GA3 in extracts of *Caulerpa racemosa* (green algae), and by Shoubaky et al. [49] in *U. lactuca* extracts, who identified two groups of PGRs, namely ABA and eight gibberellins (GA7 methyl ester, GA8, GA13, GA19, GA23, GA44, and GA75). However, the gibberellins identified in extracts of *P. durvillaei* and *U. lactuca* are considered to be biologically active in plants, contrary to those reported by Shoubaky et al. [49]; this difference is relevant, as PGRs should be in the active form for use in plant crops [50] Note that the identification of PGRs in *P. durvillaei* extracts has not been reported in the literature previously; thus, the data in this paper support the potential use of brown algae or extracts thereof as growth promoters in plants of agricultural importance.

5. Conclusions

The present study focused on the analysis of the main PGRs and the biochemical characterization of their aqueous extracts in two macroalgae involved in massive arrivals. The chromatographic analysis revealed the presence of abcisic acid (AB), auxins (IAA), cytokinins (tZ, IP, and DHZ), gibberellins (AG1 and AG4), jasmonates (AJ), and salicylates (AS); of these, AS attained the highest levels in both extracts. On the other hand, the phytochemical analysis revealed the presence of soluble compounds such as carbohydrates and uronic acids, as well as bioactive compounds such as polyphenols and flavonoids that confer antioxidant activity to extracts. The identification of PGRs in algal extracts opens the possibilities for use of algae involved in massive arrivals as potential environmentally-friendly organic biofertilizers that serve as growth promoters in agricultural crops.

Author Contributions: Conceptualization, I.B.G. and I.O.R.; methodology, A.K.D.L., E.C., I.O.R., and I.B.G.; validation, J.A.S.L., E.C., I.B.G., and I.O.R.; formal analysis, E.M.M., E.C., and I.O.R.; investigation, A.K.D.L., I.B.G., and I.O.R.; resources, E.C., I.O.R., and I.B.G.; data curation, I.B.G. and I.O.R.; writing—original draft preparation, I.B.G. and I.O.R.; writing—review and editing, I.B.G., E.C., I.O.R., E.M.M., and J.A.S.L.; visualization, E.M.M. and J.A.S.L.; supervision, I.O.R., I.B.G., E.M.M., and J.A.S.L.; project administration, I.B.G. and I.O.R.; and funding acquisition, I.O.R. and I.B.G. All authors have read and agreed to the published version of the manuscript.

Funding: This research received no external funding.

Acknowledgments: The authors are grateful to the Universidad Politécnica de Sinaloa and the Instituto de Biología Molecular y Celular de Plantas (IBMCP) CSIC-Universidad Politécnica de Valencia for providing the necessary facilities to carry out the research.

Conflicts of Interest: The authors declare no conflict of interest.

References

1. Ben Salah, I.; Aghrouss, S.; Douira, A.; Aissam, S.; El Alaoui-Talibi, Z.; Filali-Maltouf, A.; El Modafar, C. Seaweed polysaccharides as bio-elicitors of natural defenses in olive trees against verticillium wilt of olive. *J. Plant Interact.* **2018**, *13*, 248–255. [CrossRef]

2. Vera, J.; Castro, J.; Gonzalez, A.; Moenne, A. Seaweed Polysaccharides and Derived Oligosaccharides Stimulate Defense Responses and Protection Against Pathogens in Plants. *Mar. Drugs* **2011**, *9*, 2514–2525. [CrossRef]

3. Khan, W.; Rayirath, U.P.; Subramanian, S.; Jithesh, M.N.; Rayorath, P.; Hodges, D.M.; Critchley, A.T.; Craigie, J.S.; Norrie, J.; Prithiviraj, B. Seaweed Extracts as Biostimulants of Plant Growth and Development. *J. Plant Growth Regul.* **2009**, *28*, 386–399. [CrossRef]

4. Osuna-Ruiz, I.; López-Saiz, C.-M.; Burgos-Hernández, A.; Velázquez, C.; Nieves-Soto, M.; Hurtado-Oliva, M.A. Antioxidant, antimutagenic and antiproliferative activities in selected seaweed species from Sinaloa, Mexico. *Pharm. Biol.* **2016**, *54*, 2196–2210. [CrossRef] [PubMed]

5. Osuna-Ruíz, I.; Salazar-Leyva, J.A.; López-Saiz, C.M.; Burgos-Hernández, A.; Hernández-Garibay, E.; Lizardi-Mendoza, J.; Hurtado-Oliva, M.A. Enhancing antioxidant and antimutagenic activity of the green seaweed Rhizoclonium riparium by bioassay-guided solvent partitioning. *J. Appl. Phycol.* **2019**, *31*, 3871–3881. [CrossRef]

6. Farvin, K.S.; Jacobsen, C. Phenolic compounds and antioxidant activities of selected species of seaweeds from Danish coast. *Food Chem.* **2013**, *138*, 1670–1681. [CrossRef] [PubMed]

7. Meenakshi, S.; Umayaparvathi, S.; Arumugam, M.; Balasubramanian, T. In vitro antioxidant properties and FTIR analysis of two seaweeds of Gulf of Mannar. *Asian Pac. J. Trop. Biomed.* **2011**, *1*, 66–70. [CrossRef]

8. Wang, T.; Jónsdóttir, R.; Olafsdottir, G. Total phenolic compounds, radical scavenging and metal chelation of extracts from Icelandic seaweeds. *Food Chem.* **2009**, *116*, 240–248. [CrossRef]

9. Godlewska, K.; Michalak, I.; Tuhy, Ł.; Chojnacka, K. Plant Growth Biostimulants Based on Different Methods of Seaweed Extraction with Water. *BioMed Res. Int.* **2016**, *2016*, 1–11. [CrossRef]

10. Ghaderiardakani, F.; Collas, E.; Damiano, D.K.; Tagg, K.; Graham, N.S.; Coates, J.C. Effects of green seaweed extract on Arabidopsis early development suggest roles for hormone signalling in plant responses to algal fertilisers. *Sci. Rep.* **2019**, *9*, 1983. [CrossRef]

11. Prasad, K.; Das, A.K.; Oza, M.D.; Brahmbhatt, H.; Siddhanta, A.K.; Meena, R.; Eswaran, K.; Rajyaguru, M.R.; Ghosh, P.K. Detection and Quantification of Some Plant Growth Regulators in a Seaweed-Based Foliar Spray Employing a Mass Spectrometric Technique sans Chromatographic Separation. *J. Agric. Food Chem.* **2010**, *58*, 4594–4601. [CrossRef] [PubMed]

12. Yalçın, S.; Şükran Okudan, E.; Karakaş, Ö.; Önem, A.N.; Sözgen Başkan, K. Identification and quantification of some phytohormones in seaweeds using UPLC-MS/MS. *J. Liq. Chromatogr. Relat. Technol.* **2019**, *42*, 475–484. [CrossRef]

13. Hurtado, A.Q.; Neish, I.C.; Critchley, A.T. Phyconomy: The extensive cultivation of seaweeds, their sustainability and economic value, with particular reference to important lessons to be learned and transferred from the practice of eucheumatoid farming. *Phycologia* **2019**, *58*, 472–483. [CrossRef]

14. Kumar, R.; Trivedi, K.; Anand, K.G.V.; Ghosh, A. Science behind biostimulant action of seaweed extract on growth and crop yield: Insights into transcriptional changes in roots of maize treated with Kappaphycus alvarezii seaweed extract under soil moisture stressed conditions. *J. Appl. Phycol.* **2019**, *32*, 599–613. [CrossRef]

15. Hąc-Wydro, K.; Flasiński, M. The studies on the toxicity mechanism of environmentally hazardous natural (IAA) and synthetic (NAA) auxin—The experiments on model Arabidopsis thaliana and rat liver plasma membranes. *Colloids Surf. B Biointerfaces* **2015**, *130*, 53–60. [CrossRef]

16. Arioli, T.; Mattner, S.W.; Winberg, P. Applications of seaweed extracts in Australian agriculture: Past, present and future. *J. Appl. Phycol.* **2015**, *27*, 2007–2015. [CrossRef]

17. Hayashi, L.; Reis, R.P.; Dos Santos, A.A.; Castelar, B.; Robledo, D.; De Vega, G.B.; Msuya, F.E.; Eswaran, K.; Yasir, S.M.; Ali, M.K.M.; et al. The cultivation of Kappaphycus and Eucheuma in tropical and sub-tropical waters. In *Tropical Seaweed Farming Trends, Problems and Opportunities*; Springer: Cham, Switzerland, 2017; pp. 55–90.

18. Stirk, W.A.; Van Staden, J. Plant growth regulators in seaweeds: Occurrence, regulation and functions. In *Advances in Botanical Research*; Elsevier: Amsterdam, The Netherlands, 2014; Volume 71, pp. 125–159.

19. Uthirapandi, V.; Suriya, S.; Boomibalagan, P.; Eswaran, S.; Ramya, S.S.; Vijayanand, N.; Kathiresan, D. Bio-Fertilizer potential of seaweed liquid extracts of marine macro algae on growth and biochemical parameters of Ocimum sanctum. *J. Pharm. Phytochem.* **2018**, *7*, 3528–3532.

20. Hernández-Herrera, R.M.; Santacruz-Ruvalcaba, F.; Briceño-Domínguez, D.R.; Di Filippo-Herrera, D.A.; Hernández-Carmona, G. Seaweed as potential plant growth stimulants for agriculture in Mexico. *Hidrobiológica* **2018**, *28*, 129–140. [CrossRef]

21. Sharma, H.S.; Fleming, C.; Selby, C.; Rao, J.R.; Martin, T. Plant biostimulants: A review on the processing of macroalgae and use of extracts for crop management to reduce abiotic and biotic stresses. *J. Appl. Phycol.* **2013**, *26*, 465–490. [CrossRef]

22. Yalçın, S.; Okudan, E.Ş.; Karakaş, Ö.; Önem, A.N. Determination of Major Phytohormones in Fourteen Different Seaweeds Utilizing SPE–LC–MS/MS. *J. Chromatogr. Sci.* **2019**, *58*, 98–108. [CrossRef]

23. Hernández-Herrera, R.M.; Santacruz-Ruvalcaba, F.; Ruiz-López, M.A.; Norrie, J.; Hernández-Carmona, G. Effect of liquid seaweed extracts on growth of tomato seedlings (Solanum lycopersicum L.). *J. Appl. Phycol.* **2013**, *26*, 619–628. [CrossRef]

24. Tierney, M.S.; Smyth, T.; Rai, D.K.; Soler-Vila, A.; Croft, A.K.; Brunton, N. Enrichment of polyphenol contents and antioxidant activities of Irish brown macroalgae using food-friendly techniques based on polarity and molecular size. *Food Chem.* **2013**, *139*, 753–761. [CrossRef] [PubMed]

25. Dubois, M.; Gilles, K.A.; Hamilton, J.K.; Rebers, P.A.; Smith, F. Colorimetric Method for Determination of Sugars and Related Substances. *Anal. Chem.* **1956**, *28*, 350–356. [CrossRef]

26. Bradford, M.M. A rapid and sensitive method for the quantitation of microgram quantities of protein utilizing the principle of protein-dye binding. *Anal. Biochem.* **1976**, *72*, 248–254. [CrossRef]

27. Dodgson, K.; Price, R.; Lash, J.W.; Whitehouse, M.W.; Moretti, A.; Harborne, J. A note on the determination of the ester sulphate content of sulphated polysaccharides. *Biochem. J.* **1962**, *84*, 106–110. [CrossRef] [PubMed]

28. Bitter, T.; Muir, H. A modified uronic acid carbazole reaction. *Anal. Biochem.* **1962**, *4*, 330–334. [CrossRef]

29. Marigo, G. Sur une méthode de fractionnement et d'estimation des composés phénoliques chez les végétaux. *Annalusis* **1973**, *2*, 106–110.

30. Luximon-Ramma, A.; Bahorun, T.; Soobrattee, M.A.; Aruoma, O.I. Antioxidant Activities of Phenolic, Proanthocyanidin, and Flavonoid Components in Extracts of Cassia fistula. *J. Agric. Food Chem.* **2002**, *50*, 5042–5047. [CrossRef]

31. Mensor, L.L.; Menezes, F.S.; Leitão, G.G.; Reis, A.S.; Dos Santos, T.C.; Coube, C.S.; Leitão, S.G. Screening of Brazilian plant extracts for antioxidant activity by the use of DPPH free radical method. *Phytother. Res.* **2001**, *15*, 127–130. [CrossRef]

32. Przygodzka, M.; Zielińska, D.; Ciesarová, Z.; Kukurová, K.; Zieliński, H. Comparison of methods for evaluation of the antioxidant capacity and phenolic compounds in common spices. *LWT-Food Sci. Technol.* **2014**, *58*, 321–326. [CrossRef]

33. Benzie, I.; Strain, J. The Ferric Reducing Ability of Plasma (FRAP) as a Measure of "Antioxidant Power": The FRAP Assay. *Anal. Biochem.* **1996**, *239*, 70–76. [CrossRef] [PubMed]

34. Szőllősi, R.; Varga, I.S.I. Total antioxidant power in some species of Labiatae: Adaptation of FRAP method. *Acta Biol. Szeged.* **2002**, *46*, 125–127.

35. Seo, M.; Jikumaru, Y.; Kamiya, Y. Profiling of hormones and related metabolites in seed dormancy and germination studies. In *Seed Dormancy*; Springer: Boston, MA, USA, 2011; pp. 99–111.

36. Osuna-Ruiz, I.; Nieves-Soto, M.; Manzano-Sarabia, M.M.; Hernández-Garibay, E.; Lizardi-Mendoza, J.; Burgos-Hernández, A.; Hurtado-Oliva, M.Á. Gross chemical composition, fatty acids, sterols, and pigments in tropical seaweed species off Sinaloa, Mexico. *Cienc. Mar.* **2019**, *45*, 101–120. [CrossRef]

37. Shao, P.; Chen, M.; Pei, Y.; Sun, P. In intro antioxidant activities of different sulfated polysaccharides from chlorophytan seaweeds Ulva fasciata. *Int. J. Biol. Macromol.* **2013**, *59*, 295–300. [CrossRef] [PubMed]

38. Belghit, I.; Rasinger, J.D.; Heesch, S.; Biancarosa, I.; Liland, N.S.; Torstensen, B.; Waagbø, R.; Lock, E.J.; Bruckner, C.G. In-depth metabolic profiling of marine macroalgae confirms strong biochemical differences between brown, red and green algae. *Algal Res.* **2017**, *26*, 240–249. [CrossRef]

39. Prior, R.L.; Wu, X.; Schaich, K. Standardized Methods for the Determination of Antioxidant Capacity and Phenolics in Foods and Dietary Supplements. *J. Agric. Food Chem.* **2005**, *53*, 4290–4302. [CrossRef]

40. Šimura, J.; Antoniadi, I.; Široká, J.; Tarkowská, D.; Strnad, M.; Ljung, K.; Novak, O. Plant Hormonomics: Multiple Phytohormone Profiling by Targeted Metabolomics. *Plant Physiol.* **2018**, *177*, 476–489. [CrossRef]

41. Mori, I.C.; Ikeda, Y.; Matsuura, T.; Hirayama, T.; Mikami, K. Phytohormones in red seaweeds: A technical review of methods for analysis and a consideration of genomic data. *Bot. Mar.* **2017**, *60*, 153–170. [CrossRef]

42. Briceño-Domínguez, D.; Hernández-Carmona, G.; Moyo, M.; Stirk, W.; van Staden, J. Plant growth promoting activity of seaweed liquid extracts produced from Macrocystis pyrifera under different pH and temperature conditions. *J. Appl. Phycol.* **2014**, *26*, 2203–2210. [CrossRef]

43. Castellanos-Barriga, L.G.; Santacruz-Ruvalcaba, F.; Hernández-Carmona, G.; Ramírez-Briones, E.; Hernández-Herrera, R.M. Effect of seaweed liquid extracts from Ulva lactuca on seedling growth of mung bean (Vigna radiata). *J. Appl. Phycol.* **2017**, *29*, 2479–2488. [CrossRef]

44. Hernández-Herrera, R.M.; Santacruz-Ruvalcaba, F.; Zañudo-Hernández, J.; Hernández-Carmona, G. Activity of seaweed extracts and polysaccharide-Enriched extracts from Ulva lactuca and Padina gymnospora as growth promoters of tomato and mung bean plants. *J. Appl. Phycol.* **2016**, *28*, 2549–2560. [CrossRef]

45. Silva, L.D.; Bahcevandziev, K.; Pereira, L. Production of bio-Fertilizer from Ascophyllum nodosum and Sargassum muticum (Phaeophyceae). *J. Oceanol. Limnol.* **2019**, *37*, 918–927. [CrossRef]

46. Jannin, L.; Arkoun, M.; Etienne, P.; Laîné, P.; Goux, D.; Garnica, M.; Fuentes, M.; Francisco, S.S.; Baigorri, R.; Cruz, F.; et al. Brassica napus Growth is Promoted by Ascophyllum nodosum (L.) Le Jol. Seaweed Extract: Microarray Analysis and Physiological Characterization of N, C, and S Metabolisms. *J. Plant Growth Regul.* **2012**, *32*, 31–52. [CrossRef]

47. Stirk, W.; Novak, O.; Strnad, M.; Van Staden, J. Cytokinins in macroalgae. *Plant Growth Regul.* **2003**, *41*, 13–24. [CrossRef]

48. Dumale, J.; Gamoso, G.; Manangkil, J.; Divina, C. Detection and Quantification of Auxin and Gibberellic Acid in Caulerpa racemosa. *IJAT* **2018**, *14*, 653–660.

49. El Shoubaky, G.A.; Salem, E.A. Effect of abiotic stress on endogenous phytohormones profile in some seaweeds. *IJPPR* **2016**, *8*, 124–134.

50. Hedden, P.; Sponsel, V. A century of gibberellin research. *J. Plant Growth Regul.* **2015**, *34*, 740–760. [CrossRef]

Article

Roles of Nitrogen Deep Placement on Grain Yield, Nitrogen Use Efficiency, and Antioxidant Enzyme Activities in Mechanical Pot-Seedling Transplanting Rice

Lin Li [1,2], Zheng Zhang [1,2], Hua Tian [1,2], Zhaowen Mo [1,2], Umair Ashraf [1,3], Meiyang Duan [1,2], Zaiman Wang [4], Shuli Wang [1,2], Xiangru Tang [1,2,*]  and Shenggang Pan [1,2,*]

1. State Key Laboratory for Conservation and Utilization of Subtropical Agro-bioresources, College of Agriculture, South China Agricultural University, Guangzhou 510642, China; scaulilin@126.com (L.L.); scauzhangzheng@126.com (Z.Z.); tianhua@scau.edu.cn (H.T.); scaumozhw@126.com (Z.M.); umairashraf2056@gmail.com (U.A.); meiyang@scau.edu.cn (M.D.); wangshuli@scau.edu.cn (S.W.)
2. Scientific Observing and Experimental Station of Crop Cultivation in South China, Ministry of Agriculture PR China, Guangzhou 510642, China
3. Department of Botany, University of Education (Lahore), Faisalabad-Campus, Faisalabad 38000, Punjab, Pakistan
4. Key Laboratory of Key Technology for South Agricultural Machine and Equipment, Ministry of Education, Guangzhou 510642, China; wangzaiman@scau.edu.cn
* Correspondence: tangxr@scau.edu.cn (X.T.); panshenggang@scau.edu.cn (S.P.)

Received: 2 August 2020; Accepted: 21 August 2020; Published: 25 August 2020

Abstract: Mechanical pot-seedling transplanting (PST) is an efficient transplanting method and deep nitrogen fertilization has the advantage of increasing nitrogen use efficiency. However, little information is available about the effect of PST when coupled with mechanized deep nitrogen (N) fertilization on grain yield, nitrogen use efficiency, and antioxidant enzyme activities in rice. A two-year field experiment was performed to evaluate the effect of PST coupled with deep N fertilization in both early seasons (March–July) of 2018 and 2019. All seedlings were transplanted by PST and three treatments were designed as follows. There was a mechanized deep placement of all fertilizer (MAF), broadcasting fertilizer (BF), no fertilizer (N0). MAF significantly increased grain yield by 52.7%. Total nitrogen accumulation (TNA) was enhanced by 27.7%, nitrogen partial factor productivity (NPFP) was enhanced by 51.4%. nitrogen recovery efficiency (NRE) by 123.7%, and nitrogen agronomic efficiency (NAE) was enhanced by 104.3%, compared with BF treatment. Moreover, MAF significantly improved peroxidase (POD), catalase (CAT), and notably reduced the malonic dialdehyde (MDA) content for both rice cultivars, compared to BF. Hence, the result shows that mechanical pot-seedling transplanting coupled with nitrogen deep placement is an efficient method with the increase of grain yield and nitrogen use efficiency in rice cultivation in South China.

Keywords: deep N fertilization; peroxidase activity; catalase activity; rice cultivation

---

1. Introduction

Rice (*Oryza sativa* L.) is one of the world's major crops and it provides food for over three billion people [1,2]. China is the main country of rice production, with rice planting area and yield in the forefront of the world [2,3]. Therefore, increasing rice production is essential for population growth in China and the world [4,5] Transplanted rice is the most traditional planting method. Moreover, the traditional fertilization method is manually surface broadcast [6]. This method is not suitable for the stable improvement in Chinese agricultural systems because of some serious problems such as labor scarcity and low profits [7]. Therefore, improving mechanization is the main way to solve this problem.

The mechanical pot-seedling transplanting (PST) is an innovative technology for transplanting rice seedlings in the paddy field. It has the advantage of precise row and hill spacing without injury to the rice plants [8]. The technique not only reduces planting costs but improves the quality of transplanting rice seedlings [9]. Mechanical pot-seedling transplanting can transplant seedlings without root injury. It expects to reduce the transplanting shock and maintain the root activity, resulting in increased nutrient absorption and consequently a vigorous initial growth [10]. However, the application of nitrogen fertilizer under the flooding condition can cause nitrogen fertilizer to be lost through ammonia volatilization and runoff, reducing the utilization efficiency of nitrogen fertilizer [11,12]. Thus, the seedlings transplanted by mechanical pot-seedling transplanting were unable to access the N resource, restricting their performance irrespective of enhanced root activity [4]. To solve this problem, much application of chemical fertilizer is one of the approaches, but it can lead to problems of lower nitrogen use efficiency (NUE) and environmental pollution [11,13].

Alternative is the deep N fertilization in mechanical pot-seedling transplanting. Deep fertilizer application methods can maintain the nutrient and enhance nutrient use efficiency [14]. The nitrogen fertilization at about 5 cm depth notably improves the total above-ground biomass and grain yield, compared to the manual surface broadcast [15]. Moreover, deep fertilization could reduce the amount of fertilizer applications without reducing yield [16]. Deep fertilization is to bury the fertilizer near the rice root, which is beneficial for fertilizer absorption by rice root [17]. The appropriate fertilization depth could not only promote the growth of rice root but also improve the growth of rice plants in the early stage [18]. Some studies found that deep N fertilization could reduce CH_4 emissions by 40% and NO emissions by 54% [19,20]. Pan et al. [21] found that deep fertilization significantly increased peroxidase (POD) and catalase (CAT) in direct-seeded rice. Xu et al. [22] observed that ensuring the nutrient supply at the late stage of rice was conducive to improving antioxidant enzyme activities and photosynthetic performance of rice leaves. Moreover, Shu et al. [23] reported that mechanical deep fertilization could delay rice plan senescence by enhancing antioxidant enzyme activities and reducing the malonic dialdehyde (MDA) content in direct-seeded rice. Therefore, deep nitrogen fertilization is a feasible way to lessen environmental problems because of the excess fertilization in rice production.

Mechanical pot-seedling transplanting (PST) coupled with deep nitrogen fertilization is an emerging transplanting rice technology. However, little information is available about the effects of PST coupled with mechanized deep N fertilization on grain yield, nitrogen use efficiency, and antioxidant enzyme activities in rice. The aim of this study was to assess whether PST coupled with mechanized N deep fertilization could increase grain yield, nitrogen use efficiency, and antioxidant enzyme activities in rice.

2. Materials and Methods

2.1. Mechanical Pot-Seedling Transplanting Machine

A mechanical pot-seedling transplanting machine was developed by Changzhou YaMeiKe mechanical Co., Ltd. (Changzhou, China) (Figure 1). This method realized the synchronous operation of deep fertilization and transplanting seedlings and applied fertilizer quantitatively and fixed-point deep into the soil on the seedling side.

Figure 1. Pictorial view of mechanized transplanting rice machine coupled with N deep placement at the farm of South China Agricultural University, Guangzhou city, China.

2.2. Experimental Treatments and Design

Field experiments were conducted in early seasons of 2018 and 2019, respectively, at the Experimental Research Farm, College of Agriculture, South China Agricultural University, Guangzhou City, China (23°13′ N, 113°81′ E, altitude 11 m). The soil in the experimental field was sandy loam with 1010 mg kg^{-1} total N, 1080 mg kg^{-1} total p, 20,230 mg kg^{-1} total K, 73 mg kg^{-1} available p, 104 mg kg^{-1} available K, and 21,560 mg kg^{-1} organic C.

Two rice cultivars were *Yuxiangyouzhan* (*YXYZ*) and *Wufengyou615* (*WFY615*), which are inbred and hybrid rice, respectively, and widely grown in the local area. Moreover, the two rice cultivars have growth periods of 118 and 115 days for both early seasons, respectively. Both field experiments were used in a completely randomized design with three replicates with a plot area of 132 m^2 (8 m × 16.5 m). The YaraMila compound fertilizer (total nitrogen contents TN = 15%, N: P$_2$O$_5$: K$_2$O = 15%:15%:15%) was used in our experiment, which was manufactured by YaraMila Fertilizer Company, China. The application rate was 150 kg N ha^{-1} (pure N) for the fertilizer application treatment. All seedlings were transplanted by PST and three treatments were designed as follows. The mechanized deep N fertilization was a basal fertilizer in 10 cm soil depth (MAF) and fertilizers were broadcast manually on the soil surface two days before the transplanting as a basal fertilizer (BF). No fertilizer was applied during entire growth stage (N0). Water management strategies were adopted by local farmer's advice. Some chemical reagents such as herbicide, imidacloprid, tricyclazole, and carbendazim were adopted to prevent and control weeds, insects, and diseases.

2.3. Yield and Its Components

At maturity, rice grains were recorded from the harvested-area of 6 m^2. In total, 20 rice plants were collected randomly for each treatment and the averaged values were calculated for the number of productive panicles per hill. Six hills of rice plants were taken to investigate yield components and the yield components measurement were determined according to Pan et al. [21]. To divide the filled seeds, all spikelets were submerged in tap water, apart from the rachis (by manual threshing) To calculate the total number of spikelets, we counted the spikelets in three representative subsamples of 30 g. The average weight of half-filled spikelets was determined. Spikelets per panicle, grain-filling percentage, and 1000-grain-weight were also calculated from sampled plants and averaged.

2.4. Total Above-Ground Biomass (TAB) and Leaf Area Index (LAI) at Different Growth Stages

According to the average number of tillers in the plot, six plants were taken in the plot at all critical growth stages including mid-tillering (MT), panicle initiation (PI), heading stage (HS), and the maturity stage (MS). LAI and TAB were determined according to Pan et al. [21]. The soil on the rice plants were washed thoroughly. Then, leaf sheaths plus stems, leaves, and spikes were separated from the plant after the heading stage. The leaf area for all green leaf blades was measured with the Li-Cor area meter (Li Cor Model 3100, Lincoln, NE, USA) and the leaf area per m^2 (leaf area index, LAI) was then calculated. The separated part of the rice plants was oven-dried at 70 °C to constant weight and then the above-ground biomass was calculated.

2.5. Nitrogen Use Efficiency

Six plants were collected from each treatment in the physiological maturity stage. They were then divided into leaves, stems with leaf sheath, and grains. They were finally dried at 70 °C until constant weight, then stored to analyze the total N contents. Plant samples (0.2 g) were digested using the Kjeldhal method to analyze ammonia concentrations via an Alliance-Futura NP analyzer (Alliance Instruments, France) and then the N content was measured. The nitrogen use efficiency including nitrogen recovery efficiency (NRE), nitrogen agronomic use efficiency (NAE), nitrogen partial factor productivity (NPFP), nitrogen harvest index (NHI), and nitrogen grain production efficiency (NGPE) were evaluated by the formulae below:

1. NRE = (Nup − N0 up)/FN
2. NAE = (GY − GY0)/FN
3. NPFP = GY/FN
4. NGPE = GY/Nup
5. NHI = Ng/Nup

where N0 up and GY0 represented the total nitrogen uptake of above-ground plant parts and grain yields in the N0 plot, respectively. Nup and GY are the total nitrogen uptake of above-ground plant parts and grain yields in other N-fertilized plots, respectively. FN is the applied N fertilizer rate; Ng is the total nitrogen uptake in grain.

2.6. Determination of Antioxidant Enzyme Activities

About 25 leaves from each treatment were collected during the MT, PI, and HS stage. All samples were stored in −80 °C for enzyme activity determination i.e., peroxidase (POD), catalase (CAT), and malonic dialdehyde (MDA). POD and MDA were determined by the method established by Pan et al. [24]. Fresh leaf segments (<2 mm, 0.25 g) were homogenized in an ice bath in 5 mL of 50 mm borate buffer (pH 8.7) containing 5.0 mm sodium hydrogen sulfite and 0.1 g polyvinylpyrrolidone (PVP). The homogenate was centrifuged at 9000 ×g for 15 min at 4 °C. The supernatant was used as enzyme extract. POD activity was assayed by adding 0.1 mL of the enzyme extract to a substrate mixture containing acetate buffer (0.1 mol L^{-1}, pH 5.4), ortho-dianisidine (0.25% in ethyl alcohol) and 0.1 mL 0.8% H_2O_2 was added to 0.1 mL of the enzyme extract. Absorbance change of the brown guaiacol at 460 nm was recorded for calculating POD activity. One POD unit of enzyme activity was defined as the absorbance increase because of guaiacol oxidation by 1-unit min^{-1} (U g^{-1} FW min^{-1}). Leaf samples (0.5 g) were homogenized in 5 mL of 5% trichloroacetic acid. The homogenate was centrifuged at 4000 ×g for 10 min at 25 °C and 3 mL of 2-thiobarbituric acid in 20% trichloroacetic acid was added to a 2 mL aliquot of the supernatant. The mixture was heated at 98 °C for 10 min and cooled rapidly in an ice bath. After centrifugation at 4000 ×g for 10 min, the absorbance was recorded at 532 nm. Measurements were corrected for non-specific turbidity by subtracting the absorbance at 600 nm. MDA concentration was determined by the extinction coefficient MDA (ε = 155 μm cm^{-1}). CAT activities were determined according to Dhindsa et al. [25]. The 3 mL reaction mixture contained 50 mM phosphate buffer, pH 7.0, 15 mM hydrogen peroxide, and 25 μL enzyme extract. The decrease in hydrogen peroxide was followed as a decline in A240 using a Perkin-Elmer double-beam spectrophotometer connected to a recorder. The activity was expressed in units where one unit of catalase converts one μmole of hydrogen peroxide per minute.

2.7. Data Analysis

The experimental data were analyzed using DPS3.11 (Data Processing System for Practical Statistics, Hangzhou, China). In the ANOVA model, the single effect of treatment, cultivar, year, and the interaction effect were fixed, while the replication effect in year was random. The differences amongst means of the experimental treatments were separated using the least significant difference (LSD) at 0.05 probability level (ANOVA). All figures were drawn with Origin 9.0.

3. Results

3.1. Grain Yield and Its Components

The grain yield and its components varied with different N fertilizer applications for both rice cultivars (Table 1). The highest grain yield for MAF treatment was found for the BF and N0 treatments in both years. Mean grain yields of both rice cultivars under MAF were 8.4 t ha^{-1}, which was 52.7% higher than the BF treatment. Regarding yield components, deep placement produced the highest number of productive panicles ha^{-1} and spikelet per panicle, which was 272.9 $\times$ 10^4 and 189.7, respectively. No significance was found between BF and N0 treatment in the 1000-grain-weight and grain filling. Significant differences were found in the number of productive panicles, spikelets per panicle, 1000-grain-weight, and grain yield between nitrogen treatments. Both rice cultivars differed significantly in 1000-grain-weight.

Table 1. Effects of mechanical deep placement of nitrogen fertilizer on average grain yield and its components for both rice cultivars in two-year (2018 and 2019).

Treatments	Productive Panicle (10^4 Ha^{-1})	Spikelet per Panicle	Grain Filling (%)	1000-Grain-Weight (G)	Harvested Yield (T Ha^{-1})
N0	151.4 b	150.3 b	80.9	20.1 b	3.9 c
MAF	272.9 a	189.7 a	82.7	22.2 a	8.4 a
BF	185.3 b	163.3 b	81.6	21.1 b	5.5 b
Anova					
Cultivar (C)	ns	ns	ns	**	ns
Nitrogen (N)	**	*	ns	**	**
N $\times$ C	ns	ns	ns	ns	ns

MAF: mechanized deep placement of all fertilizer; BF: broadcasting fertilizer; N0: no fertilizer. Average values followed by different letter represent LSD significant differences at $p < 0.05$. **: ($p < 0.01$); *: ($p < 0.05$); ns: not significant variance.

3.2. Nitrogen Use Efficiency

The TNA, NGPE, NHI, NAE, NPFP, and NRE varied with different N fertilizer application in both rice cultivars (Table 2). In two-years, the N fertilizer application (MAF and BF) treatments significantly increased TNA relative to N0 treatment. Moreover, MAF treatment significantly increased TNA compared to BF treatment. The TNA of both rice cultivars for MAF was 173.6 kg ha^{-1}, which was 27.7% higher than the BF treatment, respectively. NGPE was highest in MAF, followed by BF, and the lowest was the N0 treatment. MAF showed the maximum NRE and NAE, while the BF had the lowest value for NRE and NRE among all treatments, respectively. Furthermore, a significant difference was found between MAF and BF. The NPFP of both rice cultivars for MAF was 54.2 kg kg^{-1}, which was

51.4 % higher than that in the BF treatment, respectively. Moreover, no remarkable difference among all treatment was found in NHI. There were notable differences in TNA, NGPE, NPFP, NAE, and NRE between nitrogen treatments. Moreover, the N × C (Nitrogen × Cultivar) factor interactions also had an obvious impact on TNA and NHI (Table 2).

Table 2. Effects of mechanical deep placement of nitrogen fertilizer on average nitrogen use efficiency for both rice cultivars in two-year (2018 and 2019).

Treatments	TNA (kg ha^{-1})	NGPE (kg kg^{-1})	NPFP (kg kg^{-1})	NHI	NRE (%)	NAE (kg kg^{-1})
N0	100.6 c	39.2 b	26.8 c	53.2		
MAF	173.6 a	46.6 a	54.2 a	54.3	46.8 a	22.6 a
BF	135.9 b	41.4 b	35.8 b	50.8	22.9 b	10.1 b
Anova						
Cultivar(C)	**	ns	*	*	ns	*
Nitrogen(N)	**	*	**	ns	**	**
N × C	**	ns	ns	**	ns	ns

Average values followed by different letter represent LSD significant differences at $p < 0.05$. **: ($p < 0.01$); *: ($p < 0.05$); ns: not significant variance. TNA: Total nitrogen accumulation; NAE: N agronomic efficiency, NRE: N recovery efficiency, NGPE: N grain production efficiency; NHI: N harvest index, NPFP: nitrogen partial factor productivity.

3.3. Total Aboveground Biomass (TAB) and Leaf Area Index (LAI) at Different Growth Stage

The N fertilizer application remarkably affected the LAI for both rice cultivars (Figure 2). For example, during the MT stage, the LAI for BF and MAF treatments were significantly higher than N0, but no significant difference was found between MAF and BF. At the PI and HS stages, MAF was significantly larger for LAI, especially when compared to N0 and BF. The result manifested that deep placement of the N application could modulate a sustainable longer growth period than surface broadcasting.

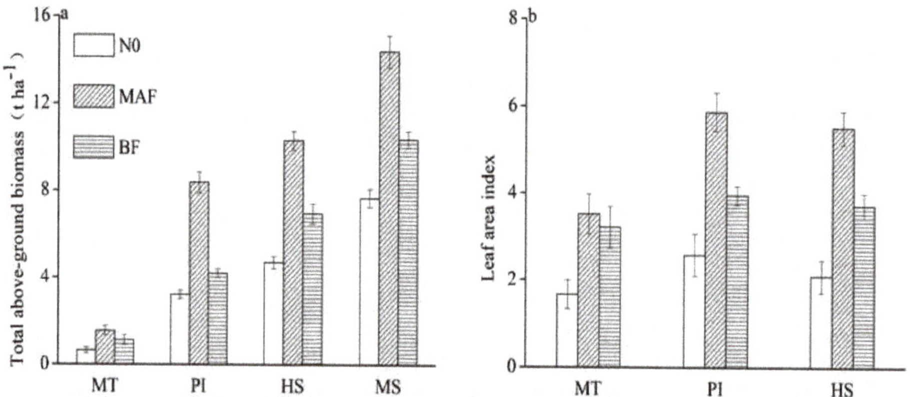

Figure 2. Effects of mechanical deep placement of nitrogen fertilizer on average for the total above-ground biomass and leaf area index for both rice cultivars (2018 and 2019). (**a**): total above-ground biomass, (**b**): leaf area index. MT: Mid-tillering stage; PI: Panicle initiation stage; HS: Heading stage; MS: Maturity stage.

The N fertilizer application remarkably affected the TAB for both rice cultivars (Figure 2). For example, at the MT stage, the TAB for BF and MAF treatments were significantly higher than N0. There were no significant differences between MAF and BF, and the TAB for MAF was higher than BF. At the HS and MS stages, N fertilizer application (MAF and BF) treatments remarkably increased TBA, especially when compared to the N0 treatment. Moreover, there were remarkable differences between

MAF and BF treatments. In the whole growth period, a similar trend for TAB of both rice cultivars was observed.

3.4. Determination of Antioxidant Enzymatic Activity

3.4.1. POD Activity

The POD activity in the leaves at all critical growth stages including the MT, PI, and HS stages were shown in Figure 3. At the MT stage, the highest POD activities were observed in the MAF treatment. However, the POD activity of the BF treatment was higher than the N0 treatment while lower than the MAF treatment. At the PI stage, MAF treatment significantly improved POD activity, especially when compared to the BF treatment. However, the POD activity did not differ significantly between BF and N0 treatment. A significant difference was found among all treatments at the HS stage. In the whole growth period, the similar trend for POD activity of both rice cultivars was observed.

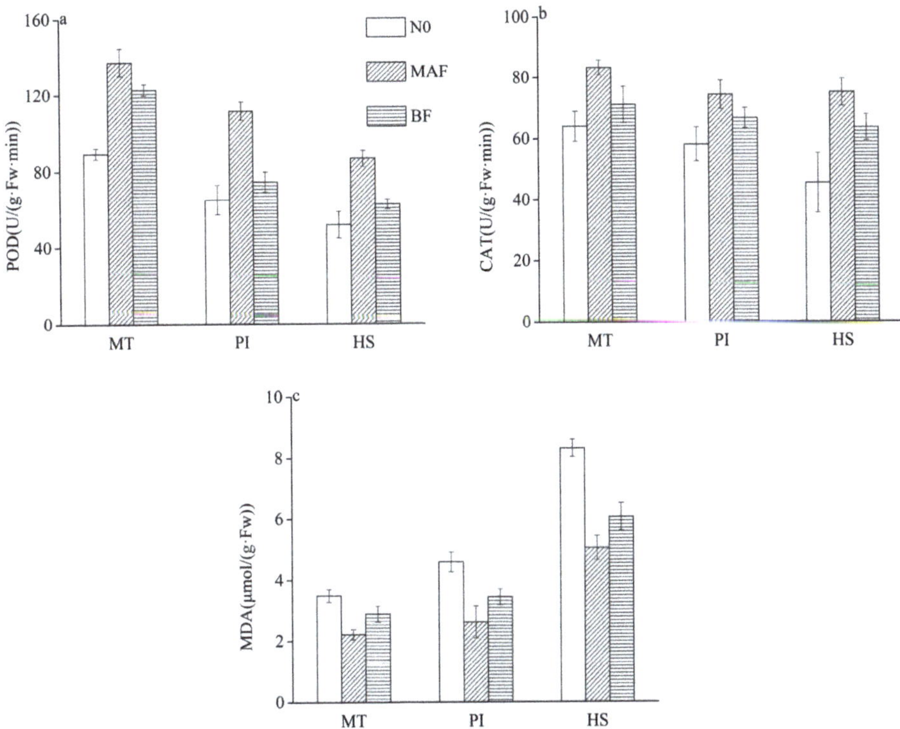

Figure 3. Effects of mechanical deep placement of nitrogen fertilizer on average for POD, CAT activity, and MDA content for both rice cultivars (2018 and 2019) (**a**): POD, (**b**); CAT, (**c**): MDA.

3.4.2. CAT Activity

The CAT activity in the leaves at all critical growth stages including the MT, PI, and HS stages were shown in Figure 3. There were significant differences in CAT activity found between the MAF and BF treatments at the MT stage. However, the BF and N0 treatments did not differ significantly. A significant difference of CAT activity was found between MAF and N0 treatment at the PI stage. At the HS stage, the highest CAT activities were observed in the MAF treatment over two years. However, CAT activity was higher in the BF treatment than the N0 treatment, while remaining lower than MAF treatment. Furthermore, a significant difference was found among all treatments.

3.4.3. MDA Content

The MDA content in the leaves at all critical growth stages including the MT, PI, and HS stages were shown in Figure 3. At the MT stage, the lowest MDA contents for both rice cultivars over two years were observed in the MAF treatment. However, the MDA content was higher in the BF treatment than the MAF treatment, while remaining lower than the N0 treatment. At the PI stage, compared to the N0 treatment, the BF and MAF treatments could significantly decrease MDA content. Marginal differences in MDA content was found between the MAF and BF treatments. Significant differences of MDA content were found among all treatments at the HS stage.

3.5. Correlation Analysis

The relationship among grain yield and its components TAB, TNA, LAI, and antioxidant enzyme activities for both rice cultivars over two years (Table 3). The variation of N treatment was focused on in the correlation analysis. Rice yield was significantly and positively correlated with the number of productive panicles ha^{-1}, spikelet per panicle, TAB, LAI, and antioxidant enzyme activities, including POD and CAT activities. The spikelet per panicle also significantly correlated with LAI, TAB, and MDA content. Furthermore, TAB at the maturity stage and LAI at the HS stage significantly correlated with antioxidant enzyme activities and TNA.

Table 3. Relationship among grain yield, and its components, TAB, TNA, LAI, and antioxidant enzyme activities for both rice cultivars (2018 and 2019).

Parameter	Productive Panicle	Spikelet per Panicle	Grain Filling	1000-Grain -Weigh (G)	LAI	TAB	POD	CAT	MDA	TNA
Productive panicle										
Spikelet per panicle	0.28									
Grain filling	−0.03	−0.18								
1000-grain-weight(g)	0.37 *	0.18	0.33 *							
LAI	0.53 **	0.47 **	0.20	0.52 **						
TAB	0.68 **	0.44 **	−0.06	0.51 **	0.85 **					
POD activity	0.55 **	0.27	0.46 **	0.45 **	0.76 **	0.60 **				
CAT activity	0.49 **	0.51 **	−0.58 **	0.03	0.47 **	0.60 **	0.28			
MDA content	−0.48 **	−0.44 **	0.54 **	−0.43 **	−0.52 **	−0.70 **	−0.18	−0.74 **		
TNA	0.65 **	0.62	−0.03	0.41 *	0.74 **	0.71 **	0.62 **	0.52 **	−0.56 **	
Yield	0.73 **	0.39	−0.06	0.44 **	0.75 **	0.79 **	0.68 **	0.62 **	−0.61 **	0.79 **

LAI: Leaf area index at HS stage; TAB: Total above-ground biomass at MS stage; POD: Peroxidase; CAT: Catalase, MDA: malonic dialdehyde at HS stage, TNA: Total nitrogen accumulation at MS stage. The same as below. **: ($p < 0.01$); *: ($p < 0.05$)

4. Discussion

4.1. Grain Yield and Its Components

Compared with broadcasting fertilizer treatment (BF), mechanized deep placement of all fertilizers at once (MAF) significantly enhanced the grain yields in mechanical pot-seedling transplanting (PST). The highest yields of MAF treatment for both rice cultivars were mainly due to the number of productive panicles ha^{-1} and spikelet per panicle, which was in agreement with Bandaogo et al. [26]. Moreover, yield increase via a mechanized deep placement in PST was 52.7% for both rice cultivars. The increase was far larger than in previous studies on mechanized deep placement in mechanical carpet-seedling transplanting (6.3–11.6%) [15], suggesting that the mechanized deep placement method could be more effective under PST conditions than non-PST conditions. Moreover, Pan et al. [21] found that deep fertilization remarkably improved spikelet number per panicle in direct-seeded rice compared to manual surface broadcast fertilizers. Tracing it to the cause, the PST was used in our experiment, which opened a fassula and then uniformly placed the N fertilizer at 10 cm depth. Finally, the fassula was covered by this applicator immediately. This method provided a continuous nitrogen supply for rice growth and the NH_4^+ absorbed by the root system remained in the soil for a longer period of time, thus promoting the growth of rice plants throughout the growth period and thereby increasing the nitrogen absorption and grain yield [27,28]. Deep placement fertilizer also maintained higher antioxidant enzyme activities at the heading stage, which was one of the reasons for the higher grain yield.

Compared with BF treatment, MAF significantly enhanced the leaf area index (LAI) and total above-ground biomass (TAB) at the PI and HS stages in PST. The main reason for this was that MAF treatment promoted rice growth in the middle and late stages by reducing nutrition loss. Broadcasting fertilizers did not meet this demand and led to insufficient nutrient supply. Moreover, LAI and TAB increased by mechanized deep placement in PST, reaching 58.5% and 40.2% for both rice cultivars, respectively. The increase was larger than in previous studies on mechanized deep placement in mechanical carpet-seedling transplanting (36.1–38.9% and 8.7–10.6%) [15,29]. In addition, a larger TAB was beneficial to the transportation of dry matter to the panicle, leading to more spikelet per panicle and 1000-grain-weight of rice, so as to improve the yield of rice. Our results showed that TAB at the MS stage remarkably correlated with LAI at the HS stage, and both of them were positively related to grain yield. The result indicated that MAF had some superiority in larger LAI and TAB, which thus resulted in a higher grain yield.

4.2. Nitrogen Use Efficiency and Antioxidant Enzyme Activities

We discovered that MAF led to a substantial increase in total N accumulation (TNA), N recovery efficiency (NRE), and agronomic efficiency (NAE) compared to BF treatment. Moreover, NAE and NRE increase by mechanized deep placement in PST was 104.3% and 123.7% for both rice cultivars, respectively. The increase was far larger than in previous studies on mechanized deep placement in mechanical carpet-seedling transplanting (17.9–43.1% and 19.6–37.4%), suggesting that the mechanized deep placement method could be more effective to improve NRE and NAE under PST conditions than non-PST conditions [29,30]. Some researchers have showed that deep fertilization could improve NUE by reducing nitrogen loss and prolonging the duration of fertilizer, which was compared with surface broadcasting [31,32]. Previous results also found that the deep N fertilization could reduce urease activity by increasing NH_4^+ concentration in soil depth, thereby reducing NH_4^+ concentration in flood water [33,34]. The reason was that the deep nitrogen fertilizer in the anaerobic soil layer caused the NH_4^+ to move slowly from the depth to soil surface, thereby the NH_4^+ content in the flood was low. Moreover, the decrease of N concentration in the flood reduced the loss of N through runoff, ammonia volatilization, and denitrification, thus improving the NUE. Deep placement of fertilizers was a concentrated application of fertilizer near the roots of rice, which was conducive to the absorption of roots and improved the NUE [18,35]. We also found that TNA content was significantly related to grain yield and antioxidant enzyme activities in a positive correlation, because the higher antioxidant enzyme activity at the full heading stage was decisive to the transfer of nutrients to grains, and increased the accumulation of nitrogen in grains, thus increasing the NAE and NRE of rice plants.

Our results showed that MAF notably increased antioxidant enzyme activities including POD and CAT and reduced the MDA content of both rice cultivars. Some reports have showed that the appropriate application of nitrogen fertilizer could maintain high antioxidant enzyme activities at the MS stage, which would be beneficial when delaying the senescence of functional leaves of rice [36,37]. The reason was that the deep application of fertilizer could fulfill nutrient requirements of rice growth and development in time, providing sufficient energy for the antioxidant enzyme activities of the plant, enhancing the activity of the plant leaf protection enzyme system, accelerating the scavenging of free radicals, and reducing membrane lipid peroxidation and the MDA content in the rice plant [38]. Moreover, the rice plant had a stronger root system and longer green leaf duration under deep N fertilization [1]. We also found a significant positive correlation between POD and CAT activities and grain yield. N-fertilized plants need such effective antioxidant machinery to cope with excessive reactive oxygen species production. In this way, it can delay the senescence of leaves in the late growth stage [23,38]. Therefore, maintaining high POD and CAT activities at the HS stage was conducive to a higher rice yield.

5. Conclusions

Compared to broadcasting fertilizer treatment, mechanized deep N fertilization enhanced rice growth under mechanical pot-seedling transplanting (PST) in terms of leaf area index and total above-ground biomass. Mechanized deep N fertilization also improved the grain yield, nitrogen use efficiency, and antioxidant enzyme activities in PST. Therefore, mechanized deep fertilization in mechanical pot-seedling transplanting should be recommended in actual production.

Author Contributions: Conceptualization, S.P. and X.T.; Methodology, S.P. and X.T.; Data curation, H.T., Z.W., M.D., and S.W.; Formal analysis, L.L., H.T., Z.Z., and M.D.; Project administration, S.W., Z.W., and Z.Z.; Supervision, L.L., H.T., S.P., and X.T.; Validation, L.L., H.T., Z.M., M.D, and U.A.; Writing—original draft, L.L., H.T., Z.M., and M.D.; Writing—review and editing, L.L., H.T., Z.M., M.D., U.A., Z.W., S.P., and X.T. All authors have read and agreed to the published version of the manuscript.

Funding: The research was supported by Key-Area Research and Development Program of Guangdong Province (No.2019 B020221003) and National Natural Science Foundation of China (31471442).

Acknowledgments: We thank Leilei Kong who provided the rice cultivars. The research was supported by Key-Area Research and Development Program of Guangdong Province (No. 2019 B020221003) and National Natural Science Foundation of China (31471442).

Conflicts of Interest: The authors declare no conflict of interest.

References

1. Liu, K.; Deng, J.; Lu, J.; Wang, X.; Lu, B.; Tian, X.; Zhang, Y. High Nitrogen Levels Alleviate Yield Loss of Super Hybrid Rice Caused by High Temperatures During the Flowering Stage. *Front. Plant Sci.* **2019**, *10*, 357. [CrossRef] [PubMed]
2. He, A.; Wang, W.; Jiang, G.; Sun, H.; Jiang, M.; Man, J.; Cui, K.; Huang, J.; Peng, S.; Nie, L. Source-sink regulation and its effects on the regeneration ability of ratoon rice. *Field Crop. Res.* **2019**, *236*, 155–164. [CrossRef]
3. Kong, L.; Ashraf, U.; Cheng, S.; Rao, G.; Mo, Z.; Tian, H.; Pan, S.; Tang, X. Short-term water management at early filling stage improves early-season rice performance under high temperature stress in South China. *Eur. J. Agron.* **2017**, *90*, 117–126. [CrossRef]
4. Chen, Y.; Fan, P.; Mo, Z.; Kong, L.; Tian, H.; Duan, M.; Li, L.; Wu, L.; Wang, Z.; Tang, X.; et al. Deep Placement of Nitrogen Fertilizer Affects Grain Yield, Nitrogen Recovery Efficiency, and Root Characteristics in Direct-Seeded Rice in South China. *J. Plant Growth Regul.* **2020**, *2*, 1–9. [CrossRef]
5. Ashraf, U.; Hussain, S.; Akbar, N.; Anjum, S.A.; Hassan, W.; Tang, X. Water management regimes alter Pb uptake and translocation in fragrant rice. *Ecotoxicol. Environ. Saf.* **2018**, *149*, 128–134. [CrossRef]
6. Tao, Y.; Chen, Q.; Peng, S.; Wang, W.; Nie, L. Lower global warming potential and higher yield of wet direct-seeded rice in Central China. *Agron. Sustain. Dev.* **2016**, *36*, 24. [CrossRef]
7. Paman, U.; Inaba, S.; Uchida, S. The mechanization of small-scale rice farming: Labor requirements and costs. *Eng. Agric. Environ. Food* **2014**, *7*, 122–126. [CrossRef]
8. Hu, Y.J.; Xing, Z.P.; Gong, J.L.; Liu, G.T.; Zhang, H.C.; Dai, Q.G.; Huo, Z.Y.; Xu, K.; Wei, H.Y.; Guo, B.W.; et al. Study on population characteristics and formation mechanisms for high yield of pot-seedling mechanical transplanting rice. *Sci. Agri. Sin.* **2014**, *47*, 865–879.
9. Jiang, X.; Li, X.; Chi, Z.; Wang, S.; Yang, F.; Zheng, J. Research on potted-tray grown rice seedling transplanting by machine. *Agri. Sci. Tech.* **2014**, *15*, 1923–1927.
10. Zhu, C.; Yang, H.J.; Guan, Y.X.; Chen, Z. Research progress on green and high efficiency cultivation technique of rice pot seedling mechanical transplanting in Jiangsu province. *China Rice* **2019**, *25*, 37–41.

11. Gaihre, Y.K.; Singh, U.; Islam, S.M.; Huda, A.; Satter, M.A.; Sanabria, J.; Islam, R.; Shah, A. Impacts of urea deep placement on nitrous oxide and nitric oxide emissions from rice fields in Bangladesh. *Geoderma* **2015**, *260*, 370–379. [CrossRef]

12. Watanabe, T.; Son, T.T.; Hung, N.N.; Van Truong, N.; Giau, T.Q.; Hayashi, K.; Ito, O. Measurement of ammonia volatilization from flooded paddy fields in Vietnam. *Soil Sci. Plant Nutr.* **2009**, *55*, 793–799. [CrossRef]

13. Yao, Y.; Zhang, M.; Tian, Y.; Zhao, M.; Zhang, B.; Zhao, M.; Zeng, K.; Yin, B. Urea deep placement for minimizing NH3 loss in an intensive rice cropping system. *Field Crop. Res.* **2018**, *218*, 254–266. [CrossRef]

14. Du, B.; Luo, H.W.; He, L.X.; Zheng, A.X.; Chen, Y.L.; Zhang, T.T.; Wang, Z.M.; Hu, L.; Tang, X.R. DEEP FERTILIZER PLACEMENT IMPROVES RICE GROWTH AND YIELD IN ZERO TILLAGE. *Appl. Ecol. Environ. Res.* **2018**, *16*, 8045–8054. [CrossRef]

15. Zhu, C.; Xiang, J.; Zhang, Y.; Zhang, Y.; Defeng, Z.; Chen, H. Mechanized transplanting with side deep fertilization increases yield and nitrogen use efficiency of rice in Eastern China. *Sci. Rep.* **2019**, *9*, 5653. [CrossRef]

16. Guo, L.; Ning, T.; Nie, L.; Li, Z.; Lal, R. Interaction of deep placed controlled-release urea and water retention agent on nitrogen and water use and maize yield. *Eur. J. Agron.* **2016**, *75*, 118–129. [CrossRef]

17. Li, X.; Chen, X.; Wang, H.; Lu, D.; Zhou, J.; Chen, Z.; Zhu, D. Effects and principle of root-zone one-time N fertilization on enhancing rice (Oryza sativa L.) N use efficiency. *Soils* **2017**, *49*, 868–875.

18. Wu, M.; Liu, M.; Liu, J.; Li, W.; Jiang, C.-Y.; Li, Z.-P. Optimize nitrogen fertilization location in root-growing zone to increase grain yield and nitrogen use efficiency of transplanted rice in subtropical China. *J. Integr. Agric.* **2017**, *16*, 2073–2081. [CrossRef]

19. Linquist, B.A.; Adviento-Borbe, M.A.; Pittelkow, C.M.; Van Kessel, C.; Van Groenigen, K.J. Fertilizer management practices and greenhouse gas emissions from rice systems: A quantitative review and analysis. *Field Crop. Res.* **2012**, *135*, 10–21. [CrossRef]

20. Yao, Z.; Zheng, X.; Zhang, Y.; Liu, C.; Wang, R.; Lin, S.; Zuo, Q.; Butterbach-Bahl, K. Urea deep placement reduces yield-scaled greenhouse gas (CH4 and N2O) and NO emissions from a ground cover rice production system. *Sci. Rep.* **2017**, *7*, 11415. [CrossRef]

21. Pan, S.; Wen, X.; Wang, Z.; Ashraf, U.; Tian, H.; Duan, M.; Mo, Z.; Fan, P.; Tang, X. Benefits of mechanized deep placement of nitrogen fertilizer in direct-seeded rice in South China. *Field Crop. Res.* **2017**, *203*, 139–149. [CrossRef]

22. Xu, Y.L.; Fu, A.B.; Liu, T.X. Effects of different nitrogen application patterns on leaf physiological characteristics and grain yield of double cropping rice. *Acta Agric. Bor. Sin.* **2019**, *34*, 155–163.

23. Shu, S.; Tang, X.R.; Luo, X.W.; Li, G.X.; Wang, Z.M.; Zheng, T.X.; Jia, X.N. Effects of deep mechanized application of slow-release fertilizers on physiological characteristics of precision hill-direct-seeding super rice. *Trans. Chin. Soc. Agric. Eng.* **2011**, *27*, 89–92.

24. Pan, S.; Rasul, F.; Li, W.; Tian, H.; Mo, Z.; Duan, M.; Tang, X. Roles of plant growth regulators on yield, grain qualities and antioxidant enzyme activities in super hybrid rice (Oryza sativa L.). *Rice* **2013**, *6*, 9. [CrossRef]

25. Plumb-Dhindsa, P.; Dhindsa, R.S.; Thorpe, T.A. Leaf Senescence: Correlated with Increased Levels of Membrane Permeability and Lipid Peroxidation, and Decreased Levels of Superoxide Dismutase and Catalase. *J. Exp. Bot.* **1981**, *32*, 93–101. [CrossRef]

26. Bandaogo, A.; Fofana, B.; Youl, S.; Safo, E.; Abaidoo, R.C.; Andrews, O. Effect of fertilizer deep placement with urea supergranule on nitrogen use efficiency of irrigated rice in Sourou Valley (Burkina Faso). *Nutr. Cycl. Agroecosystems* **2014**, *102*, 79–89. [CrossRef]

27. Kargbo, M.; Pan, S.; Mo, Z.; Wang, Z.; Luo, X.; Tian, H.; Hossain, F.; Ashraf, U.; Tang, X. Physiological Basis of Improved Performance of Super Rice (Oryza sativa) to Deep Placed Fertilizer with Precision Hill-drilling Machine. *Int. J. Agric. Boil.* **2016**, *18*, 797–804. [CrossRef]

28. Kapoor, V.; Singh, U.; Patil, S.K.; Magre, H.; Shrivastava, L.K.; Mishra, V.N.; Das, R.O.; Samadhiya, V.K.; Sanabria, J.; Diamond, R. Rice Growth, Grain Yield, and Floodwater Nutrient Dynamics as Affected by Nutrient Placement Method and Rate. *Agron. J.* **2008**, *100*, 526–536. [CrossRef]

29. Zhu, C.H.; Zhang, Y.P.; Xiang, J.; Zhang, Y.K.; Wu, H.; Wang, Y.L.; Zhu, D.F.; Chen, H.Z. Effects of side deep fertilization on yield formation and nitrogen utilization of mechanized transplanting rice. *Sci. Agric. Sin.* **2019**, *52*, 4228–4239.

30. Zhong, X.M.; Huang, T.P.; Peng, J.W.; Lu, W.L.; Kang, X.R.; Song, S.M.; Zhou, X. Effects of machine-transplanting synchronized with one-time precision fertilization on nutrient uptake and use efficiency of double cropping rice. *Chin. J. Rice Sci.* **2019**, *33*, 436–446.

31. Huda, A.; Gaihre, Y.K.; Islam, M.R.; Singh, U.; Islam, R.; Sanabria, J.; Satter, M.A.; Afroz, H.; Halder, A.; Jahiruddin, M. Floodwater ammonium, nitrogen use efficiency and rice yields with fertilizer deep placement and alternate wetting and drying under triple rice cropping systems. *Nutr. Cycl. Agroecosystems* **2016**, *104*, 53–66. [CrossRef]

32. Zhang, M.; Yao, Y.; Zhao, M.; Zhang, B.; Tian, Y.; Yin, B.; Zhu, Z. Integration of urea deep placement and organic addition for improving yield and soil properties and decreasing N loss in paddy field. *Agric. Ecosyst. Environ.* **2017**, *247*, 236–245. [CrossRef]

33. Danying, W.; Chang, Y.; Chunmei, X.; Zaiman, W.; Song, C.; Guang, C.; Zhang, X. Soil Nitrogen Distribution and Plant Nitrogen Utilization in Direct-Seeded Rice in Response to Deep Placement of Basal Fertilizer-Nitrogen. *Rice Sci.* **2019**, *26*, 404–415. [CrossRef]

34. Yao, Y.; Zhang, M.; Tian, Y.; Zhao, M.; Zhang, B.; Zeng, K.; Zhao, M.; Yin, B. Urea deep placement in combination with Azolla for reducing nitrogen loss and improving fertilizer nitrogen recovery in rice field. *Field Crop. Res.* **2018**, *218*, 141–149. [CrossRef]

35. Liu, T.; Fan, D.; Zhang, X.; Chen, J.; Li, C.; Cao, C. Deep placement of nitrogen fertilizers reduces ammonia volatilization and increases nitrogen utilization efficiency in no-tillage paddy fields in central China. *Field Crop. Res.* **2015**, *184*, 80–90. [CrossRef]

36. Zhou, J.; Meng, G.Y.; Long, J.R.; Ma, G.H.; Wan, Y.Z. Effects of application amount of slow-release nitrogen fertilizer on physiological characteristics and grain yield of early super hybrid rice zhuliangyou 02. *Hybrid Rice* **2011**, *26*, 62–68.

37. Wang, Y.; Lin, A.; Loake, G.J.; Chu, C. H2O2?induced Leaf Cell Death and the Crosstalk of Reactive Nitric/Oxygen SpeciesF. *J. Integr. Plant Boil.* **2013**, *55*, 202–208. [CrossRef]

38. Tang, H.; Xiao, X.P.; Li, C.; Tang, W.G.; Guo, L.J.; Wang, K.; Cheng, K.; Pan, X.C.; Sun, G. Effects of different long-term fertilization managements on the physiological characteristics of leaves and yield of rice in double cropping paddy field. *J. China Agric. Univ.* **2018**, *11*, 60–71.

MDPI

St. Alban-Anlage 66

4052 Basel

Switzerland

Tel. +41 61 683 77 34

Fax +41 61 302 89 18

www.mdpi.com

Agronomy Editorial Office

E-mail: agronomy@mdpi.com

www.mdpi.com/journal/agronomy